Optical Properties *of* Photonic Structures

Interplay of Order and Disorder

Optical Properties *of* Photonic Structures

Interplay of Order and Disorder

Edited by
Mikhail F. Limonov
Richard M. De La Rue

CRC Press
Taylor & Francis Group
Boca Raton London New York

CRC Press is an imprint of the
Taylor & Francis Group, an **informa** business
A TAYLOR & FRANCIS BOOK

Cover image by Diederik and Leonardo Wiersma.

CRC Press
Taylor & Francis Group
6000 Broken Sound Parkway NW, Suite 300
Boca Raton, FL 33487-2742

First issued in paperback 2019

ISBN-13: 978-1-4398-7191-1 (hbk)
ISBN-13: 978-0-367-38121-9 (pbk)

Library of Congress Cataloging-in-Publication Data

Optical properties of photonic structures : interplay of order and disorder / edited by Mikhail F. Limonov and Richard M. De La Rue.
p. cm. -- (Series in optics and optoelectronics)
Includes bibliographical references and index.
ISBN 978-1-4398-7191-1
1. Photonic crystals. 2. Crystal optics. 3. Nonlinear optics. I. Limonov, Mikhail F. II. De La Rue, Richard M.

QD941.O675 2012
548'.83--dc23 2011052600

Visit the Taylor & Francis Web site at
http://www.taylorandfrancis.com

and the CRC Press Web site at
http://www.crcpress.com

Contents

Preface

Confusion heard his voice, and wild uproar
Stood ruled, stood vast infinitude confined;
Till at his second bidding darkness fled,
Light shone, and order from disorder sprung.

—*Paradise Lost* **by John Milton**

"Disorder is the new order"—this title of an article in *Nature* (**432,** 285, 2004) announced that the era of pure, perfectly regular, photonic crystals went out some time ago. At the end of the last century, regular photonic crystal structures were believed to be essential for the effective operation of various optical devices—and regularity remains a critical issue, for example, for specific devices that rely on slow light behavior. Now it is understood that disordered materials may actually perform better—for example, in tasks such as efficient and directed light extraction from incoherent and partially coherent sources. Moreover, all the key applications of photonic crystal structures are based on deliberately introduced disorder—as is realized in channel waveguides, microcavities, embedded quantum dots, additional resonant layers, and so on. In fact, we face a dialectical problem of the unity and conflict of opposites, and there is a need to find new ways to extract order from disorder.

The volume in your hands is, we believe, the first monograph specifically concerned with disorder in photonic crystals. The leaders of research activity in this area have come together to write a comprehensive textbook. By uniting the efforts of a substantial number of experts in this field, we have been able to cover almost all the important directions, methods, and approaches; various different structures are discussed—for example, opals, aperiodic Fibonacci-type one-dimensional photonic structures, photonic glasses, and hypersonic, magnetophotonic, and plasmonic–photonic crystals.

We thank all the authors for their valuable contributions to this book. We are grateful, in particular, to Dr. Hernán Míguez for his assistance in the early stages of the compilation of the book and to Dr. Mikhail Rybin for his general help during the preparation of this book. Special thanks are also due to those Japanese authors who finished their work after the cataclysmic earthquake and tsunami on March 11, 2011.

M.F. Limonov would like to thank his wife Elena and daughter Valeria for their support during the troublesome time of the preparation of this book. R.M. De La Rue thanks his wife Barbara for her enduring, but not unlimited, patience, and for her efforts in unearthing the words of the great English poet John Milton that form the epigram for the book.

Mikhail F. Limonov
Richard M. De La Rue

Editors

Mikhail F. Limonov is a leading research scientist at the Ioffe Physical-Technical Institute of the Russian Academy of Sciences and has served as a research fellow at the National Research University of Information Technologies, Mechanics and Optics (St. Petersburg). He has more than 30 years of wide-ranging experience in optical research concerned with various dielectrics, Fullerenes, zeolites, superconductors, and photonic crystals.

Richard M. De La Rue was appointed as a lecturer at Glasgow University in 1971, and became Professor of Optoelectronics in 1986 from which he retired in 2010 and became Honorary Senior Research Fellow. He is currently visiting professor at the University of Malaya in Kuala Lumpur, Malaysia. His recent research has been concerned with photonic crystal and photonic wire structures, with waveguide microcavities and with metamaterials. Silicon, silica, ferroelectric single crystal, polymers, metals, and compound semiconductors have all been involved in this nano- and microphotonics research activity. His research in the area of photonic crystals evolved to cover compact lasers, planar microcavities, photonic-crystal LEDs, photonic integrated circuits, synthetic opal, and biosensors. He is a fellow of the Institute of Electrical and Electronics Engineers (IEEE), Optical Society of America (OSA), the Institution of Engineering and Technology (IET), the Royal Society of Edinburgh (FRSE), and the Royal Academy of Engineering (FREng).

Contributors

Richard A. Abram
Department of Physics
Durham University
Durham, United Kingdom

Andrey V. Akimov
Ioffe Physical-Technical Institute of the Russian Academy of Sciences
St. Petersburg, Russia
and
School of Physics and Astronomy
The University of Nottingham
Nottingham, United Kingdom

Lucio Claudio Andreani
Department of Physics "A. Volta"
University of Pavia
Pavia, Italy

Paolo Andrich
Department of Physics "A. Volta"
University of Pavia
Pavia, Italy

Pierre Barthelemy
Kavli Institute of Nanoscience
Delft University of Technology
Delft, the Netherlands

Alexander V. Baryshev
Ioffe Physical-Technical Institute of the Russian Academy of Sciences
St. Petersburg, Russia
and
Electronics-Inspired Interdisciplinary Research Institute
Toyohashi University of Technology
Toyohashi, Japan

Jacopo Bertolotti
Department of Physics
University of Florence
Sesto Fiorentino, Italy
and
Complex Photonic Systems
University of Twente
Enschede, the Netherlands

Konstantin Y. Bliokh
School of Physics
National University of Ireland
Galway, Ireland

Yury Bliokh
Department of Physics
Technion—Israel Institute of Technology
Haifa, Israel

Stuart Brand
Department of Physics
Durham University
Durham, United Kingdom

Hui Cao
Department of Applied Physics
Yale University
New Haven, Connecticut

Richard M. De La Rue
Photonics Research Centre
Physics Department
University of Malaya
Kuala Lumpur, Malaysia
and
School of Engineering
University of Glasgow
Glasgow, United Kingdom

Ricardo A. Depine
Department of Physics
University of Buenos Aires
Buenos Aires, Argentina

Luis A. Dorado
Department of Physics
University of Buenos Aires
Buenos Aires, Argentina

Keiichi Edagawa
Institute of Industrial Science
The University of Tokyo
Tokyo, Japan

Valentin Freilikher
Department of Physics
Bar-Ilan University
Ramat-Gan, Israel

Matteo Galli
Department of Physics "A. Volta"
University of Pavia
Pavia, Italy

Pedro David Garcia-Fernández
Niels Bohr Institute
University of Copenhagen
Copenhagen, Denmark

Dario Gerace
Department of Physics "A. Volta"
University of Pavia
Pavia, Italy

Valery G. Golubev
Ioffe Physical-Technical Institute of the Russian Academy of Sciences
St. Petersburg, Russia

Sergey V. Grigoriev
Konstantinov St. Petersburg Nuclear Physics Institute
Gatchina, Russia

Natalia A. Grigoryeva
Department of Quantum Magnetic Phenomena
St. Petersburg State University
St. Petersburg, Russia

Stephen Hughes
Department of Physics
Queen's University
Kingston, Ontario, Canada

Mitsuteru Inoue
Department of Electrical and Electronic Engineering
Toyohashi University of Technology
Toyohashi, Japan

Eugeniyus L. Ivchenko
Ioffe Physical-Technical Institute of the Russian Academy of Sciences
St. Petersburg, Russia

Sajeev John
Department of Physics
University of Toronto
Toronto, Ontario, Canada

Mikhail A. Kaliteevski
Ioffe Physical-Technical Institute of the Russian Academy of Sciences
and
Nanotechnology Research and Education Centre
Academic University
St. Petersburg, Russia

Alexander A. Kaplyanskii
Ioffe Physical-Technical Institute of the Russian Academy of Sciences
St. Petersburg, Russia

Alexander B. Khanikaev
Department of Physics
The University of Texas at Austin
Austin, Texas

Yuri S. Kivshar
Research School of Physics and Engineering
Australian National University
Canberra, Australia

Thomas F. Krauss
School of Physics and Astronomy
University of St. Andrews
St. Andrews, United Kingdom

Alexander Kuznetsov
NanoTech Institute
The University of Texas at Dallas
Dallas, Texas

Yoav Lahini
Department of Physics of Complex Systems
Weizmann Institute of Science
Rehovot, Israel

Mikhail F. Limonov
Ioffe Physical-Technical Institute of the Russian Academy of Sciences
and
National Research University of Information Technologies, Mechanics and Optics
St. Petersburg, Russia

Cefe López
Materials Science Institute of Madrid
Spanish National Research Council (CSIC)
Madrid, Spain

Juan F. Galisteo López
Materials Science Institute of Madrid
Spanish National Research Council (CSIC)
Madrid, Spain

Gabriel Lozano
Materials Science Institute of Seville
Spanish National Research Council (CSIC)–University of Seville
Seville, Spain

Alexander M. Merzlikin
Institute for Theoretical and Applied Electromagnetics
Moscow, Russia

Hernán Míguez
Materials Science Institute of Seville
Spanish National Research Council (CSIC)–University of Seville
Seville, Spain

Andrey E. Miroshnichenko
Research School of Physics and Engineering
Australian National University
Canberra, Australia

Rabia Moussa
NanoTech Institute
The University of Texas at Dallas
Dallas, Texas

Franco Nori
Digital Materials Laboratory
Advanced Science Institute, RIKEN
Wako, Japan
and
Department of Physics
University of Michigan
Ann Arbor, Michigan

Masaya Notomi
NTT Basic Research Laboratories
Atsugi, Japan

Liam O'Faolain
School of Physics and Astronomy
University of St. Andrews
St. Andrews, United Kingdom

Ryotaro Ozaki
Department of Electrical and Electronic Engineering
National Defense Academy
Kanagawa, Japan

Mark Patterson
Department of Physics
Queen's University
Kingston, Ontario, Canada

Andrei V. Petukhov
Debye Institute for Nanomaterials Science
Utrecht University
Utrecht, the Netherlands

Alexander B. Pevtsov
Ioffe Physical-Technical Institute of the Russian Academy of Sciences
St. Petersburg, Russia

Alexander N. Poddubny
Ioffe Physical-Technical Institute of the Russian Academy of Sciences
St. Petersburg, Russia

Randy C. Polson
Department of Physics and Astronomy
University of Utah
Salt Lake City, Utah

Sergei G. Romanov
Institute of Optics
University of Erlangen-Nuremberg
Erlangen, Germany
and
Ioffe Physical-Technical Institute of the Russian Academy of Sciences
St. Petersburg, Russia

Mikhail V. Rybin
Ioffe Physical-Technical Institute of the Russian Academy of Sciences
and
National Research University of Information Technologies, Mechanics and Optics
St. Petersburg, Russia

Anton K. Samusev
Ioffe Physical-Technical Institute of the Russian Academy of Sciences
and
National Research University of Information Technologies, Mechanics and Optics
St. Petersburg, Russia

Kirill B. Samusev
Ioffe Physical-Technical Institute of the Russian Academy of Sciences
and
National Research University of Information Technologies, Mechanics and Optics
St. Petersburg, Russia

Tal Schwartz
Physics Department and the Solid State Institute
Technion—Israel Institute of Technology
Haifa, Israel
and
ISIS
University of Strasbourg
Strasbourg, France

Mordechai Segev
Physics Department and the Solid State Institute
Technion—Israel Institute of Technology
Haifa, Israel

Alexander V. Sel'kin
Ioffe Physical-Technical Institute of the Russian Academy of Sciences
St. Petersburg, Russia

Yaron Silberberg
Department of Physics of Complex Systems
Weizmann Institute of Science
Rehovot, Israel

Ivan S. Sinev
Ioffe Physical-Technical Institute of the Russian Academy of Sciences
and
National Research University of Information Technologies, Mechanics and Optics
St. Petersburg, Russia

Costas M. Soukoulis
Department of Physics and Astronomy
Iowa State University
Ames, Iowa

Hironaga Uchida
Department of Electronics and Intelligent Systems
Tohoku Institute of Technology
Sendai, Japan

Z. Valy Vardeny
Department of Physics and Astronomy
University of Utah
Salt Lake City, Utah

Kevin Vynck
European Laboratory for Non-Linear Spectroscopy
University of Florence
Sesto Fiorentino, Italy

Diederik S. Wiersma
European Laboratory for Non-Linear Spectroscopy
University of Florence
Sesto Fiorentino, Italy
and
National Institute of Optics
Italian National Research Council
Florence, Italy

Alexey Yamilov
Department of Physics
Missouri University of Science and Technology
Rolla, Missouri

Anvar A. Zakhidov
NanoTech Institute
The University of Texas at Dallas
Dallas, Texas

Section 1

Introduction

1

Introduction

Sajeev John
University of Toronto

The interplay between order and disorder is of fundamental importance to the phenomenon of wave propagation and is the central mechanism for the trapping and localization of light in dielectric materials. Although the propagation of electromagnetic waves has been understood since the 1860s, only since 1984 it was realized that electromagnetic waves may have localized modes in a fully three-dimensional sense. This localization of light has far-reaching implications, both in fundamental science and in the practical applications of photonics. In fundamental science, optics is an invaluable testing ground for novel wave transport phenomena that occur near the transition between propagating and localized states. The strong focusing of light intensity in localized modes enables an unexplored regime of strong coupling between light and matter. Practical applications include circuits of light on an optical microchip for optical communications and optical computing, new types of optical fibers for laser surgery, new types of lasers and optical sensors, and a possible new generation of solar cells that are much more efficient at trapping and absorbing sunlight.

Periodic order in a dielectric microstructure provides coherent scattering of light from a precise geometrical arrangement of nearly identical scatterers. Bragg scattering resonances lead to the formation of stop gaps, where destructive wave interference prevents light propagation in specific directions over specific frequency ranges. These wave interference effects lead to a redistribution of the electromagnetic density of states as a function of frequency. In some frequency intervals this can result in a very "high" density of states, consisting of "slow" modes with a greatly reduced velocity for energy transport. In other frequency intervals, if the scattering is strong enough, this can lead to the complete absence of electromagnetic modes propagating in any direction. Such a frequency interval is referred to as a photonic band gap (PBG). The attainability of a PBG for waves obeying Maxwell's equations, using nonabsorbing dielectric materials, was first proposed in 1987. The occurrence of a PBG in certain artificial periodic microstructures has been the starting point of a broad and interdisciplinary field of research.

Disordered dielectric microstructures, unlike artificial PBG materials, are common to everyday experience. Random disorder and multiple light scattering are associated with the phenomenon of diffusion of light. This diffusion is characterized by a transport mean free path, the length scale on which light propagates before its direction is completely randomized by scattering. We all experience the effects of light diffusion when it becomes dark on a cloudy day. Another example is light propagation through human tissue where, for light of a wavelength of 1 μm, the transport mean free path is about 1 mm. However, naturally occurring systems do not scatter light so strongly that the transport mean free path becomes as short as the wavelength of light. Under such a circumstance, the classical diffusion picture of light propagation breaks down, owing to the wave interference of neighboring diffusion paths. If the scattering is sufficiently strong, transport of optical energy is completely inhibited and light localization is predicted. However, this effect is very challenging to realize in a completely random scattering medium.

Interplay between order and disorder provides the key to the systematic localization and trapping of light. From the standpoint of a photonic crystal with a complete PBG, engineered defects in the otherwise periodic microstructure produce strongly localized states of light within the gap. The escape pathways for light away from the defect are eliminated as a result of completely destructive wave interference in all directions. One remarkable feature of a strongly localized mode of light is that it can occupy a volume that is a small fraction of a cubic wavelength and yet have an arbitrarily long confinement time (i.e., large quality factor). This is very different from conventional optical resonators (that confine light by total internal reflection) where typically a very large quality factor requires a very large mode volume. This opens a new frontier in basic science where the strength of coupling between light and matter (resonant atoms, molecules, quantum dots, etc.) can reach unprecedented levels. Another distinguishing characteristic of localized light is that (unlike total internal reflection-based optical cavities) the optical energy can be confined either in the low refractive index region (such as air) or in the high refractive index region (such as silicon). This leads to an entirely new mechanism for guiding light along desired paths. One practical application of wave-interference-based guiding of light appears in hollow-core PBG fibers. Such fibers can deliver high-intensity laser light through a flexible endoscope without incurring damage to the walls of the fiber itself, enabling some laser surgeries to be less invasive.

From the standpoint of a random system, small amounts of geometrical order or correlation among nearby scatterers can make the optical diffusion coefficient vanish much more readily than when the scatterers are completely uncorrelated. Short-range order of this type can suppress the electromagnetic density of states in a manner analogous to the way that long-range periodic order leads to a PBG. Even a moderate reduction in the number of available optical pathways is enough to trigger the interference of optical diffusion paths and lead to light localization. An interesting outgrowth of this physics occurs through the addition of optical gain and loss in the medium, which leads to the phenomenon of "random lasing." Light scattering has traditionally been considered a hindrance to lasing as it reduces the optical quality factor of the lasing mode and reduces stimulated emission into the lasing mode. However, with strong multiple light scattering, a very large collection of bad optical cavities can pump one another, giving rise to a new type of laser that requires no mirrors and can literally be painted onto a surface. These random lasers are typically activated through pumping by another laser. However, if light scattering becomes very strong and the light localization threshold can be reached, stronger feedback occurs and the lasing threshold can be greatly reduced. A particularly tantalizing target is to achieve random lasing by pumping from an incoherent light source. Here again, the interplay between order and disorder may play a crucial role in achieving the target.

The collection of articles in this book offers a penetrating shaft into the still burgeoning subject of light propagation and localization in photonic crystals and disordered media. Although the subject has its origins in physics, it has broad significance and applicability in disciplines such as engineering, chemistry, mathematics, and medicine. Unlike other branches of physics, where the phenomena under consideration require extreme conditions of temperature, pressure, energy, or isolation from competing effects, the phenomena related to light localization survive under the most ordinary of conditions. This provides the science described in this book with broad applicability and vitality. However, the greatest challenge to the further development of this field is in the reliable and inexpensive synthesis of materials of the required composition, architecture, and length scale, where the proper balance between order and disorder is realized. Similar challenges have been faced and overcome in fields such as semiconductor science and technology. The challenge of photonic crystal synthesis has inspired a variety of novel fabrication protocols such as self-assembly and optical interference lithography that offer much less expensive approaches than conventional semiconductor microlithography. Once these challenges are fully met, it is likely that light propagation and localization in photonic microstructures will be at the heart of a twenty-first-century revolution in science and technology.

Section 2

Theory

2.1

Optical Properties of One-Dimensional Disordered Photonic Structures

Mikhail V. Rybin and Mikhail F. Limonov
Ioffe Physical-Technical Institute of the Russian Academy of Sciences
National Research University of Information Technologies, Mechanics and Optics

Alexander B. Khanikaev
The University of Texas at Austin

Costas M. Soukoulis
Iowa State University

CONTENTS

2.1.1 Introduction

The fundamental difference between regular crystals and photonic crystals (PhCs) is the similarity or dissimilarity of the crystal-building elements (particles). Atoms in regular crystals are absolutely identical, whereas the artificially produced or naturally grown particles (or other constituting elements) in PhCs always exhibit a certain degree of dissimilarity in size and dielectric properties. This dissimilarity leads to additional disorder-induced scattering, the character of which is determined by the shape of the particles constituting the PhC. Fabry–Pérot scattering occurs in the case of slabs, Mie scattering in the case of spheres, and so on.

Bragg scattering is responsible for the formation of stop-bands and other fundamental optical phenomena in PhCs, most of which have been studied quite extensively. Classical objects for the study of Bragg scattering are one-dimensional (1D) PhCs, which in a simplest case are composed of an ordered sequence of two (A and B) different dielectric slabs. One of the most interesting aspects of Bragg scattering is related to its interplay with disorder, which can give rise to localization of light.[1] By the time the concept of PhC was introduced in 1987,[1,2] significant advances had been made in understanding classical wave propagation in disordered structures.[3] In the case of infinite 1D structures, any solution of the corresponding wave equation is localized. That means the solution decreases on average exponentially with the distance from a bounded area of disordered structure. In the case of a long-enough disordered 1D sample, a transmission spectrum consists of a set of narrow bands with transmittivity up to one, which is superimposed on a background with vanishing transmission (Figure 2.1.1). Such a quasi-discrete state with high transparency corresponds to a large concentration of energy in randomly arranged areas inside a disordered structure. The formal analogy between problems of electron localization in random structures and bound state in a shallow potential well has been revealed in the study by Economou and Soukoulis.[4] Following this approach, one can describe a disordered structure as a well (area of wave localization) bounded on both sides by potential barriers[5] (for details see Chapter 2.4).

Yablonovitch[2] has suggested the use of PhCs to obtain control over spontaneous emission. This problem and other light manipulation tasks require a perfect structure. The stability of photonic band gaps in disordered structures has been studied extensively in various systems.[6] The studies have revealed nonlinear dependence on the disorder degree, that is, the photonic band gap is quite stable when disorder degree is below a certain value. Also, it was found that high-energy band gaps are less robust than low-energy band gaps. In contrast to Yablonovitch, John[1] considers periodicity as a mechanism for enhancing localization effects on band edges due to high density of states. A 1D disordered PhC can be described as a disordered layered structure that is periodic on average. As was noted, a typical transmission spectrum of random 1D PhC realization consists of a set of narrow peaks (as in any random structure) superimposed on a relatively broad stop-band. So it is convenient to average transmissivity or transmissivity logarithm over a large number of random realizations to analyze localization properties. In particular, the localization length becomes shorter in the vicinity of a band gap.[7] Also, localized states cannot be described in the single parameter scaling theory when the disorder degree is small enough;[8] however, when the disorder degree is high, such a description is

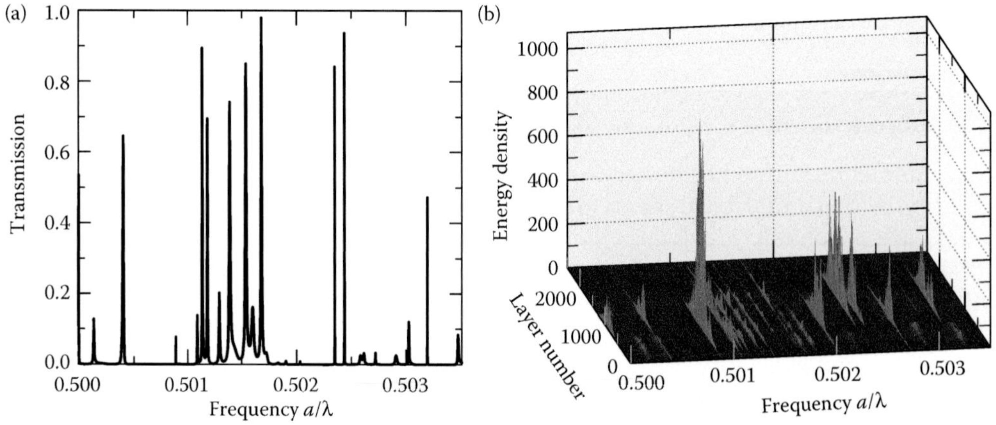

FIGURE 2.1.1
(a) Typical transmission spectrum of one-dimensional (1D) disordered sample and (b) electromagnetic energy density distribution along sample layers. Amplitude of fields inside the sample, normalized to the incident wave amplitude. Parameters of 1D photonic crystals are following $N = 2500$, $\bar{\varepsilon} = 6.0$, $\sigma_\varepsilon = 30$, and $\sigma_w = 50$.

possible. In a disordered structure formed by alternating layers of a material with a positive refraction index and a metamaterial with a negative refraction index, localization effects are suppressed.[9]

In this chapter, we develop a new picture of optical spectra transformation in disordered 1D PhCs. The main emphasis is on our recent results obtained with the use of numerical calculations. We demonstrate that the optical phenomena in 1D PhCs dramatically depend on the character of the disorder. Variation in the permittivity of the constituents of the 1D PhC gives rise to the disorder-induced Fabry–Pérot background in transmission spectra. We reveal the existence of two different types of scattering mechanisms that result in *formation of a Fano-type resonance.*[10] One can observe a distinct transformation picture of the spectrally narrow band specific for the Fano resonance, which manifests as a reversal of the photonic band gap in transmission spectra, that is, *its transformation from the stop-band to the pass-band.* Note that the Fano-type resonance involving the photonic Bragg band was previously discovered in opals[11]—representatives of the three-dimensional (3D) PhC—and is described in detail in Chapter 3.5. Experimental results lead us to suggest that the effects described in this chapter for the case of 1D PhC can also be observed in disordered two-dimensional (2D) and 3D PhCs.

In this chapter, we also consider a novel class of optical materials—structures composed of periodical arrangements of ordered components and components with fluctuating permittivity, that is, "disordered" components, whereas for disordered components, the average permittivity is equal to that of ordered components, that is, *dielectric contrast on average vanishes.* We show that such materials possess photonic band structures that can be considered as an inversion of those found in PhCs, that is, instead of Bragg stop-bands we observe Bragg pass-bands. A detailed study shows that the main mechanism behind these extraordinary optical properties is localization of light. It is worth mentioning that here we limit our consideration to the case of lossless structures and the vast majority of the results are given for the case of small dielectric contrast.

2.1.2 ε-Disorder versus *w*-Disorder

In this section, we consider the main features of the transmission spectra of 1D PhCs with different types of disorder. We consider 1D PhCs consisting of the sequence of A and B slabs (layers). For simplicity, we assume all the A slabs are identical and the B slabs are different. All the A slabs have constant permittivity ε_A and thickness w_A. In contrast, the B slabs are allowed to possess some degree of disorder both in permittivity (ε_B^i for ith layer, mean value $\bar{\varepsilon}_B$) referred to as "ε-disorder" and in thickness (w_B^i for ith layer, mean value $\bar{w}_B$) referred to as "*w*-disorder." The variations are described by the normal distribution with half-width normalized to maximum position as σ_ε and σ_w (expressed in percents below); that is, it is square of the second moment of the distribution (dispersion). At this stage, it is convenient to define an average lattice constant of 1D PhCs as $a = w_A + \bar{w}_B$. Transmission coefficients of 1D PhCs, for the case of light propagation along the structure axis, are evaluated with the use of the transfer matrix technique.[12] Then spectra are averaged for a sufficient number of realizations M.

A cornerstone of our discussion will be the results of the Bragg-band transformation with an increase of the degree of disorder of two different types, as demonstrated in Figure 2.1.2. Let us analyze this plot in detail. A classical Bragg stop-band (as shown in

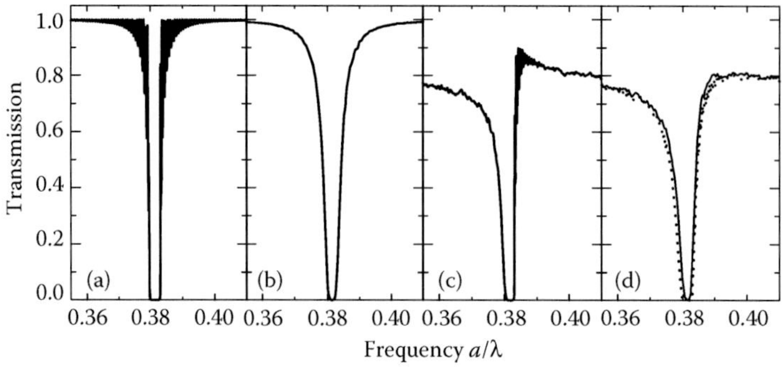

FIGURE 2.1.2
Averaged transmission spectra of one-dimensional photonic crystals (PhCs) for different types of disorder. (a) Perfect PhC; (b) PhC with w-disorder ($\sigma_w = 10$); (c) PhC with ε-disorder ($\sigma_\varepsilon = 10$); (d) PhC with both w-disorder ($\sigma_w = 10$) and ε-disorder ($\sigma_\varepsilon = 10$)—solid curve; product of two transmission spectra form (b) and (c)—dotted curve. Other parameters were: $\varepsilon_A = 1.70$, $\varepsilon_B = 1.75$, $\bar{w}_B = 0.38a$, $N = 500$, and $M = 10^3$.

Figure 2.1.2a) is observed in a conventionally ordered 1D PhC composed of, for instance, $N = 500$ unit cells of a stack of A and B layers AB with parameters $w_A = 0.62a$, $\varepsilon_A = 1.7$, $w_B = 0.38a$, and $\varepsilon_B = 1.75$. Such a Bragg stop-band is characterized by the nearly vanishing transmission, which is surrounded by the fringes associated with Fabry–Pérot-type interference on PhC boundaries.

We consider first disordered structures in which only one type of disorder, for example, w-disorder or ε-disorder, is present. The effect of w-disorder on the transmission spectra of 1D PhCs gives rise to the well-known broadening and degradation of the stop-bands and fringes (Figure 2.1.2b). At the same time, outside the Bragg bands the transmittance can reach nearly 100%. This result was revealed in multiple theoretical and experimental studies.[6] Introduction of the ε-disorder gives rise to a different transformation picture of the transmission spectra of 1D PhCs. This new picture is characterized by the asymmetric shape of the Bragg band (Figure 2.1.2c), and by the appearance of additional "background" scattering outside the Bragg bands. Consequences of the ε-disorder are discussed in Section 2.1.3.

In reality, PhC disorder involves both types of disorder—imperfection of geometry of the structure as well as variations in the dielectric constant from cell to cell. However, the results of numerical studies show that the contribution of these two types of disorder in the first approximation in disorder intensities σ_w and σ_ε can be considered as independent. Thus, a direct product of two transmission spectra for the two types of disorder calculated separately appears to be very close to the spectrum calculated for the structure in which both w-disorder and ε-disorder coexist (Figure 2.1.2d).

2.1.3 Fano Resonance between Bragg Scattering and Disorder-Induced Fabry–Pérot Scattering in 1D PhCs

In this section, we consider the main features of the transmission spectra for the case of ε-disorder. We will highlight the spectral features due to three distinct mechanisms: (i) Fabry–Pérot interference in every dielectric layer constituting a 1D PhC; (ii) Bragg diffraction in periodically arranged AB cells of 1D PhCs; and (iii) Fano-type resonance that originates from interference between narrow Bragg bands and a disorder-induced Fabry–Pérot continuum (i.e., background).

First, we focus on spectral features of an ε-disordered 1D PhC that are caused by Fabry–Pérot and Bragg diffraction. We conduct comparative analysis of the transmission spectra of *a set of single dielectric slabs* with variations of permittivity and *a set of 1D PhCs* with ε-disorder.

An averaged transmission spectrum of an ensemble of single dielectric slabs B having random permittivity ε_B^i normally distributed with a semiwidth of $\sigma_\varepsilon = 10$ is shown in Figure 2.1.3b. Each slab B is surrounded by material of permittivity ε_A. Thickness of slabs constituting the statistical ensemble B is taken to be constant. The averaged spectrum represents a sine-like function with a period corresponding to that of Fabry–Pérot resonance $T_{FP} = a/\left(2w_B\sqrt{\bar{\varepsilon}_B}\right)$ for a slab of thickness w_B with the dielectric constant $\bar{\varepsilon}_B$. Note that this sine-like function has an additional superimposed modulation of amplitude having a significantly larger period ~$10T_{FP}$. The transmission spectrum of ε-disordered 1D PhCs with the same disorder parameter σ_ε for the whole ensemble of slabs B is presented in Figure 2.1.3a. Comparison of spectra unambiguously proves that transmission of ε-disordered 1D PhCs is determined by disorder-induced Fabry–Pérot scattering with a period of $T_{FP} = a/\left(2w_B\sqrt{\bar{\varepsilon}_B}\right)$, which occurs as a result of uncompensated scattering by every particular slab B. In addition, Fabry–Pérot scattering calculations reveal a superimposed set of equidistant h-order Bragg bands, distributed with a period of $T_{\text{Bragg}} = \left(2\sqrt{(\varepsilon_A w_A + \bar{\varepsilon}_B w_B)/a}\right)^{-1}$. The Bragg bands broaden, getting shallow with the increase of index h, that is, decreasing in amplitude, which can be clearly observed in spectra up to $h \sim 15$ (this corresponds to dimensionless frequency $a/\lambda \sim 6$; see Figure 2.1.3).

An instructive example of the ε-disorder-induced transformation of the 1D PhC transmission spectra is shown in Figure 2.1.4 and is considered in more detail below. An ordered

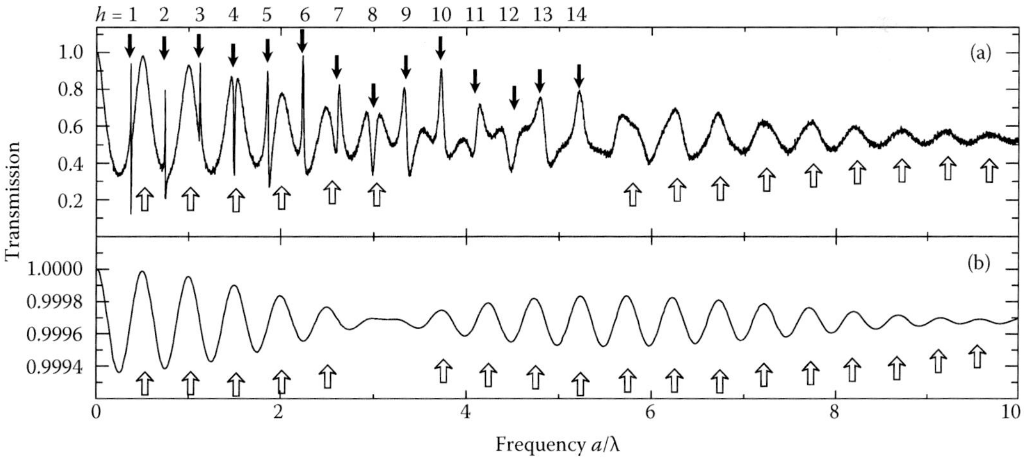

FIGURE 2.1.3
(a) Averaged transmission spectrum of one-dimensional photonic crystals (ABAB . . .) with ε-disorder ($\sigma_\varepsilon = 10$, $N = 1000$). (b) Averaged transmission spectrum of an ensemble of *single* dielectric slabs with varying permittivity ($\sigma_\varepsilon = 10$). Fringes corresponding to Fabry–Pérot resonance are shown by white arrows from bottom and those corresponding to h-order Bragg bands are shown by black arrows from top. $\varepsilon_A = 1.800$, $\bar{\varepsilon}_B = 1.793$, $w_B = 3/4a$, $M = 10^3$.

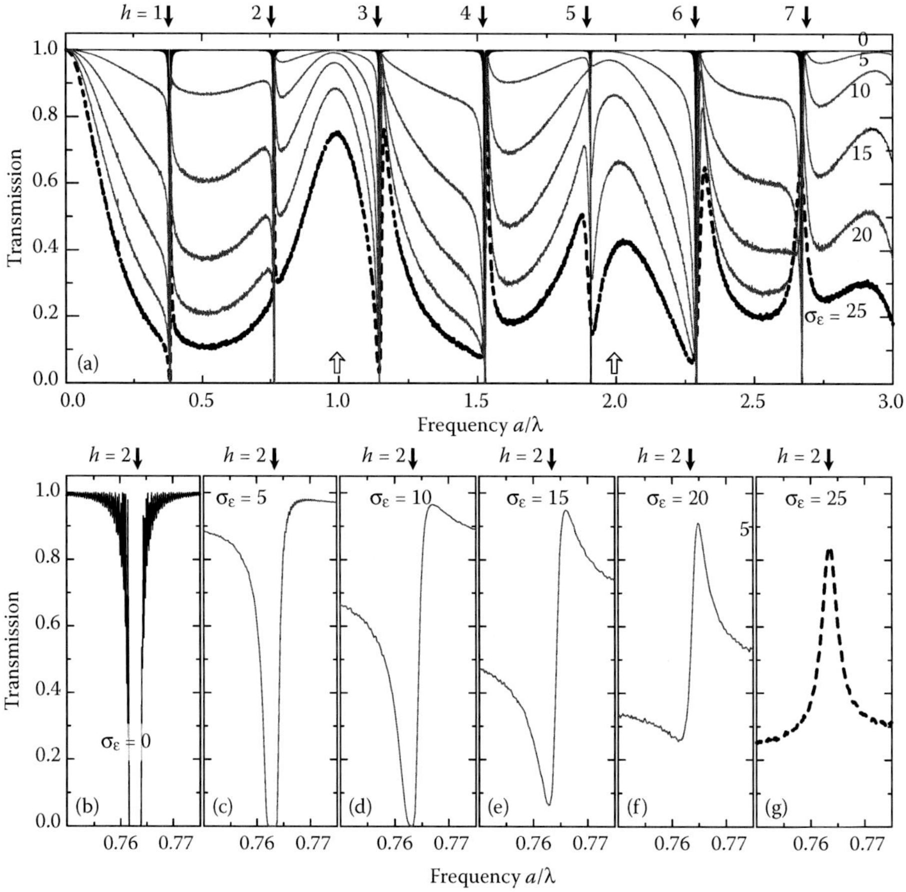

FIGURE 2.1.4
(**See color insert.**) (a) Transmission spectra of one-dimensional (1D) photonic crystal (PhCs) with ε-disorder as a function of degree of disorder σ_ε. Fringes corresponding to Fabry–Pérot resonance are shown by white arrows from bottom and those corresponding to *h*-order Bragg bands are shown by black arrows from top. (b)–(g) Transmission spectra of 1D PhCs given in a larger scale in the proximity of second-order Bragg bands for different values of disorder σ_ε. $\varepsilon_A = 1.70$, $\varepsilon_B = 1.75$, $w_B = 0.38a$, $N = 1000$, and $M = 5000$.

1D PhC (ε_A, ε_B, w_A, w_B are constants, $\sigma_\varepsilon = 10$) is optically transparent ($T = 100\%$) in the whole spectral range except for narrow *h*-order Bragg stop-bands with zero transparency ($T = 0$). Although a transmission spectrum is affected by interfaces between the PhC and the surrounding material that manifests as fringes, we will neglect that issue. Spectral areas outside the Bragg bands will be referred to as "background." Disorder destroys perfect transmittance within the background regions, transforming it into a relatively slow-changing Fabry–Pérot spectrum. Increase in σ_ε results in a gradual increase in modulation of a spectrum with a period of Fabry–Pérot resonance T_{FP}. However, all Bragg bands show an even more unexpected evolution as the ε-disorder varies from zero to rather large values, as illustrated in Figure 2.1.4b–g. First, the symmetric contour of Bragg stop-bands of perfect 1D PhCs ($\sigma_\varepsilon = 0$) becomes asymmetric when weak disorder is introduced ($\sigma_\varepsilon < 10$). Next, further growth of the degree of disorder ($\sigma_\varepsilon = 10 - 20$) causes an increasingly asymmetric shape of the spectrum. The spectrum, which is marked with a thick dashed line in Figure 2.1.4 ($\sigma_\varepsilon = 25$), illustrates the most impressive effect: reversal of the band and transformation of the *Bragg stop-band* into a *Bragg pass-band*. Bragg pass-bands have different shapes and intensity in the transmission spectrum depending on the index h. Thus, the spectrum of ε-disordered 1D PhCs can be considered in some sense as inverted with respect to the spectrum of perfectly ordered 1D PhCs: ordered 1D PhCs are completely transparent in the whole spectral range except for Bragg bands, whereas ε-disordered 1D PhCs, in contrast, are more or less transparent in the proximity of some Bragg bands.

Let us now consider transmission of ε-disordered 1D PhCs depending on structural parameters. As an example, transmission spectra were calculated depending on the thickness of B layers w_B with fixed parameters a and $\sigma_\varepsilon = 10$ (see Figure 2.1.5). In this case, both the period of the Fabry–Pérot background T_{FP} and the shape of the Bragg bands are affected. If we let $w_B \rightarrow a$, we will obtain approximately equal periodicity of Bragg and Fabry–Pérot resonances $T_{FP} \approx T_{\text{Bragg}}$ (Figure 2.1.5d). In this case, all Bragg bands positioned in the vicinity of maxima in intensity of Fabry–Pérot background become symmetrically shaped stop-bands.

What can we expect if a spectrally narrow Bragg band interacts with a broad spectrum originating from disorder-induced Fabry–Pérot scattering? If a narrow band of virtually any origin interacts with a continuum spectrum (featureless background) through an interference effect constructively or destructively, we can expect an interaction of the Fano-type.[10] Fano resonance has been observed across many different branches of physics, including studies of magnetization[13] and electronic polarization[14] phenomena, semiconductor[15] and superconductor[16] optics, to name a few. Fano resonance

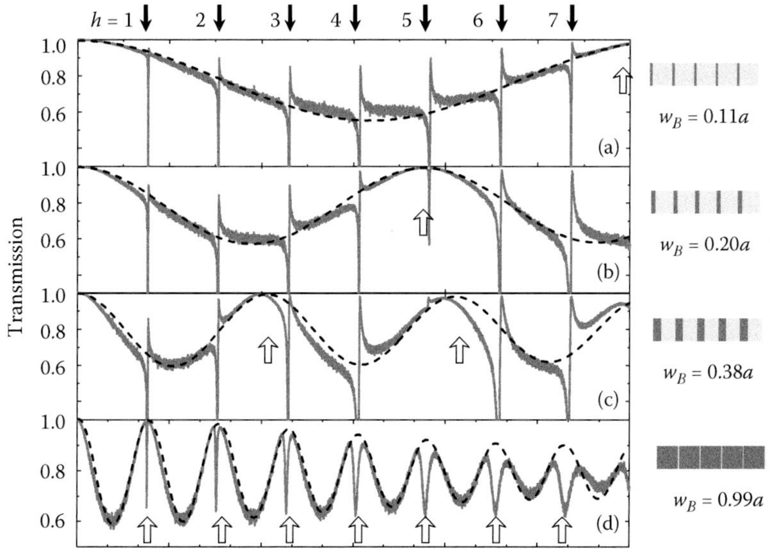

FIGURE 2.1.5
Transmission spectra of ε-disordered one-dimensional photonic crystals ($\sigma_\varepsilon = 10$) with variation of thickness of B layers w_B: (a) 0.11a, (b) 0.20a, (c) 0.38a, (d) 0.99a—gray curves. Averaged transmission spectrum of *an ensemble of single dielectric slabs* with varying permittivity ($\sigma_\varepsilon = 10$)—black dashed curves. Fringes corresponding to Fabry–Pérot resonance are shown by white arrows from the bottom and those corresponding to h-order Bragg bands are shown by black arrows from the top. $\varepsilon_A = 1.70$, $\varepsilon_B = 1.75$, $N = 1000$. Averaging was done over $M = 500$ realizations.

arises when a scattered projectile exhibiting a wave-like nature reaches the same final state via two different paths. The first scattering path corresponds to the formation of a narrow band, where the wave phase changes by ~π. The second scattering path corresponds to a broad background, where the wave phase and amplitude are nearly constant in the spectrum range of interest. To describe the resonance, which manifests as an asymmetric profile of narrow band in the observed spectrum, Ugo Fano derived a simple expression:

$$F(\Omega) = \frac{(\Omega + q)^2}{\Omega^2 + 1}, \tag{2.1.1}$$

where q is the Fano asymmetry parameter, $\Omega = (\omega - \omega_B)/(\gamma_B/2)$ is the dimensionless frequency, ω_B is a central frequency, and γ_B is the width of the narrow band. Equation 2.1.1 is often shown normalized by a coefficient $1/1 + q^2$:

$$F'(\Omega) = \frac{1}{1+q^2}\frac{(\Omega + q)^2}{\Omega^2 + 1}. \tag{2.1.2}$$

Equations 2.1.1 and 2.1.2 demonstrate that depending on the sign and value of q, for the general case where $-\infty < q < 0$ or $0 < q < +\infty$ a narrow band exhibits an asymmetric profile. Only for special values of q is the band symmetrical: $F'(\Omega) = \Omega^2/\Omega^2 + 1$ for $q = 0$ and $F'(\Omega) = 1/\Omega^2 + 1$ for $q \to \pm\infty$. The four characteristic shapes of the corresponding spectra are shown in Figure 2.1.6b.

Fano resonance has been used to describe optical properties of various nanometer-scale objects, including PhCs. For example, experimental[17] and theoretical[18] studies of PhC slab structures showed that the interference between the guided modes (narrow band) and

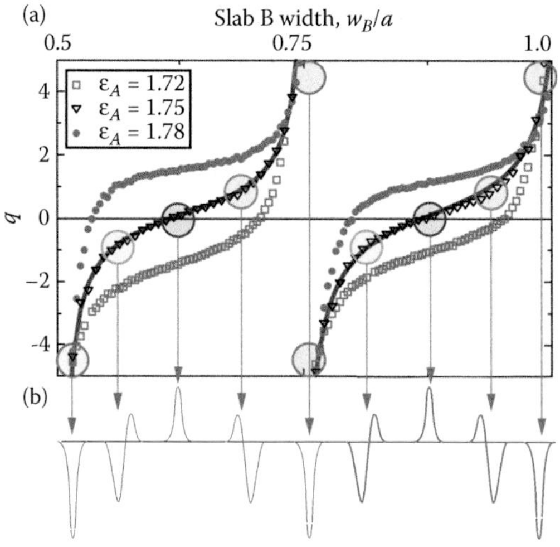

FIGURE 2.1.6
(a) Fano parameter q for $h = 4$ Bragg band in one-dimensional photonic crystals as a function of the B layer thickness (w_B/a). Calculation was done for parameters $\varepsilon_A = 1.72$, 1.75, and 1.78, $\sigma_\varepsilon = 10$, $\varepsilon_B = 1.75$, and $N = 1000$. Solid lines—$q \propto \cot(w_B/a)$. (b) Schematic view of the shape of spectral line of $h = 4$ Bragg band, corresponding to different thicknesses of layer B. The thin line shows the background level.

Fabry–Pérot scattering on the slab's boundaries results in a Fano resonance. In other studies, a microcavity mode[19] or luminescence line[20] were shown to serve as a narrow band in Fano resonance. Additionally, Fano resonance in PhCs was considered in a framework of optical bistability.[21] See also review articles by Miroshnichenko et al.[22] and Luk'yanchuk et al.[23] as well as Chapters 3.5 and 4.6.

Recently, Fano-type resonance was discovered in 3D PhCs based on synthetic opals between disorder-induced continuum (background) Mie scattering and a narrow Bragg band (see Chapter 3.5 for a detailed description). Recent works[11] suggest that a similar Fano-type resonance can be observed in 1D PhCs between narrow Bragg bands and a disorder-induced Fabry–Pérot background. The picture of light scattering in ε-disordered 1D PhCs, shown in Figures 2.1.4 and 2.1.5, possesses all specific features of Fano resonances: (i) the asymmetry of the Bragg band (the slope of either the long-wavelength wing or the short-wavelength wing of the Bragg band is much sharper) and, more importantly, (ii) the sequence of both transmission Bragg stop-bands and transmission Bragg pass-bands.

To verify the existence of Fano-type interference between narrow Bragg bands and the Fabry–Pérot background, the calculated transmission spectra (from Figure 2.1.5) were approximated with the use of Fano formula 2.1.1. We obtained an excellent fitting of the calculated spectra, and this confirms the presence of Fano resonance. As a result of this calculation, we present the dependence of Fano parameter q as a function of thickness w_B for various values of dielectric contrast $\eta = 100a(\varepsilon_A - \bar{\varepsilon}_B)/(\varepsilon_A w_A + \bar{\varepsilon}_B w_B)$ shown in Figure 2.1.6. With a variation of w_B, the shape of the Bragg band is periodically transformed in the following way: symmetric stop-band at $q \to -\infty \Rightarrow$ asymmetric line (the long-wavelength wing of the band is flat relative to a steep short-wavelength wing) at $q < 0 \Rightarrow$ symmetric pass-band at $q = 0 \Rightarrow$ asymmetric line (the short-wavelength wing of the band is flat relative to a steep long-wavelength wing) at $q > 0 \Rightarrow$ symmetric stop-band at $q \to +\infty$, and so on. Note that in the classical problem[10] the Fano parameter was expressed in terms of phase shifts as $q = \cot \delta$, where δ is the phase shift of the background continuum.[24] In this respect, the Fano parameter can be quite accurately fitted by function $q \propto \cot(w_B/a)$. At the same time, change in $q(w_B/a)$ in fact corresponds to a "scanning" by Bragg band over a sine-like Fabry–Pérot contour (Figure 2.1.5), that is, "scanning" by Bragg band along the background continuum with a varying phase shift.

2.1.4 Disorder and Inverted Photonic Band Structure

In this section, we attempt to determine whether the dielectric contrast between components constituting disordered PhCs is needed, or whether it is sufficient to have periodicity to build a material with photonic band structure. To answer this question we consider a novel class of optical materials that are referred to here as structures with ordered disorder (ODSs) and are composed of periodical alternation of components with constant and fluctuating permittivity, as illustrated in Figure 2.1.7. We consider the simplest case of a 1D ODS, which is formed from an ordered in-space sequence of two (A and B) different dielectric slabs. We define a perfectly matched 1D ODS as a structure having equal average dielectric constants of both components $\bar{\varepsilon}_A = \bar{\varepsilon}_B$, that is, the structure in which macroscopically averaged dielectric contrast vanishes. A key parameter defining optical properties of the ODS is variation of the permittivity of A and B layers (σ_ε^A and σ_ε^B). In fact,

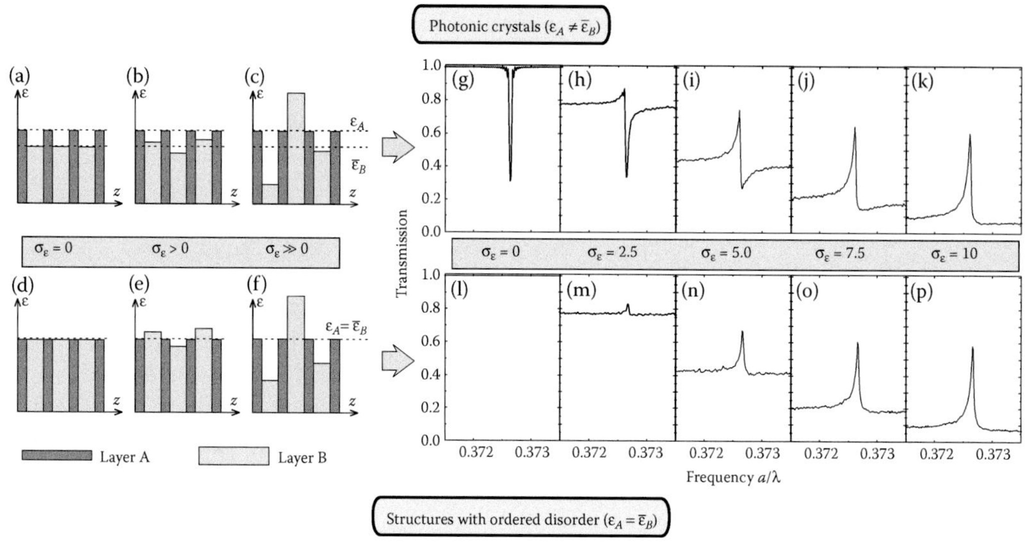

FIGURE 2.1.7
(**See color insert.**) (Left) Schematic of permittivity profile for disordered one-dimensional (1D) photonic crystals (PhCs) (a)–(c) and for 1D structures with ordered disorder (ODSs) without dielectric contrast between layers A and B permittivity in average (d)–(f) when disorder degree is increased. (Right) Transmission spectra of PhCs (g)–(k) and ODSs (l)–(p) with fixed number of unit cells $N = 10^4$ as a function of the degree of disorder $\sigma_\varepsilon = 0.0$, 2.5, 5.0, 7.5, and 10. $\varepsilon_A = 1.8$, $\bar{\varepsilon}_B = 1.8005$ (for ODSs) or $\bar{\varepsilon}_B = 1.8005$ (for PhCs), $w_B = 1/3a$, $N = 10^4$, $M = 3000$.

it will play a role as significant as that played by dielectric contrast η in the case of perfect PhCs. We demonstrate that transport of light in the ODS exhibits an unexpected behavior. Here, we present the results of numerical calculations that demonstrate the formation of the band structure for a 1D ODS, its main features, and a relation between two types of structures—the PhC and the ODS.

To investigate the transmission spectra and the photonic band structure of the ODS, we consider structures consisting of the sequence of A and B slabs (see Figure 2.1.7). As in Section 2.1.3, all the A slabs have constant permittivity ε_A and thickness w_A. The B slabs have constant thickness w_B and are allowed to possess some degree of disorder in permittivity (ε_B^i for ith layer). The variations in ε_B are described by the normal distribution with average permittivity $\bar{\varepsilon}_B$ and the second moment of the distribution (dispersion) σ_ε^2. For the ODS, we assume $\varepsilon_A = \bar{\varepsilon}_B$. Transmission coefficients of the 1D ODS for the case of light propagation along the structure axis were evaluated using a transfer matrix technique. For comparison, we also consider 1D PhCs with structural parameters ε_A, w_A, ε_B, w_B, and N with dielectric contrast $\eta \neq 0$. All calculated spectra were averaged for a sufficient number of realizations M of ODS or PhC.

The transformation of the Bragg bands in transmission spectra of ODS and *low-contrast* PhC with introduction of disorder in the permittivity σ_ε is shown in Figure 2.1.7 and is considered in more detail below. To perform this analysis we choose a narrow spectral area in the vicinity of the first Bragg band, $h = 1$. An ordered ODS ($\sigma_\varepsilon = 0$) demonstrates an optically homogeneous medium being optically transparent ($T = 100\%$) in the whole spectral range (Figure 2.1.7l), whereas an ordered PhC ($\sigma_\varepsilon = 0$) is optically transparent except for narrow Bragg stop-bands (Figure 2.1.7g). Introducing disorder destroys constant 100%—background in both cases (Figure 2.1.7h and m). For ODSs, Bragg pass-bands appear with increase in σ_ε. In the case of PhCs, the symmetric contour of a Bragg stop-band of a PhC at $\sigma_\varepsilon = 0$ becomes asymmetric when weak disorder is introduced ($\sigma_\varepsilon < 5$), as expected in the case of Fano resonance. Further increase in the degree of disorder ($\sigma_\varepsilon > 5$) transforms the stop-band into a pass-band. The transformation of the stop-band into the pass-band starts when fluctuations in ε_B become comparable with the dielectric contrast η. The complete transformation of the transmission spectra of PhCs into the transmission spectra of ODSs (Figure 2.1.7k and p) will be defined by the relation $\sigma_\varepsilon >> \eta$ when fluctuation in ε_B becomes the key parameter defining the dielectric properties of a PhC.

Thus, we start from an optically homogeneous medium (Figure 2.1.7d) that is fully transparent in the whole spectral range (Figure 2.1.7g). We move in two directions: first, introducing spatially periodic disorder (Figure 2.1.7b) and, second, introducing spatially periodic dielectric contrast (Figure 2.1.7d). Concluding, we arrive at the same result (Figure 2.1.7k and p). This result represents an inverted photonic structure, which is characterized by the transmission pass-bands surrounded by the opaque regions where transmission is vanishingly small and almost all incident light is reflected.

2.1.5 Extra-Narrow Bragg Bands Created by Disorder

Let us now consider the formation of the stop-band in the transmission spectrum of a 1D PhC and of the pass-band in the transmission spectrum of a 1D ODS with increases in the number of unit cells N (see Figure 2.1.8). In the case of the perfect 1D PhC, the photonic band structure consists of a set of narrow equidistant h-order bands with Bragg wavelengths. Figure 2.1.8a shows formation of a Bragg stop-band, namely, the narrowing of the band and the increase in dip density with increases in the number of unit cells N. In addition, Fabry–Pérot oscillations became unresolvable with increase in N. Such a pattern was demonstrated experimentally and theoretically in a variety of papers. The new result is presented in Figure 2.1.8b, which in fact in some features is similar to an inverted image of Figure 2.1.8a. The transmissivity of a 1D ODS in the spectral region of the Bragg wavelength initially decreases with increasing numbers of cells, and further ($N > 10^4$) becomes practically *unchangeable*. At the same time, outside the Bragg band the background decreases strongly to form a *pronounced Bragg pass-band*.

We treated the spectra calculated as a function of the main parameters of the structures, namely on degree of disorder $\sigma_\varepsilon = 10$ (Figure 2.1.7) and number of unit cells N (Figure 2.1.8) to demonstrate the main characteristics of the Bragg band in PhCs and ODSs. The results are presented in Figure 2.1.8c and d for most characteristic cases. The intensity of the Bragg pass-band in the transmission spectra of the ODS initially ($0 < N < 10^4$) increases and further tends to saturate (Figure 2.1.8c). With increase in N the linewidth of the pass-band strongly decreases and becomes much less than the width of the stop-band that becomes *finite and unchangeable* at $N > 10^4$ (Figure 2.1.8d).

We see that the "ordered" disorder gives rise to formation of ultra-narrow bands at Bragg frequency. In the limit $N \to \infty$ these bands become very narrow, whereas their intensity remains the same. It is worth noting that this picture resembles the resonant tunneling of an electron through a barrier with a quantum well. As shown in the studies by Economou and Soukoulis[4] and Bliokh et al.,[5] there is an analogy between an electron in a well and a localized state in a random system. Localized states arise in effective wells distributed in space and in frequency domains with some probability (Figure 2.1.1). Bragg bands are seen to affect such probability so

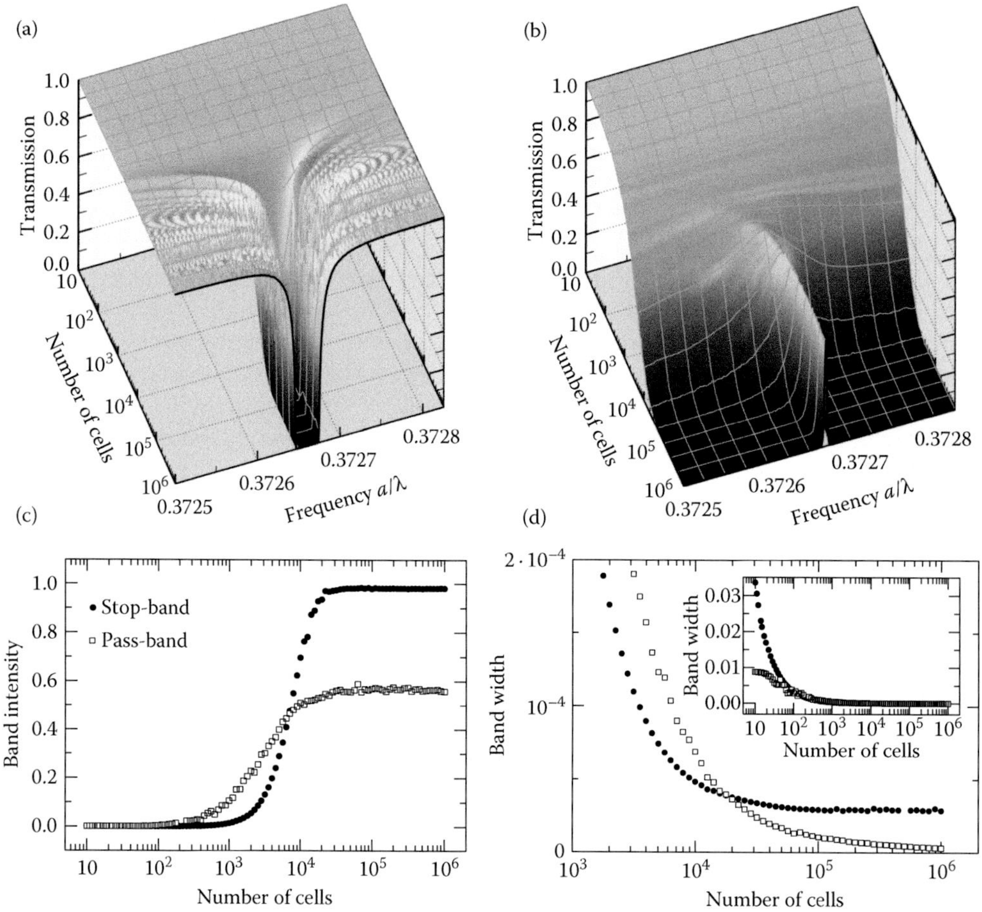

FIGURE 2.1.8
(**See color insert.**) Stop-band in transmission spectra of perfect photonic crystals (PhCs) (a) and pass-band in transmission spectra of structures with ordered disorder (ODSs) (b) as a function of the number of unit cells N. Intensity (c) and width (d) of Bragg pass-band and Bragg stop-band (ordered PhCs) obtained from transmission spectra shown at (a) and (b). $\varepsilon_A = 1.8$, $\bar{\varepsilon}_B = 1.8000$ (for ODS) or $\bar{\varepsilon}_B = 1.8005$ (for PhCs), $w_B = 1/3a$, $\sigma_\varepsilon = 10$, $M = 3000–5000$ (for ODSs).

that, on pass-band frequencies, localized states distribute very densely. As a result, a photon can freely travel from one localized state to another through a disordered sample.

2.1.6 Conclusions

The influence of disorder on optical properties of 1D PhCs was considered in a number of works, in particular in the context of broadening and degradation of stop-bands in transmission spectra. In this chapter, we demonstrated that several different scenarios are possible, including Fano-type resonance between disorder-induced Fabry–Pérot background and narrow Bragg bands and inversion of the photonic band structure. It is well known that the condition for the appearance of Fano resonance is rather simple: there should be two competing mechanisms of light scattering with dissimilar spectral characteristics (broad and narrow, respectively) that give rise to interference between the broad continuum and the narrow band. As we show, the same situation is realized in PhCs, where Fano-type resonance between narrow Bragg bands and disorder-induced slow-changing continuum was found. Thus, effects presented in this chapter demonstrate the unique transmutability of the Bragg band, including transformation of Bragg stop-bands into Bragg pass-bands and vice versa. Such effects should be observed in virtually any type of PhC of any dimensionality.

In John's pioneering work[1] on light scattering in disordered PhCs, it was predicted that localization of photons may occur in the vicinity of band edges. Since then, such interplay between Bragg diffraction and disorder-induced scattering was considered as a way to localize light and as a mechanism capable of suppressing transmittance and increasing back-scattering. We demonstrate that it is not always so and, at some particular conditions, Fano interference between scattered background radiation and Bloch waves gives rise to anomalous increase in the optical transmittance near photonic band edges or when the Bragg condition is satisfied.

Acknowledgments

We thank A. K. Samusev for his help in numerical calculation, Yu. S. Kivshar, A. E. Miroshnichenko, and A. N. Poddubny for fruitful discussions. This work was supported by the Russian Foundation for Basic Research (RFBR), Russia, Grant No. 10-02-01094.

References

1. S. John, Phys. Rev. Lett. **58**, 2486 (1987).
2. E. Yablonovitch, Phys. Rev. Lett. **58**, 2059 (1987).
3. I. M. Lifshits, S. A. Gredeskul, and L. A. Pastur, *Introduction to the Theory of Disordered Systems* (Wiley-Interscience, New York, 1988); P. Sheng, *Introduction to Wave Scattering, Localization and Mesoscopic Phenomena* (Springer, Berlin, 2010), 2nd ed.
4. E. N. Economou and C. M. Soukoulis, Phys. Rev. B **28**, 1093 (1983).
5. K. Y. Bliokh, Y. P. Bliokh, and V. D. Freilikher, J. Opt. Soc. Am. B **21**, 113 (2004).
6. S. F. Liew and H. Cao, J. Opt. **12**, 024011 (2010); M. M. Sigalas, C. M. Soukoulis, C.-T. Chan, and D. Turner, Phys. Rev. B **53**, 8340 (1996); M. M. Sigalas, C. M. Soukoulis, C.-T. Chan, R. Biswas, and K. M. Ho, *ibid.* **59**, 12767 (1999); R. Biswas, M. M. Sigalas, G. Subramania, C. M. Soukoulis, and K.-M. Ho, *ibid.* **61**, 4549 (2000); Z.-Y. Li and Z.-Q. Zhang, *ibid.* **62**, 1516 (2000); A. F. Koenderink, A. Lagendijk, and W. L. Vos, *ibid.* **72**, 153102 (2005); M. A. Kaliteevski, D. M. Beggs, S. Brand, R. A. Abram, and V. V. Nikolaev, *ibid.* **73**, 033106 (2006); E. Palacios-Lidón, B. H. Juárez, E. Castillo-Martínez, and C. López, J. Appl. Phys. **97**, 63502 (2005); M. A. Kaliteevski, D. M. Beggs, S. Brand, R. A. Abram, J. R. Fletcher, G. P. Swift, and J. M. Chamberlain, J. Mod. Opt. **53**, 2089 (2006); R. Rengarajan, D. Mittleman, C. Rich, and V. Colvin, Phys. Rev. E **71**, 16615 (2005); M. Allard and E. H. Sargent, Appl. Phys. Lett. **85**, 5887 (2004).
7. A. R. McGurn, K. T. Christensen, F. M. Mueller, and A. A. Maradudin, Phys. Rev. B **47**, 13120 (1993).
8. L. I. Deych, D. Zaslavsky, and A. A. Lisyansky, Phys. Rev. Lett. **81**, 5390 (1998).

9. A. A. Asatryan, L. C. Botten, M. A. Byrne, V. D. Freilikher, S. A. Gredeskul, I. V. Shadrivov, R. C. McPhedran, and Y. S. Kivshar, Phys. Rev. Lett. **99**, 193902 (2007).
10. U. Fano, Phys. Rev. **124**, 1866 (1961).
11. M. V. Rybin, A. B. Khanikaev, M. Inoue, K. B. Samusev, M. J. Steel, G. Yushin, and M. F. Limonov, Phys. Rev. Lett. **103**, 023901 (2009); Photonics Nanostruct. Fundam. Appl. **8**, 86 (2010).
12. M. Born and E. Wolf, *Principles of Optics* (Cambridge University Press, Cambridge, 2006), 7th ed.; P. Yeh, A. Yariv, and C.-S. Hong, J. Opt. Soc. Am. **67**, 423 (1977).
13. V. Madhavan, W. Chen, T. Jamneala, M. F. Crommie, and N. S. Wingreen, Science **280**, 567 (1998).
14. N. M. Kabachnik and I. P. Sazhina, J. Phys. B **9**, 1681 (1976).
15. J. J. Hopfield, P. J. Dean, and D. G. Thomas, Phys. Rev. **158**, 748 (1967); F. Cerdeira, T. A. Fjeldly, and M. Cardona, *ibid.* **8**, 4734 (1973).
16. B. Friedl, C. Thomsen, and M. Cardona, Phys. Rev. Lett. **65**, 915 (1990); M. F. Limonov, A. I. Rykov, S. Tajima, and A. Yamanaka, *ibid.* **80**, 825 (1998).
17. G. Levy-Yurista and A. A. Friesem, Appl. Phys. Lett. **77**, 1596 (2000); C. Grillet, D. Freeman, B. Luther-Davies, S. Madden, R. McPhedran, D. J. Moss, M. J. Steel, and B. J. Eggleton, Opt. Express **14**, 369 (2006).
18. S. Fan and J. D. Joannopoulos, Phys. Rev. B **65**, 235112 (2002); J. Song, R. P. Zaccaria, M. B. Yu, and X. W. Sun, Opt. Express **14**, 8812 (2007).
19. S. Fan, Appl. Phys. Lett. **80**, 908 (2002); M. Galli, S. L. Portalupi, M. Belotti, L. C. Andreani, L. O'Faolain, and T. F. Krauss, *ibid.* **94**, 071101 (2009); M. Notomi, E. Kuramochi, and T. Tanabe, Nat. Photonics **2**, 741 (2008).
20. T. Baba, H. Makino, T. Mori, T. Hanada, T. Yao, and H.-Y. Lee, *ibid.* **87**, 171106 (2005); R. Harbers, S. Jochim, N. Moll, R. F. Mahrt, D. Erni, J. A. Hoffnagle, and W. D. Hinsberg, Appl. Phys. Lett. **90**, 201105 (2007); S. A. Blokhin, O. A. Usov, A. V. Nashchekin, E. M. Arakcheeva, E. M. Tanklevskaya, S. G. Konnikov, A. E. Zhukov, M. V. Maksimov, N. N. Ledentsov, and V. M. Ustinov, Semiconductors **40**, 812 (2006).
21. A. R. Cowan and J. F. Young, Phys. Rev. E **68**, 046606 (2003); S. F. Mingaleev, A. E. Miroshnichenko, Y. S. Kivshar, and K. Busch, *ibid.* **74**, 046603 (2006); V. Lousse and J. P. Vigneron, Phys. Rev. B **69**, 155106 (2004).
22. A. E. Miroshnichenko, S. Flach, and Y. S. Kivshar, Rev. Mod. Phys. **82**, 2257 (2010).
23. B. Luk'yanchuk, N. I. Zheludev, S. A. Maier, N. J. Halas, P. Nordlander, H. Giessen, and C. T. Chong, Nat. Mater. **9**, 707 (2010).
24. J.-P. Connerade and A. M. Lane, Rep. Prog. Phys. **51**, 1439 (1988).

2.2

Propagation and Localization of Light in Two-Dimensional Photonic Crystals

Mikhail A. Kaliteevski
Ioffe Physical-Technical Institute of the Russian Academy of Sciences
Academic University

Stuart Brand and Richard A. Abram
Durham University

CONTENTS

2.2.1 Introduction

Since Yablonovitch[1] pointed out that photonic band gaps (PBGs) originating from the periodic modulation of the dielectric constant in photonic crystals (PhCs) could lead to the suppression of spontaneous emission from embedded sources, there has been much research effort directed toward studies of the basic properties of PhCs and a wide range of possible applications. However, the interaction of electromagnetic waves with periodic media is not a new problem in physics. For example, the atomic structure of crystals has been studied by x-ray diffraction since the early twentieth century, and crystalline materials are essentially natural PhCs for x-rays. Even within optics, multilayered structures have been used as dielectric mirrors for decades. Furthermore, all the theory necessary to describe electromagnetic waves in periodic dielectric media has existed since the nineteenth century.

The advent of PhCs as a major field of research in the late twentieth century can be largely attributed to two main developments. One was the increased availability of methods to fabricate satisfactory periodic dielectric structures in two dimensions with features on the same scale as the wavelength of visible and near-infrared light. The other was the emergence of optics and optoelectronics as key elements in information technology, resulting in an interest in new ways to control the propagation of light and its interaction with matter.

While methods to fabricate PhCs have continued to develop, those for use at near-visible frequencies require feature sizes of the order of a fraction of a micrometer and at that scale there can still be significant challenges in producing satisfactory structures. One-dimensional (1D) PhCs (defined as PhCs, such as multilayer structures, that are periodic in one dimension and have essentially uniform properties in the other two dimensions) for optical applications can be produced with very high fidelity, and atomic monolayer accuracy in the case of molecular beam epitaxy. However, two-dimensional (2D) and three-dimensional (3D) PhCs are more demanding to fabricate and the imperfections are more significant with the present level of technology. For example, synthetic opals are made from silica spheres, which are grown from a supersaturated solution and fluctuate in size and shape.[2] Also synthetic opals and other self-assembled structures can contain localized defects such as vacancies and stacking faults.[3] 2D PhCs for the optical band can be fabricated by lithography and subsequent etching or by focused ion beam etching but their detailed features can deviate from the desired shape, size, or position and such structures can also suffer from the effects of surface roughness.

The wide basic and applied interest in PhCs stems from the significant effect they can have on the propagation of light. In particular, the phase and group velocities of light within a PhC are dependent on frequency, and, most strikingly, there are ranges of frequency in which propagation does not occur—the PBGs. Classical electrodynamics shows that the electromagnetic field has the form of a Bloch wave propagating without attenuation for modes within the allowed bands and decaying exponentially within the PBGs.

A particularly valuable property of a PBG is the way in which it can be used to provide full, 3D localization of photons in a limited region of space, which is generally rather more difficult than for electrons. For electrons, the potential $V(\mathbf{r})$ in the Schrödinger equation for the wave function,

$$-\frac{\hbar^2}{2m}\nabla^2\psi + V(\mathbf{r})\psi = U\psi, \tag{2.2.1}$$

can in principle have any positive or negative values and it is therefore always possible to design simple model systems based on potential wells and barriers to confine an electron with any energy U in a chosen region of space. However, the localization of photons is not so straightforward if we restrict ourselves to conventional dielectric materials, which have real, positive values of the spatially varying dielectric constant $\varepsilon(\mathbf{r})$, because, in the associated equation

$$-\nabla^2\mathbf{E} + \nabla(\nabla\cdot\mathbf{E}) + [1-\varepsilon(\mathbf{r})]\frac{\omega^2}{c^2}\mathbf{E} = \frac{\omega^2}{c^2}\mathbf{E}, \tag{2.2.2}$$

the quantity ω^2/c^2, which is analogous to the electron energy U in the Schrödinger equation, is always greater than the analogue of the potential $[1-\varepsilon(\mathbf{r})]\omega^2/c^2$, and that is the opposite of what is required for confinement.[4] However, with a PBG it is possible to design a PhC and a cavity embedded within it that supports a mode at the required frequency within the band gap, having the form of a standing wave in the cavity and a decaying wave outside.

The use of a PBG for 3D light confinement described above can also be used to create waveguides in PhCs. Although conventional waveguides based on the total internal reflection of light at dielectric interfaces are well established in integrated optics and are entirely appropriate for many applications, the PhC approach can provide better performance, such as the retention of confinement even when there are sharp bends and other demanding features in a waveguide. However, the performance of structures that are dependent on PBG properties is critically dependent on the quality of the PBG, which in turn depends on the quality of the PhC structure.[5] As pointed out earlier, PhC structures are potentially susceptible to various kinds of disorder, and therefore it is important to have a thorough physical understanding of the possible effects of disorder on the optical states they can support and the optical properties they exhibit.

When considering the effect of disorder on the PBG and the associated optical properties of PhCs, there is a natural inclination to turn to concepts that have been developed for the apparently analogous electronic problem of the semiconductor band gap. Study of the properties of electrons in disordered semiconductors is a mature field having been the subject of sustained basic and applied research for over half a century and many useful physical pictures and associated theoretical methods have been developed that provide a good understanding of the properties of the materials and associated devices.[6] Many of the ideas in the electronic case do have relevance to the description of disorder effects in PhCs, and there are certainly clear similarities in the way that disorder creates Anderson-localized states[7] in the band gap, and as a result the density of states function exhibits disorder-induced band tails that encroach on the band gap of the perfect structure, and eventually fill the gap altogether. However, important differences emerge when the effects of disorder on system properties are measured.

As well as being scattered by static disorder (due, for example, to lattice defects, impurities, and alloy substitution), electrons in semiconductors interact with each other and with phonons. As a result, quantum mechanical coherence effects are only significant on relatively short length and time scales, and in electrical measurements are also obscured by transport through the contacts to the sample. In contrast, photons in the linear regime do not interact with each other, their interaction with phonons is weak, and they can be observed directly as they emerge from a sample. Therefore, in a typical nonabsorbing, macroscopic structure, the coherence of the photonic states is conserved. As a consequence, there is a stark contrast in what is observed in the two systems: if electrons are localized, low field transport does not occur and the electrical conductivity vanishes, but for photons the transmission coefficient of a sample at the frequency of a localized state can be as high as unity.

To illustrate how localized photonic states can actually enhance the transmission coefficient in certain ranges of frequency, consider a symmetric microcavity[8] formed by two quarter-wave Bragg reflectors (1D PhCs) with a half-wavelength layer between them, which can be considered to be a defect in a larger 1D PhC. If a photonic state is localized at the defect, it can be considered as analogous to a mode of a Fabry–Pérot cavity, and it follows from the well-known theory of that device that the transmission coefficient will be unity at its eigenfrequency. Note that the transmission is much greater than would occur in the absence of the defect, when it would be determined by the exponential attenuation of a wave at a frequency in the PBG of the Bragg reflector. However, in structures that are only weakly absorbing, the photonic state will be characterized by a finite coherence time and the increase of transmission will not occur. It is also the case that outside the PBGs, propagating Bloch waves in the PhC can be scattered by disorder resulting in a reduction of the transmission coefficient. If the disorder is sufficiently strong the propagation of light can be considered as the diffusion of photons, and the transmitted intensity T decreases inversely with sample thickness. Finally, it should be made clear that while previously we contrasted the transport of electrons in bulk disordered semiconductors with that of light in disordered PhCs, so-called resonant tunneling of electrons can be observed in suitably designed semiconductor heterostructures, where wave-function coherence across the structure can be maintained.

As localized modes in the band gap of PhCs can have such a marked effect on their transmission (and reflection) properties, it is important to have a thorough understanding of what modes can be induced by disorder and how they influence the spectral properties of a given system. In particular, we need to know the frequency distribution of modes in a typical sample of PhC and the spatial dependence of their field profiles, and how features in the transmission and reflection spectra can be related to mode properties. The latter point is particularly important for the interpretation of experiment. This article has been written with the intention of providing the reader with an appreciation of how disorder affects the mode structure and optical properties of disordered 2D PhCs at frequencies within and close to the first band gap. However, it is convenient to first look briefly at 1D PhCs as they do have some properties in common with 2D PhCs but also avoid some of the theoretical complications that come with greater dimensionality, and in doing so allow a rigorous analytic solution for some simple models of disorder.

Consider a 1D PhC composed of alternate layers of two types (labeled A and B), each of thickness $D/2$ and with refractive indices given by $n_{A,B} = n_0(1 \pm \eta)$, where n_0 is the mean refractive index of the system and η the index modulation. Basic theory shows that the PBG is centered at frequency $\omega_B = \pi c/(n_0 D)$ and has width $\Delta\omega \approx 4\eta\omega_B/\pi$. The transmission spectrum around the first PBG has a dip with a parabolic shape on a logarithmic scale. If now the parameters of the 1D PhC have random fluctuations characterized by the number δ, which is the *relative fluctuation of the optical length of the period of the structure*, then tails appear in the density of states function $\rho(\omega)$ of the bands on each side of the band gap which have a nearly Gaussian decay $\rho \sim \exp(-(\omega - \omega_0)^2/\Omega^2)$ away from the band edge ω_0 characterized by a penetration depth $\Omega = \delta\omega_B\sqrt{\eta}$. It is interesting to note that the band tailing is characterized by the numerical value of Ω irrespective of whether the disorder derives from the refractive indices of the layers, their lengths, or both. The properties of systems of finite size (typically 200-layer pairs) have been studied using the transfer matrix method, and Figure 2.2.1 shows the eigenmode frequencies and their lifetimes around the first PBG for a particular system.[9] All the modes have finite lifetimes as a result of the leakage of the light through the edges of the finite structure, but the Thouless criterion for localization[10] suggests that most of the modes within the PBG would experience strong (or Anderson) localization in an infinite

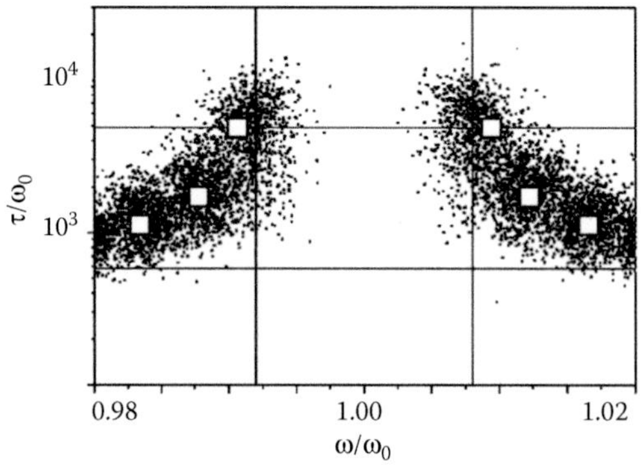

FIGURE 2.2.1
The frequencies and lifetimes of the eigenmodes of an ideal one-dimensional photonic crystal (open squares) with thickness $200D$ and refractive index modulation $\eta = 0.025$, and for 10^3 disordered structures with $\delta = 0.07$ (dots). Vertical lines mark the edges of PBG. (Adapted from D. M. Beggs et al., J. Phys. Rev. B **73**, 033106, 2006.)

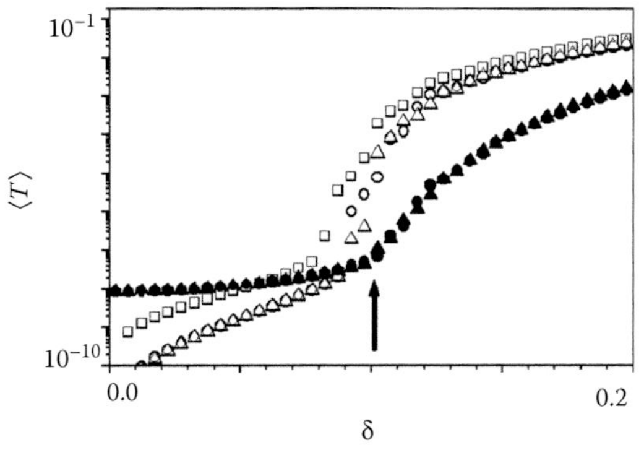

FIGURE 2.2.2
The transmission coefficient averaged over 10^6 structures (solid symbols) and its standard deviation (open symbols) as a function of disorder parameter δ for photonic crystals with $n_0 = 2$ and $\eta = 0.05$. The circles and triangles refer respectively to refractive index and layer thickness disorder with top hat distributions, and the squares refer to a Gaussian distribution. The vertical arrow indicates the "threshold level of disorder." (Adapted from M. A. Kaliteevski et al., J. Phys. Rev. E **73**, 056616, 2006.)

structure. Such localized states manifest themselves as sharp peaks in the transmission spectrum of an individual disordered structure. The transmission spectrum averaged over an ensemble of disordered structures demonstrates a smooth dip corresponding to the PBG, but the width of the dip is reduced (owing to penetration of the band tails into gap) and the transmission is increased (owing to propagation through the localized states that act like the eigenmodes of a Fabry–Pérot cavity). Transmission in this case falls exponentially with sample thickness for frequencies both inside and outside the PBG.[2]

The dependence on δ of the ensemble-averaged transmission coefficient $T(\delta)$ evaluated at the center of the PBG exhibits threshold-like behavior. When $\Omega < \Delta\omega/2$, the ensemble-averaged transmission increases only very slowly with increasing δ as shown in Figure 2.2.2, but when $\Omega \approx \Delta\omega/2$, which corresponds to $\delta = \sqrt{(\Delta\omega/\omega)/\pi}$, the band tails encroach significantly on all of the PBG and the transmission begins to increase rapidly with increasing δ. It is also significant that below the same value of δ, the ensemble standard deviation of the transmission is less than its mean value, but the opposite is the case for greater disorder. In summary, there is a threshold level of disorder defined by $\delta_{th} = \sqrt{(\Delta\omega/\omega)/\pi}$ that marks the onset of rapidly increasing transmission at the center-PBG frequency and reflects the fact that the band gap has become seriously degraded. The results discussed above were first obtained numerically, but the analytical theory developed by Greshnov et al.[11] lead to the same conclusions.

2.2.2 Mode Structure and Transmission in Disordered 2D PhCs

2.2.2.1 E-Polarized and H-Polarized Modes: Ballistic and Scattered Transmission

In the case of the 2D photonic structure shown in Figure 2.2.3, where the dielectric constant varies only in the plane perpendicular to the z-axis, the electromagnetic field can be decomposed into two independent polarizations: an H-polarization with components (H_z, E_x, E_y) and an E-polarization with components (E_z, H_x, H_y). The electromagnetic dispersion

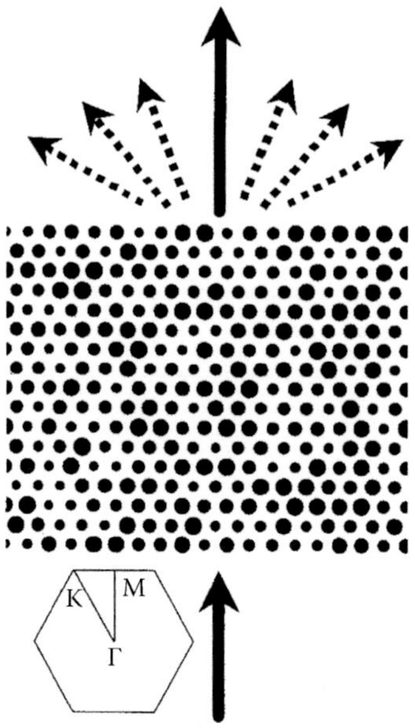

FIGURE 2.2.3
A schematic illustration of the transmission modeling undertaken. A plane wave is incident on a disordered hexagonal two-dimensional photonic crystal from the right. The light emerging from the opposite (left) side of the sample can be considered as the sum of ballistic transmission (solid arrow) and scattered transmission (dashed arrows).

relation $\omega(\mathbf{k})$ for a periodic structure is different for the two polarizations. For example, Figure 2.2.4 shows the photonic band structure for a hexagonal PhC formed by air cylinders of radius $r = 0.4a$ etched in a medium with refractive index 3.6 (which corresponds to GaAs). It can be seen that for the H-polarization a *complete* PBG exists centered at $\omega_0 \approx 2\pi \times 0.32\ d/c$ with relative width $\Delta\omega/\omega_0 \approx 0.45$, whereas for the E-polarization there is an *incomplete* PBG with a relative width that is largest at the M symmetry point in the Brillouin zone, where it has the value $\Delta\omega/\omega_0 \approx 0.14$.

The introduction of disorder into 2D PhCs leads to a wide range of phenomena, but here we will concentrate on just two aspects: the filling of the PBG with localized modes and the associated modifications in the transmission and reflection spectra. We consider the

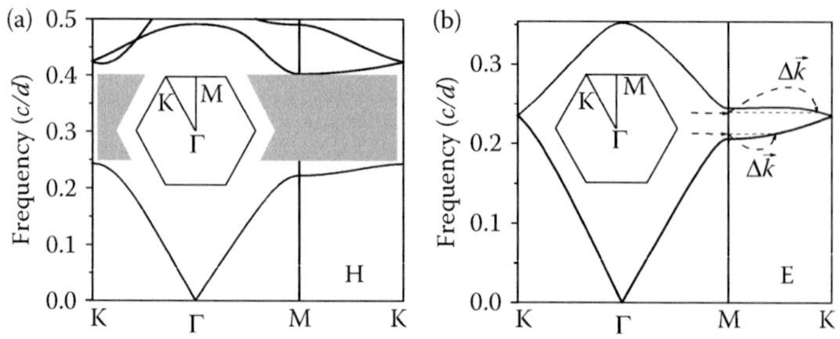

FIGURE 2.2.4
The band structure of a two-dimensional hexagonal photonic crystal formed by air cylinders etched in a material with a refractive index of 3.6 for H-polarization (a) and E-polarization (b).

effects of disorder on the specific PhC described above (which has the band structure shown in Figure 2.2.4) and by studying its optical properties for both the H- and E-polarizations we are able to compare and contrast the cases of complete and incomplete PBGs. We have carried out the calculations of the PhC band structure and of the modes in disordered structures using the plane-wave expansion method, and have modeled the transmission properties using the method described by Bell et al.[12] and using finite-difference time domain. The approach to the transmission modeling is illustrated in Figure 2.2.3. A plane wave is incident normally on one face of a slab of the disordered 2D PhC under study and the light emerging from the opposite face is calculated (giving the transmission coefficient T) and, when appropriate, analyzed into two different contributions T_b and T_s. The light propagating parallel to the direction of the incident light is considered to be the result of ballistic transmission (T_b), and that propagating in other directions is the result of scattering (and in total is referred to as the scattered transmission (T_s)).

2.2.2.2 Disordered 2D PhC with a Complete Band Gap

2D PhCs with a complete band gap are generally more useful for device applications, such as waveguides and optical sources. For a 2D PhC with a complete PBG there is a dip in the transmission spectrum for any direction of propagation (in the 2D plane) that is parabolic on a logarithmic scale. Within the PBG, the transmission coefficient falls exponentially with increasing sample thickness.

It is interesting to begin by considering the simplest type of defect in a 2D PhC, which is a vacancy formed by a "missing" cylindrical element.[13] Such a defect is not part of the kind of disorder normally produced by limitations in the quality of current fabrication technology, but consideration of its properties is a good starting point for the subsequent analysis of more typical features. Figure 2.2.5 shows that such vacancies lead to the appearance of certain spatially localized states that are discrete in frequency.

Figure 2.2.6 shows that modes localized at vacancies lead to the appearance of spikes in the transmission spectrum. In structures with a small concentration of vacancies, the associated modes are weakly coupled and result in photonic mini-bands in the band gap of the PhC. The positions of the mini-bands in the band gap are determined by the frequencies of the individual vacancy modes, and their width by the strength of the coupling, which is dependent on the concentration of vacancies. As a result, smooth, well-pronounced peaks appear in place of the narrow spikes corresponding to the localized modes of an isolated vacancy. Coupling is greater if the defects are closer to each other, as can be seen in the case when the defects are arranged in a line. That structure is essentially a waveguide, and the transparency windows of the waveguide correspond to the frequencies of the localized photonic states of a single vacancy. However, if the vacancies are in a cluster, creating a relatively large 2D cavity, the individual character of the vacancies is lost, and the transmission spectrum is determined only by the localized modes of the cavity, which will have frequencies different from those of an individual vacancy.

In a real 2D PhC it is not normally vacancies that create disorder but the random fluctuations of the position, shape, and size of individual cylinders. In contrast to the 1D case where a universal quantitative measure of disorder (the relative fluctuation of the optical length of the period of the PhC) facilitates a unified quantitative description of the effect of various kinds of disorder, in the 2D case no such parameter has yet been found, and researchers usually use random fluctuations of either the size of the cylinders (or the corresponding basic element of the PC), of their position or size and position simultaneously.

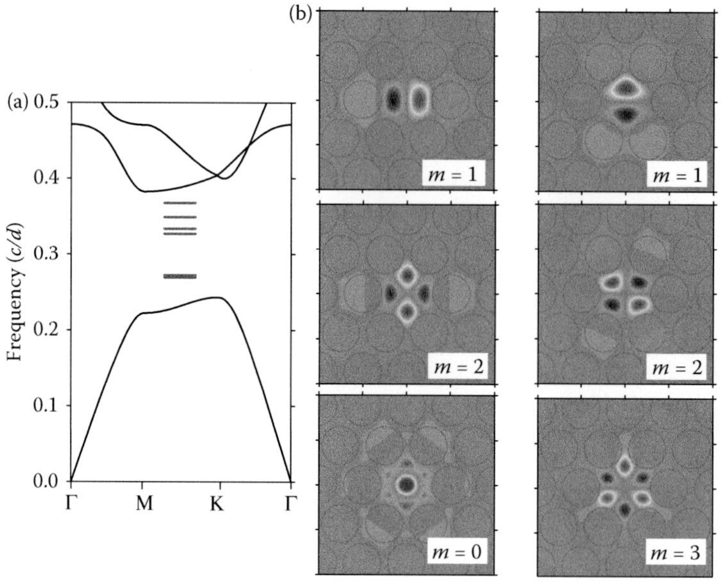

FIGURE 2.2.5
(**See color insert.**) The frequencies of eigenmodes localized at a single vacancy in a hexagonal photonic crystal (a) and the profiles of z components of the magnetic field corresponding to those modes (b).

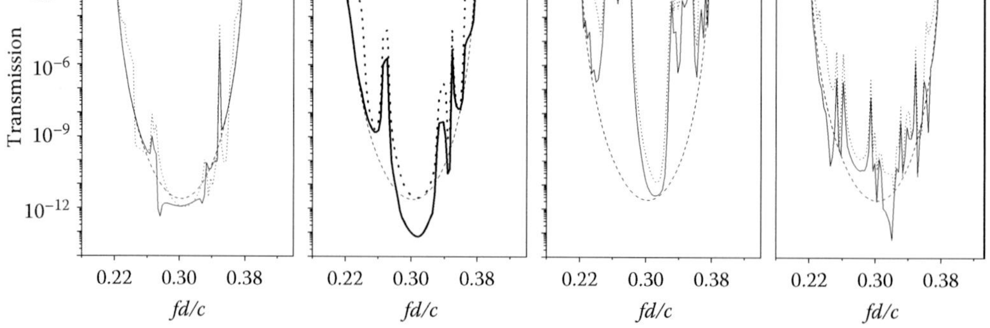

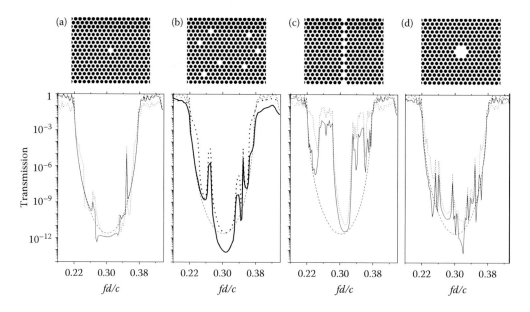

FIGURE 2.2.6
Ballistic (solid lines) and scattered (dotted lines) transmission spectra for structures, where disorder is represented by a "vacancy" of the air cylinder. The cell used for each type of disorder is shown above the spectra. Ballistic transmission for the ideal structure is shown by dashed lines. (Adapted from M. A. Kaliteevski et al., J. Phys. Condens. Matter **15**(6), 785, 2003.)

Qualitatively different kinds of random fluctuations lead to the same general behavior of the density of states and the transmission, but quantitative relationships between the various types of disorder are yet to be established.

We consider disorder in the form of a random shift of the individual cylinders away from their ideal lattice positions: the center of the air cylinders in the disordered structure is located at a random position within a circle centered on the ideal lattice points and of radius a^{TM}, where a is the lattice constant and, as before, the number δ is a measure of the degree of disorder.[14] The modification to the density of states of the 2D PhC with increasing δ is illustrated in Figure 2.2.7. The behavior is similar to that of a 1D PhC: band tails encroach into the PBG, and increase of δ leads to deeper penetration until the PBG effectively disappears (in this example, when $\delta \sim 0.2$). Penetration of the band tails into the PBG is accompanied by a modification of the transmission spectrum as shown in Figure 2.2.8, where the thin curves are the spectra for one individual configuration of disorder, shown as examples, and the thick curves are the mean spectra, averaged over many random configurations. The ballistic transmission of the ideal crystal is also shown (as a dotted black curve) for comparison.

The ideal crystal does not display any scattering of light[15] as the band gap under study is below the diffraction cut-off of the crystal. It can be seen that for small amounts of

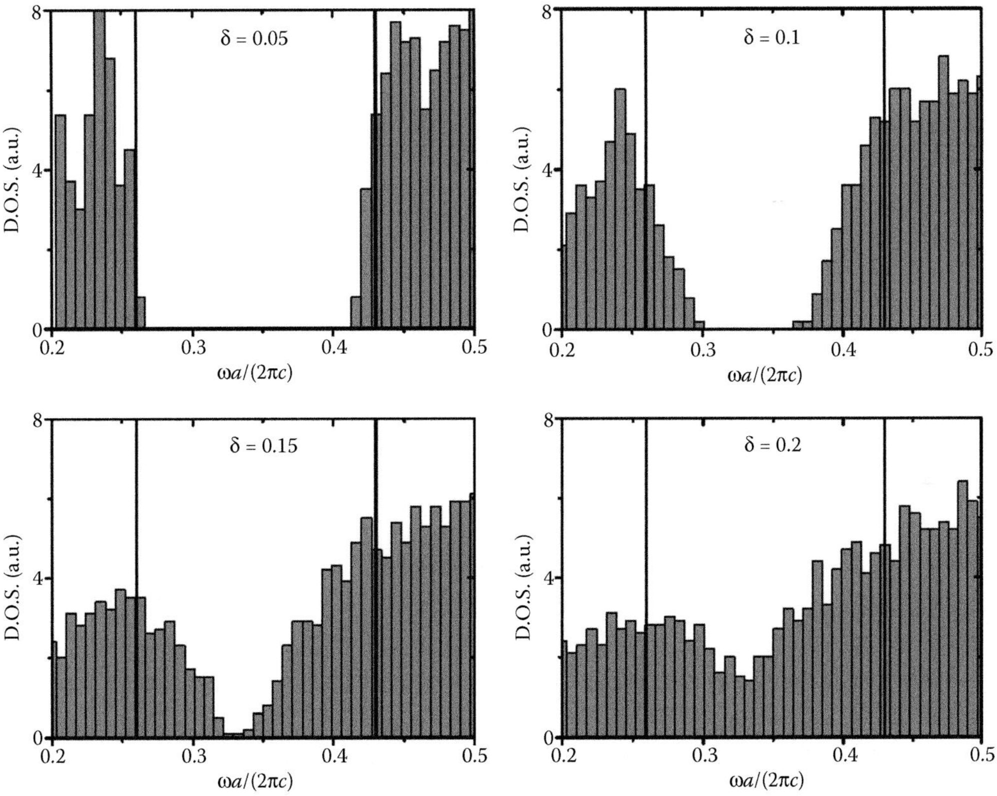

FIGURE 2.2.7
Density of states for a disordered two-dimensional photonic crystal for different degrees of displacement disorder characterized by $\delta = 0.05$, 0.1, 0.15, and 0.2. (Adapted from D. M. Beggs et al., J. Phys. Condens. Matter **17**(12), 1781, 2005.)

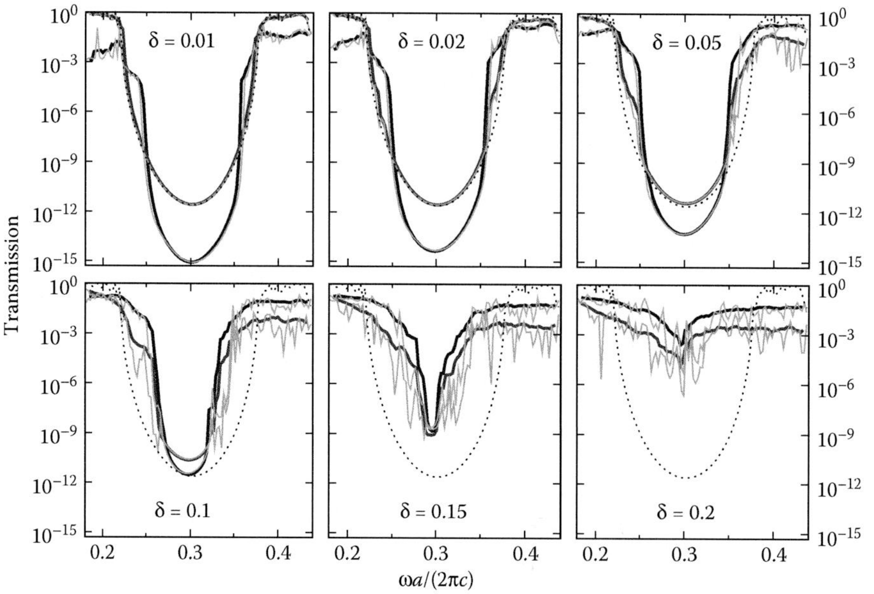

FIGURE 2.2.8
(**See color insert.**) Transmission spectra for disordered photonic crystal with δ = 0.01, 0.02, 0.05, 0.1, 0.15, and 0.2. Solid (red) curves are the ballistic transmission, dashed (blue) curves are the scattered transmission—the thin curves for an individual configuration of disorder, and the heavy curves for the mean averaged over 10 random configurations of disorder. The ballistic transmission spectrum of the ideal photonic crystal is also shown (black dotted curve) for comparison. (Adapted from D. M. Beggs et al., J. Phys. Condens. Matter **17**(12), 1781, 2005.)

disorder (δ ≤ 0.05) there is very little modification of the ballistic transmission relative to that for the ideal structure. However, with larger amounts of disorder (δ ≥ 0.1) the modification of the ballistic transmission spectrum is much more pronounced. It is also apparent that the changes with increasing δ occur first at the edges of the PBG. For individual configurations of disorder, sharp spikes are present in the spectra, which correspond to localized modes introduced into the PBG by the disorder. These sharp spikes then contribute to an increase of the average transmission in the frequency region of the PBG where they occur. Also the width of the ballistic transmission dip decreases with increasing disorder. This is due to the fact that small amounts of disorder can only introduce states into the PBG near its upper and lower edges (so-called edge states) and thus act to narrow the band gap of the PhC. Only when the disorder becomes sufficiently large can it introduce states into the center of the PBG, and influence the transmission in that region.

As in the 1D case there is evidence of a threshold value of δ above which the effect of disorder on the ballistic transmission increases rapidly. However, disorder has a marked effect on the scattered transmission even for small values of δ. Moreover, the scattered transmission increases with δ at a faster rate than the ballistic transmission, so that for large values of δ, scattered light is dominant in the transmission spectrum. Also present in the transmission spectrum is evidence of Rayleigh scattering. The ballistic transmission on the upper frequency edge of the transmission dip is smaller than that on the lower edge when disorder is present, and that is because Rayleigh scattering due to disorder is stronger at the higher frequencies.

As expected, disorder-induced states within the PGB also result in changes to the reflection properties of the PhC. The reflection spectrum of the ideal PhC has a square-shaped band corresponding to the frequency of the PBG, and light is reflected ballistically. The appearance of states within the PBG results in the reflection band becoming Gaussian in shape and the introduction of a scattered component into the reflected light (for details, see the study by Beggs et al.[14]) (Figure 2.2.9).

Analysis of the dependence of the transmission coefficient on the thickness of the disordered 2D PhC indicates that electromagnetic waves decay exponentially both within and near the PBG. The decay is characterized by an attenuation length $\xi(\delta)$ with a dependence on δ at the center of the PBG that shows a threshold-like behavior similar to the 1D case. Although δ is below a certain threshold value, $T_b > T_s$ and standard deviation of ξ for ballistic and scattered transmission is less than its mean value. Above the threshold value ($\delta \approx 0.15$ for the model 2D PhC considered here), the standard deviation exceeds the mean value. The threshold corresponds to a transition from an evanescent Bloch mode at the center of a largely preserved PBG despite the presence of disorder, to a situation in which the original PBG is filled with states.

It is interesting to consider the spatial dependence of the modes associated with spikes in the transmission spectrum. The field profiles of some modes are shown in Figure 2.2.10 and have a localized appearance, which in some cases is not so different from those associated with vacancies. However, here there are no simple vacancies and the

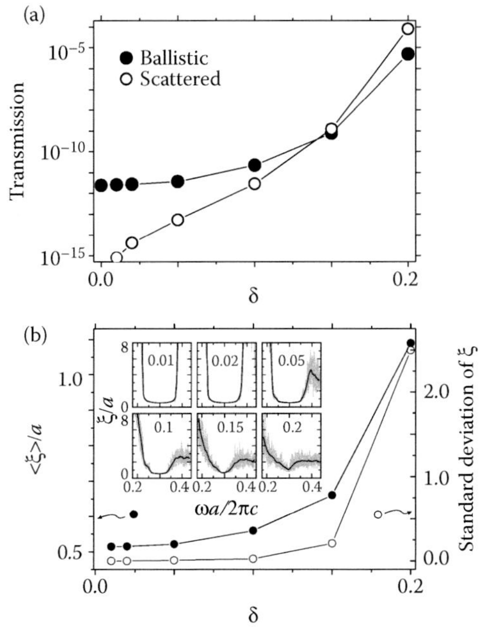

FIGURE 2.2.9
(a) The transmission coefficient at the minimum of the spectral dip for the averaged ballistic transmission (closed circles) and averaged scattered transmission (open circles) as a function of the "amount of disorder," δ; (b) The mean attenuation length (closed circles, left axis) and its standard deviation (open circles, right axis) at the center of the ideal photonic crystal band gap as a function of δ. Inset: the spectra of the attenuation lengths for $\delta = 0.01, 0.02, 0.05, 0.1, 0.15$, and 0.2 (gray lines). Black curves are the mean over the 10 random configurations, and gray curves are the 10 individual configurations. (Adapted from D. M. Beggs et al., J. Phys. Condens. Matter **17**(12), 1781, 2005.)

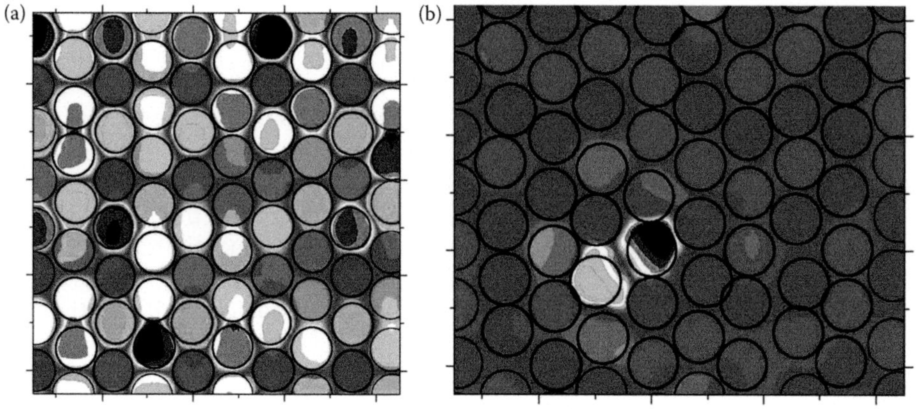

FIGURE 2.2.10
(**See color insert.**) Delocalized state near edge of PBG in the case of weak disorder (a) and localized state in the case of strong disorder (b). (Adapted from D. M. Beggs et al., J. Phys. Condens. Matter **17**(12), 1781, 2005.)

light is localized in effective microcavities formed by the structural randomness. Recently, patterns of the field associated to localized modes were observed experimentally.[16]

2.2.2.3 Disordered 2D PhC with an Incomplete Band Gap

Figure 2.2.11 shows the calculated transmission spectra of E-polarized waves for a range of levels of disorder. Light incident on the ideal PhC in the Γ–M direction in the spectral region of the incomplete PBG suffers attenuation, and hence a transmission dip is observed in the ballistic transmission that is symmetrical about the center of the PBG. When disorder is introduced, light can be scattered and for the smallest amount of disorder considered $\delta = 0.01$, the scattered transmission is comparable to the ballistic transmission, and for larger levels of disorder scattered light is dominant.[17]

The transmission spectra of Figure 2.2.11 display many similar features to the case of a complete PBG, including threshold-like behavior of the ballistic transmission. However, an interesting new feature is that for small values of δ the ballistic transmission can be smaller than for the ideal crystal counterpart, despite the fact that one might expect disorder to decrease the attenuation, and therefore lead to more transmission. However, for the narrow PBG under study, where the total attenuation of light through the crystal is small, the effect of Rayleigh scattering in the disordered crystal (which serves to remove light from the ballistic transmission) can be large enough to cause the minimum transmission to be reduced when disorder is introduced.

Another new property is that the averaged scattered transmission spectra display an asymmetric, triangular shape. The reasons for this asymmetry can be understood in terms of the following simple physical arguments[18] based on the band structure in Figure 2.2.12. Assume light is incident on the disordered crystal in the Γ–M direction at a frequency corresponding to the incomplete PBG. In this case, an evanescent wave will be excited in the crystal, characterized by some attenuation length as implicit in Figure 2.2.12a. As a result of the disorder, the wave can be scattered into a propagating state in a direction closer to Γ–K. As a result, the wave can then propagate to the rear of the sample without further attenuation (Figure 2.2.12b).

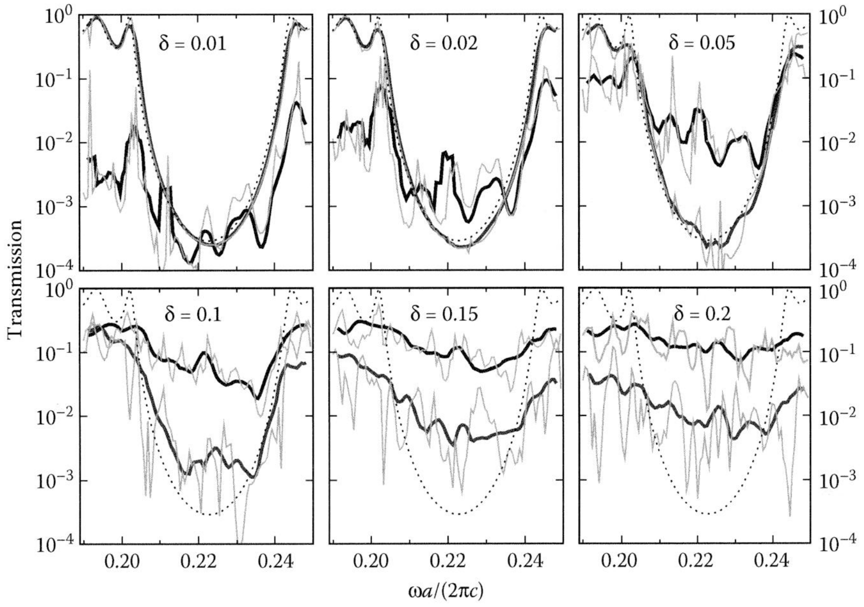

FIGURE 2.2.11
(See color insert.) Transmission spectra for disordered photonic crystals with δ = 0.01, 0.02, 0.05, 0.1, 0.15, and 0.2. The red lines are the ballistic transmission and the blue lines are the scattered transmission, with the thin lines denoting an individual configuration of disorder and the heavy lines an average over many random configurations of disorder. The ballistic transmission spectrum of the ideal photonic crystal is also shown (black dotted line) for comparison. (Adapted from D. M. Beggs et al., J. Phys. Condens. Matter **17**(26), 4049, 2005.)

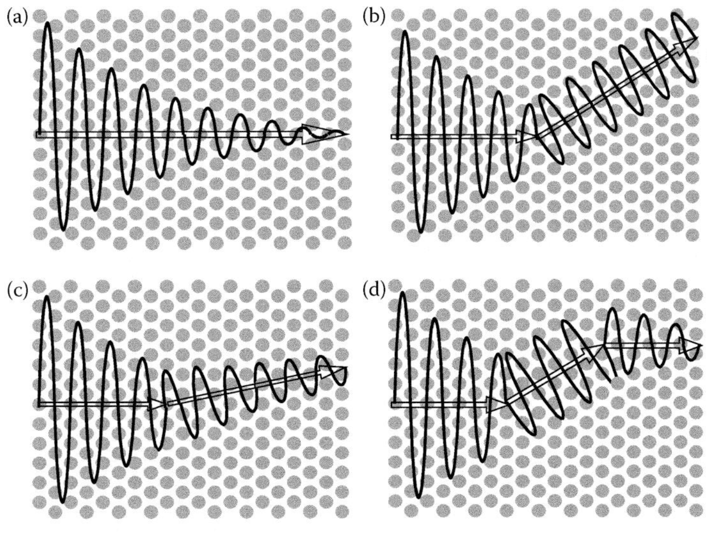

FIGURE 2.2.12
Illustration of wave propagation and scattering in disordered a photonic crystal with anisotropic incomplete bad gap. A wave at a frequency in the PBG will decay exponentially without scattering (a), but it could be scattered into a propagating mode in another direction (b), or to a direction in which attenuation is reduced (c), and multiple scattering is also possible (d). (Adapted from M. A. Kaliteevski et al., Phys. Rev. B **66**(11), 113101, 2002.)

Also possible is the scattering of the incident wave into another evanescent state that is closer to the Γ–K direction, and thus characterized by a larger attenuation length (Figure 2.2.12c). The nature of the modification to the transmission spectrum depends on the details of the curvature of the band structure between the Γ–M and Γ–K directions. For the case considered here, the change in wave vector needed to scatter to a propagating state is larger for higher frequencies (in the second photonic band) than for lower frequencies (in the first photonic band), and so there is a shift of the minimum of the transmission to higher frequency.

To a lesser degree, the averaged ballistic transmission also displays a shape asymmetry, which can be explained by similar reasoning. Once the incident wave has been scattered by the disorder, it can then be scattered a second time, occasionally back into the Γ–M direction where it will contribute to the ballistic transmission. This second-order effect leads to a shift of the minimum in the ballistic transmission away from the center of the PBG. Usually experimental studies assign the center of the PBG of real PhCs by the frequency of the minimum of the transmission spectrum. However, when scattered light is dominant in the total transmission, the shape asymmetry discussed above means that the minimum of the total transmission will depend on the details of the band structure of the PhC, and does not correspond to the center of the PBG. Simply assigning the center of the PBG to the minimum of the transmission through the crystal can thus lead to a significant error in the determination of the band structure of a disordered PhC with an incomplete PBG.

A wave scattered into propagating modes away from the incident Γ–M direction can experience multiple reflections in the crystal, and this leads to Fabry–Pérot-type oscillations in the transmission spectrum for scattered light in the spectral region of the incomplete PBG. In Figure 2.2.11, the transmission spectra for individual configurations of disorder exhibit Fabry–Pérot-type oscillations with a period similar to that for ballistic transmission at frequencies outside the PBG, and should be taken into account in the interpretation of experimental measurements. Finally, the reflection spectra experience disorder-induced modifications similar to the case of a complete PBG.

2.2.3 Conclusion

In 2D PhC with complete band gap, disorder leads to the appearance of the tail of the density of states in PBG, and some of the states that appear in PBG are localized. Appearance of localized states in PBG leads to an increase of transmission and to shrinking of observable width in PBG. A threshold amount of disorder exists: below threshold, disorder does not modify transmission properties of PhC, density of states within PBG remains negligible, and ballistic transmission dominates over scattered transmission. Above the threshold, PBG becomes filled with photonic states, and increasing the disorder has a significant effect on the spectra throughout the former PBG.

In 2D PhC with incomplete band gap, the ballistic and scattered light show different behavior as a function of the disorder parameter. The ballistic transmission is not significantly modified for small disorder, but the scattered transmission grows quickly even for small disorder. Scattered light dominates the transmission spectra within the former incomplete PBG, even for small amounts of disorder. The transmission spectrum has dips in the PBG, which becomes asymmetric for disordered PhCs, such that the minimum transmission does not coincide with the center of the PBG.

Acknowledgments

Authors are grateful to Dr. D. M. Beggs for useful discussion.

References

1. E. Yablonovitch, Phys. Rev. Lett. **58**, 2059 (1987).
2. Yu. A. Vlasov, M. A. Kaliteevski, and V. V. Nikolaev, Phys. Rev. B **60**(3), 1555 (1999).
3. V. Yannopapas, N. Stefanou, and A. Modinos, Phys. Rev. Lett. **86**, 4811 (2001).
4. S. John, in *Confined Electrons and Photons*, edited by E. Burstein and C. Weisbuch (Plenum Press, New York, 1995), p. 523.
5. S. Mazoyer, P. Lalanne, J. C. Rodier, J. P. Hugonin, M. Spasenović, L. Kuipers, D. M. Beggs, and T. F. Krauss, Opt. Express **18**, 14654 (2010).
6. R. A. Abram, G. J. Rees, and B. L. H. Wilson, Adv. Phys., **27**, 799 (1978); I. M. Lifshits, S. A. Gredeskul, and L. A. Pastur, *Introduction to the Theory of Disordered Systems* (Wiley, New York, 1988).
7. P. W. Anderson, Phys. Rev. **109**, 1492 (1958).
8. A. V. Kavokin, J. J. Baumberg, G. Malpuech, and F. P. Laussy, *Microcavities* (Oxford University Press, Oxford, 2007).
9. D. M. Beggs, M. A. Kaliteevski, S. Brand, R. A. Abram, and V. V. Nikolaev, Phys. Rev. B **73**, 033106 (2006); M. A. Kaliteevski, D. M. Beggs, S. Brand, R. A. Abram, and V. V. Nikolaev, Phys. Rev. E **73**, 056616 (2006).
10. D. J. Thouless, Phys. Rev. Lett. **39**, 1167 (1977).
11. A. A. Greshnov, M. A. Kaliteevski, R. A. Abram, S. Brand, and G.G. Zegrya, Solid State Commun. **144**, 413 (2008).
12. P. M. Bell, J. B. Pendry, L. Martin Moreno, and A. J. Ward, Comput. Phys. Commun. **85**, 306 (1995).
13. M. A. Kaliteevski, J. M. Martinez, D. Cassagne, J. P. Albert, S. Brand, and R. A. Abram, J. Phys. Condens. Matter **15**(6), 785 (2003).
14. D. M. Beggs, M. A. Kaliteevski, S. Brand, R. A. Abram, D. Cassagne, and J. P. Albert, J. Phys. Condens. Matter **17**(12), 1781 (2005).
15. Waves can be scattered by the ideal PhC due to diffraction (see E. Lidorikis, M. M. Sigalas, E. N. Economou, and C. M. Soukoulis, Phys. Rev. B **61**, 13 458 (2000)), but appropriate choice refractive indices of surrounding media allows to avoid diffraction.
16. T. Schwartz, G. Bartal, S. Fishman, and M. Segev, Nature **446**, 52 (2007).
17. D. M. Beggs, M. A. Kaliteevski, S. Brand, R. A. Abram, D. Cassagne, and J. P. Albert, J. Phys. Condens. Matter **17**(26), 4049 (2005).
18. M. A. Kaliteevski, J. M. Martinez, D. Cassagne, and J. P. Albert, Phys. Rev. B **66**(11), 113101 (2002).

2.3

Modeling the Optical Response of Three-Dimensional Disordered Structures Using the Korringa–Kohn–Rostoker Method

Gabriel Lozano and Hernán Míguez
Materials Science Institute of Seville

Luis A. Dorado and Ricardo A. Depine
University of Buenos Aires

CONTENTS

2.3.1 Introduction

Three-dimensional (3D) photonic crystals (PCs) have attracted much attention in the last decades mainly because of the possibility of an intelligent control of light and the potential applications in optoelectronics and microwave devices.[1] Spatial periodicity of the dielectric function along three independent directions is essential to fabricate structures that exhibit a complete photonic band gap, that is, the propagation of light along any direction is forbidden for those photon energies within the gap.[2] Improvements in the fabrication processes of these materials have made possible the observation of high-quality optical spectra in higher-order band frequencies.[1–18] Although new fabrication techniques have alleviated the effect of disorder in the so-called low-energy range, where the lattice constant is less than the wavelength of light, it has recently been demonstrated that extinction due to intrinsic defects, which are unintentional disruptions of the spatial periodicity of the dielectric function arising during the fabrication process, strongly affects the shape of the experimental spectra in the high-energy range, where the lattice constant is greater than the incident wavelength.[3] For this reason, the calculation of the optical response through any theoretical approach implies the modeling of disorder to establish a valid comparison between theory and measurements.

Deviations from the perfectly ordered structure cause scattering and hence exponential attenuation of the coherent radiation propagating through a PC.[4] In the low-energy range, where the size of imperfections is small compared to the incident wavelength, the degree of disorder has been evaluated using the concept of extinction length or assigning an imaginary part to an effective refractive index.[5,6] However, a rather different approach has been recently used to reproduce the optical features observed for colloidal crystals in the high-energy region. In this approach, disorder effects observed in artificial opals made of lossless materials have been satisfactorily modeled by keeping the hypothesis that the spheres are arranged in a perfectly ordered lattice while introducing some losses through the imaginary part of the dielectric constant of the spheres. Excellent agreement with experimental results obtained by different research groups worldwide has been attained.[3] Moreover, the imaginary part of the dielectric constant introduced in the model has been shown to be dependent on the crystal size,[7] which indicates that this single parameter can be identified as the most sensitive guide to accurately evaluate the progress toward the actual realization of defect-free colloidal crystals.[8] This approach is also useful to explain the physical origin of reflectance and transmittance features observed in the high-energy range and to analyze the near-field pattern and the optical diffraction in artificial opals.[9,10]

The results obtained with the extinction method indicate that the introduction of losses to model the optical response of PCs made of lossless building blocks can be regarded as a convenient theoretical tool to account for diffusely scattered light produced by disorder and imperfections, a mechanism that removes energy from the specularly reflected and forward-transmitted beams. However, this approximation does not seem to be completely satisfactory from a physical point of view as many quantitative and qualitative questions remain unanswered, such as whether the effect of imperfections can be distinguished from actual absorption, how different types of imperfections contribute to the imaginary part of the dielectric constant, or whether there is a unique way to represent a variety of imperfections by a single extinction parameter. Herein we address some of these questions using the electromagnetic counterpart of one of the simplest approximations used in the electron-scattering problem:[11] the average *T*-matrix approximation (ATA). In the ATA method—generally valid when the disorder is weak—the scattering matrix (*T*-matrix) of a single scatterer is averaged over particle states obtaining a sort of average scatterer. Thus, the optical response of a disordered structure is calculated using a perfectly ordered lattice made of average scatterers as building blocks. Particle states refer to distributions of size, shape, position, orientation, and so on, of the scatterers, so the ATA is calculated by using statistical methods.[12] Some previous uses of the ATA for the evaluation of disorder effects in PCs include systems of metallic spheres[13,14] and studies on negative refraction in random photonic alloys of polaritonic and plasmonic microspheres.[15] ATA results have also been used as a reasonable initial input for iterative schemes obtained in the frame of improved treatments of disorder based on the so-called coherent-potential approximation.[16] Here we present the ATA method to evaluate disorder effects in 3D PCs constructed from *intrinsically transparent* building blocks.[17] We exemplify the case of artificial opals, but this analysis could also apply to different PCs.

In this chapter, we use the Korringa–Kohn–Rostoker (KKR) method, a rigorous method valid for perfectly periodic structures, to simulate disorder effects in 3D PCs. Two approaches are considered: extinction methods and statistical methods. In extinction methods, energy losses are artificially added to perfectly periodic crystals made of transparent materials. These artificial losses take into account the light scattered by imperfections in the crystalline structure. In statistical methods, however, deviations from perfect periodicity are modeled by statistical distributions of sizes, shapes, and vacancies of the PC building blocks. Examples will be given for the case of synthetic opals made of dielectric spheres, a case for which several measured optical spectra are available. Thus, the chapter is organized as follows. In Section 2.3.2, we review the framework of the KKR multiple scattering approach for spherical particles. In Section 2.3.3, we present a description of the optical properties of defect-free artificial opals. We start by describing the band structure and predicting the optical response of a lattice with no losses. In Section 2.3.4, we gradually introduce defects in the model by an extinction approach and another one based on a statistical distribution of imperfections. In particular, the focus of the formulation is placed on disorder effects related to polydispersity or variations in the size of particles as well as sphere vacancies. At those energies for which very low-dispersion propagation modes are attained, we predict that perfect lattices should present a strongly fluctuating optical response that rapidly smoothes out as the amount of imperfections gradually increases. The comparison between the results obtained using the two different approaches is addressed in Section 2.3.5. Finally, in Section 2.3.6, the more outstanding results are summarized and discussed.

2.3.2 KKR Multiple Scattering Approach

Electromagnetic interactions between the scatterers arranged in a periodic lattice will be calculated by means of the layer-multiple-scattering method for spherical scatterers,[13,18–20] usually known as the vector version of the KKR method.[21] The problem of calculating the frequency band structure of a PC and of the reflection and transmission coefficients of light incident on a slab of the crystal, parallel to a given crystallographic plane, attracted a lot of attention since the very beginning of the field and was the focus of activity of many research groups. For a general introduction to the physics of this problem, the reader is referred to the study by Stefanou et al.[19] and references therein. To calculate the optical response of artificial opals, we assume spherical scatterers, although KKR methods can be, in principle, extended to lattices made of nonspherical particles as the scattering properties of the individual building block enter in the calculation method only through the *T*-matrix.[22]

Herein we present a brief description of the scattering by a single sphere and the calculations involved in the vector KKR method.[13,18–20] An incident electromagnetic wave on a sphere centered at the origin of coordinates ($\vec{r} = 0$) has an electric field with a spherical wave expansion given by[23]

$$\vec{E}(\vec{r}) = \sum_{l=1}^{\infty}\sum_{m=-l}^{l}\left[\frac{i}{k}a_{lm}^{0E}\vec{\nabla}\times j_l(kr)\vec{X}_{lm}(\theta,\phi) + a_{lm}^{0H}j_l(kr)\vec{X}_{lm}(\theta,\phi)\right] \tag{2.3.1}$$

where (r, θ, ϕ) are the usual spherical coordinates of an evaluation point $\vec{r}$, ε, and μ are the dielectric permittivity and magnetic permeability, respectively, of the homogeneous medium outside the sphere, ω is the angular frequency, c is the velocity of light in vacuum, $k = \sqrt{\mu\varepsilon}\omega/c$, $j_l(kr)$ are the spherical Bessel functions of the first kind, and $\vec{X}_{lm}(\theta,\phi)$ are the vector spherical harmonics. The magnetic field can be obtained from Maxwell's equations and has a similar series expansion. In this multipole expansion, the integers (l, m), where $l \geq 1$ and $-l \leq m \leq l$, represent a spherical partial wave of order (l, m), which can be either electric (with complex amplitude a_{lm}^{0E}) or magnetic (with complex amplitude a_{lm}^{0H}). Dipole fields correspond to $l = 1$, quadrupole fields correspond to $l = 2$, and so on. By applying the orthogonality property of the vector spherical harmonics, it is straightforward to prove that electric multipoles radiate transverse magnetic fields, whereas magnetic multipoles radiate transverse electric fields. Thus, the multipole expansion in Equation 2.3.1 can also be viewed as a linear combination of transverse magnetic and transverse electric spherical waves. In a numerical calculation, a cut-off LMAX must be introduced in the multipole expansion, retaining the first 2LMAX(LMAX + 2) terms. The electromagnetic field scattered by the sphere at the origin is given by replacing $j_l(kr)$ with the spherical Hankel function of the first kind $h_l^+(kr)$, which corresponds to outgoing spherical waves. Also, the coefficients a_{lm}^{0P} must be replaced by the new ones a_{lm}^{+P} of the scattered wave, where $P = E, H$. The T-matrix (**T**) relating the incident and scattered wave coefficients, a_{lm}^{0P} and a_{lm}^{+P}, can be found by enforcing the boundary conditions on the surface of the sphere. It can be proved that this matrix is diagonal, $\mathbf{T} = T_l^P \delta_{PP'} \delta_{ll'} \delta_{mm'}$,[20] and that its elements are given by

$$T_l^E = \left[\frac{j_l(k_s r)\dfrac{\partial}{\partial r}\left[rj_l(kr)\right]\varepsilon_s - j_l(kr)\dfrac{\partial}{\partial r}\left[rj_l(k_s r)\right]\varepsilon_s}{h_l^+(kr)\dfrac{\partial}{\partial r}\left[rj_l(k_s r)\right]\varepsilon - j_l(k_s r)\dfrac{\partial}{\partial r}\left[rh_l^+(kr)\right]\varepsilon_s} \right]_{r=S} \tag{2.3.2}$$

where S is the sphere radius, ε_s and μ_s are the dielectric permittivity and magnetic permeability of the sphere, respectively, and $k_s = \sqrt{\mu_s\varepsilon_s}\omega/c$. The matrix elements T_l^H can be obtained from Equation 2.3.2 by means of the transformations $\varepsilon_s \leftrightarrow \mu_s$ and $\varepsilon \leftrightarrow \mu$.

Once the T-matrix of a single sphere has been obtained, the multiple scattering between identical spheres arranged in a given periodic lattice can be calculated by means of the vector KKR method. In the layer version of this method,[19] the 3D PC slab is first divided into layers parallel to a given crystallographic plane and the multipole expansion in spherical waves is used to account for the multiple scattering process inside a layer. Next, a series expansion in a plane-wave basis is used to calculate the electromagnetic interactions between layers and to compute the power flux on both sides of the crystal slab.

To study the convergence of the fields involved in the optical response of the artificial opals presented in the examples, the values of the code parameters that control the number of terms in both the spherical-wave and plane-wave expansions of the electric field were increased. For the calculations shown in this chapter, we used 41 two-dimensional (2D) reciprocal lattice vectors in the plane-wave expansions and spherical waves with angular momentum up to LMAX = 9.[19,20] These are the minimum values providing good convergence in that range. Lower values of LMAX and fewer reciprocal lattice vectors still give a correct result for lower-energy bands, enough to describe the optical response around the stop band, although not sufficient to provide trustworthy information on the whole region we are interested in.

2.3.3 Optical Response of a Perfectly Ordered Artificial Opal

The model structure used herein is a close-packed face centered cubic (FCC) lattice of spheres of dielectric constant ε_s in a medium of $\varepsilon = 1$. In the case of artificial opals, the largest accessible surface is determined by the crystal growth direction, typically the (111).[24] We will focus on the optical properties when light impinges in that particular direction as they are best known.

The calculated band structure along the ΓL direction for an FCC ensemble of spheres made of a material of dielectric constant $\varepsilon_s = 2.4964$ along with the reflectance spectrum in a selected range are shown in Figure 2.3.1. This value of the dielectric constant corresponds to latex spheres in air, a structure that has been repeatedly fabricated and typically considered of high structural quality in the field of 3D PCs.[25] Energy is expressed in reduced units a/λ, where a is the lattice constant. In Figure 2.3.1c, the reflectance for a crystal made of 21 layers is shown. It can be observed that a strongly fluctuating reflectance is expected for low-dielectric-contrast FCC lattices in which extinction can be neglected for energies above the pseudogap. Reflectance may reach very high values in the absence of any gap, whereas total transmission may occur for some other narrow spectral ranges. Our results reproduce the abrupt changes in the specular reflectance already predicted by Modinos et al.[26] for perfect lattices without absorption. None of the FCC structures reported so far have shown any of the features in Figure 2.3.1 for photon energies $a/\lambda > 1$, but much smoother ones. The reasons are addressed in what follows.

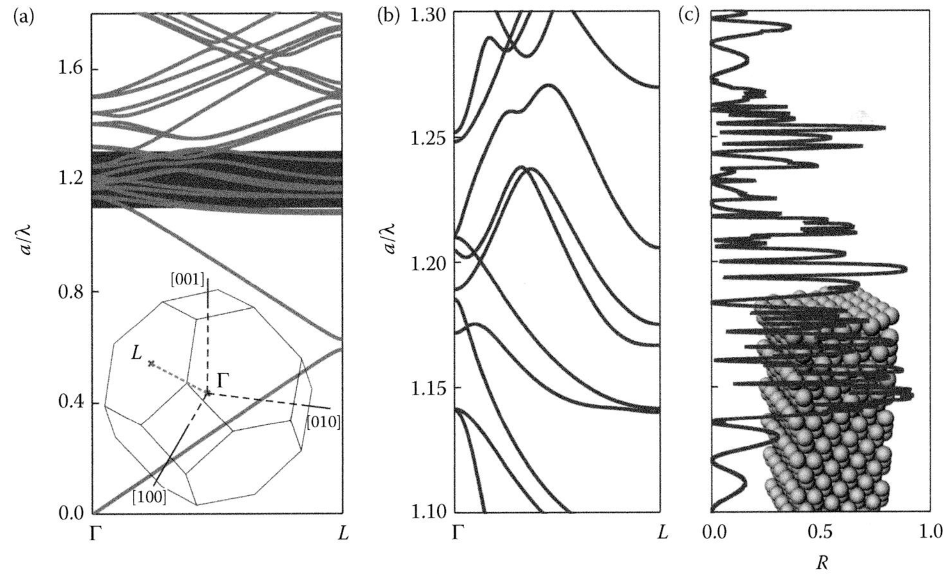

FIGURE 2.3.1
(**See color insert.**) (a) Band structure calculated along the ΓL direction of a perfect face centered cubic crystal consisting of spheres of dielectric constant $\varepsilon_s = 2.4964$ in air. (b) Zoom of the band structure in the range $1.1 \le a/\lambda \le 1.3$. (c) Specularly reflected light spectrum for $\varepsilon_s = 2.4964$. (Adapted from L. A. Dorado, R. A. Depine, and H. Míguez, Phys. Rev. B **75**, 241101(R), 2007.)

2.3.4 KKR Method for Modeling Imperfections

It must be remarked that real structures present imperfections[5,27] that cause light to be scattered diffusely, removing part of the energy from the specularly reflected or forward-transmitted beams. This extinction is here modeled following two strategies. On the one hand, the effect of disorder is considered by adding a small imaginary part in the ε_s of the spheres that form the artificial opal, which accounts for all possible sources of losses in the lattice. On the other hand, we apply a method based on a statistical distribution of sphere sizes and vacancy density in the colloidal crystal slab to take into account the diffuse light scattering produced by imperfections in the crystalline structure.

2.3.4.1 Extinction Approximation

In this approach, all possible sources of losses present in actual measurements are represented by adding an imaginary part ε_i to the dielectric constant of the spheres. Energy losses due to power dissipation are negligible in latex spheres in the optical range analyzed,[28] so in this range ε_i represents structure imperfections and disorder, responsible for the experimental deviation of the spectrum shape with respect to the theoretical optical response without extinction.[3] The parameter ε_i was varied to analyze the optical effect of different amounts of structural disorder by using the extinction approximation. Figure 2.3.2a and b show the specularly reflected and the forward-transmitted spectra for 21 sphere layers. As expected, in the low-energy region $a/\lambda < 0.9$, the effect of such losses is weak, although it has to be taken into account if a precise description of the measurements is needed.[29]

The maximum reflectance (minimun transmittance) observed around $a/\lambda \sim 0.61$ is the result of the opening of a stop band between the lowest-energy photonic bands along the ΓL direction in reciprocal space, which corresponds to the [111] direction in the real space of an FCC structure as can be observed in Figure 2.3.1a. Secondary lobes account for the

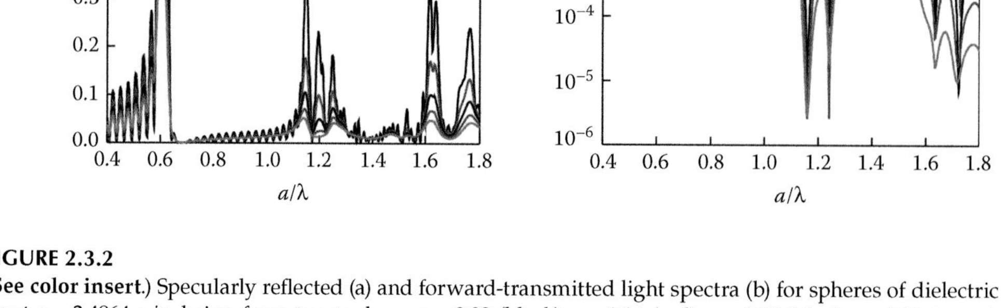

FIGURE 2.3.2

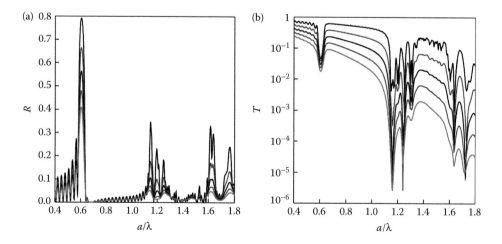

(See color insert.) Specularly reflected (a) and forward-transmitted light spectra (b) for spheres of dielectric constant $\varepsilon_s = 2.4964 + i\varepsilon_i$, being from top to down $\varepsilon_i = 0.02$ (black), $\varepsilon_i = 0.04$ (red), $\varepsilon_i = 0.06$ (blue), $\varepsilon_i = 0.08$ (green), and $\varepsilon_i = 0.10$ (magenta). (Adapted from L. A. Dorado, R. A. Depine, and H. Míguez, Phys. Rev. B **75**, 241101(R), 2007.)

coherent effects arising from the interference of beams reflected at the top and the bottom surfaces of the film, and therefore provide information on its thickness. Besides this, for a very low ε_i, two sets of reflectance peaks (deep transmission dips) are attained in the high-energy range, that is, for $1.1 \le a/\lambda \le 1.3$ and $1.55 \le a/\lambda \le 1.8$. As we increase ε_i, the reflectance peaks decrease, whereas the dips in transmission become less pronounced.

Our results confirm that the optical properties of a PC at wavelengths similar to or shorter than the lattice parameter ($a/\lambda > 1$) are much more sensitive to any modification of the structure, which is in good agreement with previous studies.[30] The extinction we are introducing has little effect on the optical properties at lower band energies, but gives rise to huge distortion of the spectra at higher frequencies. Similarly, any small structural modification can give rise to large variations of the optical response at higher frequencies. A similar absorption amplification effect has been previously described for PC waveguides, where it was found that losses scale inversely with group velocity.[31]

2.3.4.2 Average *T*-Matrix Approximation

In this approach, the characteristics of the statistical distribution of defects is explicitly considered by means of an average *T*-matrix that can be obtained within the context of the KKR method for spherical scatterers. Among all the sources of unavoidable imperfections, here we focus on those originating from a statistical distribution of sphere sizes and vacancy density, as schematically shown in Figure 2.3.3. The scatterers composing a real artificial opal are generally not identical, nor are they located at the sites of a perfectly periodic lattice. All these imperfections in the crystalline structure of a real PC produce light scattering that removes energy from the coherently scattered waves. If there are N types of scatterers with different *T*-matrices $\mathbf{T}_i$, where $i = 1, 2, \ldots N$, in the ATA method each of these scatterers is replaced by a single average scatterer described by an average *T*-matrix given by

$$\langle \mathbf{T} \rangle = \sum_{i=1}^{N} C_i T_i \tag{2.3.3}$$

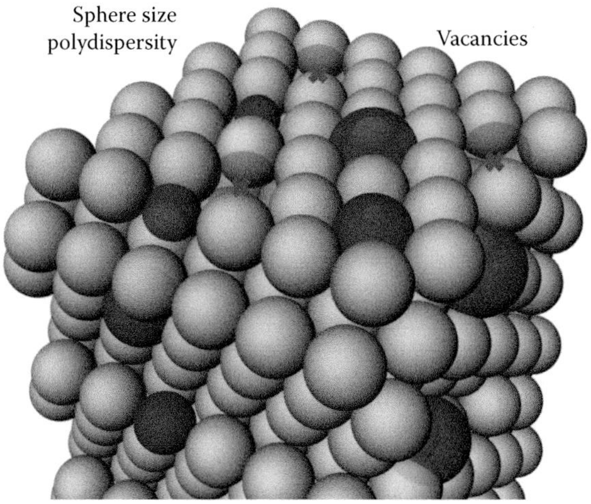

FIGURE 2.3.3
(**See color insert.**) Sketch of an artificial opal that exhibits the type of imperfections (red) that we are considering.

where C_i is the concentration of the ith scatterer and $\Sigma_{i=1}^{N} C_i = 1$. As an average scatterer occupies each lattice site, the application of the KKR method proceeds in the same fashion as in the case of ordered PC slabs. The ATA method can also be applied for the modeling of vacancies.[16] For example, if $C \leq 1$ denotes the fraction of lattice sites occupied by identical scatterers described by a T-matrix $\mathbf{T}$ and the remaining $1 - C$ sites are empty, then Equation 2.3.3 reduces to $\langle T \rangle = C\mathbf{T}$.

The ATA method can be extended for the case of a continuous distribution of scatterer types. As a first approximation, we could consider that the distributions of size, shape, position, orientation, refractive index, and so on, of the scatterers are statistically independent from each other when their deviations from the mean values are small enough.[12] Under this condition, we will focus our interest on the effect of the size distribution of spheres centered at lattice sites. Assuming that the radii s are distributed according to a random process with probability density function $P(s)$, the average T-matrix in Equation 2.3.3 is given by

$$\langle T \rangle = \int \mathbf{T}(s)P(s)\,\mathrm{d}s \tag{2.3.4}$$

Besides, if vacancies (concentration $1 - C$) are included, the spheres of the statistical ensemble can only be centered at a fraction C of the lattice sites. In this case, we see that Equation 2.3.3 becomes

$$\langle T \rangle = C\int \mathbf{T}(s)P(s)\,\mathrm{d}s \tag{2.3.5}$$

Similar expressions can be obtained whenever disorder effects associated with vacancies and with a continuous size distribution of scatterers are separated from other sources of disorder. In this study, we will assume a Gaussian distribution $P(s)$ of sphere radii with mean S_{av} (average radius) and standard deviation σ as the theoretical results obtained through this approach are compatible with measurements reported in the literature. Also, in the limit of small degrees of disorder, that is, when $\sigma \ll S_{av}$, the results and conclusions obtained from the Gaussian distribution are basically the same as those obtained from other probability density functions, for example, logarithmic normal, gamma, and modified gamma distributions.[12]

The main multipole field scattered by the spheres in the lattice is the dipole field ($l = 1$), then the T-matrix elements corresponding to electric and magnetic dipoles are T_1^E and T_1^H, respectively. In Figure 2.3.4, the real and imaginary parts of these dipolar elements are plotted as functions of the reduced photon energy. Solid lines are the results obtained directly from Equation 2.3.2 for a single sphere with radius $0.5a$, whereas dashed lines correspond to the ATA matrix obtained for an ensemble of spheres characterized by $S_{av} = 0.5a$ and $\sigma/S_{av} = 0.05$, that is, a standard deviation of 5% from the mean radius. As a result of the low standard deviation introduced, the single sphere and the average T-matrices exhibit relatively small differences. However, we will see below that even smaller amounts of disorder have a tremendous effect on the ATA predicted response, especially in the high-energy range.

In Figure 2.3.5a, we show the calculated reflectance spectrum of a close-packed perfectly ordered crystal, that is, $S_{av} = 0.5a$ and $\tilde{\sigma}/S_{av} = 0$. The spectrum exhibits a peak at $a/\lambda = 0.61$ as a result of the existence of a pseudogap in the photonic band structure and the usual

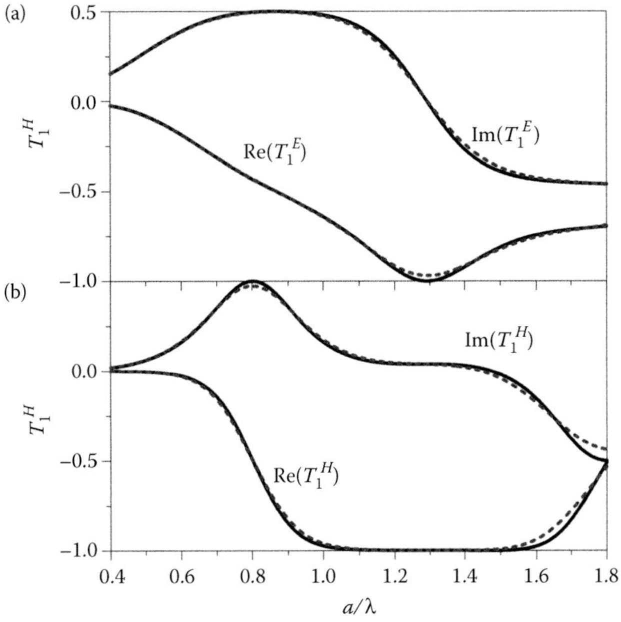

FIGURE 2.3.4
Real and imaginary parts of the single-sphere scattering coefficients corresponding to: (a) electric dipoles (T_1^E) and (b) magnetic dipoles (T_1^H) for a sphere dielectric constant $\varepsilon_s = 2.5$. Solid lines correspond to a sphere radius $S = 0.5a$ and dashed lines correspond to average *T*-matrix approximation with an average radius $S_{av} = 0.5a$ and standard deviation $\sigma = 0.05S_{av}$. (Reprinted with permission from L. A. Dorado and R. A. Depine, Phys. Rev. B **79**, 045124, 2009. Copyright 2009, by the American Physical Society.)

Fabry–Pérot oscillations related to the finite thickness of the slab. In the high-energy range, an abruptly fluctuating behavior of the reflectance can be seen with several high peaks of unitary amplitude. As in the case of sphere monolayers,[32] these peaks arise from internal resonances of the PC that can be classified according to the main multipole interactions involved in the response of the system.[9] In Figure 2.3.5b, the specular reflectance spectrum has been recalculated for $\sigma/S_{av} = 0.025$ and two different concentrations of vacancies. In the case of zero vacancies ($C = 1$, gray dashed line) and within the low-energy range, the reflectance is basically the same as that shown in Figure 2.3.5a, which means that a small degree of disorder has little effect on the optical response for these low energies. However, the spectrum changes dramatically in the high-energy range, where the reflectance values have decreased significantly and several peaks have practically disappeared. By adding 5% of vacancies ($C = 0.95$, black solid line), the reflectance decreases even more and several peaks are smoothed out. In this case, only three peaks around $a/\lambda \sim 1.2$ and two peaks above $a/\lambda \sim 1.6$ can be observed. The transmittance spectrum plotted in Figure 2.3.5c shows the same trend.

Although very fine-featured reflection spectra as that shown in Figure 2.3.5a have not been experimentally reported yet to the authors' knowledge, the optical spectra in Figure 2.3.5b and c are quite similar to measurements published in the literature. From these results we realize that the introduction of vacancies in the model seems to be essential to accurately calculate the optical response of real artificial opals in the high-energy range. Whereas the inclusion of dispersion in sphere radii attenuates several sharp resonances of the perfect lattice, the inclusion of empty sites in the lattice seems to produce a smoothing

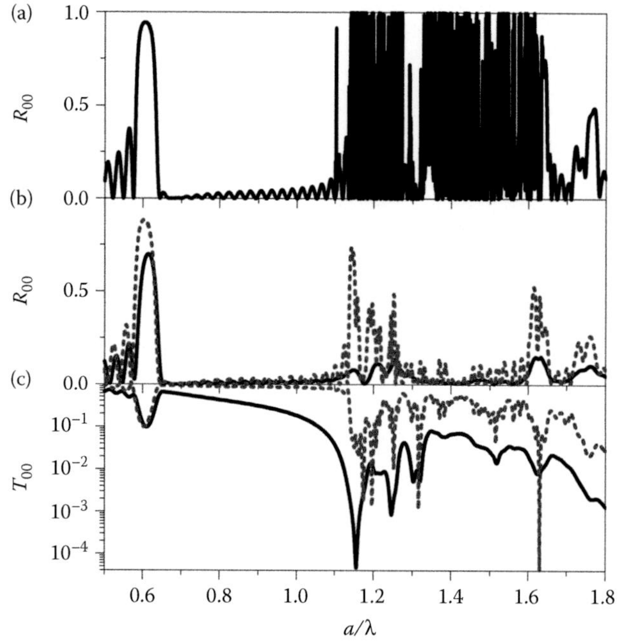

FIGURE 2.3.5
(a) Specular reflectance spectrum for a perfect 18-layer-thick crystal made of spheres of dielectric constant $\varepsilon_s = 2.5$ in air. (b) Specular reflectance and (c) forward transmittance spectra obtained using the follwing average T-matrix approximation parameters: $S_{av} = 0.5a$, $\sigma = 0.05S_{av}$, $C = 0.95$ (black solid lines), and $C = 1$ (gray dashed lines). (Reprinted with permission from L. A. Dorado and R. A. Depine, Phys. Rev. B **79**, 045124, 2009. Copyright 2009, by the American Physical Society.)

effect on the optical spectra. In Figure 2.3.6, we show the optical spectra obtained with the ATA for $C = 0.95$ and various values of σ/S_{av}. We can observe that reflectance peaks are attenuated as σ/S_{av} increases, whereas transmittance dips become more pronounced.

2.3.5 Comparison between the Results Obtained Using the Extinction and ATA Methods

In the inner extinction approximation (IEA) method, a great variety of imperfections is represented by a single parameter through the imaginary part of the dielectric constant of the latex spheres. Analogous to the IEA, another single parameter extinction approach to account for the diffusely scattered light that is removed from coherent beams could be envisaged. In this case, the energy-loss mechanism is provided by a nonzero imaginary part of the dielectric constant ε_i of the medium surrounding the spheres, instead of being provided by a nonzero imaginary part of the dielectric constant ε_s of the latex spheres.[17] Although for the case of latex spheres in air the reader could feel more comfortable when latex and not air is to be considered the lossy medium, both extinction approximations are equivalent from a modeling point of view, and we have checked that the extinction method provides similar fittings of experimental curves. On the other hand, and in contrast with

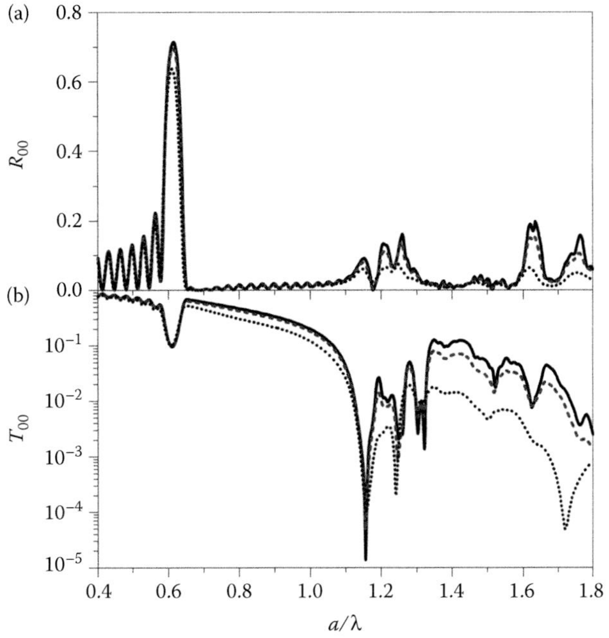

FIGURE 2.3.6
(**See color insert.**) (a) Specular reflectance and (b) forward transmittance spectra for an 18-layer-thick crystal made of spheres of dielectric constant $\varepsilon_s = 2.5$ in air. The average *T*-matrix approximation parameters are: $S_{av} = 0.5a$, $C = 0.95$, and the spectra are shown for different values of σ/S_{av}. In particular, $\sigma/S_{av} = 0.010$ (black solid line), $\sigma/S_{av} = 0.025$ (red dashed line), and $\sigma/S_{av} = 0.050$ (blue dotted line). (Reprinted with permission from L. A. Dorado and R. A. Depine, Phys. Rev. B **79**, 045124, 2009. Copyright 2009, by the American Physical Society.)

these single-parameter approaches, we have seen that the use of the ATA method allowed us to include different sources of deviation from perfect periodicity, represented in our previous examples by different parameters such as σ, S_{av}, or C. Taking into account that (i) the extinction approximations are computationally less demanding than the ATA and (ii) vacancies seem to play an important role in the optical response of self-assembled opal samples that are being produced nowadays, we have sought an improved method that, as in the ATA approach, incorporates the parameter $1 - C$ representing the concentration of vacancies while avoiding, as in the extinction approaches, direct computation of the average *T*-matrix demanded by the ATA method. The improvement can be achieved by noting that when vacancies are considered to be the only source of disorder, that is, when we assume that only a fraction $C < 1$ of lattice sites is occupied by identical scatterers described by the same *T*-matrix **T**, then the average *T*-matrix adopts the very simple form,

$$\langle \mathbf{T} \rangle = C\mathbf{T} \tag{2.3.6}$$

If we are using the ATA method to obtain the optical response of a PC affected by vacancies as the only source of disorder, the calculation of the average *T*-matrix $\langle \mathbf{T} \rangle$ does not require further explicit statistical calculations. Unfortunately, the convenient form of Equation 2.3.6 must be abandoned and explicit statistical calculations—as those performed to obtain the results presented in Section 2.3.2—are required for real samples affected not only by vacancies but also by other deviations from perfect periodicity. It is clear that for a

PC constructed from lossless building blocks, real dielectric constants are to be used for the calculation of the scattering matrix **T** in Equation 2.3.5 or Equation 2.3.6. However, we could keep the convenient form of Equation 2.3.6 for a PC affected only by vacancies if all the remaining deviations from perfect periodicity are assigned to extinction through the introduction, as in the extinction method, of a nonzero imaginary part of the dielectric constant of the PC building blocks. By doing so, we can obtain more flexible versions of the simple extinction approach where the effect of imperfections is managed through the use of two, instead of one, parameters. Although it is not our purpose to investigate here how each parameter affects the reflectance and transmittance spectra, some hints can be obtained by the ATA results presented in Figures 2.3.5 and 2.3.6, where the inclusion of dispersion in sphere radii seemed to attenuate several sharp resonances of the perfect lattice, while the introduction of vacancies seemed to produce a smoothing effect on the optical spectra, but the question remains whether the effect of vacancies can be distinguished from the effect of other kind of imperfections by only measuring reflectance and transmittance spectra.

IEA and ATA approaches are useful theoretical tools for reproducing the optical response of weakly disordered PCs. They all provide very good fittings of experimental data, as we will discuss in depth in Chapter 3.8, and essentially they lead to the same results when the disorder parameters are properly chosen. This can be observed in Figure 2.3.7 where we have compared IEA and ATA methods for the close-packed FCC lattice of spheres of dielectric constant $\varepsilon_s = 2.5$ embedded in a medium of $\varepsilon = 1$. To simulate disorder effects with the IEA method we have added an imaginary part to the dielectric constant of the

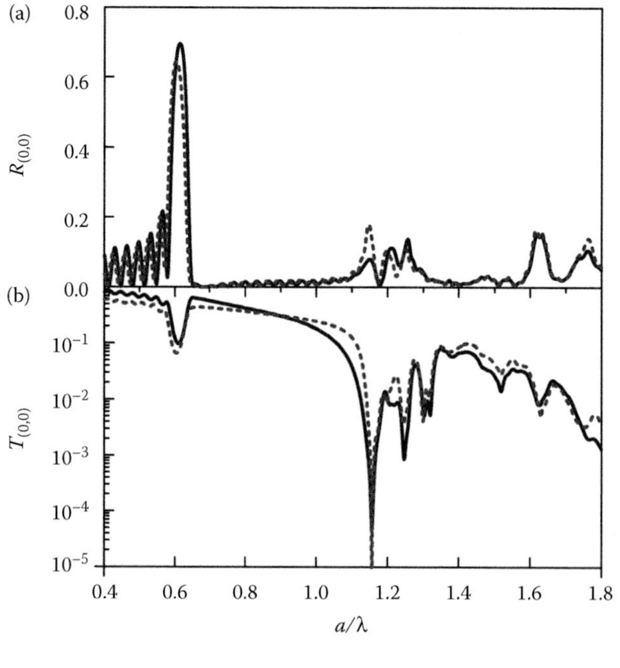

FIGURE 2.3.7
Comparison between the average *T*-matrix approximation (ATA) (solid lines) and inner extinction approximation (IEA) (dashed lines) methods. (a) Specular reflectance and (b) forward transmittance spectra for an 18-layer-thick crystal. ATA parameters are: $\varepsilon_s = 2.5$, $\varepsilon = 1$, $S_{av} = 0.5a$, $\sigma = 0.025S_{av}$, and $C = 0.95$, whereas IEA parameters are $\varepsilon_s = 2.5 + 0.04i$, $s = 1$, and $C = 1$. (Adapted from L. A. Dorado and R. A. Depine, Phys. Rev. B **79**, 045124, 2009.)

spheres ($\varepsilon_s = 2.5 + 0.04i$), and no vacancies ($C = 1$) have been introduced. For the ATA method we have used $S_{av} = 0.5a$, $\sigma = 0.025S_{av}$, and $C = 0.95$.

The close agreement between the ATA method and the extinction approach indicates that the reflectance and transmittance spectra of a 3D colloidal PC in the high-energy range are almost determined by the optical extinction due to imperfections and defects in the crystalline structure of the slab.[3] These results would also indicate that a perfect PC with material absorption generates almost the same reflectance and transmittance as a disordered PC without material loss; however, more experimental work is required in this direction. Besides, we must point out that these results are valid when no diffracted beams emerge from the PC slab,[33] which would provide a means to distinguish between different losses and disorder sources in higher-energy ranges.

2.3.6 Equivalence between IEA and ATA Approaches

The equivalence between the extinction and the average *T*-matrix approaches can easily be proved in the long wavelength limit, that is, for small values of a/λ. To do so, we consider a perfectly periodic lattice of identical lossless spheres with radius S_{av} and real dielectric constant ε_r. In this case, the average *T*-matrix is identical to the *T*-matrix of a single sphere. In the next step, we consider two physically different systems that can be regarded as perturbations of the perfectly periodic lattice of identical lossless spheres. In the first perturbed system, the spheres are not identical but their radii are randomly distributed following a Gaussian process with average S_{av} and standard deviation σ. In this case, the perturbation of the the ATA matrix is approximately given by

$$\langle \mathbf{T} \rangle = \int_{-\infty}^{+\infty} \mathbf{T}(s)P(s)ds \cong \mathbf{T}(S_{av}, \varepsilon_r) + \frac{\sigma^2}{2} \frac{\partial^2 \mathbf{T}(S_{av}, \varepsilon_r)}{\partial s^2} \tag{2.3.7}$$

which holds for small values of σ/S_{av}. In the second perturbed system, the spheres are identical but their dielectric constant is changed from the real value to a generally complex value $\varepsilon_s = \varepsilon_r + \Delta\varepsilon_s$, with $|\Delta\varepsilon_s| \ll \varepsilon_r$. Under this assumption, the ATA matrix of this system can be written as

$$\langle \mathbf{T} \rangle \cong \mathbf{T}(S_{av}, \varepsilon_r) + \frac{\partial \mathbf{T}(S_{av}, \varepsilon_r)}{\partial \varepsilon_s} \Delta\varepsilon_s \tag{2.3.8}$$

For both perturbed systems to exhibit the same response, the ATA matrices given by Equations 2.3.7 and 2.3.8 should coincide, a requirement that allows us to find an equivalence relation between $\Delta\varepsilon_s$ and σ. To obtain the explicit form of this relation, we note that in the long wavelength limit each sphere behaves essentially as electric and magnetic dipoles. In this case,

$$T_1^E \cong -\frac{4}{9}(ks)^6 \left(\frac{\varepsilon_s - \varepsilon}{\varepsilon_s + 2\varepsilon} \right)^2 + i\frac{2}{3}(ks)^3 \frac{\varepsilon_s - \varepsilon}{\varepsilon_s + 2\varepsilon} \tag{2.3.9}$$

Besides, in the long wavelength limit and for a nonmagnetic dielectric material it is $|T_1^H| << |T_1^E|$. Thus, the following equivalence between $\Delta\varepsilon_s$ and σ can be obtained:

$$\Delta\varepsilon_s = \left(\frac{\sigma}{S_{av}}\right)^2 \left(\frac{\varepsilon_r - \varepsilon}{\varepsilon}\right)\left[(\varepsilon_r + 2\varepsilon) + i2(kS_{av})^3(\varepsilon_r - \varepsilon)\right] \quad (2.3.10)$$

This relation shows the equivalence between the two physically different systems that we have considered as perturbations of the perfectly periodic lattice of identical lossless spheres and it establishes a relationship between disorder effects and energy losses.

2.3.7 Conclusions

In summary, we have presented two different approches on the basis of the KKR method to evaluate disorder effects in the optical response of PCs, that is, the extinction approximation method and the ATA method. Our numerical results show that disorder has little effect on the optical spectra in the low-energy range but affects dramatically the response of the system in the high-energy range. Both methods provide similar results and, as shown in Chapter 3.8, a very good agreement with experimental results attained from actual PCs with negligible material loss. However, more experimental work on weakly disordered crystals with a certain amount of material loss is needed to demonstrate the usefulness of the ATA and extinction methods in these cases.

Acknowledgments

This work has been partially supported by the Spanish Ministry of Science and Innovation under Grant MAT2007-02166 and Consolider HOPE CSD2007-00007, Junta de Andalucía under Grant FQM3579, Consejo Nacional de Investigaciones Científicas y Técnicas (CONICET) and Agencia Nacional de Promoción Científica y Tecnológica (PICT-11–1785).

References

1. E. Yablonovitch, Phys. Rev. Lett. **58**, 2059 (1987); K. Sakoda, *Optical Properties of Photonic Crystals* (Springer-Verlag, Berlin/Heidelberg, 2001).
2. J. D. Joannopoulos, R. D. Meade, and J. N. Winn, *Photonic Crystals* (Princeton University Press, Princeton, 1995); K. M. Ho, C. T. Chan, and C. M. Soukoulis, Phys. Rev. Lett. **65**, 3152 (1990).
3. L. A. Dorado, R. A. Depine, and H. Míguez, Phys. Rev. B **75**, 241101(R) (2007).
4. P. D. García, R. Sapienza, L. S. Froufe-Pérez, and C. López, Phys. Rev. B **79**, 241109(R) (2009).
5. A. F. Koenderink, A. Lagendijk, and W. L. Vos, Phys. Rev. B **72**, 153102 (2005).
6. J. F. Galisteo-López, M. Galli, M. Patrini, A. Balestreri, L. C. Andreani, and C. López, Phys. Rev. B **73**, 125103 (2006).

7. L. A. Dorado, G. Lozano, H. Míguez, and R. A. Depine, Phys. Rev. B **76**, 245103 (2007).
8. G. Lozano, L. A. Dorado, R. A. Depine, and H. Míguez, J. Mater. Chem. **19**, 185 (2009).
9. L. A. Dorado, G. Lozano, H. Míguez, and R. A. Depine, Opt. Express **15**, 17754 (2007).
10. J. Barrio, G. Lozano, J. Lamela, G. Lifante, L. A. Dorado, R. A. Depine, F. Jaque, and H. Míguez, J. Appl. Phys. 109, 083514 (2011); L. A. Dorado, G. Lozano, D. Shinca, R. A. Depine, and H. Míguez, Phys. Rev. B **78**, 075102 (2008); G. Lozano, J. E. Mazzaferri, L. A. Dorado, S. Ledesma, R. A. Depine, and H. Míguez, J. Opt. Soc. Am. B **27**, 1394 (2010).
11. J. L. Beeby, Proc. R. Soc. London **279**, 82 (1964); J. Phys. C **1**, 82 (1968).
12. M. I. Mishchenko, L. D. Travis, and A. A. Lacis, *Multiple Scattering of Light by Particles* (Cambridge University Press, Cambridge, 2006).
13. A. Modinos, Physica A **141**, 575 (1987).
14. N. Stefanou and A. Modinos, J. Phys. Condens. Matter **3**, 8149 (1991).
15. V. Yannopapas, Phys. Rev. B **75**, 035112 (2007).
16. A. Modinos, V. Yannopapas, and N. Stefanou, Phys. Rev. B **61**, 8099 (2000).
17. L. A. Dorado and R. A. Depine, Phys. Rev. B **79**, 045124 (2009).
18. K. Ohtaka, J. Phys. C **13**, 667 (1980).
19. N. Stefanou, V. Karathanos, and A. Modinos, J. Phys. Condens. Matter **4**, 7389 (1992).
20. N. Stefanou, V. Yannopapas, and A. Modinos, Comput. Phys. Commun. **113**, 49 (1998); **132**, 189 (2000).
21. J. Korringa, Physica **13**, 392 (1947); W. Kohn and N. Rostoker, Phys. Rev. **94**, 1111 (1954).
22. K. P. Velikov, A. Moroz, and A. van Blaaderen, Appl. Phys. Lett. **80**, 49 (2002); G. Gantzounis and N. Stefanou, Phys. Rev. B **73**, 035115 (2006).
23. J. D. Jackson, *Classical Electrodynamics* (Wiley, New York, 1975).
24. P. Jiang, J. F. Bertone, K. S. Hwang, and V. L. Colvin, Chem. Mater. **11**, 2132 (1999).
25. M. Allard, E. Sargent, E. Kumacheva, and O. Kalinina, Opt. Quantum Electron. **34**, 27 (2002).
26. A. Modinos, N.Stefanou, and V. Yannopapas, Opt. Express **8**, 197 (2001).
27. Y. A. Vlasov, V. N. Astratov, A. V. Baryshev, A. A. Kaplyanskii, O. Z. Karimov, and M. F. Limonov, Phys. Rev. E **61**, 5784 (2000); R. Rengarajan, D. Mittleman, C. Rich, and V. Colvin, Phys. Rev. E **71**, 016615 (2005).
28. T. Inagaki, E.T. Arakawa, R. N. Hamm, and M. W. Williams, Phys. Rev. B **15**, 3243 (1977).
29. J. F. Galisteo-López and W. L. Vos, Phys. Rev. E **66**, 036616 (2002); N. Tetrault, A. C. Arsenault, A. Mihi, S. Wong, V. Kitaev, I. Manners, H. Míguez, and G. A. Ozin, Adv. Mater. **17**, 1912 (2005); P. V. Braun, S. A. Rinne, and F. García-Santamaría, *ibid*. **18**, 2665 (2006); G. von Freymann, S. John, S. Wong, V. Kitaev, and G. A. Ozin, Appl. Phys. Lett. **86**, 053108 (2005); G. Lozano and H. Míguez, *ibid*. **92**, 091904 (2008).
30. K. Busch and S. John, Phys. Rev. Lett. **83**, 967 (1999).
31. D. Gerace and L. C. Andreani, Opt. Lett. **29**, 1897 (2004); S. Hughes, L. Ramunno, J. F. Young, and J. E. Sipe, Phys. Rev. Lett. **94**, 033903 (2005).
32. M. Inoue, Phys. Rev. B **36**, 2852 (1987); Y. Kurokawa, Y. Jimba, and H. Miyazaki, *ibid*. **70**, 155107 (2004).
33. F. García-Santamaría, J.F. Galisteo-López, P.V. Braun, and C. López, Phys. Rev. B **71**, 195112 (2005).

2.4

Anderson Localization of Light in Layered Dielectric Structures

Yury Bliokh
Technion—Israel Institute of Technology

Konstantin Y. Bliokh
National University of Ireland

Valentin Freilikher
Bar-Ilan University

Franco Nori
Advanced Science Institute, RIKEN
University of Michigan

CONTENTS

2.4.1 Introduction

Wave propagation in disordered media is a rich and long-standing problem that attracts many efforts, both theoretical and experimental. After almost a century of a complete sway of radiative transport and diffusion approaches, it recently became clear that the interference of multiply-scattered fields (which is neglected in classical diffusion theory) dramatically affects all wave processes, especially in systems with fluctuations. The most startling manifestation of this effect is the strong localization of electromagnetic radiation in weakly-disordered random media. Since Anderson's seminal paper,[1] localization has attracted ever-increasing attention from physicists and engineers. Without attempting an exhaustive review of available literature we note that the bibliography related to this rather young area already numbers in hundreds of original articles, reviews, and books (see, e.g., the review,[2] the monographs,[3] and references therein). Worthy of notice are the great number of meetings dedicated to the 50 years' anniversary of Anderson localization (half-a-year non-stop workshop in Cambridge, conferences in Paris, Dresden, Santa Barbara, etc.) where physicists, mathematicians, chemists, engineers, biologists, and even economists not only presented a plethora of new results but also formulated a great many challenging questions.

A boost to the studies of Anderson localization in disordered optical and quantum systems was given recently by the creation of new materials with unique properties that have spurred the rise of new conceptual challenges and high-tech applications. The most impressive latest examples include photonic crystals, plasmonics, left-handed metamaterials, Bose–Einstein condensates, and graphene. Yet it should be remembered that most of the prospects for potential technological use of these materials rest on the predicted properties of ideal (e.g., perfectly periodic) systems. Even a small amount of disorder, however, which is inevitably present in any real sample, could affect its properties dramatically (see Figure 2.4.1). Therefore, when it comes to real applications, a comprehensive study of the effects of disorder is a must. Moreover, these investigations are of interest by itself because strongly disordered (with no periodic component) systems possess further unexpected physical properties, which make them potentially useful as an alternative to the pure periodic configurations.

One-dimensional (1D) strong localization has received the most study, both analytically and numerically. In particular, the localization of the eigenstates in closed 1D disordered systems and the exponentially small (with respect to the length) transparency of open systems with 1D disorder have been scrutinized with mathematical rigor (e.g., see Ref. 4 and references therein).

The most common physical manifestation of localization is the fact that sheets of perfectly transparent paper stacked together in large numbers reflect light as a good mirror.[5] Much less evident (though long predicted theoretically[6]) is that for each sufficiently long disordered 1D sample, there exists a random set of frequencies that go all the way through the sample almost unreflected, that is, with the transmission coefficient close to unit. High transparency is always accompanied by a relatively large concentration (localization) of energy around randomly located points inside the system. Along with these "classical" trademarks of strong localization there is a plethora of not-less-amazing effects that disorder can set up in one dimension. Examples are: random lasing,[7,8] critical coupling,[9] necklace states,[10,11] level crossing and repulsion,[9] slow light and superluminal group velocities,[12] bistability and nonreciprocity of resonant transmission in nonlinear random media,[13] delocalization in metamaterials,[14–16] and in graphene superlatices,[17] and so on.

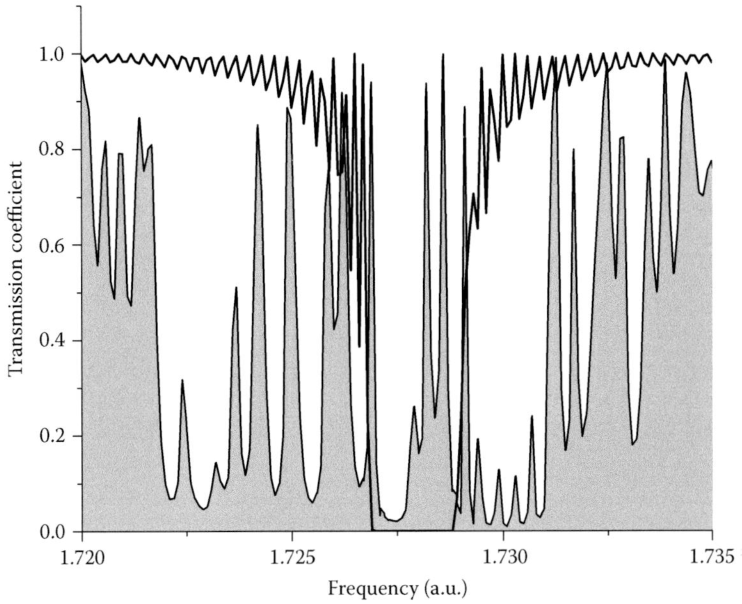

FIGURE 2.4.1
Transmission spectra of a regular periodic sample (thick black line) and of a sample whose period fluctuates in the range of 1%.

Although, generally, a disordered 1D system is a mathematical abstraction, it can provide an adequate model for many actual physical objects. For example, randomly stratified media are found in numerous geological and biological settings, as well as in fabricated materials. Interference of waves in such systems determines the transport of seismic waves in the earth's crust and sonic waves in the oceans; reflection and transmission from multilayer dielectric stacks used as optical reflectors, filters, and lasers; propagation and localization in single-mode optical fibers and microwave waveguides, etc. Even more important, it may be possible to utilize highly disordered samples for many applications. For instance, tunable switches or narrow-line laser sources can be created using randomly stacked systems.

Although the strong localization of waves in 1D random media has been well studied theoretically, most of the analytical results have been obtained for values averaged over ensembles of random realizations. These results are physically meaningful for the self-averaging Lyapunov exponent (inverse localization length), which becomes nonrandom in the macroscopic limit. For non-self-averaging quantities (field amplitude and phase, intensity, transmission and reflection coefficients, etc.), a system of any size is always mesoscopic, and, therefore, mean values have little to do with the measurements at individual (usually small in number) samples. This is most pronounced when it comes to disorder-induced resonances whose parameters are extremely sensitive to the fine structure of a particular sample and strongly fluctuate from realization to realization. In particular, the ensemble averaging wipes out all information about the frequencies and locations of individual localized states within a particular sample, even though just this set of data is essential for applications. Another frustrating inconsistency between most of the existing theories and measurements is that in real systems, losses (absorption and leakage) are inevitably present, whereas mathematicians and theoreticians usually prefer lossless (Hermitian) models that are much easier to deal with.

In this chapter, we present a brief overview of methods and results regarding the transport and localization in disordered 1D systems, followed by a detailed description of the current state-of-the-art in theoretical and experimental studies of the resonant properties of randomly layered media.

2.4.2 Lyapunov Exponent, Localization, and Transmission Lengths

Consider the 1D Helmholtz equation

$$u''(x) + k^2[\varepsilon_0 + \delta\varepsilon(x)]u(x) = 0 \tag{2.4.1}$$

with the self-adjoined (currentless) boundary condition at a point x_0

$$u(x_0) + au'(x_0) = 0, \tag{2.4.2}$$

where a is any real number. It is easy to show that Equation 2.4.2 means that the modulus of the reflection coefficient from the point x_0 equals to one. The functions ξ and φ determined as

$$\begin{aligned} u(x) &= e^{\xi(x,k)}\sin\varphi(x,k), \\ u'(x) &= ke^{\xi(x,k)}\cos\varphi(x,k) \end{aligned} \tag{2.4.3}$$

satisfy the following (nonlinear) equations:

$$\begin{aligned} \varphi'(x,k) &= k[1 - \delta\varepsilon(x)\sin^2\varphi(x,k)] \\ \xi(x,k) &= \xi(0,k) + \frac{k}{2}\int_0^x \delta\varepsilon(y)\sin 2\varphi(y,k)\,dy. \end{aligned} \tag{2.4.4}$$

Assuming that $\delta\varepsilon(x)$ is a statistically homogeneous random function with zero average, $\langle\delta\varepsilon(x)\rangle = 0$, and disappearing at infinity correlations, $W(x) = \langle\delta\varepsilon(0)\delta\varepsilon(x)\rangle_{x\to\infty} \to 0$, the following statement is true:[4] in the limit $|x| \to \infty$, the ratio $\xi(x,k)/x$ approaches a non-random limit that is positive for all k:

$$\lim_{|x|\to\infty}\frac{\xi(x,k)}{x} = \lim_{|x|\to\infty}\left\langle\frac{\xi(x,k)}{x}\right\rangle \equiv \gamma(k) > 0. \tag{2.4.5}$$

In principle, this result follows from the Furstenberg theorem,[18] which holds that, under some conditions, the specific logarithm of the product of N transfer matrices M_j tends to a positive limit as N goes to infinity:

$$\lim_{N\to\infty}\left(\frac{1}{N}\ln\prod_{j=1}^{N} M_j\right) = \mu > 0.$$

Equation 2.4.5 presents two fundamental properties of a 1D random system satisfying the above-listed conditions:

1. The Lyapunov exponent, $\gamma(k)$, is a self-averaging quantity; that is, at any single random realization it tends to the ensemble-averaged nonrandom mean value when the size of the realization increases infinitely.
2. For a single random realization, the amplitude of the wave function increases exponentially with nonrandom increment $\gamma(k)$ on both sides of the point, at which the currentless boundary condition (2.4.2) holds.

The inverse of the Lyapunov exponent

$$l_{\text{loc}} = \frac{1}{2\gamma} \tag{2.4.6}$$

is called the localization length. The meaning of this term becomes clear if one considers two solutions of Equation 2.4.1, $u_1(x)$ and $u_2(x)$, in the interval $0 \le x \le L$ (closed 1D system), each satisfying the boundary condition (2.4.2) at $x = 0$ and $x = L$, respectively, and therefore each increasing exponentially away from these points. An eigenfunction, $\psi_n(x)$, of this system can be constructed from $u_1(x)$ and $u_2(x)$ under the condition that these functions and their derivatives match at some point x_n inside the interval. Obviously, an eigenfunction, $\psi_n(x)$, obtained in this way is localized; that is, its envelop, $A^2(x)$, decreases exponentially on both sides of x_n:

$$A^2(x) \equiv \left[|u(x)|^2 + \frac{1}{k^2} |u'(x)|^2 \right] \sim \exp\left(-\frac{|x - x_n|}{l_{\text{loc}}} \right) \tag{2.4.7}$$

In the case of the white-noise disorder [$\delta\varepsilon(x)$ is a δ-correlated random process], the closed Fokker–Planck equation for the probability density distribution, $P(\xi, z)$, of the quantity $\xi(x)$ can be derived and solved using a averaging over rapid random phase.[4,19] This is impossible if $\delta\varepsilon(x)$ has a finite correlation radius, and the random-phase approximation breaks down (as well as the single-parameter scaling theory of localization). In this case, the ordered cummulant method of Van-Kampen can be used to obtain the weak disorder expansion of the Lyapunov exponent $\gamma(k)$.[20] The first term of this expansion can be obtained from Equation 2.4.4 by solving the equation for the phase perturbatively and substituting the result into the integral for $\xi(x,k)$. The limit $|x| \to \infty$ yields the famous result

$$\gamma(k) = \frac{k^2}{4} \int_0^\infty \mathrm{d}x\, W(x) \cos 2kx, \tag{2.4.8}$$

which means that the localization is due to the resonant Brag backscattering provided by the $2k$ Fourier component of the random potential. Higher orders of the weak disorder expansion have been calculated and can be found in the literature.[20,21]

The notion of the Lyapunov exponent is related to the eigenvalues boundary problem, and therefore is well defined only for closed disordered systems. From the physical point of view, not less relevant is the scattering problem that addresses the transmission, reflection, and propagation in open structures with fluctuating parameters.

Transport properties of a 1D system of a finite length L can be described by the transfer matrix $\hat{M}$, which relates the amplitudes of the incident (A_L) and outgoing (B_L) waves on one side of the sample to those on the other side (A_R and B_R):

$$\begin{pmatrix} B_R \\ A_R \end{pmatrix} = \hat{M} \begin{pmatrix} B_L \\ A_L \end{pmatrix} \tag{2.4.9}$$

Assuming time-reversal invariance, the transfer matrix $\hat{M}$ can be written as

$$\hat{M} = \left\| \begin{matrix} \frac{1}{t^*} & -\left(\frac{r}{t}\right)^* \\ -\left(\frac{r}{t}\right) & \frac{1}{t} \end{matrix} \right\| = \frac{1}{\sqrt{T}} \left\| \begin{matrix} e^{i\phi_t} & -\sqrt{1-T}e^{-i(\phi_r-\phi_t)} \\ -\sqrt{1-T}e^{i(\phi_r-\phi_t)} & e^{-i\phi_t} \end{matrix} \right\|, \tag{2.4.10}$$

where $t = \sqrt{T}\exp(i\phi_t)$, and $r = \sqrt{1-T}\exp(i\phi_r)$ are the transmission and reflection amplitudes, respectively; $T \equiv |t|^2$ is the transmission coefficient. In what follows, the quantity $l_{\mathrm{tr}}(L) = \left[-\frac{1}{2L}\ln T(L)\right]^{-1}$ is called the transmission length.

Evidently, the solution of the boundary value problem (2.4.1), (2.4.2) and the solution of the scattering problem in the limit $L \to \infty$ are uniquely related; that is,

$$\begin{pmatrix} u(L) \\ u'(L) \end{pmatrix} = \hat{M}' \begin{pmatrix} u(0) \\ u'(0) \end{pmatrix}$$

where $\hat{M}'$ is an elementary linear transformation of the transfer matrix $\hat{M}$.

When the size L of the system is much larger than the localization length l_{loc}, the transmission coefficient T becomes exponentially small (with the probability exponentially close to one), and statistically independent of the phases ϕ_t and ϕ_r. As a result, each matrix $\hat{M}$ and $\hat{M}'$ factorizes into a product of a large factor $1/\sqrt{T}$ and a matrix of the order of unity, which is statistically independent of T. This means that, asymptotically for large L, one can write

$$\gamma(k) = \frac{1}{2L}\ln\left[|u(L)|^2 + \frac{1}{k^2}|u'(L)|^2\right] = -\frac{1}{2L}\ln T(L) + O(L^0), \tag{2.4.11}$$

where the second term in the asymptotic expansion is independent of the first one, and their cummulants are of the order of unity. Apparently, in the limit $L \to \infty$, the transmission length coincides with the inverse Lyapunov exponent (localization length):

$$l_{\mathrm{tr}} = l_{\mathrm{loc}}. \tag{2.4.12}$$

This fact is generally recognized. However, it was shown recently[14] that, surprisingly enough, l_{loc} and l_{tr} can be different in stacks made of alternating right- and left-handed dielectric layers with random refractive indices and thicknesses.

The equality Equation 2.4.12 means, in particular, that the localization length can be probed noninvasively from the transmission coefficient, without measurements of the field amplitude inside random samples.

2.4.3 Statistics of the Transmission in 1D Disordered Systems

2.4.3.1 Transport and Localization in Randomly Layered Media

2.4.3.1.1 Normal Incidence: Transfer Matrix Method

One of the most efficient theoretical methods of studying general properties of transmission in 1D disordered systems is based on the composition rule for a chain of statistically identical and independent random scatterers.[22] For stratified media, the method is straightforward and involves the calculation of the transfer matrix $\hat{M}$ using the following exact recurrence relations for the transmission coefficients (for details, see Ref. 14 and references therein):

$$T_n = \frac{T_{n-1}t_n}{1 - R_{n-1}r_n}, \tag{2.4.13}$$

$$R_n = r_n + \frac{R_{n-1}t_n^2}{1 - R_{n-1}r_n} \tag{2.4.14}$$

Here T_n and R_n are, respectively, the total transmission and reflection coefficients of a stack of n layers, t_n and r_n are the (complex) transmission and the reflection amplitudes of a single layer. Equations 2.4.13 and 2.4.14 are general and, taking into account all multiply-scattered fields, present exact solutions that can be used for direct numerical simulations.

For a sample composed of N statistically identical and independent random layers of normal (with positive refractive index) dielectrics, the following expression for the average inverse transmission length can be derived from Equations 2.4.13 and 2.4.14:[14]

$$\left\langle \frac{1}{l_{\text{tr}}(N)} \right\rangle = \frac{1}{l_{\text{loc}}} + \frac{1}{Nd}\text{Re}\left[\langle r \rangle^2 \frac{1-\langle t^2 \rangle^N}{\left(1-\langle t^2 \rangle\right)^2} \right], \tag{2.4.15}$$

where d is the average width of the layers, and the inverse localization length is

$$\frac{1}{l_{\text{loc}}} = \frac{1}{l_{\text{tr}}(\infty)} = -\langle \ln |t| \rangle - \text{Re}\frac{\langle r \rangle^2}{1-\langle t^2 \rangle}. \tag{2.4.16}$$

Note that when $N \to \infty$, Equation 2.4.15 transforms into Equation 2.4.12, that is, the localization and transmission lengths become equal.

In the case of weak scattering, the reflection from a single layer is small, $|r_n| \ll 1$ and Equation 2.4.16 yields:

$$l_{\text{loc}} = \begin{cases} \dfrac{12d}{\delta^2}, & \lambda = 2\pi/k \to 0, \\ \dfrac{3\lambda^2 d}{2\pi^2\delta^2}, & \lambda = 2\pi/k \to \infty. \end{cases} \tag{2.4.17}$$

In the derivation of Equation 2.4.17, it was assumed that the width of each layer was distributed uniformly over the interval $[d - \delta,\ d + \delta]$, and the refractive index did not fluctuate. Nevertheless, the functional λ-dependences, Equation 2.4.17, of the localization length in weakly disordered systems are rather general (e.g., see Ref. 23).

2.4.3.1.2 Oblique Incidence: Reduction to the Oscillatory Problem

An original efficient method of calculating the localization length was developed in Ref. 24. It uses the fact that the reflection from an adequately long, randomly layered sample differs from unity by an exponentially small number, $1 - R(L) \sim \exp(-L/l_{\text{loc}})$, and, therefore, the flux along the system is also exponentially small. This *a priori* information enables one to assume (with an exponential accuracy) that the field in each layer inside the sample is a standing wave, and to reduce the wave propagation problem to the oscillatory one, with the real-valued wave amplitude being a single unknown. This simplifies the problem significantly as compared to the conventional transfer matrix method, where the evolution of two independent waves in each layer is considered. Using this method, the oblique incidence of electromagnetic waves on a randomly layered medium was studied.[24] Two effects not found at normal incidence were predicted: dependence of the localization length on the polarization and the decrease of the localization length as a result of the internal reflections from layers with small refractive indices. The attenuation rate for *p*-polarized radiation is shown to be always smaller than that of *s*-polarized waves, which is to say that an adequately long, randomly layered sample polarizes transmitted radiation. The localization length for *p*-polarization depends nonmonotonically on the angle of propagation and, under certain conditions, turns to infinity at some angle, which means that typical (nonresonant) random realizations become transparent at this angle of incidence (stochastic Brewster effect).

2.4.3.2 Transport and Localization in Continuous Active Media

2.4.3.2.1 Invariant Embedding Method

An alternative approach to the 1D random scattering problem is the invariant embedding method,[25] which amounts to finding the solution of the following system of (exact) first-order Langevin-type equations for the reflection and transmission coefficients:

$$\frac{\mathrm{d}r(L)}{\mathrm{d}L} = \frac{i}{2}k\varepsilon(L)\left[e^{-ikL} + r(L)e^{ikL}\right]^2 \tag{2.4.18}$$

$$\frac{\mathrm{d}t(L)}{\mathrm{d}L} = \frac{i}{2}k\varepsilon(L)t(L)\left[e^{-ikL} + r(L)e^{2ikL}\right] \tag{2.4.19}$$

Equations 2.4.18 and 2.4.19 can be treated statistically by means of exact numerical calculations and approximate analytical methods as well. A Fokker–Planck equation can be derived, which, in the case of the white-noise disorder, yields the distribution function for the reflection coefficient $R(L) = |r(L)|^2$. In the absence of absorption or amplification, this distribution function provides complete information on the transmission coefficient $T = |t(L)|^2$.

The problem becomes much more complicated in the case of lossy media where the energy conservation law not only connects T and R but also involves a random amount of the absorbed intensity. In the paper,[26] the asymptotically exact expressions for all moments of the transmission coefficient have been obtained by mapping the Fokker–Planck problem onto a Shrödinger equation with imaginary time. In particular, it has been shown that in the case of small absorption, $l_{\text{loc}} << l_a$ (l_a is the absorption length),

$$-\frac{\ln\langle T(L)\rangle}{L} = \left(\frac{1}{4l_{\text{loc}}} + \frac{1}{l_{\text{in}}}\right), \qquad (2.4.20)$$

where

$$l_{\text{in}} = l_{\text{loc}} \ln^2\left(\frac{l_a}{l_{\text{loc}}}\right) \qquad (2.4.21)$$

is a disorder-induced absorption length, which lies between the localization and absorption lengths, see Ref. 26

$$l_{\text{loc}} \ll l_{\text{in}} \ll l_a.$$

This means that in the localized regime, the disorder causes drastic enhancement of the attenuation of the average transmission coefficient as compared to that in the corresponding pure ($\delta\varepsilon = 0$) sample ($l_{\text{in}} << l_a$). Note that the disorder-induced absorption length for the localized waves, Equation 2.4.21, is also significantly smaller (i.e., the effect of absorption is much stronger) than that in the diffusive regime: $l_{\text{in}} \ll \sqrt{l_{\text{loc}} l_a}$.

In contrast to Equation 2.4.20, the contributions from scattering and absorption to the average decrement of the transmission coefficient (or to the Lyapunov exponent) are additive:

$$-\left\langle\frac{\ln T(L)}{L}\right\rangle = \left(\frac{1}{l_{\text{loc}}} + \frac{1}{l_a}\right) > -\frac{\ln\langle T(L)\rangle}{L}. \qquad (2.4.22)$$

It can be shown that $\ln\langle T(L)\rangle$ and $\langle \ln T(L)\rangle$ are different not only in lossy media (compare Equations 2.4.20 and 2.4.22) but also in nonabsorbing ($l_a \to \infty$) systems. This is because $\ln T(L)/L$ is a self-averaging quantity (see Equations 2.4.8 and 2.4.11), with very narrow distribution (δ function at $L \to \infty$) centered at its mean value. This means that at a randomly chosen realization, $\ln T(L)$ will be found in a small vicinity of its average value with a probability exponentially close to one, and therefore the value of a function $F[\ln T(L)]$ will be close to $F[\langle \ln T(L)\rangle]$ with the same probability. In particular, the transmission coefficient typically is exponentially small:

$$T_{\text{typ}}(L) = \exp\left[\frac{\ln\langle T(L)\rangle}{L}\right] = \exp\left[-L\left(\frac{1}{l_{\text{loc}}} + \frac{1}{l_a}\right)\right]. \qquad (2.4.23)$$

On the other hand, $T(L)$ itself is a strongly fluctuating random variable with broad distribution. It turns out that the main contribution to its average value comes not from the typical (nontransparent) realizations but from low-probable ones, so-called resonant realizations (see Section 2.4.4), corresponding to the non-Gaussian tail of the distribution of $\ln T$, where the transmission coefficient is of the order of unity. This is due to these resonantly transparent realizations that the average transmission is much larger than the typical one. For example,[4] in the lossless media with delta-correlated disorder

$$\frac{\langle T(L)\rangle}{T_{\text{typ}}(L)} = \exp\left(\frac{3}{4}\frac{L}{l_{\text{loc}}}\right) \gg 1.$$

while both $T_{\text{typ}}(L)$ and $\langle T(L)\rangle$ are exponentially decaying functions of L (the difference is in the attenuation rate). More than that, there are quantities for which the typical and mean values have completely different functional dependencies. An example of how misleading a formally calculated mean value can be is the energy flux, $J = 2\text{Im}(u^*u')$, created by a point source located at the perfectly reflecting edge ($x = 0$) of a disordered sample of length L. As shown in Ref. 2, the mean flux does not interact with disorder and is equal to its value in the homogeneous sample: $\langle J(L)\rangle = 2/k$. This result is physically meaningless because to obtain it experimentally, averaging over an exponentially large number of realizations is necessary. At the same time, the measurement at a single random sample will give (with a probability exponentially close to unity) the typical value, which is exponentially small as a result of the localization effect: $J_{\text{typ}} = \exp(\langle \ln J\rangle) \sim \exp(-L/l_{\text{loc}})$. This is because $\ln J$ is an additive self-averaging quantity.

The typical-medium approach in the theory of Mott–Anderson localization in electron systems is discussed in Ref. 27.

An outstanding distinction between the transmission at typical and resonant configurations of amplifying random media has been found in Ref. 28. It has been shown that in random systems with complex dielectric permittivity, $\varepsilon(x) = \varepsilon_0 + \delta\varepsilon(x) + i\Gamma$, the inverse Lyapunov exponent is always negative, independent of the sign of Γ;[29] that is, the typical transmission through a finite disordered dielectric sample is exponentially small for both absorbing and amplifying disordered samples. To the contrary, the mean value of the transmission coefficient in random media with gain ($\Gamma > 0$) diverges (because of the infinitely increasing resonant intensity) even at samples of finite size. To obtain physically meaningful finite values of the transmission, the nonlinear effect of saturation should be included in the model.

2.4.3.3 Transport and Localization in 1D Periodic Structures with Disorder

The study of the effects of disorder on the wave properties of periodic structures is essential for better understanding the physics of the interplay between periodicity and disorder, and also for practical applications. Indeed, though considerable effort has been expended to develop highly periodic structures, deviations from periodicity inevitably present in any manufactured photonic crystal can significantly modify its optical characteristics. To reveal the most general transport properties of disordered 1D structures that are periodic on average (1D photonic crystals), the Helmholtz Equation 2.4.1 can be used, in which ε_0 is a periodic function of the coordinate x, and $\langle\delta\varepsilon(x)\rangle = 0$. Three types of periodic systems with weakly-perturbed periodicity were studied in Ref. 30: (i) stacks of alternating discrete dielectric layers with constant permittivities, $\varepsilon^{(1)}$ and $\varepsilon^{(2)}$, and fluctuating width of each layer, $d_i = d + \delta d_i$; (ii) samples of the same geometry but with constant $d_i = d$ and

$\varepsilon_{2i-1}^{(1)} = \varepsilon_0^{(1)} + \delta\varepsilon_{2i-1}$, $\varepsilon_{2i}^{(2)} = \varepsilon_0^{(2)} + \delta\varepsilon_{2i}$; (iii) continuous periodic media with $\varepsilon(x) = A\cos qx + \delta\varepsilon(x)$. The quantities δd_i, $\delta\varepsilon_j$, and $\delta\varepsilon$ are assumed to be random variables with known statistics. For numerical simulations of the propagation in discrete systems (i) and (ii), the transfer matrix approach (Section 2.4.3.1) is appropriate, whereas for the continuous model (iii), the invariant embedding method (Section 2.4.3.2) is best suited.

The following property is found to be universal, independent of the geometry of the system and of the type of disorder for the frequencies of the incident wave belonging to a band gap of the underlying periodic structure, weak disorder enhances (in contrast to homogeneous in average random 1D systems) the transparency. Moreover, the localization length and the transmission coefficient grow when the strength of the disorder increases. This is because in the presence of disorder, the channels of the propagation that are closed in the perfectly periodic system open up as a result of the partial filling of the density of photonic states in the gap by the tails of this density from the transparency zones bordering the gap.

In contrast to the band gap, the features of the transmission for frequencies in the transparency zone depend on the type of disorder. For these frequencies, the surprising non-monotonic dependence of the localization length on the strength of disorder was observed in stratified media with geometrical disorder and constant dielectric permittivities [type (i)]. In such a medium, the initial decrease of the transmission coefficient is a classical manifestation of Anderson localization, which is usually stronger for larger fluctuations (see previous sections). When the fluctuations of the width become adequately large, the decrease gives way to the enhancement of the transmission for increasing disorder. To explain this rather counterintuitive result, we note that in the strong localization regime, the inverse localization length is approximately equal to[30]

$$\frac{1}{l_{\mathrm{loc}}} \approx \langle R_1 \rangle, \tag{2.4.24}$$

where R_1 is the reflection coefficient of a single layer. This means that, in this case, the total transmission of a stack is completely determined by the mean value of the reflection coefficient of a single element. For a dielectric layer, R_1 is a periodic function of the width d of the layer, and the averaging in Equation 2.4.24 means the integration of R_1 over d in an interval Δ_d, in which the fluctuations δd_i are distributed. Evidently, if the disorder is strong, $\Delta_d \simeq d$, the increase of the interval of the integration of the periodic function causes a decrease of $\langle R_1 \rangle$.

In a system of the second type (random ε), R_1 is proportional to $\delta\varepsilon$ and the growth of its variance enhances the strength of a single scattering, leading to a monotonic increase of $\langle R_1 \rangle$. On further increase of the disorder, all three types of random systems finally lose all traces of the underlying periodicity, the band structure disappears, and waves of all frequencies experience the same disordered medium that becomes homogeneous in average.

2.4.4 Disorder-Induced Resonances in 1D Systems

2.4.4.1 Exploration by Analogy: Deterministic Model of Random Resonances

As shown in Section 2.4.1, disorder can strongly affect the transport properties of periodic systems, sometimes to the point where the photonic band structure is completely destroyed.

As fluctuations of the dielectric and geometrical parameters are inevitably present in any manufactured periodic sample, this could create a serious obstacle in the efficient practical use of photonic crystals. Therefore, nowadays considerable efforts of researches and producers go into the control of fluctuations. However, if rather than combating imperfections of periodicity, one fabricated highly disordered samples, they could be equally well harnessed, for example, for creating tunable resonant elements. This is because 1D random configurations have a unique band structure that for some applications has obvious advantages over those of photonic crystals.

The transparency spectrum of a typical 1D random sample consists of very narrow bands separated by broad gaps (Figure 2.4.2). For adequately long structures, the bands are so narrow that they can be treated as (quasi)-resonances that are well pronounced; that is, their widths are much smaller than the distances between them, both in the frequency domain and in real space. Physically, at each resonant frequency, an open random 1D configuration can be considered as an open resonator with high quality factor. An important advantage of application features of such a system is that, in contrast to a regular resonator whose modes occupy all inner space, in a 1D random structure, each eigenfrequency (mode) is localized inside its own effective "cavity" whose size is much smaller than that of the sample. Figure 2.4.3 shows the intensities of the fields generated by a resonant frequency (central curve), and by two off-resonance, typical frequencies (side curves) with exponentially small transmission coefficients. Another important advantage of disordered samples is that they are much easier to fabricate as they do not require precise periodicity.

The existence of disorder-induced resonances in 1D random media was predicted a while ago.[6] The random set of resonant frequencies is a sort of 1D optical "speckle pattern," which is individual for each random configuration and represents its unique "fingerprint."

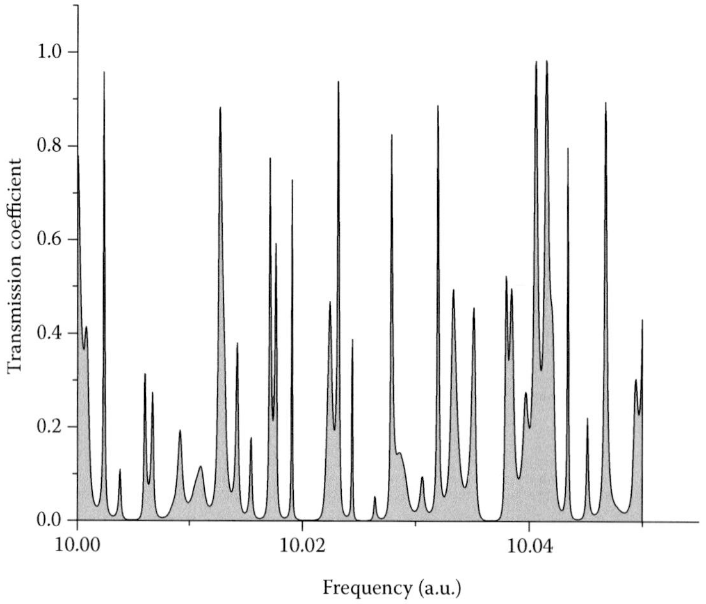

FIGURE 2.4.2
Transmission coefficient T as a function of the wave frequency.

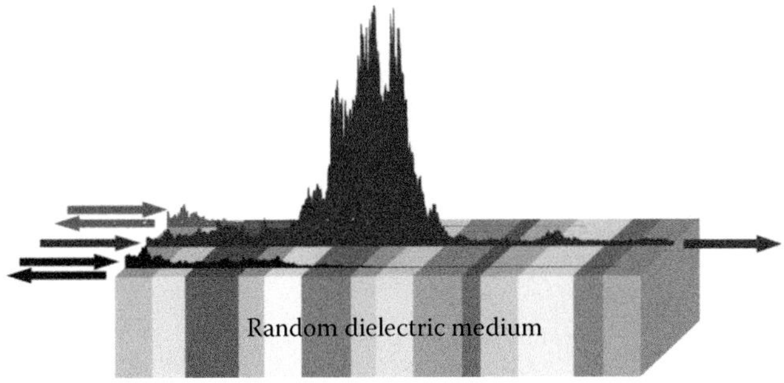

FIGURE 2.4.3
(**See color insert.**) Amplitude of the field inside the sample as a function of the coordinate for three different wavelengths. (Reprinted with permission from K. Bliokh et al., Rev. Mod. Phys. **80**, 1201, 2008. Copyright 2008 by the American Physical Society.)

Although the resonances indirectly manifest themselves in the dominant contribution to the mean transmission coefficient (see Section 2.4.3.2), the ensemble-averaged quantities do not provide any information about the frequencies, locations, and spatial intensity distribution of the individual localized states, yet just this set of data is essential for applications (e.g., random lasing). Unfortunately, an explicit, general analytical solution of Equation 2.4.1 with arbitrary function $\delta\varepsilon(x)$ does not exist, and standard approximate methods (e.g., small perturbations expansion) are of little help, because the adequate description of strong localization calls for the summation of infinite number of multiply-scattered fields. On the other hand, direct optical measurements of the field inside a given disordered sample are generally not feasible.

In such a situation, the question arises as to whether the outgoing radiation bears the necessary information on what happens inside the sample or, more specifically, whether the parameters and internal structure of individual resonances can be retrieved from the standard external measurements of the transmission and reflection amplitudes.

The positive answer to this question is given by means of the approach developed by Bliokh et al.,[31] which is based on the concept that the fundamental properties of resonances are universal and independent of the physical nature of the system, regular or random, whether it is a quantum-mechanical potential well, an optical or microwave resonator, or a 1D random medium.[32] The distinguishing feature of a random structure is that there are no regular walls in it, and the strong reflection that locks the radiation in an effective resonant cavity is the result of Anderson localization. Moreover, different segments of the sample turn out to be transparent for different frequencies; that is, each localized mode is associated with its own resonator.

In the framework of approach,[31] the problem of the transport and localization in a 1D random medium is mapped onto an exactly solvable quantum mechanical problem of tunneling and resonant transmission through an effective deterministic potential. With the formulae derived on the basis of this mapping, the parameters of the individual resonances, in particular, their spectral and spatial widths, field amplitudes, and transmission coefficients, can be calculated. These results also enable solving inverse problems, namely, to find (with some accuracy) the absorption rate of the medium and the positions of the effective cavities, using the measured transmission and reflection coefficients as the input data. The analytical results deduced from quantum-mechanical analogy are in close agreement with the results of direct numerical simulations.

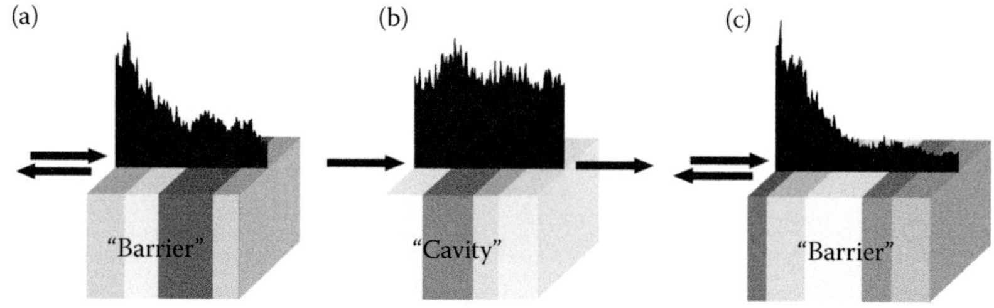

FIGURE 2.4.4
(**See color insert.**) Amplitude of the field as a function of the coordinate inside the whole sample, in the (a), (b), and (c) parts taken as a separate sample each. (Reprinted with permission from K. Bliokh et al., Rev. Mod. Phys. **80**, 1201, 2008. Copyright 2008 by the American Physical Society.)

The idea of this mapping was inspired by the well-pronounced similarity between Figures 2.4.3 and 2.4.4 obtained for a random stack of layers, and the corresponding dependencies calculated for a regular potential well bounded at both sides by two potential barriers (e.g., see Ref. 33). Although the physics of the propagation in each system is totally different (interference of the multiply-scattered random fields in a randomly layered medium, and tunneling through a regular two-humped potential), the affinity between them stands out. Indeed, in both cases, the transmission coefficients are exponentially small for most of the frequencies (energies) and have well-pronounced resonant maxima (sometimes of the order of unity) at discrete points corresponding to the eigen levels of each system. The energy at resonant frequencies is localized, and the transmission depends drastically on the position of the area of localization.

Figure 2.4.4 presents the transparencies of three segments of the sample cut in accordance with the central curve (resonant wave) in Figure 2.4.3 and separately illuminated by the same resonant (for the whole sample) wave. It is seen that at the resonant frequency, the middle part where the energy is concentrated is almost transparent, whereas the side sections are practically opaque. This qualitatively corresponds to what happens to quantum particles in a potential well. Moreover, even quantitatively, the intensity distributions presented in Figure 2.4.3 and the corresponding values of the transmission coefficients compare favorably with those calculated quantum-mechanically by solving Equation 2.4.1, with the random function $\delta\varepsilon(x)$ being replaced by a regular potential profile with properly chosen parameters.

For any (not too high) energy, the transmission coefficient of a two-humped potential profile (potential well) can be calculated in the Wentzel–Kramers–Brillouin (WKB) approximation,[33] which yields expressions independent of the "fine structure" of the profile and uniquely determined by the size of the well and by the tunneling transparencies of the bounding barriers. To use these formulae for the quantitative description of the wave transmission through a random sample, it is necessary to express the parameters of the effective potential through the parameters of the disordered system in hand. For an ensemble of 1D random realizations, those parameters are the length L of the samples and the (self-averaging) localization length l_{loc} (Equation 2.4.6). Note that in the localized regime, $L \gg l_{\mathrm{loc}} \gg \lambda$. This inequality justifies the validity of the WKB approximation. To estimate the length of the effective well, we note that the appearance of a transparent segment (effective well) inside a random sample is the result of a very specific (and therefore low-probable) combination of phases of the multiply-scattered fields. Obviously, the longer such segment the less the probability of its occurrence. On the other hand, the typical scale in the localized regime is l_{loc}. Hence, the minimal and thus the most probable size of the effective well, l_{res}, is of the order of the localization length, which we assume (as a result of self-averaging) to be the same in all realizations. Under this assumption, different values of the resonant transmission coefficients and different intensities can be reproduced by variations of the location of the well in the corresponding quantum-mechanical formulae.

If the center of the transparent segment of a resonant realization is shifted a distance d from the center of the sample, the lengths of the nontransparent parts of the resonant realizations become

$$L_{1,2} = \frac{L - l_{res}}{2} \pm d, \tag{2.4.25}$$

and the transmission coefficients of the confining barriers are

$$T_{1,2} = \exp\left(-\frac{L_{1,2}}{l_{loc}}\right). \tag{2.4.26}$$

Substituting Equations 2.4.25 and 2.4.26 into the corresponding WKB formulae, we obtain

$$T_{res}(d) = \frac{4T_1T_2}{(T_1 + T_2)^2} = \frac{4}{[\exp(d/l_{loc}) + \exp(-d/l_{loc})]^2}, \tag{2.4.27}$$

$$|A_{res}(d)|^2 = \frac{8T_1}{(T_1 + T_2)^2} = \frac{8\exp(L/l_{loc} - 1 - d/l_{loc})}{\left[\exp(d/l_{loc}) + \exp(-d/l_{loc})\right]^2}, \tag{2.4.28}$$

where $T_{res}(d)$ and $A_{res}(d)$ are, respectively, the transmission coefficient and the peak amplitude of the field pumped into the cavity located at a point d by an incident resonant monochromatic wave of unit amplitude. Note that $|A_{res}|^2$ is asymmetric with respect to T_1 and T_2, which means that the intensity induced in an effective cavity by the resonant incident wave depends on the direction of incidence.[34] The width of high ($T_{res} \sim 1$) resonances in a long sample is exponentially small:

$$\delta k_{res} \sim \frac{1}{l_{loc}} \exp\left(-\frac{L}{l_{loc}}\right). \tag{2.4.29}$$

Equation 2.4.27 shows that the transmission coefficient of a disordered sample at a resonance is independent of the length of the sample and is governed exclusively by the location of the effective cavity. This rather counterintuitive result is totally different from that for the typical (nonresonant) transmission, which decays exponentially with the increase of the length L.

From Equation 2.4.28 it follows that the amplitudes of the centrally located ($d \ll l_{loc}$) strong resonances are an exponentially increasing functions of the length of the sample:

$$|A|^2 \simeq 2\exp\left(\frac{L}{l_{loc}}\right) \gg 1. \tag{2.4.30}$$

Whereas the transmission is maximal when the effective cavity is located precisely in the middle of the sample, $T_{res}(d = 0) = 1$, the largest amplitude can be pumped in a resonator shifted from the center toward the input by the distance

$$d_A = -0.27 l_{loc}, \tag{2.4.31}$$

which is independent of the length of the sample.

It should be emphasized that in a disordered sample, for each resonant frequency, ω_{res}, there exists its own "strange resonator," which (unlike regular potential wells) is transparent only for this particular ω_{res}. These resonators are located at different random points; therefore, it should be remembered that the coordinate d in Equations 2.4.25 through 2.4.28 is frequency-dependent: $d = d(\omega_{res})$. The structure of the effective cavities is described below.

To test the ability of the above-introduced deterministic model to describe quantitatively the resonances in randomly layered samples, extensive numerical experiments were carried out. The comparison of the results of direct numerical simulations with those given by Equations 2.4.27 through 2.4.31 demonstrates that the seemingly rough analogy based on just one fitting parameter (localization length) performs surprisingly well.[31] For example, not only the coordinate of the effective cavity with the highest resonant amplitude is independent of the total length of the sample and proportional to l_{loc}, as predicted by Equation 2.4.31, but also the coefficient in Equation 2.4.31 coincides with that obtained in numerical simulations with an accuracy ~10%.

To provide an adequate description of the resonances in real dielectric structures, the absorption should be incorporated in the model. Although in quantum-mechanical problems losses are rather uncommon, the absorption of light in a dielectric medium can be taken into account by formally adding to the corresponding effective potential a negative imaginary part proportional to the imaginary part of the permittivity. If the spatial decrement of the wave energy due to loss, $\Gamma = 1/l_a \sim \mathrm{Im}\varepsilon_0$, is small compared to the inverse localization length, $l_a \gg l_{loc}$, calculations of the resonant transmittance and intensity yield[35] (compare with Equations 2.4.27 and 2.4.28):

$$T_{res}(d) = \frac{4T_1T_2}{(\Gamma l_{res}\sqrt{\mathrm{Re}\,\varepsilon_0} + T_1 + T_2)^2}, \tag{2.4.32}$$

$$|A_{res}(d)|^2 = \frac{2T_{res}}{T_2}. \tag{2.4.33}$$

The connection between the resonant reflection and transmission coefficients follows from the energy conservation law and in, a case of small losses, takes the form

$$R_{res} = 1 - T_{res} - \Gamma l_{res} |A_{res}|^2 \sqrt{\mathrm{Re}\,\varepsilon_0}, \tag{2.4.34}$$

where the last term is due to absorption in the medium. The width of the resonant peak in the reflection is determined by the absorption in the medium and by the transmittance of the effective walls that form the effective wells, and is equal to

$$\delta k_{res} = \Gamma + \frac{T_1 + T_2}{l_{res}\sqrt{\mathrm{Re}\,\varepsilon_0}}. \tag{2.4.35}$$

When the parameter

$$b = 2\Gamma l_{\text{res}}\sqrt{\operatorname{Re}\varepsilon_0}\exp\left(\frac{L}{2l_{\text{loc}}}\right) \gg 1, \tag{2.4.36}$$

even small absorption ($l_a \gg l_{\text{loc}}$ or $l_a \gg L$), which practically does not affect the typical transmission, suppresses dramatically both the reflected and transmitted fluxes at resonances. Interestingly, the resonant reflection coefficient, Equation 2.4.34, is a nonmonotonic function of the dissipation rate Γ, and turns to zero when $b = -2\sinh(d/l_{\text{loc}})$. This effect is known in optics and microwave electronics as critical coupling.[36] As no energy is reflected from the sample at critical coupling, it corresponds to the resonance with the highest intensity.

2.4.4.2 How Is an Effective Cavity Built Up?

The model presented above gives a more penetrating insight into the mechanism of formation of localized modes and allows an explanation of how and why the characteristics of a resonance (Q-factor, transmission coefficient, linewidth, etc.) depend on the parameters of the effective cavity (location, size, and absorption rate) associated with it. It is now clear that the resonant transmission through a random sample takes place at a frequency ω_{res} if around some (random) point inside this sample there exists an area of length $l_{\text{res}} \gtrsim l_{\text{loc}}$ which is transparent for this frequency.

Here another question arises: why do effective resonant cavities exist in 1D disordered samples? In other words, why are different segments of a random configuration transparent or opaque for the wave with a given wave number k? To answer this question recall that in the case of weak scattering, the so-called resonance reflection takes place; that is, the reflection coefficient of an adequately long segment $[x - a, x + a]$ of a random medium is proportional to the amplitude of the $2k$-harmonic in the power spectrum, $\tilde{\varepsilon}(2k, x)$, of its dielectric constant $\varepsilon(x)$.[37] This amplitude can be calculated as

$$\tilde{\varepsilon}(2k, x) = \int_{x-a}^{x+a} \varepsilon(y)e^{-i2ky}dy. \tag{2.4.37}$$

This expression is known as a window Fourier transformation with a rectangular window function. To investigate the structure of a resonant configuration, we first determined the resonant wave numbers for which the transparency $T(k)$ of this configuration exceeded, for example, 0.5. For each of those waves, the spatial distributions of the amplitude and of the corresponding local power spectrum (2.4.37) were juxtaposed. An example is shown in Figure 2.4.5. One can see that in the areas where the field is localized, the function $\tilde{\varepsilon}(2k, x)$ is strongly suppressed, whereas in the nontransparent parts it has well-pronounced maxima. Therefore, the resonant cavity for a frequency ω_{res} arises in the area of a disordered sample where, accidentally, the harmonic with wave number $q_{\text{res}} = 2k_{\text{res}}$ in the power spectrum of the (random) permittivity has small amplitude.

This result is of profound importance for the correct understanding of the resonant transparency of 1D random systems. It has been commonly accepted that, in accordance with the celebrated formula (2.4.8), high (not "typical," exponentially small) transparency at a given frequency requires the suppression of the corresponding spatial spectral component *in the whole sample*. A fundamentally new outcome of the above consideration is that for arising of a localized eigenmode, which provides resonant transmission, it suffices for the $2k_{\text{res}}$-harmonic to be suppressed in *a much smaller area of size* $l_{\text{res}} \gtrsim l_{\text{loc}}$.

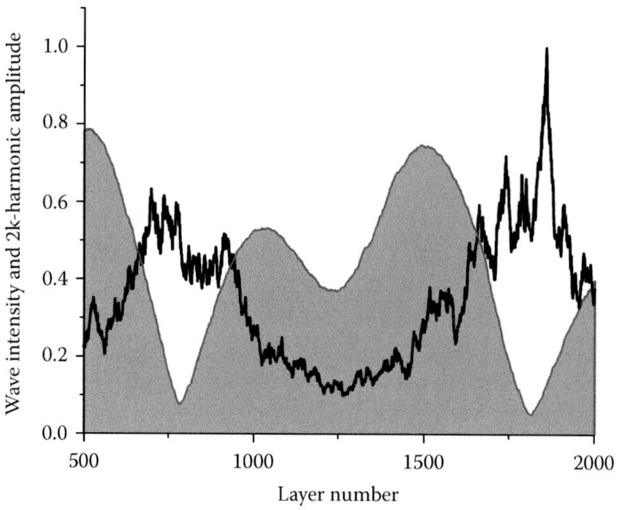

FIGURE 2.4.5
Normalized wave (black line) and $2k$-harmonic (gray area) amplitudes as functions of the coordinate (layer number).

Obviously, the probability of such a random event is exponentially (with respect to $L/l_{res} >> 1$) larger than that of the same to occur simultaneously in the whole random configuration. This not only explains the observed high spectral density of the eigenstate (2.4.3) but also provides a background for harnessing disordered 1D systems in designing tunable light-tailoring devices (see Section 2.4.7).

2.4.4.3 Coupling and Level Repulsion

An adequately long, disordered 1D sample can contain several isolated regions where the spectral harmonics with wave numbers close to q_{res} are suppressed. The spatial overlap of the wave functions localized in such regions couples these modes and leads to the formation of the so-called necklace states, which have been predicted theoretically in the studies by Lifshits and Kirpichenkov[38] and by Pendry[11] and observed and studied experimentally in Refs. 9 and 39. These states have broadened spectral lines[7] and contribute substantially to the overall transmission in localized regime. An example of a necklace state is shown in Figure 2.4.6.

Necklace states can be easily incorporated in the modeling scheme as two or more potential wells coupled by the evanescent fields that tunnel through barriers separating the wells.[9,32] The temporal dynamics of the field in such a chain of coupled resonators is described by a system of oscillator equations with an external force, damping, and coupling coefficient, which account for the incident wave, the finite Q-factors, and the spatial overlap of the modes, respectively. In the simplest case of two cavities with coordinates d_1 and d_2, the equations that provided an effective description of coupled modes can be written as:[32]

$$\frac{d^2\psi_1}{d\tau^2} + Q_1^{-1}\frac{d\psi_1}{d\tau} + (1-\Delta_1)^2\psi_1 = q\psi_2 + f, \tag{2.4.38}$$

$$\frac{d^2\psi_2}{d\tau^2} + Q_2^{-1}\frac{d\psi_2}{d\tau} + (1-\Delta_2)^2\psi_2 = q\psi_1. \tag{2.4.39}$$

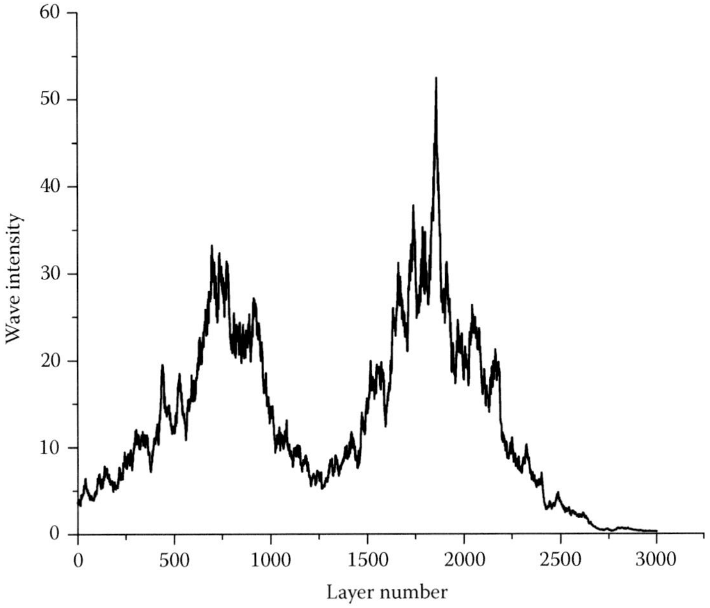

FIGURE 2.4.6
Necklace state. Normalized wave intensity as a function of the coordinate (layer number).

Here $\psi_i(\tau)$ is the field in the ith effective resonator, $\tau = \omega_0 t$ is the dimensionless time (ω_0 is a characteristic central frequency of the problem), and $1 - \Delta_i$ ($|\Delta_i| \ll 1$) is the dimensionless (in units of ω_0) eigenfrequency of the ith resonator. The effective external force f is

$$f = \psi_0 \exp\left(\frac{-d_1}{l_{\text{loc}}}\right) e^{-i\nu\tau}, \tag{2.4.40}$$

where ψ_0 and ν, ($|\nu - 1| \ll 1$), are the amplitude and the frequency of the external field, exciting the first (close to the input) resonator. The coupling coefficient $q \ll 1$ of two cavities, which is due to the spatial overlap of modes, is equal to

$$q = \exp\left(\frac{-d}{l_{\text{loc}}}\right) \tag{2.4.41}$$

with d being the distance between the effective cavities.

The Q-factors describing the losses of energy in the ith resonator are:[40]

$$Q_i^{-1} = \Gamma_i + \frac{v_g T_i}{2 l_i \omega_0} \ll 1, \tag{2.4.42}$$

where Γ_i is the dissipation rate in the ith resonator, v_g is the wave group velocity inside the resonator, and l_i is the cavity length. The last term in Equation 2.4.42 accounts for the leakage of the energy from the system; therefore, analogously to Equation 2.4.26, the transmission coefficients $T_{1,2}$ are given by[32]

$$T_1 = \exp\left(\frac{-d_1}{l_{\text{loc}}}\right),$$
$$T_2 = \exp\left[-\left(\frac{L - d_2}{l_{\text{loc}}}\right)\right]. \tag{2.4.43}$$

Substitution $\psi_i = A_i \exp(-i\nu\tau)$ reduces Equations 2.4.38 and 2.4.39 to a couple of algebraic equations, from which the eigenfrequencies of two complex independent eigenmodes of the system can be found:

$$\nu^{\pm} = 1 - \frac{\Delta_1 + \Delta_2}{2} - i\frac{Q^{-1}}{2} \pm \frac{1}{2}\sqrt{(\Delta_1 - \Delta_2)^2 + q^2}. \tag{2.4.44}$$

We have assumed here that $Q_1 = Q_2 = Q$. Depending on the parameters, Equation 2.4.44 describes either level anticrossing (level repulsion) or the coupling of isolated resonators to form collective eigenmodes. The anticrossing takes place when the eigenfrequencies of two cavities come close and are transformed into double-peaked, more extended modes. The gap between eigenfrequencies is equal to

$$G_e = \sqrt{(\Delta_1 - \Delta_2)^2 + q^2}, \tag{2.4.45}$$

and is minimal at the resonance, $\Delta_1 = \Delta_2$. Note that the level repulsion of modes, that is generally agreed to be an inherent feature of diffusion, arises here in the regime of strong localization. Away from resonance, $|\Delta_1 - \Delta_2| \gg q$, the eigenmodes tend to the modes of isolated resonators. The shapes of the modes are exchanged when passing through the resonance; that is, + (–) eigenmodes correspond to the first (second) resonator at $\Delta_1 \ll \Delta_2$, and to the second (first) resonator when $\Delta_1 \gg \Delta_2$.

The frequency (ν) dependences of the amplitudes A_1 and A_2 of the oscillations induced in the cavities by incident monochromatic waves are determined by the ratio between the coupling coefficient q and losses Q^{-1}. Both amplitudes are at maximum at two frequencies

$$\nu^{\pm}_{\max} = 1 - \frac{\Delta_1 + \Delta_2}{2} \pm \frac{1}{2}\sqrt{q^2 - Q^{-2}}. \tag{2.4.46}$$

There are two different regimes of the excitation of coupled resonators, depending on the value of the qQ factor, as shown in Figure 2.4.7 for the case of identical ($\Delta_1 = \Delta_2 = 0$) cavities.[32] When losses are small, so that the condition $qQ > 1$ holds, there are two collective anticrossing resonant modes with the frequency gap $G_A = \sqrt{q^2 - Q^{-2}}$ and equal field amplitudes in both resonators. As qQ decreases, the resonant peaks in the spectra are located near each other and meet when $qQ = 1$. In the regime $qQ < 1$, there is one peak at $\nu = 1$.

The parameter qQ that appears in the model has a simple physical meaning: it determines whether the two resonators should be considered as essentially coupled or isolated. When $Q^{-1} \ll q$, the losses are negligible and the field characteristics are

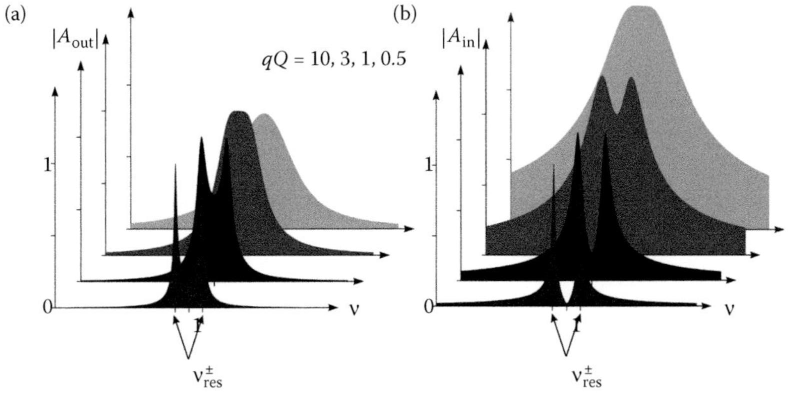

FIGURE 2.4.7
(**See color insert.**) Near-resonant transmission of an incident wave through two coupled open resonators at different values of qQ. The normalized (i.e., multiplied by the factor $2Q^{-1}$) absolute values of the field amplitudes in two resonators, $|A_{\text{out}}|$ (a) and $|A_{\text{in}}|$ (b), are shown. (Reprinted with permission from K. Bliokh et al., Rev. Mod. Phys. **80**, 1201, 2008. Copyright 2008 by the American Physical Society.)

essentially determined by the coupling. Remarkably, in this case the field intensity in the first (incoming) resonator is negligible at $\nu = 1$, and almost all the energy is concentrated in the second resonator: $A_2 >> A_1$. On the contrary, when the losses prevail over the coupling, $Q^{-1} >> q$, the incident wave only excites the first resonator, and the energy is concentrated mostly in it: $A_1 >> A_2$.

2.4.4.4 Bistability of Anderson Localized States in Nonlinear Random Media

A combination of disorder and nonlinearity offers a multitude of striking physical phenomena, some of which still remain enigmatic and call for further investigation. In particular, nonlinear interactions between electromagnetic radiation and disorder influences the interference of the multiply-scattered waves and can affect localization in rather unusual ways. This area has long been the subject of keen scientific interest that has quickened recently, mostly as a result of the creation of high-power lasers and of the latest advancement in studies of Bose–Einstein condensates. Although publications on the transport and localization in nonlinear random media are numerous, the overwhelming majority of the analytical results have a common shortcoming: they are related to ensemble-averaged characteristics, which are of a little use when individual localized states are concerned. A breakthrough in the theoretical study of the disorder-induced resonances has been made possible with the above-presented quantum-mechanical deterministic model, in which the nonlinearity was incorporated. Surprisingly, this rather simple approach not only offered a clearer insight into the physics of the resonances in nonlinear random media but also performed well in their quantitative description.[13]

According to the model, the transmittance spectrum $T(k)$ in the vicinity of a resonant wavelength, $|k - k_{\text{res}}| << k_{\text{res}}$ is given by the Lorentzian dependence

$$T(k) = \frac{I_{\text{out}}}{I_{\text{in}}} = \frac{T_{\text{res}}}{1 + \left[2Q(k/k_{\text{res}} - 1)\right]^2}, \tag{2.4.47}$$

where I_{in} and I_{out} are the intensities of the incident and outgoing waves, respectively. The resonant transmission coefficient $T_{res} = T(d)$ is given by Equation 2.4.32, where $d = d(k_{res})$.

Obviously, the nonlinearity becomes most noticeable at the points where the resonances are located and the intensity is maximal, $I = I_{res}$. It changes the effective refractive index of the medium leading to the intensity-dependent shift of the resonant wave number: $k_{res} \to \tilde{k}_{res}(I_{res})$ (Figure 2.4.8).

As the values of I_{res} and I_{out} are unambiguously connected, the resonant wave number is a function of the output intensity, and Equation 2.4.47 establishes a relation between the input and output wave intensities, which in the case of weak Kerr-type nonlinearity is cubic with respect to I_{out}.[13] It has a universal form typical of nonlinear resonators with optical bistability.[41] The ultimate dependence $I_{out}(I_{in})$ is of the S-type, and, in some range of parameters, the stationary transmission spectrum $T(k)$ is a three-valued function. Typically, one of the solutions is unstable, whereas the other two form a hysteresis loop in the $I_{out}(I_{in})$ dependence.[13] Figure 2.4.8 shows nonlinear deformations of the resonant transmission spectra $T(k)$, which at large values of the parameter χI_{in} (χ is the Kerr coefficient) exhibit transitions to bistability. The analytical dependence $T(k)$, derived from Equation 2.4.47, with parameters found from the numerical experiments, are in excellent agreement with the direct numerical simulations.

The adequacy of the model has also been substantiated by numerical modeling in time domain. In these simulations, the transitional oscillations and reshaping of the transmitted pulse that typically accompany switching between two regimes of transmission in deterministic bistable nonlinear structures[42] have been found in disordered nonlinear samples. Nonreciprocity (diode-like unidirectional propagation) of resonant tunneling

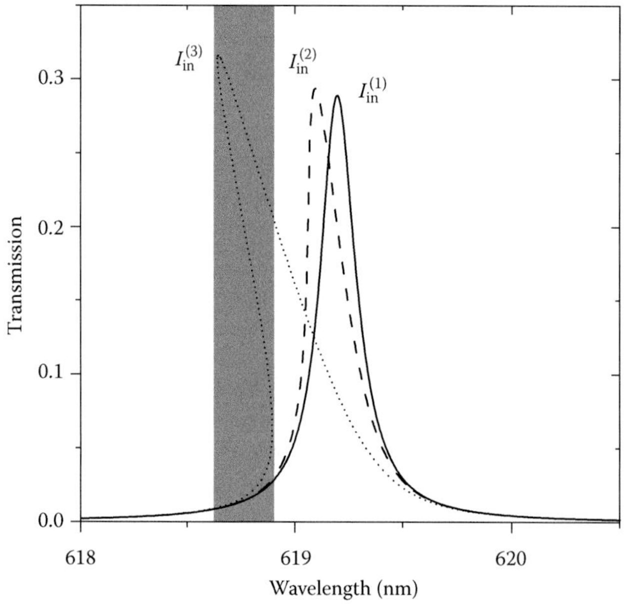

FIGURE 2.4.8
Nonlinear deformations of the transmission spectrum of the resonance at different intensities $I_{in}^{(1)} < I_{in}^{(2)} < I_{in}^{(3)}$ of the incident wave. The light-gray stripe indicates the three-valued region. Only the lower and upper branches of the transmission spectrum are stable. (Reprinted with permission from I. Shadrivov et al., Phys. Rev. Lett. **104**, 123902, 2010. Copyright 2010 by the American Physical Society.)

through a nonlinear random structure that stems from the intrinsic asymmetry of disorder has also been observed.

2.4.5 Experimental Studies of Resonances

Comprehensive experimental studies of localized states and disorder-induced resonances were carried out in the microwave frequency range (14 GHz $\leq f \leq$ 20 GHz).[9,35] A long metallic single-mode (at these frequencies) waveguide filled with randomly arranged, weakly absorbing dielectric slabs were used as 1D disordered system. The experimental setup allowed measurements of the complex transmission and reflection amplitudes and the complex field inside the waveguide for different random configurations. Figure 2.4.9 depicts the intensity $I(x,f) \equiv |A(x,f)|^2$ generated inside a sample by an incident monochromatic wave with frequency f, as a function of coordinate x and frequency f.[35] Although the "fine structure" of the field changes dramatically from sample to sample, the general features intrinsic in all 1D disordered systems are clearly recognized in the results of a single measurement presented, as an example, in Figure 2.4.9. Localized states (resonances) excited by the incident wave are clearly seen in Figure 2.4.9. When $b \gg 1$ (Equation 2.4.36), the highest of them are located in the left (close to the input) part of the sample. The transmitted signal is suppressed by losses below the experimental noise and is indiscernible in Figure 2.4.9 even at resonant frequencies. At the same time, the resonances manifest themselves (and can be easily detected) by sharp dips in the frequency dependence of the reflection coefficient (see Figure 2.4.10). Moreover, at different values of b (i.e., of Γ) localized eigenstates are excited and detectable in reflection in different regions of the system, thus providing a possibility for scanning the sample through variations of losses. This means that dissipation, which usually impairs the excitation and observation of resonances in 1D random samples improves essentially the "observability" of the localized states.

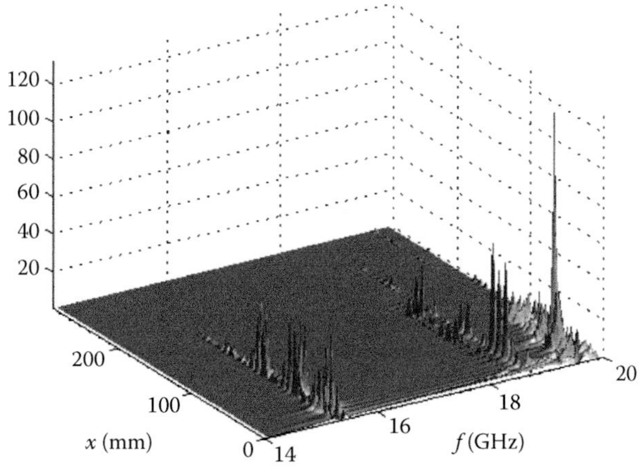

FIGURE 2.4.9
Intensity versus frequency and position inside the sample. (Reprinted with permission from K. Bliokh et al., Phys. Rev. Lett. **97**, 2439094, 2006. Copyright 2006 by the American Physical Society.)

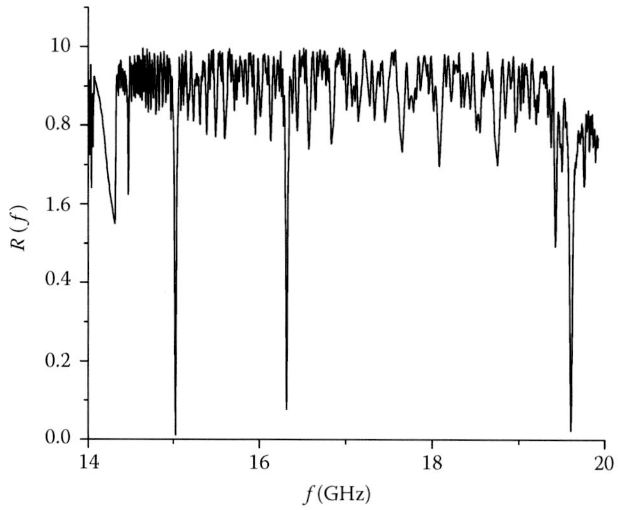

FIGURE 2.4.10
Spectrum of reflection for Figure 2.4.9. (Reprinted with permission from K. Bliokh et al., Phys. Rev. Lett. **97**, 2439094, 2006. Copyright 2006 by the American Physical Society.)

In the study by Blikoh et al.,[9] the same setup was used to explore experimentally the dynamics of formation of the necklace states and to study their spectral and transport properties. Measurements were made in a sequence of configurations in which the spacing between two randomly located scatterers could change steps in a controlled way. The position at which the air gap was introduced was chosen to correspond to the peak of a single Anderson localized mode of the unperturbed random sample. This allowed to manipulate the frequency of the selected mode in a manner similar to the tuning of a defect state through a band gap in a periodic structure. In doing so, the mode frequency shifted and crossed the frequencies of other localized states, which made it possible to study the coupling of modes. Changing the air spacing at points where other states have been localized, allowed to couple several localized modes, thereby creating necklace states extended throughout the sample. The spectral positions of the localized states as functions of the air gap introduced into the sample are plotted in Figure 2.4.11. The frequencies of modes either cross or anticross, in reasonable compliance with Equation 2.4.44. Direct measurements of the electromagnetic field inside the samples have revealed that in the case of anticrossing (regions 1,2,4,5 in Figure 2.4.11), the coupling within the sample was accompanied by the theoretically predicted exchange of shapes. In contrast, when modes crossed (region 3 in Figure 2.4.11) they did not exchange shapes and remained practically independent of each other. When the frequencies of the modes were closest, the two localized states coupled into double-peaked modes signifying the formation of quasi-extended necklace states.

The minimum frequency differences were calculated for the interacting pairs 1,2,4,5 in Figure 2.4.11, as $G_A = \sqrt{q^2 - Q^{-2}}$ with q and Q found from Equations 2.4.41 and 2.4.42, and then compared to the measured values of the gap. A comparison of the measured and calculated data is presented in Figure 2.4.12 for the following parameters of the system:[9] $f_0 = \omega_0/2\pi = 15.5$ GHz, $l_{\mathrm{loc}} = 12$ mm, $\omega_0\Gamma = 7 \times 10^7 \mathrm{s}^{-1}$, and $v_g = c/2.4$. Good agreement exists between the experiment and the model.

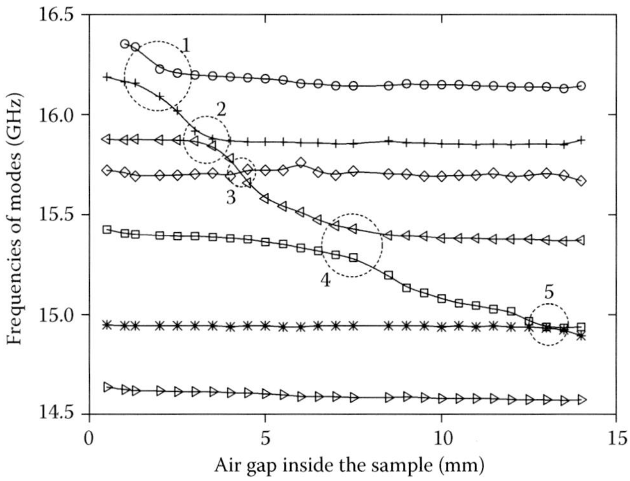

FIGURE 2.4.11
Resonant frequencies of excited localized modes versus the driving parameter—the air gap inside the sample. Five pair-interaction regions are circled. (Reprinted with permission from K. Bliokh et al., Phys. Rev. Lett. **101**, 133901, 2008. Copyright 2008 by the American Physical Society.)

Anderson localization of millimeter electromagnetic waves (75–110 GHz) has been studied experimentally,[12] utilizing 100-layer dielectric stacks of randomly shuffled quartz and Teflon wafers. Exponentially small transmission at typical frequencies, resonant transmission at eigenfrequencies, and enhanced absorption have been observed. Slow light and superluminal group velocities, which in contrast to photonic crystals were not associated with any periodicity in the system, have also been discovered.

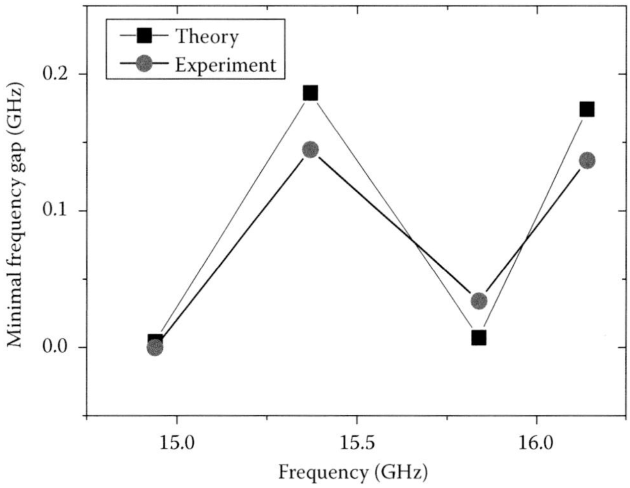

FIGURE 2.4.12
Experimentally measured and theoretically calculated minimal frequency gaps $G_A = \sqrt{q^2 - Q^{-2}}$ for pairs of interacting modes 1,2,4,5 presented in Figure 2.4.11. (Reprinted with permission from K. Bliokh et al., Phys. Rev. Lett. **101**, 133901, 2008. Copyright 2008 by the American Physical Society.)

2.4.5.1 Inverse Scattering Problems and Remote Sensing of Disordered Samples

Microwave experiments[9,12,35] made it possible to test immediately the validity of the model introduced in Section 2.4.4. Theoretical predictions based on the model have been checked against the results of measurements carried out at a large number of random configurations and in a wide range of parameters. It turned out that Equations 2.4.23 and 2.4.32 through 2.4.35 not only provided a new insight into the physics of the experimentally observed features presented in the previous subsection but also were in a good quantitative agreement with the measured data. This was made possible using these equations as a basis for formulation and solving a classical wave inverse problem: retrieval of internal characteristics of a medium of propagation from parameters of the external fields. In practice, an algorithm of remote sensing of random samples have been developed, which has enabled nonintrusive detection and monitoring of the disorder-induced resonances and determination of the absorption and localization lengths by measuring the reflected and transmitted fields. The algorithm is very simple. Indeed, Equations 2.4.23 and 2.4.32 through 2.4.35 can be treated as four algebraic equations for four unknowns, and by solving them one can find the location, d, and the size, l_{res}, of an effective cavity for each resonant frequency, and the localization and absorption lengths of the sample. Then, the intensity, A_{res}, pumped in a cavity by the incident wave can be calculated by Equation 2.4.30.

In such a manner, these parameters had been retrieved in Ref. 12 for many disordered configurations, using the directly measured values of T_{res}, R_{res}, $\delta f = c\delta k_{res}$, and l_{loc}. An example is presented in Table 2.4.1. Shown in columns 2, 3, and 4 are $1 - R_{res}$, T_{res}, and δf, respectively, measured for the resonances indicated in column 1. The loss tangent, $\tan(\alpha) = \Gamma c/f\sqrt{\mathrm{Re}\varepsilon_0}$, is given in column 6. The value of the loss tangent averaged over the five resonances equals 8.35×10^{-4}. The genuine weighted loss tangent for the disordered quartz/Teflon system was 5.2×10^{-4}, so that the measured and retrieved values of the absorption agreed to within the accuracy of the experiment. Similar experiments in the centimeter-wavelength range[35] also yielded retrieved data consistent with the true values.

The remote sensing procedure can be also applied for monitoring nonlinear disordered samples. In this instance, T_{res}, k_{res}, and Q are determined from the transmission spectrum in the linear regime as it was described above, and the additional external parameter of the medium—Kerr coefficient—is retrieved from the measured shift of the transmission spectral line when the intensity changes. This enables one to obtain the whole dependence $I_{out}(I_{in}, k)$ for any given resonance performing external measurements of $T(k)$ at only two different intensities of the incident wave.

TABLE 2.4.1

Measured and Calculated Parameters Associated with Five Resonances

Frequency (GHz)	$1 - R_{res}$	T_{res}	δf (GHz)	$\Gamma/\sqrt{\varepsilon}\cdot 10^2$	$\tan\alpha \cdot 10^4$
$f_1 = 83.5$	0.978	0.75	0.40	0.83	4.77
$f_2 = 92.0$	0.998	0.33	0.39	2.6	13.45
$f_3 = 105.7$	0.993	0.31	0.34	2.25	10.14
$f_4 = 101.8$	0.87	0.18	0.25	1.33	6.22
$f_5 = 99.8$	0.77	0.30	0.45	1.5	7.16

Source: Reprinted with permission from J. Scales et al., Phys. Rev. B **76**, 085118, 2007. Copyright 2007 by the American Physical Society.

Note: The localization length is 1 cm (obtained from the nonresonant transmission coefficient).

2.4.6 Anderson Localization in Exotic Materials

2.4.6.1 Suppression of Anderson Localization in Disordered Metamaterials

The theoretical study of any newly discovered physical phenomenon or laboratory-created material always starts from a simplified, ideal model, which makes it possible to understand the underlying principles and to explain the basic features observed in the pioneering experiments. More in-depth investigations call for more realistic models. Sooner or latter, in particular when it comes to applications, taking account of disorder becomes necessary. Such is indeed the case in the current status of research on metamaterials and graphene.

Unusual physical properties of metamaterials open up unique possibilities for numerous applications in modern optics and microelectronics. As all real metamaterials are always disordered (mostly as a result of inevitable fabrication errors), the investigation of the effects of random scattering on their transport properties is not only a fundamental academic problem but is also of significant practical importance.

The analytical and numerical analyses based on the transfer matrix method presented in Section 2.4.3 show that in stratified media with alternating layers of right- and left-handed materials (mixed stacks), the localization properties differ dramatically from those exhibited by conventional disordered materials. In particular, at long wavelengths, the localization length of mixed stacks with random refractive indices and nonfluctuating thicknesses is proportional to the sixth power of the wavelength, the result that has been neither predicted nor observed in conventional 1D random media.[14,15] It means that left-handed metamaterials can substantially suppress Anderson localization in 1D disordered systems. The suppression reveals itself also in the vanishing of the disorder-induced resonances when left-handed layers are added to a random stack of normal dielectrics. This is attributable to the lack of phase accumulation over a mixed sample, due to the cancellation of the phase across alternating left- and right-handed layers. When both refractive index and thickness of the layers constituting a mixed stack fluctuate the transmission length in the long-wave range of the localized regime exhibits the well-known quadratic power wavelength dependence with different coefficients of proportionality for mixed and homogeneous (only metalayers) random stacks. However, the transmission length of a mixed stack differs from the reciprocal of the Lyapunov exponent of the corresponding infinite stack, presenting a unique example of a 1D disordered system, in which the localization and transmission lengths are different. In contrast to normal disordered materials, the characteristic ballistic and localization lengths of mixed stacks are also different, at least in the weak scattering limit. The crossover region from localization to the ballistic regime is relatively narrow for both mixed and homogeneous stacks.

Polarization effects have been considered in Ref. 36. It is shown that the transport length strongly depends on the angle of incidence for both vertical (p) and horizontal (s) polarizations of the incident wave. In particular, when the angle of incidence exceeds a critical angle, an additional exponential decay arises due to the internal reflection from the individual layers. In mixed stacks with only refractive-index disorder, p-polarized waves are strongly localized, whereas for the s-polarization the localization is substantially suppressed at all angles of incidence. The Brewster anomaly angle depends on both the polarization and the nature of disorder, that is, disorder in either the permittivity or the permeability. For incidence at the Brewster angle, localization is suppressed, and, in contrast to the case of normal incidence, the localization length is proportional to the square of the variance of the fluctuations rather than to the variance itself.

The effects of absorption on 1D transport and localization have been studied both analytically and numerically.[15] It turns out that the crossover region is particularly sensitive to losses, so that even small absorption noticeably suppresses frequency-dependent oscillations in the transmission length. The disorder-induced resonances, which present an important signature of the localization regime, are also strongly affected (suppressed) by absorption.

The frequency dependence (dispersion) of the permittivity, ε, and/or permeability, μ, has a profound effect on Anderson localization leading to rather unusual, sometimes counterintuitive phenomena. The most exotic behavior is observed in mixed stacks at the frequencies, at which ε or μ turns to zero. In this instance, the waves are delocalized for normal incidence, whereas the localization is enhanced when disorder is present in both permittivity and permeability of the layers.

2.4.6.2 Transport and Localization in Disordered Graphene Superlattices

Shortly after the discovery of highly unusual physical properties of graphene, it was realized that the electron transport in this material had many common features with the propagation of light in dielectrics. In mathematical terms, under some (rather general) conditions, Dirac equations describing the charge transport in a graphene superlattice created by applying an inhomogeneous external electric potential could be reduced to Maxwell equations for the propagation of light in a dielectric medium. The role of the refractive index of this effective medium is played by the quantity $n_{\text{eff}} = E - U$, where E and U are, respectively, the dimensionless energy of the charge carrier and the scalar potential of the external electric field. It is easy to see that if the potential is a piecewise constant function of one coordinate, the corresponding graphene superlattice reproduces a layered dielectric structure.[17] In particular, a layer, in which the potential exceeds the energy of the particle, $U > E$, is similar to a slab with negative refractive index (metamaterial). It is because of this similarity that a junction of two regions having opposite signs of $E - U$ (so-called p–n junction) focuses Dirac electrons in graphene in the same way as an interface between left- and right-handed dielectrics focuses electromagnetic waves.[43] However, the analogy is not complete: although the equations are akin, the boundary conditions are, generally, different. Comparing these conditions one can infer that in the particular case of normal incidence, the transmission of Dirac electrons through a junction is similar to the transmission of light through an interface between two media with different refractive indices but equal impedances. Such an interface is absolutely transparent to light and therefore to the Dirac electrons in graphene as well. This explains the Klein paradox (perfect transmission through a high potential barrier) in graphene systems, and leads to the surprising conclusion that Dirac electrons are delocalized in a disordered 1D graphene structure, providing a minimal nonzero overall conductivity, which cannot be destroyed by fluctuations, no matter how strong they are.[44] Nevertheless, many features of Anderson localization can be found in random graphene systems.[17] There exist a discrete random set of angles (or a discrete random set of energies for each given angle) for which the corresponding wave functions are exponentially localized. Depending on the type of unperturbed system, the disorder could either suppress or enhance the transmission. The transmission of a graphene system built of alternating p–n and n–p junctions has an anomalously narrow angular spectrum and, in some range of directions, it is practically independent of the amplitude of the fluctuations of the potential.

Disorder manifests itself in various other situations, including graphene devices considered in Ref. 48, and the localization of acoustics waves in disordered and partially disordered one-dimensional structures.[49–53]

2.4.7 Conclusion

The term "disorder" usually bears a negative connotation. It is deemed obvious (both in everyday life and in physics and engineering as well) that irregularities are always injurious and detrimental. Contrary to this widely held view, here we argue that if treated properly, disorder can be taken advantage of in numerous technical applications. The unique spectral properties of wave transport in the localized regime presented in this chapter lead one to believe that exploiting randomness can be an effective strategy for creating light-tailoring devices, in particular switchable mirrors and tunable resonant micro- and nano-cavities.

Nowadays photonic crystals are the most extensively used for these purposes. The ability of perfectly periodic structures to manipulate light have long been demonstrated with regard to high-Q cavity resonances, spontaneous emission control, cavity quantum electrodynamical effects, and so on.[45] However, although in laboratory studies photonic crystals perform wholly satisfactorily, practical applications are frequently problematic because of the heavy demands on the accuracy of manufacturing: even small deviations from periodicity could modify dramatically the optical characteristics and hinder the performance of crystal-based devices. That is why significant effort and financial resources are expended to eliminate disorder and to develop pure, ideally regular structures. Yet, a different approach is a possibility: rather than combat the imperfections in periodicity, one can attempt harnessing highly disordered samples as high-Q resonators in optical and microwave switches, filters, and amplifiers. Despite the random character of Anderson modes, their behavior and evolution are rather deterministic, and, therefore, these modes can be used for efficient control of light similar to regular cavity modes.

To suit the optical device designer's requirements, the transmission should be fast-tunable. Resonant cavities in photonic crystals are created by implanting specially designed defects. Then, the transmission can be controlled by varying the spacing between the given frequency of the incident radiation and the resonant frequency of the cavity, which has to be easily tunable. However, any shift of the spectral line usually requires structural changes of the whole sample,[45,46] which makes such methods practically unusable.

As the localization length (and therefore the typical transmission coefficient) of a 1D random configuration is determined by the power spectrum of disorder, Equation 2.4.8, it is obvious that the frequency spectrum of the transmission can be tailored by varying the spatial structure of the correlation function.[47] Although the physical idea is trivial, its implementation for designing fast-tunable optical devices is problematic because, just as it is in the case of photonic crystals, it needs a rearrangement of the sample as a whole. This however does not mean that unique transport properties of disordered systems cannot be utilized. More sophisticated analysis of the nature of the disorder-induced resonances leads to the conclusion that they are extremely sensitive to changes of the parameters of the medium only inside the effective cavities where eigenmodes are localized. Figure 2.4.13 presents the numerically calculated dependence of the resonant transmission coefficient on the variations of the dielectric constant, $T(\delta\varepsilon)$, in an area occupying 1/50 of the total length of a random

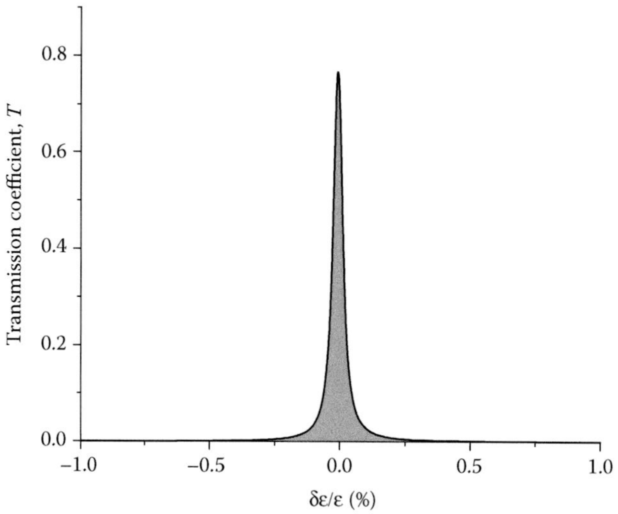

FIGURE 2.4.13
Dependence of the resonant transmission coefficient on the variations of the dielectric constant, $T(\delta\varepsilon)$, in the area occupying 1/50 of the total length of a random stack of layers.

stack of layers. It is seen that 1% change in the permittivity only in this area already results in a decrease in T by the factor 10^3. This example gives good grounds to believe that one can switch a sample from reflection to transmission or tune the emission of a source located inside the sample by external actions; for example, illuminating it by electromagnetic radiation that changes the dielectric constant of the material due to nonlinear effects.

Acknowledgments

V.F. acknowledges partial support from the Israel Science Foundation (Grant No. 894/10). K.Y.B. acknowledges partial support from the European Commission (Marie Curie Action) and Science Foundation Ireland (Grant No. 07/IN.1/I906). F.N. acknowledges partial support from the National Security Agency (NSA), Laboratory Physical Sciences (LPS), Army Research Office (ARO), and National Science Foundation (NSF) (Grant No. 0726909), Grant-in-Aid for Scientific Research (S), MEXT Kakenhi on Quantum Cybernetics, and Funding Program for Innovative R&D on S&T (FIRST).

References

1 P.W. Anderson, Phys. Rev. **109**, 1492 (1958).
2. V. Freilikher and S. Gredeskul, Prog. Opt. **30**, 37 (1992).
3. P. Cheng, *Introduction to Wave Scattering, Localization and Mesoscopic Phenomena* (Springer-Verlag, Heidelberg, 2006); E. Akkermans and G. Montambaux, *Mesoscopic Physics of Electrons and*

Photons (Cambridge University Press, Cambridge, 2007); J.-P. Fouque, J. Garnier, G. Papanicolaou, and K. Solna, *Wave Propagatin and Time Reversal in Randomly Layered Media* (Springer Science + Business Media, LLC, New York, 2007); E. Abrahams (ed.), *50 Years of Anderson Localization* (World Scientific, New Jersey/London/Singapore, 2010).
4. I. M. Lifshits, S. A. Gredeskul, and L. A. Pastur, *Introduction to the Theory of Disordered Systems* (Wiley, New York, 1988).
5. M. V. Berry and S. Klein, Eur. J. Phys. **18**, 222 (1997).
6. U. Frisch, C. Froeschle, J.-P. Scheidecker, and P.-L. Sulem, Phys. Rev. A **8**, 1416 (1973); M. Ya. Azbel, Phys. Rev. B **28**, 4106 (1983); M. Ya. Azbel and P. Soven, *ibid.* **27**, 831 (1983).
7. D. S. Wiersma, Nature **406**, 132 (2000); V. Milner and A. Genack, Phys. Rev. Lett. **94**, 073901 (2005).
8. H. Cao, Y. G. Zhao, S. T. Ho, E. W. Seelig, Q. H. Wang, and R. P. H. Chang, Phys. Rev. Lett. **82**, 2278 (1999).
9. K. Bliokh, Y. Bliokh, V. Freilikher, A. Genack, and P. Sebbah, Phys. Rev. Lett. **101**, 133901 (2008).
10. N. F. Mott, Philos. Mag. **22**, 7 (1970); I. Lifshits, V. Kirpichenkov, Zh. Eksp. Teor. Fiz. **77**, 989 (1979) [Sov. Phys. JETP **50**, 499 (1979)]; J. Pendry, Adv. Phys. **43**, 461 (1994).
11. J. Pendry, J. Phys. C **20**, 733 (1987).
12. J. Scales, L. Carr, D. McIntosh, V. Freilikher, and Y. Bliokh, Phys. Rev. B **76**, 085118 (2007).
13. I. Shadrivov, K. Bliokh, Y. Bliokh, V. Freilikher, and Y. Kivshar, Phys. Rev. Lett. **104**, 123902 (2010).
14. A. Asatryan, L. Botten, M. Byrne, V. Freilikher, S. Gredeskul, I. Shadrivov, R. McPhedran, and Y. Kivshar, Phys. Rev. Lett. **99**, 193902 (2007).
15. A. Asatryan, L. Botten, M. Byrne, V. Freilikher, S. Gredeskul, I. Shadrivov, R. McPhedran, and Y. Kivshar, Phys. Rev. B **81**, 075124 (2010).
16. A. Asatryan, L. Botten, M. Byrne, V. Freilikher, S. Gredeskul, I. Shadrivov, R. McPhedran, and Y. Kivshar, Phys. Rev. B **82**, 205124 (2010).
17. Y. Bliokh, V. Freilikher, S. Savel'ev, and F. Nori, Phys. Rev. B **79**, 075123 (2009).
18. H. Furstenberg, Trans. Am. Math. Soc. **108**, 377 (1963).
19. V. Kliatskin, *Stochastic Equations and Waves in Randomly Layered Media* (Nauka, Moscow, 1980).
20. E. Gurevich, Ph.D. thesis, Technion, Haifa, 2011.
21. E. Gurevich and A. Iomin, Phys. Rev. E **83**, 011128 (2011).
22. P. Anderson, D. Thouless, E. Abrahams, and D. Fisher, Phys. Rev. B **22**, 3519 (1980).
23. V. Baluni and J. Willemsen, Phys. Rev. A **31**, 3358 (1985).
24. K. Bliokh and V. Freilikher, Phys. Rev. B **70**, 245121 (2004).
25. R. Wing, *An Introduction to Invariant Imbedding* (Wiley, New York, 1976).
26. V. Freilikher, M. Pustilnik, and I. Yurkevich, Phys. Rev. B **73**, 810 (1994); **50**, 6017 (1994).
27. V. Dobrosavljevic, in *50 Years of Anderson Localization*, edited by E. Abrahams (World Scientific, New Jersey/London/Singapore, 2010).
28. V. Freilikher, M. Pustilnik, and I. Yurkevich, Phys. Rev. B **56**, 5974 (1957).
29. J. Paasschens, T. Misirpashaev, and C. Beenakker, Phys. Rev. B **54**, 11887 (1996).
30. V. Freilikher, B. Lianskii, I. Yurkevich, A. Maradudin, and A. McGurn, Phys. Rev. E **51**, 6301 (1995).
31. K. Bliokh, Y. Bliokh, and V. Freilikher, J. Opt. Soc. Am. B **21**, 113 (2004).
32. K. Bliokh, Y. Bliokh, V. Freilikher, S. Savel'ev, and F. Nori, Rev. Mod. Phys. **80**, 1201 (2008).
33. D. Bohm, *Quantum Theory* (Prentice-Hall, New York, 1952).
34. B. Payne, J. Andreasen, H. Cao, and A. Yamilov, Phys. Rev. B **82**, 104204 (2000).
35. K. Bliokh, Y. Bliokh, V. Freilikher, A. Genack, B. Hu, and P. Sebbah, Phys. Rev. Lett. **97**, 2439094 (2006).
36. J. C. Slater, *Microwave Electronics* (Van Nostrand, Princeton, 1950).
37. S. Rytov, Y. Kravtsov, and V. Tatarskii, *Principles of Statistical Radiophysics IV: Wave Propagation through Random Media* (Springer-Verlag, Berlin, 1989).
38. I. Lifshits and V. Kirpichenkov, Sov. Phys. JETP **50**, 499 (1979).
39. J. Bertolotti, S. Gottardo, and D. Wiersma, Phys. Rev. Lett. **94**, 113903 (2005).

40. Y. Bliokh, J. Felsteiner, and Y. Slutsker, Phys. Rev. Lett. **95**, 165003 (2005).
41. H. Gibbs, S. McCall, and T. Venkatesan, Phys. Rev. Lett. **36**, 1135 (1976).
42. M. Feise, I. Shadrivov, Y. Kivshar, and M. Feise, Phys. Rev. E **71**, 037602 (2005).
43. V. Cheianov, V. Fal'ko, and B. L. Altshuler, Science **315**, 1252 (2007).
44. M. Titov, Europhys. Lett. **79**, 17004 (2007).
45. Soon-Hong Kwon, T. Sunner, M. Kamp, and A. Forchel, Opt. Express **16**, 11709 (2008).
46. P. Kohli, J. Chatterton, D. Stieler, G. Tuttle, M. Li, X. Hu, Z. Ye, and K. Ho, Opt. Express **16**, 1984424 (2008); K. Aoki, D. Guimard, M. Nishioka, M. Nomura, S. Iwamotoi, and Y. Arakawa, Nat. Photonics **2**, 688 (2008).
47. F. Izrailev and A. Krokhin, Phys. Rev. Lett. **82**, 4062 (1999); U. Kuhl, F. Izrailev, and A. Krokhin, *ibid.* **100**, 126402 (2008); U. Kuhl, F. M. Izrailev, A. Krokhin, and H.-J. Stockmann, Appl. Phys. Lett. **77**, 633 (2000); O. Dietz, U. Kuhl, H.-J. Stockmann, N. Makarov, and F. Izrailev, Phys. Rev. B **83**, 134203 (2011).
48. A.V. Rozhkov, G. Giavaras, Y. P. Bliokh, V. Freilikher, and F. Nori, Phys. Reports **503**, 77 (2011).
49. S. Tamura and F. Nori, Phys. Rev. B **40**, 9790 (1989).
50. S. Tamura and F. Nori, Phys. Rev. B **41**, 7941 (1990).
51. M. Kolar, M. K. Ali, and F. Nori, Phys. Rev. B **43**, 1034 (1991).
52. N. Nishiguchi, S. Tamura, and F. Nori, Phys. Rev. B **48**, 2515 (1993).
53. N. Nishiguchi, S. Tamura, and F. Nori, Phys. Rev. B **48**, 14426 (1993).

2.5

*The Disorder Problem for Slow-Light Photonic Crystal Waveguides**

Mark Patterson and Stephen Hughes
Queen's University

CONTENTS

* This chapter is partly based on a thesis submitted by Patterson [1] to Queen's University, Canada, in partial fulfilment of the requirements for the M.Sc. (Eng.) degree.

2.5.1 Introduction

Photonic crystals (PhCs)* are structured dielectric composites that are designed and fabricated to have periodic optical properties that strongly alter the properties and propagation of light. Structures with one-dimensional periodicity such as Bragg mirrors were investigated as early as the nineteenth century. After the seminal works of John [2] and Yablonovitch [3], interest in structures with periodicity in two or three dimensions grew dramatically and the field of PhCs was born. The inspiration for PhCs comes from solid-state physics and the continuing success of the semiconductor industry. Electrons and holes in an atomic crystal, with a periodic potential due to the lattice of atoms, exhibit complex band structures that facilitate the control and manipulation of charge carriers. Understanding and exploiting these electronic "lattices" have revolutionized the development of transistors and computers. It is hoped that by using photonic lattices, an analogous revolution in optical systems will be possible [4]. One of the defining properties of PhCs are photonic band gaps (PBGs)—frequency ranges where light cannot propagate because of destructive interference between coherent scattering paths. A one-dimensional PhC, namely, a material that is periodic only along one spatial direction, will exhibit a *stop gap* for normally incident light, and, in this forbidden frequency range, acts as a perfect mirror. Two- and three-dimensional PhCs also exhibit PBGs for any given angle of incidence,† but careful engineering may be required to create a complete band gap—a frequency range where a PBG exists, regardless of angle of incidence.

PhC waveguides are structures typically formed by a line defect in an otherwise nominally perfect (disorder-free) PhC lattice. Waveguides are key components in many optical devices, as they allow one to route light from one location to another. Any platform technology involving PhCs would require a waveguide. Semiconductor-based PhC slab waveguides are of particular interest because they can be fabricated using high-quality etching and lithography techniques, and the high refractive index results in guided modes with very strong transverse spatial confinement, essentially trapping light to a spatial scale that is much smaller than the wavelength of the light. PhC waveguides have the intriguing characteristic of often exhibiting a region of slow-light propagation [5], which has potential applications as an optical delay line [6] and for tailoring the emission properties of embedded light sources such as quantum dots [2,7,8]. Slow light also leads to enhanced nonlinearities [9].

Slow-light waveguides are fundamentally interesting, as slowing down light can cause local fields to enhance proportionally. Formally, at the mode edge, the group velocity of a propagation mode becomes zero. The slope of the dispersion gives the group velocity, v_g, of the mode,

$$v_g(k) = \frac{d\omega(k)}{dk} = \frac{c}{n_g(k)}, \tag{2.5.1}$$

where k is the wave vector, $\omega(k)$ is the angular frequency as shown in the band structure, and n_g is the group index that can be a convenient way of expressing the group velocity.

* Although "PC" is a more natural abbreviation for photonic crystal, many PC researchers prefer the somewhat strange convention of PhC, including the Editors of this book, so we too use the latter in this chapter.

† Provided, for the two-dimensional case, that the light is incident in the plane of periodicity.

The group velocity is the speed at which an optical pulse centered at that frequency will propagate. Stopped light implies that the local fields can be infinite, which already flags a problem; indeed there is a *problem*: slowing down light in a photonic nanostructure comes at a price! It is now widely accepted that slow-light propagation in PhC waveguides enhances scattering from *structural disorder*, leading to large propagation losses [10,11], and, at best, slow group velocities of around $c/100$ can be measured, with extremely large propagation loss. Similar slow-light problems apply to other photonic structures as well; for example, slow-light metamaterial waveguides have massive propagation losses [12]. This "disorder problem" has often been ignored by many research groups in predicting possible applications of photonic nanostructures. However, disorder-induced scattering is an intrinsic part of photonic nanostructures, which is especially important and significant for slow-light PhC waveguides. Thus, the effects of disorder-induced scattering cannot be ignored in these structures. On the one hand, disorder is the number one *Achilles' Heel* preventing commercial development of PhC devices, and, on the other hand, disorder effects in PhCs are fundamentally interesting to study rich *extrinsic* scattering phenomena. The complexity of modeling planar PhC waveguide devices, and the need to account for fabrication tolerances, has been known for some time [13]. General predictions about the dramatic impact of disorder-induced scattering on the performance of PhC devices and PhC waveguides were made in the studies by Hughes et al. [14] and Koenderink et al. [15].

In this chapter, we give a thorough introduction to the disorder problem for slow-light PhC waveguides. We introduce detailed theoretical derivations and models of extrinsic scattering from an injected waveguide mode due to fabrication imperfections. Using an incoherent scattering theory, we show how slow-light propagation enhances extrinsic scattering from unavoidable manufacturing imperfections leading to backscattering and radiation loss that scale with the group velocity v_g, roughly as v_g^{-2} and v_g^{-1}, respectively. We also discuss when and why these simple scaling rules break down. We then improve the modeling of scattering using a coherent, multiple scattering approach to better explain the experimental observation of disordered resonances in slow-light waveguide modes. Finally, we demonstrate the influence of local-field effects that are specific to disordered high-index-contrast structures. We connect our theory and calculations to experimental measurements. Some of the general techniques also apply, and have been applied, to understanding and modeling the effect of disorder in other periodic structures, such as PhC cavities [16] and coupled cavity resonators [17]. To help set the scene for this chapter, an example schematic to show multiple scattering due to disorder in a PhC waveguide is illustrated in Figure 2.5.1a, with an example disordered hole shown in Figure 2.5.1b.

This chapter is organized as follows. In Section 2.5.2, we introduce the basic concepts of planar PhC waveguides and discuss their underlying electromagnetic properties. We introduce concepts such as band structure, Bloch modes, and Green functions, all of which will be useful for calculating the effects of disorder-induced scattering. In Section 2.5.3, we describe an *incoherent* scattering theory of disorder-induced scattering losses in PhC waveguides, and derive useful analytical formulas for backscatter and radiation mode (*out-of-plane*) scatter. This theory is then improved to account for multiple scattering processes, and we demonstrate the breakdown of the Beer–Lambert law. In Section 2.5.4, we extend the theory to account for multiple *coherent* scattering effects between disorder sites. This allows us to model various disordered instances and connect to disorder-induced localization effects. We directly connect this theory to experimental observations. In Section 2.5.5, we briefly address the problem of how to include local-field effects into the disorder theory, which results in broadening of the band structure and a systematic frequency shift of the entire band structure.

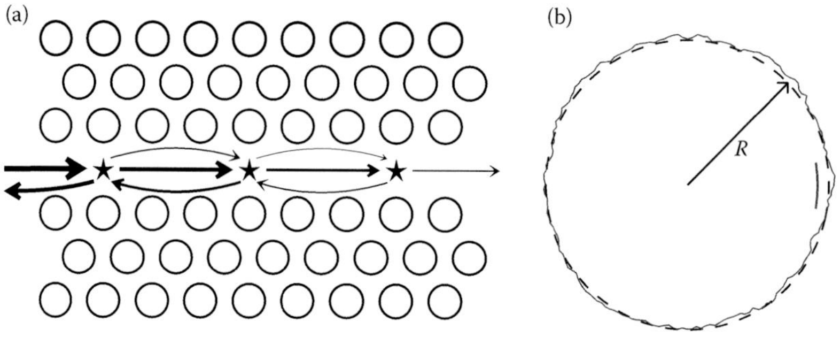

FIGURE 2.5.1
(a) Schematic of the "disorder problem" for a PhC waveguide. We show a plan view of a PhC slab waveguide with scattering sites (stars) whose strength is enhanced by the smaller group velocities. The PhC waveguide is formed by omitting a row of holes in an otherwise perfect photonic lattice. Incident light from the left, indicated by the arrows, undergoes scattering at each of the disorder sites leading to a complex interplay of forward and backward propagating waves. In practice, the scattering sites are continuously distributed throughout the system and one must account for the three-dimensional nature of the structure. (b) A realistic disordered hole that we will consider in our modeling, which is the source of the various scattering sites; the disorder perturbs the perimeter of the holes, with straight walls, where the disorder profile depends on statistical parameters such as the root mean square roughness σ and the correlation length l_p around the circumference (short arc).

2.5.2 Electromagnetic Properties of Planar PhC Waveguides

In this section, we introduce some of the key electromagnetic properties of PhC slab waveguides. Two-dimensional—or quasi two-dimensional—slab PhCs have periodicity in a plane. These *planar* dielectric structures are popular as they can be fabricated with conventional semiconductor growth techniques that are optimized for imprinting a two-dimensional pattern onto a substrate. One of the most common fabrication techniques is to pattern a periodic array of holes onto a thin layer of semiconductor, for example, silicon (Si), gallium–arsenide (GaAs), and indium–phosphide (InP), using electron beam lithography. The air holes are etched through the top layer that is then undercut to create a semiconductor membrane with periodic patterning. Figure 2.5.2 shows such a planar PhC

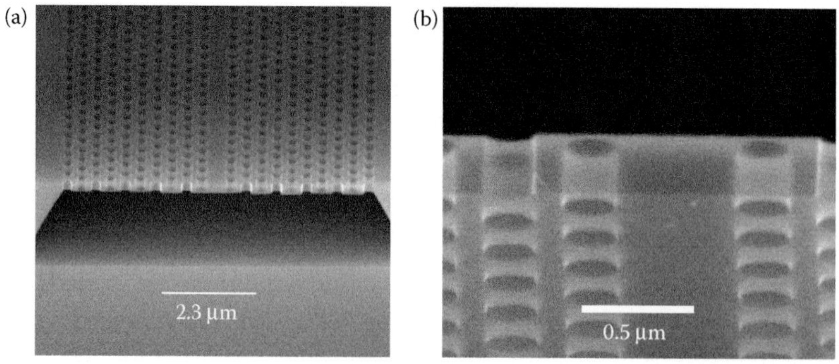

FIGURE 2.5.2
Two example scanning electron microscope images of a photonic crystal slab waveguide formed by omitting a row of holes in a photonic crystal membrane. Images kindly supplied by our collaborators, Alfredo De Rossi and Sylvain Combrié. (Adapted from S. Combrié et al., Opt. Express **14**(16), 7353, 2006.)

membrane looking down at an angle on the membrane edge [18]. The region of missing holes creates a waveguide, which will be discussed below. Although there are always finite-size effects, generally one tries to have enough unit cells to build up the *infinite-size* response. In this regard, the high-index-contrast nature of these semiconductor PhCs means that the 10 rows of holes (five unit cells) flanking the waveguide shown in Figure 2.5.2a are sufficient to build up a bulk-like band gap.

2.5.2.1 PhC Slabs

In this chapter, we will deal exclusively with air-clad semiconductor membranes or two-dimensional PhC slabs, such as those shown in Figure 2.5.2. The key advantage of these structures is that vertical confinement can be achieved by the high index-contrast between the semiconductor and air. Light with an in-plane wave vector* larger than the free space value, ω/c, will be trapped by total internal reflection. This critical value defines the "light line" or "light cone" and modes below the light line are confined to the slab. Above the light line, there is a continuum of modes that diffract out of the slab, termed radiation modes (c.f. bound or evanescent modes).

In two- and three-dimensional PhCs, one thinks about the Bloch modes of the structure (see Section 2.5.2.3) where the mode is parameterized by a Bloch wave vector **k**. Using techniques that will be discussed below, one can solve for the band structure or frequency as a function of wave vector, as shown in Figure 2.5.3 for a semiconductor membrane with a triangular array of holes. The frequency is given in normalized units that will be discussed later. Here, the band gap is visible as the range of frequencies without a propagating mode (marked by the horizontal dashes). The gray shaded regions indicate the continuum of radiation modes above the light line that are not confined by total internal reflection.

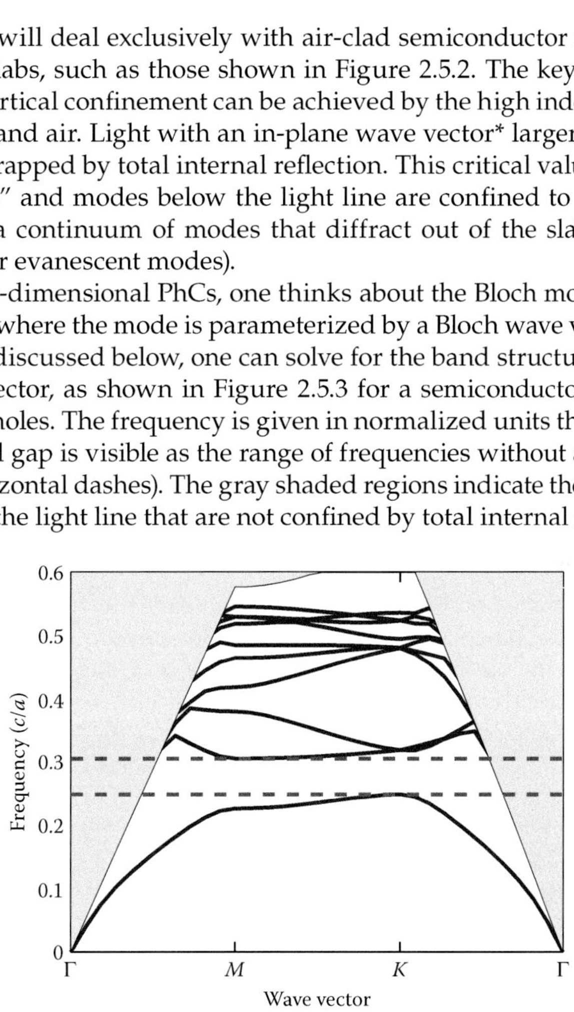

FIGURE 2.5.3
Band structure, that is, the dependence of frequency on wave vector, for the pseudo-transverse electric or quasi-transverse electric modes of a triangular lattice of air holes in a semiconductor membrane. The gray regions represent the continuum of radiation modes above the light line and the band gap is marked by the horizontal dashes. The Γ, *M*, and *K* mark high-symmetry points in wave vector space. The frequency is given in normalized units of the speed of light *c*, divided by the periodicity of the photonic crystal *a*. The structure parameters are the same as those for the W1.1 waveguide listed in Table 2.5.1 (but without the waveguide width).

* Properly, the Bloch wave vector.

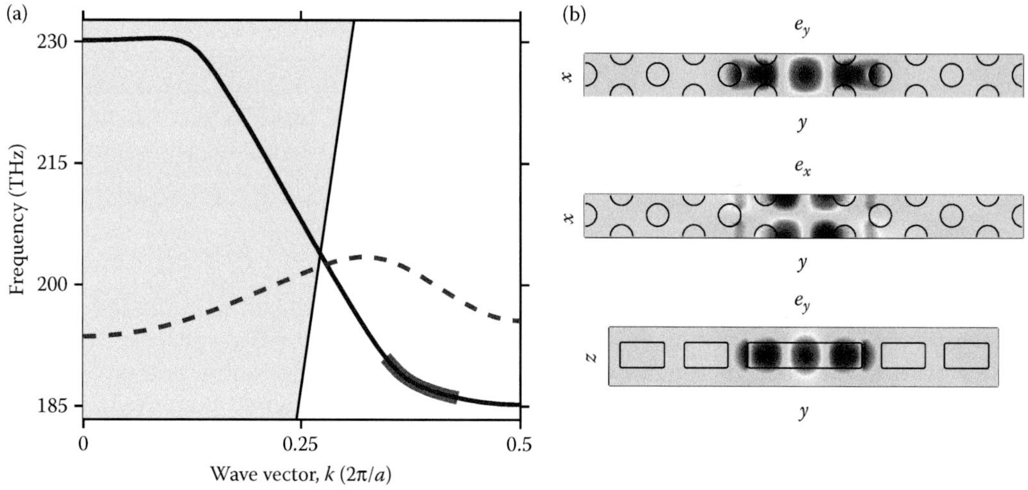

FIGURE 2.5.4
(See color insert.) (a) Band structure for the waveguide shown in Figure 2.5.2. The horizontal axis is the propagation (Bloch) wave vector. This waveguide supports two modes in the photonic band gap of the surrounding material, with the fundamental waveguide mode depicted by the solid curve and the secondary mode by the dashed curve. A typical region of operation for a slow-light waveguide is highlighted and does not extend all the way to the band edge because, although the velocity continues to decrease, scattering losses become significant (see Sections 2.5.3 and 2.5.4). The gray region to the left of the figure represents the continuum of radiation modes above the light line. (b) Components of the Bloch mode electric field (absolute value) for the fundamental waveguide mode near the band edge. The x- and y-components are shown on planes that bisect the unit cell. The z-component is negligible. The parameters for this "W1.1" waveguide are given in Table 2.5.1.

When discussing modes in PhC slabs, it can be useful to describe them in terms of pseudo-transverse electric (pseudo-TE) and pseudo-transverse magnetic (pseudo-TM) modes in analogy to the classifications of the modes of a dielectric stack. A pseudo-TE mode has the electric field polarized in the plane of the PhC slab and approximately perpendicular to the in-plane propagation direction. Owing to the lack of translational invariance of the structure along the propagation direction, the electric field in-plane polarization will vary throughout the unit cell. As the normal component of the electric field is discontinuous at dielectric interfaces, it is strongly influenced by the constituent holes of the PhC. In contrast, a pseudo-TM mode has the electric field polarized generally perpendicular to the slab surface and the magnetic field polarized in-plane. As the electric field is now mostly tangential to the holes, the pseudo-TM modes are less affected by the PhC* (see Figure 2.5.4 for the pseudo-TE modes).

2.5.2.2 PhC Slab Waveguides

PhC slabs can readily be made to produce waveguides, where quasi-one-dimensional channels are introduced for transporting light. For example, a PhC waveguide can be made by omitting a row of holes in a two-dimensional PhC slab as shown in Figure 2.5.2. Light in this local defect cannot spread out in the plane as a result of the perfectly reflective

* At least for the structures under consideration. For a more detailed discussion, see the study by Joannopoulos et al. [4] (pp. 128–130).

TABLE 2.5.1

Material Parameters

Parameter	W1 Waveguide	W1.1 Waveguide
Waveguide width	$\sqrt{3}a$	$1.1\sqrt{3}a$
Lattice period, a	480 nm	400 nm
Slab thickness, h	160 nm	265 nm
Hole radius, r	95 nm	0.27 a
Length, L	1.5 mm	250 μm
Index of refraction, n	3.18	3.38

PhC and cannot escape the slab because of total internal reflection. The waveguide is periodic along the propagation direction and so will have Bloch modes just like the perfect PhC. The waveguide modes occur at frequencies inside the PBG of the host crystal (or they would not be confined to the waveguide). The band structure or dispersion relation can be calculated for the waveguide, just like for the perfect PhC, and an example is shown in Figure 2.5.4a. The electric field distribution near the band edge for the fundamental waveguide mode is shown in Figure 2.5.4b, showing that the Bloch mode changes as a function of wave vector.

A ubiquitous class of PhC waveguides is "W1 waveguides" that are formed by omitting a single row of holes from a triangular lattice. There are numerous variations on this design such as a W3 waveguide where three rows of holes are omitted or a W1.1 waveguide where the waveguide width is 10% larger than for a W1 waveguide. For any given width, the hole radius r, slab thickness h, and periodicity a can be tuned to alter the properties of the waveguide. Values of $r = 0.27\,a$ and $h = 0.5\,a$ are typical of many designs. The results presented in this work are for W1 and W1.1 waveguides, although the formalisms are general and alternate structures can easily be studied [19]. Waveguide parameters that we will use in this chapter are listed in Table 2.5.1 and correspond to values in the study by Patterson et al. [20].

In Figure 2.5.4a the slow-light region occurs near 185 THz. Frequencies close to 200 THz, corresponding to a wavelength of about 1.5 μm, are used extensively in telecommunications. This is owing to the low loss properties of optical fibers and other components at this frequency. The frequency at which the device operates is determined by the periodicity a, so most of our scaling arguments and techniques can easily be scaled to any frequency. One may first observe that, in the case of Figure 2.5.4a, the group velocity is negative. This is interesting but not of concern as there is an equivalent point with negative wave vector with positive group velocity. As discussed in the introduction, one also observes that at the mode edge (the high k limit), the slope of the curve and the group velocity go to zero. The vanishing group velocity can also be interpreted as a divergence in the density of modes (proportional to v_g^{-1}), and we will use this interpretation to aid in the explanation of disorder-induced scattering and losses.

2.5.2.3 Electromagnetic Theory for PhC Waveguides

We now describe some of the useful analytic methods used to model light propagation in PhC structures. We introduce the fundamental Maxwell equations, Bloch symmetry for periodic systems, and photon Green function techniques.

2.5.2.3.1 Maxwell Equations

The electromagnetic fields are well described by the Maxwell equations. The two primary vector fields are the electric field $\mathbf{E}(\mathbf{r}; t)$ and the magnetic field $\mathbf{B}(\mathbf{r}; t)$. The two auxiliary fields, the electric displacement $\mathbf{D}(\mathbf{r}; t)$ and its magnetic equivalent $\mathbf{H}(\mathbf{r}; t)$, describe the effect of the fields being in a material such as a dielectric or air. In a volume with no free charges or currents on the macroscopic scale, the four Maxwell equations are

$$\nabla \cdot \mathbf{D}(\mathbf{r}; t) = 0, \tag{2.5.2}$$

$$\nabla \cdot \mathbf{B}(\mathbf{r}; t) = 0, \tag{2.5.3}$$

$$\nabla \times \mathbf{E}(\mathbf{r}; t) = -\frac{\partial}{\partial t}\mathbf{B}(\mathbf{r}; t), \tag{2.5.4}$$

$$\nabla \times \mathbf{H}(\mathbf{r}; t) = \frac{\partial}{\partial t}\mathbf{D}(\mathbf{r}; t). \tag{2.5.5}$$

It proves convenient to transform to the frequency domain and work with phasors for the fields. For example, the electric field phasor $\mathbf{E}(\mathbf{r}; \omega)$ is related to the time domain field by $\mathbf{E}(\mathbf{r}; t) = \mathrm{Re}[\mathbf{E}(\mathbf{r}; \omega)e^{-i\omega t}]$. Under this transform, the curl equations become

$$\nabla \times \mathbf{E}(\mathbf{r}; \omega) = i\omega\mathbf{B}(\mathbf{r}; \omega), \tag{2.5.6}$$

$$\nabla \times \mathbf{H}(\mathbf{r}; \omega) = -i\omega\mathbf{D}(\mathbf{r}; \omega). \tag{2.5.7}$$

In Equations 2.5.2 and 2.5.3, ω is simply substituted for t.

For normal dielectrics, that is, linear and isotropic, the primary and auxiliary fields are related by

$$\mathbf{D}(\mathbf{r}; \omega) = \varepsilon_0\, \varepsilon(\mathbf{r})\mathbf{E}(\mathbf{r}; \omega) \tag{2.5.8}$$

$$\mathbf{B}(\mathbf{r}; \omega) = \mu_0\, \mu(\mathbf{r})\mathbf{H}(\mathbf{r}; \omega), \tag{2.5.9}$$

where ε_0 and μ_0 are the vacuum permittivity and permeability, respectively. Their relative counterparts, $\varepsilon(\mathbf{r})$ and $\mu(\mathbf{r})$, are material-dependent and equal to 1 for vacuum. In this work, we only consider nonmagnetic ($\mu(\mathbf{r}) = 1$), lossless ($\mathrm{Im}[\varepsilon(\mathbf{r})] = 0$) materials.

Equations 2.5.6 and 2.5.7 can be combined, using Equations 2.5.8, 2.5.9, and $c^{-2} = \varepsilon_0\mu_0$, to form the well-known electromagnetic wave equation,

$$\frac{1}{\varepsilon(\mathbf{r})}\nabla \times \nabla \times \mathbf{E}(\mathbf{r}; \omega) = \left(\frac{\omega}{c}\right)^2 \mathbf{E}(\mathbf{r}; \omega). \tag{2.5.10}$$

To find a solution to the Maxwell equations, it is sufficient to solve Equation 2.5.10 while enforcing Equation 2.5.2. The magnetic field is then recovered using Equation 2.5.6.

Equation 2.5.10 is an eigenvalue equation; the left-hand side consists of an operator acting on the eigenvector $\mathbf{E}(\mathbf{r}; \omega)$ and the right-hand side gives the associated eigenvalue $(\omega/c)^2$

(with eigenfrequency ω). As written, the operator is non-Hermitian, but if the equation is rearranged in terms of the modified eigenvectors $\sqrt{\varepsilon(\mathbf{r})}\mathbf{E}(\mathbf{r};\omega)$, the operator becomes Hermitian. We can then invoke the properties of a Hermitian eigenvalue problem to show that the set of solutions for the electric field is both orthogonal and complete [21]. The orthogonality relationship for two modes l and l' is

$$\int_{\text{all space}} d\mathbf{r}\,\varepsilon(\mathbf{r})\mathbf{E}_l^*(\mathbf{r};\omega)\cdot\mathbf{E}_{l'}(\mathbf{r};\omega) = \delta_{l,l'}. \tag{2.5.11}$$

The completeness relation is

$$\delta(\mathbf{r}-\mathbf{r}')\vec{\mathbf{1}} = \varepsilon(\mathbf{r})\left[\sum_l \mathbf{E}_l^T(\mathbf{r})\otimes\mathbf{E}_l^{T*}(\mathbf{r}') + \sum_l \mathbf{E}_l^L(\mathbf{r})\otimes\mathbf{E}_l^{L*}(\mathbf{r}')\right], \tag{2.5.12}$$

where $\vec{\mathbf{1}}$ is the unit tensor, $\otimes$ denotes an outer (tensor) product, and we have explicitly characterized the modes as transverse (T) or longitudinal (L). Both types of modes are solutions to Equation 2.5.10, but the longitudinal modes have an eigenfrequency of 0 and do not satisfy Equation 2.5.2. Although they are unphysical, the longitudinal modes are required to construct the completeness relation.

2.5.2.3.2 Bloch Modes

Because the dielectric constant is periodic, one expects the intensity of the mode to be periodic too. However, the electromagnetic fields may have a phase component that is not periodic with the PhC. Such fields are called Bloch states or Bloch modes and can be written as a purely periodic component, $\mathbf{e}_\mathbf{k}(\mathbf{r})$, modulated by a plane-wave phase factor that describes propagation:

$$\mathbf{E}(\mathbf{r};\omega) = \sqrt{\frac{a}{L}}\mathbf{e}_\mathbf{k}(\mathbf{r})e^{i\mathbf{k}\cdot\mathbf{r}}, \tag{2.5.13}$$

where $L \to \infty$ is the length of the waveguide. The Bloch wave vector, $\mathbf{k}$, and angular frequency, ω, are inextricably linked and their relation is the band structure of the system. Bloch states exist for PhCs and also for PhC waveguides, which are periodic along the waveguide propagation direction. Wave vectors that differ by $2\pi/a$, where a is the periodicity along a basis direction, are equivalent. For this reason, we work within the reduced zone scheme where the wave vector is restricted to the first Brillouin zone. This requires, for a one-dimensional PhC with periodicity a, $-\pi/a \le k \le \pi/a$. For any given wave vector, there will be multiple solutions with distinct electric-field profiles and frequencies that form the multiple bands or modes of the structure.

The Bloch modes have a number of convenient properties. Modes with a negative wave vector are related to their positive wave vector counterpart by

$$\mathbf{e}_{-k}(\mathbf{r}) = \mathbf{e}_k^*(\mathbf{r}). \tag{2.5.14}$$

Following from Equation 2.5.11, the modes are orthogonal by

$$\int_{\text{cell}} d\mathbf{r}\;\varepsilon(\mathbf{r})\mathbf{e}_k^*(\mathbf{r})e^{-ikx}\cdot\mathbf{e}_{k'}(\mathbf{r})e^{ik'x} = \delta_{k,k'}, \tag{2.5.15}$$

where we have been able, through symmetry, to reduce the integration volume to a single unit cell.

2.5.2.3.3 Green Function

A convenient mathematical tool for describing the electromagnetic properties of a system is the Green function. The electric field Green function $\overleftrightarrow{G}(\mathbf{r},\mathbf{r}';\omega)$ satisfies the Maxwell wave equation under a dipole excitation:

$$\nabla \times \nabla \times \overleftrightarrow{G}(\mathbf{r},\mathbf{r}';\omega) - \left(\frac{\omega}{c}\right)^2 \varepsilon(\mathbf{r})\overleftrightarrow{G}(\mathbf{r},\mathbf{r}';\omega) = \left(\frac{\omega}{c}\right)^2 \delta(\mathbf{r}-\mathbf{r}')\overleftrightarrow{\mathbf{1}}. \tag{2.5.16}$$

The photon Green function is a tensor (or 3×3 matrix) whose components depend on two space points and frequency. The $G_{ij}(\mathbf{r}, \mathbf{r}'; \omega)$ component can be physically interpreted as giving the *i*-component of the electric field at position $\mathbf{r}$ as a result of the introduction of a *j*-polarized dipole at position $\mathbf{r}'$ with angular frequency ω.

An excellent treatment of the properties of PhCs using Green function techniques is given by Sakoda [21]. The key property of the Green function is that the effect of adding a macroscopic polarization density $\mathbf{P}(\mathbf{r}'; \omega)$ to a system described by $\overleftrightarrow{G}(\mathbf{r},\mathbf{r}';\omega)$ is given by the well-known integral equation:

$$\mathbf{E}(\mathbf{r};\omega) = \mathbf{E}_i(\mathbf{r};\omega) + \int \overleftrightarrow{G}(\mathbf{r},\mathbf{r}';\omega)\cdot\frac{\mathbf{P}(\mathbf{r}';\omega)}{\varepsilon_0}d\mathbf{r}', \tag{2.5.17}$$

where $\mathbf{E}_i(\mathbf{r};\omega)$ is the field before the source is introduced. This is a fully self-consistent, nonperturbative solution for the new electric field and forms the basis of all the scattering calculations we present.

In simple geometries, the Green function can be computed analytically, for example, for a homogeneous dielectric [22]. One of the physical properties accessible through the Green function is the (electric) local density of states, proportional to the trace of $\mathrm{Im}[\overleftrightarrow{G}(\mathbf{r},\mathbf{r}';\omega)]$ [23]. For a homogeneous dielectric, with dielectric constant ε, one has

$$\mathrm{Im}\left[\overleftrightarrow{G}_0(\mathbf{r},\mathbf{r}';\omega)\right] = \frac{\omega^3\sqrt{\varepsilon}}{6\pi c^3}\overleftrightarrow{\mathbf{1}}. \tag{2.5.18}$$

In inhomogeneous structures, it can be frequently easier to work with the scattered Green function $\overleftrightarrow{G}^S(\mathbf{r},\mathbf{r}';\omega)$ that expresses the difference between the inhomogeneous material and a homogeneous dielectric as

$$\overleftrightarrow{G}^S(\mathbf{r},\mathbf{r}';\omega) = \overleftrightarrow{G}(\mathbf{r},\mathbf{r}';\omega) - \overleftrightarrow{G}_0(\mathbf{r},\mathbf{r}';\omega), \tag{2.5.19}$$

where the homogeneous Green function is evaluated for the material in which the source is located. This can be especially useful when evaluating the Green function for $\mathbf{r} \to \mathbf{r}'$. In this limit, the full Green function diverges (for valid physical reasons), but the scattered Green function may be finite.

To aid in its calculation in an *inhomogeneous* system, the Green function can be decomposed into transverse and longitudinal components as

$$\overleftrightarrow{\mathbf{G}}(\mathbf{r},\mathbf{r}';\omega) = \overleftrightarrow{\mathbf{G}}^T(\mathbf{r},\mathbf{r}';\omega) + \overleftrightarrow{\mathbf{G}}^L(\mathbf{r},\mathbf{r}';\omega), \tag{2.5.20}$$

and then further decomposed into a sum over the electromagnetic mode as

$$= \sum_l \frac{\omega^2}{\omega_l^2 - \omega^2} \mathbf{E}_l^T(\mathbf{r}) \otimes \mathbf{E}_l^{T*}(\mathbf{r}') - \sum_l \mathbf{E}_l^L(\mathbf{r}) \otimes \mathbf{E}_l^{L*}(\mathbf{r}'). \tag{2.5.21}$$

Using the completeness relation (Equation 2.5.12), the explicit longitudinal mode contribution can be eliminated,

$$\overleftrightarrow{\mathbf{G}}(\mathbf{r},\mathbf{r}';\omega) = \sum_l \frac{\omega_l^2}{\omega_l^2 - \omega^2} \mathbf{E}_l^T(\mathbf{r}) \otimes \mathbf{E}_l^{T*}(\mathbf{r}') - \frac{\delta(\mathbf{r}-\mathbf{r}')}{\varepsilon(\mathbf{r})} \overleftrightarrow{\mathbf{1}}. \tag{2.5.22}$$

Note that the δ-function contains contributions from both the transverse and the longitudinal modes. For convenience, we partition the Green function into contributions from the bound waveguide mode, radiation modes, and "other" modes as

$$\overleftrightarrow{\mathbf{G}}(\mathbf{r},\mathbf{r}';\omega) = \overleftrightarrow{\mathbf{G}}_{\text{bound}}(\mathbf{r},\mathbf{r}';\omega) + \overleftrightarrow{\mathbf{G}}_{\text{rad}}(\mathbf{r},\mathbf{r}';\omega) + \overleftrightarrow{\mathbf{G}}_{\text{other}}(\mathbf{r},\mathbf{r}';\omega), \tag{2.5.23}$$

where $\overleftrightarrow{\mathbf{G}}_{\text{bound}}(\mathbf{r},\mathbf{r}';\omega)$ and $\overleftrightarrow{\mathbf{G}}_{\text{rad}}(\mathbf{r},\mathbf{r}';\omega)$ are purely transverse.

The bound mode Green function is given by a sum over modes from Equation 2.5.22 as

$$\overleftrightarrow{\mathbf{G}}_{\text{bound}}(\mathbf{r},\mathbf{r}';\omega_0) = \sum_k \frac{\omega_k^2}{\omega_k^2 - (\omega_0 + i\delta)^2} \mathbf{E}_k^T(\mathbf{r}) \otimes \mathbf{E}_k^{T*}(\mathbf{r}'), \tag{2.5.24}$$

where the sum is now over the wave vector of the mode k, and δ is an infinitesimal positive quantity we introduce to ensure that we obtain a causal solution to the Maxwell equations. The electric field solutions for the entire structure $\mathbf{E}_k^T(\mathbf{r})$ are related to the Bloch modes by Equation 2.5.13. The discrete summation is converted to an integration by $\Sigma_k \to \frac{L}{2\pi}\int dk$, to yield

$$\overleftrightarrow{\mathbf{G}}_{\text{bound}}(\mathbf{r},\mathbf{r}';\omega_0) = \frac{a}{4\pi} \int_{-\pi/a}^{\pi/a} dk \left(\frac{\omega(k)}{\omega(k) + \omega_0 + i\delta} + \frac{\omega(k)}{\omega(k) - \omega_0 - i\delta} \right) \mathbf{e}_k(\mathbf{r}) \otimes \mathbf{e}_k^*(\mathbf{r}') e^{ik(x-x')}, \tag{2.5.25}$$

where ω_0 is a parameter of the Green function and $\omega(k)$ is a function of the integration variable. To carry out the integration, it is useful to note that the eigenfrequency $\omega(k)$ is positive everywhere and so only the second term in the bracket will have a pole. We denote the wave vector with positive group velocity near the pole as k_0. Note that $\omega(k)$ is a function of the complex variable k and for arbitrary k, it too may be complex (but not on the real k-axis for lossless materials). Exploiting Cauchy's residue theorem, one can readily derive

$$\overleftrightarrow{\mathbf{G}}_{\text{bound}}(\mathbf{r},\mathbf{r}';\omega) = i\frac{a\omega}{2v_g}\Big[\mathbf{e}_{k_0}(\mathbf{r}) \otimes \mathbf{e}^*_{k_0}(\mathbf{r}')e^{ik_0(x-x')}\Theta(x-x')$$
$$+ \mathbf{e}^*_{k_0}(\mathbf{r}) \otimes \mathbf{e}_{k_0}(\mathbf{r}')e^{-ik_0(x-x')}\Theta(x'-x)\Big], \tag{2.5.26}$$

where k_0 must be chosen so that $v_g(k) > 0$, and $\Theta(x)$ is the Heaviside step function, equal to 1 if $x > 0$ and 0 if $x < 0$. If the eigenmode data is calculated at a wave vector with $v_g(k_0) < 0$, mirror it to the other side of the Brillouin zone by taking $k_0 \to -k_0$ and $\mathbf{e}_{k_0}(\mathbf{r}) \to [\mathbf{e}_{k_0}(\mathbf{r})]^*$ (by Equation 2.5.26). This is one of the key results of this section and will be used repeatedly in further derivations. It allows one to efficiently calculate the bound mode contribution to the Green function, which dominates the total Green function, using easily obtained properties of the waveguide mode.

The radiation mode Green function is also written as a sum over modes as

$$\overleftrightarrow{\mathbf{G}}_{\text{rad}}(\mathbf{r},\mathbf{r}';\omega) = \sum_l \frac{\omega_l^2}{\omega_l^2 - (\omega + i\delta)^2}\mathbf{E}_l^T(\mathbf{r}) \otimes \mathbf{E}_l^{T*}(\mathbf{r}'), \tag{2.5.27}$$

where l does not include the waveguide modes, nor modes with a wave vector greater than $\sqrt{\varepsilon}\omega/c$. The radiation modes are less tractable and can only be evaluated exactly for simple geometries, for example, a dielectric tube [24]. In Section 2.5.3.3, we will describe an efficient numerical technique for approximating the radiation Green function of an effective PhC slab.

The remaining contributions to the Green function, lumped into $\overleftrightarrow{\mathbf{G}}_{\text{other}}(\mathbf{r},\mathbf{r}';\omega)$, include the δ-function from Equation 2.5.22, evanescent modes, and surface plasmons (if they exist). These contributions ensure well-known physical properties of the Green function such as the divergence of the real part of the Green function as $\mathbf{r} \to \mathbf{r}'$ (due to the evanescent modes) and an additional divergence at $\mathbf{r} = \mathbf{r}'$ (due to the δ-function). For PhC waveguides, it is a reasonable approximation to neglect these contributions to the Green function:

$$\overleftrightarrow{\mathbf{G}}(\mathbf{r},\mathbf{r}';\omega) \simeq \overleftrightarrow{\mathbf{G}}_{\text{bound}}(\mathbf{r},\mathbf{r}';\omega) + \overleftrightarrow{\mathbf{G}}_{\text{rad}}(\mathbf{r},\mathbf{r}';\omega). \tag{2.5.28}$$

Although this is an approximation, Equation 2.5.28 is exact for the imaginary part of the Green function. The dominant effect that is neglected in this approximation is *local-field corrections*. In Sections 2.5.3 and 2.5.4, we will essentially ignore $\overleftrightarrow{\mathbf{G}}_{\text{other}}$. The role of local-field effects on disorder-induced scattering will be addressed in Section 2.5.5.

2.5.2.4 Numerical Methods for Obtaining Photonic Bloch Modes and Band Structure

For all but the simplest structures, the Maxwell equations cannot be solved analytically and one resorts to numerical methods. In this chapter, we utilize two complementary approaches, the finite-difference time-domain (FDTD) method and the frequency-domain plane-wave expansion method (PWEM). We highlight each of these approaches below; however, we keep our discussions brief, as these techniques are well known to the PhC community and further details are well explained in the cited references.

The FDTD method was pioneered by Yee [25] and has been studied extensively. Two excellent resources are the works by Sullivan [26], which focuses on writing a functional FDTD code, and Taflove and Hagness [27], which provides in-depth background on the method. The method discretizes both $\mathbf{E}(\mathbf{r}; t)$ and $\mathbf{H}(\mathbf{r}; t)$ in space and then iterates through time, solving the Maxwell curl Equations 2.5.4 and 2.5.5. The heart of the method is the realization that, since $\partial\mathbf{E}(\mathbf{r}; t)/\partial t$ depends only on $\mathbf{H}(\mathbf{r}; t)$ and vice versa, the two grids may be interleaved to achieve much greater accuracy. This distribution of the discretized field components throughout the unit cell is called the Yee cell. We make use of Lumerical's FDTD Solutions [28], a commercial FDTD code. There are also typically in-house FDTD codes within PhC research groups (including our own), and there are free FDTD codes available to use within the optics community [29]. To calculate the eigenfrequencies of a structure using FDTD, a simulation volume is designed with the appropriate geometry and boundary conditions (absorbing for isolated cavities or Bloch for periodic systems). A broadband dipole is used to excite the system and the field as a function of time is then analyzed, using, for example, Fourier transform techniques, to determine the mode frequencies. If the eigenmode is also desired, a separate simulation must be run with a narrowband excitation. The convergence of eigenfrequencies is inversely proportional to the length of the simulation and so achieving a high accuracy can be computationally intensive, although some improvements can be made [30,31]. FDTD can also be used to directly calculate the Green function for a complex structure. The system is excited with a broadband electric dipole at $\mathbf{r}'$ for a short period of time and the electric field at other points of interest $\mathbf{r}$ is recorded.

For determining eigenmodes of an infinite periodic system, frequency-domain techniques are significantly more efficient. In particular, the PhC and waveguide properties presented in this work are almost exclusively calculated using a freely available implementation of the PWEM by Johnson and Joannopoulos [32]. For systems with lossless eigenmodes, PWEM is orders of magnitude more efficient than FDTD. For example, to calculate the band structure in Figure 2.5.4, we performed simulations of the structure using both numerical methods. FDTD required 400 cpu-minutes per wave vector and PWEM needed only 18 cpu-minutes per wave vector. In addition, we were only interested in calculating the eigenfrequencies; PWEM also returned the eigenmodes but FDTD would have required a second round of simulations to calculate them. The weakness of PWEM is the requirement for periodicity in all spatial dimensions. Thus, when modeling, for example, a quasi two-dimensional air-clad PhC slab waveguide, the simulated system is actually an array of parallel waveguides. By making the simulation supercell large in both transverse directions, evanescent coupling between the waveguides is reduced and the simulation converges to the results for a single isolated waveguide. PWEM cannot correctly resolve modes above the light line (not confined to the PhC slab by total internal reflection) which are intrinsically lossy. In the case of Figure 2.5.4, it was necessary to use FDTD as the band structure is shown above the light line.

2.5.3 Disorder-Induced Incoherent Scattering in PhC Waveguides

Light in a planar PhC waveguide mode, propagating below the light line, is trapped in-plane by the PBG of the cladding PhC and vertically by total internal reflection. Thus, if the structure is made of lossless dielectric, light in the mode should propagate indefinitely without loss for all frequencies from the mode edge to the intersection of the mode's

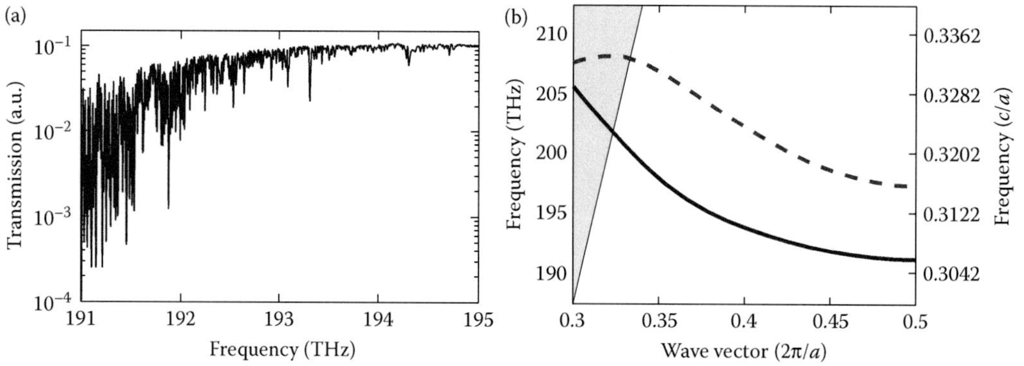

FIGURE 2.5.5
(a) Typical experimental transmission spectrum for a 1.5-mm-long waveguide provided by Sylvain Combrié and Alfredo De Rossi, Thales Research and Technology, France. (Adapted from M. Patterson et al., Phys. Rev. Lett. **102**, 253903, 2009.) Approaching the band edge, the transmission rolls off as a result of *extrinsic* scattering. (b) For reference, the calculated band structure of the waveguide with the mode of interest is shown with the solid curve. Owing to uncertainty in the membrane thickness, there may be a systematic frequency shift between the calculated and experimental band structures. The structure is a W1 waveguide with parameters given in Table 2.5.1.

dispersion with the light line. Instead, measurements observe a gradual roll-off of transmission near the band edge as the group velocity decreases (see Figure 2.5.5). This phenomenon can be explained by considering small imperfections in the fabricated devices that act as extrinsic scattering sites. As the light slows down, both the local electric field strength and the interaction time increase, leading to an amplification of extrinsic scattering. As can be seen, the waveguide transmission loss increases by several orders of magnitude as the light slows down, and the very-slow-light regime ($v_g \sim< c/100$) is not really accessible—in fact, the losses are so large that most of the light is completely reflected. This is a common and typical example of the *disorder problem*.

In this section, we present a formalism for calculating the incoherent ensemble average loss in a disordered PhC waveguide. The formalism used here is based on the earlier work of Hughes et al. [14], but is extended through the efficient calculation of radiation losses and the extrapolation of the loss per unit cell to experimental length scales. There has been significant recent interest in disorder theories for PhC waveguides and there are a number of other formalisms with complementary results, notably by Povinelli et al. [33] and Gerace and Andreani [34,35]. The general results of Hughes et al. [14] show that there are two dominant scattering processes: backscattering in which light scatters from a forward propagating mode into a backward propagating mode, and radiative scattering in which light scatters into a radiation mode above the light line that quickly leaks out of the slab. We will show why these scattering processes depend on the group velocity v_g, scaling *approximately* as v_g^{-2} for backscattering and v_g^{-1} for radiation scattering. These approximate loss-scaling relations have been confirmed experimentally by a number of groups [10,36], but they may break down at low group velocities where multiple disorder-induced scattering becomes significant.

In Section 2.5.3.1, we present the theory of incoherent scattering in PhC waveguides with the main results given in Equations 2.5.35 and 2.5.39. We then introduce a statistical model for disorder in these structures (Section 2.5.3.2) and a method for approximating the radiation modes (Section 2.5.3.3). In Section 2.5.3.4, we outline a method for connecting the loss per unit cell to experimental length scales before finally, in Section

2.5.3.5, providing sample calculations included transmission spectra, the dependence of the loss on the sample disorder, and a comparison with the experimental results of Figure 2.5.5. The failure of the Beer–Lambert law, which assumes that light transmission $T \propto e^{-\alpha x}$ (with α the loss coefficient), is shown to break down for typical experimental sample lengths.

2.5.3.1 Theory and Analytical Expressions for Disorder-Induced Losses in Optically Thin Samples

An ideal infinite PhC waveguide with spatially dependent dielectric constant $\varepsilon_i(\mathbf{r})$ has an initial solution consisting of a forward propagating wave given by

$$\mathbf{E}_0(\mathbf{r};\omega) = \mathcal{E}_0\mathbf{e}_k(\mathbf{r})e^{ikx}, \tag{2.5.29}$$

where $\mathcal{E}_0$ is an amplitude, and $\mathbf{e}_k(\mathbf{r})$ is the Bloch mode electric field with wave vector k. If a region of disorder is introduced in the center of the waveguide, described by $\Delta\varepsilon(\mathbf{r}) = \varepsilon(\mathbf{r}) - \varepsilon_i(\mathbf{r})$, the new solution is related to the ideal solution by an integral Equation 2.5.17

$$\mathbf{E}(\mathbf{r};\omega) = \mathbf{E}_0(\mathbf{r};\omega) + \int_{\text{all space}} d\mathbf{r}'\overleftrightarrow{\mathbf{G}}(\mathbf{r},\mathbf{r}';\omega)\cdot[\Delta\varepsilon(\mathbf{r}')\mathbf{E}(\mathbf{r}';\omega)], \tag{2.5.30}$$

where the field we wish to solve for $\mathbf{E}(\mathbf{r};\omega)$ appears on both sides of the equation and $\overleftrightarrow{\mathbf{G}}(\mathbf{r},\mathbf{r}';\omega)$ is the photon Green function for the ideal structure. The integral is over all space but will be nonzero only in the disordered region owing to the inclusion of $\Delta\varepsilon(\mathbf{r}')$. For compactness, the subscript on the integral will be omitted and the integration domain implied by $\Delta\varepsilon(\mathbf{r}')$. In a PhC waveguide system, $\overleftrightarrow{\mathbf{G}}(\mathbf{r},\mathbf{r}';\omega) \simeq \overleftrightarrow{\mathbf{G}}_{\text{bound}}(\mathbf{r},\mathbf{r}';\omega) + \overleftrightarrow{\mathbf{G}}_{\text{rad}}(\mathbf{r},\mathbf{r}';\omega)$ (Equation 2.5.28). As briefly mentioned in Section 2.5.2, the dominant omission is local-field effects, which arise from the δ-function in Equation 2.5.22 and which have been studied in part by Johnson et al. [37] for extremely limited disorder configurations and by Wang et al. [38] for a simple two-dimensional geometry. The main effect of the local fields is a frequency shift of the mode (see the study by Ramunno and Hughes [16] for an example of a PhC cavity), which does not significantly alter the roll-off of the spectrum. In Section 2.5.5, we will describe the effects of local-field corrections in more detail, which is based on the work by Patterson and Hughes [39].

We focus on three-dimensional modeling, with reasonable approximations, using general and realistic disorder models such as those seen and measured in experiments. Equation 2.5.30 provides a fully self-consistent, nonperturbative solution to the disordered system but is difficult to solve. In this section, we take a Born expansion by iteratively substituting the equation into itself to give

$$\begin{aligned}\mathbf{E}(\mathbf{r};\omega) &= \mathbf{E}_0(\mathbf{r};\omega) + \int d\mathbf{r}'\overleftrightarrow{\mathbf{G}}(\mathbf{r},\mathbf{r}';\omega)\cdot[\Delta\varepsilon(\mathbf{r}')\mathbf{E}_0(\mathbf{r}';\omega)]\\ &\quad+\int d\mathbf{r}'\overleftrightarrow{\mathbf{G}}(\mathbf{r},\mathbf{r}';\omega)\cdot\left[\Delta\varepsilon(\mathbf{r}')\int d\mathbf{r}''\overleftrightarrow{\mathbf{G}}(\mathbf{r}',\mathbf{r}'';\omega)\cdot[\Delta\varepsilon(\mathbf{r}'')\mathbf{E}_0(\mathbf{r}'';\omega)\right]\\ &\quad+O(\overleftrightarrow{\mathbf{G}}(\mathbf{r},\mathbf{r}';\omega)^3),\end{aligned} \tag{2.5.31}$$

which expands to

$$\mathbf{E}(\mathbf{r};\omega) \simeq \mathbf{E}_0(\mathbf{r};\omega) \tag{2.5.32a}$$

$$+\int d\mathbf{r}'\Delta\varepsilon(\mathbf{r}')\overleftrightarrow{\mathbf{G}}_{\mathrm{bound}}(\mathbf{r},\mathbf{r}';\omega)\cdot\mathbf{E}_0(\mathbf{r}';\omega) \tag{2.5.32b}$$

$$+\int d\mathbf{r}'\Delta\varepsilon(\mathbf{r}')\overleftrightarrow{\mathbf{G}}_{\mathrm{rad}}(\mathbf{r},\mathbf{r}';\omega)\cdot\mathbf{E}_0(\mathbf{r}';\omega) \tag{2.5.32c}$$

$$+\iint d\mathbf{r}'d\mathbf{r}''\Delta\varepsilon(\mathbf{r}')\Delta\varepsilon(\mathbf{r}'')\overleftrightarrow{\mathbf{G}}_{\mathrm{bound}}(\mathbf{r},\mathbf{r}';\omega)\cdot\overleftrightarrow{\mathbf{G}}_{\mathrm{bound}}(\mathbf{r}',\mathbf{r}'';\omega)\cdot\mathbf{E}_0(\mathbf{r}'';\omega) \tag{2.5.32d}$$

$$+\iint d\mathbf{r}'d\mathbf{r}''\Delta\varepsilon(\mathbf{r}')\Delta\varepsilon(\mathbf{r}'')\overleftrightarrow{\mathbf{G}}_{\mathrm{bound}}(\mathbf{r},\mathbf{r}';\omega)\cdot\overleftrightarrow{\mathbf{G}}_{\mathrm{rad}}(\mathbf{r}',\mathbf{r}'';\omega)\cdot\mathbf{E}_0(\mathbf{r}'';\omega) \tag{2.5.32e}$$

$$+\iint d\mathbf{r}'d\mathbf{r}''\Delta\varepsilon(\mathbf{r}')\Delta\varepsilon(\mathbf{r}'')\overleftrightarrow{\mathbf{G}}_{\mathrm{rad}}(\mathbf{r},\mathbf{r}';\omega)\cdot\overleftrightarrow{\mathbf{G}}_{\mathrm{bound}}(\mathbf{r}',\mathbf{r}'';\omega)\cdot\mathbf{E}_0(\mathbf{r}'';\omega) \tag{2.5.32f}$$

$$+\iint d\mathbf{r}'d\mathbf{r}''\Delta\varepsilon(\mathbf{r}')\Delta\varepsilon(\mathbf{r}'')\overleftrightarrow{\mathbf{G}}_{\mathrm{rad}}(\mathbf{r},\mathbf{r}';\omega)\cdot\overleftrightarrow{\mathbf{G}}_{\mathrm{rad}}(\mathbf{r}',\mathbf{r}'';\omega)\cdot\mathbf{E}_0(\mathbf{r}'';\omega) \tag{2.5.32g}$$

$$+O(\overleftrightarrow{\mathbf{G}}(\mathbf{r},\mathbf{r}';\omega)^3). \tag{2.5.32h}$$

In Section 2.5.4, we will solve Equation 2.5.30 without resorting to a Born expansion. Equation 2.5.32 includes all terms up to second order in $\overleftrightarrow{\mathbf{G}}(\mathbf{r},\mathbf{r}';\omega)$. We will calculate the backscatter loss by examining the solution at $x \to -\infty$, where only terms to first order in $\overleftrightarrow{\mathbf{G}}(\mathbf{r},\mathbf{r}';\omega)$ are necessary (one scattering event). To calculate the radiation loss, we will examine the solution at $x \to \infty$ and associate the reductions in transmission with backscatter and radiation loss. In transmission, it is necessary to retain terms up to second order in $\overleftrightarrow{\mathbf{G}}(\mathbf{r},\mathbf{r}';\omega)$ for the first order scattering to attenuate the forward wave.

At $x \to -\infty$,* the solution will consist of the initial solution and a reflected term $\mathbf{E}(x \to -\infty;\omega) = \mathbf{E}_0(\mathbf{r};\omega) + r\mathcal{E}_0\mathbf{e}_k^*(\mathbf{r})e^{-ikx}$, where r is the reflected amplitude. We isolate the reflected term using the orthogonality relation (Equation 2.5.15) by multiplying from the left by $\mathcal{E}_0^{-1}\varepsilon_i(\mathbf{r})\mathbf{e}_k(\mathbf{r})e^{ikx}$ and integrating over a unit cell near $x \to -\infty$. Keeping terms only to first order in $\overleftrightarrow{\mathbf{G}}(\mathbf{r},\mathbf{r}';\omega)$ from Equation 2.5.32 (i.e., 2.5.32a through c), only Equation 2.5.32b will have a nonzero projection and the reflection coefficient is

$$\begin{aligned}
r &= \int_{\mathrm{cell}} d\mathbf{r}\,\mathcal{E}_0^{-1}\varepsilon_i(\mathbf{r})\mathbf{e}_k(\mathbf{r})e^{ikx}\cdot\int d\mathbf{r}'\Delta\varepsilon(\mathbf{r}')\overleftrightarrow{\mathbf{G}}_{\mathrm{bound}}(\mathbf{r},\mathbf{r}';\omega)\cdot\mathbf{E}_0(\mathbf{r}';\omega),\\
&= i\frac{a\omega}{2v_g}\int_{\mathrm{cell}} d\mathbf{r}\,\varepsilon_i(\mathbf{r})\mathbf{e}_k(\mathbf{r})e^{ikx}\cdot\mathbf{e}_k^*(\mathbf{r})e^{-ikx}\\
&\quad\times\int d\mathbf{r}'\Delta\varepsilon(\mathbf{r}')\Theta(x'-x)\mathbf{e}_k(\mathbf{r}')e^{ikx'}\cdot\mathbf{e}_k(\mathbf{r}')e^{ikx'},\\
&= i\frac{a\omega}{2v_g}\int d\mathbf{r}'\Delta\varepsilon(\mathbf{r}')\mathbf{e}_k(\mathbf{r}')e^{ikx'}\cdot\mathbf{e}_k(\mathbf{r}')e^{ikx'},
\end{aligned} \tag{2.5.33}$$

where the form of $\overleftrightarrow{\mathbf{G}}_{\mathrm{bound}}(\mathbf{r},\mathbf{r}';\omega)$ is given in Equation 2.5.26. The power reflection coefficient α_{back}, is given by

* It is not actually necessary to take $x \to -\infty$. It is sufficient that there exists an undisordered (perfect) unit cell on the input side of the disorder region over which the orthogonality integration may be performed. Similarly, instead of $x \to \infty$, there need only be a unit cell with no disorder on the transmission side of the disorder region.

$$\alpha_{\text{back}} = \left| \frac{r\mathcal{E}_0 \mathbf{e}_k^*(\mathbf{r})e^{-ikx}}{\mathbf{E}_0(\mathbf{r};\omega)} \right|^2,$$

$$= r r^*,$$

$$= \left(\frac{a\omega}{2v_g}\right)^2 \left[\int d\mathbf{r}\, \Delta\varepsilon(\mathbf{r})\mathbf{e}_k(\mathbf{r})e^{ikx} \cdot \mathbf{e}_k(\mathbf{r})e^{ikx}\right]$$

$$\times \left[\int d\mathbf{r}' \Delta\varepsilon(\mathbf{r}')\mathbf{e}_k^*(\mathbf{r}')e^{-ikx'} \cdot \mathbf{e}_k^*(\mathbf{r}')e^{-ikx'}\right],$$

$$= \left(\frac{a\omega}{2v_g}\right)^2 \iint d\mathbf{r}\, d\mathbf{r}' \Delta\varepsilon(\mathbf{r})\Delta\varepsilon(\mathbf{r}')[\mathbf{e}_k(\mathbf{r}) \cdot \mathbf{e}_k(\mathbf{r})][\mathbf{e}_k^*(\mathbf{r}') \cdot \mathbf{e}_k^*(\mathbf{r}')]e^{i2k(x-x')}. \qquad (2.5.34)$$

The only *stochastic* term on the right-hand side is $\Delta\varepsilon$ and so the expectation value of the power reflection coefficient is

$$\langle \alpha_{\text{back}} \rangle = \left(\frac{a\omega}{2v_g}\right)^2 \iint d\mathbf{r}\, d\mathbf{r}' \langle \Delta\varepsilon(\mathbf{r})\Delta\varepsilon(\mathbf{r}')\rangle [\mathbf{e}_k(\mathbf{r}) \cdot \mathbf{e}_k(\mathbf{r})][\mathbf{e}_k^*(\mathbf{r}') \cdot \mathbf{e}_k^*(\mathbf{r}')]e^{i2k(x-x')}, \qquad (2.5.35)$$

where $\langle \cdots \rangle$ indicates the expectation value when incoherently averaged over a set of nominally identical disordered waveguides. Equation 2.5.35 is one of the primary results of this section. It shows that the backscatter loss scales with the group velocity as v_g^{-2}, leading to high losses at low group velocity. The dependence on v_g is only approximate as the Bloch mode electric field is also frequency-dependent and so the integrand changes simultaneously with v_g.

At $x \rightarrow +\infty$, the solution will consist of the initial forward propagating wave less backscatter losses (Equation 2.5.34) and less losses to radiation modes; $\mathbf{E}(x \rightarrow \infty;\omega) = t\,\mathcal{E}_0 \mathbf{e}_k(\mathbf{r})e^{ikx}$, where t is the transmitted amplitude. Retaining terms from Equation 2.5.32 up to second order in $\overleftrightarrow{\mathbf{G}}(\mathbf{r},\mathbf{r}';\omega)$ and projecting onto a forward propagating wave, only terms with $\overleftrightarrow{\mathbf{G}}_{\text{bound}}$ in the leftmost position, namely 2.5.32b, d, and e, will have nonzero contributions. The transmission amplitude is

$$t = \int_{\text{cell}} d\mathbf{r}\, \mathcal{E}_0^{-1} \varepsilon_i(\mathbf{r})\mathbf{e}_k^*(\mathbf{r})e^{-ikx} \cdot \mathbf{E}_0(\mathbf{r};\omega)$$

$$+ \int_{\text{cell}} d\mathbf{r}\, \mathcal{E}_0^{-1} \varepsilon_i(\mathbf{r})\mathbf{e}_k^*(\mathbf{r})e^{-ikx} \cdot \int d\mathbf{r}' \Delta\varepsilon(\mathbf{r}')\overleftrightarrow{\mathbf{G}}_{\text{bound}}(\mathbf{r},\mathbf{r}';\omega) \cdot \mathbf{E}_0(\mathbf{r}';\omega)$$

$$+ \int_{\text{cell}} d\mathbf{r}\, \mathcal{E}_0^{-1} \varepsilon_i(\mathbf{r})\mathbf{e}_k^*(\mathbf{r})e^{-ikx} \cdot \iint d\mathbf{r}' d\mathbf{r}'' \Delta\varepsilon(\mathbf{r}')\Delta\varepsilon(\mathbf{r}'')\overleftrightarrow{\mathbf{G}}_{\text{bound}}(\mathbf{r},\mathbf{r}';\omega) \cdot \overleftrightarrow{\mathbf{G}}_{\text{rad}}(\mathbf{r}',\mathbf{r}'';\omega) \cdot \mathbf{E}_0(\mathbf{r}'';\omega)$$

$$+ \int_{\text{cell}} d\mathbf{r}\, \mathcal{E}_0^{-1} \varepsilon_i(\mathbf{r})\mathbf{e}_k^*(\mathbf{r})e^{-ikx} \cdot \iint d\mathbf{r}' d\mathbf{r}'' \Delta\varepsilon(\mathbf{r}')\Delta\varepsilon(\mathbf{r}'')\overleftrightarrow{\mathbf{G}}_{\text{bound}}(\mathbf{r},\mathbf{r}';\omega) \cdot \overleftrightarrow{\mathbf{G}}_{\text{bound}}(\mathbf{r}',\mathbf{r}'';\omega) \cdot \mathbf{E}_0(\mathbf{r}'';\omega),$$

$$= 1 + i\frac{a\omega}{2v_g}\int d\mathbf{r}'\Delta\varepsilon(\mathbf{r}')|\mathbf{e}_k(\mathbf{r}')|^2$$

$$+ i\frac{a\omega}{2v_g}\iint d\mathbf{r}'d\mathbf{r}''\Delta\varepsilon(\mathbf{r}')\Delta\varepsilon(\mathbf{r}'')\mathbf{e}_k^*(\mathbf{r}')e^{-ikx'}\cdot\overleftrightarrow{\mathbf{G}}_{\text{rad}}(\mathbf{r}',\mathbf{r}'';\omega)\cdot\mathbf{e}_k(\mathbf{r}'')e^{ikx''}$$

$$-\left(\frac{a\omega}{2v_g}\right)^2\iint d\mathbf{r}'d\mathbf{r}''\Delta\varepsilon(\mathbf{r}')\Delta\varepsilon(\mathbf{r}'')|\mathbf{e}_k(\mathbf{r}')|^2|\mathbf{e}_k(\mathbf{r}'')|^2\Theta(x'-x'')$$

$$-\left(\frac{a\omega}{2v_g}\right)^2\iint d\mathbf{r}'d\mathbf{r}''\Delta\varepsilon(\mathbf{r}')\Delta\varepsilon(\mathbf{r}'')[\mathbf{e}_k^*(\mathbf{r}')\cdot\mathbf{e}_k^*(\mathbf{r}')][\mathbf{e}_k(\mathbf{r}'')\cdot\mathbf{e}_k(\mathbf{r}'')]e^{i2k(x''-x')}\Theta(x''-x'), \qquad (2.5.36)$$

where the last two terms come from the $\overleftrightarrow{\mathbf{G}}_{\text{bound}}\cdot\overleftrightarrow{\mathbf{G}}_{\text{bound}}$ term. The transmitted power $T = |t|^2$, in the limit that the losses given by $t - 1$ are small, expands to

$$T \simeq 1 + 2\,\text{Re}[t-1],$$

$$= 1 + 0$$

$$-2\,\text{Im}\left[\frac{a\omega}{2v_g}\iint d\mathbf{r}'d\mathbf{r}''\Delta\varepsilon(\mathbf{r}')\Delta\varepsilon(\mathbf{r}'')\mathbf{e}_k^*(\mathbf{r}')e^{-ikx'}\cdot\overleftrightarrow{\mathbf{G}}_{\text{rad}}(\mathbf{r}',\mathbf{r}'';\omega)\cdot\mathbf{e}_k(\mathbf{r}'')e^{ikx''}\right]$$

$$-2\,\text{Re}\left[\left(\frac{a\omega}{2v_g}\right)^2\iint d\mathbf{r}'d\mathbf{r}''\Delta\varepsilon(\mathbf{r}')\Delta\varepsilon(\mathbf{r}'')|\mathbf{e}_k(\mathbf{r}')|^2|\mathbf{e}_k(\mathbf{r}'')|^2\Theta(x'-x'')\right]$$

$$-2\,\text{Re}\left[\left(\frac{a\omega}{2v_g}\right)^2\iint d\mathbf{r}'d\mathbf{r}''\Delta\varepsilon(\mathbf{r}')\Delta\varepsilon(\mathbf{r}'')[\mathbf{e}_k^*(\mathbf{r}')\cdot\mathbf{e}_k^*(\mathbf{r}')][\mathbf{e}_k(\mathbf{r}'')\cdot\mathbf{e}_k(\mathbf{r}'')]e^{i2k(x''-x')}\Theta(x''-x')\right],$$

and using $2\,\text{Re}\left[\iint d\mathbf{r}'d\mathbf{r}''f(\mathbf{r}')f^*(\mathbf{r}'')\Theta(x'-x'')\right] = \iint d\mathbf{r}'d\mathbf{r}''f(\mathbf{r}')f^*(\mathbf{r}'')$, one finds

$$T \simeq 1 \qquad (2.5.37a)$$

$$-2\frac{a\omega}{2v_g}\iint d\mathbf{r}'d\mathbf{r}''\Delta\varepsilon(\mathbf{r}')\Delta\varepsilon(\mathbf{r}'')\,\text{Im}\left[\mathbf{e}_k^*(\mathbf{r}')e^{-ikx'}\cdot\overleftrightarrow{\mathbf{G}}_{\text{rad}}(\mathbf{r}',\mathbf{r}'';\omega)\cdot\mathbf{e}_k(\mathbf{r}'')e^{ikx''}\right] \qquad (2.5.37b)$$

$$-\left(\frac{a\omega}{2v_g}\right)^2\iint d\mathbf{r}'d\mathbf{r}''\Delta\varepsilon(\mathbf{r}')\Delta\varepsilon(\mathbf{r}'')|\mathbf{e}_k(\mathbf{r}')|^2\,|\mathbf{e}_k(\mathbf{r}'')|^2 \qquad (2.5.37c)$$

$$-\alpha_{\text{back}}. \qquad (2.5.37d)$$

As expected, the loss in transmission is due to backscattering (Equation 2.5.37d) and radiation scattering (Equation 2.5.37b). The term 2.5.37c arises from light that is initially forward scattered and which, to first order, must be transmitted. For consistency with the calculation of α_{back} to first order, this term is omitted. The radiation term 2.5.37b can be further simplified. The integral runs over all $\mathbf{r}'$ and $\mathbf{r}''$ so for the integration point $(\mathbf{r}', \mathbf{r}'') = (\mathbf{r}_a, \mathbf{r}_b)$, there will always be a corresponding point $(\mathbf{r}', \mathbf{r}'') = (\mathbf{r}_b, \mathbf{r}_a)$. Combining these two integrand terms, and exploiting $\mathbf{f}\cdot\overleftrightarrow{\mathbf{G}}(\mathbf{r}',\mathbf{r}'';\omega)\cdot\mathbf{g}=\mathbf{g}\cdot\overleftrightarrow{\mathbf{G}}(\mathbf{r}'',\mathbf{r}';\omega)\cdot\mathbf{f}$,‡ we can write the radiation loss $\alpha_{\text{rad}} = 1 - T - \alpha_{\text{back}}$ as

$$\alpha_{\text{rad}} = \frac{a\omega}{v_g}\iint d\mathbf{r}'d\mathbf{r}''\Delta\varepsilon(\mathbf{r}')\Delta\varepsilon(\mathbf{r}'')\mathbf{e}_k^*(\mathbf{r}')e^{-ikx'}\cdot\text{Im}\left[\overleftrightarrow{\mathbf{G}}_{\text{rad}}(\mathbf{r}',\mathbf{r}'';\omega)\right]\cdot\mathbf{e}_k(\mathbf{r}'')e^{ikx''}, \quad (2.5.38)$$

and the expectation value as

$$\langle\alpha_{\text{rad}}\rangle = \frac{a\omega}{v_g}\iint d\mathbf{r}'d\mathbf{r}''\langle\Delta\varepsilon(\mathbf{r}')\Delta\varepsilon(\mathbf{r}'')\rangle\mathbf{e}_k^*(\mathbf{r}')e^{-ikx'}\cdot\text{Im}\left[\overleftrightarrow{\mathbf{G}}_{\text{rad}}(\mathbf{r}',\mathbf{r}'';\omega)\right]\cdot\mathbf{e}_k(\mathbf{r}'')e^{ikx''}. \quad (2.5.39)$$

Equation 2.5.39 is the second key result of this section. As an *ensemble* average, it should be compared with the average loss in a set of nominally identical disordered waveguides. The equation shows that the radiation loss can be calculated if the radiation Green function is known and that the loss scales approximately as v_g^{-1}—a less severe scaling than the backscatter loss.

Evaluating Equations 2.5.35 and 2.5.39 requires two components that we have not yet discussed in detail: $\Delta\varepsilon(\mathbf{r}')$ and $\overleftrightarrow{\mathbf{G}}_{\text{rad}}(\mathbf{r},\mathbf{r}';\omega)$. A model for disorder in PhC slabs is given in Section 2.5.3.2, which gives an analytic expression for $\langle\Delta\varepsilon(\mathbf{r}')\Delta\varepsilon(\mathbf{r}'')\rangle$. Methods to calculate $\overleftrightarrow{\mathbf{G}}_{\text{rad}}(\mathbf{r}',\mathbf{r}'';\omega)$ are given in Section 2.5.3.3.

2.5.3.2 Fabrication Process and Disorder Model

PhC slab structures are fabricated using lithographic techniques that were pioneered by the semiconductor electronics industry. Continuous advances have enabled excruciating tolerances to be achieved as demonstrated by cavities with quality factors exceeding 10^6 [41]. Nevertheless, even minute, nanometer-scale imperfections and disorder in the structure can be enhanced by slow-light propagation and cause significant scattering losses. The semiconductor membrane is grown epitaxially and so the top and bottom surfaces are crystallographic facets with roughness, if any, on the order of the atomic lattice constant (0.5–0.6 nm). The key PhC features, that is, the holes, are defined by electron beam lithography that can be used to position features with an accuracy of approximately 1 nm. The features are then imprinted into the membrane using an etch mask and etching process. The etch mask is usually the largest source of disorder and the holes are found to have straight side walls but rough perimeters, as schematically illustrated in Figure 2.5.1b.

A seminal investigation of the statistical properties of disorder in PhC slabs is given by Skorobogatiy et al. [42] who extracted the disordered device geometry from scanning electron microscope images and performed detailed statistical analysis. On the basis of their analysis (and the analysis of other PhC slab images), we take the disorder in our PhC structures to be well described by small perturbations of the radius around the hole perimeter $\Delta r(\phi_\alpha)$, that follow the distribution

$$\langle\Delta r(\phi_\alpha)\Delta r(\phi'_{\alpha'})\rangle = \sigma^2 e^{-r|\phi_\alpha-\phi'_{\alpha'}|/l_p}\delta_{\alpha,\alpha'}, \quad (2.5.40)$$

‡ See, for example, Appendix A.1 in the study by Vogel and Welsch [40].

where α indexes the holes, and ϕ_α is the angular coordinate of the point relative to the hole center. The root mean square (RMS) roughness σ, and correlation length l_p, are typically on the order of 2–3 nm and 40–50 nm, respectively. The disorder used in Equations 2.5.35 and 2.5.39 is then given as (assuming $\Delta r(\phi_\alpha) \ll r$)

$$\Delta\varepsilon(\mathbf{r}) = (\varepsilon_2 - \varepsilon_1)\Theta(h/2 - |z|)\sum_{\alpha}\Delta r(\phi_\alpha(x,y))\delta\left(\sqrt{(x-x_\alpha)^2 + (y-y_\alpha)^2} - r\right), \tag{2.5.41}$$

where ε_1 (ε_2) is the dielectric constant of the membrane (holes), h is the slab thickness, and hole α is centered at (x_α, y_α). This disorder model allows for calculations of the loss in PhC systems to be performed using only two parameters that can be directly related to experiments. Although it would be significantly easier to assume randomly varying perfect holes, as done commonly in the disorder literature, such models are far removed from the actual disorder that occurs in fabricated PhC waveguides. A numerical example of a disordered hole using the statistics of Equation 2.5.40 is shown in Figure 2.5.1b.

2.5.3.3 Effective-Slab Radiation Green Function

Earlier attempts to numerically evaluate the radiation loss utilized FDTD to calculate the Green function [10,14]. Specifically, the total loss α_{total} was calculated through an equation analogous to Equation 2.5.39 but with $\overleftrightarrow{\mathbf{G}}_{\text{rad}}(\mathbf{r}',\mathbf{r}'';\omega)$ replaced with the total Green function that was obtained using FDTD. The properties of the bound modes could also be determined from FDTD, and so Equation 2.5.35 could be evaluated and the radiation loss calculated as $\alpha_{\text{rad}} = \alpha_{\text{total}} - \alpha_{\text{back}}$. This method has the advantage of rigorously including the radiation modes of the complex geometry. The disadvantage is that FDTD simulations are very time-consuming and a separate simulation must be run for each possible point $\mathbf{r}''$ in the integral. Moreover, the features of $\overleftrightarrow{\mathbf{G}}_{\text{rad}}$ are considerably less interesting than those of $\overleftrightarrow{\mathbf{G}}_{\text{bound}}$, and the contribution is much smaller than bound mode scattering. This is why backscatter typically dominates the transmission loss in PhC waveguides. This property can also be an advantage for certain applications, for example, for emitting "single photons" into a PhC waveguide mode [7,43].

Instead of using the brute-force FDTD approach, we calculate an approximation of $\overleftrightarrow{\mathbf{G}}_{\text{rad}}(\mathbf{r}',\mathbf{r}'';\omega)$ directly. The bound waveguide mode will be strongly dependent on the features of the PhC slab; but the radiation modes, which quickly escape and propagate away, should have a much weaker dependence. To calculate the radiation Green function, we replace the PhC slab with a homogeneous slab with an effective dielectric constant that will be determined through comparison with rigorous FDTD results. Slabs with an effective dielectric constant of 6–7 are typical for PhCs with $\varepsilon \simeq 12$. Gerace and Andreani [34] use a similar approach of an effective dielectric slab; however, they work with the field patterns for the individual modes; the Green function provides a more natural treatment. In this way, both contributions to the Green function can be obtained analytically, offering considerable insight into the key contributions to disorder-induced scattering.

This approximation is only useful if the radiation Green function of a homogeneous dielectric slab can be calculated efficiently. We make use of the work by Paulus et al. [22], who give the slab Green function analytically in terms of a single integral that must be performed numerically as*

$$\overleftrightarrow{\mathbf{G}}_{\text{slab}}(\mathbf{r},\mathbf{r}';\omega) = \frac{\hat{\mathbf{z}}\hat{\mathbf{z}}}{\varepsilon^2(\mathbf{r})}\delta(\mathbf{r}-\mathbf{r}') + \frac{i\omega^2}{4\pi c^2}\int_0^\infty dk_\rho \overleftrightarrow{\mathbf{f}}(k_\rho;\mathbf{r},\mathbf{r}'), \quad (2.5.42)$$

where k_ρ is the radial (in-slab) wave vector that we are integrating over, and the components of $\overleftrightarrow{\mathbf{f}}(k_\rho;\mathbf{r},\mathbf{r}')$ are listed in the work of Paulus et al. [22]. The total slab Green function will contain contributions from the radiation modes of the slab and also spurious bound slab modes. Bound slab modes occur naturally in a dielectric slab but, as we use the effective slab only to approximate the radiation modes of the patterned PhC slab, their contribution must be suppressed. We adapt the work of Paulus et al. [22] to partition their total Green function into contributions from radiation modes, bound slab modes, and evanescent modes. Evanescent modes are nonpropagating and account for the expected divergence of the real part of $\overleftrightarrow{\mathbf{G}}(\mathbf{r},\mathbf{r}';\omega)$ as $\mathbf{r}\to\mathbf{r}'$. However they do not contribute to the imaginary part of $\overleftrightarrow{\mathbf{G}}(\mathbf{r},\mathbf{r}';\omega)$ (refer to Equation 2.5.39) and so are safely ignored.

The bound and radiation components of the Green functions for a sample effective slab are shown in Figure 2.5.6 in units such that a value of 1 for the imaginary part is equivalent to the spontaneous emission rate (proportional to the local density of states) in a homogeneous dielectric. Although the bound mode contribution (dark gray) is dominant, we are able to isolate it from the radiation component (light gray). By virtue of the slab geometry, the strength of radiative emission is strongly decreased relative to a homogeneous system.

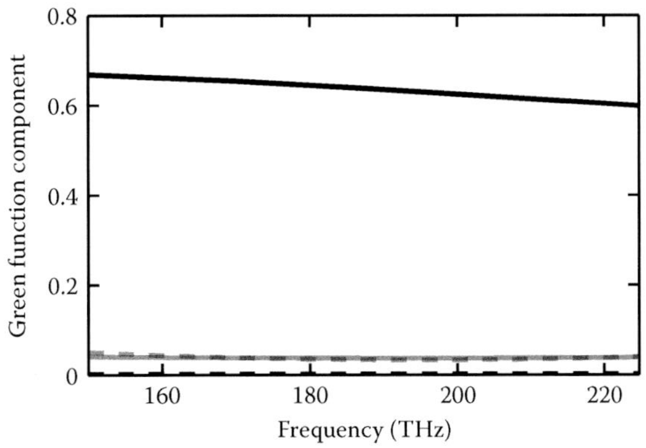

FIGURE 2.5.6
Radiation (light—middle curves) and bound mode (dark—upper and lower curves) Green functions of the effective slab with the source and sensor point in the center of the slab. The real (dashed) and imaginary (solid) parts are shown. The slab is 265-nm thick with an effective dielectric constant of $\varepsilon = 7$. The values are normalized by the imaginary part for the Green function for a homogeneous dielectric with $\varepsilon = 12$. The radiative decay rate of the slab is clearly much less than that of a corresponding bulk structure, by about a factor of ~0.05.

* Note that Paulus et al. [21] use a different definition of the Green function and their results should be multiplied by $(\omega/c)^2$ to be compatible with our convention. Equation 2.5.42 uses our definition of the Green function.

In the loss calculation, we build a look-up table for the radiation Green function that spans 10 possible values each of frequency, the radial separation of **r** and **r**′, the z-component of **r**, and the z-component of **r**′. Using symmetry, this is sufficient to interpolate the radiation Green function for all parameters of interest. The radiation Green function is calculated separately for each parameter combination and the computation of the full look-up table requires 40 cpu-minutes. To produce the same look-up table with FDTD, a separate simulation need only be run for each value of the source position **r**′. However, 10 such FDTD simulations require approximately 288 cpu-hours.

2.5.3.4 Extrapolation of Disorder-Induced Loss to Long Waveguides, and Breakdown of the Beer–Lambert Law

Equations 2.5.35 and 2.5.39 give the loss per unit cell but care must be taken when comparing with experimental loss measurements that may be for waveguides that are millimeters long (thousands of periods). At low group velocities where multiple disorder-induced scattering becomes significant, the approximate loss scaling rules naturally breakdown; Le Thomas et al. [44] discuss this breakdown in terms of a crossover from a *dispersive* to a *diffusive* transport regime. There are three extrapolation methods that we will discuss which are, in decreasing degree of approximation, the small loss limit (or optically thin sample), the Beer–Lambert model, and the multiple scattering model. In all three cases, we assume that light scattered from different disorder sites adds incoherently.

The most obvious method to extrapolate to longer waveguides is to simply extend the integration volume when calculating the loss coefficients. Because the calculation is an expectation calculation, it can be shown that the loss over a waveguide of length L is simply $\alpha L/a$ where $\alpha = \langle\alpha_{\text{back}}\rangle + \langle\alpha_{\text{rad}}\rangle$ is the loss per unit cell. This treatment works if the total loss is a small fraction of the forward wave but breaks down if the loss is on the order of the forward wave. One can easily see this by taking $L > a/\alpha$ that causes the transmission $T = 1 - \alpha L/a$ to be negative—clearly nonsensical. This failure arises since α gives the loss per period under *unit* excitation; once the initial wave is partially depleted, α overestimates the loss. The loss as a function of waveguide length using this model is plotted in Figure 2.5.7 (short-dashed curve). The length where the transmission goes negative is marked with the dashed vertical line in this figure.

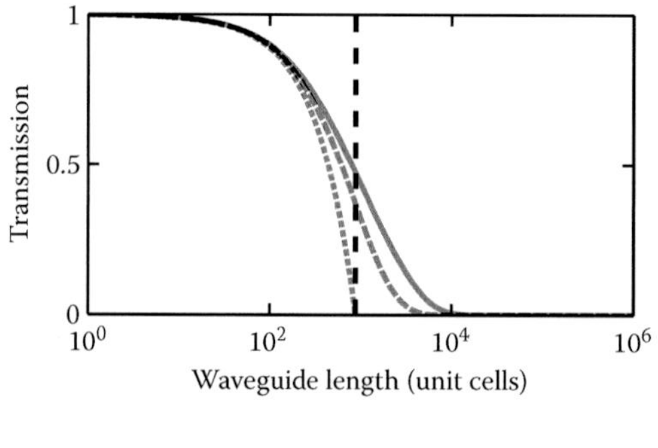

FIGURE 2.5.7
Transmission as a function of length through a waveguide with loss per unit cell $\alpha_{\text{back}} = 10^{-3}$ and $\alpha_{\text{rad}} = 10^{-4}$. The transmission is shown using three different extrapolation models: weak loss limit (short-dashed curve), Beer–Lambert model (long-dashed curve), and multiple scattering model (solid curve). The value $(\alpha_{\text{back}} + \alpha_{\text{rad}})\,L/a = 1$ is marked with the vertical dashed black line.

The resolution to the unphysical negative transmission is the Beer–Lambert extrapolation model, named in analogy to the famous law for the attenuation of light in an absorbing medium. The loss at a given position along the waveguide should be proportional to the intensity in the forward wave at that point. The intensity $P(x)$ is then described by

$$\frac{\mathrm{d}P(x)}{\mathrm{d}x} = -\frac{\alpha}{a}P(x), \tag{2.5.43}$$

which has the solution

$$T = P(L) = e^{-\alpha L/a}. \tag{2.5.44}$$

It is easy to see that the weak loss limit simply solved Equation 2.5.43 under the assumption that the $P(x)$ on the right-hand side could be approximated by 1. The Beer–Lambert solution extrapolates to arbitrary lengths without obvious unphysical behavior like the weak loss limit. This relation is typically used in the literature to analyze PhC waveguide results and is shown with the long-dashed curve in Figure 2.5.7.

There is a subtle failure to the Beer–Lambert model that is particular to the slow-light PhC waveguide systems under consideration. In these systems, backscattering can dominate radiative losses. Equation 2.5.43 is not representative of a system with strong backscattering as it implicitly assumes that light which scatters once will never reach the output port. In a system with strong backscattering, it is quite possible for light to backscatter multiple times before being transmitted. Wang et al. [38] show this for a two-dimensional system, which is expected as there is no radiative (out of plane) loss. One resolution is to self-consistently model light propagation along an entire disordered waveguide [20], not just a single period, as described in Section 2.5.4. Although this technique provides a wealth of information, the method is somewhat computationally expensive. Instead, we can extrapolate the loss per unit cell to long waveguides using an incoherently averaged coupled mode approach [19]. With $\Psi_f(x)$ and $\Psi_b(x)$ giving the intensity in the forward and backward modes, the propagation is governed by

$$\frac{\mathrm{d}}{\mathrm{d}x}\Psi_f(x) = -\left(\langle\alpha_{\mathrm{back}}\rangle + \langle\alpha_{\mathrm{rad}}\rangle\right)\Psi_f(x) + \langle\alpha_{\mathrm{back}}\rangle\Psi_b(x), \tag{2.5.45}$$

$$\frac{\mathrm{d}}{\mathrm{d}x}\Psi_b(x) = \left(\langle\alpha_{\mathrm{back}}\rangle + \langle\alpha_{\mathrm{rad}}\rangle\right)\Psi_b(x) - \langle\alpha_{\mathrm{back}}\rangle\Psi_f(x). \tag{2.5.46}$$

These equations are straightforward to solve analytically and have the following solution* for the transmitted power T and reflected power R:

$$T = \Psi_f(L) = \frac{2\gamma e^{-\gamma L}}{\left(\gamma + \langle\alpha_{\mathrm{back}}\rangle + \langle\alpha_{\mathrm{rad}}\rangle\right) + \left(\gamma - \langle\alpha_{\mathrm{back}}\rangle - \langle\alpha_{\mathrm{rad}}\rangle\right)e^{-2\gamma L}}, \tag{2.5.47}$$

$$R = \Psi_b(0) = \frac{\langle\alpha_{\mathrm{back}}\rangle\left(1 - e^{-2\gamma L}\right)}{\left(\gamma + \langle\alpha_{\mathrm{back}}\rangle + \langle\alpha_{\mathrm{rad}}\rangle\right) + \left(\gamma - \langle\alpha_{\mathrm{back}}\rangle - \langle\alpha_{\mathrm{rad}}\rangle\right)e^{-2\gamma L}}, \tag{2.5.48}$$

* Provided neither $\langle\alpha_{\mathrm{back}}\rangle$ nor $\langle\alpha_{\mathrm{rad}}\rangle$ is 0.

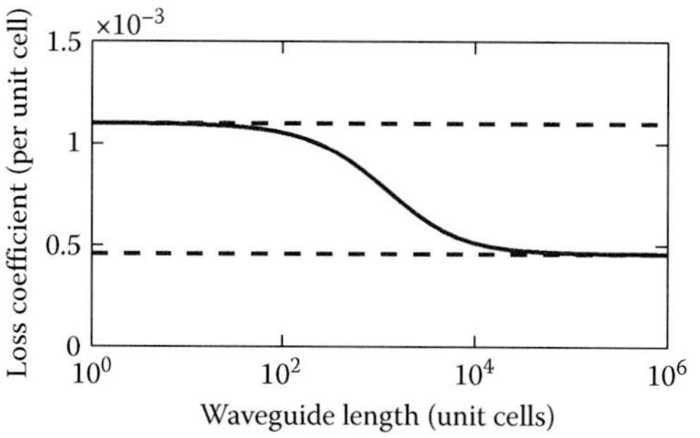

FIGURE 2.5.8
Effective loss coefficient using the multiple scattering extrapolation model (solid). In the short waveguide limit, the loss approaches $\langle\alpha_{back}\rangle + \langle\alpha_{rad}\rangle$ (top dashed line), in agreement with the weak and Beer–Lambert models. In the long waveguide limit, the effective loss coefficient decreases to $\sqrt{\langle\alpha_{rad}\rangle^2 + 2\langle\alpha_{back}\rangle\langle\alpha_{back}\rangle}$ (bottom dashed line).

where $\gamma = \sqrt{\alpha_{rad}^2 + 2\alpha_{rad}\alpha_{back}}$. The *extrapolated* loss using this model is shown with the solid curve in Figure 2.5.7, and, once the waveguide is sufficiently long, predicts a lower loss than the Beer–Lambert model.

To further examine the multiple scattering model, we plot the effective loss coefficient per unit cell $[\alpha_{eff} = -(a/L)\ln(T)]$ as a function of waveguide length in Figure 2.5.8 (solid), where T is calculated from Equation 2.5.47. In the short waveguide limit, the loss is equal to $\langle\alpha_{back}\rangle + \langle\alpha_{rad}\rangle$, in agreement with the Beer–Lambert model (top dashed line). In the long waveguide limit, where multiple scattering is more significant, the loss coefficient decreases to $\sqrt{\langle\alpha_{rad}\rangle^2 + 2\langle\alpha_{back}\rangle\langle\alpha_{back}\rangle}$ (bottom dashed limit) and disagrees with the Beer–Lambert model. Note that if $\langle\alpha_{back}\rangle \ll \langle\alpha_{rad}\rangle$, the two limits are degenerate and the multiple scattering method agrees with the Beer–Lambert method for all waveguide lengths.

2.5.3.5 Example Disorder-Induced Loss Calculations

Using the disorder model and effective radiation Green function, we evaluate Equations 2.5.35 and 2.5.39 for the loss per unit cell in a disordered waveguide. Figure 2.5.9a shows the radiation loss $\langle\alpha_{rad}\rangle$ (short dashed), backscatter loss $\langle\alpha_{back}\rangle$ (long dashed), and total loss $\langle\alpha_{back}\rangle + \langle\alpha_{rad}\rangle$ (solid), as a function of frequency for a W1 waveguide (whose parameters are in Table 2.5.1). For reference, the group index is plotted in the bottom plot on a matched frequency scale. In the fast light regime (high frequencies), backscattering and radiation loss are of comparable magnitudes and both are needed to correctly calculate the total loss. As the group velocity decreases, moving to lower frequencies, the v_g^{-2} scaling of backscattering (vs. v_g^{-1} for radiation scattering) causes it to dominate.

We examine the loss more closely at the frequency marked by the dashed black line in Figure 2.5.9 (~193.6 THz). Figure 2.5.10 plots the dependence of the loss at this fixed frequency as a function of the disorder parameters: the RMS roughness σ (left plot) and the correlation length l_p (right plot). The dependence on σ is quadratic as can be seen by simple inspection of the loss equations. The dependence on l_p is more complex, with the loss starting at zero for $l_p = 0$, increasing rapidly at first and then slowing and actually decreasing for very large l_p. Similar behavior has been found using a numerical supercell approach by Gerace and Andreani [45]. The limit $l_p \to \infty$ corresponds to a random change in the radius

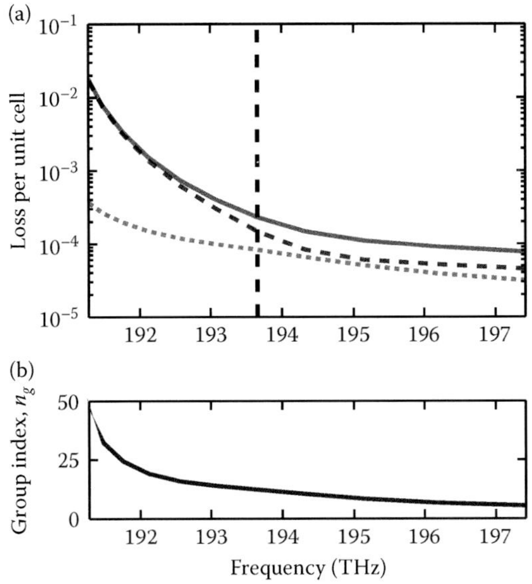

FIGURE 2.5.9
(a) Loss in a W1 photonic crystal waveguide as a function of frequency. The total loss (solid) is divided into its contributions from backscattering (long-dashed) and radiation loss (short-dashed). The vertical dashed black line indicates the frequency of Figure 2.5.10. (b) Group index for the waveguide mode on a matched frequency scale. The calculation uses $\sigma = 3$ nm and $l_p = 40$ nm, and the same geometry as Figure 2.5.5.

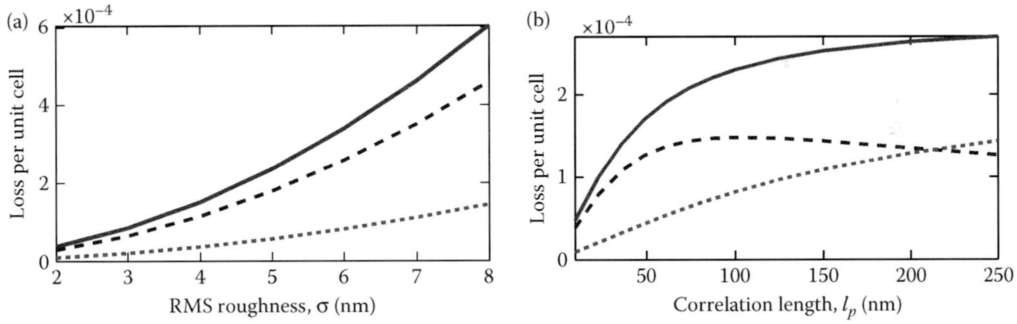

FIGURE 2.5.10
Scaling of the average loss per unit cell with the disorder parameters: root mean square roughness σ (a) and correlation length l_p (b). Both plots are calculated at the frequency marked by the vertical dashed black line in Figure 2.5.9. The radiation (short-dashed), backscatter (long-dashed), and total loss (solid) are all shown.

of the holes but no deviation in their circularity. We can see that such models will overestimate the radiation loss but could actually underestimate the back loss.

2.5.3.6 Comparison with Experiment

The calculated loss can now be compared with the original transmission spectrum as shown in Figure 2.5.11. The calculated loss is shown extrapolated to the 1.5 mm length using both the Beer–Lambert model (light gray—lower) and the multiple scattering model (dark gray—upper). It is clear that the multiple scattering model is a much better fit and that the scattering calculation reproduces the experimentally observed roll-off. The glaring

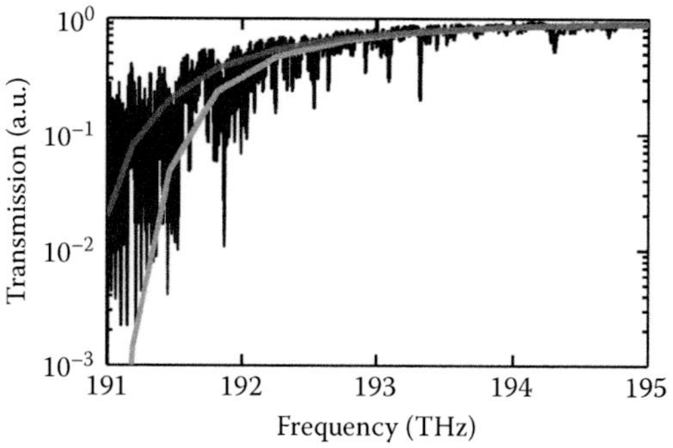

FIGURE 2.5.11
(**See color insert.**) The experimental transmission spectrum for a W1 waveguide with the calculated loss spectrum superimposed. The calculated loss per period is extrapolated to the 1.5-mm waveguide length using the Beer–Lambert model (light gray—lower) and the multiple scattering model (dark gray—upper). The disorder parameters used in the calculation are $\sigma = 3$ nm and $l_p = 40$ nm. A frequency shift of −0.3 THz has been applied to the calculated spectrum to account for uncertainty in the experimental slab thickness. (Based on Patterson, M. et al., Phys. Rev. Lett. **102**, 253903, 2009, with experimental measurement from Sylvain Combrié and Alfredo De Rossi, Thales Research and Technology.)

discrepancy is the strong variations in the transmission that intensify near the band edge. As this calculation gives the incoherent average transmission, it should be compared with the transmission averaged over many nominally identical disordered waveguides; it should not be surprising that any single experimental transmission spectrum has features that deviate from the predicted averages loss. The resonances in the experimental data will be explored in detail in the next section. Further comparisons with experimental results and discussions about the breakdown of the Beer–Lambert law are given in a joint paper with the Thomas Krauss group [19], where a different waveguide design was used.

The models discussed here show that one can easily extend the previous models of incoherent loss [14] to optically long samples, which is evidently necessary for the slow-light regime. This retains many of the analytical benefits of the thin-sample-approximation formulas, and adds essentially no additional complexities to the numerical computations. Importantly, our presented techniques and calculations are fully three-dimensional and use a realistic disorder model for the imperfect holes.

We also highlight that the analytical formulas allow one to design lower loss waveguides. As pointed out by Hughes et al. [14], Equations 2.5.35 and 2.5.39 clearly show the way to design waveguides with lower losses. Simply stated, one should pull the Bloch mode fields away from the region of disorder, that is, the overlap of the Bloch modes with the hole interfaces should be minimized. Although one may find that a particular design has lower losses than another (e.g., by experimental trial and error), using the above prescriptions can easily enable the deterministic design of lower loss PhC waveguides.

2.5.4 Disorder-Induced Coherent Scattering

The incoherent extrinsic scattering calculation of Section 2.5.3 captures the average transmission roll-off due to disorder-induced scattering; however, it is only useful for understanding measurements that are incoherently averaged over a large set of nominally identical samples. There is a wealth of interesting behavior present in individual disordered waveguides that disappears when averaged. Figure 2.5.11 shows the possible discrepancy between the incoherent theory and the transmission through a particular disordered waveguide. The experimental data, like that shown by Topolancik et al. [46], contain sharp spectral resonances near the band edge where the transmission may be enhanced by orders of magnitude.

In this section, we develop a coupled mode theory [47] for *coherent* extrinsic scattering in slow-light PhC waveguides, which successfully explains the band-edge resonances observed in experiments [20]. In contrast to Section 2.5.3, we include unlimited forward- and backscattering events and the phase of the reflected light, and thus develop a coherent scattering theory. These methods can also be employed to model time-frequency reflectometry maps [11] that can be used to visualize the frequency-dependent impulse response and reveal a number of interesting features such as disorder-induced scattering and facet reflections.*

2.5.4.1 Disorder-Mediated Coupled Mode Equations

The electric field in the ideal waveguide can be decomposed into the complete Bloch-mode basis consisting of the target bound waveguide modes $\mathbf{e}_{\pm k}(\mathbf{r})$, and the set of radiation modes $\{\mathbf{q}(\mathbf{r})\}$ as

$$\mathbf{E}(\mathbf{r};\omega) = \mathcal{E}_0[\mathbf{e}_k(\mathbf{r})e^{ikx}\psi_f(x) + \mathbf{e}_k^*(\mathbf{r})e^{-ikx}\psi_b(x) + \sum_{\mathbf{q}} \mathbf{q}(\mathbf{r})e^{ik_q x}\psi_q(x)], \quad (2.5.49)$$

where $\mathcal{E}_0$ is an amplitude and $\psi_f(x)$, $\psi_b(x)$, and $\{\psi_q(x)\}$ are the envelopes for the forward, backward, and radiation modes, respectively. We use envelopes only for convenience and do not require that they are slowly varying. We are only interested in the envelopes for the bound waveguide modes but we initially track the radiation modes to include radiation scattering.

The field in a disordered waveguide can be calculated analytically from Equation 2.5.17, using the effective PhC waveguide Green function and the disorder polarization density $\mathbf{P}(\mathbf{r};\omega) = \varepsilon_0\Delta\varepsilon(\mathbf{r})\,\mathbf{E}(\mathbf{r};\omega)$, as

$$\mathbf{E}(\mathbf{r};\omega) \simeq \mathbf{E}_i(\mathbf{r};\omega) + \int d\mathbf{r}'\left[\overleftrightarrow{\mathbf{G}}_{\text{bound}}(\mathbf{r},\mathbf{r}';\omega) + \overleftrightarrow{\mathbf{G}}_{\text{rad}}(\mathbf{r},\mathbf{r}';\omega)\right]\cdot[\Delta\varepsilon(\mathbf{r}')\mathbf{E}(\mathbf{r}';\omega)], \quad (2.5.50)$$

Where we assume an initial electric field $\mathbf{E}_i(\mathbf{r};\omega) = \mathcal{E}_0\mathbf{e}_k(\mathbf{r})e^{ikx}$, and a total field including scattering $\mathbf{E}(\mathbf{r};\omega)$ given by Equation 2.5.49.

For this section, as we are interested in developing subunit-cell propagation equations, we prefer the Bloch mode normalization integration to be over only the plane perpendicular to the propagation direction (cf. Equation 2.5.15, where the Bloch modes are normalized

* Although we do not show any refrectometry maps in this chapter, the reader is referred to the study by Patterson et al. [20].

in three dimensions). Using the electric and magnetic field orthogonality relations, the Maxwell constitutive relations, and the divergence theorem, one can derive

$$0 = \frac{i}{\omega\varepsilon_0}\left(1 - e^{i(k'-k)a}\right) \times \iint_{x=x_0} \mathrm{d}y\,\mathrm{d}z\,\hat{\mathbf{x}} \cdot \left(\mathbf{h}_k^*(\mathbf{r})e^{-ikx} \times \mathbf{e}_{k'}(\mathbf{r})e^{ikx}\right), \tag{2.5.51}$$

where the integration here is performed over a single plane transverse to the propagation direction, and $\mathbf{h}_k(\mathbf{r})$ is the magnetic Bloch mode formed in an analogous fashion to $\mathbf{e}_k(\mathbf{r})$. For $k \neq k'$, the term in brackets is nonzero and the integral must evaluate to zero. For $k = k'$, the integral can be recognized as the power flux at the transverse plane that is clearly nonzero (except for a radiation mode propagating perpendicular to the slab). Thus, a new projection (orthogonality) operator can be defined as [1]

$$\mathcal{P}_k\mathbf{E}^{\mathrm{P}}(\mathbf{r}) = \frac{\iint_{x=x_0} \mathrm{d}y\,\mathrm{d}z\,\hat{\mathbf{x}} \cdot \left(\mathbf{h}_k^*(\mathbf{r})e^{-ikx} \times \mathbf{E}^{\mathrm{P}}(\mathbf{r})\right)}{\iint_{x=x_0} \mathrm{d}y\,\mathrm{d}z\,\hat{\mathbf{x}} \cdot \left(\mathbf{h}_k^*(\mathbf{r})e^{-ikx} \times \mathbf{e}_k(\mathbf{r})e^{ikx}\right)}, \tag{2.5.52}$$

where $\mathbf{E}^{\mathrm{P}}(\mathbf{r})$ is the field being projected and $x = x_0$ is an arbitrary plane. This result is in agreement with that of Marcuse [47] and the standard form for overlap integrals [48]. The projection operator $\mathcal{P}_k$ has the useful property that $\mathcal{P}_k\mathbf{e}_{k'}(\mathbf{r})e^{ik'x} = \delta_{k,k'}$.

We begin by projecting Equation 2.5.50 onto a forward propagating wave by operating with $\mathcal{P}_k$. We then multiply by $\mathcal{E}_0^{-1}$ and differentiate with respect to x. The left-hand side becomes simply $\mathrm{d}\Psi_f(x)/\mathrm{d}x$. The projection of $\mathbf{E}_i(\mathbf{r};\omega)$ equals 1 and differentiating eliminates the contribution of the field in the ideal structure. This derivation will transform the integral description of the total electric field into a set of coupled propagation equations and the electric field in the ideal structure will be included as a wave injected from the input port. Equation 2.5.50 for the forward wave becomes

$$\frac{\mathrm{d}}{\mathrm{d}x}\psi_{\mathrm{f}}(x) = \frac{i}{v_g}\left[c_{\mathrm{ff}}(x)\psi_{\mathrm{f}}(x) + c_{\mathrm{fb}}(x)e^{-i2kx}\psi_{\mathrm{b}}(x) + \sum_{\mathrm{q}} c_{\mathrm{fq}}(x)\psi_{\mathrm{q}}(x)\right]. \tag{2.5.53}$$

The terms on the right-hand side all arise from the projection of the $\overleftrightarrow{\mathbf{G}}_{\mathrm{bound}}(\mathbf{r},\mathbf{r}';\omega)$ term; the projection of the $\overleftrightarrow{\mathbf{G}}_{\mathrm{rad}}(\mathbf{r},\mathbf{r}';\omega)$ term is 0 since the constituent radiation modes are orthogonal to the chosen bound mode. The volume integral has been converted to an integral over the transverse plane by the derivative of the Heaviside function in $\overleftrightarrow{\mathbf{G}}_{\mathrm{bound}}(\mathbf{r},\mathbf{r}';\omega)$ (Equation 2.5.26). The scattering coefficients, corresponding to *forward-forward*, *forward-backward*, and *forward-radiation* scatter, are

$$c_{\mathrm{ff}}(x) = \frac{a\omega}{2}\iint \mathrm{d}y\,\mathrm{d}z\,\mathbf{e}_k^*(\mathbf{r}) \cdot \mathbf{e}_k(\mathbf{r})\,\Delta\varepsilon(\mathbf{r}), \tag{2.5.54}$$

$$c_{\mathrm{fb}}(x) = \frac{a\omega}{2}\iint \mathrm{d}y\,\mathrm{d}z\,\mathbf{e}_k^*(\mathbf{r}) \cdot \mathbf{e}_k^*(\mathbf{r})\,\Delta\varepsilon(\mathbf{r}), \tag{2.5.55}$$

$$c_{\mathrm{fq}}(x) = \frac{a\omega}{2}\iint \mathrm{d}y\,\mathrm{d}z\,\mathbf{e}_k^*(\mathbf{r})e^{-ikx}\cdot\mathbf{q}(\mathbf{r})e^{ik_{\mathrm{q}}x}\Delta\varepsilon(\mathbf{r}). \tag{2.5.56}$$

An analogous equation to Equation 2.5.53 for $\mathrm{d}\psi_{\mathrm{b}}(x)/\mathrm{d}x$ is formed by projecting Equation 2.5.50 onto a backward propagating wave. One has

$$\frac{\mathrm{d}}{\mathrm{d}x}\psi_{\mathrm{b}}(x) = \frac{-i}{v_g}\left[c_{\mathrm{bb}}(x)\psi_{\mathrm{b}}(x) + c_{\mathrm{bf}}(x)e^{i2kx}\psi_{\mathrm{f}}(x) + \sum_{\mathrm{q}} c_{\mathrm{bq}}(x)\psi_{\mathrm{q}}(x)\right], \tag{2.5.57}$$

where the negative sign arises from the Heaviside function in Equation 2.5.26, $c_{\mathrm{bb}}(x) = c_{\mathrm{ff}}(x)$, $c_{\mathrm{bf}}(x) = c_{\mathrm{fb}}^*(x)$, and

$$c_{\mathrm{bq}}(x) = \frac{a\omega}{2}\iint \mathrm{d}y\,\mathrm{d}z\,\mathbf{e}_k(\mathbf{r})e^{ikx}\cdot\mathbf{q}(\mathbf{r})e^{ik_{\mathrm{q}}x}\Delta\varepsilon(\mathbf{r}).$$

Next, we seek to eliminate the $\psi_{\mathrm{q}}(x)$ from the equation as there are a large (infinite) number of radiation modes. We project Equation 2.5.50 onto any one of the radiation modes to derive a radiation mode envelope equation. The left-hand side becomes simply $\psi_{\mathrm{q}}(x)$. Only the $\overleftrightarrow{\mathbf{G}}_{\mathrm{rad}}(\mathbf{r},\mathbf{r}';\omega)$ term on the right-hand side will have a nonzero projection as any chosen radiation mode will be orthogonal to the bound waveguide modes. Thus, we obtain a set of equations, one for each of the radiation modes $\mathbf{q}$,

$$\begin{aligned}
\psi_{\mathrm{q}}(x) &= \mathcal{E}_0^{-1}\mathcal{P}_{\mathrm{q}}\int \mathrm{d}\mathbf{r}'\overleftrightarrow{\mathbf{G}}_{\mathrm{rad}}(\mathbf{r},\mathbf{r}';\omega)\cdot\left[\mathbf{E}(\mathbf{r}';\omega)\,\Delta\varepsilon(\mathbf{r}')\right] \\
&= \mathcal{P}_{\mathrm{q}}\int \mathrm{d}\mathbf{r}'\overleftrightarrow{\mathbf{G}}_{\mathrm{rad}}(\mathbf{r},\mathbf{r}';\omega)\cdot\mathbf{e}_k(\mathbf{r})e^{ikx}\psi_{\mathrm{f}}(x)\,\Delta\varepsilon(\mathbf{r}') && \text{(2.5.58a)} \\
&\quad + \mathcal{P}_{\mathrm{q}}\int \mathrm{d}\mathbf{r}'\overleftrightarrow{\mathbf{G}}_{\mathrm{rad}}(\mathbf{r},\mathbf{r}';\omega)\cdot\mathbf{e}_k^*(\mathbf{r})e^{-ikx}\psi_{\mathrm{b}}(x)\,\Delta\varepsilon(\mathbf{r}') && \text{(2.5.58b)} \\
&\quad + \mathcal{P}_{\mathrm{q}}\int \mathrm{d}\mathbf{r}'\overleftrightarrow{\mathbf{G}}_{\mathrm{rad}}(\mathbf{r},\mathbf{r}';\omega)\cdot\sum_{\mathrm{q}}\mathbf{q}(\mathbf{r})e^{-ik_{\mathrm{q}}x}\psi_{\mathrm{q}}(x)\,\Delta\varepsilon(\mathbf{r}'). && \text{(2.5.58c)}
\end{aligned}$$

The three sources of energy for the radiation modes are scattering from the forward wave (Equation 2.5.58a), scattering from the backward wave (Equation 2.5.58b), and scattering from all the radiation modes (including self-scattering from the current radiation mode into itself) (Equation 2.5.58c).

To eliminate the ψ_{q} from Equation 2.5.53, first we omit Equation 2.5.58c, as we assume that radiation-mode scattering is just a loss mechanism and inter-radiation-mode scattering will not feed back into the waveguide modes. We also neglect Equation 2.5.58b; this would give rise to radiation-assisted backscattering where light from the backward mode scatters into a radiation mode and then the forward mode. These assumptions are reasonable because the radiation modes quickly leak from the slab and so do not interact with the scattering regions for very long. This leaves only Equation 2.5.58a that accounts for loss from the forward mode into the radiation modes. The $\mathcal{P}_{\mathrm{q}}$ prefix in Equation 2.5.58 is a projection operator acting on the radiation Green function. In Equation 2.5.53, the projected Green function (in ψ_{q}) is multiplied by the basis vector (in c_{fq}). As the set $\{\mathbf{q}(\mathbf{r})\}$ spans all radiation modes included in $\overleftrightarrow{\mathbf{G}}_{\mathrm{rad}}(\mathbf{r},\mathbf{r}';\omega)$, this is an identity transform of $\overleftrightarrow{\mathbf{G}}_{\mathrm{rad}}(\mathbf{r},\mathbf{r}';\omega)$ and Equation 2.5.53, under substitution by Equation 2.5.58, becomes

$$v_g \frac{\mathrm{d}}{\mathrm{d}x}\psi_{\mathrm{f}}(x) = i\, c_{\mathrm{ff}}(x)\,\psi_{\mathrm{f}}(x) + i\, c_{\mathrm{fb}}(x)e^{-i2kx}\,\psi_{\mathrm{b}}(x) + i\, c_{\mathrm{fr}}(x)\,\psi_{\mathrm{f}}(x), \tag{2.5.59}$$

where the radiation coupling coefficient c_{fr} is given in Equation 2.5.61 (which is further simplified below). Note that we now have conveniently eliminated the sum over **q**.

For the backward wave, Equation 2.5.57 is transformed using Equation 2.5.58 with only term 2.5.58b retained. The backward wave equation is

$$-v_g \frac{\mathrm{d}\psi_{\mathrm{b}}(x)}{\mathrm{d}x} = i\, c_{\mathrm{bb}}(x)\,\psi_{\mathrm{b}}(x) + i\, c_{\mathrm{bf}}(x)e^{i2kx}\psi_{\mathrm{f}}(x) + i\, c_{\mathrm{br}}(x)\,\psi_{\mathrm{b}}(x). \tag{2.5.60}$$

The final coupled mode equations are Equations 2.5.59 and 2.5.60. The coupling coefficients can be physically interpreted as $c_{\mathrm{ff}} = c_{\mathrm{bb}}$ (Equation 2.5.54) driving scattering from a mode into itself, $c_{\mathrm{bf}} = c_{\mathrm{fb}}^*$ (Equation 2.5.55) driving scattering into the counter-propagating mode, and c_{fr} and c_{br} driving scattering from the waveguide mode into radiation modes above the light line. With the elimination of the radiation mode envelopes, the coupling coefficients into radiation modes (e.g., Equation 2.5.56) become

$$c_{\mathrm{fr}}(x) = \frac{a\omega}{2}\iint \mathrm{d}y\,\mathrm{d}z \int_{\text{all space}} \mathrm{d}\mathbf{r}'\Delta\varepsilon(\mathbf{r})\,\Delta\varepsilon(\mathbf{r}')e^{-ikx}\mathbf{e}_k^*(\mathbf{r})\cdot\overleftrightarrow{\mathbf{G}}_{\mathrm{R}}(\mathbf{r},\mathbf{r}';\omega)\cdot\mathbf{e}_k(\mathbf{r}')e^{ikx'}, \tag{2.5.61}$$

$$c_{\mathrm{br}}(x) = \frac{a\omega}{2}\iint \mathrm{d}y\,\mathrm{d}z \int_{\text{all space}} \mathrm{d}\mathbf{r}'\,\Delta\varepsilon(\mathbf{r})\Delta\varepsilon(\mathbf{r}')e^{ikx}\mathbf{e}_k(\mathbf{r})\cdot\overleftrightarrow{\mathbf{G}}_{\mathrm{R}}(\mathbf{r},\mathbf{r}';\omega)\cdot\mathbf{e}_k^*(\mathbf{r}')e^{-ikx'}. \tag{2.5.62}$$

The radiation scattering coefficients of Equations 2.5.61 and 2.5.62 are difficult to evaluate because of the integral over the entire waveguide. Although we assume disorder between holes is uncorrelated in the expectation sense, for any instance of disorder, there may be a nonzero correlation between holes mediated by radiation modes. However, we are primarily interested in coherent scattering that is contained within the waveguide, and can reasonably assume that any field scattered out of a bound mode will not be scattered back into a bound mode; this is justified as the bound mode scattering channel is by far the dominant one. Therefore, we can simplify the radiation loss by using $c_{\mathrm{fr}} = i\langle\alpha_{\mathrm{rad}}\rangle\, v_g/2a$ where $\langle\alpha_{\mathrm{rad}}\rangle$ is the incoherent average radiation loss derived earlier (Equation 2.5.39). Comparing Equations 2.5.39 and 2.5.56, the former is just the expectation value of the imaginary part of the latter integrated over a unit cell. A factor of 2 is necessary to convert from a power loss to an amplitude loss.

For modeling an incident field at one end of the waveguide, the boundary conditions for a wave injected into the waveguide [and consistent with $\mathbf{E}_i(\mathbf{r};\omega)$] are

$$\psi_{\mathrm{f}}(x_{\mathrm{start}}) = 1, \tag{2.5.63}$$

$$\psi_{\mathrm{b}}(x_{\mathrm{end}}) = 0, \tag{2.5.64}$$

where x_{start} and x_{end} are the positions of the input and output ports, respectively. The propagating envelopes are subsequently computed at all spatial position within the waveguide using the presented coupled mode Equations 2.5.59 and 2.5.60.

We stress that the full three-dimensional Bloch mode and disordered holes are self-consistently included in these final coupled-mode equations.

A similar theoretical work by Mazoyer et al. [49] obtains similar results of the present theory [20], with the main difference being the formalism. The technique of Mazoyer et al. [49] uses a powerful three-dimensional Fourier–Bloch-mode method, and it is similar to our own approach in the sense that (i) multiple-scattering events due to disorder are included, and (ii) light that is scattered out (by radiation modes) is not rescattered back into a bound mode. The method of Mazoyer et al. [49] is a full numerical method, where it can be difficult to have analytical insight into the scattering phenomena; the approach also assumes unit-cell averaged quantities, which, as we will show below (c.f. Figure 2.5.13), can lead to inaccuracies. From a general modeling perspective, it is not clear (at least, not to us) how to adapt the Fourier–Bloch-mode method to incorporate other effects such as nonlinear and quantum optical interactions, which we believe are ideally described in the language of Green functions and coupled modes. Nonlinear coupled mode equations for PhC waveguides have been described, for example, by Pereira and Sipe [50], and it would be interesting to combine these effects with disorder. Further, as with the incoherent loss section (see final paragraph of Section 2.5.3), the main advantage of having analytical formulas is that one can then design for manufactured waveguides with particular properties, for example, with lower or higher losses. Both approaches should be seen as complimentary, however, and it is encouraging that similar predictions are obtained using different numerical methods.

2.5.4.2 Numerical Implementation

This calculation requires, as inputs, the ideal waveguide mode dispersion and spatial field distribution. As a representative example, we again consider a W1 semiconductor waveguide, with parameters give in Table 2.5.1. The dispersion of the waveguide mode is shown in Figure 2.5.12a (solid, left scale) along with the group index (dashed, right scale). An example of the spatial distribution of the electric field in the center of the slab is shown in Figure 2.5.12b.

To solve Equations 2.5.59 and 2.5.60 numerically, the coupling coefficients are assumed to be constant over a short ($\Delta x << a$) interval in x and are integrated analytically. This yields a

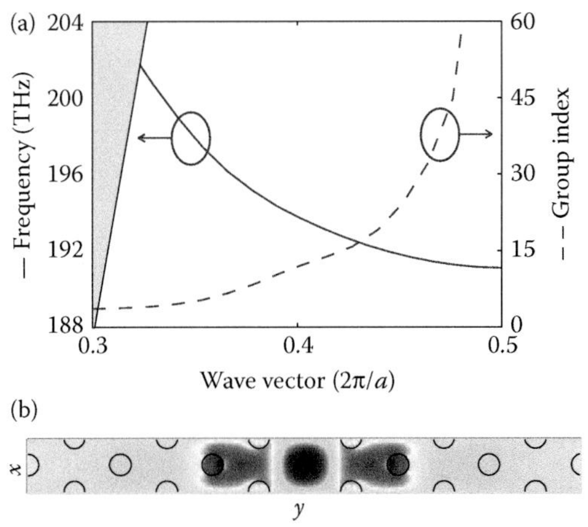

FIGURE 2.5.12
Properties of the nominal structure (W1 waveguide). (a) Dispersion of the waveguide mode (solid, left scale). The continuum of radiation mode above the light line is indicated by the shading on the left side of the figure. The group index (dashed, right scale) is also shown and diverges at the band edge ($k = 2\pi/a$). (b) Distribution of the transverse component of the electric field Bloch mode at the middle of the slab near the mode edge.

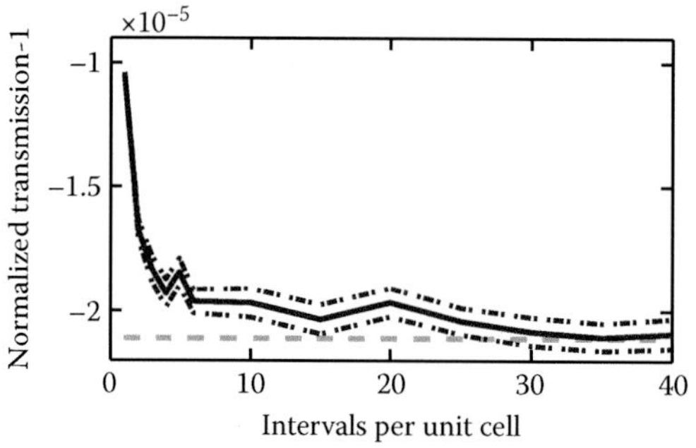

FIGURE 2.5.13
Mean transmission through 500 disordered waveguides (dark, solid) as a function of the number of intervals each unit cell is divided into. The error in the mean is marked by the dash-dotted limits and the mean agrees well with the prediction of the incoherent calculation [14] (light, dashed), except for very coarse discretizations.

pair of transfer equations linking the envelopes on either side of the chosen interval. In this way, a set of transfer equations that span the entire waveguide length can be built, and then solved using linear algebra techniques. This approach is particularly amenable to adding reflective facets and other features by simply including an appropriate transfer matrix.

The average coupling constants for each interval are calculated for each hole, generating an *instance* of a disordered profile from the statistical distribution of Equation 2.5.40. The coupling coefficients are calculated at multiple points within the interval and then averaged. Typically, there are 20 intervals per unit cell to satisfy the assumption that the coefficients are relatively constant. As shown in Figure 2.5.13, if the discretization of the unit cell is too coarse, the loss is underestimated. Thus, one must include subunit-cell propagation effects in general.

We highlight that the calculation is orders of magnitude more efficient than standard brute-force numerical techniques, for example, FDTD. We also note that we only need to calculate the coupled mode coefficients wherever disorder has an influence, namely at the hole interfaces. Further numerical details are discussed in the study by Patterson and Hughes [51]. For now, we mention that this semianalytic treatment can solve for thousands of frequency points and thousands of spatial grid points (e.g., 50,000 grid points for 2500 unit cells) on a desktop computer. The speed up required in computational time compared to a numerical technique like FDTD is literally tens of thousands.

2.5.4.3 Computed Transmission Spectra

Figure 2.5.14 shows transmission spectra for four disordered waveguides calculated by solving Equations 2.5.59 and 2.5.60 (solid). For reference, previous incoherent scattering results, computed within a second-order Born approximation and with multiple scattering extrapolation (Section 2.5.3), are also shown (dashed). Each row of plots is for a different waveguide with the left plot showing a broad frequency range and the right plot showing a narrow frequency range near the band edge. The top row is for a disordered 1.5 mm (3125 unit cells). The second row is for a different disorder instance of the same 1.5 mm. Experimentally, this would be similar to carrying out measurements on a second waveguide fabricated with nominally identical parameters. It has the same qualitative shape but the particular disordered resonances are substantially different. This is important if it was desired to take advantage of these sharp resonances as their resonant

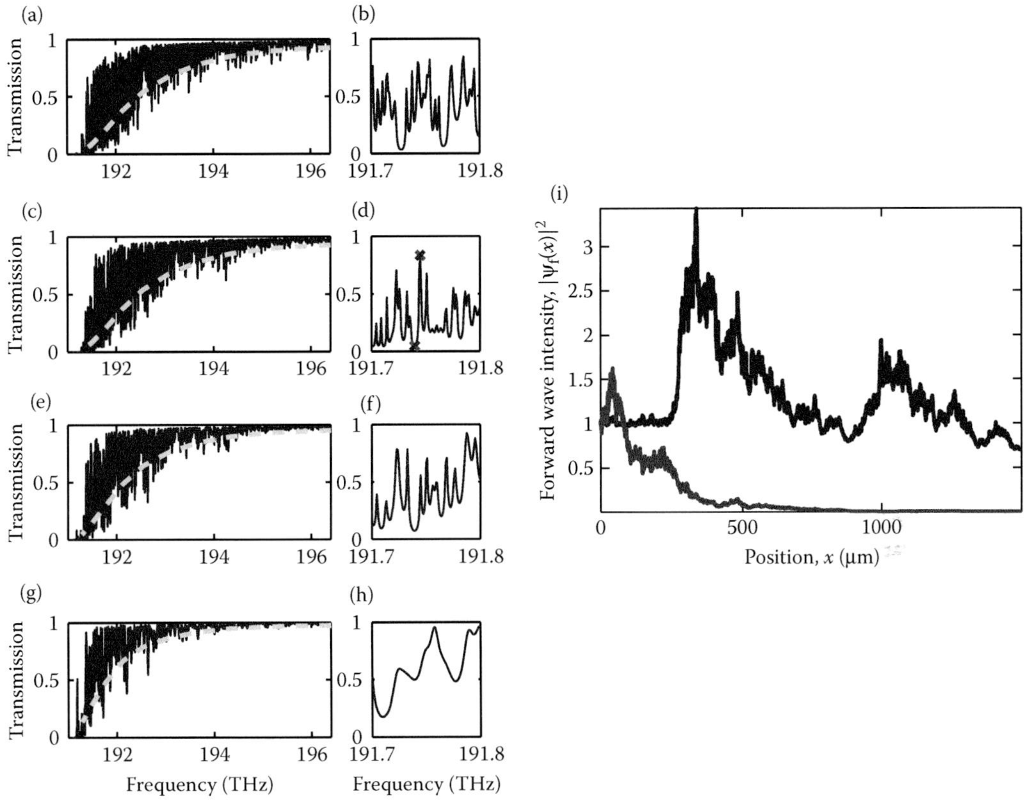

FIGURE 2.5.14
(**See color insert.**) Simulated transmission spectra of four disordered W1 waveguides using the coherent scattering theory (solid) and the first- and second-order Born incoherent theory (dashed) [14]. Each row of plots is for a different waveguide with the left plot showing a broad frequency range and the right plot showing a narrow frequency range near the band edge. Plots (a) and (b) are for a disordered 1.5 mm. Plots (c) and (d) are for a different disorder instance of the same 1.5 mm. Plots (e) and (f) and plots (g) and (h) are for the same disorder instance as plots (c) and (d) but with the length reduced to 1.0 and 0.5 mm, respectively. The calculation uses a root mean square roughness of $\sigma = 3$ nm, and a disorder correlation length of $l_p = 40$ nm. (i) Forward wave intensity in a disordered waveguide at two wave vectors. The upper curve ($n_g = 24.96$) corresponds to a local transmission maximum and the lower curve ($n_g = 25.11$) is a neighboring transmission minimum. These two curves correspond to the crosses shown in plot (d).

frequency cannot be easily designed. The third and fourth rows are for the same disorder instance as the second but with the length reduced to 1.0 and 0.5 mm respectively. Here the qualitative roll-off changes because of the length reduction but disordered resonances can be found at similar frequencies across the three lengths, especially between 1.5 and 1.0 mm.

We can examine the position-dependent distribution of energy in the waveguide under continuous wave illumination. In Figure 2.5.14d, a neighboring transmission minimum and maximum are marked with crosses. The forward wave intensity at these frequencies is plotted in Figure 2.5.14i. Although the points are very close in frequency, the minute difference in group index ($n_g = 25.11$ compared to $n_g = 24.96$) creates a difference in the accumulated phase and a dramatic change in the transmission.

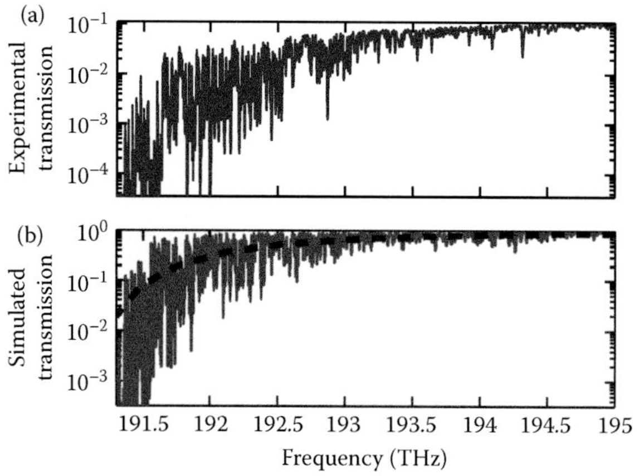

FIGURE 2.5.15
(a) Experimental transmission spectra for a 1.5-mm W1 waveguide showing resonances near the band edge. (Adapted from M. Patterson et al., Phys. Rev. Lett. **102**, 253903, 2009.) (b) Theoretical transmission spectra calculated using the incoherent (dashed) and coherent (solid) scattering theories. The incoherent scattering calculation uses the multiple scattering extrapolation technique described in Section 2.5.3.

2.5.4.4 Comparison to Experiment

An example of the simulated and experimental transmission plot versus frequency is shown in Figure 2.5.15, for a 1.5-mm W1 waveguide. As well as the expected transmission loss roll-off behavior that scales approximately with $\propto 1/v_g^2$, we also observe numerous sharp resonances where the transmission varies by orders of magnitude. These resonances are associated with multiple scattering events that are necessary to build up Fabry–Pérot-like resonances between disorder sites. Excellent agreement between the simulations and measurements are obtained over more than three orders of magnitude, without using any fit parameters.* Moreover, one can easily apply our technique to simulate time-frequency reflectance maps, and model intensity plots of the reflected signal as a function of time [11,20].

2.5.4.5 Light Localization

This coherent scattering formalism naturally provides insights into localization phenomena in PhC structures. Strong localization or Anderson localization was first proposed by Anderson [52] for electrons propagating in a periodic atomic potential. Anderson envisioned that as the lattice became disordered, an electron wave-packet would transition from a regime of diffusive propagation to a regime of exponential confinement. The transition occurs when the mean free path l, of a propagating Bloch mode is reduced to the order of the Bloch wave vector k; formally $kl < 1$ [2]. To the best of our knowledge, strong localization has not yet been conclusively observed for electrons owing to the difficulty of such experiments and complications due to electron–electron interactions and a lack of time-invariance in disordered atomic lattices (due to thermal vibrations). Strong localization has been shown to be a general wave phenomenon and so there is hope that it may be observed in other systems, such as engineered optical structures.

* The nominal sample parameters and disorder parameters are obtained from the experiments and statistical analysis of the fabricated waveguides.

Strong localization of light was proposed by both John [53] and Anderson [54] at about the same time. Since 1987, there has been significant progress toward observing strong localization of light. In 1997, Wiersma et al. [55] reported strong optical localization in a random powder; however, this does not satisfy Anderson's original formulation of a disordered *periodic* media with Bloch modes that become localized. Vlasov et al. [56] observed the thickness dependence of transmission through a disordered face-centered-cubic lattice of silica spheres near the band edge and argued that their observed exponential decrease in transmission was due to the interplay of Bragg scattering and incoherent scattering. A number of experiments [57] have shown strong localization in periodic structures transverse to the direction of propagation. This is an easier experimental feat as the transverse component of the wave vector can be significantly smaller than the wave vector magnitude. Topolancik et al. [46] reported the observation of resonances at the band edge of an artificially roughened PhC waveguide and argued these were due to strong localization. Similar observations have been reported by Garcia et al. [58].

The transmission spectra of PhC waveguides (e.g., Figure 2.5.14) exhibit sharp resonances near the band edge, similar to those reported by Topolancik et al. [46]. These features can also be resolved in high-resolution time-frequency reflection maps [20]. To address the question of whether these features are indicative of localization in a three-dimensional structure, a localization length l can be defined as $l^{-1} = \langle \ln T \rangle / L$ where T is the transmitted power and L is the sample length [56]. Using the PhC waveguide parameters above, $k = 0.45 \times 2\pi/a$ where $v_g = c/45$, the localization length is calculated to be $l \simeq 100$ μm, yielding $kl \simeq 700$, far from the criteria for strong localization. Thus, we advocate describing these features as Fabry–Pérot-like resonances between scattering sites and not strong localization, in agreement with the interpretation of Vlasov et al. [56]. Measurements [20] also determine localized modes spread throughout the entire waveguide, supporting the view of weak localization for these naturally disordered waveguides.

On the other hand, if one chooses to adopt a simpler one-dimensional analysis for the planar PhC waveguide, then Anderson localization will likely always occur for sufficiently long samples, as no one had yet manufactured such a structure without disorder. The same is also true for two-dimensional models and analysis. Certainly from a modeling perspective, these PhC slab waveguides require a full three-dimensional model, and it will be interesting in future work to quantitatively address the regime of *deliberately introduced strong disorder*. The deliberate disorder will act in addition to the intrinsic disorder that we focus on here. Probably more care is needed in assessing the degree of localization as radiation mode scatter introduces effective absorption channels, and there do exist other approaches to probing the statistical signatures of strong localization, for example, by measuring the relative size of *fluctuations* of certain transmission quantities [59]. More electromagnetic modes may also be needed in the theoretical analysis.

2.5.5 Local-Field Effects

In this final section, we introduce an approximate theory to describe disorder-induced scattering, including the influence of local-field effects—which were essentially ignored in Sections 2.5.3 and 2.5.4. Local-field effects will be shown to increase the predicted disorder-induced scattering loss and result in resonance shifts of the propagating waveguide mode. Performing an incoherent averaging calculation, we demonstrate that two types of frequency shifts can be expected, a mean frequency shift and an RMS frequency shift, both acting to blue-shift and broaden the nominal band structure. For the representative W1 waveguide, at telecommunication operating frequencies, we predict substantial millielectron volt frequency shifts and band structure broadening. The disorder-induced broadening is also found to increase as the propagation frequency approaches the slow-light regime (mode edge) due to restructuring of the electric field distribution.

2.5.5.1 The Local-Field Problem for High-Index-Contrast Dielectric Perturbations

Although many experiments have confirmed the slow-light loss behavior described in Sections 2.5.3 and 2.5.4 [10,36], an open question that still remains is: what is the effect of disorder on the band structure? The answer to this question is complicated and involves a complex interplay between local-field corrections and enforcing boundary conditions between parallel and perpendicular field components. Furthermore, experimental evidence for a more dramatic reduction of transmission near the mode edge is observed to take place much sooner than that predicted for the scaling of scattering loss. This suggests that some unknown effect is either shifting the mode edge or broadening the band structure (see Figure 2.5.12a). For PhC waveguides, this question was partially addressed by Patterson and Hughes [39], and we summarize that work below.

The usual theoretical approaches to modeling disorder-induced scattering use the standard perturbation theory where the nominal (disorder-free) electric field is used with the dielectric index change to model polarization scatterers, through $\mathbf{P}_{\text{dis}} = \varepsilon_0 \Delta\varepsilon \mathbf{E}_0$. For high-index-contrast perturbations (e.g., air to silicon gives $\Delta\varepsilon \approx 12$), as shown by Johnson et al. [37], this polarization scatterer is problematic for several reasons: (i) the parallel components of the electric field and the perpendicular components of the displacement field must be continuous across the surface, and (ii) the index change results in local-field corrections. Andreani and Garcie [60] estimated the magnitude of this error in their theory by comparing with a simple numerical supercell calculation of perfect hole shapes with different radii, and they concluded that the effects of local fields was perhaps not important. Wang et al. [38] used the *slowly varying surface approximation* [37], valid only for smooth bumps, and demonstrated the impact on increasing the scattering losses. For PhC cavity systems, Ramunno and Hughes [16] showed that a *quickly varying surface perturbation* can cause mean resonance shifts of a strongly confined cavity mode. RMS ensemble average shifts were computed for intrinsically lossy coupled-resonances optical waveguide (CROW) structures [17,61], though local fields were not considered and only band-edge broadening was predicted.

2.5.5.2 Theory and Impact on Incoherent Scattering Calculations

From Section 2.5.2, we know that the total electric field, in a system with disorder, can be written as (Equation 2.5.17)

$$\mathbf{E}(\mathbf{r};\omega) = \mathbf{E}_i(\mathbf{r};\omega) + \int_{\text{all space}} d\mathbf{r}' \overleftrightarrow{\mathbf{G}}(\mathbf{r},\mathbf{r}';\omega) \cdot \frac{\mathbf{P}(\mathbf{r}';\omega)}{\varepsilon_0},$$

where $\overleftrightarrow{\mathbf{G}}(\mathbf{r},\mathbf{r}';\omega) = \overleftrightarrow{\mathbf{G}}_{\text{bound}}(\mathbf{r},\mathbf{r}';\omega) + \overleftrightarrow{\mathbf{G}}_{\text{rad}}(\mathbf{r},\mathbf{r}';\omega) + \overleftrightarrow{\mathbf{G}}_{\text{other}}(\mathbf{r},\mathbf{r}';\omega)$. In Sections 2.5.3 and 2.5.4, we essentially neglected the influence of $\overleftrightarrow{\mathbf{G}}_{\text{other}}$. We now consider the role of this term, which is connected to local-field effects and high-index-contrast perturbations. We again consider the W1 waveguide and show the band structure in Figure 2.5.16, with and without disorder,

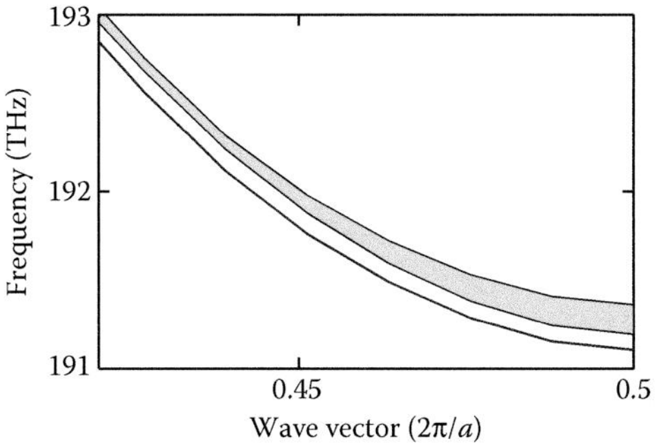

FIGURE 2.5.16
Nominal dispersion of a W1 photonic crystal waveguide mode (thick solid), and the broadened disorder-induced band structure due to local-field effects (gray shading, shifted up in frequency—see text).

where the latter effect is discussed below. To help clarify the physics of local-field effects, we refer back to Figure 2.5.1b, which shows a numerically generated disordered hole (Equation 2.5.40). We stress that this is an *instance*; each hole will look quantitatively different because of disorder.

In our disorder models before (Sections 2.5.4 and 2.5.3), we used $\mathbf{P}_{\mathrm{dis}} = \varepsilon_0 \Delta\varepsilon \mathbf{E}_0$ as the polarization perturbation, which may be termed the *weak index-contrast model*. For high index-contrast structures, such as with air holes on semiconductor slabs, this model fails in general, except for specific polarization directions of the fields. The failure of the weak index-contrast model is most obvious by considering a small dielectric sphere introduced in a homogeneous background with dielectric constant ε_2. It is well known that the weak contrast polarizability $\Delta\varepsilon$ must be replaced with the corrected polarizability $3\Delta\varepsilon/(3\varepsilon_2 + \Delta\varepsilon)$; this can easily be proven from Equation 2.5.50 (see also the study by Ramunno and Hughes [16]).

Therefore, because of local-field effects, the macroscopic scattering depends on the microscopic geometry of the scatter, and this is also true for disordered PhC waveguides. This issue has been partly investigated by Johnson et al. [37], who give the corrected disorder polarization density due to a disorder element at $\mathbf{r}'$ as

$$\mathbf{P}(\mathbf{r};\omega) = \left(\frac{\varepsilon_1 + \varepsilon_2}{2} \alpha_{\parallel} \varepsilon_0 \mathbf{E}_{\parallel}(\mathbf{r};\omega) + \varepsilon(\mathbf{r}) \gamma_{\perp} \mathbf{D}_{\perp}(\mathbf{r};\omega) \right) \Delta V \delta(\mathbf{r} - \mathbf{r}'), \tag{2.5.65}$$

where $\alpha_{\parallel}$ and $\gamma_{\perp}$ are polarizabilities for the disorder element, $\varepsilon(\mathbf{r})$ takes a different value depending on which side of the interface $\mathbf{r}'$ is located, and ΔV is the volume of the disorder element. Owing to the $\varepsilon(\mathbf{r})$ factor, $\mathbf{E}^*(\mathbf{r}) \cdot \mathbf{P}(\mathbf{r})$ will have terms proportional to $|\mathbf{E}_{\parallel}(\mathbf{r})|^2$ and $|\mathbf{D}_{\perp}(\mathbf{r})|^2$ and these fields are well defined at a dielectric interface. This improved disorder model hides a great deal of complexity in the polarizabilities $\alpha_{\parallel}$ and $\gamma_{\perp}$. The polarizabilities are different for a positive bump (ε_1 extending into ε_2, $\alpha_{\parallel} = \alpha_{\parallel}^{+}$) and a negative bump ($\alpha_{\parallel}^{-}$). Further, in general, the polarizabilities are asymmetric so that $\alpha_{\parallel}^{+} \neq -\alpha_{\parallel}^{-}$. The exact polarizabilities also depend on the precise geometry of the disorder element and must be calculated numerically. This is presumably why only a single isolated bump was treated in the study by Johnson et al. [37].

One method to simplify the treatment of the polarizabilities is to assume a structure for the disorder where the polarizability is known. In the study by Wang et al. [38], a polarization density of the form $\mathbf{P}(\mathbf{r}) = \Delta\varepsilon(\mathbf{r}) \left(\varepsilon_0 \mathbf{E}_{\parallel}(\mathbf{r}) + \varepsilon(\mathbf{r}) \frac{\mathbf{D}_{\perp}(\mathbf{r})}{\varepsilon_1 \varepsilon_2} \right) \delta(\mathbf{r} - \mathbf{r}')$ was used. This takes care of the issue of the **E**(**r**) being ill-defined at the interface but it has some subtle and questionable assumptions. Implicit in the derivation of this term is the assumption that the disordered surface is smooth; effectively that the hole radius changes but remains nearly circular. As a model for the disorder in real fabricated samples, this is rather suspect, and, as we have discussed earlier, it is known that the interface fluctuations vary rapidly (c.f. Figure 2.5.1b, and the study by Skorobogatiy et al. [42]). One solution to this general problem is to use a representative model for the polarizability, namely Equation 2.5.65, with the numerically calculated polarizabilities for a cylindrical bump [37], in combination with our numerically generated disordered profile.

It is easiest to compare these polarization density models above using our incoherent scattering calculation where the band-edge resonances are suppressed. The incoherent average power loss, which is dominated by backscatter loss for slow-light PhC waveguides, is (c.f. Equation 2.5.35)

$$\langle \alpha_{\text{back}}(\omega) \rangle = \left(\frac{a\omega}{2v_g}\right)^2 \iint d\mathbf{r}\, d\mathbf{r}' \left\langle \left[\mathbf{E}(\mathbf{r};\omega)\cdot\frac{\mathbf{P}(\mathbf{r};\omega)}{\varepsilon_0}\right]\left[\mathbf{E}^*(\mathbf{r}';\omega)\cdot\frac{\mathbf{P}^*(\mathbf{r}';\omega)}{\varepsilon_0}\right]\right\rangle, \tag{2.5.66}$$

where the integration is performed over a single unit cell and $\mathbf{E}(\mathbf{r};\omega) = \mathbf{e}_k(\mathbf{r})e^{ikx}$. The incoherent averaged backscatter loss is plotted in Figure 2.5.17 for the weak contrast (light solid), smooth surface (medium gray solid), and cylindrical bump (dark solid) models. Although the various loss predictions differ, all the models predict similar loss trends and their magnitudes are similar. For the roughness statistics typical of PhC waveguide systems, we find reasonable agreement between the backscatter loss predictions of the three models, though the smooth-surface and weak-contrast models underestimate the scattering loss. This by itself, although an improvement, is evidently not a drastic revision of previous disorder models. In terms of predicting loss trends versus frequency, all three models qualitatively agree.

However, one of the subtle effects neglected when one ignores the $\overleftrightarrow{\mathbf{G}}_{\text{other}}(\mathbf{r},\mathbf{r}';\omega)$ contribution of the Green function is disorder-induced frequency shifts. This phenomenon for PhC waveguides is closely related to the selection of a suitable polarization model, as the

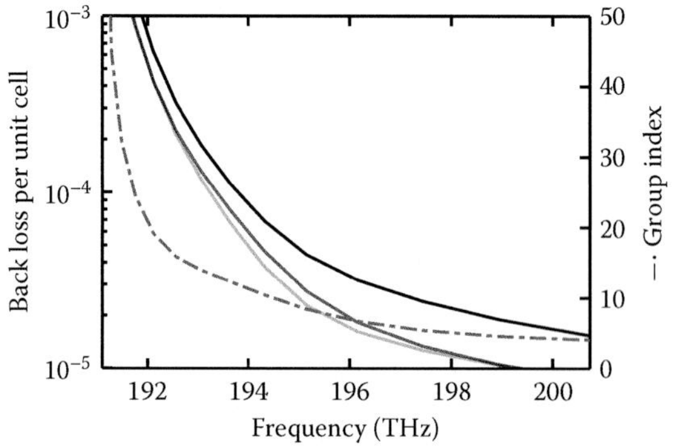

FIGURE 2.5.17
Incoherent averaged backscatter loss for a single unit cell using the weak contrast (lower—light solid), smooth surface (middle—medium gray solid), and cylindrical bump (upper—dark solid) polarization density models. For reference, the group index (dot-dash) is shown on the right scale.

predicted frequency shift is sensitive to this term. The first-order *mean frequency shift* due to disorder is given as [21]

$$\langle \Delta\omega \rangle = -\frac{\omega}{2}\int d\mathbf{r}\left\langle \mathbf{E}^*(\mathbf{r};\omega)\cdot\frac{\mathbf{P}(\mathbf{r};\omega)}{\varepsilon_0}\right\rangle, \tag{2.5.67}$$

with $\mathbf{E}(\mathbf{r};\,\omega)$ normalized as before (Equation 2.5.11). The frequency shift is usually taken to be zero for zero-mean surface perturbations, but as shown previously for a PhC cavity [16], correctly treating local-field effects yields a nonzero first-order frequency shift. Considering the above polarization densities, both the weak-contrast and smooth-surface models predict $\langle\Delta\omega\rangle = 0$ due to the symmetry in the polarization for positive and negative bumps. The cylindrical bump polarizability predicts a nonzero $\langle\Delta\omega\rangle$, but the details differ from those of Ramunno and Hughes [16] (the simpler PhC cavity case) owing to differing disorder models; for the waveguide, we are also dealing with a continuous mode rather than a discrete resonance. When using the polarization density of Equation 2.5.65, care must be taken when evaluating the expectation value as the value of the polarizabilities depends on the direction of the bump. In addition to the mean frequency shift, which in general may or may not be zero, the RMS frequency shift $\langle\Delta\omega\rangle_{RMS} = \sqrt{\langle\Delta\omega^2\rangle}$ is not. It is calculated in a similar way to the corrected backscatter loss, from

$$\langle \Delta\omega^2 \rangle = \frac{\omega^2}{4}\iint d\mathbf{r}\,d\mathbf{r}'\left\langle \left[\mathbf{E}^*(\mathbf{r};\omega)\cdot\frac{\mathbf{P}(\mathbf{r};\omega)}{\varepsilon_0}\right]\left[\mathbf{E}^*(\mathbf{r}';\omega)\cdot\frac{\mathbf{P}(\mathbf{r}';\omega)}{\varepsilon_0}\right]\right\rangle. \tag{2.5.68}$$

Figure 2.5.18 plots the mean (dashed) and RMS (solid) frequency shifts for the weak contrast (medium gray—solid), smooth surface (light gray), and cylindrical bump (dark gray) polarization density models. The weak-contrast and smooth-surface models always predict zero mean frequency shifts and thus are not shown. The disorder-induced frequency shifts are particularly important for understanding experimental transmission spectra. Typically, very near the band edge, there will be an abrupt drop in the transmission associated with a local frequency shift causing the band to be shifted such that the injected

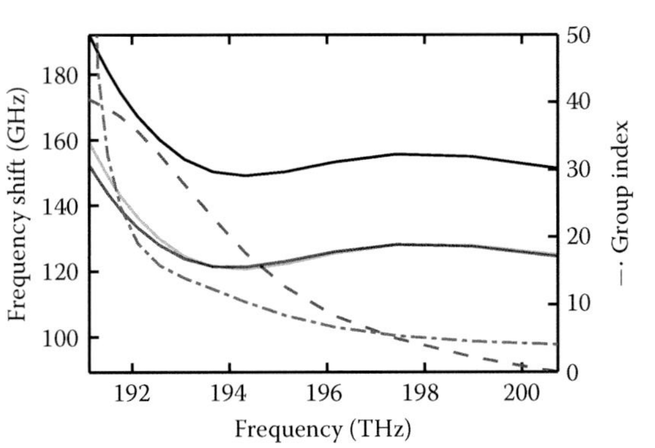

FIGURE 2.5.18
Mean frequency shift (dashed) and root mean square frequency shift (solid) using the weak contrast (light—middle solid), smooth surface (medium gray—lower solid), and cylindrical bump (dark—upper solid) polarization density models. Only the cylindrical bump model has a nonzero mean frequency shift. For reference, the group index (dot-dash) is shown on the right scale.

frequency is below the new band edge. As can be seen from the RMS frequency shifts, this will be an issue regardless of the polarization model, but can be further exasperated by a nonzero mean frequency shift.

To help summarize these local-field effects, Figure 2.5.16 depicts the computed band structure with disorder* (gray shaded band), showing that it is actually impossible to reach the nominal band edge. Qualitatively, for waveguide mode frequencies within a few standard deviations of the mode edge (slow-light regime), we expect that somewhere along the waveguide, total reflection will inevitably occur. This will enhance the increasing loss roll-off effects described in the previous sections.

2.5.6 Summary

In Section 2.5.2, we described the concept of PhCs, PhC slabs, and the PhC slab waveguide, and presented some of the theory and numerical techniques to describe them. We introduced the band structure, Bloch modes, and Green function theory.

In Section 2.5.3, we showed how the incoherent average loss in a single period of a disordered PhC waveguide can be calculated from properties of the bound mode, a disorder model, and an effective radiation Green function. We also described an extrapolation method for connecting these results with meaningful experimental length scales. The primary limitation of this calculation is the assumption that there are no long-range phase effects such as interference between reflections from difference disorder sites, a limitation that is overcome in Section 2.5.4. The key findings of this section were that by neglecting contributions due to the changing Bloch mode electric field, the backscatter loss scales as v_g^{-2} and the radiation loss scales as v_g^{-1}. This raises serious concerns for the possibility of operating devices in the slow-light regime since, for any degree of disorder, there will always be a limit where the slow-light propagation amplifies the scattering sufficiently so that propagation is significantly disrupted. Deviations from these simple scaling rules can be caused by a changing Bloch mode as a function of frequency, and by the assumption of the Beer–Lambert law, which was shown to clearly breakdown for a 1.5-mm waveguide. The general breakdown in these simple loss scaling rules has also been shown and discussed for so-called dispersion-engineered guides [19]. These predictions and simulations agree very well with recent experiments. The presented physically intuitive analytical formulas also make clear how to design waveguides with lower losses, for example, by minimizing the overlap of the Bloch modes with the hole interfaces.

In Section 2.5.4, we described and applied a theory for self-consistently modeling coherent scattering in a disordered PhC waveguide instance, allowing one to map directly onto a realistic experimental situation. As with the incoherent scattering case, slow-light propagation enhances backscattering (and, to a lesser extent, radiation scattering) leading to high losses near the band edge. The key feature of disorder-induced coherent scattering is formation of sharp spectral resonances near the band edge, which is mediated by Fabry–Pérot-like resonances between disorder sites. This theory is computationally efficient, making the analysis of very long waveguides (thousands of periods using the full three-dimensional structure) feasible on a desktop computer. This approach has been used to

* The standard deviation of the frequency shift about the new mean value is defined as $\sqrt{\langle \Delta\omega^2 \rangle - \langle \Delta\omega \rangle^2}$.

explain a rich range of experimental features without introducing any fitting parameters [20]. The method also allows one to directly probe light localization phenomena. The spirit is similar to Section 2.5.3, in that the formalisms have been kept semianalytic, where the analytical formulas clarify the underlying physics of light scattering; all one requires is the Bloch mode and a model for the disordered PhC waveguide.

Finally, in Section 2.5.5, we have described an approximate theory of disorder-induced scattering that includes the influence of local-field effects and high-index-contrast perturbations [39]. The model calculations of incoherent backscatter loss are shown to increase the predicted waveguide losses and result in significant and complex (asymmetric) disorder-induced resonance shifts. The band structure broadening also offers further insights into the fabrication limits of slow-light propagation in PhC waveguides and serves as a further warning that propagation modes near the mode edge will have an increasingly better chance of being completely reflected. Although we have specialized in this local-field study to incoherent averages, these general results can also be used to guide and improve other models that have been used to describe disorder-induced coherent scattering, for example, see the models of Patterson et al. [19] and Patterson and Hughes [51]; also see Section 2.5.4.

Acknowledgments

This work was supported by the National Sciences and Engineering Research Council of Canada, and the Canadian Foundation for Innovation. We thank many colleagues and collaborators for useful discussion over the years, including Lora Ramunno, Sylvain Combrié, Alfredo De Rossi, Thomas Krauss, John Sipe, Jeff Young, Steven Johnson, Michelle Povinelli, Lucio Andreani, Marc Dignam, and Masaya Notomi. In particular, we acknowledge Alfredo De Rossi for forcing us to dig much deeper into the underlying mechanisms of disorder-induced scattering, which helped in the development of our coherent scattering theory (Section 2.5.4); we also acknowledge Alfredo De Rossi and Sylvain Combrié for use of their data shown in Figures 2.5.5, 2.5.11, and 2.5.15.

References

1. M. Patterson, Master's thesis, Queen's University, Canada, 2009. http://hdl.handle.net/1974/5122.
2. S. John, Phys. Rev. Lett. **58**, 2486 (1987).
3. E. Yablonovitch, Phys. Rev. Lett. **58**, 2059 (1987).
4. J. D. Joannopoulos, S. G. Johnson, J. N. Winn, and R. D. Meade, *Photonic Crystals: Molding the Flow of Light* (Princeton University Press, Princeton, NJ, 2008), 2nd ed.
5. M. Notomi, K. Yamada, A. Shinya, J. Takahashi, C. Takahashi, and I. Yokohama, Phys. Rev. Lett. **87**, 253902 (2001); Y. A. Vlasov, M. O'Boyle, H. F. Hamann, and S. J. McNab, Nature **438**, 65 (2005).
6. T. Baba, Nat. Photonics **2**, 465 (2008).
7. S. Hughes, Opt. Lett. **29**, 2659 (2004); V. S. C. Manga Rao and S. Hughes, Phys. Rev. Lett. **99**, 193901 (2007); Phys. Rev. B **75**, 205437 (2007); G. Lecamp, P. Lalanne, and J. P. Hugonin, Phys. Rev. Lett. **99**, 023902 (2007); T. Lund-Hansen, S. Stobbe, B. Julsgaard, H. Thyrrestrup, T. Sunner, M. Kamp, A. Forchel, and P. Lodahl, *ibid*. **101**, 113903 (2008).

8. E. Viasnoff-Schwoob, C. Weisbuch, H. Benisty, S. Olivier, S. Varoutsis, I. Robert-Philip, R. Houdré, and C. J. M. Smith, Phys. Rev. Lett. **95**, 183901 (2005).
9. Y. Hamachi, S. Kubo, and T. Baba, Opt. Lett. **34** 1072 (2009); B. Corcoran, C. Monat, C. Grillet, D. J. Moss, B. J. Eggleton, T. P. White, L. O'Faolain, and T. F. Krauss, Nat. Photonics **3**, 2006 (2009); A. Baron, A. Ryasnyanskiy, N. Dubreuil, P. Delaye, Q. V. Tran, S. Combrié, A. de Rossi, R. Frey, and G. Roosen, Opt. Express **17**, 552 (2009).
10. E. Kuramochi, M. Notomi, S. Hughes, A. Shinya, T. Watanabe, and L. Ramunno, Phys. Rev. B **72**, 161318(R) (2005).
11. A. Parini, P. Hamel, A. D. Rossi, S. Combrié, N.-V.-Q. Tran, Y. Gottesman, R. Gabet, A. Talneau, Y. Jaouën, and G. Vadalà, IEEE J. Lightwave Technol. **26**, 3794 (2008).
12. A. Reza, M. M. Dignam, and S. Hughes, Nature **23**, E10 (2008).
13. R. De La Rue, Opt. Quantum Electron. **34**, 417 (2002).
14. S. Hughes, L. Ramunno, J. F. Young, and J. E. Sipe, Phys. Rev. Lett. **94**, 033903 (2005).
15. A. F. Koenderink, Ad Lagendijk, and W. L. Vos, Phys. Rev. B **72**, 153102 (2005).
16. L. Ramunno and S. Hughes, Phys. Rev. B **79**, 161303(R) (2009).
17. D. P. Fussell, S. Hughes, and M. M. Dignam, Phys. Rev. B **78**, 144201 (2008).
18. S. Combrié, E. Weidner, A. De Rossi, S. Bansropun, S. Cassette, A. Talneau, and H. Benisty, Opt. Express **14**, 7353 (2006).
19. M. Patterson, S. Hughes, S. Schulz, D. M. Beggs, T. P. White, L. O'Faolain, and T. F. Krauss, Phys. Rev. B **80**, 195305 (2009).
20. M. Patterson, S. Hughes, S. Combrié, N.-V.-Q. Tran, A. De Rossi, R. Gabet, and Y. Jaouën, Phys. Rev. Lett. **102**, 253903 (2009).
21. K. Sakoda, *Optical Properties of Photonic Crystals* (Springer, Berlin/Heidelberg, 2005), 2nd ed.
22. M. Paulus, P. Gay-Balmaz, and O. J. F. Martin, Phys. Rev. E **62**, 5797 (2000).
23. K. Joulain, R. Carminati, J.-P. Mulet, and J.-J. Greffet, Phys. Rev. B **68**, 245405 (2003).
24. C. G. Poulton, C. Koos, M. Fujii, A. Pfrang, T. Schimmel, J. Leuthold, and W. Freude, IEEE J. Sel. Top. Quantum Electron. **12**, 1306 (2006).
25. K. Yee, IEEE Trans Antennas Propag. **14**, 302 (1966).
26. D. M. Sullivan, *Electromagnetic Simulation Using the FDTD Method* (IEEE Press, Piscataway, NJ, 2000.
27. A. Taflove and S. C. Hagness, *Computational Electrodynamics: The Finite-Difference Time-Domain Method* (Artech House, Norwood, MA, 2005), 3rd ed.
28. FDTD Solutions is a product of Lumerical Solutions, Inc. http://www.lumerical.com
29. A. F. Oskooi, D. Roundy, M. Ibanescu, P. Bermel, J. D. Joannopoulos, and S. G. Johnson, Comput. Phys. Commun. **181**, 687 (2010).
30. V. A. Mandelshtam and H. S. Taylor, J. Chem. Phys. **107**, 6756 (1997).
31. S. G. Johnson, http://ab-initio.mit.edu/wiki/index.php/Harminv. A free implementation of the harmonic inversion analysis by Mandelshtam and Taylor [30].
32. S. G. Johnson and J. D. Joannopoulos, Opt. Express **8**, 173 (2001).
33. M. L. Povinelli, S. G. Johnson, E. Lidorikis, J. D. Joannopoulos, and M. Soljacic, Appl. Phys. Lett. **84**, 3639 (2004).
34. D. Gerace and L. C. Andreani, Opt. Lett. **29**, 1897 (2004).
35. D. Gerace and L. C. Andreani, Opt. Express **13**, 4939 (2005).
36. L. O'Faolain, T. P. White, D. O'Brien, X. Yuan, M. D. Settle, and T. F. Krauss, Opt. Express **15**, 13129 (2007); R. J. P. Engelen, D. Mori, T. Baba, and L. Kuipers, Phys. Rev. Lett. **101**, 103901 (2008).
37. S. G. Johnson, M. L. Povinelli, M. Soljačić, A. Karalis, S. Jacobs, and J. D. Joannopoulos, Appl. Phys. B **81**, 283 (2005).
38. B. Wang, S. Mazoyer, J. P. Hugonin, and P. Lalanne, Phys. Rev. B **78**, 245108 (2008).
39. M. Patterson and S. Hughes, Phys. Rev. B **81**, 245321 (2010).
40. W. Vogel and D.-G. Welsch, *Quantum Optics* (Wiley-VCH, Berlin, 2006), 3rd ed.
41. E. Kuramochi, M. Notomi, S. Mitsugi, A. Shinya, T. Tanabe, and T. Watanabe, Appl. Phys. Lett. **88**, 041112 (2006); S. Combrié, A. D. Rossi, Q. V. Tran, and H. Benisty, Opt. Lett. **33**, 1908 (2008).
42. M. Skorobogatiy, G. Bégin, and A. Talneau, Opt. Express **13**, 2487 (2005).

43. P. Yao, V. S. C. Manga Rao, and S. Hughes. Laser Photonics Rev. **4**, 499 (2010).
44. N. Le Thomas, H. Zhang, J. Jágerská, V. Zabelin, R. Houdré, I. Sagnes, and A. Talneau, Phys. Rev. B **80**, 125332 (2009).
45. D. Gerace and L. C. Andreani, Photonics Nanostruct. Fundam. Appl. **3**, 120 (2005).
46. J. Topolancik, B. Ilic, and F. Vollmer, Phys. Rev. Lett. **99**, 253901 (2007).
47. D. Marcuse, *Theory of Dielectic Optical Waveguides* (Academic Press, New York, 1974).
48. M. Palamaru and P. Lalanne, Appl. Phys. Lett. **78**, 1466 (2001).
49. S. Mazoyer, J. P. Hugonin, and P. Lalanne, Phys. Rev. Lett. **103**, 063903 (2009).
50. S. Pereira and J. E. Sipe, Phys. Rev. E **66**, 026606 (2002).
51. M. Patterson and S. Hughes, J. Opt. (Special Issue on "slow light") **12**, 104013 (2010).
52. P. W. Anderson, Phys. Rev. **109**, 1492 (1958).
53. S. John, Phys. Rev. Lett. **53**, 2169 (1984).
54. P. W. Anderson, Philos. Mag. B **52**, 505 (1985).
55. D. S. Wiersma, P. Bartolini, A. Lagendijk, and R. Righini, Nature **390**, 671 (1997).
56. Y. A. Vlasov, M. A. Kaliteevski, and V. V. Nikolaev, Phys. Rev. B 60, 1555 (1999).
57. T. Schwartz, G. Bartal, S. Fishman, and M. Segev, Nature **446**, 52 (2007); Y. Lahini, A. Avidan, F. Pozzi, M. Sorel, R. Morandotti, D. N. Christodoulides, and Y. Silberberg, Phys. Rev. Lett. **100**, 013906 (2008).
58. P. D. Garcia, S. Smolka, S. Stobbe, and P. Lodahl, Phys. Rev. B **82**, 165103 (2010).
59. A. A. Chabanov, M. Stoytchev, and A. Z. Genack, Nature **404**, 850 (2000).
60. L. C. Andreani and D. Gerace, Phys. Status Solidi B **244**, 3528 (2007).
61. S. Mookherjea and A. Oh, Opt. Lett. **32**, 289 (2007).

2.6

Quasicrystalline Photonic Structures: Between Order and Disorder

Alexander N. Poddubny and Eugeniyus L. Ivchenko
Ioffe Physical-Technical Institute of the Russian Academy of Sciences

CONTENTS

The concept of quasicrystal as a nonperiodic structure with perfect long-ranged order was brought in solid-state physics by Levine and Steinhardt.[1] At present, it has become clear that, in addition to crystalline and amorphous materials, there exists a third form of solids that unexpectedly fills the gap between the two well-defined condensed-matter states. Moreover, this intermediate class called aperiodic deterministic structures includes the famous Fibonacci sequence $\mathcal{ABAAB}\ldots$ and other quasicrystals that can be described by a projection onto the n-dimensional (nD) space with $n = 1, 2,$ or 3 of an mD periodic lattice with dimensionality $m > n$. Examples of aperiodic structures different from quasicrystals are Thue–Morse and period-doubling sequences. Discovery of quasicrystals and other deterministic aperiodic structures initiated new fields of research in photonics. The studies of aperiodic long-range-ordered systems were extended to optics in the work by Kohmoto et al.,[2] where a 1D quasicrystal constructed of dielectric layers forming the Fibonacci sequence was proposed. Since then, photonic quasicrystals and other artificial long-range-ordered aperiodic objects have aroused an increasing interest in optical spectroscopy of solids.[3–5]

In this chapter, we first define the quasicrystals and present their structure factors. Then we consider light propagation in aperiodic photonic structures and pay particular attention to application of the two-wave approximation (TWA). The latter allows one to interpret the optical spectra of aperiodic structures in terms of the periodic objects and underline the specific features arising as a result of the nonperiodicity. To illustrate, we analyze not only the binary Fibonacci optical superlattices built of two constituent layers A and B but also the recently proposed artificial objects, namely, Fibonacci multiple quantum-well (QW) structures. An important point is that the regimes where TWA is invalid demonstrate the properties of optical spectra that are forbidden for periodic structures: (i) the localization characteristic for disordered systems and (ii) scaling and self-similarity that are absent in both conventional crystals and disordered materials. In the final part of this chapter, we briefly enumerate other proposed and studied photonic aperiodic long-range-ordered photonic structures—1D, 2D, and 3D.

2.6.1 Definition and Structure Factor

An nD quasicrystal is defined as a partial projection of points of a periodic lattice in the mD hyperspace onto nD space ($n = 1, 2, 3$; $m > n$). Here, in the first two subsections, we present, in addition to this cut-and-project method, two other methods to classify the 1D quasicrystals. In the third subsection, we give brief information concerning the structure factor of quasicrystals.

2.6.1.1 One-Dimensional Quasicrystals

There are three equivalent definitions of quasicrystalline systems: (i) the incommensurate chains, (ii) the substitution rules, and (iii) the cut-and-project method. The *incommensurate chains* and related structures[6,7] have been studied since the 1960s, even before the term "quasicrystal" was introduced by Levine and Steinhardt.[1] For 1D quasicrystals, the coordinates of the chain sites are written in the form

$$z_j = z_0 + j\bar{d} + r(j), \tag{2.6.1}$$

where $\bar{d}$ is the mean period of the 1D lattice of sites, z_0 is an arbitrary shift of the lattice as a whole, and the modulation $r(j)$ is the periodic function

$$r(j) = \Delta\left\{\frac{j}{t} + \varphi\right\}, \tag{2.6.2}$$

$\{x\}$ stands for the fractional part of x; Δ, t, and φ are the structure parameters, with t being irrational and φ being a noninteger. At vanishing Δ, Equations 2.6.1 and 2.6.2 specify a simple periodic lattice with the period $\bar{d}$. In case of rational t, the structure is still periodic but has a compound supercell, whereas for irrational values of t Equation 2.6.1 leads to a deterministic aperiodic chain known as a "modulated crystal."[8] The parameter Δ describes the modulation strength and the value of φ specifies the initial phase of the function $r(j)$. For z_j defined according to Equations 2.6.1 and 2.6.2 the spacings $z_{j+1} - z_j$ take one of the two values

$$a = \bar{d} + \Delta\left(\frac{1}{t} - 1\right) \quad \text{and} \quad b = \bar{d} + \frac{\Delta}{t}. \tag{2.6.3}$$

Therefore, a and b satisfy the conditions $\Delta = b - a$ so that the values of Δ lie in the interval between $-a$ and b. Excluding Δ in Equation 2.6.3 one can find the relation $\bar{d} = [(t-1)b + a]/t$. Moreover, the ratio N_B/N_A of the number of spacings a and b in an infinite lattice is related to t by $N_B/N_A = t - 1$.

Under certain conditions imposed on the values t and φ,[9,10] the arrangement of points (Equation 2.6.1) can be also obtained by the *substitution rules* acting on the building segments $\mathcal{A}$ and $\mathcal{B}$ as follows:

$$\begin{aligned} \mathcal{A} &\rightarrow \sigma(\mathcal{A}) = \mathcal{M}_1\mathcal{M}_2\ldots\mathcal{M}_{\alpha+\beta}, \\ \mathcal{B} &\rightarrow \sigma(\mathcal{B}) = \mathcal{N}_1\mathcal{N}_2\ldots\mathcal{N}_{\gamma+\delta}. \end{aligned} \tag{2.6.4}$$

Each of the symbols $\mathcal{M}_k$ and $\mathcal{N}_k$ in the right-hand side of Equation 2.6.4 stands for $\mathcal{A}$ or $\mathcal{B}$, α and β denote the numbers of letters $\mathcal{A}$ and $\mathcal{B}$ in $\sigma(\mathcal{A})$, whereas γ and δ are the numbers of $\mathcal{A}$ and $\mathcal{B}$ in $\sigma(\mathcal{B})$.[11] The correspondence between the two definitions is established by the relation $t = 1 + (\lambda_1 - \alpha)/\gamma$ between a value of t and indices α, β, γ, and δ, where $\lambda_1 = (v + \sqrt{v^2 + 4w})/2$, $v = \alpha + \delta$ and $w = \beta\gamma - \alpha\delta$. For the quasicrystals w must be equal to ± 1.[12]

2.6.1.2 Fibonacci Structures

The 1D Fibonacci lattice, being one of the most studied quasicrystals, is determined by the substitution rule:[8] $\mathcal{A} \rightarrow \mathcal{AB}$, $\mathcal{B} \rightarrow \mathcal{A}$. Equivalently, it can be defined as an infinite sequence $\lim_{m\to\infty} \mathcal{F}_m$, where $\mathcal{F}_m$ is the finite Fibonacci sequence of the mth order ($m = 1, 2, 3 \ldots$) satisfying the recurrence relations $\mathcal{F}_m = \mathcal{F}_{m-1}\mathcal{F}_{m-2}$ with the initial conditions $\mathcal{F}_1 = \mathcal{B}$ and $\mathcal{F}_1 = \mathcal{A}$. For example, one has for the seventh-order sequence $\mathcal{F}_7 = (\mathcal{ABAABABA})(\mathcal{ABAAB})$, where the brackets separate the sequences $\mathcal{F}_6$ and $\mathcal{F}_5$. The chain $\mathcal{F}_m$ shares the name with the Fibonacci numbers F_m owing to the similar recursion rule, $F_1 = F_2 = 1$, $F_{m+1} = F_m + F_{m-1}$.

In terms of the incommensurate chains, see Equations 2.6.1 and 2.6.2, and the substitution rules 2.6.4, parameters of the Fibonacci structure are given by

$$\begin{aligned} &t = \tau \equiv (\sqrt{5} + 1)/2, \quad \varphi = 0, \quad N_A/N_B = \tau, \\ &\alpha = \beta = \gamma = 1, \quad \delta = 0; \quad w = v = 1, \lambda_1 = \tau. \end{aligned} \tag{2.6.5}$$

If the first site is chosen at the plane $z = 0$ then $z_0 = -b$. Here, the ratio a/b is arbitrary. For $a = b$, the structure becomes periodic.[13] For a/b equal to the golden mean τ, it becomes the *canonical* Fibonacci chain.[14] In the noncanonical Fibonacci structures, this ratio is different from 1 and τ.

The structure described by Equations 2.6.1 and 2.6.2 can be equivalently defined by the *cut-and-project method*.[15] In this method, the Fibonacci chain is generated by projection of a stripe in the auxiliary 2D space under an irrational slope.

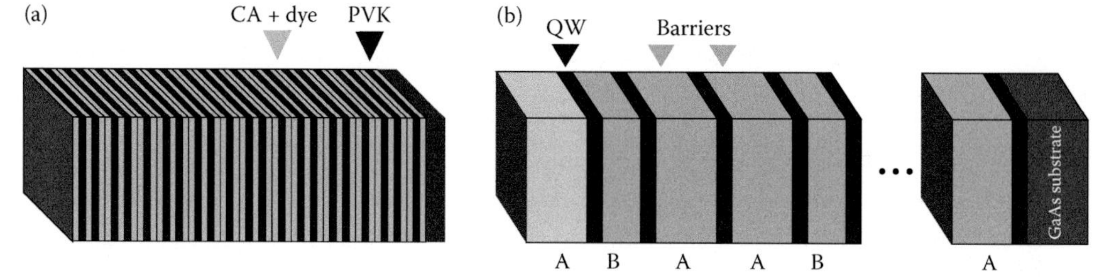

FIGURE 2.6.1
Examples of one-dimensional Fibonacci quasicrystalline structures. (a) The sample is composed of polymer layers made from cellulose acetate (CA) and poly vinyl carbazole (PVK) grown by spin-coating on the glass substrate. (b) Fibonacci structure containing GaAs quantum wells sandwiched between the long (A) and short (B) AlGaAs barrier layers. (Panel (b) adapted from J. Hendrickson et al., Opt. Express **16**, 15382, 2008; M. Werchner et al., Opt. Express **17**, 6813, 2009.)

To the best of our knowledge, deterministic aperiodic photonic structures do not exist in nature. All the available systems are artificially fabricated. Since the pioneering work by Hattori et al.,[16] where the first experiment on photonic quasicrystal based on the Fibonacci sequence had been reported, technology has substantially developed and many different structures are now realized experimentally. To illustrate, Figure 2.6.1a schematically shows a structure fabricated using two polymer materials, cellulose acetate and poly vinyl carbazole, stacked in the Fibonacci sequence.[17] The reflectivity of such a Fibonacci structure is discussed in Section 2.6.3. Another realization is presented in Figure 2.6.1b. The structure consists of thin GaAs QWs separated by AlGaAs barriers. The barriers are of two types: long (A) and short (B), with the interwell distances arranged into the Fibonacci sequence.[18–20] This is an example of the *resonant* photonic quasicrystal owing to a possibility of optical excitation of the quasi-2D excitons confined in the QWs. Its optical properties are considered in Section 2.6.3.2.

2.6.1.3 Structure Factor

We start by writing the wave equation for the electric field **E** of the electromagnetic wave

$$\Delta \mathbf{E}(\mathbf{r}) - \text{grad}\,\text{div}\,\mathbf{E}(\mathbf{r}) = -\left(\frac{\omega}{c}\right)^2 \mathbf{D}(\mathbf{r}), \tag{2.6.6}$$

where ω is the light frequency, c is the light velocity in vacuum and **D** is the displacement vector. The material relation between **D** and **E** is taken in the form $\mathbf{D}(\mathbf{r}) = \varepsilon(\mathbf{r})\mathbf{E}(\mathbf{r}) + 4\pi\mathbf{P}_{\text{exc}}(\mathbf{r})$. Here, $\varepsilon(\mathbf{r})$ is the local nonresonant permittivity and $\mathbf{P}_{\text{exc}}(\mathbf{r})$ is the exciton contribution to the dielectric polarization. For a system with the dielectric response modulated only along one direction z and for the light propagating in this direction, Equation 2.6.6 reduces to

$$\frac{d^2E(z)}{dz^2} = -\left(\frac{\omega}{c}\right)^2 [\varepsilon(z)E(z) + 4\pi P_{\text{exc}}(z)]. \tag{2.6.7}$$

In the following, we consider separately two particular cases of 1D photonic aperiodic systems, namely, (I) binary aperiodic optical superlattice, consisting of two types of uniform, isotropic and nonabsorbing layers $\mathcal{A}$ and $\mathcal{B}$ of the thicknesses a and b with the dielectric constants ε_A and ε_B, respectively, and arranged along the z direction according to a deterministic rule, and (II) multiple QW structure consisting of N identical QWs embedded in a matrix with the dielectric constant ε_b, with the interwell distances taking two values a and b. In the former case, the excitonic polarization is absent, $P_{\text{exc}}(z) \equiv 0$, and the dielectric function is given by $\varepsilon(z) = \varepsilon_A$ if $z \in \mathcal{A}$ and $\varepsilon(z) = \varepsilon_B$ if $z \in \mathcal{B}$. In the latter case, the function $\varepsilon(z)$ is a constant ε_b and, for thin enough QWs, the excitonic dielectric response is given by

$$4\pi P_{\text{exc}}(z)\left(\frac{\omega}{c}\right)^2 = \frac{2q\Gamma_0}{\omega_0 - \omega - i\Gamma} \sum_{j=1}^{N} \delta(z - z_j)E(z). \qquad (2.6.8)$$

Here $q = \omega\sqrt{\varepsilon_b}/c$; ω_0, Γ_0, and Γ are the 2D exciton resonance frequency and radiative and nonradiative damping rates in a single QW structure; and z_j is the center of the jth QW.

It follows from Equation 2.6.8 that the optical properties of the resonant quasicrystalline chain are described by its structure factor defined by

$$f(q) = \lim_{N\to\infty} f(q,N), \quad f(q,N) = \frac{1}{N}\sum_{j=1}^{N} e^{2iqz_j}. \qquad (2.6.9)$$

In the limit $N \to \infty$, the structure factor of a quasicrystal[21] consists of δ-peaks corresponding to the Bragg diffraction and characterized by two integer numbers h and h' (see, e.g., the studies by Janot[8] and Luck et al.[10]):

$$f(q) = \sum_{h,h'=-\infty}^{\infty} \delta_{2q,G_{hh'}} f_{hh'}, \qquad (2.6.10)$$

with the diffraction vectors

$$G_{hh'} = \frac{2\pi}{\bar{d}}\left(h + \frac{h'}{t}\right) \qquad (2.6.11)$$

filling the wavevector axis in a dense quasicontinuous way. We recall that in the periodic lattice ($\Delta = 0$) the structure factor has nonzero peaks at the single-integer diffraction vectors $G_h = 2\pi h/\bar{d}$ with $|f_h| = 1$, as shown in Figure 2.6.2, at integer values of the dimensionless diffraction vector. For arbitrary values of t, φ, and Δ, the structure-factor coefficients $f_{hh'}$ are given in the studies by Werchner et al.[19] and Aviram.[22] For the particular case of the canonic Fibonacci lattice, one has

$$f_{hh'} = \frac{\sin S_{hh'}}{S_{hh'}} e^{i\theta_{hh'}}, \quad S_{hh'} = \pi\frac{\tau(\tau h' - h)}{1+\tau^2}, \qquad (2.6.12)$$

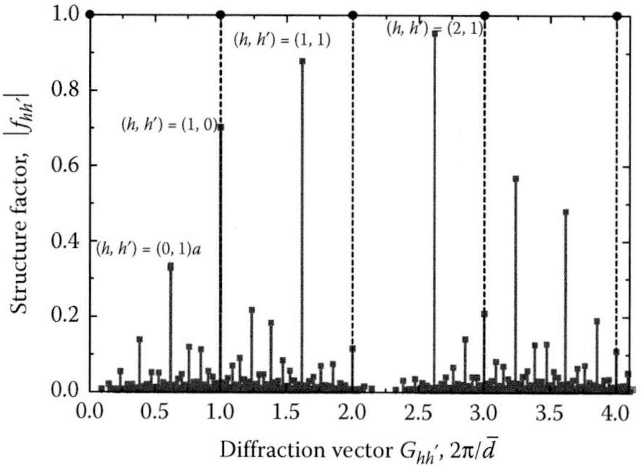

FIGURE 2.6.2
Coefficients of the structure factor as a function of the diffraction vector. The lengths of dashed and solid vertical lines indicate values of $|f_h|$ and $|f_{hh'}|$ for periodic and canonic Fibonacci chains, respectively.

$\theta_{hh'} = (1 - 2\tau^{-1})S_{hh'}$ and the mean period $\bar{d} = b(3 - \tau)$. The largest values of $|f_{hh'}| \approx 1$ correspond to h and h' equal to the subsequent Fibonacci numbers: $(h, h') = (F_j, F_{j-1})$.

For a nonresonant canonical Fibonacci binary chain, the role of structure factor is played by the Fourier transform ε_G of the dielectric constant defined by

$$\varepsilon(z) = \sum_G \varepsilon_G\, e^{iGz}, \quad \varepsilon_G = \frac{1}{L}\int_0^L dz\, \varepsilon(z)e^{-iGz}, \tag{2.6.13}$$

where L is the chain length. For the Fibonacci sequence, the straightforward derivation leads to the following results:

$$\varepsilon_{-G} = \varepsilon_A \frac{e^{iGa} - 1}{iG\bar{d}} f^{(A)}(G) + \varepsilon_B \frac{e^{iGb} - 1}{iG\bar{d}} f^{(B)}(G), \tag{2.6.14}$$

where, for $G \equiv G_{hh'}$,

$$\begin{aligned} f^{(A)}(G) &= \frac{\sin(S_{hh'}/\tau)}{S_{hh'}} \exp\left[i\frac{S_{hh'}}{\tau}\left(\frac{2}{\tau} - 1\right)\right], \\ f^{(B)}(G) &= \frac{\sin(S_{hh'}/\tau^2)}{S_{hh'}} \exp\left[i\frac{S_{hh'}}{\tau}\left(\frac{1}{\tau} - 2\right)\right]. \end{aligned} \tag{2.6.15}$$

In fact, these equations agree with the result obtained by Dharma-wardana et al.[23] for the Fourier transform of the local photoelastic coefficient in the Fibonacci superlattices. Note that the sum $f^{(A)}(G) + f^{(B)}(G)$ reduces to the structure-factor coefficient $f_{hh'}$ in Equation 2.6.12.

2.6.2 Two-Wave Approximation

In this section, we apply TWA to derive the light dispersion and the reflectivity spectra of the aperiodic multilayered structures. The electric field of the light wave propagating in the multilayered structure satisfies the following wave equation:

$$\left(-\frac{d^2}{dz^2} - q^2\right)E(z) = 2q\sum_G e^{iGz}P_G E(z), \tag{2.6.16}$$

where $E(z)$ is the electric field of the light wave at the frequency ω, G is the reciprocal vector G_h in the case of a periodic structure and the diffraction vector $G_{hh'}$ in the case of a photonic quasicrystal, and the wave vector q and coefficients P_G are defined by

$$q = \frac{\omega}{c}\sqrt{\bar{\varepsilon}}, \quad P_G = \frac{q}{2}\frac{\varepsilon_G}{\bar{\varepsilon}} \tag{2.6.17}$$

for the nonresonant Fibonacci lattice (case I) and by

$$q = \frac{\omega}{c}\sqrt{\varepsilon_b}, \quad P_G = \xi f^*_{hh'}, \quad P_{-G} = \xi f_{hh'}, \quad \xi = \frac{1}{d}\frac{\Gamma_0}{\omega_0 - \omega - i\Gamma} \tag{2.6.18}$$

for case II, where $f_{hh'}$ is given by Equation 2.6.12 and $\bar{\varepsilon}$ is the average dielectric constant

$$\bar{\varepsilon} = \frac{\varepsilon_A N_A a + \varepsilon_B N_B b}{N_A a + N_B b} = \frac{\varepsilon_A(\tau+1) + \varepsilon_B}{\tau + 2}. \tag{2.6.19}$$

Note that, in the absence of dissipation, $P_{-G} = P^*_G$. Moreover, in case I the coefficient P_0 reduces to zero. Equation 2.6.16 allows one to write the electromagnetic field in the structure as a superposition of the "Bloch-like" waves

$$E_K(z) = \sum_G e^{i(K-G)z}E_{K-G}. \tag{2.6.20}$$

The difference $K_1 - K_2$ for two different waves $E_{K_1}(z)$ and $E_{K_2}(z)$ in Equation 2.6.20 cannot be an integer number of any diffraction vector G. Moreover, for unambiguity the dispersion $\omega_{K'}$ with K' being the real part of K is defined in the extended-zone scheme where, for positive K', $\omega_{K'}$ is a monotonously increasing function of K'.

In TWA, only two space harmonics K and $K - G$ are taken into consideration in the superposition 2.6.16 and the electric field is approximated as

$$E(z) = E_K e^{iKz} + E_{K-G}e^{i(K-G)z}. \tag{2.6.21}$$

We analyze a narrow frequency region near the Bragg frequency ω_G satisfying the condition

$$q(\omega_G) = G/2 \tag{2.6.22}$$

and assume $K \approx G - K \approx q(\omega_G)$. Then, the exact wave equation 2.6.16 is reduced to

$$\begin{aligned} (q - K + P_0)E_K + P_G E_{K-G} &= 0, \\ P_{-G}E_K + (q + K - G + P_0)E_{K-G} &= 0. \end{aligned} \tag{2.6.23}$$

The two eigenvalues K satisfying Equation 2.6.23 are given by

$$K = \frac{G}{2} \pm Q, \quad Q = \sqrt{\left(q - \frac{G}{2} + P_0\right)^2 - P_G P_{-G}}. \tag{2.6.24}$$

First, we take the coefficients P_G to be frequency-independent and assume $P_0 = 0$. Then, the energy spectrum has a band gap around ω_G with the edges ω_+ and ω_- spaced by $\omega^+ - \omega^- = 2c\,|\,P_G\,|/\sqrt{\bar{\varepsilon}}$ and centered at frequency $(\omega_+ + \omega_-)/2 = (c/\sqrt{\bar{\varepsilon}})(G/2)$.

If the Fourier components have a pole at frequency ω_0 and, in addition, the resonant Bragg condition

$$\sqrt{\varepsilon_b}\,(\omega_0/c) = G/2 \tag{2.6.25}$$

is satisfied, the structure is characterized by two symmetrical exciton–polariton band gaps. Particularly, in a resonant Bragg QW quasicrystal with the coefficients P_G given by Equation 2.6.18, the edges $\omega^{\pm}_{\text{out}}$ and $\omega^{\pm}_{\text{in}}$ of these band gaps are obtained from Equation 2.6.23 by setting $K = G/2$.

The result reads (for $\Gamma \to +0$)

$$\omega^{\pm}_{\text{out}} = \omega_0 \pm \Delta_0\sqrt{\frac{1 + |f_{hh'}|}{2(h + h'/\tau)}}, \quad \omega^{\pm}_{\text{in}} = \omega_0 \pm \Delta_0\sqrt{\frac{1 - |f_{hh'}|}{2(h + h'/\tau)}}, \tag{2.6.26}$$

where $\Delta_0 = \sqrt{2\omega_0\Gamma_0/\pi}$. In the resonant periodic Bragg structures, the values of P_0, P_G, and P_{-G} coincide with ξ and the system has a single gap of width $2\Delta_0$.[24] In a quasicrystal, $|P_G|$ with $G \neq 0$ is smaller than P_0 and the system has two gaps of equal widths separated by an allowed band of the width $\omega^{+}_{\text{in}} - \omega^{-}_{\text{in}} \propto \sqrt{1 - |f_{hh'}|}$.

2.6.3 Reflection and Transmission from Fibonacci Structures

We consider the light reflection from a finite multilayered structure formed by layers $\mathcal{A}$ and $\mathcal{B}$ sandwiched between the semi-infinite layers of the material A (for QW structures) or the material with the dielectric constant $\bar{\varepsilon}$ (for nonresonant structures).

For an aperiodic structure of thickness L, the electric field is described as follows

$$E(z) = \begin{cases} E_0 e^{iqz} + E_r e^{-iqz} & (z < 0), \\ \sum_{\pm} E_{K_\pm} e^{\pm iQz} (e^{iGz/2} + \zeta_\pm e^{-iGz/2}) & (0 < z < L), \\ E_t e^{iq(z-L)} & (L < z). \end{cases} \tag{2.6.27}$$

Here E_0, E_r, and E_t are the amplitudes of the incident, reflected, and transmitted wave, $E_{K_\pm}$ is the amplitude E_K in Equation 2.6.21 for the wavevector $K_\pm = (G/2) \pm Q$, and

$$\zeta_\pm = \frac{E_{K_\pm - G}}{E_{K_\pm}} = -\frac{P_{-G}}{q + K_\pm - G + P_0}.$$

The boundary conditions of electric-field continuity at $z = 0$ and $z = L$ relate the amplitudes E_0, E_r, $E_{K_\pm}$, and E_t. If number of wells N in the Fibonacci QW structure coincides with $F_j + 1$, where F_j is one of the Fibonacci numbers, then the phase factor $\exp(iG_{hh'}L)$ and the length L can be replaced, respectively, by unity and the product $N\bar{d}$. The straightforward derivation results in comparatively simple expressions for the reflection and transmission coefficients from a Bragg Fibonacci QW structure,

$$\begin{aligned} r_N &\equiv \frac{E_r}{E_0} = \frac{\xi f_{hh'}}{(G/2) - q - \xi - iQ\cot(QN\bar{d})}, \\ t_N &\equiv \frac{E_t}{E_0} = -\frac{iQ}{\xi f_{hh'} \sin(QN\bar{d})} r_N. \end{aligned} \tag{2.6.28}$$

For nonresonant Fibonacci chain the Bragg condition 2.6.22 reduces to

$$\frac{\omega_G \sqrt{\bar{\varepsilon}}}{c} \bar{d} = \pi\left(h + \frac{h'}{\tau}\right), \tag{2.6.29}$$

where $\bar{\varepsilon}$ is the average dielectric constant (Equation 2.6.19). The solution results in the following expressions for the reflection and transmission coefficients, respectively:

$$r_N = \frac{P_G}{(G/2) - q - iQ\cot(QN\bar{d})}, \quad t_N = -\frac{iQ}{P_G \sin(QN\bar{d})} r_N. \tag{2.6.30}$$

2.6.3.1 Nonresonant Fibonacci Binary Chains

Let us use TWA to analyze light reflection from the Fibonacci binary chains. Figure 2.6.3 shows the reflectivity and transmittivity of the structure containing a finite number, $F_{11} = 89$, of layers $\mathcal{A}$ and $\mathcal{B}$ arranged in the Fibonacci sequence. The parameters of the structure are as follows: $n_A = 1$, $n_B = \tau$. The positions of maxima in the reflection spectrum and minima in the transmission spectrum correlate with the gaps obtained in the 12-wave

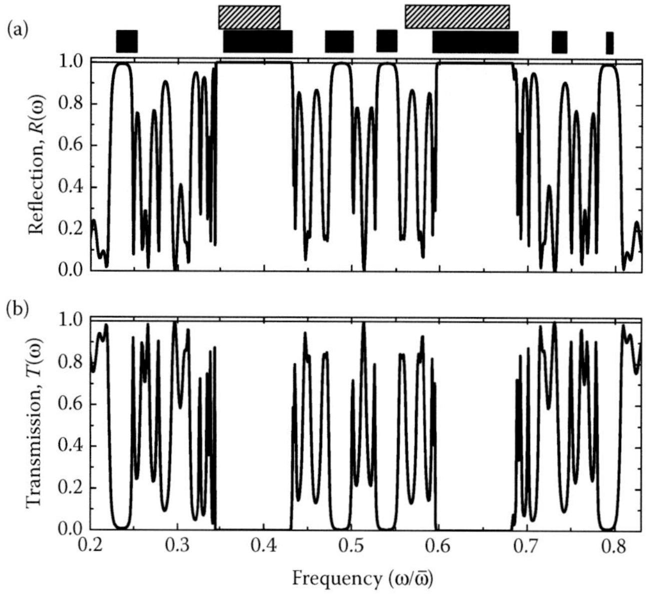

FIGURE 2.6.3
The reflection (a) and transmission (b) spectra calculated for the finite Fibonacci sequence. The shaded and filled rectangles indicate the band gaps obtained in the two-wave approximation and the 12-wave calculation, respectively. The light frequency ω is scaled to the frequency $\bar{\omega} = \pi c/(\bar{d}\sqrt{\bar{\varepsilon}})$. The parameters of calculation are indicated in the text.

calculation and shown by filled rectangles. The shaded rectangles correspond to the largest gaps found in analytical TWA calculations taking into account plane waves K, $K - G_{1,-1}$ or K, $K - G_{1,0}$.

2.6.3.2 Fibonacci QW Structures

Periodic structures in which the dielectric response of at least one composite material as a function of frequency ω has a pole at a resonance frequency are grouped into a specific class of resonant photonic crystals whose normal waves are polaritons. Similarly, one can define resonant photonic quasicrystals and other aperiodic deterministic sequences. This subsection concerns their optical spectroscopy near the exciton resonant frequency.

If the light frequency lies in the vicinity of the exciton resonant frequency then, in general, a photon and an exciton mix and form a combined quasiexcitation called the exciton–polariton. De Medeiros et al.[25] were the first to investigate theoretically the propagation of exciton–polaritons in quasiperiodic binary superlattices of Fibonacci type. Layer A was represented by a spatially dispersive medium modeled by a semiconductor from the nitride family (GaN) which alternated with a typical dielectric medium B (sapphire). The layer thicknesses were $a = b = 500$ Å, which means that the exciton Bohr radius was small as compared with a. Therefore, the exciton can be considered as a bulk-like particle in the same way as it is treated in thin films[26] and optical superlattices.[27] The distribution of the allowed and forbidden energy bandwidths of the exciton–polaritons was calculated by De Medeiros et al.[25] for the periodic approximants as a function of the Fibonacci generation index m. Scaling and localization of exciton–polariton modes propagating in these quasiperiodic structures are also described.

Poddubny et al.[14] proposed a new kind of resonant photonic quasicrystal, namely, a multiple QW structure with the exciton resonant frequency tuned to the resonant Bragg condition 2.6.25. The allowed diffraction vectors $G_{hh'}$ form a dense pseudocontinuous set. We recall that the largest values of $|f_{hh'}|$ are reached for the diffraction vectors $G_{hh'}$ with pairs h, h' coinciding with two successive Fibonacci numbers F_m, F_{m-1}. Thus, for $(h, h') = (F_m, F_{m-1}) = (1,0), (1,1), (2,1), (3,2)$, and $(5,3)$ corresponding to $m = 1 \dots 5$, the modulus of $f_{hh'}$ equals to ≈0.70, 0.88, 0.95, 0.98, and 0.99, respectively. It follows then that if the exciton resonance frequency satisfies the condition

$$\frac{\omega_0 n_b}{c}\bar{d} = \pi\left(F_m + \frac{F_{m-1}}{\tau}\right), \quad m = 1,2\ldots \tag{2.6.31}$$

one can expect the structure to show super-radiative behavior, similar to the super-radiance of a periodic multiple QW structure with the period d tuned to the resonance condition:[13]

$$\frac{\omega_0 n_b}{c} d = \pi. \tag{2.6.32}$$

Figure 2.6.4a presents reflection spectra calculated for four structures containing $N = 54$ QWs. The exciton parameters used are as follows: $\hbar\omega_0 = 1.533$ eV, $\hbar\Gamma_0 = 50$ µeV, $\hbar\Gamma = 100$ µeV, and $n_b = 3.55$. Curve 1 is calculated for the resonant canonic Fibonacci QW structure satisfying the exact Bragg condition 2.6.31 with $m = 2$. Curves 2 and 3 correspond

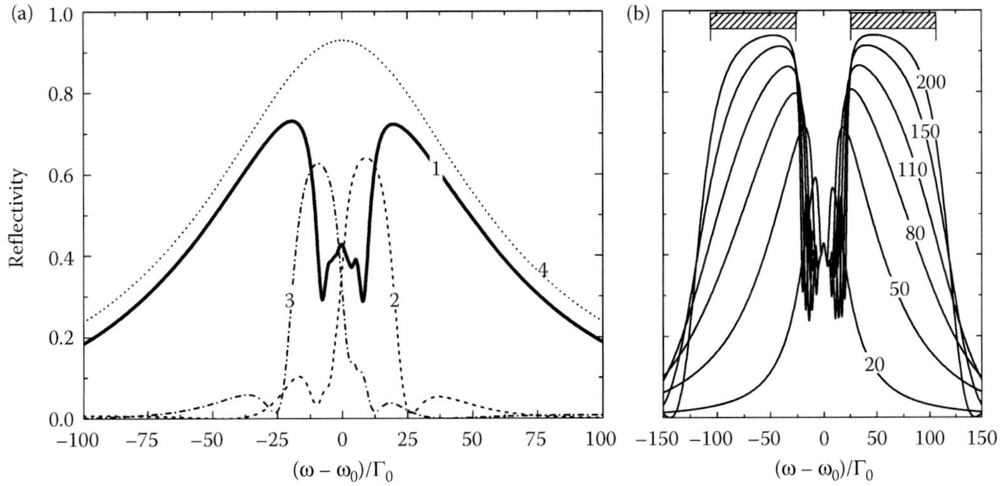

FIGURE 2.6.4
(a) Reflection spectra calculated for three Fibonacci structures satisfying the resonant Bragg condition 2.6.31 for $m = 2$ (curve 1) and detuned by ±2% from this condition (curves 2 and 3) in comparison with the reflection spectrum from the periodic resonant Bragg quantum-well structure (curve 4). The values of parameters are indicated in the text. (b) Reflectivity from Bragg Fibonacci structures satisfying the condition 2.6.31 with $m = 2$. Six curves are calculated for the structures with $N = 20, 50, 80, 110, 150$, and 200. The number of wells is indicated near each corresponding curve. Shaded areas indicate two exciton–polariton band gaps given by Equation 2.6.26.

to the Fibonacci structures also with $a/b = \tau$, but the barrier thicknesses slightly detuned so that the average period differs from the exactly tuned period by a factor of 1.02 for curve 2 and 0.98 for curve 3. Curve 4 describes the reflection from the periodic Bragg structure with the same exciton parameters and period $d = \pi/q(\omega_0)$, satisfying Equation 2.6.32. By comparing curves 1 and 4 we conclude that the reflection spectra from the resonant Fibonacci and periodic structures tuned to Bragg conditions 2.6.31 and 2.6.32 are close to each other outside the frequency region around ω_0. Moreover, it follows from curves 2 and 3 that a slight deviation from condition 2.6.31 results in a radical decrease in the effective spectral half-width. Thus, the sensitivity to the resonance condition, the characteristic of periodic Bragg QW systems, holds also for aperiodic QW systems such as the Fibonacci structures. The remarkable structured dip in the middle of spectrum 1 is the only qualitative difference from the periodic structures, this dip is a consequence of an allowed exciton–polariton band lying between the frequencies $\omega_{in}^{\pm}$ defined in Equation 2.6.26. We recall that for a periodic resonant Bragg structure an inner allowed band is absent, $\omega_{in}^{+} = \omega_{in}^{-}$.

Evolution of the reflection spectra with the increasing QW number is illustrated in Figure 2.6.4b. The spectral envelope shows a behavior similar to that of the conventional Bragg QW structure. Indeed, for small N the envelope is a Lorentzian with the half-width increasing as a linear function of N. This is a straightforward manifestation of superradiant regime, which, as one can see here, does not necessarily require periodicity even if the interwell distances are comparable to the light wavelength. The saturation of the spectral half-width (photonic quasicrystal regime) begins at large N of the order of $\sqrt{\omega_0/\Gamma_0}$, in a similar way as for the periodic Bragg structures. The shape of the spectra for large N confirms existence of two wide symmetrical stop bands in the energy spectrum of the structure with an allowed band between them, as described by Equation 2.6.26. Of course, the terms "allowed" and "stop" bands are applicable to an aperiodic structure in a reasonable sense assuming TWA.

In the frequency region around ω_0, the reflection spectra from the nonperiodic structures show wide dips where the reflection coefficient oscillates with the period of oscillations decreasing as ω approaches ω_0. The spectral dip naturally appears for a multilayered deterministic system tuned to a Bragg diffraction vector with the structure-factor coefficient f_G smaller than unity, and it widens as the difference $1 - |f_G|$ increases.

In realistic semiconductor QWs, the nonradiative decay rate Γ is larger than or comparable to Γ_0, and the majority of spectral fine-structure features are smoothed.[18] For small values of Γ lying beyond experimentally available values, an intricate fine structure of optical spectra is developed in this region. The spectral properties in the frequency range $|\omega - \omega_0| \sim \Gamma_0$ for $\Gamma \ll \Gamma_0$ are discussed in Section 2.6.4 in more detail.

Figure 2.6.5 compares the measured reflectivity for two graded samples with $N = 21$: (a) Fibonacci QW sequence and (b) periodically spaced multiple QWs.[18] A reflectivity maximum with broad linewidth occurs at the Bragg condition in both samples for both the heavy-hole (HH) and the light-hole (LH) exciton resonances. Notably, however, the spectral peak is smooth for the crystalline structure, whereas it reveals a pronounced fine structure (dips, see arrows in Figure 2.6.5) for the Fibonacci quasicrystals. This fine structure is visible for both the HH and the LH exciton resonances, strongly suggesting that this signature is a fingerprint of the quasicrystal (see Figure 2.6.4).

Figure 2.6.5c presents the calculated and measured reflection for the 54 Fibonacci QWs. Owing to Fibonacci spacing, the spectrum displays multiple features that are well reproduced by the theory.[19] To take into account explicitly the disorder in the studied samples the nonradiative damping rate is assumed to be frequency-dependent as $\Gamma(\omega) = \Gamma\{1 + \exp$

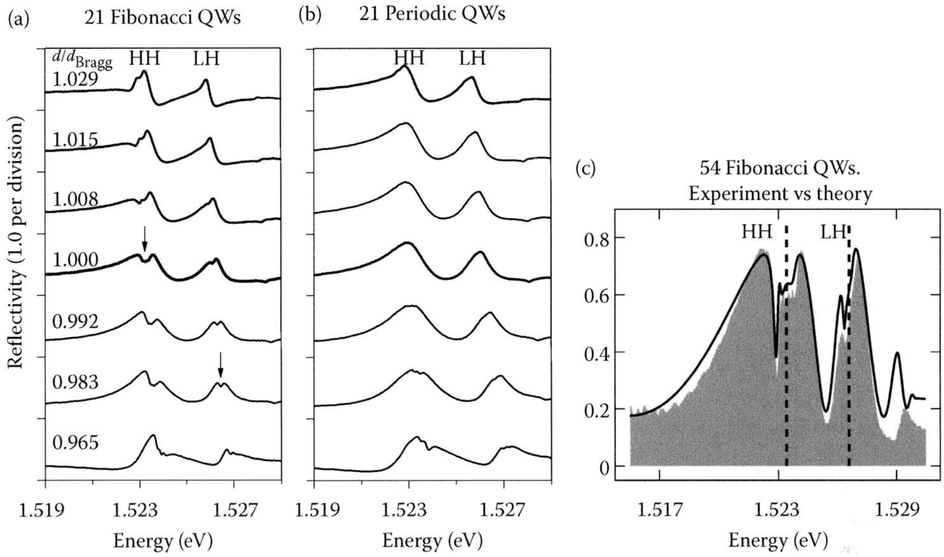

FIGURE 2.6.5
(a) and (b) Comparison of the thickness dependence of the measured reflectivity. The spacings of 21 quantum wells (QWs) either (a) form a Fibonacci sequence with optical thicknesses of $b = 0.36\lambda(\omega_0)$ and $a = 0.59\lambda(\omega_0)$ or (b) are equal to $\lambda(\omega_0)/2$. A reflectivity dip is present at the Bragg resonance only in (a). Curves are labeled by the ratio d/d_{Bragg}, where $d_{Bragg} = \lambda(\omega_0)/2$, $\lambda(\omega_0)$ is the wavelength of the light inside the sample at the exciton resonance frequency, and d is the average interwell distance. Each curve of larger d/d_{Bragg} is shifted up by one unit of reflectivity from the preceding one. (c) Fit (solid curve) to measured reflection spectrum (shaded area) for 54 Fibonacci-spaced QWs using a frequency-dependent dephasing $\Gamma(\omega)$. Labels HH and LH indicate the spectral positions of the heavy-hole and light-hole exciton resonances, respectively. (Panel (a) adapted from J. Hendrickson et al., Opt. Express **16**, 15382, 2008; panel (b) adapted from M. Werchner et al., Opt. Express **17**, 6813, 2009.)

$[-\hbar(\omega - \omega_0 + \Omega_{cut})/E_0]\}^{-1}$. For the HH resonance, the best fit is obtained for the following set of constants: 0.16 meV for $\hbar\Gamma$, 0.25 meV for the cut-off energy $\hbar\Omega_{cut}$, and 0.1 meV for E_0. The refractive indices of the layers are fitted using a single small set of parameters such that the calculations reproduce both the single-QW and the 54-QW measured reflection spectra.

2.6.4 Scaling Features and Localization of Excitonic Polaritons

In this section, we focus on the features of Fibonacci QW structures lying beyond TWA. The resonant structures under consideration have two distinct differences from the nonresonant systems widely studied before,[2,28,29] which considerably affect their optical properties: (i) the resonant spectral behavior of the dielectric response of the QW excitons (Equation 2.6.8), and (ii) the possibility of light absorption, characterized by the exciton nonradiative decay rate Γ. In our analysis, we will first assume $\Gamma = 0$ and study the density of states (DOS) of the periodic approximants of the Fibonacci chains. We will demonstrate how this system can be compared to the nonresonant systems known in the literature.

Then, we will take into account nonradiative decay and analyze the interplay between the absorption and localization.

We concentrate on the narrow frequency region $|\omega - \omega_0| \sim \Gamma_0$ and neglect the nonradiative decay of excitonic polaritons, assuming $\Gamma = 0$. In this regime, the optical spectra and the exciton–polariton dispersion demonstrate scaling invariance and self-similarity.[30] To reveal the behavior of exciton polaritons it is instructive to calculate the polariton dispersion for the approximants[7] of the aperiodic chains containing the periodically repeating sequences $\mathcal{F}_m$. Optical properties of the elementary supercell $\mathcal{F}_m$ of the structure can be characterized by transfer matrix $T_m(\omega)$. The transfer matrix depends on the light frequency ω, the geometrical parameters of the system, and the excitonic parameters. The details of transfer matrix calculation are given in the literature.[5,24] For the periodic approximant one can introduce Bloch wavevector $K_m(\omega)$ defined by $K_m L_m = \arccos[\mathrm{Tr}(T_m)/2]$, where L_m is the approximant period. The integrated DOS is given by[5]

$$\rho(\omega) = \frac{\mathrm{Re}\, K_m(\omega) L_m}{2\pi}, \tag{2.6.33}$$

where the values of K_m are defined in an extended Brillouin zone scheme. The DOS is constant inside the band gaps of the approximant and monotonously increases in the allowed bands. Owing to the resonant character of the dielectric response of the exciton (Equation 2.6.8), the sequence of the allowed bands and forbidden gaps is very intricate. It can be more efficiently visualized as a function of the auxiliary variable

$$S(\omega) = \frac{\Gamma_0}{\omega - \omega_0} \tag{2.6.34}$$

proportional to the QW susceptibility. After transformation 2.6.34 is made, the resonant effects do not complicate the analysis of the band structure. Thus, instead of the DOS $\rho(\omega)$ we introduce below the effective DOS

$$\tilde{\rho}(S) = -\frac{L_m\, \mathrm{Re}\, K_m[\omega(S)]}{\pi} + C \tag{2.6.35}$$

depending on $S(\omega)$. Here, the minus sign is introduced to make $\rho(S)$ an increasing function of S, and for the sake of convenience a constant C is added and the denominator 2π is replaced by π.

Solid and dashed curves in Figure 2.6.6 present the results of calculation for the approximants $\mathcal{F}_{13}$ and $\mathcal{F}_7$ with 233 and 13 wells. The dotted curve shows the imaginary part of the wavevector in the approximant with 233 wells. The structure was tuned to Bragg condition with $(h, h') = (1, 0)$. The constant C in Equation 2.6.35 was chosen to make $\tilde{\rho}(0) = 0$. The overall shape of the curves $\tilde{\rho}(S)$ is generally the same as for the nonresonant dielectric and phononic Fibonacci structures:[2,28] the DOS resembles the devil's staircase. We note, that one can obtain similar DOS for negative S, that is, $\omega < \omega_0$. The curves clearly demonstrate the presence of the band gaps, where the DOS is constant and $\mathrm{Im}\, K > 0$. The leftmost gap at $0.008 \le S \le 0.02$ (solid arrow in Figure 2.6.6) is just the TWA band gap $(\omega_{\mathrm{in}}^{+}, \omega_{\mathrm{out}}^{+})$ (see Equation 2.6.26). All the other band gaps are absent in TWA and cannot be described by including the single peak in the structure factor. The rightmost gap at $S \ge 1.07$ corresponds

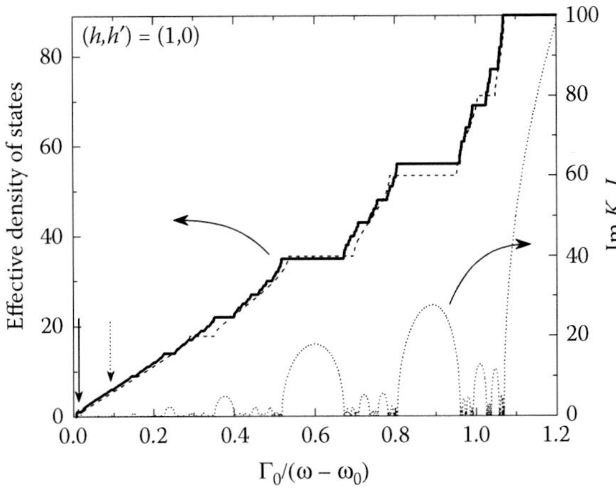

FIGURE 2.6.6
Integrated density of states of the periodic approximants of Fibonacci quantum-well structures as a function of the variable $S = \Gamma_0/(\omega - \omega_0)$. Solid and dashed curves correspond to the approximants $\mathcal{F}_{13}$ and $\mathcal{F}_7$ with 233 and 13 wells, respectively. The dashed curve has been multiplied by 89/5 for better presentation. The solid vertical arrow indicates the band gap at $(\omega_{in}^+, \omega_{out}^+)$ given in two-wave approximation by Equation 2.6.26. The dotted curve presents the imaginary part Im $K_{13}L_{13}$ of the Bloch wavevector for the approximant $\mathcal{F}_{13}$. The dotted vertical arrow indicates the narrow band gap centered at $S^* \approx 0.096$ and corresponding to $\omega^* \approx \omega_0 + 10.42\Gamma_0$. Other parameters are given in the text.

to the frequency region $0 \le (\omega - \omega_0)/\Gamma_0 \le 0.93$. It is the narrow central band gap around ω_0 found in the study by Poddubny et al.[30] In the intermediate region $0.02 \le S \le 1.07$ [corresponding to $0.93 \le (\omega - \omega_0)/\Gamma_0 \le 54$], the sequence of the allowed and forbidden gaps is very complex. It behaves with the order of the approximant m similarly to the Cantor set. In particular, the number of band gaps increases with m, although the spectral position of each given gap saturates with m. One can see from Figure 2.6.6 that the largest band gaps for approximant $\mathcal{F}_{13}$ (solid curve), for example, the central one at $0.52 \le S \le 0.67$, are manifested already for approximant $\mathcal{F}_7$ (dashed curve). The total width of the allowed bands decreases with m. In the limit $m \to \infty$ the Lebesque measure of the set of allowed bands is zero.[31] It means that at almost all frequencies the structure transmission coefficient decays exponentially with the structure length L as follows: $T(L) \propto \exp(-L/L_{dec})$. The length L_{dec} has the meaning of localization length in the quasiperiodic structure. It is related to the imaginary part of the wavevector K_m by $L_{dec} = 2(\text{Im } K_m)^{-1}$ and saturates as a function of m. One can see from Figure 2.6.6 that Im $K_m(\omega)$ strongly oscillates on frequency. Peak values of this function, corresponding to minima of L_{dec}, increase when ω approaches ω_0.

Let us now analyze the impact of the excitonic absorption, that is, assume $\Gamma \neq 0$. Naturally, the damping of the excitonic resonance smears the spectral features in Figure 2.6.6 which are narrower than Γ. The decay length L_{dec} in turn decreases with Γ owing to a possibility of absorption. The dependence of L_{dec} on Γ is presented by a solid line in Figure 2.6.7. The calculation was done for $(h, h') = (1, 0)$ at frequency $\omega^* = \omega_0 + 10.42\Gamma_0$, indicated by a dotted vertical arrow in Figure 2.6.6. It is instructive to compare an exact numerical result for L_{dec} with the absorption length $L_{abs} = 2(\text{Im } K)^{-1}$ where K is given by TWA (see Equation 2.6.24). The TWA result is shown by the dotted line in Figure 2.6.7. At $\Gamma = 0$, TWA is invalid: the decay length diverges at $\Gamma \to 0$ as, in TWA, frequency ω^* lies in the allowed band. Thus, in TWA the decay length is finite only because of the absorption. If the absorption is high, the TWA becomes applicable, so difference between the two curves in Figure 2.6.7 tends to

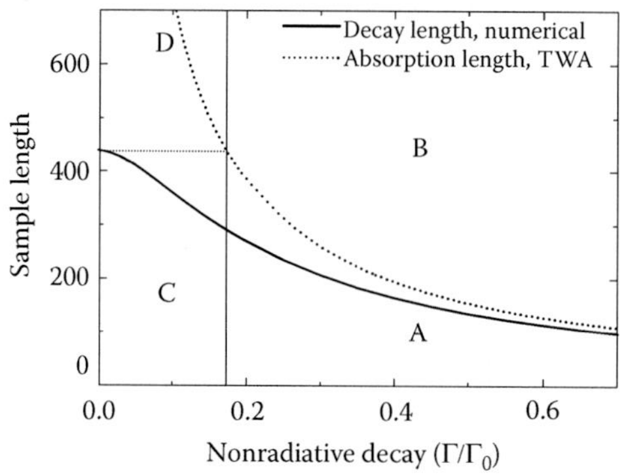

FIGURE 2.6.7
Decay length (solid curve) and two-wave approximation absorption length (dotted curve) as functions of the nonradiative damping rate Γ. Letters A–D indicate the qualitatively different regions of values of the sample length L and the damping Γ and are explained in the text.

zero. Consequently, at any given value of Γ the spectral properties of the Fibonacci QW structure are determined by the relation between the three characteristic lengths: (i) sample length L, (ii) decay length L_{dec} that includes both effects of localization and absorption, and (iii) the TWA absorption length L_{abs}. As a result, Figure 2.6.7 can be treated as the phase diagram where letters A, B, C, and D mark qualitatively different regions of parameters L and Γ, explained below. In region A, TWA is applicable as $L_{abs} \sim L_{dec}$ and the sample is thin as compared to the absorption length, $L < L_{abs}$. The Fabry–Pérot interference of excitonic polaritons is possible. The reflection coefficient is sensitive to the structure length and well approximated by Equation 2.6.28. In region B, TWA is still applicable but the reflection coefficient is independent of the structure length L because L is larger than the absorption length L_{abs}. In region C, TWA is invalid and the effects of localization are important because $L_{dec} < L_{abs}$. As the sample length L is smaller than the localization length the reflection spectrum is strongly sensitive to L. Scaling and self-similarity effects are also demonstrated in this region.[2,30] In region D, TWA is also invalid and the reflection spectrum is saturated as a function of L because $L > L_{dec}$.

2.6.5 Variety of Aperiodic Long-Range-Order Photonic Structures

2.6.5.1 Non-Fibonacci Aperiodic Sequences

The generalized two-component Fibonacci lattices can be generated from seed $\mathcal{A}$ by the substitution rule $\sigma_n(\mathcal{A}) \to \mathcal{A}^n\mathcal{B}$, $\sigma_n(\mathcal{B}) \to \mathcal{A}$ $(n \geq 1)$ with $\alpha = n$, $\beta = \gamma = 1$, and $\delta = 0$ (see Equation 2.6.4). As $\omega = \beta\gamma - \alpha\delta = 1$ they are quasicrystals. The structure with $n = 2$ is called the silver mean lattice.[32] Another generalization is the substitution rule $\mathcal{A} \to \mathcal{B}^{n-1}\mathcal{A}\mathcal{B}$, $\mathcal{B} \to \mathcal{B}^{n-1}\mathcal{A}$. It is called the Fibonacci class sequence of the nth order and labeled as FC(n).[33] As for this sequence a value of ω equals to unity it also represents a quasicrystal.

The substitution rule $\sigma(\mathcal{A}) = \mathcal{A}\mathcal{B}$, $\sigma(\mathcal{B}) = \mathcal{A}^2$ generates the lattice known as the period-doubling sequence.[4] For the generalized substitutional rule $\mathcal{A} \to \mathcal{A}^n\mathcal{B}^m$, $\mathcal{B} \to \mathcal{A}$ with $m > 1$, the corresponding aperiodic lattice[34] does not belong as well to quasicrystals because in this case a value of $\omega = m$ differs from ±1. Another example of nonquasicrystalline aperiodic deterministic chain is the Thue–Morse sequence introduced at the beginning of the twentieth century.[35,36] It has recently attracted much attention as a result of its rich and complex spectral properties revealed in electronic,[37] phononic,[38] and photonic systems.[39] This lattice is based on the substitution rule $\sigma(\mathcal{A}) = \mathcal{AB}$, $\sigma(\mathcal{B}) = \mathcal{BA}$, starting from the initial chain $\mathcal{TM}_0 = \mathcal{A}$, and characterized by the parameters $\alpha = \beta = \gamma = \delta = 1$ and $\omega = 0 \neq \pm 1$. Three Thue–Morse sequences with generation numbers $m = 1 \ldots 3$ are given below: $\mathcal{TM}_1 = \mathcal{AB}$, $\mathcal{TM}_2 = \mathcal{ABBA}$, $\mathcal{TM}_3 = \mathcal{ABBABAAB}$. The chain $\mathcal{TM}_m$ contains $N = 2^m$ layers and has the length $L_m = 2^m \bar{d}(m \geq 1)$, where $\bar{d} = (a + b)/2$.

2.6.5.2 Two-Dimensional Structures

The Penrose tiling, illustrated by Figure 2.6.8a, is one of the most famous and charming 2D quasicrystalline lattices.[1,40–42] Indeed, the pioneering paper by Sir Roger Penrose[43] is even

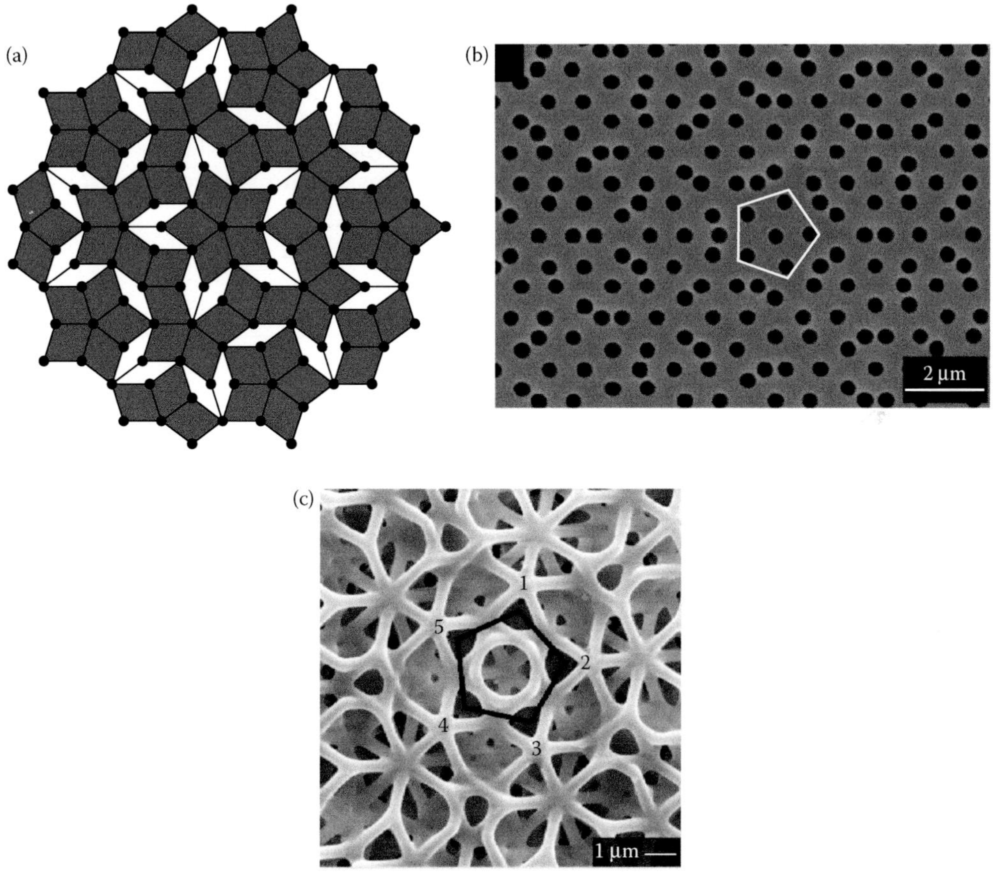

FIGURE 2.6.8
(**See color insert.**) Examples of two-dimensional (2D) and three-dimensional (3D) quasicrystalline structures. (a) Schematic illustration of the 2D Penrose tiling. (b) Scanning electron microscope image of the 2D Penrose quasicrystal. (c) Electron micrographs of silicon 3D icosahedral quasicrystal structure. (Panel (b) from D. Shir et al., Nano Lett. **8**, 2236, 2008. With permission; panel (c) from A. Ledermann, G. von Freymann, and M. Wegener, Phys. Unserer Zeit **38**, 300, 2007. With permission.)

titled "The role of aesthetics in pure and applied mathematical research." The common feature of all different kinds of Penrose tilings termed as *decagonal phase*[44] is a 10-fold orientational symmetry of the diffraction images. In fact, the Penrose tiling can be thought of as a generalization of the Fibonacci lattice to the 2D case.

The Penrose photonic quasicrystals were first fabricated by electron beam lithography.[42] Another approach is to use photolithography[40,45] (see Figure 2.6.8b). One can also utilize the optical interference holographic lithography technique.[46] This approach can be straightforwardly extended to produce structures with arbitrarily rotational symmetry. The allowance for high rotational symmetry leads to the large variety of 2D and 3D quasicrystals and, favoring the full photonic band gap, presents their advantage over traditional photonic crystals. Resonant light reflection from the 2D Penrose grating composed of quantum dots has been theoretically analyzed by Poddubny.[47]

2.6.5.3 Three-Dimensional Structures

Three-dimensional icosahedral structures with the quasiperiodic order arranged in all three spatial dimensions have been fabricated only recently by Man et al.[48] with a stereolithography machine (in microwave spectral range), by Ledermann et al.[49] using the direct laser writing technique (see Figure 2.6.8b and c), and by Xu et al.[50] using the optical interference holography technique. Icosahedral quasicrystals present a generalization of Penrose tiling to the 3D space. Moreover, 3D Penrose tiling with icosahedral symmetry can be obtained by taking the 6D lattice vectors that lie inside a unit triacontahedron in the pseudospace, and finding their partial projections on the physical space.

2.6.6 Summary

The discovery of quasicrystals and other deterministic aperiodic structures have initiated new fields of research in solid-state electronics, phononics, and photonics. In this chapter, we have presented recent achievements in fabrication and analysis of photonic quasicrystals. The studies of quasicrystalline structures have widened the existing concepts of wave diffraction. Emphasis has been placed on similarities and differences between optical properties of this third form of solid matter and those of conventional crystals and amorphous materials. Similarly to periodic systems, TWA is often efficient to describe, within narrow frequency intervals, light propagation in quasicrystalline medium. Moreover, sometimes it is even instructive to introduce effective allowed bands and forbidden gaps (or stop-bands) for light waves propagating in such medium. An important difference with periodic systems with a simple (undecorated) unit cell is that, in an aperiodic system, the coefficients of the Bragg structure factor differ from unity, which has an influence on the gap width and, in the Fibonacci QW structures, leads to formation of an allowed band inside the forbidden gap. The resemblance to disordered materials reveals properties of photonic crystalline approximants of quasicrystals. By increasing the thickness of the approximant supercell, the optical spectra provide evidence for the localization of light waves. Important specific features of the approximants are a scaling and self-similarity of optical spectra that are completely absent in crystalline and amorphous materials.

References

1. D. Levine and P. J. Steinhardt, Phys. Rev. Lett. **53**, 2477 (1984).
2. M. Kohmoto, B. Sutherland, and K. Iguchi, Phys. Rev. Lett. **58**, 2436 (1987).
3. E. L. Albuquerque and M. G. Cottam, Phys. Rep. **376**, 225 (2003).
4. W. Steurer and D. Sutter-Widmer, J. Phys. D **40**, R229 (2007).
5. A. N. Poddubny and E. L. Ivchenko, Physica E **42**, 1871 (2010).
6. M. Y. Azbel, Sov. Phys. JETP **19**, 634 (1964).
7. M. Y. Azbel, Phys. Rev. Lett. **43**, 1954 (1979).
8. C. Janot, *Quasicrystals. A Primer* (Clarendon Press, Oxford, UK, 1994).
9. Z. Lin, M. Goda, and H. Kubo, J. Phys. A **28**, 853 (1995).
10. J. M. Luck, C. Godreche, A. Janner, and T. Janssen, J. Phys. A **26**, 1951 (1993).
11. X. Fu, Y. Liu, P. Zhou, and W. Sritrakool, Phys. Rev. B **55**, 2882 (1997).
12. M. Kolář, Phys. Rev. B **47**, 5489 (1993).
13. E. L. Ivchenko, A. I. Nesvizhskii, and S. Jorda, Phys. Solid State **36**, 1156 (1994).
14. A. N. Poddubny, L. Pilozzi, M. M. Voronov, and E. L. Ivchenko, Phys. Rev. B **77**, 113306 (2008).
15. M. C. Valsakumar and V. Kumar, Pramana **26**, 215 (1986); Z. Lin, H. Kubo, and M. Goda, Z. Phys. B **98**, 111 (1995).
16. T. Hattori, N. Tsurumachi, S. Kawato, and H. Nakatsuka, Phys. Rev. B **50**, 4220 (1994).
17. V. Passias, N. V. Valappil, Z. Shi, L. Deych, A. A. Lisyansky, and V. M. Menon, Opt. Express **17**, 6636 (2009).
18. J. Hendrickson, B. C. Richards, J. Sweet, G. Khitrova, A. N. Poddubny, E. L. Ivchenko, M. Wegener, and H. M. Gibbs, Opt. Express **16**, 15382 (2008).
19. M. Werchner, M. Schafer, M. Kira, S. W. Koch, J. Sweet, J. D. Olitzky, J. Hendrickson, B. C. Richards, G. Khitrova, H. M. Gibbs, A. N. Poddubny, E. L. Ivchenko, M. Voronov, and M. Wegener, Opt. Express **17**, 6813 (2009).
20. B. C. Richards, J. Hendrickson, J. Sweet, G. Khitrova, D. Litvinov, D. Gerthsen, B. Myer et al., Opt. Express **16**, 21512 (2008).
21. D. Levine and P. J. Steinhardt, Phys. Rev. B **34**, 596 (1986).
22. I. Aviram, J. Phys. A **19**, 3299 (1986).
23. M. W. C. Dharma-wardana, A. H. MacDonald, D. J. Lockwood, J.-M. Baribeau, Houghton, and D. C., Phys. Rev. Lett. **58**, 1761 (1987).
24. E. L. Ivchenko, *Optical Spectroscopy of Semiconductor Nanostructures* (Alpha Science International, Harrow, UK, 2005).
25. F. de Medeiros, E. Albuquerque, M. Vasconcelos, and G. Farias, Surf. Sci. **600**, 4337 (2006).
26. E. Rashba and M. Sturge, eds., *Excitons* (North-Holland, Amsterdam, 1982).
27. E. L. Ivchenko and G. Pikus, *Superlattices and Other Heterostructures: Symmetry and Optical Phenomena* (Springer-Verlag, Berlin, 1997).
28. D. Würtz, T. Schneider, and A. Politi, Phys. Lett. A **129**, 88 (1988).
29. M. A. Kaliteevski, V. V. Nikolaev, R. A. Abram, and S. Brand, Opt. Spectrosc. **81**, 109 (2001).
30. A. N. Poddubny, L. Pilozzi, M. M. Voronov, and E. L. Ivchenko, Phys. Rev. B **80**, 115314 (2009).
31. B. Sutherland and M. Kohmoto, Phys. Rev. B **36**, 5877 (1987).
32. S. Chattopadhyay and A. Chakrabarti, Phys. Rev. B **65**, 184204 (2002); P. Tong, Phys. Rev. B **53**, 1795 (1996).
33. X. Q. Huang, S. S. Jiang, R. W. Peng, and A. Hu, Phys. Rev. B **63**, 245104 (2001).
34. A. Chakrabarti and S. N. Karmakar, Phys. Rev. B **44**, 896 (1991).
35. A. Thue, Norske Vididensk. Selsk. Skr. **7**, 1 (1906).
36. M. Morse, Am. J. Math. **43**, 35 (1921).
37. Z. Cheng, R. Savit, and R. Merlin, Phys. Rev. B **37**, 4375 (1988); C. S. Ryu, G. Y. Oh, and M. H. Lee, *ibid.* **46**, 5162 (1992); **48**, 132 (1993).

38. S. Tamura and F. Nori, Phys. Rev. B **40**, 9790 (1989).
39. N.-H. Liu, Phys. Rev. B **55**, 3543 (1997); S. Chattopadhyay, A. Ghosh, and A. Chakrabarti, *ibid.* **63**, 064201 (2001); H. Lei, J. Chen, G. Nouet, S. Feng, Q. Gong, and X. Jiang, *ibid.* **75**, 205109 (2007); M. S. Vasconcelos, P. W. Mauriz, F. F. de Medeiros, and E. L. Albuquerque, *ibid.* **76**, 165117 (2007); L. Moretti and V. Mocella, Opt. Express **15**, 15314 (2007); L. D. Negro, M. Stolfi, Y. Yi, J. Michel, X. Duan, L. C. Kimerling, J. LeBlanc, and J. Haavisto, Appl. Phys. Lett. **84**, 5186 (2004); V. Agarwal, J. Soto-Urueta, D. Becerra, and M. E. Mora-Ramos, Photonics Nanostruct. Fundam. Appl. **3**, 155 (2005).
40. D. Shir, H. Liao, S. Jeon, D. Xiao, H. T. Johnson, G. R. Bogart, K. H. A. Bogart, and J. A. Rogers, Nano Lett. **8**, 2236 (2008).
41. M. A. Kaliteevski, S. Brand, R. A. Abram, T. F. Krauss, P. Millar, and R. M. DeLa Rue, J. Phys. Condens. Matter **13**, 10459 (2001).
42. M. A. Kaliteevski, S. Brand, R. A. Abram, T. F. Krauss, R. De La Rue, and P. Millar, Nanotechnology **11**, 274 (2000).
43. R. Penrose, Bull. Inst. Math. Appl. **10**, 266 (1974).
44. W. Steurera and T. Haibacha, *International Tables for Crystallography Volume B. Chapter 4.6. Reciprocal-Space Images of Aperiodic Crystals* (International Union of Crystallography, Kluwer Academic Publishers, Dordrecht, The Netherlands, 2006).
45. A. Ledermann, G. von Freymann, and M. Wegener, Phys. Unserer Zeit **38**, 300 (2007).
46. X. Wang, C. Ng, W. Tam, C. Chan, and P. Sheng, Adv. Mater. **15**, 1526 (2003).
47. A. N. Poddubny, Phys. Rev. B **83**, 075106 (2011).
48. W. Man, M. Megens, P. J. Steinhardt, and P. M. Chaikin, Nature **436**, 993 (2005).
49. A. Ledermann, L. Cademartiri, M. Hermatschweiler, C. Toninelli, G. A. Ozin, D. S. Wiersma, M. Wegener, and G. von Freymann, Nat. Mater. **5**, 942 (2006).
50. J. Xu, R. Ma, X. Wang, and W. Y. Tam, Opt. Express **15**, 4287 (2007).

2.7

Multicomponent Photonic Crystals with Inhomogeneous Scatterers

Alexander B. Khanikaev
The University of Texas at Austin

Mikhail V. Rybin and Mikhail F. Limonov
Ioffe Physical-Technical Institute of the Russian Academy of Sciences
National Research University of Information Technologies, Mechanics and Optics

CONTENTS

2.7.1 Introduction

The design and architecture of photonic crystals (PhCs) may be defined as the science and art of selecting discrete scatterers and assembling them into the structures that meet some functional and performance goals. The first step in the design process is the selection of the shape and dielectric properties of scatterers. In general, dielectric function $\varepsilon(\mathbf{r}, \omega)$ of scatterers, also called permittivity, can have any prescribed spatial dependence, that is, scatterers can be inhomogeneous. The next step is to arrange scatterers spatially in some definite way in space of one, two, or three dimensions (1D, 2D, or 3D). After building in such a way a lattice of PhCs, one then characterizes photonic properties of the designed structure with the use of modern computational techniques.[1]

Although much effort has been made to fabricate and study different kinds of artificial PhCs, most PhCs considered both theoretically and experimentally were mainly composed of two homogeneous materials with different dielectric constants. These structures will hereafter be referred to as *two-component* photonic crystals (2c-PhCs). However, in many circumstances one has to consider materials with *multicomponent structures* formed from three or more homogeneous components with different dielectric constants or even composed of inhomogeneous components. Such structures[2,3] will be referred to as *multicomponent* photonic crystals (MPhCs).

Many PhCs known today represent multicomponent structures. For example, various semiconductor-based 2D PhCs have a complex space-periodic morphology and a defect surface formed during their fabrication by lithographic techniques, which in reality represents a third component. In addition, in the process of fabricating 2D PhCs, following etching (both ion and electrochemical), samples are coated with a so-called passivation layer[4] that represents an additional component of the structure. One more instructive example is the well-known 3D opal-based PhC, which, until 2007, in most papers devoted to their spectroscopic studies, were considered as two-component structures composed of homogeneous spheres and homogeneous filler. However, there are no natural or synthetic opals made up of homogeneous spheres; they are just theoretical models. Inhomogeneity of a-SiO_2 spheres that may seem inessential at first sight, in reality gives rise to remarkable optical effects,[2,3,5] some of which will be discussed in this chapter as well as in Chapters 3.5 and 3.6 of this book.

This chapter describes an original theoretical approach to the study of optical properties of low-contrast MPhC.[2,3,6,7] MPhCs have been considered in theoretical[8–10] and experimental[11,12] works. The convergence problem with plane-wave expansion method (PWEM) that arises from the step-like permittivity variation on the boundary of spherical scatterers was discussed by Sözüer et al.[8] Authors have compared such 2c-PhCs with structures formed from spheres with the Gaussian permittivity profile, which are essentially MPhCs with the inhomogeneous component. Also with the use of PWEM the influence of the surface oxide layer on the band structure of macroporous silicon[9] [i.e., a three-component photonic crystal (3c-PhC)] has been considered. Photonic stop-band intensity dependences in opal-based 3c-PhCs composed from coated spheres have been studied numerically by PWEM and within two-wave approximation.[10] Authors have revealed that the lower stop-band width in two different directions $\Gamma \to L$ (i.e., {111} stop-bands) and $\Gamma \to X$ (i.e., {200} stop-bands) vanishes at different coat-to-nucleolus radii ratios. Similar results have been experimentally obtained from reflection spectra of polystyrene opals infiltrated by Si[11] and ZnO.[12]

To analyze photonic band structures of 1D, 2D, and 3D MPhCs within the framework of the same theoretical model, we unify description of their structures in the following way. We will consider PhCs composed of "scatterers" embedded into a homogeneous medium with the dielectric constant ε_f which will be referred to as "filler." Besides, we will consider scatterers of a certain symmetry that is specific to a wide class of real PhCs; scatterers are centrosymmetric with respect to the Wigner–Seitz cell for 1D, cylindrically symmetric for 2D, and spherically symmetric for 3D MPhCs (Figure 2.7.1). Taking symmetry into account, it is convenient to define permittivity as a function $\varepsilon_s(r)$ of radial coordinate r. However, results obtained below can be generalized to an MPhC with a scatterer of any symmetry.

A 2c-PhC formed from slabs (1D), cylinders (2D), and spheres (3D) consisting of two compounds characterized by their permittivities ε_s and ε_f represents the most trivial case, whereas a 3c-PhC is the simplest example of MPhC and is schematically shown in a separate row in Figure 2.7.1. In the latter case, the scatterer consists of a homogeneous nucleus with the permittivity ε_n and a homogeneous coat with the permittivity ε_c, whereas a homogeneous filler with the permittivity ε_f represents the third component. The case of 3c-PhC is of special interest because it provides a relatively simple analytical solution and representative results, which will be described in Sections 2.7.4 and 2.7.5.

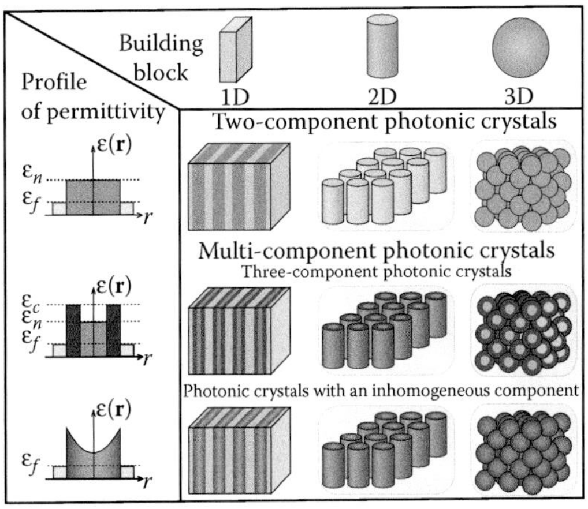

FIGURE 2.7.1
The conceptual design of photonic crystals. Simple examples of two-component and multicomponent photonic crystals for one, two, and three dimensions.

All of the MPhCs described above with different dimensions will be considered in the framework of an analytical approach intended to describe specific optical properties of photonic stop-bands.[2,3,6,7] Within this approach the photonic properties of the MPhCs have been studied as depending on various dielectric and structural parameters. As a particular example allowing description of all of the main effects, we will treat the filler permittivity ε_f as a continuous variable. Such ε_f adjustment can be realized in practice by some external means (e.g., electromagnetic field, temperature, applied stress, etc.) or simply by changing the filler material itself by embedding the "matrix" constituting the PhC into a liquid or gaseous medium. Such a procedure is referred to as "immersion" (after the Latin *immersio*, meaning *embedding*). In this chapter, primary attention will be given to the effect of the vanishing of photonic stop-bands in optical spectra of low-contrast PhCs when the parameter ε_f is varied, and this *spectroscopic* effect will be referred to as *photonic band immersion.*

The problem of photonic band structure in MPhCs was first formulated in papers by Baryshev et al.[2] and Rybin et al.,[3] in which immersion of the photonic bands in synthetic opals was studied both theoretically and experimentally. The analytical approach proposed by Baryshev et al.[2] was based on the well-known fact that the appearance or disappearance of Bragg diffraction due to the scattering by certain systems of crystallographic planes in PhCs is the mechanism responsible for the formation (and therefore existence or absence) of the stop-bands in low-contrast PhCs. The developed analytical approach was intended to demonstrate the effects of immersion of various Bragg stop-bands associated with certain reciprocal lattice vectors **g** when the structural inhomogeneity of scatterers a-SiO_2 in opals was taken into account. It was shown that Bragg band intensities as a function of $|\mathbf{g}|$ in MPhCs exhibit *quasiperiodic resonant behavior,* thus providing the possibility to *separately immerse different Bragg bands* and thus simultaneously and independently manipulate light at various Bragg wavelengths.

Note that when we refer to photonic *stop-bands,* we mean energy bands with practically *zero light transmission* in certain directions in the crystal lattice. However, if our object under study is a finite low-contrast PhC near the photonic band immersion condition, we are far from zero light transmission. For this reason, in this chapter we will use the more appropriate term "photonic Bragg bands" instead of the term "photonic stop-band."

2.7.2 General Approach for Analytical Model

In this section, we describe a general approach that allows a simple description of the Bragg band immersion effects in low-contrast MPhCs. Immersion effects such as the disappearance of Bragg bands can be considered as the vanishing of their bandwidth. We have to mention here that evaluation of the spectral width of a Bragg band in the simplest case of a 1D PhC at normal light incidence is given in the monograph by Yariv and Yeh[13] and can be summarized as follows. First, as a result of the periodicity of the structure the electric field can be expanded in the Fourier series on the basis of plane waves, and then one straightforwardly obtains an approximate secular equation limiting the basis to the two plane waves. Such two-wave approximation allows one to determine photonic dispersion relation in the proximity of a particular isolated Bragg band with quite good precision. In the general case of 3D MPhCs, however, this problem results in a non-Hermitian eigenvalue problem,[1] and the non-Hermiticity does not admit a straightforward application of a mathematical apparatus well-developed for the Hermitian operators.

However, such a problem does not arise when a similar eigenvalue problem is formulated for the magnetic field, which in the case of constant magnetic permeability results in a Hermitian eigenvalue problem. We briefly describe the algorithm for calculation of the spectral width of Bragg bands for the case of 3D PhCs in the two-wave approximation. We write down the corresponding eigenvalue problem in a way analogous to that given by Ho et al.[14] formulated for the column vector $\mathbf{h}_{\mathbf{k}}^{3D}$, with each element representing a Fourier component of the magnetic field corresponding to a particular quasi-wave vector $(\mathbf{k}+\mathbf{g})$

$$\hat{\mathbf{M}}\mathbf{h}_{\mathbf{k}}^{3D} = \frac{\omega^2}{c^2}\mathbf{h}_{\mathbf{k}}^{3D}, \tag{2.7.1}$$

where $\hat{\mathbf{M}}$ is an operator with the matrix elements $\mathbf{M}_{\mathbf{g}',\mathbf{g}} = -(\mathbf{g}'+\mathbf{k})\times\tilde{\varepsilon}_{\mathbf{g}'-\mathbf{g}}(\mathbf{g}+\mathbf{k})\times$, $\mathbf{g}$, $\mathbf{g}'$ are reciprocal lattice vectors of the PhC, ω is frequency of light, c is speed of light in vacuum, and $\tilde{\varepsilon}_{\mathbf{g}} = \frac{1}{V}\iiint d\mathbf{r}\frac{1}{\varepsilon(\mathbf{r})}e^{i\mathbf{g}\mathbf{r}}$ is Fourier coefficients of the inverse permittivity with the integration performed over the unit cell volume V. We focus here on transparent materials, which implies that we can treat $\varepsilon(\mathbf{r})$ as purely real and positive.

For every particular reciprocal lattice vector $\mathbf{g}$ we choose two auxiliary unit vectors $\mathbf{u}_{\mathbf{g}}$ and $\mathbf{v}_{\mathbf{g}}$ in such a way that together with the vector $(\mathbf{g}+\mathbf{k})$ they form a right-handed triple of vectors. Taking into account the divergenceless character of the magnetic field, we can write it down as a constraint $(\mathbf{g}+\mathbf{k})\mathbf{h}_{\mathbf{k}}^{3D}{}_{\mathbf{g}} = 0$. Then, it follows that $\mathbf{h}_{\mathbf{k}}^{3D}{}_{\mathbf{g}}$ can be expressed as $\mathbf{h}_{\mathbf{k}}^{3D}{}_{\mathbf{g}} = \mathbf{U}_{\mathbf{k}\mathbf{g},\mathbf{g}}\mathbf{h}_{\mathbf{k}}^{2D}{}_{\mathbf{g}}$, where $\mathbf{U}_{\mathbf{k}}$ is a block diagonal matrix with each block equal to

$$\mathbf{U}_{\mathbf{k}\mathbf{g},\mathbf{g}} = \begin{pmatrix} u_{\mathbf{g},x} & v_{\mathbf{g},x} \\ u_{\mathbf{g},y} & v_{\mathbf{g},y} \\ u_{\mathbf{g},z} & v_{\mathbf{g},z} \end{pmatrix}. \tag{2.7.2}$$

As a result, one obtains a Hermitian eigenvalue problem

$$\mathbf{U}_{\mathbf{k}}^{T}\hat{\mathbf{M}}\mathbf{U}_{\mathbf{k}}\mathbf{h}_{\mathbf{k}}^{2D} = \frac{\omega^2}{c^2}\mathbf{h}_{\mathbf{k}}^{2D}. \tag{2.7.3}$$

It is worth mentioning here that the action of the matrix $\mathbf{U}_{\mathbf{k}\mathbf{g},\mathbf{g}}$ on a particular Fourier component $\mathbf{h}^{2D}_{\mathbf{k}\ \mathbf{g}}$ can be represented as a sum $h^{\mathrm{u}}_{\mathbf{k}\ \mathbf{g}}\mathbf{u} + h^{\mathrm{v}}_{\mathbf{k}\ \mathbf{g}}\mathbf{v}$ (in what follows, to simplify notations, we will omit indices $\mathbf{g}$ of vectors $\mathbf{u}_{\mathbf{g}}$ and $\mathbf{v}_{\mathbf{g}}$). Expression 2.7.3 allows one to calculate Fourier component $\mathbf{h}^{2D}_{\mathbf{k}\ \mathbf{g}}$ through other components that can be expressed as

$$\frac{\omega^2}{c^2}(h^{\mathrm{u}}_{\mathbf{k}\ \mathbf{g}}\mathbf{u} + h^{\mathrm{v}}_{\mathbf{k}\ \mathbf{g}}\mathbf{v}) = \tilde{\varepsilon}_0 \mid \mathbf{g} + \mathbf{k} \mid^2 (h^{\mathrm{u}}_{\mathbf{k}\ \mathbf{g}}\mathbf{u} + h^{\mathrm{v}}_{\mathbf{k}\ \mathbf{g}}\mathbf{v}) + R^{\mathrm{u}}\mathbf{u} + R^{\mathrm{v}}\mathbf{v}, \tag{2.7.4}$$

where terms of the expansion not containing $\mathbf{h}_{\mathbf{k}\mathbf{g}}$ are referred to as $R^{\mathrm{u,v}}$. From expression 2.7.4 one can see that $h^{\mathrm{u,v}}_{\mathbf{k}\ \mathbf{g}} = R^{\mathrm{u,v}}\left(\frac{\omega^2}{c^2} - \tilde{\varepsilon}_0 \mid \mathbf{g} + \mathbf{k} \mid^2\right)^{-1}$, and from this we can conclude that coefficients $h^{\mathrm{u,v}}_{\mathbf{k}\ \mathbf{g}}$ appear to be dominant as compared to other Fourier coefficients of the series when the condition $\mid \mathbf{g} + \mathbf{k} \mid \approx \frac{\omega}{c\sqrt{\tilde{\varepsilon}_0}}$ is satisfied. The Bragg band is known to appear when the Bragg condition $|\mathbf{g}+\mathbf{k}| = |\mathbf{k}|$ is satisfied, and, in this case, plane waves determined by the coefficients $h^{\mathrm{u,v}}_{\mathbf{k}\ \mathbf{g}}$ and $h^{\mathrm{u,v}}_{\mathbf{k}\ 0}$ appear to be dominant. In what follows, we assume that orientation of the quasi-wave vector does not correspond to the regime of the multiple Bragg wave diffraction;[15] that is, we assume that there are no two distinct reciprocal lattice vectors $\mathbf{g}$ and $\mathbf{g}'$ for which the condition $|\mathbf{g}+\mathbf{k}| \approx |\mathbf{k}| \approx |\mathbf{g}'+\mathbf{k}|$ is satisfied. This allows us to argue that only two plane waves dominate in the particular frequency range and in the orientation of the crystal with respect to Bloch wave vector $\mathbf{k}$. In the case of such two-wave approximation, by vanishing amplitudes of all the other plane waves, Equation 2.7.3 reduces to the system

$$\begin{cases} \tilde{\varepsilon}_0 k^2(\mathbf{u}h^{\mathrm{u}}_{\mathbf{k}\,0} + \mathbf{v}h^{\mathrm{v}}_{\mathbf{k}\,0}) + \tilde{\varepsilon}_{-\mathbf{g}} \mid \mathbf{g} + \mathbf{k} \mid (\mid k \times t \mid \mathbf{u}h^{\mathrm{u}}_{\mathbf{k}\,\mathbf{g}} + k\mathbf{v}h^{\mathrm{t}}_{\mathbf{k}\,\mathbf{g}}) = \dfrac{\omega^2}{c^2}(\mathbf{u}h^{\mathrm{u}}_{\mathbf{k}\,0} + \mathbf{v}h^{\mathrm{v}}_{\mathbf{k}\,0}) \\ \tilde{\varepsilon}_{\mathbf{g}} k(\mid (\mathbf{g} \times \mathbf{v} - \mathbf{u}k) \mid \mathbf{u}h^{\mathrm{u}}_{\mathbf{k}\,0} + k\mathbf{t}h^{\mathrm{v}}_{\mathbf{k}\,0}) + \tilde{\varepsilon}_0 \mid \mathbf{g} + \mathbf{k} \mid^2 (\mathbf{u}h^{\mathrm{u}}_{\mathbf{k}\,\mathbf{g}} + \mathbf{t}h^{\mathrm{t}}_{\mathbf{k}\,\mathbf{g}}) = \dfrac{\omega^2}{c^2}(\mathbf{u}h^{\mathrm{u}}_{\mathbf{k}\,\mathbf{g}} + \mathbf{t}h^{\mathrm{t}}_{\mathbf{k}\,\mathbf{g}}) \end{cases}, \tag{2.7.5}$$

where three unit vectors $\mathbf{u}$, $\mathbf{v}$ and $\mathbf{t}$ are such that sets $\mathbf{k}$, $\mathbf{u}$, $\mathbf{v}$ and $(\mathbf{k}+\mathbf{g})$, $\mathbf{u}$, $\mathbf{t}$ form right-handed triple vectors. System 2.7.5 defines a set of four seqular equations. The first pair of equations formed from coefficients at the vectors $\mathbf{v}$ and $\mathbf{t}$ defines a two-dimensional eigenvalue problem for *s*-polarization [magnetic field in the $(\mathbf{k}, \mathbf{g})$ plane]. The second pair (coefficients at the vector $\mathbf{u}$) defines an eigenvalue problem for *p*-polarization. Looking for nontrivial solutions, we are setting determinants of the corresponding matrices to be equal to zero and obtain the following equations:

$$\left(\tilde{\varepsilon}_0 k^2 - \frac{\omega^2}{c^2}\right)\left(\tilde{\varepsilon}_0 \mid \mathbf{g} + \mathbf{k} \mid^2 - \frac{\omega^2}{c^2}\right) - \tilde{\varepsilon}_{-\mathbf{g}}\tilde{\varepsilon}_{\mathbf{g}} k^3 |\mathbf{g} + \mathbf{k}| = 0, \tag{2.7.6a}$$

$$\left(\tilde{\varepsilon}_0 k^2 - \frac{\omega^2}{c^2}\right)\left(\tilde{\varepsilon}_0 \mid \mathbf{g} + \mathbf{k} \mid^2 - \frac{\omega^2}{c^2}\right) - \tilde{\varepsilon}_{-\mathbf{g}}\tilde{\varepsilon}_{\mathbf{g}} k \mid \mathbf{g} + \mathbf{k} \mid\mid \mathbf{k} \times \mathbf{t} \mid\mid \mathbf{g} \times \mathbf{v} - \mathbf{u}k \mid = 0, \tag{2.7.6b}$$

for *s*- and *p*-polarizations, respectively. Let us define the angle between vectors $\mathbf{k}$ and $\mathbf{g}$ as $\theta_{\mathbf{g}}$. With this definition we can determine the value of splitting between eigenvalues, that is, squared eigenfrequencies ω^2 by the following expressions: $2|\tilde{\varepsilon}_{\mathbf{g}}|c^2k^2$ for *s*-polarization and $2c^2\left|\tilde{\varepsilon}_g\right|k^2\left|\cos 2\theta_g\right|$ for p-polarization. And, finally, with the use of these expressions we can estimate the value of splitting as:

$$\Delta\omega_s = \omega_0 \frac{\left|\tilde{\varepsilon}_g\right|}{\left|\tilde{\varepsilon}_0\right|}, \quad \text{for } s\text{-polarization,} \tag{2.7.7a}$$

$$\Delta\omega_p = \omega_0 \frac{\left|\tilde{\varepsilon}_g\right|}{\left|\tilde{\varepsilon}_0\right|}\left|\cos 2\theta_g\right|, \quad \text{for } p\text{-polarization,} \tag{2.7.7b}$$

where $\omega_0 = ck/\sqrt{\tilde{\varepsilon}_0}$ is the Bragg frequency.

From this result we can conclude that spectral width of the Bragg band is proportional to the Fourier component of the inverse permittivity $|\tilde{\varepsilon}_g|$. Besides, in the case of p-polarization, spectral width of the Bragg band is proportional to the factor $|\cos 2\theta_g|$. Thereby, there are two mechanisms resulting in the disappearance (vanishing) of the Bragg band in spectra of *perfectly ordered* PhCs. The first mechanism has polarization-related origin and is inherently related to the Brewster effect.[16] It gives rise to the vanishing of the Bragg bands in p-polarized spectra when $\theta_g = 45°$ (Brewster angle). This effect was experimentally observed and noted in several works.[17]

The second mechanism resulting from the immersion of the Bragg bands is related to a specific distribution of permittivity within the unit cell of the structure and can be realized by varying permittivity of one of the MPhC components (this effect was experimentally observed[2,3]). The corresponding condition has the form

$$\tilde{\varepsilon}_g = 0. \tag{2.7.8}$$

In terms of x-ray scattering the parameter analogous to $\tilde{\varepsilon}_g$ is referred to as the scattering form factor and is usually denoted as $F(\mathbf{g})$. Thus, the condition $\tilde{\varepsilon}_g \equiv F(\mathbf{g}) = 0$ in many circumstances is equivalent to the condition of the disappearance of the corresponding Bragg band. One should remember, however, that this is true only in the case of validity of the two-wave approximation, that is, in the absence of the multi-wave diffraction that is closely related to the need of strong band separation and therefore to the condition of small dielectriccontrast ($\eta = ((\max(\varepsilon(r)) - \min(\varepsilon(r)))/\langle\varepsilon(r)\rangle) \ll 1$).

2.7.3 MPhCs with Centrosymmetric Inhomogeneous Scatterers

In this section, we will focus on the effects of selective immersion of the photonic Bragg bands. To analyze the photonic properties of MPhCs, we will follow the approach adopted by Baryshev et al.[2] and Rybin et al.[3] Let us consider the Fourier coefficient of the inverse permittivity, that is, the scattering form factor $F(\mathbf{g})$:

$$F(\mathbf{g}) = \frac{1}{V_0}\int_{V_0} d\mathbf{r}\frac{1}{\varepsilon(\mathbf{r})}\exp(-i\mathbf{g}\cdot\mathbf{r}), \tag{2.7.9}$$

which describes the intensity of Bragg diffraction from a family of crystal planes as a function of the reciprocal lattice vector **g**. The reciprocal lattice vectors for 1D, 2D, and 3D PhCs are defined as $\mathbf{g}_h = h\mathbf{b}_1$, $\mathbf{g}_{hk} = h\mathbf{b}_1 + k\mathbf{b}_2$, and $\mathbf{g}_{hkl} = h\mathbf{b}_1 + k\mathbf{b}_2 + l\mathbf{b}_3$, respectively, where $\{\mathbf{b}_i;\ i = 1, 2, 3\}$ are the unit reciprocal lattice vectors. V_0 represents the length (1D PhC), area (2D PhC), or volume (3D PhC) of the primitive cell.

Let us consider the case of MPhCs with the centrosymmetric inhomogeneous scatterers. The inverse permittivity of a Wigner–Seitz cell of the crystal is defined as

$$\frac{1}{\varepsilon(r)} = \frac{1}{\varepsilon_f} + \left(\frac{1}{\varepsilon_s(r)} - \frac{1}{\varepsilon_f}\right)\Theta(r_s - r), \tag{2.7.10}$$

where r is the distance from the Wigner–Seitz cell center, and $\varepsilon_s(r)$ is an arbitrary profile of the permittivity of the scatterer of radius (or length in the case of 1D PhC) r_s and possessing the proper symmetry illustrated in Figure 2.7.1 for the structures of different dimensionality, 1D, 2D, or 3D. The unit step function $\Theta(r)$ in Equation 2.7.10 is defined as: $\Theta(r) = 1$ at $r \geq 0$ and $\Theta(r) = 0$ at $r < 0$. It is worth noting that Equation 2.7.10 implies that the structure under study represents a crystal with the only scatterer within the Wigner–Seitz cell. Such limitation allows us to obtain a simple and instructive expression; however, it is straightforward to show that an arbitrary distribution of the permittivity over the unit cell of the MPhC results in analogous results.

We will analyze the immersion conditions when the scattering form factor turns to zero: $F(\mathbf{g}) = 0$. At $|\mathbf{g}| \neq 0$, the relation $F(\mathbf{g}) = 0$ can be rewritten as

$$F(\mathbf{g}) = \frac{1}{V_0}\int_{V_0} d\mathbf{r}\left(\frac{1}{\varepsilon_s(r)} - \frac{1}{\varepsilon_f}\right)\Theta(r_s - r)\exp(-i\mathbf{g}\cdot\mathbf{r}) = 0. \tag{2.7.11}$$

As mentioned above, we take the filler to be a homogeneous material (ε_f = constant). Thereby Equation 2.7.11 defines the *value of the filler permittivity yielding the immersion condition for the particular photonic Bragg band determined by the reciprocal lattice vector* **g**. Such value of the filler permittivity will be denoted by $\varepsilon_f^0(g)$, where $g = |\mathbf{g}|$. Integration of the appropriate expressions provides $\varepsilon_f^0(g)$ for MPhCs of different dimensionality:

$$\text{for 1D MPhC:}\quad \varepsilon_f^0(g) = \frac{\sin(gr_s)}{g\displaystyle\int_0^{r_s}\frac{1}{\varepsilon_s(r)}\cos(gr)\,dr}, \tag{2.7.12a}$$

$$\text{for 2D MPhC:}\quad \varepsilon_f^0(g) = \frac{r_s J_1(gr_s)}{g\displaystyle\int_0^{r_s}\frac{1}{\varepsilon_s(r)}rJ_0(gr)\,dr}, \tag{2.7.12b}$$

$$\text{for 3D MPhC:}\quad \varepsilon_f^0(g) = \frac{R(gr_s)}{g^2\displaystyle\int_0^{r_s}\frac{1}{\varepsilon_s(r)}r\sin(gr)\,dr}. \tag{2.7.12c}$$

In expression 2.7.12b functions $J_0(x)$ and $J_1(x)$ are the zero- and first-order Bessel functions of the first kind, whereas in expression 2.7.12c $R(x) \equiv \sin(x) - x\cos(x)$ is the Rayleigh–Hans function.

The domain of functions 2.7.12a–c is limited by the discrete set of values given by the absolute value of the reciprocal lattice vectors; for example, g_h for 1D MPhC, g_{hk} for 2D MPhC, and g_{hkl} for 3D MPhC. Nevertheless, it is convenient for the subsequent analysis to extend the domain of functions 2.7.12a–c to the whole positive semiaxis. At the same time, one should remember that all the expressions are meaningful only for certain discrete values of g corresponding to the lengths of the reciprocal lattice vectors. We do not limit our analysis to the discrete values of g but consider it as a continuous function g and omit the *hkl* indices.

It is clear that, in general, the quantity $\varepsilon_f^0(g)$ is a function of the reciprocal lattice vector modulus g; therefore, it differs for different Bragg bands corresponding to different g. As a consequence, *different Bragg bands in MPhCs can be immersed selectively*. It is worth mentioning that denominators in Equations 2.7.12a–c may change sign and for certain values of g become equal to zero. Thus, for some Bragg bands we can find the immersion condition $\varepsilon_f^0(g) \to \infty$. Such a Bragg band cannot be immersed by any realistic filler medium and will be referred to as *the resonant Bragg band*.

2.7.4 Three-Component Photonic Crystals

As a particular example, let us consider the case of Bragg band immersion in the 3c-PhC of different dimensionality. In this case, the scatterers consist of a homogeneous nucleus and a homogeneous coat (see Figure 2.7.1).

We consider a 1D 3c-PhC structure having the lattice period a, in which inhomogeneous scatterers are composed of homogeneous slabs (representing nucleus) characterized by their thickness $2r_n$ and permittivity ε_n. The slabs are covered by homogeneous layers (referred to as a coat) having thickness $r_c - r_n$ and permittivity ε_c. These scatterers are separated by layers of thickness $(a - 2r_c)$ with homogeneous filler characterized by permittivity ε_f.

To analyze 2D 3c-PhCs, we consider a structure having inhomogeneous scatterers composed of homogeneous cylinders of the infinite extent (nucleus) and radius r_n and characterized by permittivity ε_n, coated by the homogeneous coat of radius $r_c \geq r_n$ and thickness $(r_c - r_n)$ and characterized by the permittivity ε_c. Centers of these infinite cylinders are positioned in the nodes of an arbitrary "monoatomic" Bravais lattice. More specifically, we will discuss the square lattice of period a. Reciprocal lattice vectors $\mathbf{g}$ and radius vectors r are defined to lie in the plane perpendicular to the cylinders' axes. The space between cylinders is filled by a homogeneous medium having permittivity ε_f.

Immersion effects in 3D 3c-PhCs are considered for the case of a face-centered cubic (FCC) structure, in which inhomogeneous scatterers are formed by spheres, each consisting of homogeneous nucleus of radius r_n with permittivity ε_n, a homogeneous coat ε_c, and the outer radius $r_c \geq r_n$ and therefore having the thickness $r_c - r_n$. The intersphere space is filled with a homogeneous material of permittivity ε_f. The distance between the nearby spheres centers in the FCC lattice $a_{00} \geq 2r_c$ (here $a_{00} = \frac{1}{\sqrt{2}}a$), where a is a cubic lattice constant.

Note that in all three cases of 1D, 2D, and 3D 3c-PhCs, definitions of the parameters characterizing internal structure of the scatterers are unified and the same notations are used; for example, thickness of the coat in all three cases is given by $r_c - r_n$ whereas thickness of the nucleus is equal to $2r_n$.

In the case of 3c-PhC of arbitrary dimensionality, formula 2.7.10 for the inverse permittivity of a Wigner–Seitz cell can be written as

$$\frac{1}{\varepsilon(r)} = \frac{1}{\varepsilon_f} + \left(\frac{1}{\varepsilon_c} - \frac{1}{\varepsilon_f}\right)\Theta(r_c - r) + \left(\frac{1}{\varepsilon_n} - \frac{1}{\varepsilon_c}\right)\Theta(r_n - r). \tag{2.7.13}$$

In general, Bragg band immersion effects in MPhCs can be considered as a parametric problem, with the parameter chosen to be any of the structure parameters ε_f, ε_n, ε_c, r_n, or r_c. As an example we will consider Bragg band immersion problems in 1D, 2D, and 3D 3c-PhCs by varying parameter ε_f and fixing the other parameters. The corresponding expressions for $\varepsilon_f^0(g)$ now can be written in the following form that is convenient for further analysis:

$$\text{for 1D MPhC:} \quad \frac{1}{\varepsilon_f^0(g)} = \frac{1}{\varepsilon_c} + \left(\frac{1}{\varepsilon_n} - \frac{1}{\varepsilon_c}\right)\frac{\sin(gr_n)}{\sin(gr_c)}, \tag{2.7.14a}$$

$$\text{for 2D MPhC:} \quad \frac{1}{\varepsilon_f^0(g)} = \frac{1}{\varepsilon_c} + \left(\frac{1}{\varepsilon_n} - \frac{1}{\varepsilon_c}\right)\frac{r_n}{r_c}\frac{J_1(gr_n)}{J_1(gr_c)}, \tag{2.7.14b}$$

$$\text{for 3D MPhC:} \quad \frac{1}{\varepsilon_f^0(g)} = \frac{1}{\varepsilon_c} + \left(\frac{1}{\varepsilon_n} - \frac{1}{\varepsilon_c}\right)\frac{R(gr_n)}{R(gr_c)}. \tag{2.7.14c}$$

As can be seen from expressions 2.7.14a–c, the value of ε_f^0 in determining the immersion conditions depends on the modulus of the reciprocal lattice vector g. Therefore in 3c-PhCs of any dimensionality, immersion of photonic Bragg bands is selective.

To analyze Bragg band immersion effects, we have to define values of the reciprocal lattice vector modulus for actual systems of crystallographic planes for each of the three structures under discussion. In Figure 2.7.2a–c, vertical lines define values of g_h for low-index (low in h) systems of planes of 1D lattice $\{h\}$, values g_{hk} for low-index $\{hk\}$ systems of planes of square 2D lattice, and values g_{hkl} for low-index $\{hkl\}$ systems of planes of 3D FCC lattice, respectively. For 1D lattice the expression $g_h = 2\pi h a^{-1}$ is used and length of the low-index reciprocal lattice vectors ($g_h < 20$) take values $g_1 = 6.28$, $g_2 = 12.57$, $g_3 = 18.85$, and so on. For 2D square lattice the corresponding expression has a form $g_{hk} = 2\pi\sqrt{(h^2 + k^2)}a^{-1}$ and defines values $g_{10} = 6.28$, $g_{11} = 8.89$, $g_{20} = 12.57$, $g_{21} = 14.05$, $g_{22} = 17.77$, $g_{30} = 18.85$, $g_{31} = 19.87$, and so on. And, finally, for 3D FCC lattice it is convenient to measure radii and reciprocal vector modulus in the distance between the nearby spheres centers a_{00} and a_{00}^{-1}, respectively (in 1D and 2D cases, a_{00} coincides with lattice constant a). In these units, $g_{hkl} = \pi\sqrt{2(h^2 + k^2 + l^2)}a_{00}^{-1}$, $g_{111} = 7.70$, $g_{200} = 8.89$, $g_{220} = 12.57$, $g_{311} = 14.74$, $g_{222} = 2g_{111} = 15.40$, $g_{400} = 2g_{200} = 17.78$, $g_{331} = 19.37$, $g_{420} = 19.87$, and so on. The values of g_h, g_{hk}, and g_{hkl} are taken in the units of the inverse of distances between the nearby scatterer centers a_{00}^{-1}.

Now we can consider dependence of the filler permittivity ε_f^0 corresponding to the immersion conditions as a function of the reciprocal lattice vector modulus g. Parameters

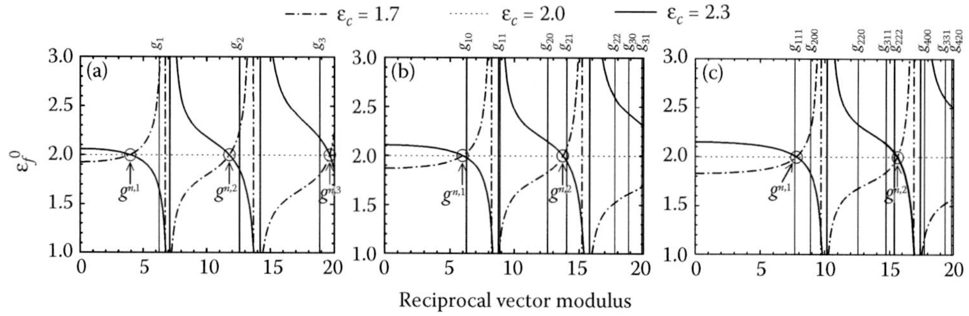

FIGURE 2.7.2
(a–c) The filler permittivity $\varepsilon_f^0(g)$ corresponding to the immersion conditions as a function of the reciprocal lattice vector modulus g is calculated from Equation 2.7.14 for a one-dimensional (a), two-dimensional (b), and three-dimensional (c) three-component photonic crystal with the same structural parameters $\varepsilon_n = 2.0$, $r_n = 0.35$, $r_c = 0.45$, and three different values of coat permittivity, namely, less ($\varepsilon_c = 1.7$, dot-and-dashed curve), equal ($\varepsilon_c = 2.0$, doted line), and greater ($\varepsilon_c = 2.3$, solid curve) than the nucleus permittivity (ε_n). The modulus of the shortest reciprocal lattice vectors g_h, g_{hk}, and g_{hkl} are shown by vertical lines. The special points at $g = g^{n,i}$ are marked by circles.

of 1D, 2D, and 3D 3c-PhCs are given in the caption of Figure 2.7.2. In what follows we consider two major cases. Firstly, we consider the case of 2c-PhC $\varepsilon_n = \varepsilon_c$. As can be seen from expressions 2.7.14a–c, when $\varepsilon_n = \varepsilon_c$ the value of ε_f^0 does not depend on g. Therefore, the function $\varepsilon_f^0(g)$ in Figure 2.7.2 becomes the straight line and has the same value at all the intersections with the vertical dashed lines given by $g = g_h$ (1D), g_{hk} (2D), and g_{hkl} (3D). Thus, in 2c-PhCs all the Bragg bands are immersed simultaneously and the structure becomes fully transparent when $\varepsilon_f^0 = \varepsilon_c = \varepsilon_n$. The second case of $\varepsilon_n \neq \varepsilon_c$ corresponds to the 3c-PhC structure. The corresponding functions $\varepsilon_f^0(g)$, regardless of the dimensionality of the MPhC, demonstrate quite nontrivial behavior, that is, a quasiperiodic dependence as a function of g with resonant peculiarities.

2.7.5 The Resonance Bragg Band in MPhCs

We consider now the resonant quasiperiodic character of the function $\varepsilon_f^0(g)$, which follows directly from the general expressions for MPhCs of any dimensionality (Equations 2.7.12a–c), as well as from the expressions for 3c-PhCs (Equations 2.7.14a–c). The resonance condition implies such a value of $g = g^{\mathrm{res},i}$ at which $\varepsilon_f^0 \to \pm\infty$. Because in the experiment it is impossible to implement the case of a filler of infinite permittivity, the Bragg band corresponding to $g^{\mathrm{res},i}$ cannot be immersed by changing the filler permittivity; that is, the immersion condition is unfeasible. The resonant values $g = g^{\mathrm{res},i}$ for 3c-PhCs are determined from the following equations:

$$\text{for 1D MPhC:}\quad \frac{1}{\varepsilon_c} = \left(\frac{1}{\varepsilon_c} - \frac{1}{\varepsilon_n}\right)\frac{\sin(g^{\mathrm{res}} r_n)}{\sin(g^{\mathrm{res}} r_c)}, \tag{2.7.15a}$$

$$\text{for 2D MPhC:}\quad \frac{1}{\varepsilon_c} = \left(\frac{1}{\varepsilon_c} - \frac{1}{\varepsilon_n}\right)\frac{r_n}{r_c}\frac{J_1(g^{\mathrm{res}} r_n)}{J_1(g^{\mathrm{res}} r_c)}, \tag{2.7.15b}$$

$$\text{for 3D MPhC:}\quad \frac{1}{\varepsilon_c} = \left(\frac{1}{\varepsilon_c} - \frac{1}{\varepsilon_n}\right)\frac{R(g^{\text{res}} r_n)}{R(g^{\text{res}} r_c)}. \tag{2.7.15c}$$

These expressions point out that alternation of the sign of functions $\sin(gr)$, $J_1(gr)$, and $R(gr)$ (see Figure 2.7.3) is responsible for the resonance $\varepsilon_f^0 \to \pm\infty$. Resonant values of $g^{\text{res},i}$ depend on dielectric (ε_n and ε_c) and structural parameters (r_n and r_c) and for this reason the condition $g = g^{\text{res},i}$ can be satisfied by means of tuning the corresponding parameters of the MPhC.

It is important to mention here that the resonant Bragg band, which may appear to be unrealistic at first glance, was observed experimentally in studies of the transmission spectra of synthetic opals composed of *inhomogeneous* *a*-SiO_2 particles.[3] These results will be discussed in detail in Chapter 3.5. Here we mention that the intensity of the (222) Bragg band was found to remain nearly unaltered when the filler permittivity ε_f was varied within the whole experimentally accessible range, whereas other (*hkl*) bands showed quite sharp dependences.

Let us consider another specific feature of the function $\varepsilon_f^0(g)$, namely, special points at $g = g^{n,i}$, when all curves do not show any dependence from ε_c and intersect at the value $\varepsilon_f^0(g^{n,i}) = \varepsilon_n$ (Figure 2.7.2). The corresponding values $g^{n,i}$ determined with the use of Equations 2.7.14a–c for 3c-PhCs are given by the following expressions:

$$\text{for 1D MPhC:}\quad \frac{\sin(g^{n,i} r_n)}{\sin(g^{n,i} r_c)} = 1, \tag{2.7.16a}$$

$$\text{for 2D MPhC:}\quad \frac{J_1(g^{n,i} r_n)}{J_1(g^{n,i} r_c)}\frac{r_n}{r_c} = 1, \tag{2.7.16b}$$

$$\text{for 3D MPhC:}\quad \frac{R(g^{n,i} r_n)}{R(g^{n,i} r_c)} = 1. \tag{2.7.16c}$$

The first three values of $g^{n,i}$ are given by $g^{n,1} = 3.93$, $g^{n,2} = 11.78$, and $g^{n,3} = 19.63$ in the case of 1D 3c-PhCs; $g^{n,1} = 6.00$, $g^{n,2} = 13.76$, and $g^{n,3} = 21.57$ in the case of 2D 3c-PhCs; and $g^{n,1} = 7.81$, $g^{n,2} = 15.62$, and $g^{n,3} = 23.43$ in the case of 3D 3c-PhCs.

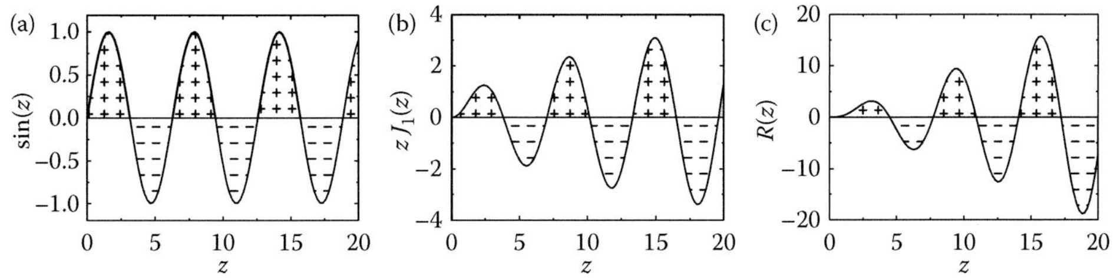

FIGURE 2.7.3
(a–c) The graph of alternating-sign functions: (a) the sine function; (b) the first-order Bessel function of the first kind multiplied by its argument; (c) the Rayleigh–Hans function. (Reprinted from A. K. Samusev et al. J. Phys. Condens. Matter 22, 115401, 2010. With permission.)

These points represent a significant interest because at $g = g^{n,i}$ the alteration of the immersion conditions is observed. At first we consider the area of the smallest values of $g < g^{n,1}$ (see Figure 2.7.2). When $\varepsilon_c > \varepsilon_n$ (solid curves) the condition $\varepsilon_c > \varepsilon_f^0 > \varepsilon_n$ is satisfied, whereas when $\varepsilon_n > \varepsilon_c$ (dot-and-dashed lines) the opposite condition is met, that is, $\varepsilon_n > \varepsilon_f^0 > \varepsilon_c$. Therefore, to implement the immersion condition when $g < g^{n,1}$ the filler has to have the intermediate dielectric permittivity lying between ε_n and ε_c. However, after passing the point $g = g^{n,1}$ the immersion condition is changing. When $\varepsilon_c > \varepsilon_n$, immersion takes place for $\varepsilon_c > \varepsilon_n > \varepsilon_f^0$, whereas when $\varepsilon_n > \varepsilon_c$ it takes place for $\varepsilon_f^0 > \varepsilon_n > \varepsilon_c$. Thus, the filler should be optically more or less dense than both of the components of the scatterers. These conditions again recur to the previous one ($\varepsilon_c > \varepsilon_f^0 > \varepsilon_n$ or $\varepsilon_n > \varepsilon_f^0 > \varepsilon_c$) when g is growing further as we pass the first resonant point $g^{\mathrm{res},1}$, and so on.

Another interesting feature of the function $\varepsilon_f^0(g)$ is demonstrated in Figure 2.7.4. The figure presents a wide range of g dependences of function $\varepsilon_f^0(g)$ for the cases of 1D (a), 2D (b), and 3D (c) 3c-PhCs, which are plotted like ones shown in Figure 2.7.2; that is, there are the three cases considered: $\varepsilon_n = \varepsilon_c$, $\varepsilon_n > \varepsilon_c$, and $\varepsilon_n < \varepsilon_c$. To analyze how the geometrical dimensionality affects the immersion of the Bragg bands, we superimposed curves from Figure 2.7.4a–c to 2.7.4d. All curves were shifted along the abscissas by values (also shown by arrows) $g^{n,1} = 3.93$ (1D), 6.00 (2D), and 7.81 (3D). A quite unexpected result is revealed by Figure 2.7.4d. Despite the fact that the immersion conditions in 3c-PhCs with different

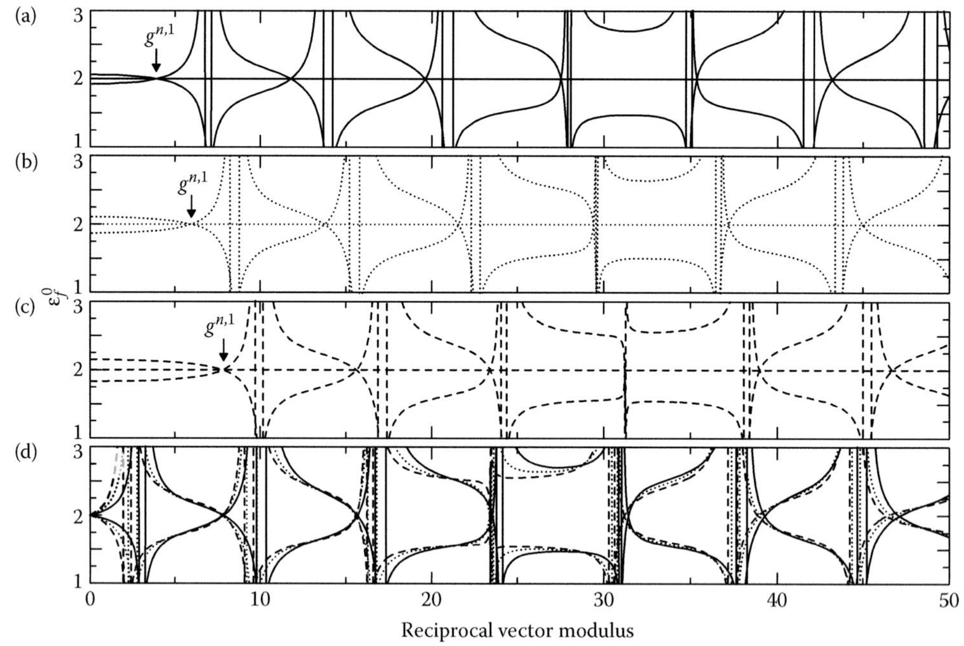

FIGURE 2.7.4
(a)–(c) The filler permittivity $\varepsilon_f^0(g)$ corresponding to the Bragg band immersion conditions as a function of the reciprocal lattice vector modulus g. The three-component photonic crystal composed of scatterers with the nucleus ($r_n = 0.35$, $\varepsilon_n = 2.0$) and different coats ($r_c = 0.45$, $\varepsilon_c = 1.7, 2.0, 2.3$) for different dimensions: (a) one dimension—solid lines, (b) two dimensions—dotted lines, (c) three dimensions—dashed lines. (d) The comparison of the $\varepsilon_f^0(g)$ dependences for three different dimensions. The curves are shifted along the abscissas by the value of the $g^{n,1}$ (shown by arrows). The styles of the curves correspond to (a)–(c). (Reprinted from A. K. Samusev et al. J. Phys. Condens. Matter **22**, 115401, 2010. With permission.)

dimensions have different mathematical definitions, after centering at $g^{n,1}$ the functions $\varepsilon_f^0(g - g^{n,1})$ virtually do not depend on the PhC dimensionality.

Thus, change of the dimensionality of 3c-PhC, 1D ⇒ 2D ⇒ 3D, which is described by the substitution of the multiplier $(\sin(gr_n)/\sin(gr_c)) \Rightarrow (J_1(gr_n)/J_1(gr_c))(r_n/r_c) \Rightarrow (R(gr_n)/R(gr_c))$ in expressions 2.7.14a–c for $1/\varepsilon_f^0(g)$, mainly results in translation of the curve of the function $\varepsilon_f^0(g)$ as a whole by the value $g^{n,1}$; that is, it is practically equivalent to the transition to the region of photonic bands with higher Miller indices.

2.7.6 Splitting Effects in the Scales of Dielectric and Structural Parameters

Let us now discuss optical properties of the PhC during the Gedanken experiment of transition from two-component structures to those with inhomogeneous scatterers. In the 2c-PhCs consisting of homogeneous scatterers, the immersion condition for all photonic Bragg bands reduces to a trivial one, $\varepsilon_f = \varepsilon_s$, that leads to complete optical homogeneity of the structure. In the case of MPhCs the picture is qualitatively different because a complete optical homogeneity of the medium cannot be achieved. However, for any given Bragg band, except the resonant case, the realistic immersion conditions can be defined.

We use Equation 2.7.14c to analyze the immersion conditions for (*hkl*) Bragg bands of the three-component FCC structure and estimate the splitting of ε_f^0. To this end, we expressed the dependence $\varepsilon_f^0(g_{hkl})$ through the 3c-PhC parameters ε_c, ε_n, and r_n by assuming that the FCC structure is closely packed and, hence, by taking $r_c = 0.5$ Note that Equation 2.7.14c is linear relative to reciprocal permittivity; therefore, the $\Delta\varepsilon_f^0(g_{hkl})$ splitting with respect to the degeneracy point $\varepsilon_{f0}^0(g_{hkl})$ $(\varepsilon_{n0} = \varepsilon_{c0})$, which corresponds to the situation of 2c-PhC, can be defined as

$$\Delta\left(\frac{1}{\varepsilon_f^0(g_{hkl})}\right) = \frac{1}{\varepsilon_f^0(g_{hkl})} - \frac{1}{\varepsilon_{f0}^0(g_{hkl})}. \tag{2.7.17}$$

Let us analyze the effect of ε_f^0 splitting in the scale of ε_c. If the coat permittivity deviates from that of the nucleus, the $\Delta\varepsilon_f^0(g_{hkl})$ splitting is

$$\Delta\left(\frac{1}{\varepsilon_f^0(g_{hkl})}\right) = \left(\frac{1}{\varepsilon_c} - \frac{1}{\varepsilon_{c0}}\right)\left(\frac{R(0.5\cdot g_{hkl}) - R(g_{hkl}r_n)}{R(0.5\cdot g_{hkl})}\right). \tag{2.7.18}$$

Note that ε_n does not enter this expression.

The splitting effect in the case of deviation of ε_n from the magnitude corresponding to the degeneracy point is found in a similar way. The splitting $\Delta\varepsilon_f^0(g_{hkl})$ is derived from the expression

$$\Delta\left(\frac{1}{\varepsilon_f^0(g_{hkl})}\right) = \left(\frac{1}{\varepsilon_n} - \frac{1}{\varepsilon_{n0}}\right)\frac{R(g_{hkl}r_n)}{R(0.5\cdot g_{hkl})}, \tag{2.7.19}$$

which does not contain ε_c.

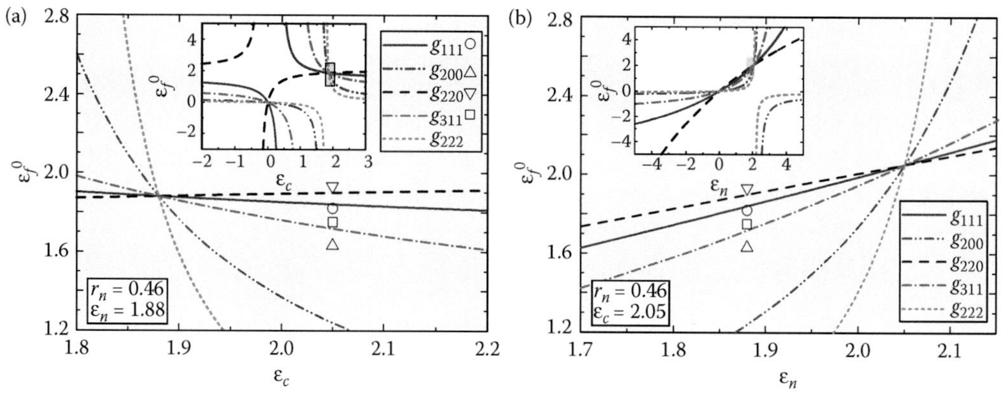

FIGURE 2.7.5
Splitting of the filler permittivity $\varepsilon_f^0(g_{hkl})$ in the scales of ε_c and ε_n. $\varepsilon_f^0(g_{hkl})$ corresponding to the immersion condition of the {111}, {200}, {220}, {311}, and {222} Bragg bands as a function of permittivity of the coat (a) and nucleus (b). The circle, triangles, and square are experimental data.[3] The inset shows the dependences of $\varepsilon_f^0(g_{hkl})$ in a reduced scale where the regions corresponding to the main plots are indicated. (Reprinted from A. V. Moroz et al. J. Opt. Soc. Am. B **26**, 176, 2009. With permission.)

The calculated dependences are shown in Figure 2.7.5. The structural parameters are $r_n = 0.46$, $r_c = 0.5$, $\varepsilon_n = 1.88$ (in the dependence on ε_c), and $\varepsilon_c = 2.05$ (in the dependence on ε_n). The symbols in the figures indicate experimental values of $\varepsilon_f^0(g_{hkl})$ from the study by Rybin et al.[3] The three-component model describes fairly well three of four magnitudes of $\varepsilon_f^0(g_{hkl})$, but there is a considerable discrepancy for the near-resonance (200) Bragg band. Nevertheless, the mutual positions of $\varepsilon_f^0(g_{hkl})$ and (which is quite important) the sign of the $\Delta\varepsilon_f^0(g_{hkl})$ splitting are described correctly. As ε_c increases with respect to the degeneracy point $\varepsilon_{c0} = 1.88$, $\varepsilon_f^0(g_{220})$ increases but $\varepsilon_f^0(g_{111})$, $\varepsilon_f^0(g_{200})$, and $\varepsilon_f^0(g_{311})$ decrease, as depicted in Figure 2.7.5a. As follows from Equation 2.7.18, such a behavior of the $\Delta\varepsilon_f^0(g_{hkl})$ splitting for different (*hkl*) Bragg bands can be attributed to the behavior of the Rayleigh–Hans function $R(gr)$. The mutual positions of $\varepsilon_f^0(g_{hkl})$ and the sign of the splitting $\Delta\varepsilon_f^0(g_{hkl})$ in the dependences on ε_n are also described correctly (see Figure 2.7.5b).

Splitting of parameter $\varepsilon_f^0(g_{hkl})$ in the scale of structure parameter r_n is described in detail by Moroz et al.[7]

2.7.7 Conclusion

In this chapter, we suggested classification of PhCs as 2c-PhCs and MPhCs. The main difference in the immersion conditions in the transmission spectra of the 2c-PhCs and MPhCs is summarized schematically in Figure 2.7.6. In the case of 2c-PhCs, all the Bragg bands vanish simultaneously when the condition $\varepsilon_f = \varepsilon_n$ is satisfied and the structure becomes optically transparent. The MPhC is formed from scatterers with inhomogeneous permittivity profile $\varepsilon_s(r)$, and hence its structure is optically inhomogeneous irrespective of the filler permittivity ε_f. Inhomogeneity of $\varepsilon_s(r)$ gives rise to the effects of "splitting" of the immersion conditions in the scale of ε_f: the filler with the permittivity $\varepsilon_f^0(g)$ can immerse, in general, only one particular Bragg band. Note, however, that the case of accidental

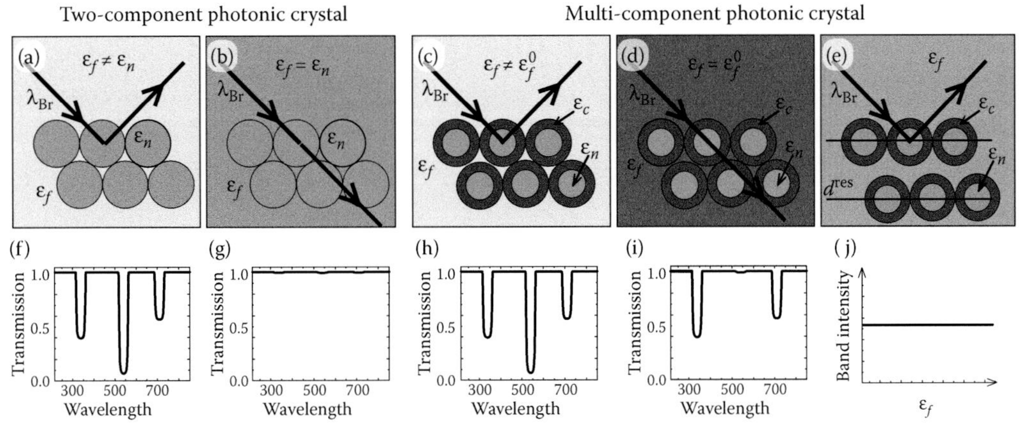

FIGURE 2.7.6
Schematic depiction of the Bragg diffraction by a two-component photonic crystal (2c-PhC) and a multicomponent PhC (MPhC) (upper row) and corresponding transmission spectra (lower row). For a 2c-PhC: the general case of the Bragg scattering (a), (f); the immersion condition (b), (g). For an MPhC: the general case of the Bragg scattering (c), (h); the immersion condition (d), (i); the resonant "nonimmersive" Bragg band defined by g^{res} that is by interlayer distance d_{res} (e), (j). (Reprinted from A. K. Samusev et al. J. Phys. Condens. Matter 22, 115401, 2010. With permission.)

degeneracy is also possible when a particular value of $\varepsilon_f^0(g)$ immerses several Bragg bands belonging to different families of planes. Nevertheless, even in this case, one will observe a number of nonimmersed Bragg bands in the spectra of MPhCs. The important exception is the resonant Bragg band, which cannot be immersed by varying the permittivity of one of the components. Resonant Bragg bands determine the Bragg wavelengths at which the MPhC is opaque regardless of the value of the filler permittivity. The described photonic properties are inherent to MPhCs of any dimension (1D, 2D, and 3D).

The main expressions describing Bragg band immersion effects in perfectly ordered 1D, 2D, and 3D MPhCs built up from centrosymmetric inhomogeneous scatterers are summarized in Table 2.7.1. The separate rows give expressions for 3c-PhCs, which are the simplest representatives of MPhCs. It is noteworthy that the quantitative results obtained in this study are applicable to *the case of low dielectric contrast only*.

The developed analytical model allows description of a variety of nontrivial experimental results obtained for synthetic opals a-SiO_2 and considers classified opals as a representative of MPhCs.[2,3] We underline that all the results of the model presented in this chapter *are valid only for the case of perfectly ordered structures consisting of identical scatterers*. As demonstrated in Chapter 3.5 (with opals as an example), if scatterers are not perfectly similar, but have some random variation in their dielectric constants, the condition of complete immersion, that is, a complete disappearance of the corresponding Bragg bands in the optical spectra of MPhCs, cannot be achieved.

Even though approximate analytical techniques are known to have their limits of applicability, they not only permit computations of optical properties from the known structural parameters of the PhC (*direct problem*) but also allow one to "tailor" the structure of PhCs to have desirable optical characteristics (*inverse design problem*,[18] for which fully numerical techniques are inapplicable or rather cumbersome for most situations). A few examples of such inverse problems demonstrating design of PhCs and their structural parameters aiming to meet certain Bragg band characteristics are presented in the literature.[3,7]

TABLE 2.7.1

Expressions Describing the Effect of Bragg Band Immersion in Perfectly Ordered Low-Contrast 1D, 2D, and 3D MPhC

	1D	2D	3D
$\frac{1}{\varepsilon(r)}$, MPhC		$\frac{1}{\varepsilon_f}+\left(\frac{1}{\varepsilon_s(r)}-\frac{1}{\varepsilon_f}\right)\Theta(r_s-r)$	
$\frac{1}{\varepsilon(r)}$, 3c-PhC		$\frac{1}{\varepsilon_f}+\left(\frac{1}{\varepsilon_c}-\frac{1}{\varepsilon_f}\right)\Theta(r_c-r)+\left(\frac{1}{\varepsilon_n}-\frac{1}{\varepsilon_c}\right)\Theta(r_n-r)$	
$\frac{1}{\varepsilon_f^0(g)}$, MPhC	$\frac{g\int_0^{r_s}\frac{1}{\varepsilon_s(r)}\cos(gr)\,dr}{\sin(gr_s)}$	$\frac{g\int_0^{r_s}\frac{1}{\varepsilon_s(r)}rJ_0(gr)\,dr}{r_sJ_1(gr_s)}$	$\frac{g^2\int_0^{r_s}\frac{1}{\varepsilon_s(r)}r\sin(gr)\,dr}{R(gr_s)}$
$\frac{1}{\varepsilon_f^0(g)}$, 3c-PhC	$\frac{1}{\varepsilon_c}+\left(\frac{1}{\varepsilon_n}-\frac{1}{\varepsilon_c}\right)\frac{\sin(gr_n)}{\sin(gr_c)}$	$\frac{1}{\varepsilon_c}+\left(\frac{1}{\varepsilon_n}-\frac{1}{\varepsilon_c}\right)\frac{r_n}{r_c}\frac{J_1(gr_n)}{J_1(gr_c)}$	$\frac{1}{\varepsilon_c}+\left(\frac{1}{\varepsilon_n}-\frac{1}{\varepsilon_c}\right)\frac{R(gr_n)}{R(gr_c)}$
g^{res}, 3c-PhC	$\frac{1}{\varepsilon_c}=\left(\frac{1}{\varepsilon_c}-\frac{1}{\varepsilon_n}\right)\frac{\sin(g^{res}r_n)}{\sin(g^{res}r_c)}$	$\frac{1}{\varepsilon_c}=\left(\frac{1}{\varepsilon_c}-\frac{1}{\varepsilon_n}\right)\frac{r_n}{r_c}\frac{J_1(g^{res}r_n)}{J_1(g^{res}r_c)}$	$\frac{1}{\varepsilon_c}=\left(\frac{1}{\varepsilon_c}-\frac{1}{\varepsilon_n}\right)\frac{R(g^{res}r_n)}{R(g^{res}r_c)}$

Note: 1D, one dimension; 2D, two dimensions; 3D, three dimensions; 3c-PhC, three-component photonic crystal; MPhC, multicomponent photonic crystal.

Acknowledgments

This work was done in collaboration with A.K. Samusev, A.V. Baryshev, and A.V. Moroz. The work is supported by the Russian Foundation for Basic Research (Grant Nos. 10-02-01094 and 11-02-00865).

References

1. J. D. Joannopoulos, S. G. Johnson, J. N. Winn, and R. D. Meade, *Photonic Crystals: Molding the Flow of Light* (Princeton University Press, Princeton, 2008), 2nd ed.; K. Inoue and K. Ohtaka (eds.), *Photonic Crystals: Physics, Fabrication and Applications* (Springer-Verlag, Berlin/Heidelberg, 2004); K. Sakoda, *Optical Properties of Photonic Crystals* (Springer-Verlag, Berlin/Heidelberg, 2004), 2nd ed.
2. A. V. Baryshev, A. B. Khanikaev, M. Inoue, P. B. Lim, A. V. Sel'kin, G. Yushin, and M. F. Limonov, Phys. Rev. Lett. **99**, 063906 (2007).
3. M. V. Rybin, A. V. Baryshev, A. B. Khanikaev, M. Inoue, K. B. Samusev, A. V. Sel'kin, G. Yushin, and M. F. Limonov, Phys. Rev. B **77**, 205106 (2008).
4. V. Mizeikis, S. Juodkazis, A. Marcinkevicius, S. Matsuo, and H. Misawa, J. Photochem. Photobiol. C **2**, 35 (2001).
5. M. V. Rybin, A. B. Khanikaev, M. Inoue, K. B. Samusev, M. Steel, G. Yushin, and M. F. Limonov, Phys. Rev. Lett. **103**, 023901 (2009); Photonics Nanostruct. Fundam. Appl. **8**, 86 (2010).
6. A. K. Samusev, M. V. Rybin, and M. F. Limonov, Phys. Solid State **51**, 518 (2009); A. K. Samusev, K. B. Samusev, I. S. Sinev, M. V. Rybin, and M. F. Limonov, *ibid.* **53**, 1415 (2011); A. K. Samusev, K. B. Samusev, M. V. Rybin, and M. F. Limonov, J. Phys. Condens. Matter **22**, 115401 (2010);

7. A. V. Moroz, M. V. Rybin, K. B. Samusev, and M. F. Limonov, J. Opt. Soc. Am. B **26**, 176 (2009).
8. H. S. Sözüer, J.W. Haus, and R. Inguva, Phys. Rev. B **45**, 13962 (1992).
9. A. Glushko and L. Karachevtseva, Photonics Nanostruct. Fundam. Appl. **4**, 141 (2006).
10. H. Takeda and K. Yoshino, Appl. Phys. Lett. **80**, 4495 (2002); J. Appl. Phys. **93**, 3188 (2003).
11. J. F. Galisteo-López, F. García-Santamaría, D. Golmayo, B. H. Juárez, C. López, and E. Palacios-Lidón, Photonics Nanostruct. Fundam. Appl. **2**, 117 (2004).
12. P. D. García, J. F. Galisteo-López, and C. López, Appl. Phys. Lett. **87**, 201109 (2005).
13. A. Yariv and P. Yeh, *Optical Waves in Crystals* (John Wiley & Sons, New York, 1984).
14. K. M. Ho, C.T. Chan, and C. M. Soukoulis, Phys. Rev. Lett. **65**, 3152 (1990).
15. H. M. van Driel and W. L. Vos, Phys. Rev. B **62**, 9872 (2000).
16. M. Born and E. Wolf, *Principles of Optics* (Cambridge University Press, Cambridge, 2006), 7th ed.
17. A. V. Baryshev, A. B. Khanikaev, H. Uchida, M. Inoue, and M. F. Limonov, Phys. Rev. B **73**, 033103 (2006); M. V. Rybin, A. V. Baryshev, M. Inoue, A. A. Kaplyanskii, V. A. Kosobukin, M. F. Limonov, A. K. Samusev, and A. V. Sel'kin, Photonics Nanostruct. Fundam. Appl. **4**, 146 (2006); A. G. Bazhenova, A. V. Sel'kin, A. Y. Menshikova, and N. N. Shevchenko, Phys. Solid State **49**, 2109 (2007).
18. M. Burger, S. Osher, and E. Yablonovitch, IEICE Trans. Electron. **87**, 258 (2003).

Section 3

Experiment

3.1

Anderson Localization of Light: Disorder-Induced Linear, Nonlinear, and Quantum Phenomena

Yoav Lahini

Weizmann Institute of Science

Tal Schwartz

Technion—Israel Institute of Technology

University of Strasbourg

Yaron Silberberg

Weizmann Institute of Science

Mordechai Segev

Technion—Israel Institute of Technology

CONTENTS

3.1.1 Introduction, General Concept, and First Observations

The physics of periodic systems, as is largely discussed in this book, is of fundamental importance and results in a multitude of phenomena governing the transport of waves. However, as it often so happens, deviations from perfect periodicity can result in higher complexity and give rise to surprising effects. Anderson localization is one of the most well-known scenarios where such behavior arises, and it describes the alteration of a conducting crystal into an insulator, as a result of disorder superimposed on the underlying periodic structure. Traditionally, disorder in crystals was modeled as a perturbation scattering the electrons in a random fashion, whereby the electrons were treated as point-like particles. This logic leads to the diffusive Brownian motion that stands behind Ohm's law. However, in his seminal paper of 1958, Phillip Anderson revisited the effect of disorder on the conduction of an otherwise-periodic crystal, taking into account the wave nature of electrons.[1] He found that, under a broad range of conditions, the classical diffusive motion of electrons breaks down as the electronic wave-functions become exponentially localized. Consequently, when an electron is initially placed on one atom, its wave function will no longer diffuse to cover the whole crystal, but will rather remain localized around its initial position. In other words, the material will cease to conduct charge and will become an insulator. This localization phenomenon is a direct consequence of interference between different paths, arising from multiple scattering of the electron by lattice defects.

Anderson's model assumed that the atomic potential is time-invariant, and it was later found that temporal variations in the potential diminish localization effects. Moreover, the presence of temporal fluctuations (e.g., phonon vibrations at finite temperature) reduces the coherence of the scattering process and destroys the interference effects and eventually leads to recovery of Ohm's law. Perhaps even more importantly, Anderson's model represents a single particle or an ensemble of noninteracting particles. But of course electrons are fermions that fundamentally interact (e.g., through Coulomb's law, spin exchange, etc.). When fermionic interactions are included the scenario changes dramatically, and generally localization does not occur. These two preconditions underlying Anderson's model, namely, time-invariance of the potential and absence of interactions, have posed great difficulties for observing Anderson localization in atomic crystals.[2] Nevertheless, as Anderson localization is in essence a wave-mechanics phenomenon, it was realized that it is universal to all wave systems, and in particular it should occur in optics.[3] In fact, random scattering of light is ubiquitous in nature, and it can be found, for example, in clouds, milk, or sugar. These materials are all microscopically transparent to light; however, they appear opaque because of multiple scattering of the light traveling inside them. Thus, optics seems an ideal framework to study localization effects, as coherence is naturally preserved and as photons are inherently noninteracting bosons. This was the logic behind experiments studying the light-transmission properties of random media, which showed exponential decay of transmittance with sample length.[4,5] In performing such experiments, one should bare in mind that any slight absorbance will also result in an exponentially decaying transmittance (as the path of the scattered light inside the medium can be much longer than the actual length of the sample) and it is therefore difficult to discriminate between the two effects, requiring more elaborate measurements.[5]

A different avenue for studying the effects of disorder on the transport of light, called the transverse localization scheme, was proposed by De Raedt, Lagendijk, and de Vries in the late 1980s,[6] but this proposition remained dormant for a long time. The progress came following another research direction: discrete solitons.[7] In 1988, Christodoulides and

Joseph proposed the existence of solitons in waveguide arrays, which, under some approximations (paraxial propagation, tight-binding, and Kerr-type nonlinearity), were modeled by a discrete cubic Schrödinger-type equation, hence the name discrete solitons.[7] It took a decade to observe discrete solitons experimentally,[8] and another five years to demonstrate them in two transverse dimensions.[9] But around 2005 this area of nonlinear waves and solitons in photonic lattices (waveguide arrays) has become a major research area in both nonlinear optics and soliton science (see recent reviews[10]). Following this progress, it became natural to search for localization effects in paraxial disordered photonic systems. However, early experiments seemed rather discouraging,[11,12] partly because some of the concepts were missing (e.g., the necessity for ensemble averaging to obtain meaningful results), and partly because Lagendijk's transverse localization scheme[6] was unknown to researchers in the area of discrete solitons. It was not until 2007 that the first successful transverse localization experiments emerged,[13,14] and many experiments have followed since.[15–17] To date, transverse localization of light has become the most convenient and direct scheme for observing localization effects with electromagnetic waves, possibly even the most important avenue for localization experiments in all areas of research.

The evolution of the optical waves in the transverse localization scheme is described by the Schrödinger-type paraxial equation for monochromatic light

$$i\frac{\partial\Psi}{\partial z} = \left[-\frac{1}{2k}\left(\frac{\partial^2}{\partial x^2}+\frac{\partial^2}{\partial y^2}\right)-\frac{k}{n_0}\Delta n(x,y,z)\right]\Psi \equiv \hat{H}\Psi \tag{3.1.1}$$

where z is the propagation coordinate, x and y are the transverse dimensions, Ψ is the slowly varying envelope of an optical field $E(\overline{r},t) = \mathrm{Re}[\Psi(x,y,z)e^{i(kz-\omega t)}]$ of frequency ω and wave number $k = \omega n_0/c$, n_0 is the bulk refractive index, and Δn is the local change in the refractive index (lattice + disorder), on the background of a much larger refractive index of the bulk n_0 (with $|\Delta n| \ll n_0$). Equation 3.1.1 has the form of the Schrödinger equation: the equivalence emerges when $z \to t$ and $-\Delta n \to V$. Hence, the evolution of a light beam behaves like the wave packet of a quantum particle in a two-dimensional (2D) potential but with the coordinate z replacing time. Recalling that Anderson localization requires a stationary potential implies that in Equation 3.1.1 the index change Δn must be propagation-invariant; that is, $\Delta n(x, y)$ has to be z-independent.

It is important to emphasize the role of ensemble averaging in the transverse localization scheme, as explained in the first experiments.[13] Namely, the observables of the analogous quantum system, being an expectation-value problem, are all of a statistical nature. In this vein, to obtain meaningful experimental results in the photonic system, one has to average over multiple realizations of the disorder (the quantum system is actually self-averaging given an adequately long evolution time, but the propagation distance in photonic systems is not large enough to exhibit self-averaging). The most important observable in the transverse localization scheme is therefore $\langle|\Psi(x, y, z = L)|^2\rangle$: the intensity emerging from the photonic system after propagating a distance L, averaged over multiple (~100) experiments with different spatial realizations of disorder (with the disorder characterized by some statistical distribution).

As stated earlier, the analogy between the Schrödinger equation and paraxial optics was used in many recent experiments, demonstrating concepts from solid-state physics using paraxial optical settings.[10] Specifically for localization experiments, this approach opened the possibility to follow the evolution of wave packets in the disordered optical structures at the microscopic level. Together with the inherent flexibility and controllability of optical systems, this approach led to the first direct observations of Anderson localization,[13] as originally described by Anderson: the expansion of an initially confined wave packet comes to a complete halt and attains exponentially decaying tails, owing to multiple scattering in a disordered crystal. In addition, this approach made it possible to directly observe other predictions of the theory, such as the observation of individual localized eigenmodes and their properties.[14] Absorption and thermal vibrations of the underlying potential do not pose problems in this system. Rather, they can be introduced in a controlled manner. Furthermore, it is possible to directly measure the wave-function profile inside the disordered system. This approach, which was originally conceived and implemented in the context of optical waves, was later echoed in experiments on Anderson localization of matter waves (Bose–Einstein condensates).[18]

An important additional aspect of the optical setups described here is that they were the first to enable the experimental study of a related new problem: the interplay between nonlinear interactions and Anderson localization. As stated above, the localization phenomenon predicted by Anderson is a noninteracting linear interference effect. It therefore corresponds to disordered systems containing a single particle, or many identical noninteracting particles. This, however, is not the general case for real systems that usually contain many particles for which mutual interactions are not always negligible. The problem of the interplay of disorder and interactions remains one of the greatest challenges in modern solid state physics.[19] In essence, the nonlinearity can couple the localized modes of the disordered potential in a nontrivial manner, resulting in effects that are much more complex than the linear ones. Such nonlinear interactions may appear in various forms in different systems—Coulomb or spin-exchange interactions among the electrons in solids, or dipole–dipole interaction between cold atoms in Bose–Einstein condensates, to name a few. In optics, nonlinear response of the disordered medium gives rise to indirect interactions between the photons, through a variety of mechanisms, for example, an intensity-dependent contribution to refractive index. This particular interaction corresponds to the case of cold (bosonics) atoms with pair-wise attractive or repulsive interactions. This kind of optical nonlinearity enters the evolution equation as a nonlinear term $\Delta n_{\mathrm{NL}} = f(|\Psi|^2)$ added to the index change Δn (the "potential" term) in Equation 3.1.1. Hence, this simple system of Anderson localization of light in the presence of nonlinearity may set a basis for a better understanding of complicated quantum many-body systems.

The configuration used to observe Anderson localization in 2D disordered lattices[13] is depicted in Figure 3.1.1a: a periodic waveguide array with a controlled level of disorder is formed in a photorefractive material using the optical induction technique,[20] by interfering several plane waves with a random speckled beam. This creates a volume hologram that is transformed into a variation in the refractive index within the material. The optical induction technique enables fine control over the level of disorder by varying the intensity ratio between the speckled beam and the plane waves creating the periodic lattice, going from a perfectly periodic structure up to high levels of disorder, and eventually even to a completely random potential. More importantly, this technique facilitates the real-time creation of many different statistically identical realizations of the disorder and ensemble averaging, which is crucial in such a statistical system. After the disordered lattice is formed, a narrow light beam (of width comparable to that of a single waveguide) is launched into the medium. In the absence of disorder, the beam undergoes diffraction broadening, expanding during propagation (Figure 3.1.1a, top) due to coupling between adjacent waveguides. Linking to quantum mechanics, time is replaced with the

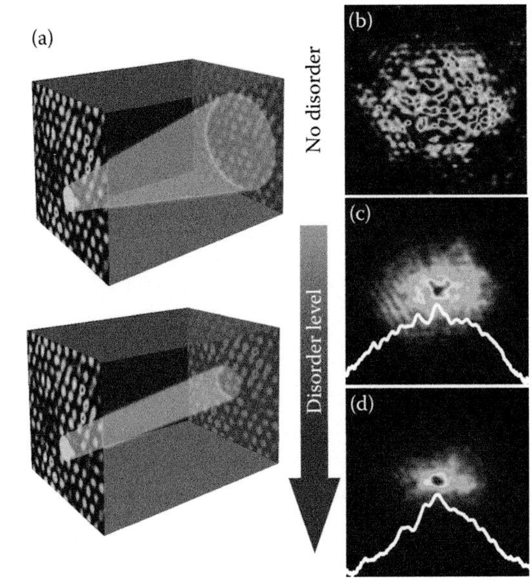

FIGURE 3.1.1
(**See color insert.**) (a) The transverse localization scheme: a probe beam propagates in photonic lattice with a controlled level of disorder. In the absence of disorder, the beam exhibits ballistic transport: its width increases linearly with propagation distance (top). Under the influence of disorder, the beam becomes exponentially localized in the transverse plane, maintaining its width throughout propagation (bottom). (b)–(d) Ensemble-averaged intensity distribution at the output face of lattice. The results show gradual transition from ballistic transport (b), where the diffraction pattern reflects the hexagonal symmetry of the lattice, to diffusion (c) in presence of disorder (designated by a Gaussian shape of the intensity profile, plotted in logarithmic scale) and, at stronger disorder, to localization with an exponentially decaying intensity profile (d). (Adapted from T. Schwartz et al., Nature **446**, 52, 2007.)

propagation distance z, whereas the spatial evolution is restricted to the transverse x–y plane normal to propagation direction. In this view, the effect of disorder is to halt the broadening of the beam, as illustrated at the bottom of Figure 3.1.1a, with the light remaining exponentially localized in that transverse plane. Hence the term "transverse localization." The experimental observation of the effect is shown in Figure 3.1.1b–d. In the absence of disorder, the beam diffracts in the periodic structure by ballistic transport, manifested in the hexagonal symmetry of the diffraction intensity pattern (Figure 3.1.1b), and in the fact that the width of the expanding beam grows in proportion to the propagation distance. When weak disorder is introduced, the lattice symmetry is lost, and the intensity tunnels randomly among the lattice sites (Figure 3.1.1c). Here, the (ensemble-averaged) transport is diffusive, as is evident by the Gaussian profile of the beam. When the level of disorder is increased, the output intensity profile narrows down and the beam acquires exponentially decaying tails (Figure 3.1.1d). This exponential decay means that the (transverse) transport of light stops: after a short propagation distance, along which the beam expands diffusively, the (ensemble-averaged) beam diameter reaches the localization regime and its diffraction broadening is arrested. The exponentially decaying tails of the ensemble-averaged beam are a direct indication of Anderson localization ("strong localization"). At the same time, the reduced transport can be also seen as a narrowing of beam at the output face as the level of disorder is increased. To characterize the beam width, the inverse participation ratio, defined as $P = \langle [\int I(x, y, L)^2 \, dx \, dy] \rangle / \langle [\int I(x, y, L) \, dx \, dy]^2 \rangle$, is used, with

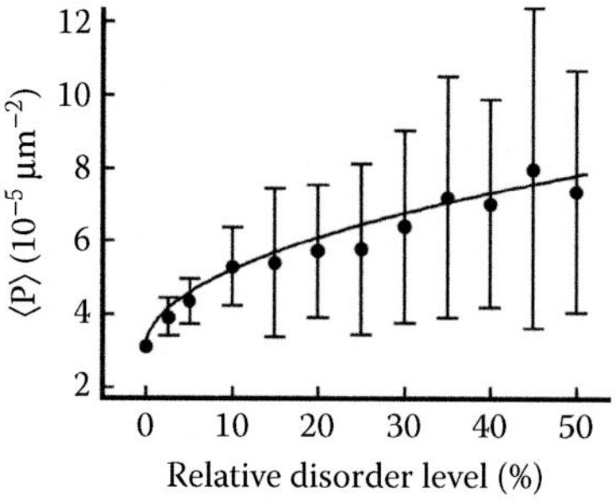

FIGURE 3.1.2
Average inverse participation ratio as a function of disorder level, showing the gradual localization of the beam. The error bars are the statistical fluctuations of different random realizations. (Adapted from T. Schwartz et al., Nature **446**, 52, 2007.)

$I(x, y) = |\Psi(x, y, z = L)|^2$ being the intensity distribution at the output plane $z = L$. This quantity gives a measure of the confinement of the beam, and it is inversely proportional to the effective beam area. Figure 3.1.2 shows the ensemble-averaged value of the inverse participation ratio P as a function of the disorder level, along with its statistical standard deviation (marked by error bars). The monotonic increase in P indicates the gradual narrowing of the beam with disorder. Interestingly, it also reveals that, when Anderson localization occurs, the relative fluctuations of the inverse participation ratio are very large—of the order of unity,[13] indicating large variations of the beam width between different realizations of the disorder, and hence a remarkable sensitivity of the transport to the fine details of the disordered potential.

As discussed above, a most significant issue related to Anderson localization is the interplay between disorder and nonlinearity. One question that immediately comes to mind is what happens to the localization process itself, under weak or strong nonlinear conditions. In the first experiments,[13] the optically induced system was also used as a well-controlled tool to study the fundamental issue of localization in the presence of nonlinearity. The simplest way to introduce nonlinearity into such a system is by increasing the intensity of the probe beam, such that it creates a nonlinear index change on top of the disordered lattice. Figure 3.1.3a compares the ensemble-averaged width of the beam with a nonlinearity of the self-focusing type to the results of linear propagation, as a function of disorder strength. The experiments showed that under self-focusing (positive, attractive) nonlinearity the localization is enhanced. Not only does the ensemble-averaged beam narrow down but also the characteristic exponential decay of localization appears at a lower level of disorder, for which the linear transport is still diffusive. This is revealed by examining the average intensity profile, shown in Figure 3.1.3b (on a logarithmic scale) for the linear case, and with various strengths of nonlinearity. Thus, self-focusing nonlinearity promotes the localization process, at least at finite propagation distances and when the strength of the nonlinearity is not too high (i.e., the nonlinear index change is smaller than the linear index change defining the disordered lattice).

The influence of nonlinearity on the transport in disordered lattices could be rather complex, if one thinks of the transport in periodic structures and their associated band structure. Recalling that dispersion can be either normal or anomalous (like the positive or negative effective mass of the electron in a crystal), one may expect that localization effects would behave differently in these two regimes, when nonlinearity combines with disorder. In the anomalous dispersion regime, close to the Bragg surface (the edge of the

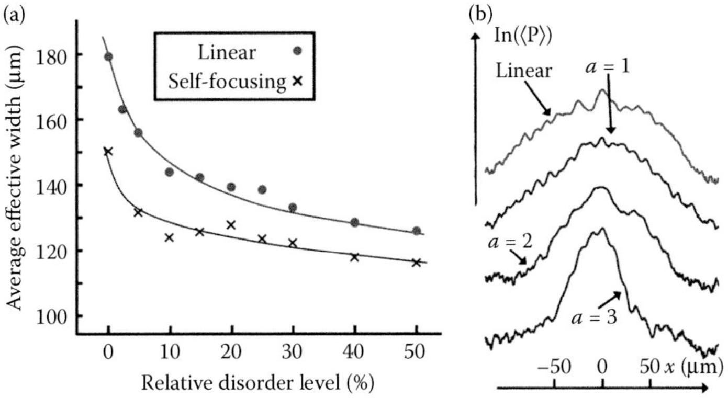

FIGURE 3.1.3
(a) Output beam width versus disorder strength with self-focusing nonlinearity (crosses), compared with results of linear propagation (dots). (b) Ensemble-averaged intensity profiles (in logarithmic scale) at the output of the lattice, showing transition from diffusive transport under linear conditions (upper curve) to localization characterized by exponentially decaying tails under nonlinear conditions, demonstrating that self-focusing nonlinearity promotes localization (α scales the nonlinearity strength). (Adapted from T. Schwartz et al., Nature **446**, 52, 2007.)

first Brillouin zone), a wave packet tends to narrow under a negative (self-defocusing, repulsive) nonlinearity, whereas a positive nonlinearity causes broadening of the wave packet. For such a wave packet in the negative effective mass regime, one would expect that negative nonlinearity would enhance the localization process in the presence of weak disorder. This is indeed the case (at least for finite propagation distances); however, for a strong enough disorder the concept of effective mass would no longer hold, and hence at some disorder level the system should change its behavior abruptly. These ideas raise intriguing questions on the interplay between the periodic structure, disorder, and nonlinearity.

Finally, these optical techniques enable one to elucidate quantum properties of localization that go beyond the wave nature of the propagation. By considering quantum statistical properties of the propagating photons, it was discovered that unique correlations develop between copropagating particles.[21] As described in the next section, such properties can be observed by studying the propagation of nonclassical states of light in disordered lattices, but they are also echoed in the intensity correlations of classical waves.

3.1.2 Anderson Localization and Nonlinearity in One Dimension

3.1.2.1 The Disordered One-Dimensional Waveguide Lattice

In this section, we describe the experimental study of the effects of disorder and the interplay with nonlinear effects in one-dimensional (1D) lattices containing disorder.

The experimental setup used to carry out the experiment described below is a 1D lattice of evanescently coupled, single-mode optical waveguides, patterned on an AlGaAs substrate.[8,10] The waveguide array is illustrated schematically in Figure 3.1.4a. In these

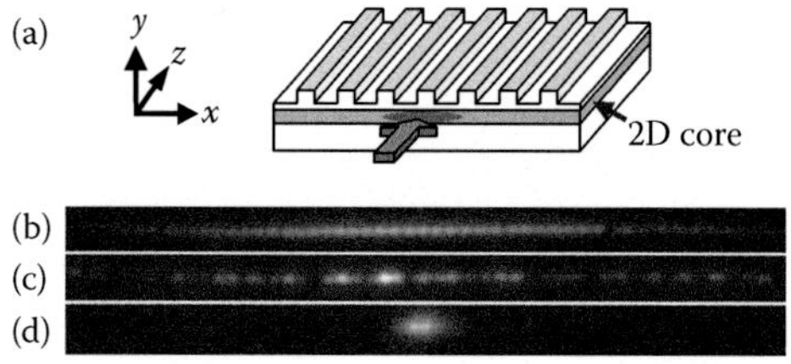

FIGURE 3.1.4
(a) Schematic view of the sample used in the experiments. The large arrow indicates the input beam. (b)–(d) Images of output light distribution, when the input beam covers a few lattice sites: (b) in a periodic lattice, (c) in a disordered lattice, when the input beam is coupled to a location that exhibits a high degree of expansion, and (d) in the same disordered lattice when the beam is coupled to a location in which localization is clearly observed. (Adapted from Y. Lahini et al., Phys. Rev. Lett. **100**, 013906, 2008.)

systems, light is injected into one waveguide (or a few waveguides) at the input facet. The light is allowed to propagate freely in the z direction (see Figure 3.1.4a), and it also tunnels to neighboring waveguides (via evanescent coupling) in the transverse direction (x axis in Figure 3.1.4a). The distribution of the light intensity is measured at the output facet (see an example in Figure 3.1.4b). In addition, at sufficiently high intensities, the waveguides behave in a nonlinear manner—that is, the index of refraction of each waveguide changes as a function of the intensity of light propagating within it.

The physical description of the propagation of light in 1D waveguide lattices can be approximated by a set of coupled nonlinear discrete Schrödinger equations:

$$-i\frac{\partial U_n}{\partial z} = \beta_n U_n + \Sigma_{<m,n>} C_{m,n} U_m + \gamma |U_n|^2 U_n \tag{3.1.2}$$

Here $n = 1 \ldots N$ where N is the number of lattice sites (waveguides), U_n is the wave amplitude at site n, β_n is the eigenvalue (propagation constant) associated with the reflection index of the nth waveguide, $C_{n,m}$ are the tunneling amplitudes between two adjacent sites, which is related to the separation between the waveguides, and z is the longitudinal space coordinate along the waveguides. The nonlinear parameter γ describes the additional dependence of the index of refraction at site n on the local light intensity.

In the linear limit, that is, for $\gamma = 0$ (and up to an inversion in sign), these equations are identical to the equations describing the time evolution of a single electron in an atomic lattice under the tight binding or the evolution of noninteracting matter waves in optical lattices, where in the optical case the time evolution is accompanied by propagation in the z direction, and U_n is the wave amplitude at site n. With nonlinearity, these equations are identical to the discrete Gross–Pitaevskii equation describing the mean field dynamics of ultra-cold atomic gases in lattices.[22]

3.1.2.1.1 *Types of Disorder*

For a perfectly periodic waveguide lattice, all waveguides are identical and equally spaced—that is, all $\beta_n = \beta$, and all $C_{n,m} = C$. In full correspondence to the tight-binding model, disorder can be introduced to the model, where the disorder type can be divided into two broad categories. One category is of "diagonal" disorder, in which the β_n parameters are random in a certain range $\beta_0 \pm \Delta/2$, but the tunneling amplitudes are fixed across the lattice. A measure of disorder in such a case is given by the ratio Δ/C.[23] Such disorder was considered by Anderson in his original work,[1] in what is now known as the Anderson model. A second type of disorder is known as "off-diagonal" disorder, in which the β_n parameters are fixed across the lattice, yet the tunneling amplitudes $C_{n,m}$ are randomized.

3.1.2.1.2 Eigenmodes and Spectrum of Disordered Lattices

To make things clearer later on, we first review the properties of the linear disordered tight-binding model. We then show how these properties, which were hard to observe directly before, can be probed experimentally in waveguide lattices. We then describe an experimental study of the effect nonlinear perturbations on localized excitations.

We start with an analysis of the Anderson model in one dimension, that is, a lattice with pure diagonal disorder. For finite size lattices and in the linear limit, the spectrum of a disordered lattice can be exactly calculated by diagonalizing the corresponding Equation 3.1.1. Figure 3.1.5 shows the result for $N = 69$ and $\Delta/C = 1$. For weak disorder, the resulting band of eigenvalue deviates only slightly from the cosine-shaped band of a perfectly ordered lattice.[10] It is instructive to consider the structure of the different eigenmodes of the system. Several eigenmodes that have their corresponding eigenvalues near both edges of the spectrum are localized in space, with exponentially decaying tails. This happens even though the density of states near the band edges is not significantly lower than the one at the band center, an indication that the localization effects are indeed a result of interference and not of trivial defect states.[23,24] However, the localized eigenmodes with eigenvalues coming from the two edges of the spectrum have a different phase structure: localized eigenmodes near the high eigenvalue edge of the band are all flat in phase—that is, their wave function's amplitude is in-phase at all sites (see inset), whereas the localized eigenmodes at the low edge of the band are all staggered, that is, their wave function's

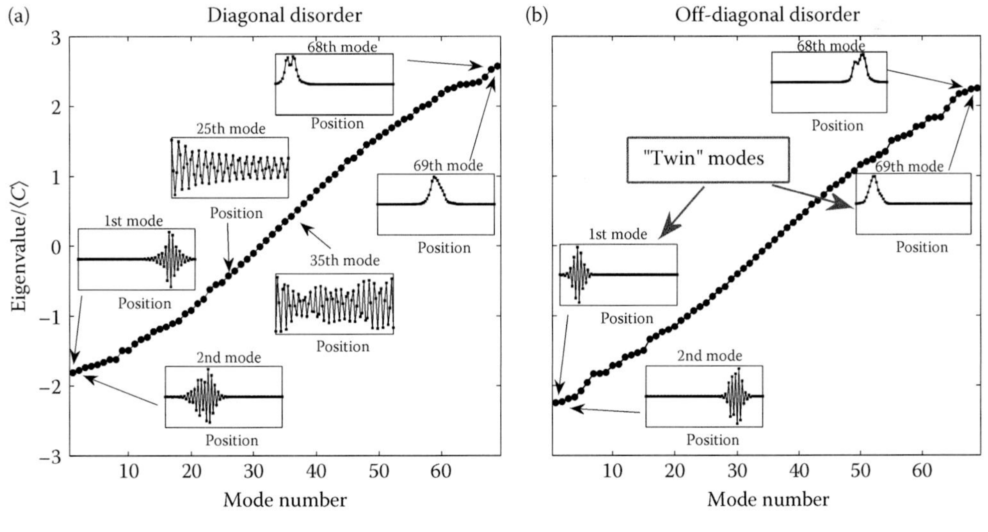

FIGURE 3.1.5
Spectrum of eigenvalues and eigenmodes of a disordered, one-dimensional, tight-binding lattice with $N = 69$. (a) For the case of a lattice with pure diagonal disorder. (b) For the case of pure off-diagonal disorder. Note how in contrast to (a), the spectrum in (b) is symmetric with respect to the middle of the band, and modes with symmetric eigenvalue (i.e., 1 and 69, 2 and 68, etc.) share the same region in space.

amplitude has a π phase flip between adjacent sites (inset). Notably, different eigenmodes from the same band edge are localized at well-separated regions in space. As for eigenmodes from the middle of the band, in finite size systems as the one considered here, these remain "extended" (inset). This does not contradict the common knowledge that for infinite 1D disordered systems all the modes are localized. Indeed, these modes will be also be localized in adequately large systems, but on a much longer length scale. As disorder is increased, a larger fraction of the modes becomes localized within the finite lattice, and the sample becomes localized.

Next, we run a similar analysis on lattices with pure off-diagonal disorder. Such lattices are known to exhibit several unique spectral properties that are different from the ones exhibited by lattices with diagonal disorder.[25] These properties and others described below will become important later on.

In Figure 3.1.5b, we show the eigenvalues and eigenmodes of a single realization of a disordered lattice with pure off-diagonal disorder. As in the diagonal disorder case, eigenmodes near the edge of the spectrum are tightly localized; modes from the high eigenvalue edge of the spectrum have a flat-phase structure, whereas modes from the low eigenvalue edge of the spectrum have a staggered phase. However, in lattices with off-diagonal disorder each eigenmode at one edge of the spectrum had a twin-eigenmode at the other edge. These twin eigenmodes occupy the same region of the lattice, have the same envelope distribution of the absolute amplitude, but differ in phase structure as shown in Figure 3.1.5b. This property does not exist in lattices with diagonal disorder.

Theoretically, this property can be understood in the following manner: without loss of generality, we can set the diagonal terms of the Hamiltonian to zero, and we denote the randomized tunneling terms as C_n. Let $(a_1, \ldots, a_N)$ be an eigenvector with an eigenvalue. Then, $C_1{*}a_2 = \lambda{*}a_1$, $C_1{*}a_1 + C_2{*}a_3 = \lambda{*}a_2$, $C_2{*}a_2 + C_3{*}a_4 = \lambda{*}a_3$, and so on. Now it is obvious that the vector $b_n = (-1)^n \times a_n$, is also an eigenvector with an eigenvalue $-\lambda$. That is, the eigenvectors are paired around the center of the band, where twin eigenmodes share the same density distribution in absolute value, but an opposite (staggered) phase structure. It is also easy to see that this property will not be exhibited by lattices with diagonal disorder.

3.1.2.1.3 Experimental Setup

To directly measure the properties of disordered systems we use the experimental system of the 1D waveguide lattice, as described previously. The salient feature of this system is that it can be perfectly controlled, and all parameters can be tuned at will. For example, diagonal disorder can be introduced to the lattice by randomly changing the width of each waveguide in a finite range $W_0 \pm \delta/2$ where W_0 is the mean value (typically 4 μm in our samples), which will result in a distribution of the β_n parameters in the range $\beta_0 \pm \Delta/2$. Off-diagonal disorder can be introduced in a similar fashion, by randomizing the distances between now-identical waveguides.

3.1.2.1.4 Effect of Beam Width

As pointed out earlier in the experimental technique described in this chapter, it is possible to tailor the initial condition (i.e., a light distribution at $z = 0$) by injecting light into one or several adjacent lattice sites, and by controlling the amplitude and phase at each site using beam-shaping techniques. The field injected into the array will now evolve according to Equation 3.1.1 until it emerges and is measured at the other end of the lattice (see Figure 3.1.4a). It is therefore possible, using different initial conditions, to directly measure different properties of disordered lattices, for example, to directly observe localized eigenmodes or to measure the Anderson localization of expanding wave packets.

As we noted above, the localized eigenmodes of disordered lattices have several interesting properties: the width of a typical eigenmode depends on the location of its corresponding eignvalue in the spectrum: modes near the edges of the spectrum are tightly localized, whereas modes near the middle of the spectrum are typically wider. This can create a situation in which in a finite size system (i.e., with N not too large) the modes from the edges of the band will be localized within the system, whereas modes from the middle of the spectrum will be localized on a length scale much larger than the system, essentially behaving as extended states. This property will become important when we discuss wave-packet expansion.

Another interesting property of the band-edge eigenmodes is their phase structure and spatial distribution: The first few lowest (highest) laying modes of the spectrum all have an almost purely flat (staggered) phase structure, and each one of these modes cover a different region in the lattice. As a result a very wide excitation beam that covers the whole lattice and is flat (staggered) in phase will excite mostly these modes (as shown, e.g., in Figure 3.1.6a and b). It can be shown that in this case the output light distribution is a sum

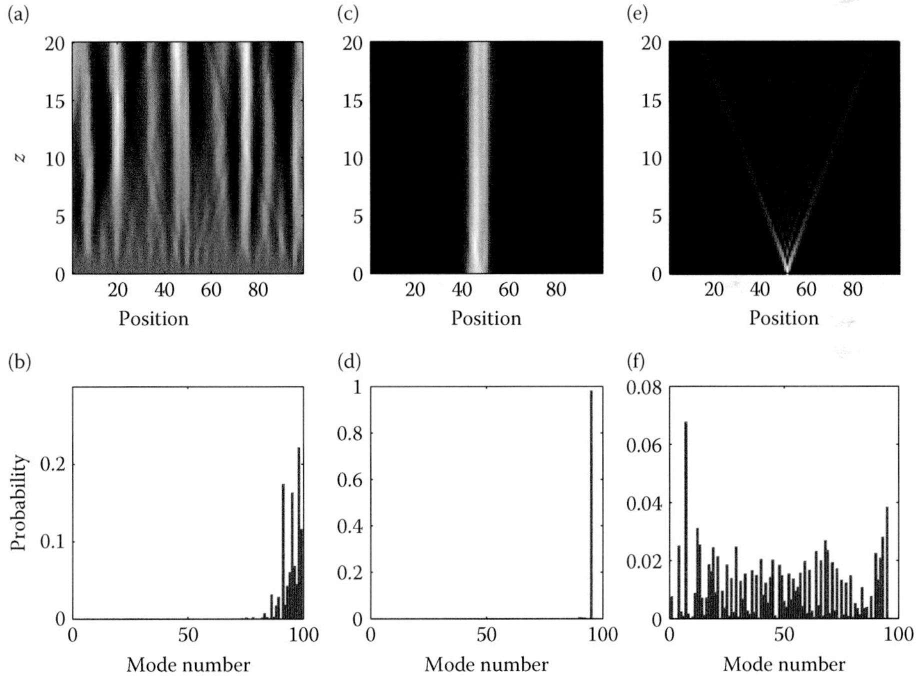

FIGURE 3.1.6
Observed propagation in a weakly disordered lattice for different beam widths. (a) For an initial excitation having a uniform excitation of all lattice sites, the beam is fragmented to few, stable localized excitations. (b) Spectral decomposition of this initial condition reveals that the localized excitations are the flat-phased eigenmodes with eigenvalues at the high-end part of the spectrum that significantly overlap such an initial condition. (c) In the same lattice, an initial excitation at the location of a flat-phased localized eigenmode, and at the correct width, excites only that mode, as can be seen also in the spectral decomposition in (d). (e) In the same lattice, an initial excitation of a single site results in a ballistically expanding wave packet, only slightly different from the one exhibited by periodic lattices. (f) The spectral decomposition shows that many lattice modes are excited, including the wide modes in the middle of the spectrum. In highly disordered lattices in which all the modes are localized within the lattice, this propagation will result in exponentially localized wave packets and smaller spectral content.

of the different, nonoverlapping modes, each occupying a different region of the lattice. If narrower beams are used, this will result in the excitation of a different combination of eigenmodes from the set of edge-localized modes, depending on the exact location and width of the beam. The excited wave packets, however, will show very little dynamics, as only tightly localized modes are excited (see Figure 3.1.6c and d). This situation changes when the width of the exciting beam becomes narrower than the width of the narrowest modes of the system (see Figure 3.1.6e and f). In this case, a wave packet of modes that also includes the wider "extended" modes is excited, and the wave packet shows transient dynamics or expansion before settling into a localized distribution.

3.1.2.2 Observation of Localized States

3.1.2.2.1 Observation of Single Localized Eigenmodes

In the experiment described below, we used a beam with a width that was adjusted to fit the width of the band-edge modes of a lattice with pure diagonal disorder. The disorder level in the sample was $\Delta/C = 1$, which yields an expected localization length of ~10. Using a beam with a width of approximately the localization length ξ, and scanning that beam across the lattice, it was possible to selectively excite the lowest lying localized eigenmodes of the disordered sample one by one. The results of this procedure are depicted in Figure 3.1.7. The observed localized output distributions are compared to calculations based on the specific realization of disorder, demonstrating the ability to excite and measure pure localized eigenmodes. A similar procedure, only with a staggered excitation beam, was used to excite the highest lying modes of the system (see Figure 3.1.7). Note that in diagonally disordered lattices the positions of the lowest eigenvalues of flat-phased eigenmodes never coincided with the position of the highest eigenvalues, staggered localized modes. Indeed, an input beam exciting a flat-phase localized eignmode was tilted to excite neighboring sites with a π phase difference. The output density showed considerable expansion (see dashed line in Figure 3.1.8b), suggesting that no staggered localized eigenmode resides in the same location.

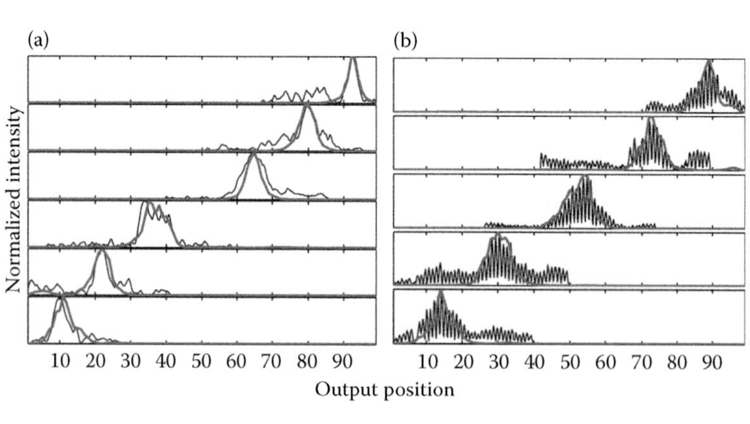

FIGURE 3.1.7
Measurements of single Anderson localized modes in a lattice with diagonal disorder. (a) Measurements of pure flat-phase Anderson localized modes from the low-eigenvalue edge of the spectrum. Panels show a comparison between measurements (thin line) and the corresponding calculated eigenmodes of the lattice (thick line). (c) Same for staggered localized eigenmodes, from the high-eigenvalue edge of the spectrum. (Adapted from Y. Lahini et al., Phys. Rev. Lett. **100**, 013906, 2008.)

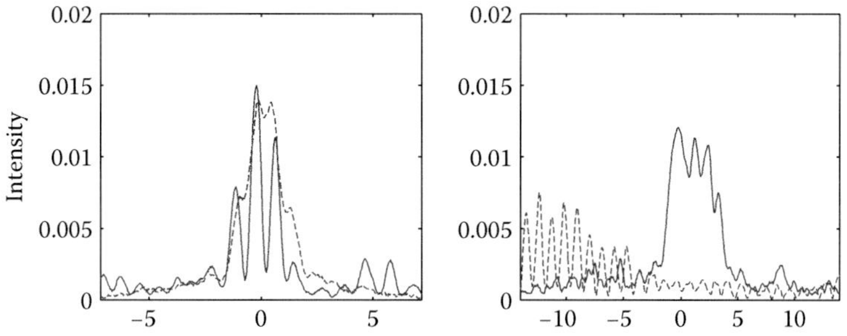

FIGURE 3.1.8
(a) Experimental measurement of two spectral-symmetric eigenmodes in a lattice with off-diagonal disorder, showing they share lattice position and envelope shape. (b) A similar measurement in a lattice with diagonal disorder cannot excite two modes sat the same position. (Adapted from Y. Lahini et al., Phys. Rev. A **84**, 041806(R), 2011.)

In contrast, Figure 3.1.8a shows the same procedure in a lattice with off-diagonal disorder. Here, the flat-phased beam excited a flat-phased localized eigenmode, and a beam with a π phase difference between adjacent sites excited a staggered localized eigenmode with the same width, the same spatial profile, and at the same location, in agreement with the properties of off-diagonal disordered lattices described above.

3.1.2.3 Effect of Nonlinear Perturbations on Localized Excitation

3.1.2.3.1 Weak Disorder Case

The effect of nonlinear perturbations on localized excitations is studied by exciting a pure localized mode in the linear regime and increasing the input beam power. The intensities are kept way below those required to exhibit self-focusing in a periodic lattice with the same average parameters[10] keeping the experiments in the weak nonlinear regime. Some localized states were found to exhibit significant response to nonlinearity. The results of two such experiments are shown in Figure 3.1.9. It is found that weak positive nonlinearity has an opposite effect on localized states from different edges of the spectral band. For the flat-phased localized states nonlinearity results in increased localization. In contrast, for the staggered localized stated, nonlinearity tends to result in decreased localization.

These results, understood on the basis of the theory developed by Kopidakis and Aubry,[26] show that nonlinear shifts of a localized eigenmode's frequency (represented here by the eigenvalues) can lead to delocalization if this frequency crosses a resonance with other modes of the lattice (see Figure 3.1.9b). This condition can be satisfied in the case of weak disorder for the staggered localized modes, as nonlinearity shifts the eigenvalue of these modes from the edge of the band to higher values, deeper into the linear spectrum. In such a case, the localized excitation can cross a resonance with the "extended" modes of the system, resulting in expansion of the wave function. In contrast, the flat-phased modes at the other edge of the band are shifted by nonlinearity out of the linear spectrum and into the gap; thus they become localized defect-like states (see also discussion in the study by Pertsch et al.[12]). Clearly, if the sign of nonlinearity would have been opposite, the reverse effect would occur; that is, staggered localized eigenmodes would become more localized, whereas flat-phased modes would become delocalized under the influence of nonlinearity. These predictions were corroborated numerically (not shown).

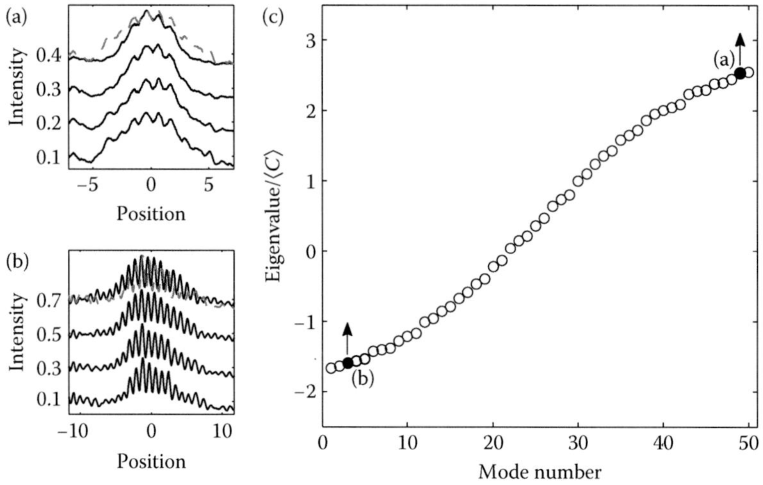

FIGURE 3.1.9
Experimental measurement of the effect of nonlinear perturbations on pure localized eigenmode excitations. (a) Measures cross section as a function of laser intensity. For the flat-phased localized modes from the high-eigenvalue edge of the spectrum, nonlinearity results in increased localization. (b) For the staggered localized modes from the low-eigenvalue edge of the spectrum, nonlinearity results in increased localization. (c) The heuristic explanation in terms of a nonlinear shift of the eigenvalue of the flat-phased modes out of the spectrum and into the gap (resulting in increased localization), whereas the eigenvalue of the staggered modes is shifted by nonlinearity into the band, where the excitation can resonantly couple to the more extended modes.

3.1.2.3.2 Strong Disorder: Nonlinearly Driven Hopping

A slightly more complicated scenario takes place in lattices with a higher level of disorder, as shown in Figure 3.1.10. In these cases, the disorder level was $\Delta/C = 3$, resulting in a lattice in which all the eigenmodes are localized within the finite sample. Here, a different nonlinear effect becomes dominant, as in most cases the narrow excitation beam excites localized states from the middle of the spectrum. Now, as the self-frequency (or eignvalue) is shifted by nonlinearity to higher values, the beam remains localized until the eigenvalue happens to cross a resonance with another near-by localized eignmode (see Figure 3.1.10). In such a case the excitation can "hop" between localized states, using a nonlinearly driven mechanism for transport. As some of the experimental examples show, the excitation can travel this way a considerable distance.

3.1.2.4 Observation of Anderson Localization in 1D Wave-Packet Expansion

Up till now we were focused on the experimental study of localized eigenmodes of a disordered system. As described earlier, a complementary view of disordered systems, which was the one adopted by Anderson in his originally work, is the study of the dynamical problem—the expansion or diffusion of a quantum particle confined to a small region. It was already known that in perfectly ordered lattices the particle's initially localized wave function is a wave packet of infinitely extended modes (the Bloch modes), and therefore the wave packet will expand indefinitely and ballistically. Anderson has shown that under certain conditions, in the presence of disorder interference effects can result in a suppression of this expansion. This can also be understood as now the particle wave function is a wave packet of localized modes. The wave packet can initially expand, but this expansion obviously cannot continue beyond the width of the widest localized mode in the wave packet.

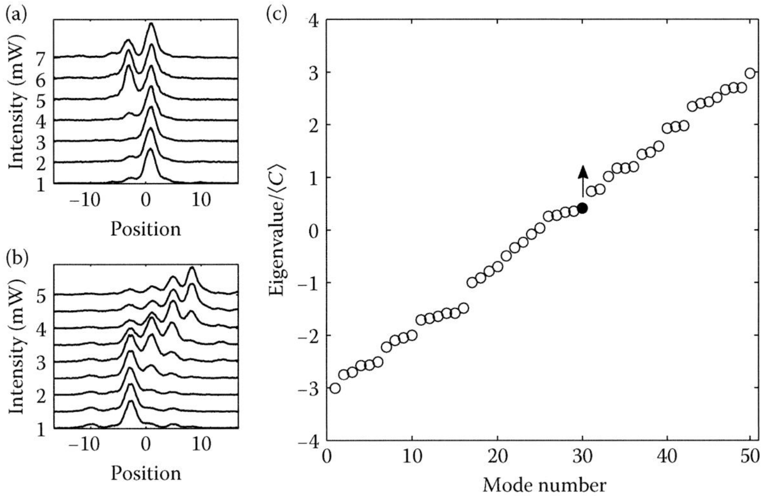

FIGURE 3.1.10
Experimental measurement of nonlinearly induced hopping in strongly disordered waveguide lattices. (a) Measures cross sections as a function of laser intensity. The linear excitation is localized, but at certain threshold intensity the energy can (a) hop to a nearby localized mode, or (b) hop several locations. (c) The heuristic explanation in terms of a nonlinear shift of the eigenvalue that pushed the eigenvalue of the excited localized state inside the band, until it crosses a resonance with a different, near-by localized state.

In the linear case, the viewpoints of eigenmodes and wave-packet expansion are of course equivalent—they are simple transforms of each other. One can add the theories of transmission through disordered samples. However, in the presence of nonlinearity this equivalence no longer holds, and one cannot simply map the nonlinear effects between the cases.

We turn then to describe a set of experiments, aimed to study the effect of disorder on wave-packet expansion in 1D in the linear and nonlinear regimes. This aspect is studied by injecting light into the waveguide lattice at $z = 0$ into a single lattice site, thus exciting a tight δ-like wave packet of all the eigenmodes having nonvanishing overlap with the excited site. The wave packet evolves in the lattice and the light distribution is measured at the output facet. In the experiments we average the output patterns obtained, to gain a statistical measure of localization. Here, we do this by separately exciting each site in the lattice while keeping the measurement window centered about the input site position.

The results of these measurements in the linear regime, taken in 5-mm-long samples with different disorder levels, are shown in Figure 3.1.10a–d. Without disorder, single-site excitation results in ballistic propagation (wave-packet width grows linearly with time), recognized by a characteristic signature of two separated lobes[10] (Figure 3.1.11a). At moderate disorder, a second component emerges, localized around the input site position (Figure 3.1.11b and c). The localized and the ballistic components coexist in this regime. At high disorder a highly localized, exponentially decaying distribution is observed (Figure 3.1.11d). This exponential decay of the expansion profile is the hallmark of Anderson localization.

These results offer a direct look of the short time evolution of wave packets in a 1D disordered system. It is known that for infinite disordered 1D systems and for long timescales, wave-packet expansion is always fully suppressed. However, on short timescales, wave packets evolve.[6,27] The results in Figure 3.1.11a–d show how localization is reached in the 1D case; localization emerges from ballistic expansion through the continuous buildup of a localized component and the suppression of a ballistic component.

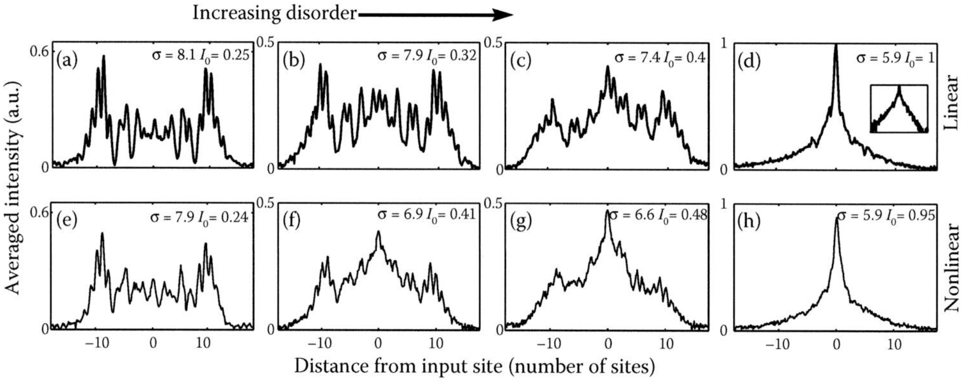

FIGURE 3.1.11
The effect of disorder on wave-packet expansion in the linear and nonlinear regimes. The results shown are normalized lattice averages of the output light distribution, when initially a single site is excited (see discussion in the text). A measure of localization is given by the mean-square displacement from the input site σ and by the intensity at the origin site I_0. (a)–(d) Measurements in the linear case as a function of disorder in: (a) $\Delta/C = 0$, (b) $\Delta/C = 1$, (c) $\Delta/C = 1.5$, and (d) $\Delta/C = 3$. Note the transition from ballistic transport in (a) to exponential localization in (d) through the buildup of a central component and the suppression of the ballistic side lobes. Inset in (d) shows the localized distribution in a semilog scale, demonstrating the exponential tails. (e)–(h): Measurements of the same lattices in the nonlinear case, showing that on average nonlinearity tends to increase localization for intermediate disorder levels (e.g., compare (b) and (f)). (Adapted from Y. Lahini et al., Phys. Rev. Lett. **100**, 013906, 2008.)

This dynamics is different from the one observed in 2D systems[12,13] or quasi-1D systems.[28] In these cases the expansion is initially diffusive, and becomes localized after much longer propagation times. In the real 1D case the diffusive dynamics is absent, as discussed for example in the study by Izrailev et al.[27] To illustrate this dynamics we present a numerical simulation of it in Figure 3.1.12, where we also compare the propagation and measures of intensity distributions for lattices with diagonal and off-diagonal disorder. We find that in this case there is no measurable difference between them.

To study the effect of nonlinearity on wave-packet expansion, we repeat these measurements at increased light intensities. Again, we remain in the weak nonlinear regime. Results are shown in Figure 3.1.11e–h. On average, the results indicate increased localization at intermediate disorder levels; the intensity in the localized (ballistic) component is higher (lower) under nonlinear conditions (e.g., compare Figure 3.1.11b and f). This suggests that the buildup of the localized component and the suppression of the ballistic component occur faster under nonlinear conditions. This description holds for short timescales in which the ballistic component is still present.

3.1.2.4.1 Simulations on Nonlinear Expansion

In the theoretical and numerical study of the interplay of nonlinearity and disorder there is a long-standing problem—the problem of the effect of nonlinearity on Anderson localization of wave packets in the long run. The question is whether in the long time limit Anderson localization survives nonlinear effects or will nonlinearity result in delocalization. At first, one can erroneously think that the answer simply depends on the sign of the nonlinear interactions: if the interaction is attractive (focusing nonlinearity) the effect of disorder (Anderson localization) and the effect of nonlinearity (soliton formation—nonlinear localization) will perhaps combine to form an even more localized wave packet,

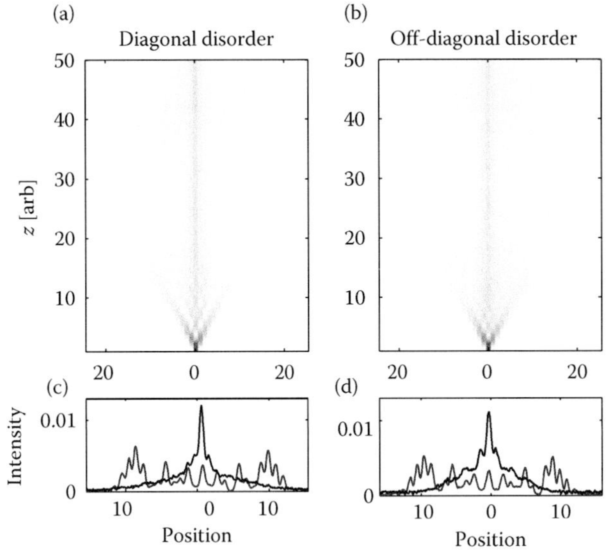

FIGURE 3.1.12
Observation of Anderson localization in lattices with diagonal and off-diagonal disorder. Top panels show simulation of the average wave-packet expansion in lattices with diagonal (a) and off-diagonal (b) disorder, when excited at a single site, showing practically the same dynamics. (c), (d) Experimental measurements of the output distributions for the two types of disorder (dark lines) as compared to the same distribution in a periodic lattice. (Adapted from Y. Lahini et al., Phys. Rev. A **84**, 041806(R), 2011.)

whereas for repulsive interaction the answer might depend on the relative strength of nonlinearity and disorder. This simple view however turns out to be incorrect. Anderson localization is an interference effect, and nonlinearly accumulated phases can modify the results. In essence, through nonlinear effects the propagating waves induce a time-dependent (or z-dependent) disordered potential. It is well established that wave packets expanding in lattices with time-dependent disorder exhibit diffusion—that is, their width grows like the square root of the propagation parameter (see the studies by Amir et al.[29] and Levi et al.,[30] and references therein]. However, in this case the driving force of the disorder fluctuations is the nonlinear effects that decrease as the wave packet expands. The final outcome depends on the interplay of these effects, and still remains unclear.[31]

The nonlinear experiment described above cannot yet address the issue as presented above, as they would require unrealistically long arrays. However, in this section we would like to point out a variant of this problem that can be (an in a way already has been) measured, and several other interesting phenomena that occur for expanding nonlinear wave packets on the short time limit. Figure 3.1.13 depicts a simulation of the width of an initially single-site wave packet expanding in a disordered lattice. In the linear case the expansion quickly saturates to its final width, related to the localization length. In this regime the wave packet remain exponentially localized throughout the propagation, with minor fluctuations in the width of the distribution. This curve corresponds to the linear experiments described earlier. When nonlinear interactions are introduced, the simulations show clear slow expansion at late times (beyond the scope of our experiments, marked in a dashed line in Figure 3.1.13). As the nonlinear interactions are increased the short time expansion of the wave packet is further suppressed, and the wave packet

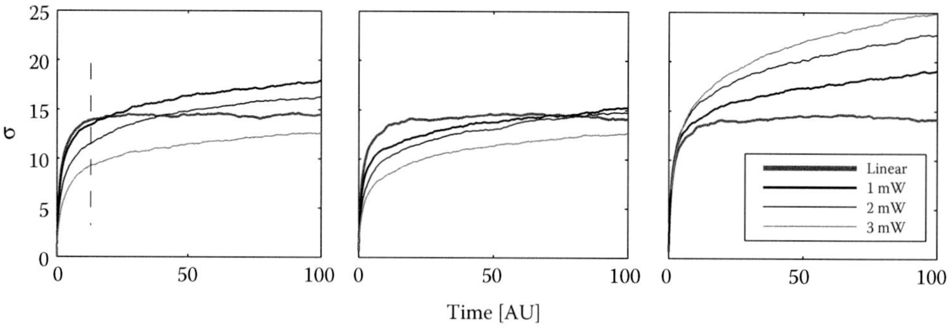

FIGURE 3.1.13
Simulations on the effect of nonlinearity on the width of an expanding wave packet. (a) When initially a single site is excited, a linear beam (solid line) initially expands and then saturates to a final width. In the nonlinear case, the beam becomes more localized on the short timescale and exhibits slow subdiffusion on long timescales. The increased localization as a function of nonlinearity does not depend on the sign of the nonlinear coefficient. The experiment described in Figure 1.1.11 corresponds to the dashed line. (b) For a two-site initial condition with positive (focusing) nonlinearity, the initial localization is even more pronounced. (c) For an out-of-phase initial condition (or exchangeably a negative sign to the nonlinear coefficient), the result on the short timescales is earlier delocalization. This effect holds a promise to observe nonlinear delocalization in experimentally accessible sample lengths.

becomes more localized on the short timescales. Still, the rate of expansion on the long timescales remains the same.

Interestingly, in the case of a single-site initial excitation, these short and long time trends are identical for both attractive and repulsive interactions, that is, for focusing and defocusing nonlinearity. In both cases the wave packet is more localized on the short timescales, and it expands at the same rate on the long timescales. This symmetry is broken when the initial excitation is wider than a single site. Now, the effect of the nonlinear interaction highly depends on the sign of the nonlinear parameter γ and on the phase structure of the initial excitation. For example, for a two-site initial condition (Figure 3.1.13b) with a flat phase and focusing nonlinearity, the wave packet becomes even more localized on the short timescales and the crossover to the subdiffusive expansion is delayed for longer propagation distanced. On the other hand, for either defocusing or focusing nonlinearity and a staggered phase structure, the short time behavior is different: the wave packet shows a tendency to an earlier crossover to subdiffusion expansion.

The effects described above can be explained in terms of the different parts of the spectrum excited in each case. In the case of a single-site initial excitation, the initial condition excites all the eignmodes that overlap with that site, equally (on average) from all regions of the spectrum. As a result, the sign of the nonlinear parameter does not change the average result. A two-site excitation is more selective: a flat-phase two-site excitation, for example, will excite localized eignmodes from the flat-phase edge of the band, but will have zero overlap with the staggered localized modes from the other edge of the band. As a result, the dominant nonlinear effect will be the one corresponding to the flat-phased modes—in the case of focusing nonlinearity this means increased localization. Similarly, for staggered initial condition (or defocusing nonlinearity) the result will be earlier delocalization.

The fact that under the right conditions delocalization sets in earlier suggests that nonlinear delocalization can perhaps be observed on shorter samples this way. In fact, the expansion of the staggered excitation in Figure 3.1.9b can be considered an observation of this effect in the limit of wide excitations.

3.1.2.5 Intensity Correlations of Localized Waves

The transverse localization scheme offers a unique viewpoint on Anderson localization, as it enables measurements of the spatial properties of the localized waves inside the lattice. In this section, we use this ability to measure another important property of localized waves that was difficult to access previously: the spatial correlations. We present results of an experimental study of the intensity correlations in 1D disordered photonic lattices exhibiting Anderson localization.

The coherent propagation of light in disordered media results in a complex and irregular interference pattern. This intensity distribution shows large fluctuations between different realizations of disorder (see Figure 3.1.2);[13] each realization generates a different speckled intensity distribution. As discussed above, to obtain a statistical measure of localization, the intensity distribution is averaged over many different realizations of disorder. In the localized regime the averaged distribution exhibits an exponential decay in space, the hallmark of Anderson localization. This procedure was also used in the recent experiments on Anderson localization of matter waves. However, in this procedure the information on the fine features of the intensity distributions is lost. This information can be recovered by measurements of other statistical properties of intensity—for example, the spatial correlation function.

3.1.2.5.1 Intensity Correlations for Single-Site Excitations

We first present results for the case of a δ-like initial condition. A measure of the spatial coherence of the generated speckle pattern was obtained by injecting the beam at different locations across the lattice. For each location, we measured the intensity at the output of the lattice and calculated its autocorrelation. We find that intensity correlations distinguish between the different types of disorder. As noted earlier, the two different types of disorder (diagonal and off-diagonal) show very similar dynamical properties for this initial condition—Anderson localization of wave packets is very similar in both cases. A significant difference was observed, however, when the density correlations of the output distribution were measured. Figure 3.1.14 presents experimental results of density-correlation measurements in disordered lattices. Here, for each realization of disorder, the density autocorrelation is measured, and then averaged over many realizations (again by shifting the input site). The result is then normalized, so that $\Gamma_{r,q} = \langle I_r I_q \rangle \langle I_r \rangle \langle I_q \rangle$, where r and q represent lattice site numbers and I_r is the intensity at site r. As can be seen in Figure 3.1.14, both types of disorder show a distinct diagonal feature in the correlation matrix. The length scale of this feature along the main diagonal ($q = r$) of the matrix is the localization length. The width of the diagonal feature is given by the correlation length, which is not represented in the ensemble-averaged density distributions. As discussed earlier, in each single realization, the density distribution is not a smooth exponentially decaying distribution; it is speckled.[13,14] The width of the diagonal feature reflects the average speckle size. As in each realization the speckle pattern varies, the features are smeared out in the averaged distribution and the information about their width is lost. However, the fact that these speckles have a characteristic length scale is recorded in the averaged correlation function.

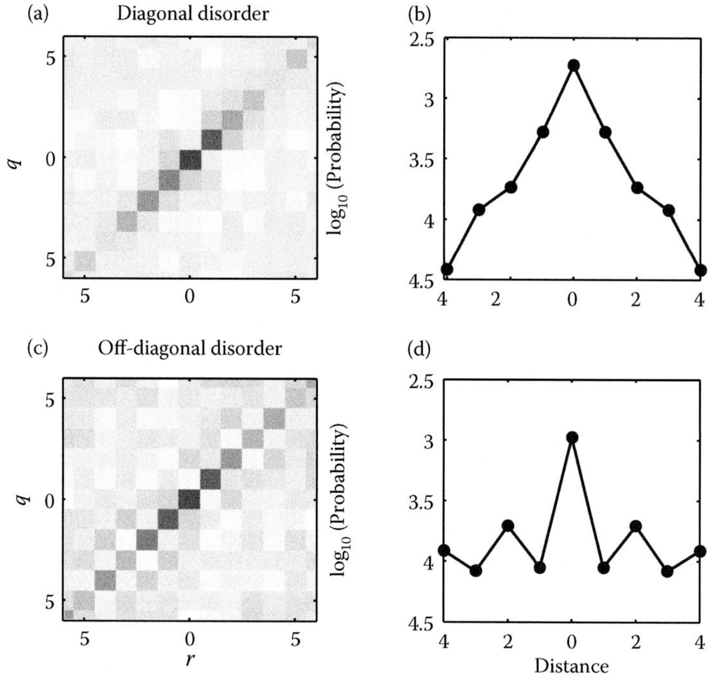

FIGURE 3.1.14
Measured density–density correlations $\Gamma_{r,q} = \langle I_r I_q \rangle \langle I_r \rangle \langle I_q \rangle$ for localized wave packets in: (a) lattices with diagonal disorder. The strong diagonal feature reflects finite coherence length of the waves Distance. (b) The extracted correlation function. (c) Density correlations in lattices with off-diagonal disorder, showing checker-like correlations. (d) The extracted correlation function, showing oscillating correlations. (Adapted from Y. Lahini et al., Phys. Rev. A **84**, 041806(R), 2011.)

We find that the correlation function carried additional information on the type of disorder in the lattice, information that is also lost when one considers the average density distributions. A closer look at the correlation matrix reveals that for lattices with off-diagonal disorder (Figure 3.1.14c) the correlations tend to form a checkered pattern. This can be better seen when looking for the correlation function $g(\Delta r)$, extracted from the correlation matrix $\Gamma_{r,q}$ by summing over the diagonals, $g(\Delta r) = \Sigma_r \Gamma_{r,q+\Delta r}$. The density correlation decays smoothly for lattices with diagonal disorder, yet it exhibits decaying oscillations for lattices with off-diagonal disorder. These results were corroborated in numerical simulations (not shown).

Now, to explain the different density–density correlations shown in Figure 3.1.14, we need to consider the effect of the different spectral properties of lattices with diagonal versus diagonal disorder on the expanding wave packets when they are excited at a single site. In lattices with off-diagonal disorder, an initial excitation of a single lattice site necessarily involves the simultaneous excitations of pairs of "twin" eigenmodes, as they have identical overlap with the initially excited site. The sum of two identical amplitude distributions with a π phase difference in each second site results in a density comb-like pattern that nulls at every second site. In the dynamic problem, several pairs could be excited simultaneously by the single-site initial condition, and the two modes of the pair accumulate phase in a different rate (according to their eigenvalues). Nevertheless, the wave packet will contain a component with an oscillating intensity pattern, with a spatial frequency of two sites. This effect is washed out in the density distribution averaged over all realizations of disorder, as in each realization the oscillations appear in a different location. However, the fact that such oscillations appear in each realization will be recorded in the averaged correlation.

3.1.2.5.2 Intensity Correlations for Nonlinear Localized Waves

We next describe the results of a similar experiment with nonlinear waves. Figure 3.1.15b presents a typical measurement of the speckle pattern generated at the output of a disordered waveguide lattice, when the input beam was a narrow Gaussian beam covering about three lattice sites. The intensity correlation function shown in Figure 3.1.15c was obtained by averaging these autocorrelations over many different input locations.

To measure the effect of nonlinear interactions on the spatial coherence, we repeated the above experiment, for the same disorder realizations, with high-intensity light. As the results show, nonlinearity results in a sharper, narrower correlation function.

To better understand these results, we conducted numerical simulations corresponding to the experimental conditions. As shown in Figure 3.1.15d, the simulation reproduces the observed effects; the correlation function becomes sharper and narrower. These results can be related to previous results on the effect of nonlinearity on pure eigenmodes of disordered lattices. A Gaussian flat-phase initial condition excites a wave packet of mostly flat-phase modes, each of which tends to become localized under the effect of nonlinearity. As a result, the bright features of the output speckle pattern tend to become narrower. In contrast, a staggered Gaussian initial condition excites a wave packet of staggered modes, and therefore should result in the opposite effect. To verify this, we checked numerically the effect of nonlinearity on staggered initial conditions. This is a case in which the input has the same envelope and width, but adjacent lattice sites have a π phase difference. As the simulations show, this results in an opposite effect; that is, the correlation functions become wider and lower.

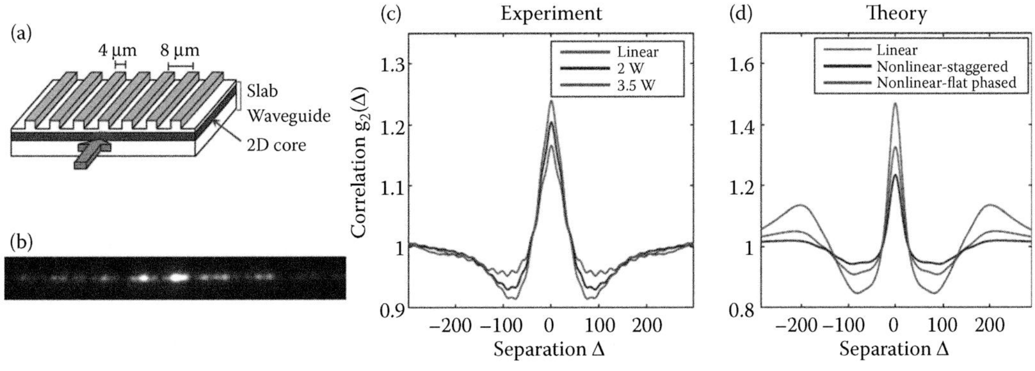

FIGURE 3.1.15
(a) A schematic view of the waveguide lattice used in the experiment. The arrow represents the input light beam. (b) Measured speckled intensity distribution after propagation in the lattice. (c) Experimentally measured intensity correlation function, averaged over many realizations of input positions. The graph compares the results for linear propagation, and increasingly nonlinear propagation. (d) Numerical simulations.

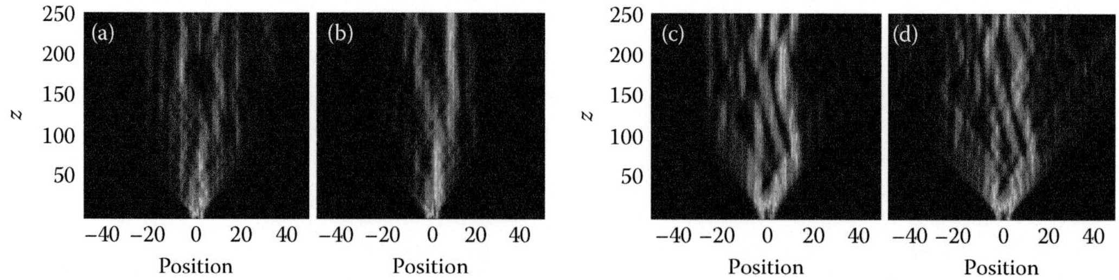

FIGURE 3.1.16
Linear and nonlinear propagation in a single realization of disorder. (a) Linear propagation of a flat-phased initial condition. (b) Nonlinear propagation of a flat-phased initial condition in the same disorder realization, showing narrowing of the bright features in the speckled field. (c) Linear propagation of a staggered initial condition. (d) Nonlinear propagation of a staggered initial condition in the same disorder realization, showing flattening of the bright features in the speckled field.

Further insights can be obtained by looking at the propagation along the lattice. In Figure 3.1.16, we show two numerical simulations of the propagation, with flat-phased and staggered initial conditions, with and without nonlinearity. Figure 3.1.16a presents the propagation in the linear case, and shows how the output speckle pattern is generated from multiple scattering inside the lattice. In the presence of nonlinearity (Figure 3.1.16b), the speckles become narrower and sharper, and this results in a sharper correlation function. For similar propagation with staggered input conditions (see Figure 3.1.16c and d), the speckles become wide and flat in the presence of nonlinearity, which results in a wider and flatter correlation function.

3.1.2.6 Toward the Anderson Localization of Nonclassical Light

Since the first prediction by Anderson, there has been an ongoing effort to observe the signature of Anderson localization experimentally. The novel approach described in this chapter enabled the direct observation in space of Anderson localization for photons and later also of ultra-cold atoms. In this final section, we show that the transverse localization approach can be used to directly measure other quantities that contain additional information not accessible otherwise. One such quantity—the spatial two-point correlation—can be used to reveal new effects related to quantum properties of light in disordered media.

The experiments described in this chapter, as well as the experiments describing the localization of ultra-cold atoms, have all reported measurements of the particle density distribution. As such, these results follow the predictions of the single-particle model described by Anderson (as long as no interactions are involved). For example, running the experiments described in this chapter with photon sources or other types of nonclassical light will result in the same photon density distribution one would get by using classical light. The following question then arises: what new phenomena can be exhibited by quantum states of light?

When several quantum particles propagate together, exchange terms can result in the formation of nonclassical correlations between their positions even in the absence of interactions—an effect known as Hanbury-Brown and Twiss interference.[32] In the study by Lahini et al.,[21] the problem of two-particle Anderson localization was analyzed by calculating the correlations between two indistinguishable quantum particles evolving simultaneously in disordered lattices. As noninteracting particles were considered, particle density follows single-particle dynamics: both particles exhibit Anderson localization.

Nevertheless, it was shown that the two particles develop nontrivial spatial correlations due to interferences of all the scattering paths that the two particles can take as a pair. On short timescales, the localization of one of the particles uniquely determines whether or not the other particle will be localized. On longer timescales, when both particles are localized, in some cases the particles exhibit oscillatory correlations within the localization length that survive multiple scattering even after very long evolution times. Anderson localization of path-entangled states sometimes exhibit fermionic-like correlations.

The experimental study of the quantum correlation properties of disordered media using optics requires the use of nonclassical light, for example, pairs of correlated photon generated by parametric down-conversion. Higher-order spatial correlations can be used to characterize the many-photon localization. Analogous experiments can be conducted using atoms in optical lattices. Interactions between few quantum particles[33] (e.g., two ultra-cold atoms) can be used to study the quantum limit of the interplay between disorder and nonlinearity described throughout this chapter.

3.1.3 Outlook

Today, merely four years after the appearance of the first papers on Anderson localization of light in disordered photonic lattices,[13,14] it is already clear that the experimental techniques and the concepts described therein have opened a plethora of experimental possibilities, and provided a new path to study many questions related to localization and nonlinear effects in disordered media. Many long-sought fundamental ideas have become experimentally accessible. Examples range from disorder-enhanced transport in quasiperiodic lattices[17] and localization transition in quasiperiodic structures[15] to localization near an interface[16] and amorphous photonic lattices exhibiting a bandgap[34] and even hypertransport in the presence of dynamically evolving disorder, where a wave packet expands by virtue of disorder much faster than ballistic expansion.[35] One of the most intriguing avenues prompted by these techniques relates to nonclassical phenomena[21,36] with the added complexity of quantum correlations revealing many new effects. Many of these concepts are now being adopted in the domain of matter waves, where the main challenges are to carry out localization experiments in domains where optics cannot provide answers, such as localization of interacting fermions[37] and localization of a Tonks–Girardeu gas.[38] We have discussed some of these ideas here and provided a contemporary review of this new emerging area. However, as often happens in science when a new experimental paradigm is invented, most probably, the best ideas are yet to be suggested, and they will reveal new information on the universal phenomena associated with the transport of waves in random media.

References

1. P. W. Anderson, Phys. Rev. **109**, 1492 (1958).
2. B. I. Shklovskii, and B. Z. Spivak. in *Hopping Conduction in Semiconductors*, edited by M. Pollak and B. I. Shklovskii (North-Holland, Amsterdam, 1990).
3. S. John, S. Phys. Rev. Lett. **53**, 2169 (1984); P. W. Anderson, Philos. Mag. B **52**, 505 (1985).

4. D. S. Wiersma, P. Bartolini, A. Lagendijk, and R. Righini, Nature **390**, 671 (1997); A. A. Chabanov, M. Stoytchev, and A. Z. Genack, *ibid.* **404**, 850 (2000).
5. M. Störzer, P. Gross, C. M. Aegerter, and G. Maret. Phys. Rev. Lett. **96**, 063904 (2006).
6. H. De Raedt, A. Lagendijk, and P. de Vries. Phys. Rev. Lett. **62**,47–50 (1989).
7. D. N. Christodoulides and R. I. Joseph, Opt. Lett. **13**, 794 (1988).
8. H. Eisenberg, Y. Silberberg, R. Morandotti, A. R. Boyd, and J. S. Aitchison, Phys. Rev. Lett. **81**, 3383 (1998).
9. J. W. Fleischer, M. Segev, N. K. Efremidis, and D. N. Christodoulides, Nature **422**, 147 (2003).
10. D. N. Christodoulides, F. Lederer, and Y. Silberberg, Nature **424**, 817 (2003); F. Lederer, G. I. Stegeman, D. N. Christodoulides, G. Assanto, M. Segev, and Y. Silberberg, Phys. Rep. **463**, 1–126 (2008).
11. H. Eisenberg, Ph.D. thesis, Weizmann Institute of Science, 2002.
12. T. Pertsch, U. Peschel, J. Kobelke, K. Schuster, H. Bartelt, S. Nolte, A. Tünnermann, and F. Lederer, Phys. Rev. Lett. **93**, 053901 (2004).
13. T. Schwartz, G. Bartal, S. Fishman, and M. Segev, Nature **446**, 52 (2007).
14. Y. Lahini, A. Avidan, F. Pozzi, M. Sorel, R. Morandotti, D. N. Christodouldies, and Y. Silberberg, Phys. Rev. Lett. **100**, 013906 (2008).
15. Y. Lahini, R. Pugatch, F. Pozzi, M. Sorel, R. Morandotti, N. Davidson and Y. Silberberg, Phys. Rev. Lett. **103**, 013901 (2009).
16. A. Szameit, Y. V. Kartashov, P. Zeil, F. Dreisow, M. Heinrich, R. Keil, S. Nolte, A. Tunnermann, V. A. Vysloukh, and L. Torner, Opt. Lett. **35** 1172 (2010).
17. L. Levi, M. Rechtsman, and B. Freedman, Science **332**, 1541 (2011).
18. J. Billy, V. Josse, Z. Zuo, A. Bernard, B. Hambrecht, P. Lugan, D. Clement, L. Sanchez-Palencia, P. Bouyer, and A. Aspect, Nature **453**, 891–894 (2008); G. Roati, C. D'Errico, L. Fallani, M. Fattori, C. Fort, M. Zaccanti, G. Modugno, M. Modugno, and M. Inguscio, *ibid.* **453**, 895–898 (2008).
19. D. Belitz and T. R. Kirkpatrick, Rev. Mod. Phys. **66**, 261–380 (1994); D. M. Basko, I. L. Aleiner, and B. L. Altshuler, Ann. Phys. **321**, 1126–1205 (2006).
20. N. K. Efremidis, S. Sears, D. N. Christodoulides, J. W. Fleischer, and M. Segev, Phys. Rev. E **66**, 046602 (2002); J. W. Fleischer, T. Carmon, and M. Segev, Phys. Rev. Lett. **90**, 023902 (2003).
21. Y. Lahini, Y. Bromberg, D. N. Christodoulides, and Y. Silberberg, Phys. Rev. Lett. **105**, 163905 (2010).
22. Lev. P. Pitaevskii and S. Stringari, *Bose–Einstein Condensation*, (Clarendon Press, Oxford, 2003).
23. P. Sheng, *Introduction to Wave Scattering, Localization, and Mesoscopic Phenomena* (Springer, Berlin, 2006).
24. A. McGurn, K. Christensen, F. Mueller, and A. Maradudin, Phys. Rev. B **47**, 13120 (1993).
25. F. J. Dyson, Phys. Rev. **92**, 1331 (1953); C. M. Soukoulis and E. N. Economou, Phys. Rev. B **24**, 5698 (1981); C. M. Soukoulis, I. Webman, G. S. Grest, and E. N. Economou, *ibid.* **26**, 1838 (1982); M. Inui, S. A. Trugman, and E. Abrahams, *ibid.* **49**, 3190 (1994); A. Komiyama, IEICE Trans. Electron. **E83**, 736 (2000).
26. G. Kopidakis and S. Aubry, Phys. Rev. Lett. **84**, 3236 (2000); Physica D **139**, 247 (2000); **130**, 155 (1999); C. Albanese and J. Frohlich, Commun. Math. Phys. **138**, 193 (1991).
27. F. M. Izrailev, T. Kottos, A. Politi, and G. P. Tsironis, Phys. Rev. E **55**, 4951 (1997); E. P. Nachmedov, Sov. Phys. JETP **65**, 1202 (1987).
28. A. A. Chabanov and A. Z. Genack, Phys. Rev. E **56**, R1338 (1997); A. García-Martín, J. A. Torres, J. J. Saíenz, and M. Nieto-Vesperinas, Appl. Phys. Lett. **71**, 1912 (1997); A. García-Martín, F. Scheffold, M. Nieto-Vesperinas, and J. J. Sáenz, Phys. Rev. Lett. **88**, 143901 (2002).
29. A. Amir, Y. Lahini, and H. B. Perets, Phys. Rev. E **79**, 050105 (2009).
30. L. Levi, T. Schwartz, M. Segev, and S. Fishman, in *Conference on Lasers and Electro-Optics/ International Quantum Electronics* (Optical Society of America, 2009), paper IThD2 (http:// www.opticsinfobase.org/abstract.cfm?URI=IQEC-2009-IThD2).
31. D. L. Shepelyansky, Phys. Rev. Lett. **70**, 1787–1790 (1993); A. S. Pikovsky and D. L. Shepelyansky, *ibid.* **100**, 94101 (2008); S. Flach, D. O. Krimer, and Ch. Skokos, *ibid.* **102**, 24101 (2009); S. Fishman, A. Iomin, and K. Mallick, Phys. Rev. E **78**, 66605 (2008).

32. R. Hanbury Brown and R. Q. Twiss, Nature **177**, 27–29 (1956).
33. Y. Lahini, M. Verbin, S. D. Huber, Y. Bromberg, R. Pugatch, and Y. Silberberg, arXiv **1105**, 2273 (2011).
34. M. Rechtsman, A. Szameit, F. Dreisow, M. Heinrich, R. Keil, S. Nolte, and M. Segev, Phys. Rev. Lett. **106**, 193904 (2011).
35. L. Levi, Y. Krivolapov, S. Fishman, and M. Segev, in *Quantum Electronics and Laser Science (QELS) Conference*, (Optical Society of America, 2011), paper QThF2 (http://www.opticsinfobase.org/abstract.cfm?URI=QELS-2011-QThF2).
36. Y. Bromberg, Y. Lahini, R. Morandotti, and Y. Silberberg, Phys. Rev. Lett. **102**, 253904-4 (2009); A. Peruzzo, M. Lobino, J. C. F. Matthews, N. Matsuda, A. Politi, K Poulios, X. Zhou et al., Science **329**, 1500–1503 (2010).
37. V. Oganesyan and D. A. Huse, Phys. Rev. B **75**, 155111 (2007).
38. J. Radic, V. Bacic, D. Jukic, M. Segev, and H. Buljan, Phys. Rev. A **81**, 063639 (2010).
39. Y. Lahini, Y. Bromberg, Y. Shechtman, A. Szameit, D. N. Christodoulides, R. Morandotti, Y. Silberberg, Phys. Rev. A **84**, 041806(R) (2011).

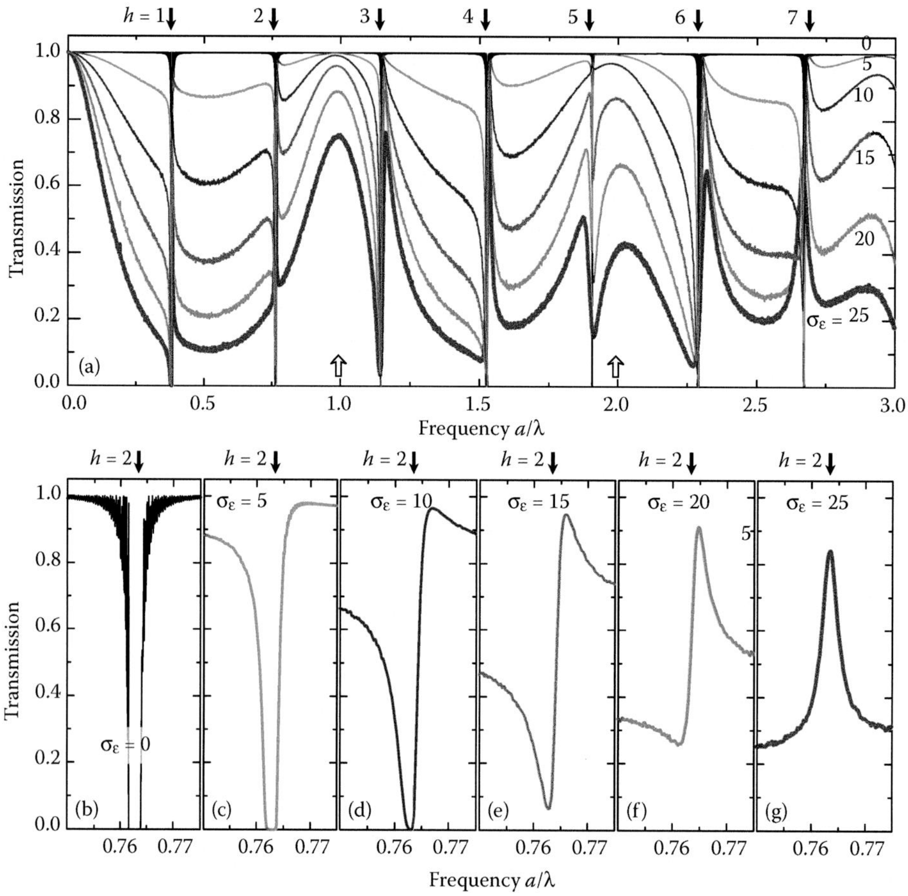

FIGURE 2.1.4
(a) Transmission spectra of one-dimensional (1D) photonic crystal (PhCs) with ε-disorder as a function of degree of disorder σ_ε. Fringes corresponding to Fabry–Pérot resonance are shown by white arrows from bottom and those corresponding to *h*-order Bragg bands are shown by black arrows from top. (b)–(g) Transmission spectra of 1D PhCs given in a larger scale in the proximity of second-order Bragg bands for different values of disorder σ_ε. $\varepsilon_A = 1.70$, $\varepsilon_B = 1.75$, $w_B = 0.38a$, $N = 1000$, and $M = 5000$.

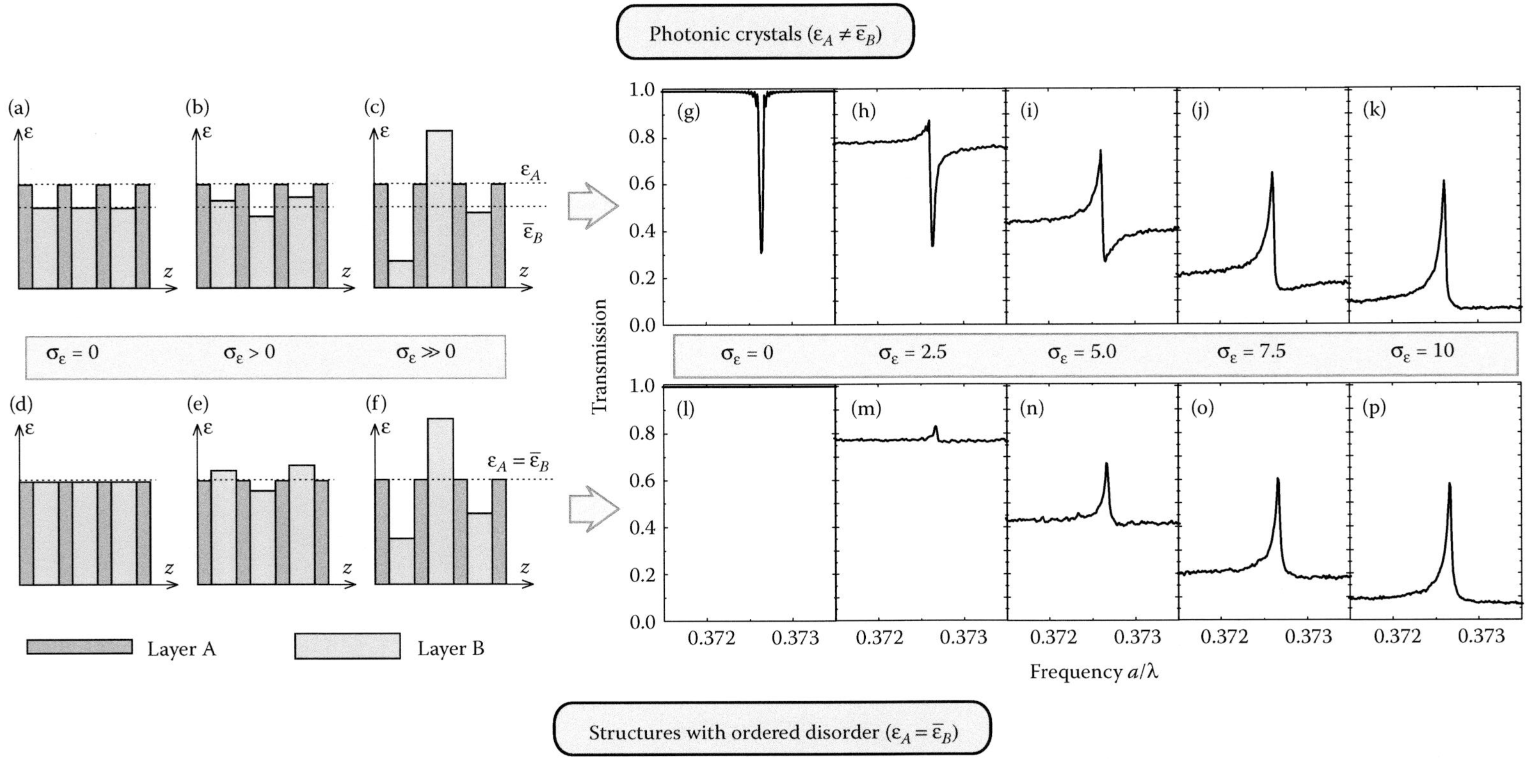

FIGURE 2.1.7
(Left) Schematic of permittivity profile for disordered one-dimensional (1D) photonic crystals (PhCs) (a)–(c) and for 1D structures with ordered disorder (ODSs) without dielectric contrast between layers A and B permittivity in average (d)–(f) when disorder degree is increased. (Right) Transmission spectra of PhCs (g)–(k) and ODSs (l)–(p) with fixed number of unit cells $N = 10^4$ as a function of the degree of disorder $\sigma_\varepsilon = 0.0$, 2.5, 5.0, 7.5, and 10. $\varepsilon_A = 1.8$, $\bar{\varepsilon}_B = 1.8005$ (for ODSs) or $\bar{\varepsilon}_B = 1.8005$ (for PhCs), $w_B = 1/3a$, $N = 10^4$, $M = 3000$.

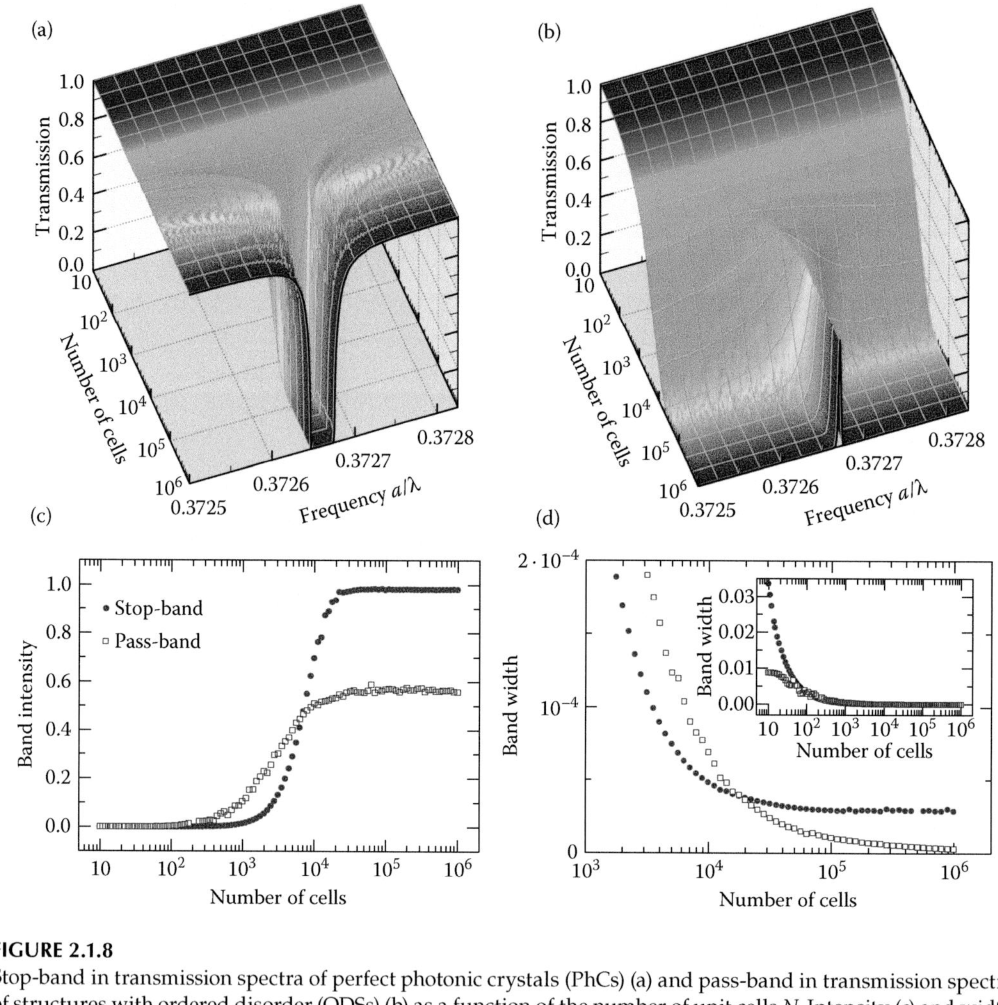

FIGURE 2.1.8
Stop-band in transmission spectra of perfect photonic crystals (PhCs) (a) and pass-band in transmission spectra of structures with ordered disorder (ODSs) (b) as a function of the number of unit cells N. Intensity (c) and width (d) of Bragg pass-band and Bragg stop-band (ordered PhCs) obtained from transmission spectra shown at (a) and (b). $\varepsilon_A = 1.8$, $\bar{\varepsilon}_B = 1.8000$ (for ODS) or $\bar{\varepsilon}_B = 1.8005$ (for PhCs), $w_B = 1/3a$, $\sigma_\varepsilon = 10$, $M = 3000$–5000 (for ODSs).

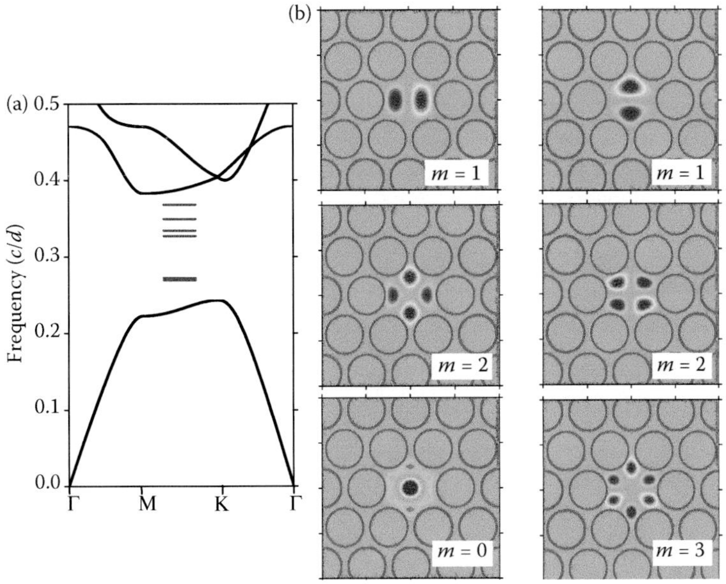

FIGURE 2.2.5
The frequencies of eigenmodes localized at a single vacancy in a hexagonal photonic crystal (a) and the profiles of z components of the magnetic field corresponding to those modes (b).

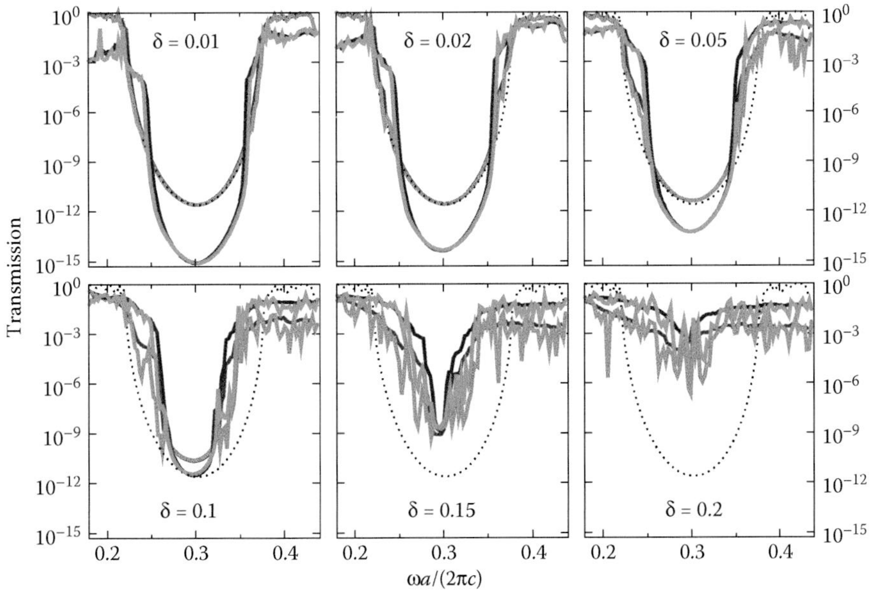

FIGURE 2.2.8
Transmission spectra for disordered photonic crystal with $\delta = 0.01$, 0.02, 0.05, 0.1, 0.15, and 0.2. Solid (red) curves are the ballistic transmission, dashed (blue) curves are the scattered transmission—the thin curves for an individual configuration of disorder, and the heavy curves for the mean averaged over 10 random configurations of disorder. The ballistic transmission spectrum of the ideal photonic crystal is also shown (black dotted curve) for comparison. (Adapted from D. M. Beggs et al., J. Phys. Condens. Matter **17**(12), 1781, 2005.)

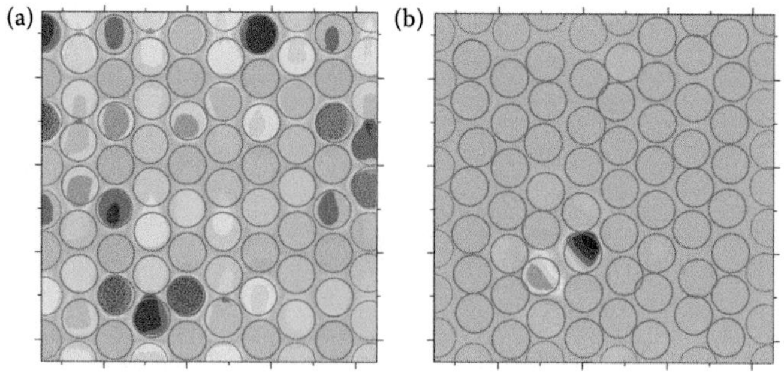

FIGURE 2.2.10
Delocalized state near edge of PBG in the case of weak disorder (a) and localized state in the case of strong disorder (b). (Adapted from D. M. Beggs et al., J. Phys. Condens. Matter **17**(12), 1781, 2005.)

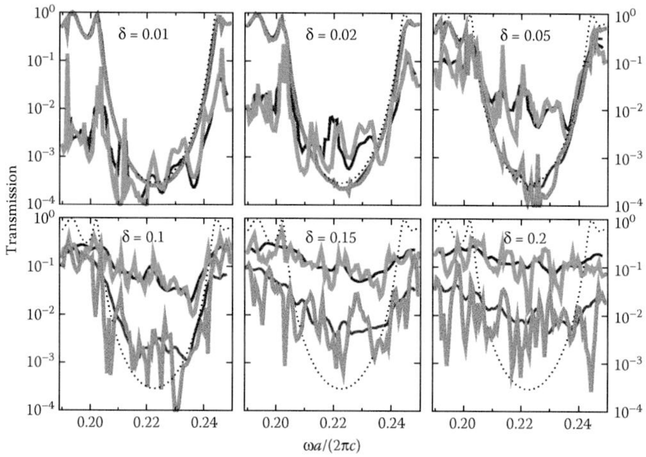

FIGURE 2.2.11
Transmission spectra for disordered photonic crystals with $\delta = 0.01, 0.02, 0.05, 0.1, 0.15$, and 0.2. The red lines are the ballistic transmission and the blue lines are the scattered transmission, with the thin lines denoting an individual configuration of disorder and the heavy lines an average over many random configurations of disorder. The ballistic transmission spectrum of the ideal photonic crystal is also shown (black dotted line) for comparison. (Adapted from D. M. Beggs et al., J. Phys. Condens. Matter **17**(26), 4049, 2005.)

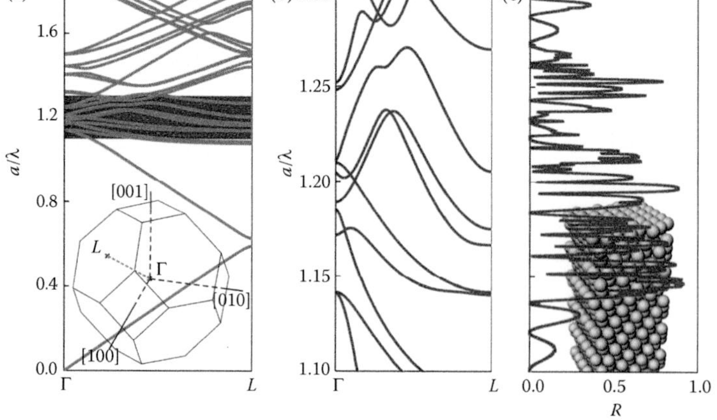

FIGURE 2.3.1
(a) Band structure calculated along the ΓL direction of a perfect face centered cubic crystal consisting of spheres of dielectric constant $\varepsilon_s = 2.4964$ in air. (b) Zoom of the band structure in the range $1.1 \le a/\lambda \le 1.3$. (c) Specularly reflected light spectrum for $\varepsilon_s = 2.4964$. (Adapted from L. A. Dorado, R. A. Depine, and H. Míguez, Phys. Rev. B **75**, 241101(R), 2007.)

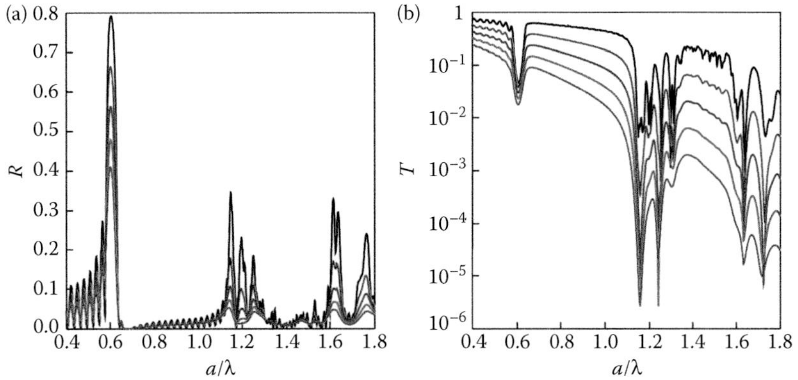

FIGURE 2.3.2
Specularly reflected (a) and forward-transmitted light spectra (b) for spheres of dielectric constant $\varepsilon_s = 2.4964 + i\varepsilon_i$, being from top to down $\varepsilon_i = 0.02$ (black), $\varepsilon_i = 0.04$ (red), $\varepsilon_i = 0.06$ (blue), $\varepsilon_i = 0.08$ (green), and $\varepsilon_i = 0.10$ (magenta). (Adapted from L. A. Dorado, R. A. Depine, and H. Míguez, Phys. Rev. B **75**, 241101(R), 2007.)

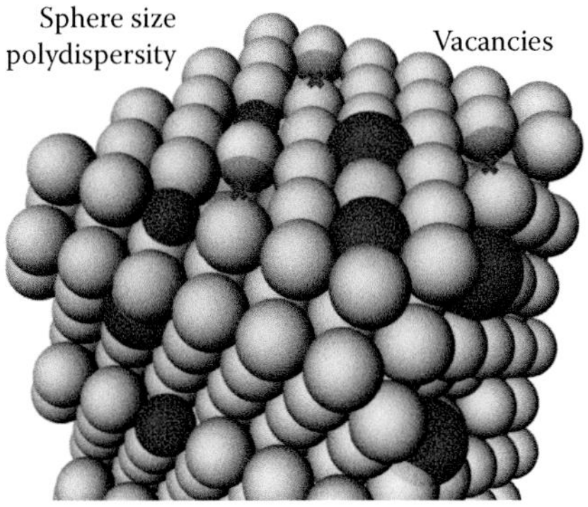

FIGURE 2.3.3
Sketch of an artificial opal that exhibits the type of imperfections (red) that we are considering.

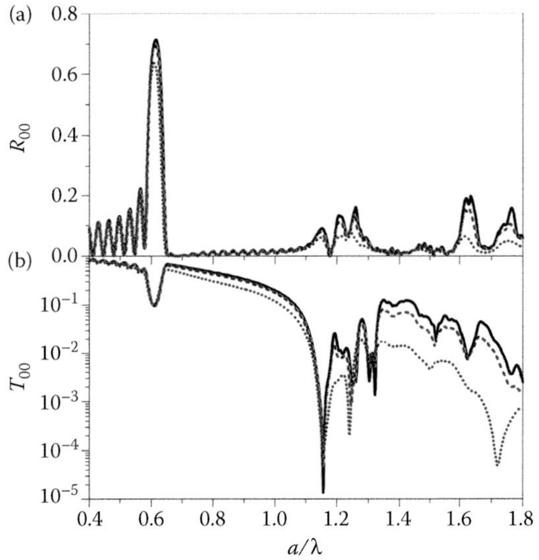

FIGURE 2.3.6
(a) Specular reflectance and (b) forward transmittance spectra for an 18-layer-thick crystal made of spheres of dielectric constant $\varepsilon_s = 2.5$ in air. The average *T*-matrix approximation parameters are: $S_{av} = 0.5a$, $C = 0.95$, and the spectra are shown for different values of σ/S_{av}. In particular, $\sigma/S_{av} = 0.010$ (black solid line), $\sigma/S_{av} = 0.025$ (red dashed line), and $\sigma/S_{av} = 0.050$ (blue dotted line). (Reprinted with permission from L. A. Dorado and R. A. Depine, Phys. Rev. B **79**, 045124, 2009. Copyright 2009, by the American Physical Society.)

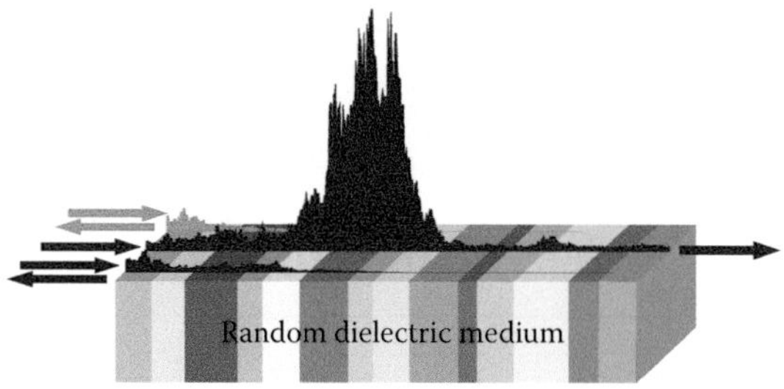

FIGURE 2.4.3
Amplitude of the field inside the sample as a function of the coordinate for three different wavelengths. (Reprinted with permission from K. Bliokh et al., Rev. Mod. Phys. **80**, 1201, 2008. Copyright 2008 by the American Physical Society.)

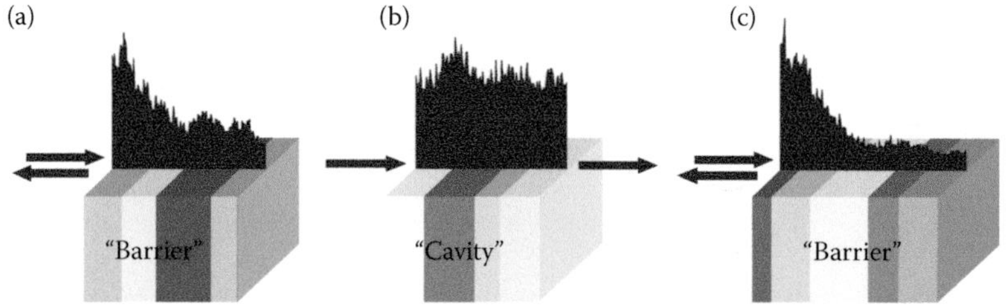

FIGURE 2.4.4
Amplitude of the field as a function of the coordinate inside the whole sample, in the (a), (b), and (c) parts taken as a separate sample each. (Reprinted with permission from K. Bliokh et al., Rev. Mod. Phys. **80**, 1201, 2008. Copyright 2008 by the American Physical Society.)

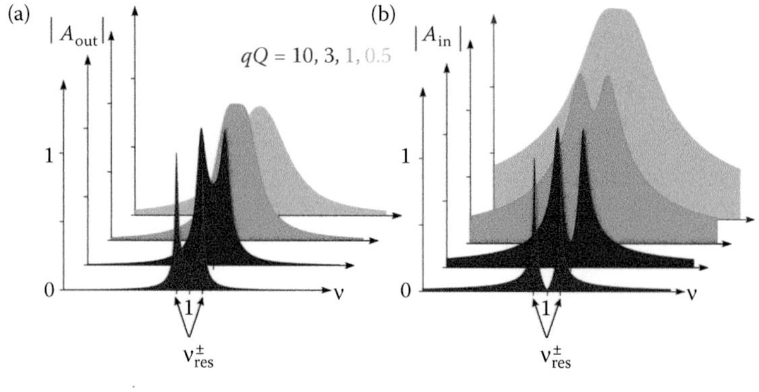

FIGURE 2.4.7
Near-resonant transmission of an incident wave through two coupled open resonators at different values of qQ. The normalized (i.e., multiplied by the factor $2Q^{-1}$) absolute values of the field amplitudes in two resonators, $|A_{out}|$ (a) and $|A_{in}|$ (b), are shown. (Reprinted with permission from K. Bliokh et al., Rev. Mod. Phys. **80**, 1201, 2008. Copyright 2008 by the American Physical Society.)

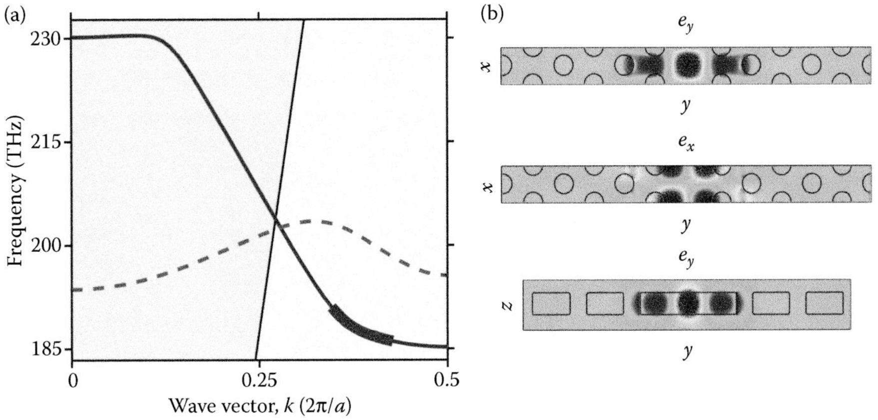

FIGURE 2.5.4
(a) Band structure for the waveguide shown in Figure 2.5.2. The horizontal axis is the propagation (Bloch) wave vector. This waveguide supports two modes in the photonic band gap of the surrounding material, with the fundamental waveguide mode depicted by the solid curve and the secondary mode by the dashed curve. A typical region of operation for a slow-light waveguide is highlighted and does not extend all the way to the band edge because, although the velocity continues to decrease, scattering losses become significant (see Sections 2.5.3 and 2.5.4). The gray region to the left of the figure represents the continuum of radiation modes above the light line. (b) Components of the Bloch mode electric field (absolute value) for the fundamental waveguide mode near the band edge. The x- and y-components are shown on planes that bisect the unit cell. The z-component is negligible. The parameters for this "W1.1" waveguide are given in Table 2.5.1.

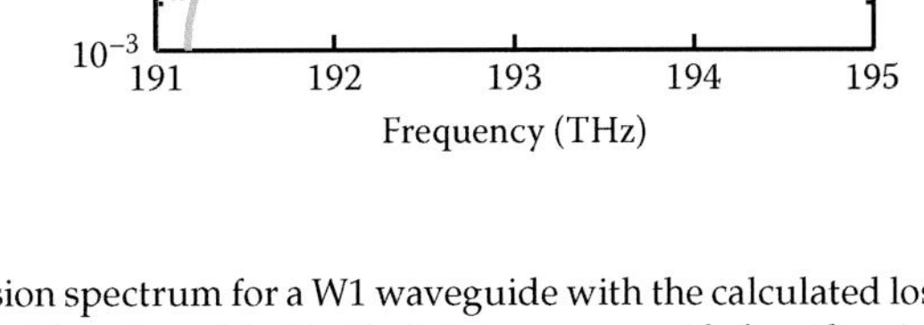

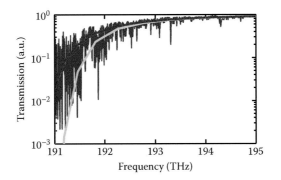

FIGURE 2.5.11
The experimental transmission spectrum for a W1 waveguide with the calculated loss spectrum superimposed. The calculated loss per period is extrapolated to the 1.5-mm waveguide length using the Beer–Lambert model (cyan—lower) and the multiple scattering model (red—upper). The disorder parameters used in the calculation are $\sigma = 3$ nm and $l_p = 40$ nm. A frequency shift of −0.3 THz has been applied to the calculated spectrum to account for uncertainty in the experimental slab thickness. (Based on Patterson, M. et al., Phys. Rev. Lett. **102**, 253903, 2009, with experimental measurement from Sylvain Combrié and Alfredo De Rossi, Thales Research and Technology.)

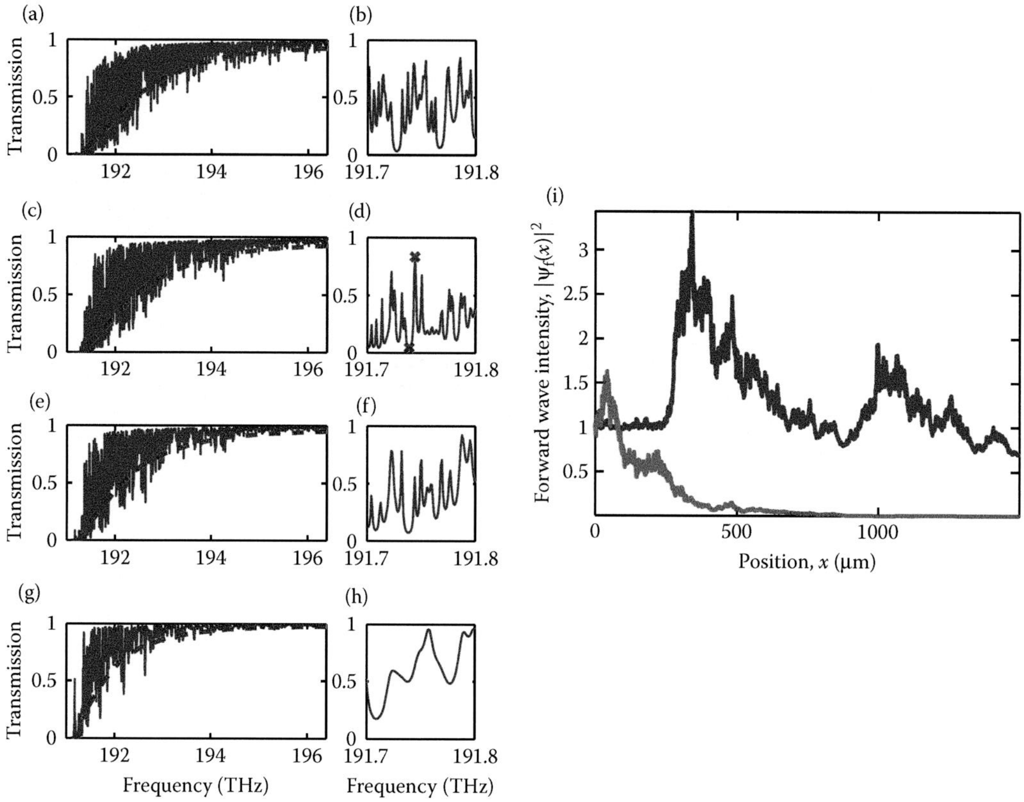

FIGURE 2.5.14
Simulated transmission spectra of four disordered W1 waveguides using the coherent scattering theory (solid) and the first- and second-order Born incoherent theory (dashed) [14]. Each row of plots is for a different waveguide with the left plot showing a broad frequency range and the right plot showing a narrow frequency range near the band edge. Plots (a) and (b) are for a disordered 1.5 mm. Plots (c) and (d) are for a different disorder instance of the same 1.5 mm. Plots (e) and (f) and plots (g) and (h) are for the same disorder instance as plots (c) and (d) but with the length reduced to 1.0 and 0.5 mm, respectively. The calculation uses a root mean square roughness of $\sigma = 3$ nm, and a disorder correlation length of $l_p = 40$ nm. (i) Forward wave intensity in a disordered waveguide at two wave vectors. The upper curve ($n_g = 24.96$) corresponds to a local transmission maximum and the lower curve ($n_g = 25.11$) is a neighboring transmission minimum. These two curves correspond to the crosses shown in plot (d).

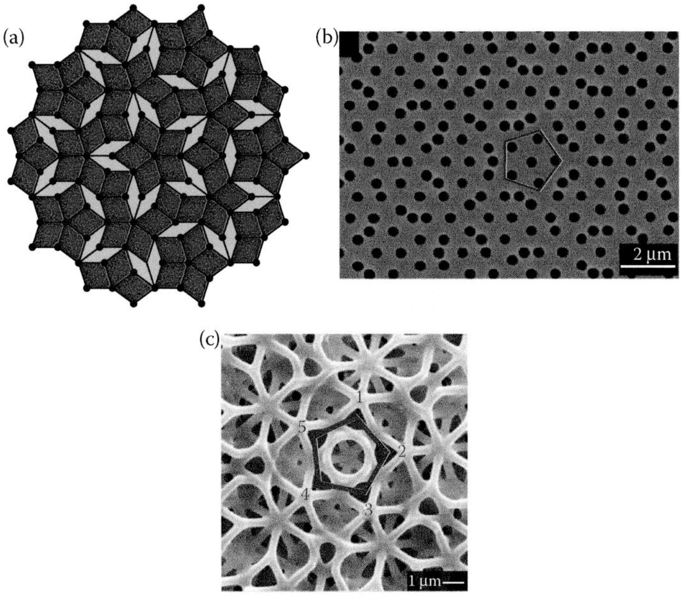

FIGURE 2.6.8
Examples of two-dimensional (2D) and three-dimensional (3D) quasicrystalline structures. (a) Schematic illustration of the 2D Penrose tiling. (b) Scanning electron microscope image of the 2D Penrose quasicrystal. (c) Electron micrographs of silicon 3D icosahedral quasicrystal structure. (Panel (b) from D. Shir et al., Nano Lett. **8**, 2236, 2008. With permission; panel (c) from A. Ledermann, G. von Freymann, and M. Wegener, Phys. Unserer Zeit **38**, 300, 2007. With permission.)

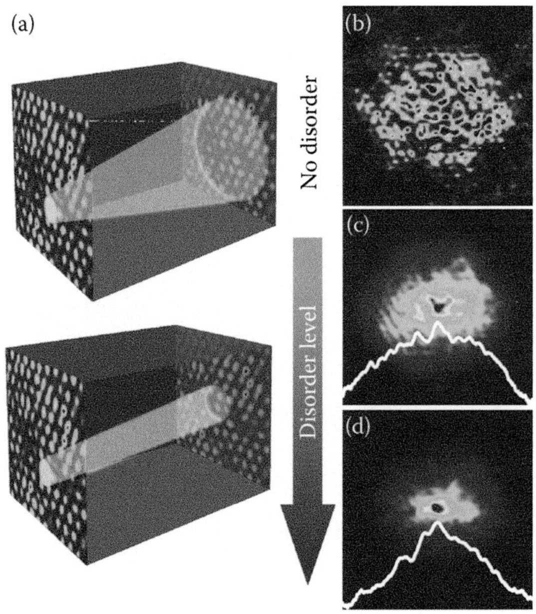

FIGURE 3.1.1
(a) The transverse localization scheme: a probe beam propagates in photonic lattice with a controlled level of disorder. In the absence of disorder, the beam exhibits ballistic transport: its width increases linearly with propagation distance (top). Under the influence of disorder, the beam becomes exponentially localized in the transverse plane, maintaining its width throughout propagation (bottom). (b)–(d) Ensemble-averaged intensity distribution at the output face of lattice. The results show gradual transition from ballistic transport (b), where the diffraction pattern reflects the hexagonal symmetry of the lattice, to diffusion (c) in presence of disorder (designated by a Gaussian shape of the intensity profile, plotted in logarithmic scale) and, at stronger disorder, to localization with an exponentially decaying intensity profile (d). (Adapted from T. Schwartz et al., Nature **446**, 52, 2007.)

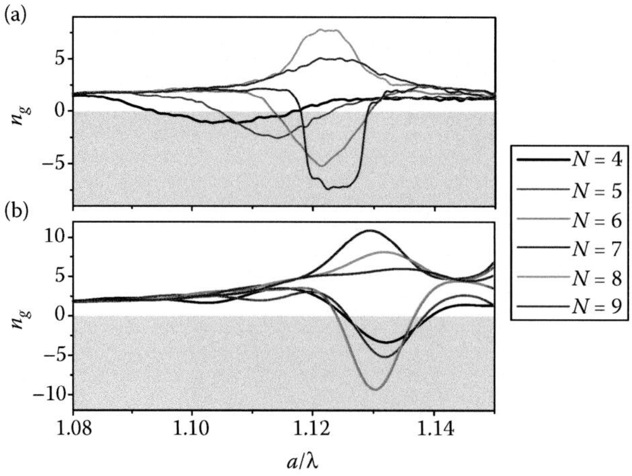

FIGURE 3.2.6
Experimental (a) and calculated (b) group indices for artificial opals grown from 705 nm polystyrene spheres with different number of layers. The gray box indicates negative values of the group velocity. (Reproduced with permission from J. F. Galisteo-López et al., Opt. Express **15**, 15342, 2007.)

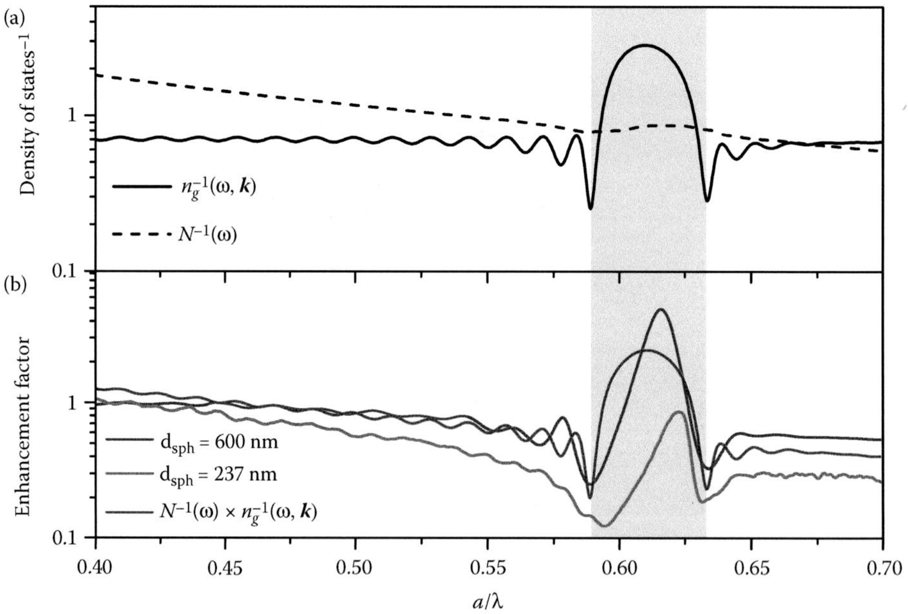

FIGURE 3.2.10
(a) Inverse of the total $N(\omega)$ density of states and group index $n_g[\omega, \boldsymbol{k} = (111)]$ along the incident (ΓL) direction are plotted. (b) Enhancement factor, $\ell_s(a/\lambda)/\ell_s(a/\lambda = 0.4)$, for two opals with no vacancy doping and the quantity $N^{-1}(\omega)/n_g^{-1}[\omega, \boldsymbol{k} = (111)]$. (Reproduced with permission from P. D. García et al., Phys. Rev. B **79**, 241109(R), 2009.)

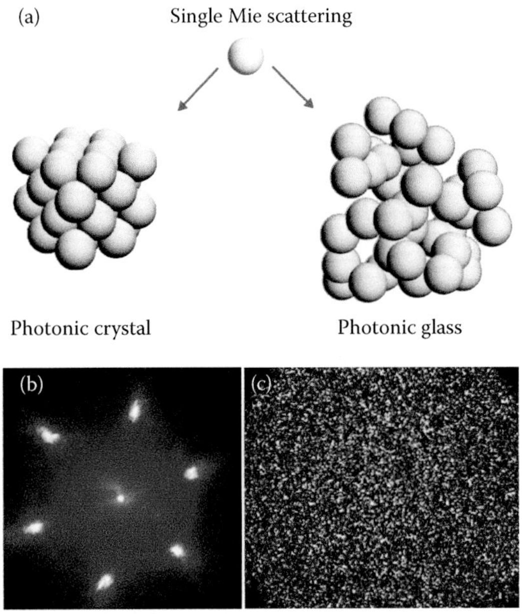

FIGURE 3.3.1
(a) Ordered and disordered packing of a single-scattering particle, a dielectric sphere in this case. (b) and (c) Light speckle pattern arising from an ordered face-centered cubic arrangement of dielectric microspheres and from a random arrangement of the same spheres, respectively, and collected on a far-field screen. (Adapted from P. D. García, R. Sapienza, and C. López, Adv. Mater. **22**, 12, 2010.)

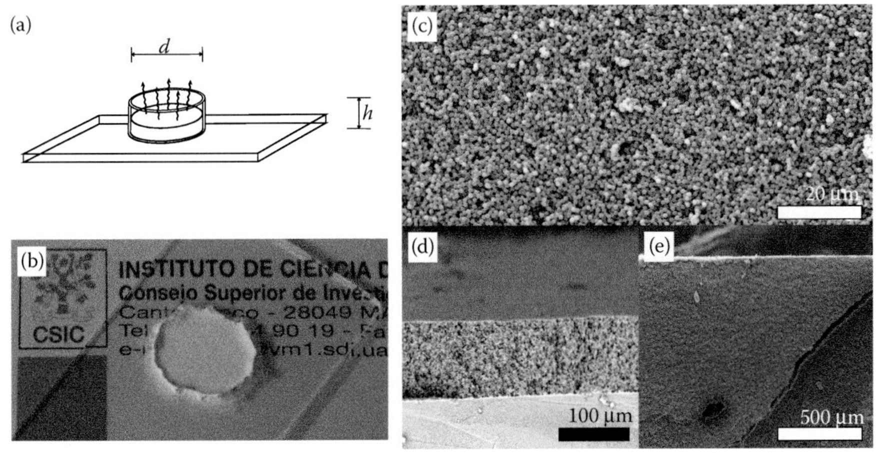

FIGURE 3.3.3
(a) Schematic of the photonic glass growth method. A methacrylate cylinder of high h ~ 1 cm is fixed with impermeable gum to a clean, hydrophilic glass substrate. It is then filled with a charged colloidal suspension previously prepared and shaken under ultrasound. Then it is placed in an oven under constant temperature (~45°C) to force the evaporation of the liquid phase. Finally, the methacrylate cylinder is removed from the glass substrate. (b) A photonic glass grown on a substrate. The sample shows a high degree of planarity, apart from the irregularity of the edge. Scanning electron microscopy images from different parts of a photonic glass made with polystyrene spheres with $d = 1220$ nm. (c) Surface of the sample. The random arrangement of the spheres is revealed here. (d) and (e) Cleaved edges of the sample. In both cases, the images reveal the planarity of the surface over millimeters. The crack visible in (e) is produced when removing the methacrylate cylinder from the glass substrate. (Adapted from P. D. García, R. Sapienza, and C. López, Adv. Mater. **22**, 12, 2010.)

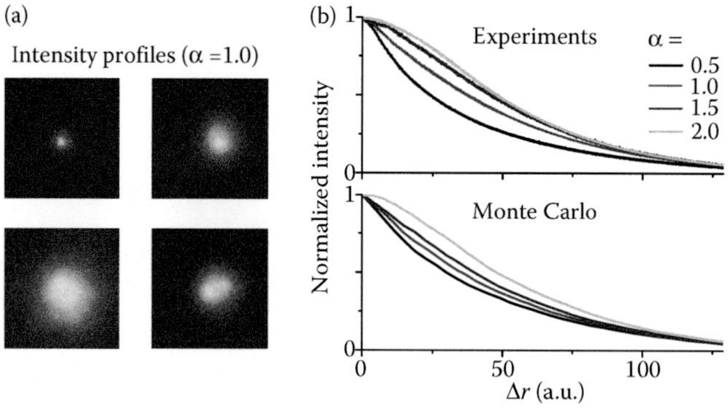

FIGURE 3.4.7
(a) Spatial distributions of the transmitted intensity for a Lévy glass with $\beta = 2.0$, obtained by focusing light on different points of the sample. The transmitted profiles strongly differ from one to the other. (Reproduced from P. Barthelemy, J. Bertolotti, and D. S. Wiersma, Nature **453**, 495, 2008. With permission.) (b) Upper panel: measured intensity profiles for Lévy glasses with different sphere diameter distributions. Lower panel: intensity profiles calculated by Monte Carlo simulations of Lévy walks with the same nominal value of α as in the experiments. Δr represents the distance from the peak of the transmission profile. (J. Bertolotti et al., Adv. Funct. Mater. **20**, 965, 2010. Copyright Wiley-VCH Verlag GmbH & Co. KgaA. Reproduced with permission.)

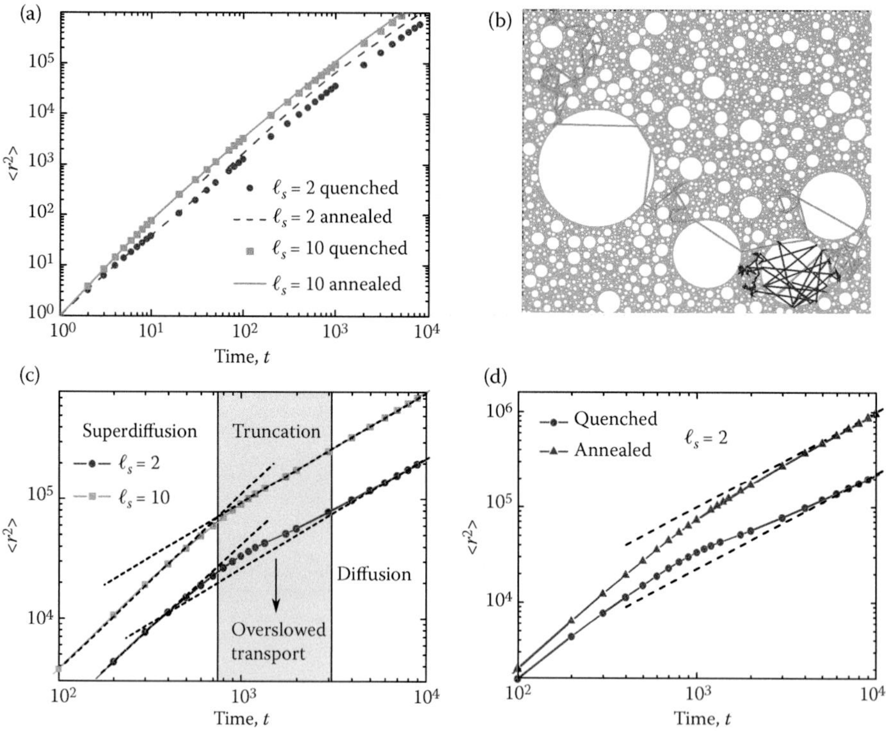

FIGURE 3.4.9
Influence of quenched disorder on transport for large mean free path ($\ell_s = 10 \approx \langle\Delta\rangle$, green) and small mean free path ($\ell_s = 2 < \langle\Delta\rangle$, red). (a) Mean-square displacement versus time in the quenched Lévy glasses (dots/squares) and annealed systems (dashed/solid lines). (b) Typical trajectory of random walkers in two-dimensional (2D) Lévy glasses. (c) Mean-square displacement at the truncation time in 2D quenched Lévy glasses with maximum diameter $\phi_m = 1000$. (d) Comparison of quenched (dots) and annealed (triangles) systems for $\ell_s = 2$. (Reproduced from P. Barthelemy et al., Phys. Rev. E **82**, 011101, 2010. With permission.)

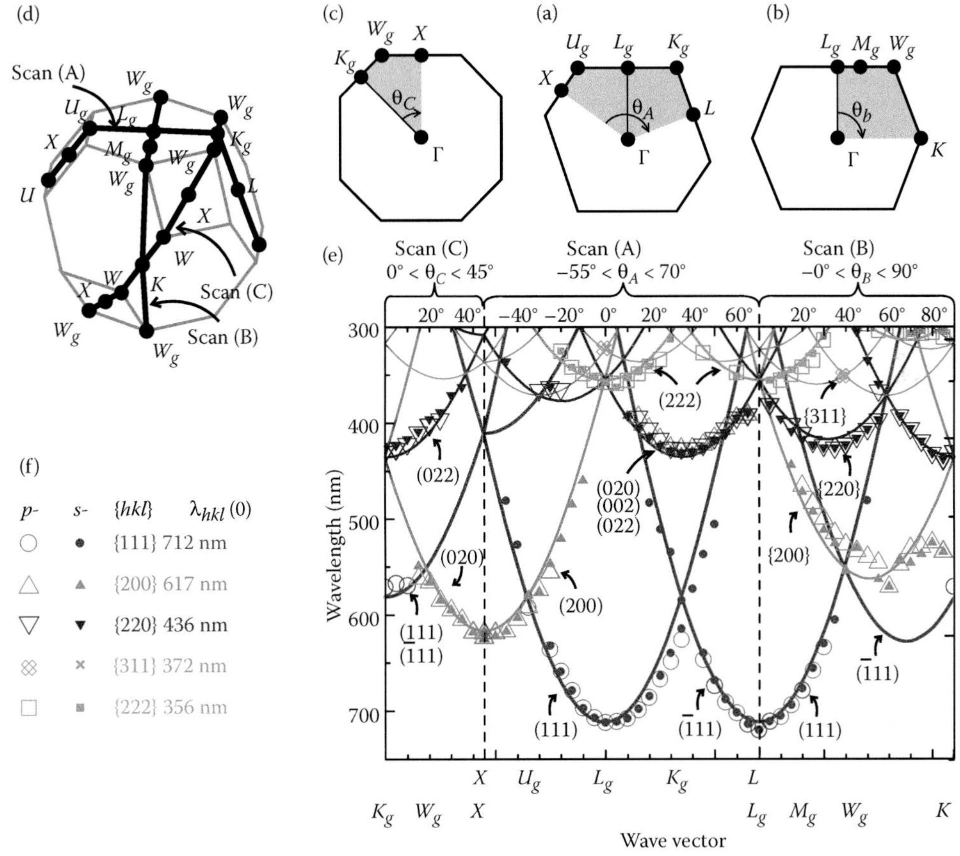

FIGURE 3.5.5

(a)–(c) Three cross sections of the Brillouin zone (BZ) of the face-centered cubic (FCC) lattice made by the scanning planes A (a), B (b), and C (c). (d) The BZ of the FCC lattice showing three scanning paths (A), (B), and (C). (e) Photonic band structure of low-contrast opaline photonic crystals. Solid curves are theoretical dependences of the Bragg wavelength versus angles of light incidence onto the *(hkl)* planes in the FCC lattice; the dependencies are calculated from Equation 3.5.3 for three scanning paths. Symbols are dip positions in the experimental transmission spectra. The symmetry points along the paths are shown in the lower abscissa. (f) Symbols used for denoting the *{hkl}* Bragg bands for *p*- and *s*-polarizations in panel (e) and the theoretical Bragg wavelength $\lambda_{hkl}(\theta)$ at $\theta_{hkl} = 0$ (see Equation 3.5.3). (Panels (a)–(c) and (e) adapted from M. V. Rybin et al., Phys. Rev. B **77**, 205106, 2008. Panels (d) and (f) adapted from M. V. Rybin, K. B. Samusev, and M. F. Limonov, Photonics Nanostruct. Fundam. Appl. **5**, 119, 2007. With permission.)

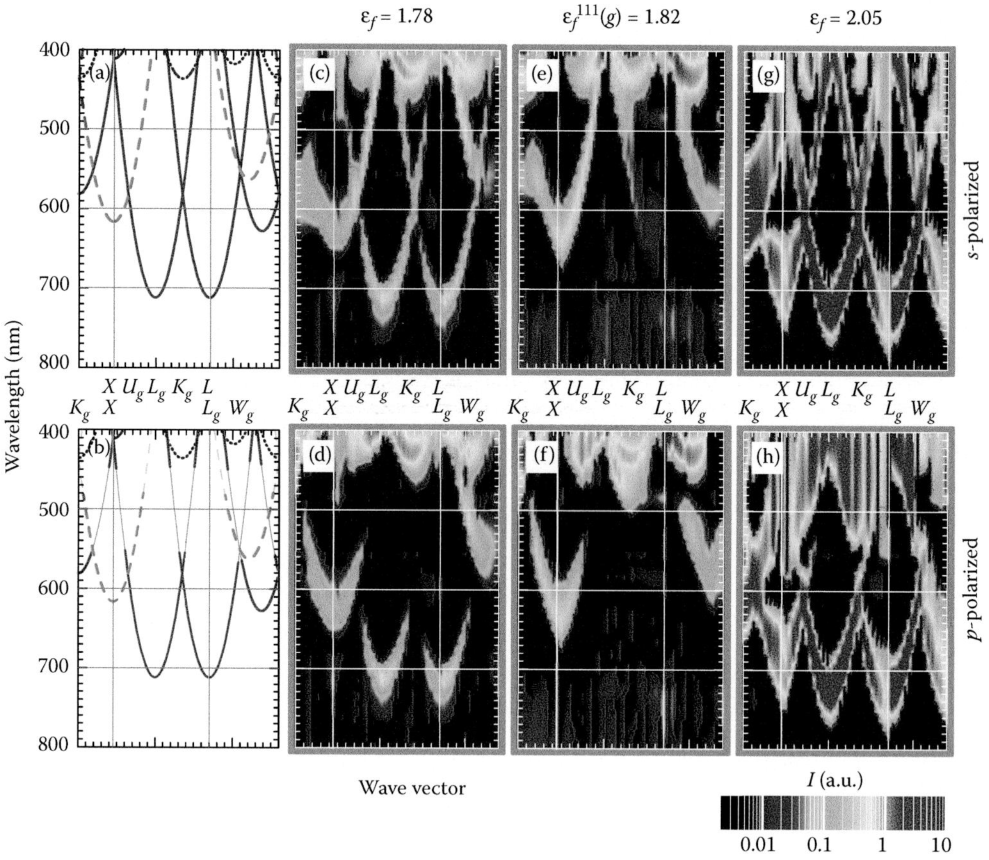

FIGURE 3.5.8
The photonic band structure of opal. (a) Bragg wavelengths $\lambda_{hkl}(\theta)$ for s-polarization versus the angle of incidence for diffraction from different $\{hkl\}$ plane families, calculated from Equation 3.5.3 for three scanning paths (A–C) (see Equation 3.5.2). Solid lines represent the {111} family of planes, dashed lines the {200} family of planes, and dotted lines the {220} family of planes. (b) Bragg wavelengths $\lambda_{hkl}(\theta)$ for p-polarization. The dispersion regions near the Brewster angle $\cos 2\theta_{hkl} = 0$ (see Equation 2.7.7 in Chapter 2.7) are represented by thin curves. (c)–(h) The photonic band structure of opals for three fillers and two linear polarizations is shown on the same scales of wavelengths and wave vectors and in the same scanning planes as the calculations in panels (a) and (b). The intensity scale of photonic bands is given in the lower right corner. (c), (d) $\varepsilon_f^0(g_{111}) = 1.78$, s- and p-polarization; (e), (f) $\varepsilon_f = 1.82$, s- and p-polarization; (g), (h) $\varepsilon_f = 2.05$, s- and p-polarization. (Adapted from M. V. Rybin et al., Phys. Rev. B **77**, 205106, 2008.)

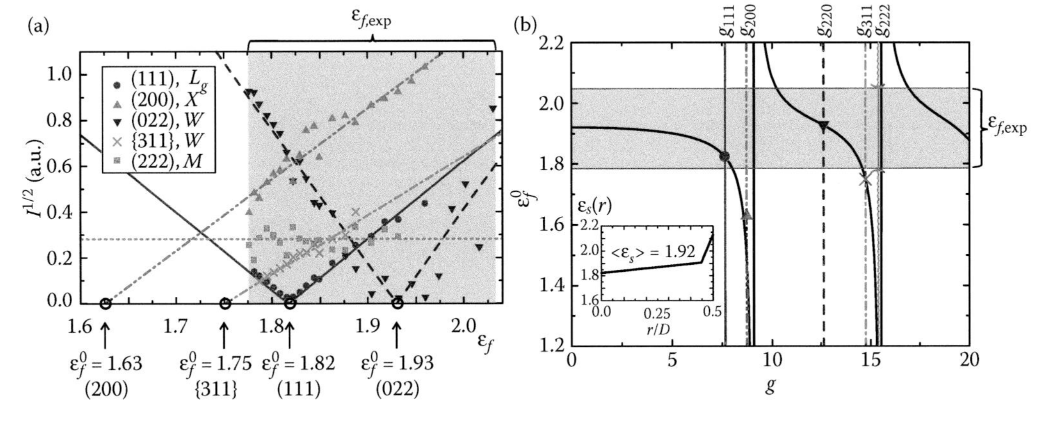

FIGURE 3.5.9
(a) The square root of the intensity of the *(hkl)* Bragg dips as a function of the filler permittivity ε_f. Straight lines represent the fitting of experimental points by a linear function. The gray region is for the experimental range of ε_f. (b) The immersion conditions for an opal structure. The filler permittivity $\varepsilon_f^0(g)$ that corresponds to the immersion condition as a function of the reciprocal lattice vector module $g = |\mathbf{g}|$ calculated from Equation 3.5.4 for the permittivity profile $\varepsilon_s(r)$ simulating the a-SiO_2 spheres. The $\varepsilon_s(r)$ profile is given in the inset. The modules of the shortest reciprocal lattice vector g_{hkl} are shown by vertical lines. Symbols represent experimental values of $\varepsilon_f^0(g_{hkl})$ for four investigated *(hkl)* Bragg bands. (Reprinted from M. V. Rybin et al., Phys. Rev. B **77**, 205106, 2008. With permission.)

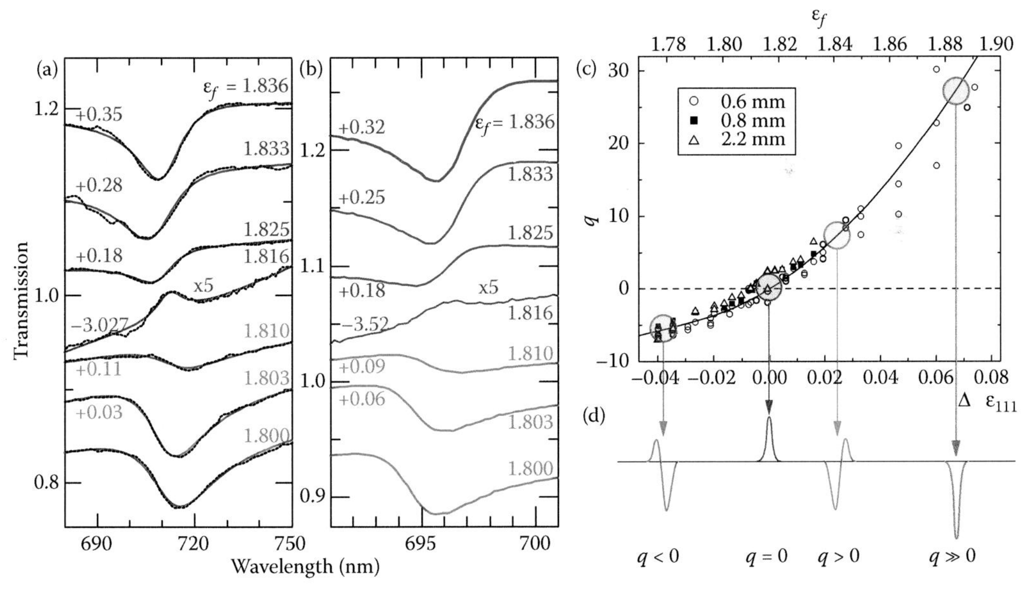

FIGURE 3.5.11
(a) The measured transmission spectra of an opal sample ($D = 316$ nm) as a function of the filler permittivity ε_f in the region of the (111) photonic band (dotted curves). The solid curves are the results of the fitting with a Fano function. (b) The results of calculations of transmission spectra using the "quasi-3D" model of disordered opal structure.[11,24] In (a) and (b) the curves are shifted vertically by the values shown. (c) The Fano asymmetry parameter q as a function of the filler permittivity ε_f, determined by fitting the transmission spectra of the three samples with different thicknesses of 0.6, 0.8, and 2.2 mm. The solid line is a guide for the eyes only. (d) Schematic of the Bragg band shape observed in the transmission spectra on changes in the filler permittivity ε_f, which, in turn depends on Fano parameter q. (Adapted from M. V. Rybin et al., Phys. Rev. Lett. **103**, 023901, 2009; M. V. Rybin et al., Photonics Nanostruct. Fundam. Appl. **8**, 86, 2010. With permission.)

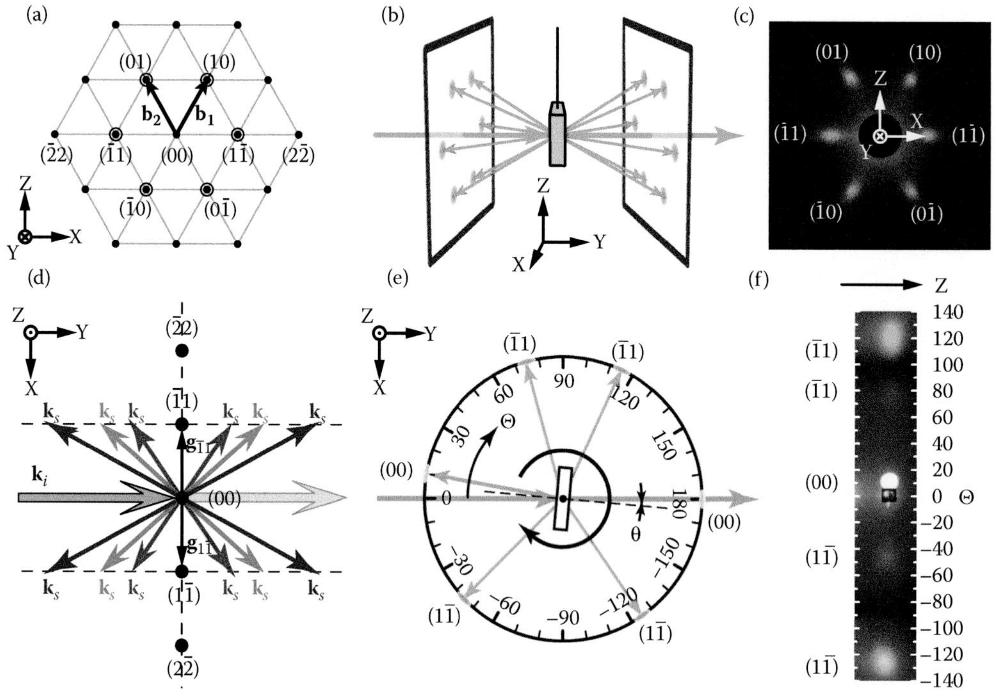

FIGURE 3.6.3

(a) The reciprocal lattice of a two-dimensional (2D) hexagonal layer placed in the XZ plane normally to the incident beam. The lattice consists of an array of nodes, the position of each determined by the 2D vector $\mathbf{g}_{hk} = h\mathbf{b}_1 + k\mathbf{b}_1$, where $\mathbf{b}_1$ and $\mathbf{b}_2$ stand for the basis vectors of the reciprocal lattice, and h and k are integers. The nodes defining the first order of diffraction for the beam incident along the film normal (along the Y axis) are marked with circles. (b) A schematic illustration of the experimental setup designed to observe transmitted or reflected light on a flat screen. (c) The experimental pattern of monochromatic ($\lambda = 532$ nm) light diffraction from a thin opal film observed on a flat screen positioned behind the sample. Six observed reflections are marked with the respective pairs of 2D diffraction indices (*hk*). (d) A schematic illustration of 2D diffraction conditions in the XY plane $\mathbf{q}_{\parallel} = \mathbf{g}_{\bar{1}1}$ and $\mathbf{q}_{\parallel} = \mathbf{g}_{1\bar{1}}$ with the incident white light beam propagating along the Y axis and the hexagonal layer positioned in the XZ plane. (e) A sketch of the experimental setup designed to observe light diffraction in the XY plane on a cylindrical screen. Six diffracted beams are shown. Three of them are scattered forward with respect to the incident beam (i.e., the Y axis), whereas the other three are scattered backward. The angle of incidence θ and the angle of scattered light registration Θ are indicated. Oblique incidence is exemplified here: $\theta = 5°$. (f) A surface development of the white light diffraction pattern from opal film observed on the cylindrical screen in the XY plane at $140° \leq \Theta \leq 140°$. (From A. K. Samusev et al., Phys. Solid State **53**(5), 1056, 2011. With permission.)

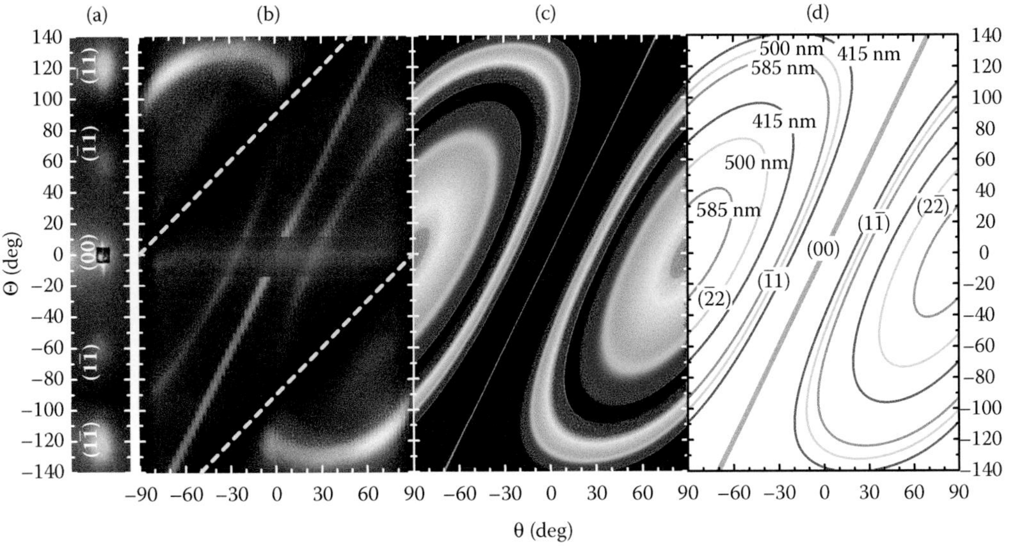

FIGURE 3.6.4
(a) The surface development of a white light diffraction pattern recorded on a cylindrical screen at the normal incidence ($\theta = 0$) on a six-layer-thick opal film consisting of *a*-SiO_2 particles with diameter 720 nm. (b) The combined experimental pattern of light diffraction in the XY plane (see Figure 3.6.3e) for the angle of light incidence varying within the range $-90° \leq \theta \leq 90°$. Dashed lines correspond to the angles $\Theta = \theta \pm 90°$. (c) The numerical simulation of the two-dimensional (2D) diffraction pattern in the Born approximation. The square of the scattering structure factor $S^2(\mathbf{q})$ was calculated using Equation 3.6.6. (d) The calculation of positions of the zeroth-, first-, and second-order 2D diffraction reflections obtained from Equation 3.6.7 for three different wavelengths: $\lambda = 415$, 500, and 585 nm. (From A. K. Samusev et al., Phys. Solid State **53**(5), 1056, 2011. With permission.)

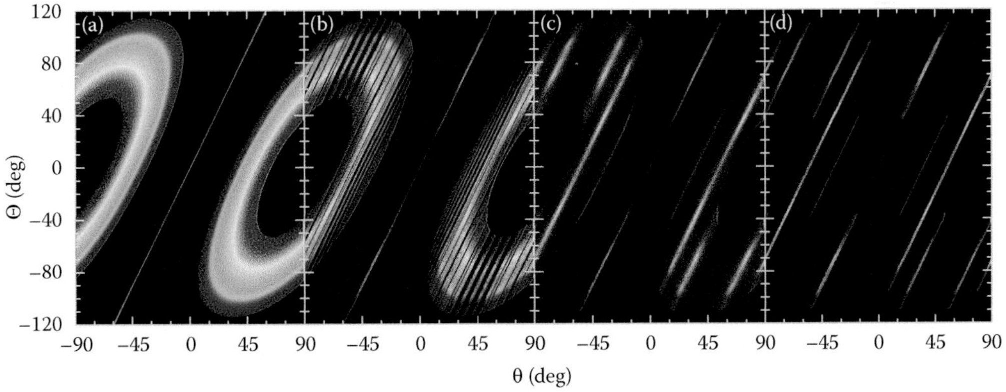

FIGURE 3.6.5
Intensity of the visible (400 ÷ 700 nm) light diffraction from the twinned face-centered cubic opal structure with different number of layers: $N = 1$ (a), $N = 20$ (b), $N = 50$ (c), and $N = 3000$ (d). θ is the angle of incidence and Θ is the angle of scattered light registration. The calculations of the square of scattering structure factor $S^2(\mathbf{q})$ were performed for the structures with *a*-SiO_2 particle diameter of 330 nm. The opal filler is water ($\varepsilon_{H_2O} = 1.78$). (From A. K. Samusev et al., Phys. Solid State **53**(7), 1415, 2011. With permission.)

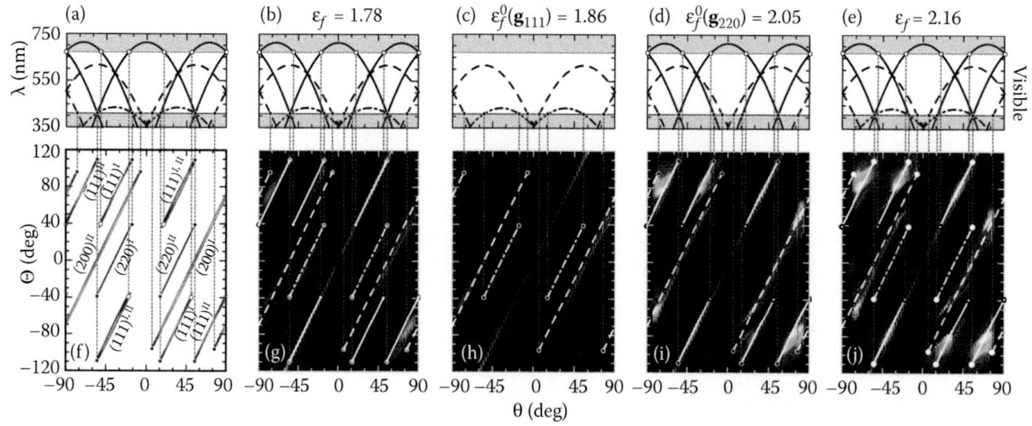

FIGURE 3.6.6
(a)–(e) The dispersion relations $\lambda_{hkl}(\theta)$ of the Bragg wavelengths for the twinned face-centered cubic (FCC) opal lattice formed by a-SiO_2 particles with $D = 330$ nm. $\lambda_{\{111\}}(\theta)$ are indicated by solid curves, $\lambda_{\{200\}}(\theta)$ by dashed curves, and $\lambda_{\{220\}}(\theta)$ by dot-dash curves (see Equation 3.6.8). Panel (c) misses the dispersion curves $\lambda_{\{111\}}(\theta)$ and panel (d) misses the dispersion curves $\lambda_{\{220\}}(\theta)$. The visible spectral range of electromagnetic waves is separated out by two horizontal lines; the ultraviolet and infrared spectral ranges are darkened. (f) The calculated intensity of visible (400 ÷ 700 nm) light diffraction from the twinned FCC opal structure consisting of 3000 × 3000 × 3000 scatterers immersed in water. (g)–(j) The experimental patterns of light diffraction from the synthetic opal sample immersed in fillers with dielectric permittivity $\varepsilon_f = 1.78$ (g), 1.86 (h), 2.05 (i), and 2.16 (j). θ is the angle of incidence counted from growth (111) plane set of opal FCC lattice and Θ is the angle of scattered light registration. (From A. K. Samusev et al., Phys. Solid State **53**(7), 1415, 2011. With permission.)

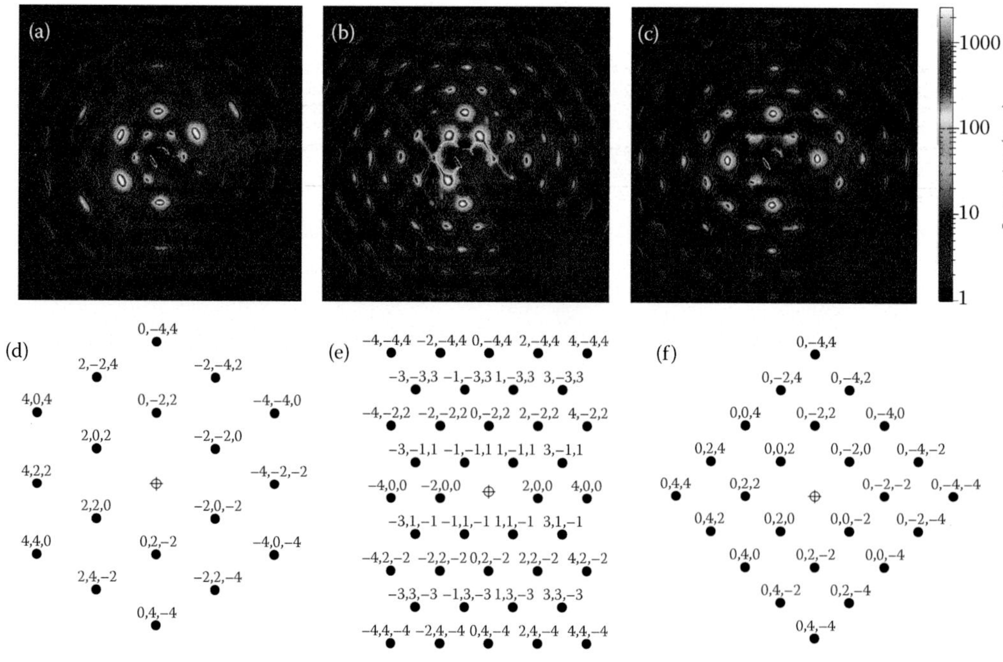

FIGURE 3.6.9
Microradian x-ray diffraction patterns of opal-like photonic films of polystyrene colloidal spheres. The film is about 20 layers thick. The patterns are measured at normal incidence θ = 0° (a) and after sample rotation by θ = −35° (b) and 55° (c). The sketches (d–f) below each pattern (a–c) present the expected Bragg reflections for a face-centered cubic crystal with corresponding orientations.

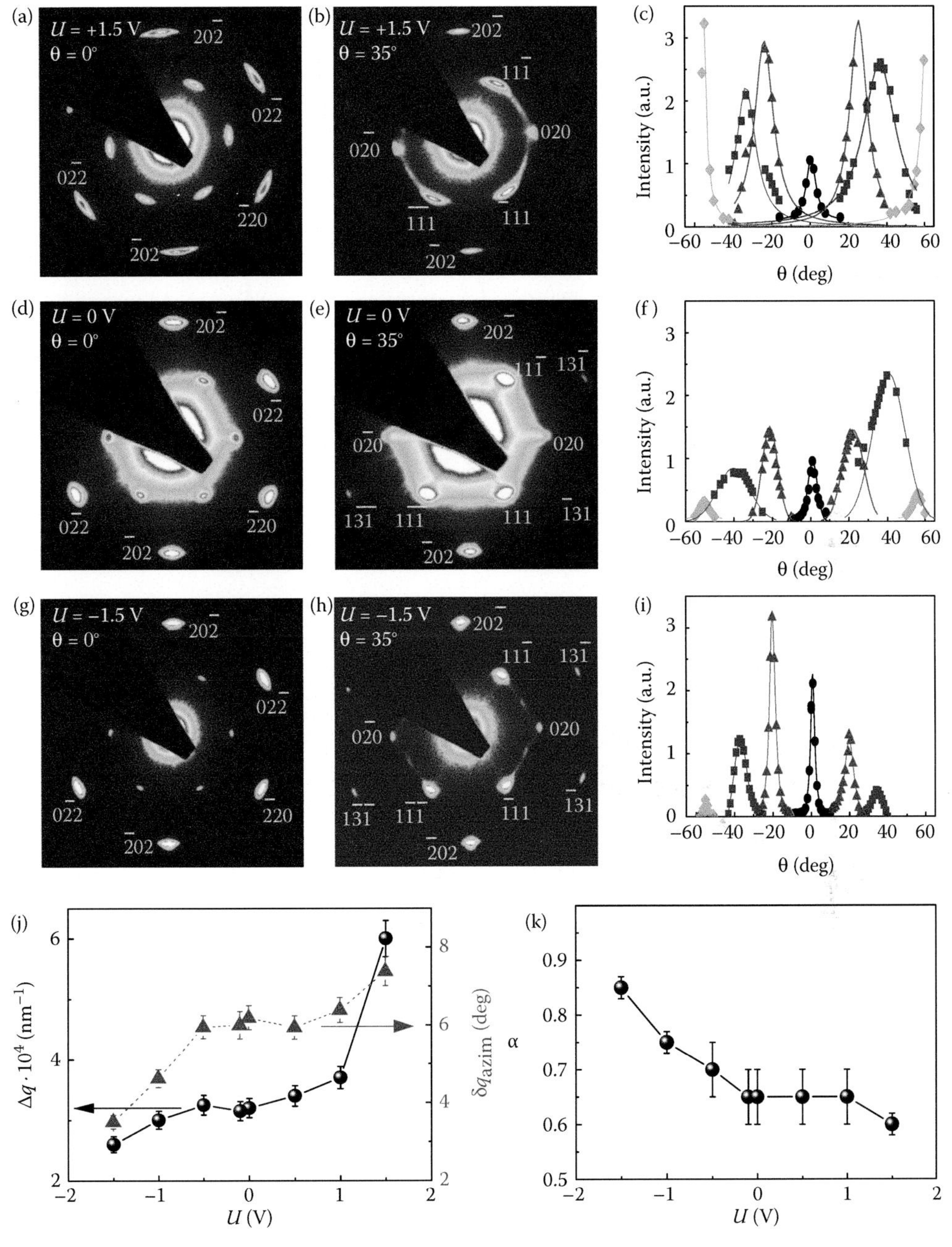

FIGURE 3.6.12
The diffraction patterns obtained at rotation angles $\theta = 0°$ (a, d, g) and 35° (b, e, h) for samples synthesized at a few selected voltage values $U = -1.5$ V (g, h), 0 V (d, e), and +1.5 V (a, b). Panels (c), (f), and (i) present corresponding rocking curves: the θ dependence of the intensity for (220) (circles), $(1\bar{1}\bar{1})$ (triangles), $(1\bar{1}1)$ (squares), and (202) (rhombs) Bragg reflections. Panel (j) displays the longitudinal (Δq) and transversal (δq_{azim}) width of $(1\bar{1}1)$ reflection (measured at $\theta = 19°$) as a function of the potential applied on colloidal film synthesis. The dependence of stacking probability α of a colloidal film on applied voltage U is given in panel (k). Negative (positive) values of U correspond to cathode (anode) polarizations.

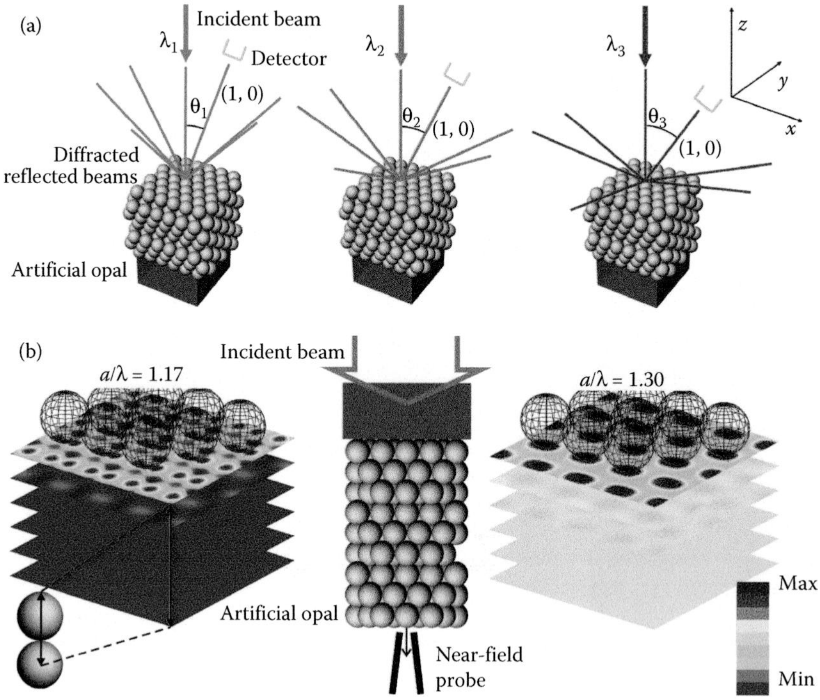

FIGURE 3.7.2
(a) Scheme of the experimental setup used to measure the spectral dependence of the intensity of the reflected diffracted beams. Note that $\lambda_1 < \lambda_2 < \lambda_3$. (b) Schematic drawings of the setup used to measure the near-field intensity pattern in the collection mode. Calculations showing the evolution of the transmitted near-field intensity profile before reconstruction of the far field takes place for two values of reduced frequency for a 10-layer-thick artificial opal made of spheres of dielectric constant $\varepsilon_s = 2.5 + 0.05i$. Profiles are calculated each 0.3*d* (*d* being the diameter of the sphere) between 0, that is, the plane tangent to the poles of the spheres, and 1.5*d*. Color scale bar is also indicated.

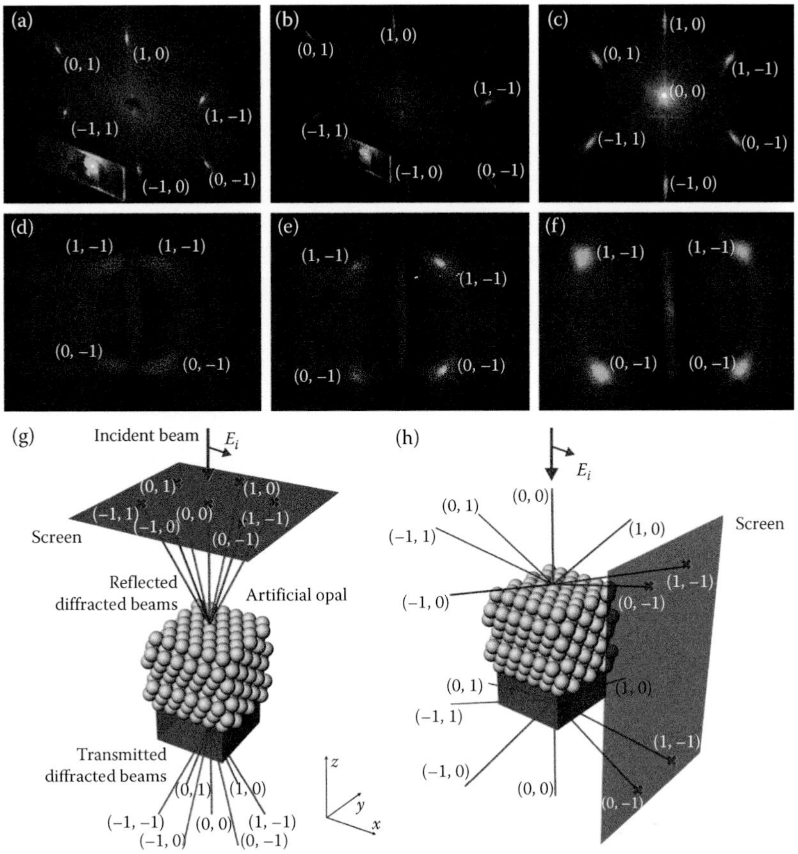

FIGURE 3.7.6
(a), (b) Diffraction patterns of reflected beams projected on a screen parallel to the xy-plane for: (a) $\lambda = 484$ nm and (b) $\lambda = 565$ nm. (c) Diffraction pattern of transmitted beams projected on a screen parallel to the xy-plane for $\lambda = 539$ nm. (d)–(f) Diffraction patterns of reflected and transmitted beams projected on a screen parallel to the yz-plane for: (d) $\lambda = 622$ nm, (e) $\lambda = 593$ nm, and (f) $\lambda = 512$ nm. Sketches depicting the reflected and transmitted diffracted beams projected on a screen parallel to the (g) xy-plane and (h) yz-plane. (Adapted from L. A. Dorado et al., Phys. Rev. B **78**, 075102, 2008.)

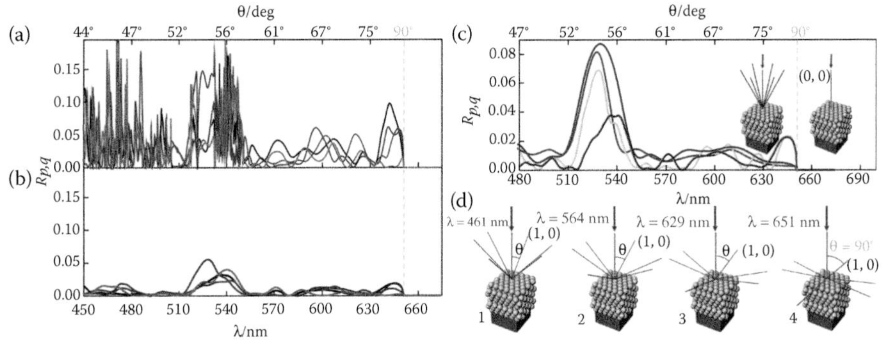

FIGURE 3.7.7
(a), (b) Calculated efficiencies of nonspecular diffracted reflected beams for a perfectly ordered ($\varepsilon_i = 0$) artificial opal (a) and the same photonic structure after introducing extinction ($\varepsilon_i = 0.04$). (c) Measured and calculated reflection efficiencies of diffracted modes (1, 0) (red and light gray, respectively) and (1, –1) (blue and dark gray line, respectively). Vertical dashed line indicates the onset of diffraction for the reflected beams diffracted by an artificial opal made of spheres of 750 nm in diameter. (d) Scheme showing the evolution of the angle (θ) that forms the wave vector of the diffracted beams ($\vec{K}_g^{\pm}$) with the z-axis for different values of the incident wavelength: that is, $\theta = 45°$ for $\lambda = 461$ nm (1), $\theta = 60°$ for $\lambda = 564$ nm (2), $\theta = 75°$ for $\lambda = 629$ nm (3), and $\theta = 90°$ for $\lambda = \lambda_\chi = 651$ nm (4). (Adapted from L. A. Dorado et al., Phys. Rev. B **78**, 075102, 2008.)

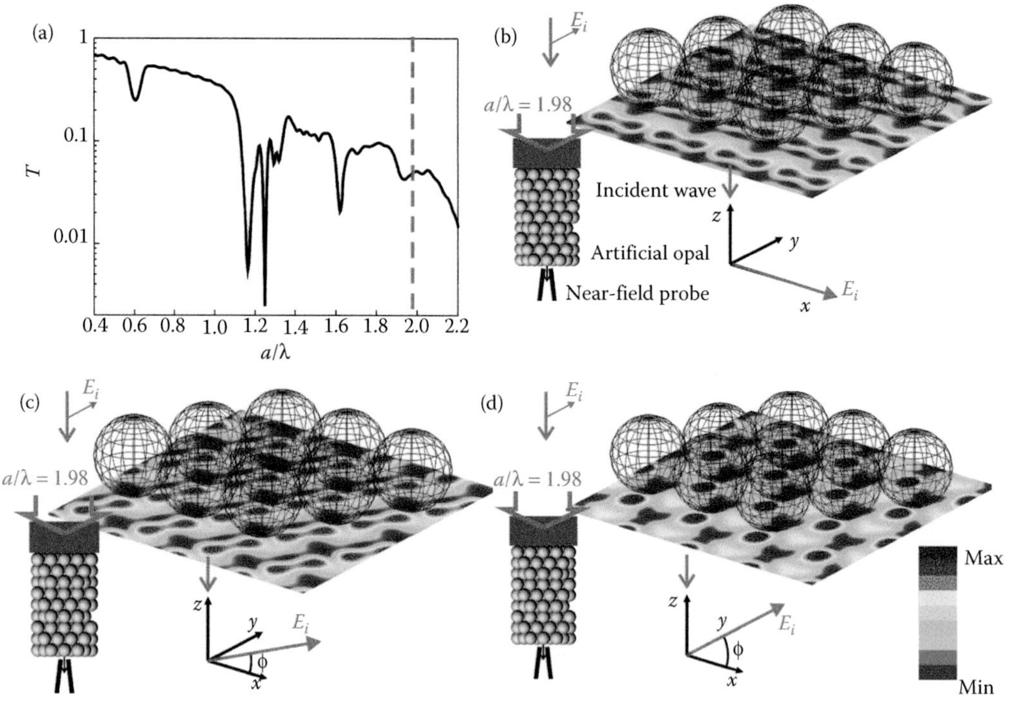

FIGURE 3.7.10
(a) Calculated forward transmittance spectrum for a 10-layer-thick glass-supported colloidal crystal made of spheres of dielectric constant $\varepsilon_s = 2.5 + 0.05i$. Green vertical dashed line indicates the experimental incident wavelength value ($\lambda = 532$). (b)–(d) Spatial distribution of the intensity of the transmitted electric field calculated in the plane tangent to the poles of the spheres that form the outer compact plane in the 10-layer stack. The angle ϕ that forms the incident electric field with respect to the normal of the outer surface of the colloidal crystal (z-axis in our reference system) is $\phi = 0°$ in (b), $\phi = 60°$ in (c), and $\phi = 90°$ in (d). Color scales are also indicated. (Adapted from J. Barrio et al., J. Appl. Phys. **109**, 083514, 2011.)

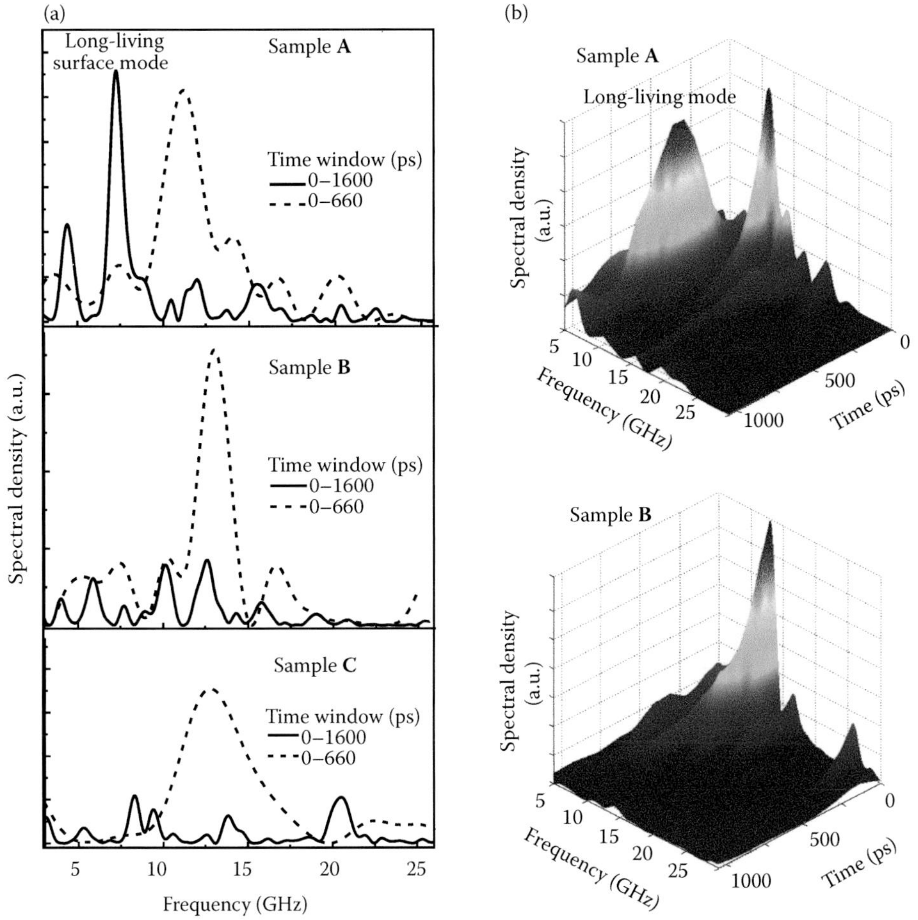

FIGURE 3.8.5
(a) Power spectra of coherent elastic vibration obtained by fast Fourier transform of the measured reflectivity signals Δt (see Figure 3.8.4) for various opal samples. The solid and dashed curves in (a) give the spectra obtained in long ($\Delta t = 1600$ ps) and short ($\Delta t = 660$ ps) time windows, respectively. (b) Temporal evolutions of the vibrational spectra for samples **A** and **B**; the time scale in (b) corresponds to the start of the time window that was fixed at $\Delta t = 660$ ps. The long-living mode is observed only in sample **A**. (From A. S. Salasyuk et al., Nano Lett. **10**, 1319, 2010. With permission.)

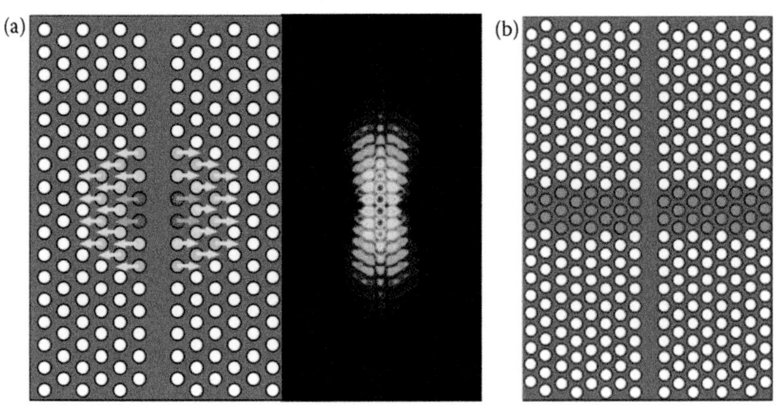

FIGURE 4.1.2
Modulated mode-gap cavities in two-dimensional photonic crystals. (a) Width-modulated line-defect cavity. Schematic and field-intensity profile. (b) Double heterostructure cavity.

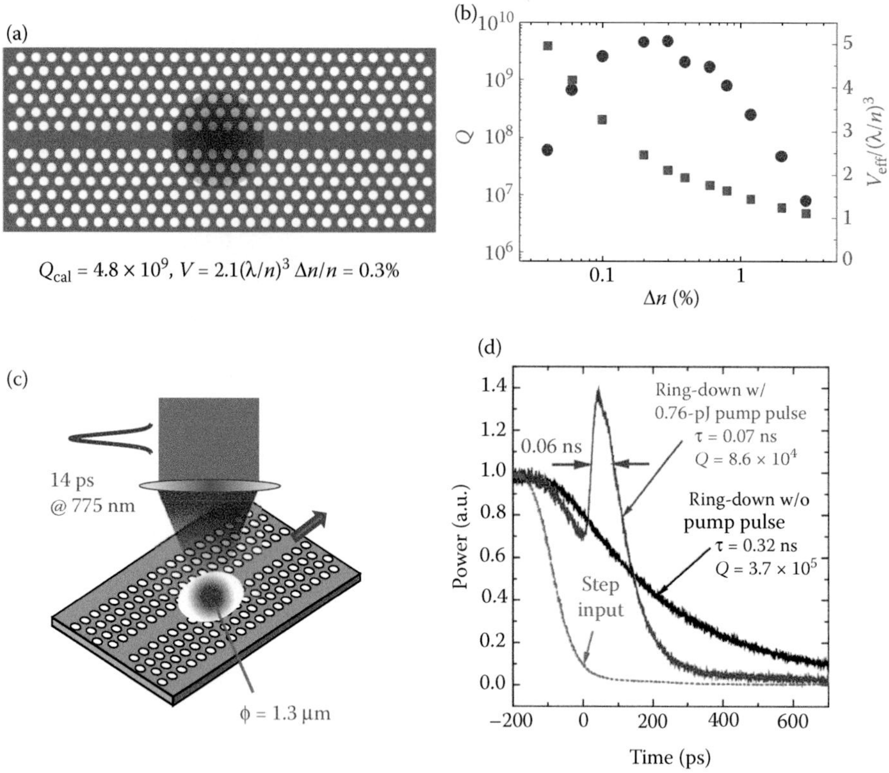

FIGURE 4.1.5
Light confinement by index modulation. (a) Ultrahigh-Q cavity formed by local refractive index modulation. (b) Calculated Q and mode volume for index-modulated cavities. (c) Experimental setup for dynamic Q tuning for a Si width-modulated cavity. (d) Experimental observation of Q tuning. A short pulse is generated in the output waveguide because the coupling Q is dynamically lowered by optical pumping. (Panels (a) and (b) adapted from M. Notomi and H. Taniyama, Opt. Express **16**, 18657, 2008. Panels (c) and (d) adapted from T. Tanabe et al., Phys. Rev. Lett. **102**, 043907, 2009.)

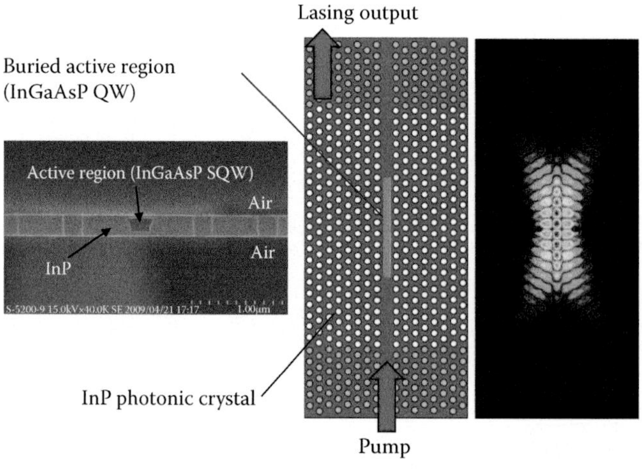

FIGURE 4.1.7
Index-modulated cavity for a photonic crystal buried heterostructure laser. (a) Scanning electron micrograph. (b) Schematic and calculated field-intensity profile.

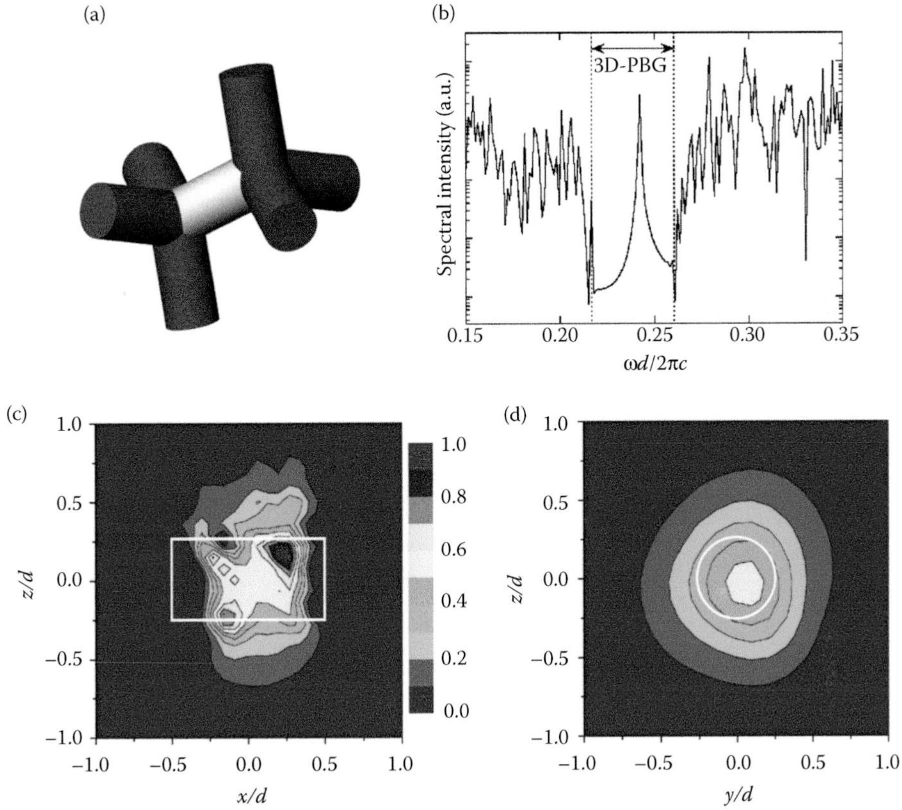

FIGURE 4.1.13
(a) A schematic illustration of the defect introduced in a photonic amorphous diamond by removing a rod. Here, the light gray rod is removed, creating two threefold junctions. (b) The spectral intensity at the center of the removed rod. (c) and (d) The defect mode profiles on two mutually orthogonal planes, where the position of the missing rod is indicated by a white rectangle or circle. (Panels (b)–(d) adapted from K. Edagawa, S. Kanoko, and M. Notomi, Phys. Rev. Lett. **100**, 013901, 2008.)

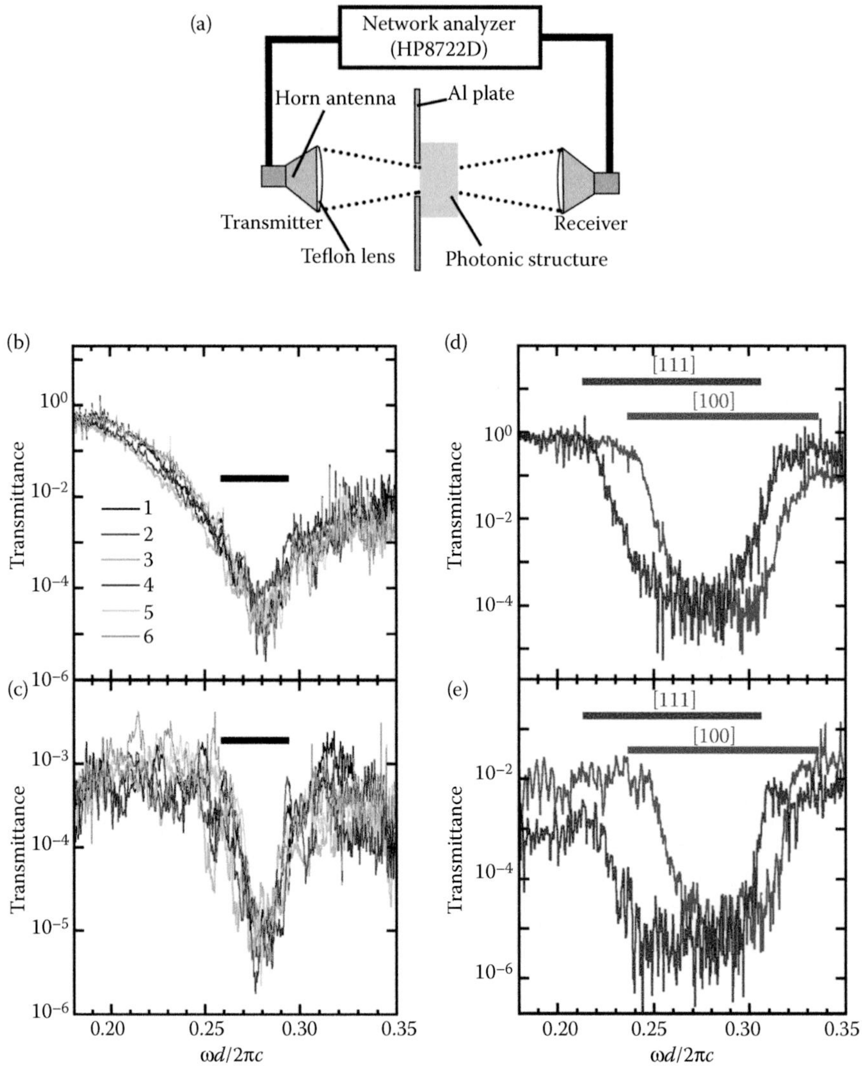

FIGURE 4.1.15
(a) A schematic illustration of the experimental setting for microwave transmission measurements. (b)–(e) Measured transmission spectra: (b) T_p for PAD, (c) T_c for PAD, (d) T_p for PCD, and (e) T_c for PCD. The six curves in (b) and (c) are for three samples with different orientations, and for two mutually orthogonal directions of the incident microwave polarizations for each sample. The blue and red spectra in (d) and (e) are for the microwave incident directions along [100] and [111] of the diamond lattice, respectively. The black bar in (b) and (c) indicates the frequency range of the three-dimensional photonic band-gap in PAD deduced by a finite-difference time-domain calculation. The blue and red bars in (d) and (e) indicate the gap positions of PCD in the ΓX and ΓL directions, respectively, deduced by a photonic band-structure calculation using a plane-wave expansion method. Note that these gap frequency ranges are different from those in Figures 4.1.11 and 4.1.12 because of the different refractive index.PAD, photonic amorphous diamond; PCD, photonic crystalline diamond. (Panels (b)–(e) adapted from S. Imagawa et al., Phys. Rev. B **82**, 115116, 2010.)

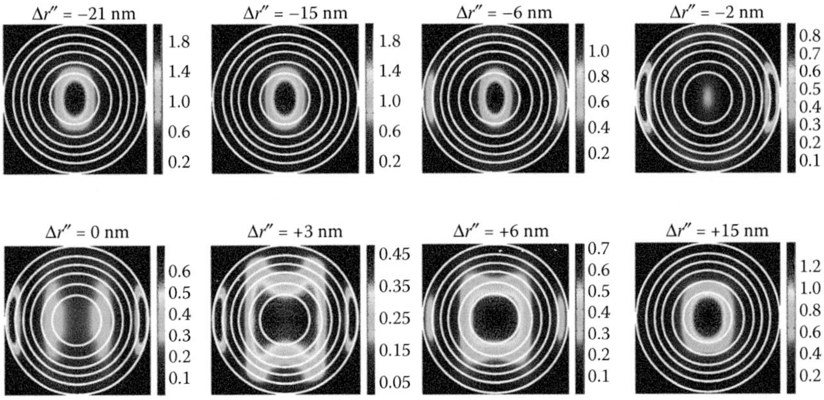

FIGURE 4.2.3
A selection of calculated far-field patterns (electric field intensity profile, $|\mathbf{E}|^2$) by three-dimensional finite-difference time-domain simulations, corresponding to the measured devices whose results are shown in Figure 4.2.2. Field intensities are normalized to the total emitted power in the vertical half-space. Concentric circles correspond to $\theta = 20°, 30°, 40°, 50°, 60°$, and $90°$ from the inner to the outer one, respectively. (Reproduced from S. L. Portalupi et al., Opt. Express **18**, 16064, 2010. With permission.)

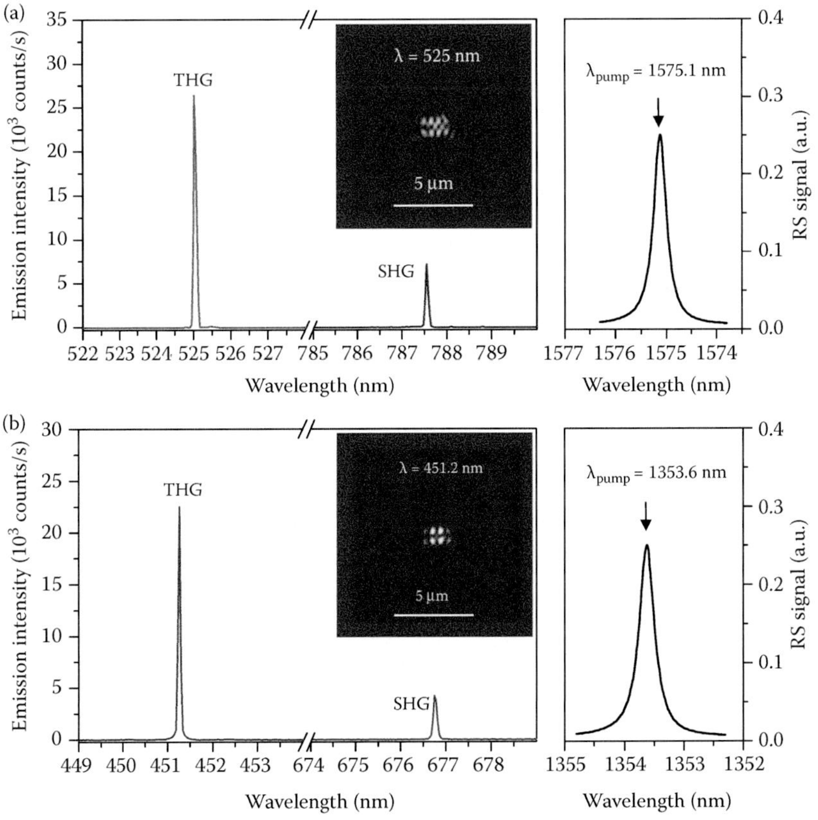

FIGURE 4.2.4
(a) Resonant scattering spectrum (right) of a PhC nanocavity with fundamental mode at $\lambda_c = 1575.1$ nm (pump wavelength). SHG and THG emission spectra (left) at deep-red and green wavelengths $\lambda_{SH} = 787.5$ nm and $\lambda_{TH} = 525$ nm, respectively. (b) Resonant scattering spectrum (right) of a PhC nanocavity with fundamental mode at $\lambda_c = 1353.6$ nm (pump wavelength). SHG and THG emission spectra (left) at red and blue wavelengths $\lambda_{SH} = 676.8$ nm and $\lambda_{TH} = 451.2$ nm, respectively. Insets show THG images taken with a commercial CCD camera.

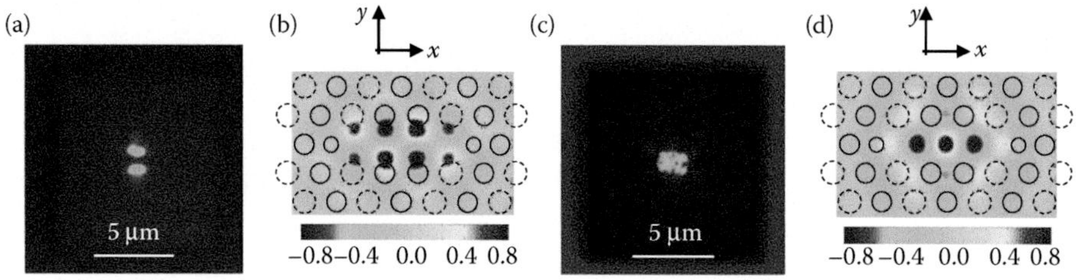

FIGURE 4.2.7
Spectrally filtered optical image of (a) second-harmonic generation and (c) third-harmonic generation emission taken with a high-sensitivity Si CCD. Calculated electric field inside the photonic crystal cavity at the resonance frequency for (b) E_x and (d) E_y components, respectively. (Reproduced from M. Galli et al., Opt. Express **18**, 26613, 2010. With permission.)

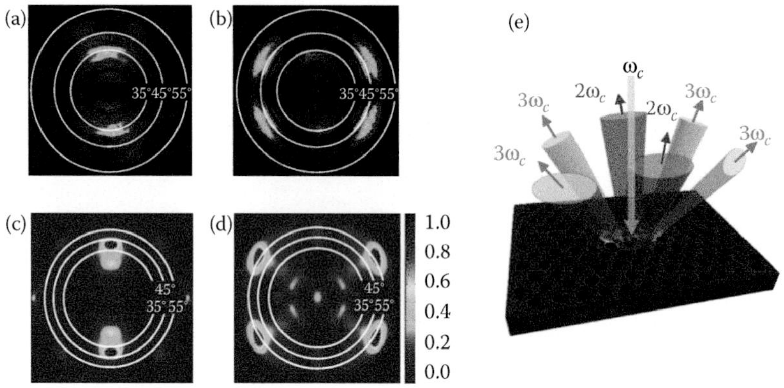

FIGURE 4.2.8
Experimental Fourier images showing the polar far-field emission profile of (a) second-harmonic generation and (b) third-harmonic generation light. The corresponding far-field intensity patterns in (c) and (d) are calculated by three-dimensional finite-difference time-domain for (a) second-harmonic and (b) third-harmonic frequencies, respectively. (e) An artist's impression of the SHG and THG emission from the photonic crystal nanocavity. (Reproduced from M. Galli et al., Opt. Express **18**, 26613, 2010. With permission.)

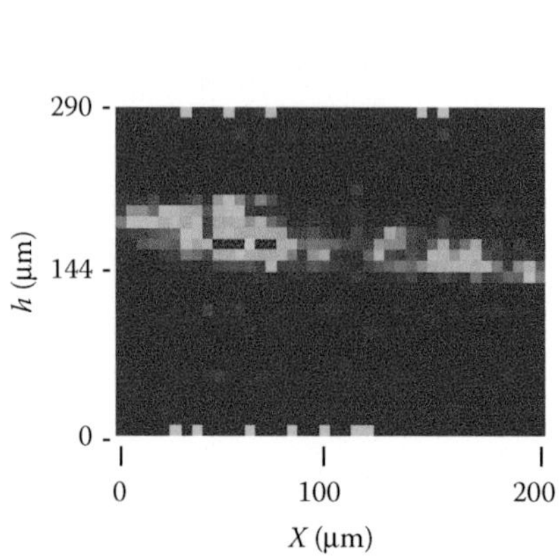

FIGURE 4.3.11
Emission intensity map from poly(2,5-dioctyloxy-p-phenylenevinylene) or DOO-PPV polymer film. (Adapted from R. C. Polson and Z. V. Vardeny, Opt. Lett. **35** (16), 2801, 2010.)

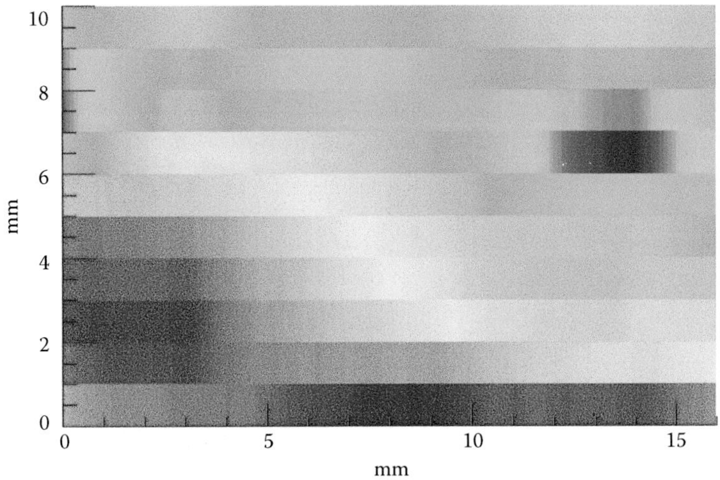

FIGURE 4.3.15
Mapping of relative health of tissue based of averaged Fourier transform of lasing tissue. (From R. C. Polson and Z. V. Vardeny, J. Opt. **12**, 24010, 2010. With permission.)

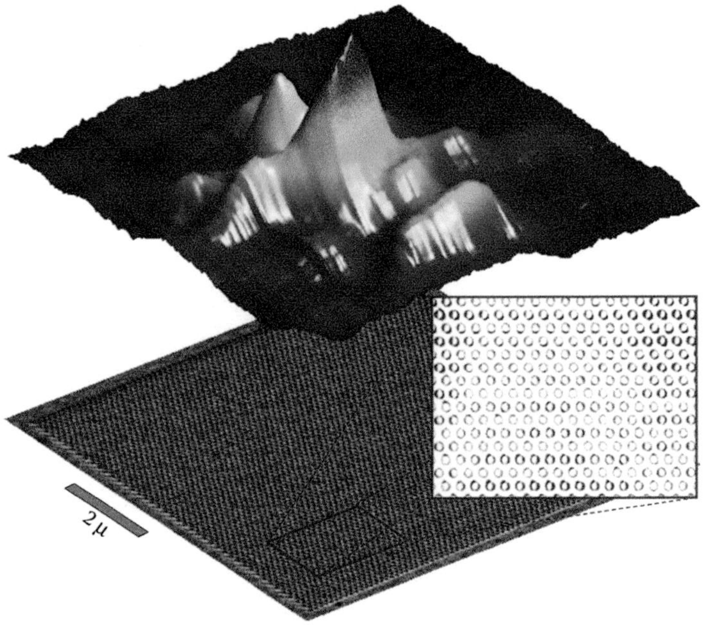

FIGURE 4.4.3
Top-view scanning electron micrograph (SEM) of a ZnO photonic crystal (PhC) slab. Difference between the digitized SEM of a real sample and the perfect honeycomb lattice (blow-out box) reveals the structural disorder. Superimposed is the measured intensity distribution of the lasing mode in a ZnO PhC slab with $a = 115$ nm and $R = 0.25a$. The pattern measures 8×8 μm.

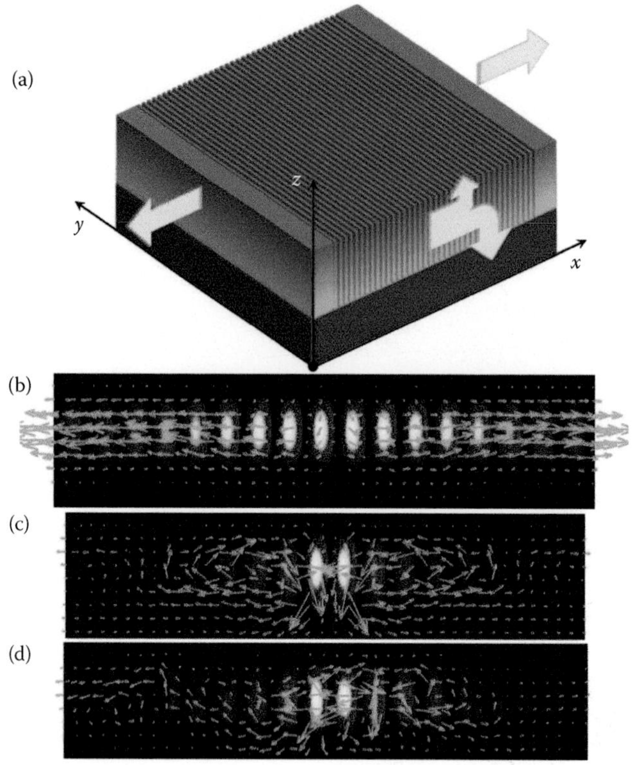

FIGURE 4.4.4
(a) Simplified model of a photonic crystal slab used in numerical simulations. Infinitely long grooves run parallel to the y-axis. Disorder only affects x position of the groves. Although such a system can be modeled in two dimensions—in the xz-plane—it already incorporates the possibilities of vertical and horizontal leakage. Intensity of the mode with the highest Q factor in (a) an ordered system (band-edge mode), (b) an ordered structure with a localized defect, and (c) a disordered structure with a localized defect. Arrows represent the amplitude and direction of the local Poynting vector. The radiative losses are dominated by in-plane leakage in (b) and by out-of-plane leakage in (c), whereas both loss mechanisms are comparable in the disordered system (d).

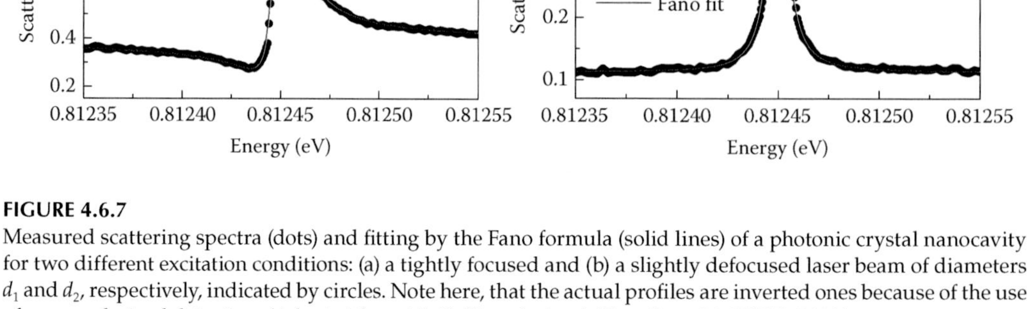

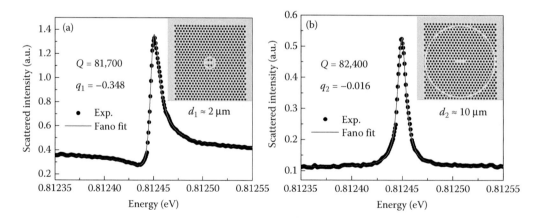

FIGURE 4.6.7
Measured scattering spectra (dots) and fitting by the Fano formula (solid lines) of a photonic crystal nanocavity for two different excitation conditions: (a) a tightly focused and (b) a slightly defocused laser beam of diameters d_1 and d_2, respectively, indicated by circles. Note here, that the actual profiles are inverted ones because of the use of cross-polarized detection. (Adapted from M. Galli et al., Appl. Phys. Lett. **94**, 071101, 2009.)

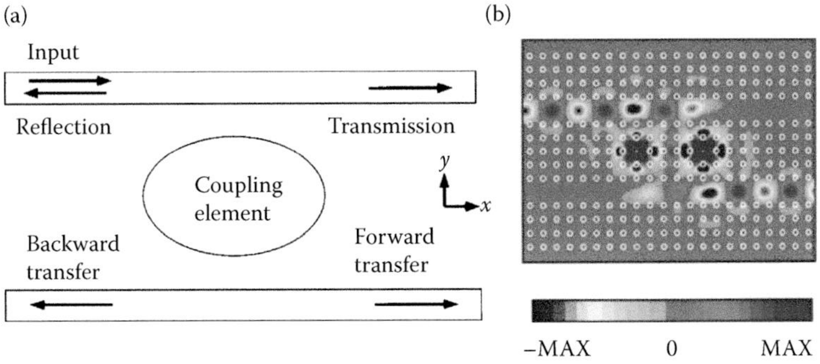

FIGURE 4.6.9
Add-drop filter. (a) Schematic diagram of two waveguides coupled through an element which supports a localized resonant state. (b) Electric field pattern of the photonic crystal at the resonant frequency. The white circles indicate the position of the rods. (Adapted from S. Fan et al., Phys. Rev. Lett. **80**, 960, 1998.)

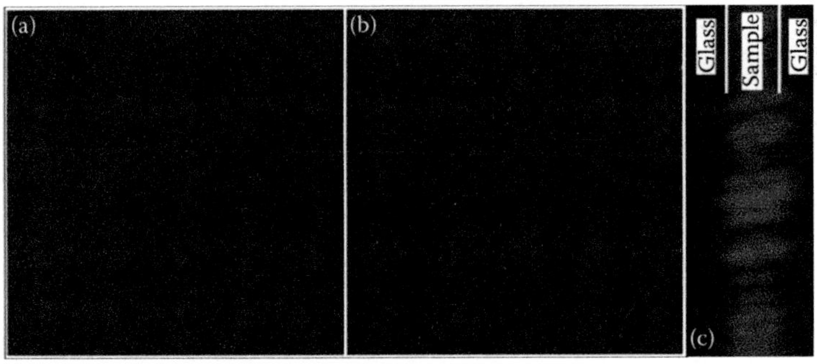

FIGURE 4.9.5
Grayscale confocal image of porous silicon (a) before infiltration and (b) after infiltration with quantum dots. (c) Side view of reconstructed three-dimensional confocal image showing porous silicon sample infiltrated with quantum dots. Image size is 0.75 mm × 0.75 mm.

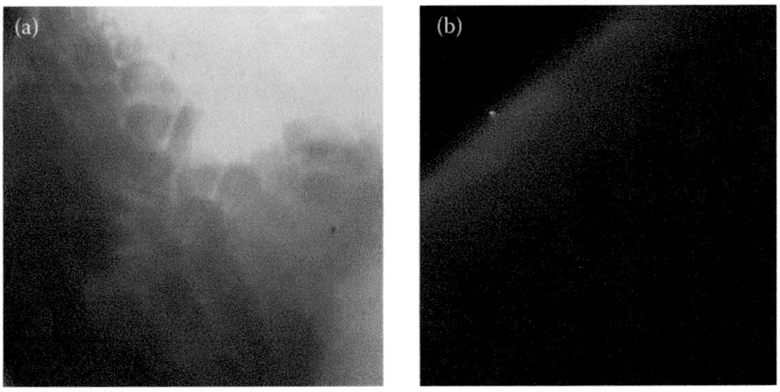

FIGURE 4.9.16
Cytoviva images of the red opal with blue quantum dots. In (a) the opal is seen under the fully optical light and in (b) it is seen under fully florescent light. (With kind permission from Springer Science+Business Media: *Nanoscale Photonics and Optoelectronics*, Vol. 9, 2010, p. 70, R. Moussa et al., Figures 4.1–4.8.)

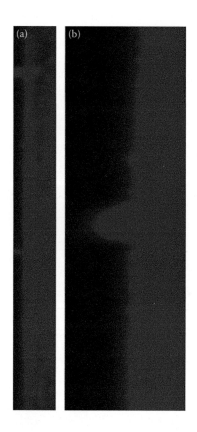

FIGURE 4.9.17
Two snapshots of the multiple generated scans. (a) Snapshot for the whole surface of the imaged red opal with blue quantum dots and (b) snapshot for the top left part of the surface of the imaged red opal with blue quantum dots. (With kind permission from Springer Science+Business Media: *Nanoscale Photonics and Optoelectronics*, Vol. 9, 2010, p. 71, R. Moussa et al., Figures 4.1–4.8.)

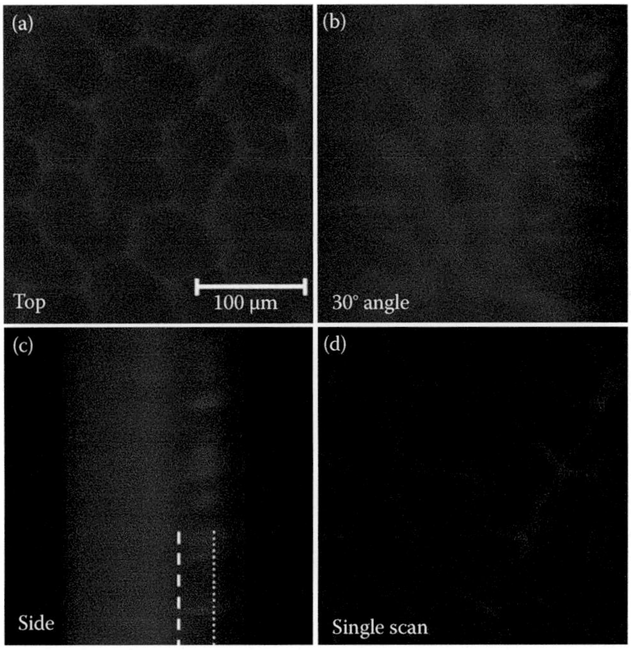

FIGURE 4.9.18
Three-dimensional reconstructed image of light propagation in red opal infiltrated with 481 nm quantum dots. (a) Top view of the opal; (b) 30° inclined from top view; (c) side view of the opal where the dashed line shows opal–glass interface with opal being on the left side and dotted line is where the single scan was taken. (d) Single scan at 30 μm above the surface of the opal. (With kind permission from Springer Science+Business Media: *Nanoscale Photonics and Optoelectronics*, Vol. 9, 2010, p. 72, R. Moussa et al., Figures 4.1–4.8.)

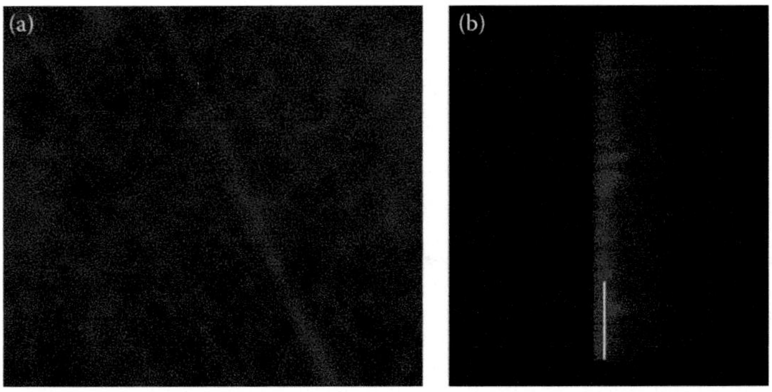

FIGURE 4.9.19
Three-dimensional reconstructed image of light propagation in red inverted opal with optical glue and infiltrated with 481 nm blue quantum dots. (a) Image seen from the top; (b) image seen from the side. The solid line indicates roughly the interface between the inverted opal and the glass. (With kind permission from Springer Science+Business Media: *Nanoscale Photonics and Optoelectronics*, Vol. 9, 2010, p. 73, R. Moussa et al., Figures 4.1–4.8.)

3.2

Optical Spectroscopy of Real Three-Dimensional Self-Assembled Photonic Crystals

Juan F. Galisteo López and Cefe López
Materials Science Institute of Madrid

CONTENTS

3.2.1 Introduction

Self-assembled structures in the form of artificial opals are arguably the most widespread approach to the fabrication of three-dimensional (3D) photonic crystals (PhCs). As a matter of fact, such systems were already present in the 1987 seminal work of Yablonovitch[1] as a probe system to study the optical properties of 3D PhCs. More than two decades after that initial proposal, artificial opals have been thoroughly studied from both experimental and theoretical points of view. Their crystalline quality has been improved by exploring several fabrication methods, their optical response has been studied using a wide variety of experimental techniques, and many applications have been proposed for them such as chemical or biological sensors, components of dye sensitized solar cells or displays to name but a few.[2]

Artificial opals tend to grow forming a face-centered cubic lattice owing to the natural tendency of colloidal particles to self-assemble under the appropriate conditions. Further, unless a patterned substrate is used to direct the growth, they tend to grow with the close-packed {111} planes parallel to the sample surface. This means that if conventional reflection or transmission spectroscopy measurements are performed to study their optical response, the ΓL direction in reciprocal space will be probed in a normal incidence configuration, as depicted in Figure 3.2.1, where the dispersion relation of an artificial opal made from polystyrene (PS) spheres is shown.

The dispersion relation along this particular direction shows two well-differentiated spectral regimes. One is what is commonly termed the low-energy regime, where a pseudogap of forbidden frequencies for light propagation lies within two energy bands that resemble the linear dispersion for a homogeneous material. For frequencies contained within such a pseudogap the available wave vectors have an imaginary component as a

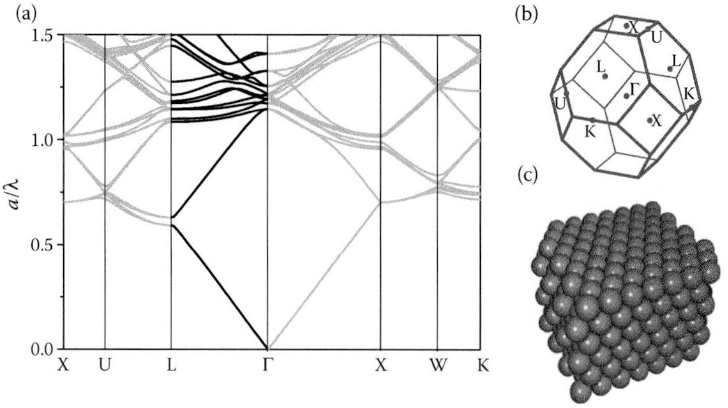

FIGURE 3.2.1
(a) Band structure for a polystyrene opal along high symmetry directions. ΓL direction is highlighted with black lines. (b) Brillouin zone for a face-centered cubic lattice. (c) Image of an artificial opal.

consequence of light extinction due to Bragg diffraction by the {111} planes. The second is the so-called high-energy regime, for reduced frequencies above ~1.1, where multiple Bragg diffraction by several crystallographic planes leads to a complex dispersion relation formed by several energy bands.

The optical study of the pseudogap is commonly used to assess the optical quality of the sample and its spectral position can be further used to monitor postprocessing stages in the fabrication procedure, such as infiltration with other materials or inversion of the original structure by removing the opaline scaffold. In studying the pseudogap spectral region, reflection and transmission spectroscopy as well as emission of broadband internal sources are routinely used. The pseudogap appears as a peak (dip) in a reflection (transmission or emission) spectrum evidencing an exponential attenuation of light due to Bragg diffraction. Its spectral position at normal incidence and its evolution as the angle of incidence is varied along high-symmetry directions have been monitored using these techniques in a number of systems, and several features such as anticrossings resulting from simultaneous Bragg diffraction by several families of planes have been observed. However, most of these techniques, though valid for exploring the existence of forbidden intervals, fail to provide in-depth information on how light propagates inside these systems. Only in samples with a reduced number of layers or a low refractive index contrast, reflection and transmission spectroscopy have been used to study the dispersion of allowed energy bands in the surroundings of the pseudogap.[3]

However, information on how light transport takes place inside these samples is extremely relevant not only from a fundamental point of view but also when envisaging applications. Issues such as how disorder affects light transport or whether slow-light regimes can be achieved where enhanced light matter interaction is expected are important in the study of 3D PhCs. In this chapter, we present results on how experimental techniques such as white-light phase-sensitive spectroscopy or the combination of reflection and transmission spectroscopy can be used to gain further insight into how light propagates inside real 3D PhCs. Although the results presented deal with PhCs in the form of artificial opals, the conclusions extracted are valid for any kind of 3D system.

3.2.2 Optical Phase and Effective Refractive Index

To study the dispersive properties of 3D PhCs, phase-sensitive measurements have been used over the last two decades. In phase-sensitive measurements, interferometric techniques are used to retrieve the optical phase of a beam traversing the sample under study from which the wave vector of light within the crystal can be extracted. Initial measurements in the visible spectral range were performed with monochromatic tunable laser sources using a modified Mach–Zehnder interferometer (MZI) and 3D PhCs in the form of colloidal crystals with[4] and without[5] the presence of intentionally added point defects. In the former case the phase delay was extracted in a narrow spectral range in the surrounding of the pseudogap edges, whereas in the latter case the phase was also extracted for frequencies close to the localized state associated with the defects. Later, artificial opals were used as reflectors of a Fabry–Pérot cavity and the phase upon reflection for those frequencies contained within the pseudogap was obtained at normal incidence.[6]

In obtaining the phase delay introduced by a sample over a broad spectral range, a combination of a modified MZI and a commercial Michelson interferometer can be used together with a tungsten lamp.[7] The probe beam is divided into two arms within the MZI, one containing the sample and the other being the reference. The output of the MZI, containing the reference signal and a delayed sample signal, is then fed into the Michelson interferometer where the two are cross-correlated and an interferogram obtained. Finally, the phase introduced by the sample can be extracted from the interferogram by means of Fourier analysis. In obtaining the phase one must be careful to properly subtract the contribution of the glass substrate on which the sample rests, which is acquired in a separate measurement. It must be mentioned that to obtain an interferogram with the Michelson interferometer coherent radiation must be used. In our case, the output of the tungsten lamp remains coherent over a temporal delay of the order of femtoseconds, enough to perform our measurement.

Figure 3.2.2 shows transmittance and absolute phase-delay measurements performed on PS artificial opals having an increasing number of layers parallel to the sample surface. When looking at the transmittance one can see how a dip evolves in the spectral region where the pseudogap is expected for this kind of sample. Increasing the number of layers increases the amplitude of the dip as expected, as Bragg attenuation becomes more efficient. At both sides of the dip the transmittance presents a background of high transmission, corresponding to the spectral regime where two energy bands are available, and the sample can be treated as a homogeneous effective medium. This background presents a monotonically decreasing trend toward high energies as a consequence of scattering due to structural imperfections. In parallel, we can see how the overall absolute phase increases with sample thickness as expected from a larger optical path within the sample. For those spectral regions where the sample behaves as a homogeneous medium, the phase presents a linear behavior characteristic of a transparent material. As sample thickness increases, a change in slope develops for those frequencies contained within the pseudogap and an overall change of absolute phase of π takes place. This value is the same as the one measured for one-dimensional (1D) PhCs[8] and is the phase-change characteristic of a resonance.

Similar measurements have been performed for samples grown along the (100) direction.[9] In these samples, in the low-energy range no pseudogap opens up and the samples become transparent (see Figure 3.2.1) so that the measured phase presents a linear behavior. Phase-sensitive measurements have also been performed on samples with planar

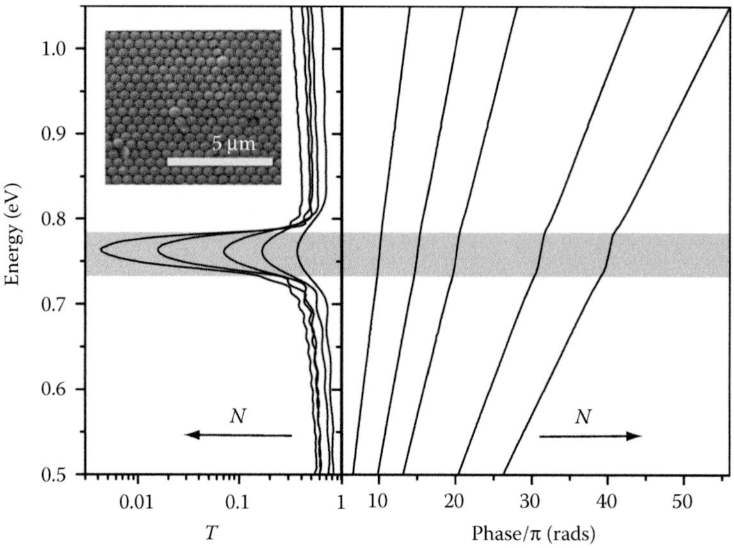

FIGURE 3.2.2
Transmittance (left panel) and absolute phase delay (right panel) for artificial opals grown from 705 nm polystyrene spheres having an increasing number of layers. Horizontal arrow indicates growing number of layers: 10, 15, 20, 31, and 40. Inset shows the surface of a sample. The gray box indicates the calculated spectral position of the pseudogap. (Reproduced with permission from J. F. Galisteo-López et al., Phys. Rev. B **73**, 125103, 2006.)

defects, providing valuable information on light propagation in these systems.[10] These results will be further discussed in the next section.

Once the transmittance and absolute phase have been experimentally obtained, the real and imaginary components of the effective refractive index of the sample can be retrieved. To do so, one can assume an effective medium approach that, for the case of 1D PhCs, has been proven to satisfy Kramers–Kronig relations.[11] This approach is size-dependent and describes the PhCs as an effective medium with the same transmittance and phase delay as the PhCs when probed along the measured direction. In this model, the complex transmission coefficient is defined as:

$$t(\omega) = |t|e^{i\phi} \tag{3.2.1}$$

where $t(\omega)$ is the square root of the transmittance $T(\omega)$ and ϕ is the optical phase in transmission. The complex refractive index can then be defined as:

$$n_{\text{eff}} = \text{Re}(n_{\text{eff}}) + i\,\text{Im}(n_{\text{eff}}) = \frac{c}{\omega D}\left(\phi(\omega) - i\ln|t|\right) \tag{3.2.2}$$

where D is the sample thickness $D = d_{111}N$, in which N is the total number of sphere planes parallel to the surface and d_{111} is the separation between {111} planes. Hence, from the measurements presented in Figure 3.2.2 one can extract the real and imaginary components of the effective refractive index. Figure 3.2.3 shows experimental results for samples having an increasing number of layers alongside numerical simulations obtained using the scattering matrix method[12] that has proved to correctly describe the optical response of artificial opals.[13] For the real part of n_{eff} we can see how a region of anomalous dispersion,

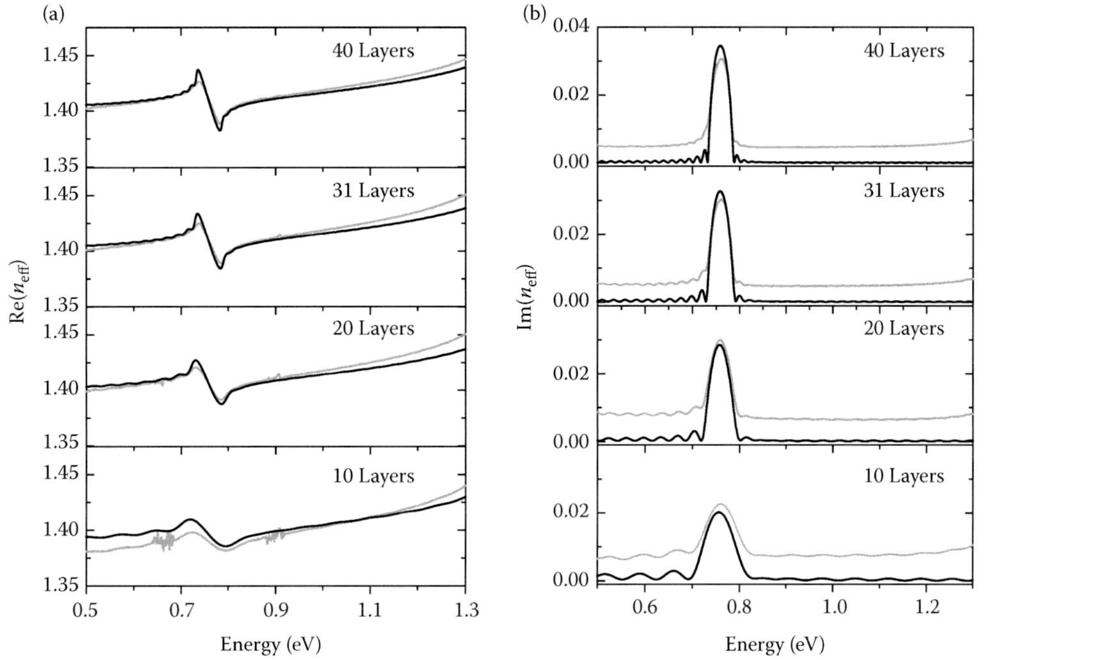

FIGURE 3.2.3
Experimental (gray lines) and calculated (black lines) real (a) and imaginary (b) components of the effective refractive index for artificial opals grown from 705 nm polystyrene spheres having an increasing number of layers. (Reproduced with permission from J. F. Galisteo-López et al., Phys. Rev. B **73**, 125103, 2006.)

where refractive index decreases with increasing frequency, occurs for those frequencies contained within the pseudogap. Although in homogeneous materials anomalous dispersion is associated with extinction due to absorption, in our case, as in the case of 1D PhCs,[11] the origin of such behavior is Bragg diffraction. This anomalous dispersion has been used as a means to achieve phase matching when employing self-assembled PhCs to increase the efficiency in nonlinear processes such as the generation of second[14] or third harmonics.[15] The results presented in Figure 3.2.3 point to the relevance of sample thickness in achieving such phenomena, as the dispersive properties of the samples present a clear dependence on the number of {111} planes parallel to the sample surface.

It is also interesting to consider the behavior of the imaginary component of n_{eff}. Here we see how for those frequencies within the pseudogap a strong extinction takes place as a consequence of Bragg diffraction by the {111} planes. The magnitude of this extinction increases with the number of layers for both theory and simulations, as expected from a more efficient diffraction process as the number of layers increases. Outside the pseudogap experimental results present a nearly constant background of extinction not present in the simulations, having its origin in scattering by structural disorder not included in the theoretical model. Here it must be noted that the value of Im(n_{eff}) for frequencies outside the pseudogap, which is linked to the amount of disorder present in the sample, decreases with sample thickness, indicating that thin samples with $N < 15$ layers present worse crystalline quality than thicker ones.

So far we have only focused on the low-energy regime where Bragg diffraction by {111} planes takes place, but phase-sensitive measurements may also be carried out in the

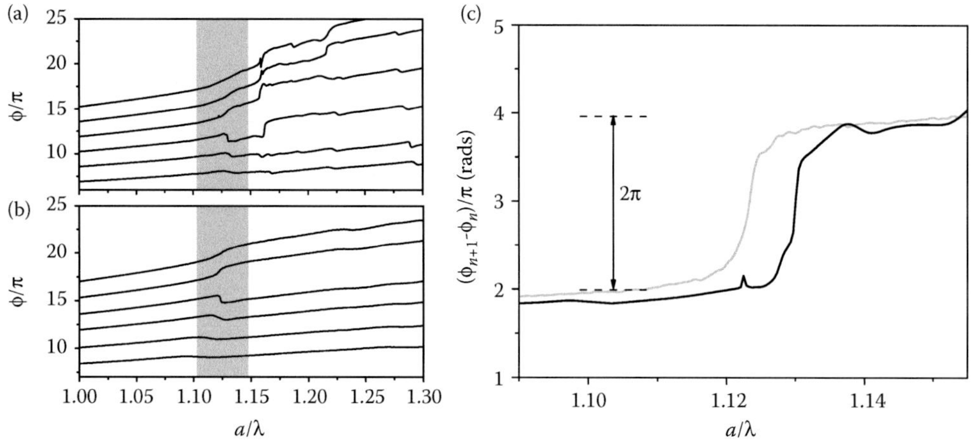

FIGURE 3.2.4
Calculated (a) and measured (b) phase delay in the high-energy spectral range for polystyrene artificial opals made from 705 nm spheres and having an increasing number of layers. The gray box marks the spectral region where the first weakly dispersive band appears. (c) Difference between the optical phase for samples with seven and eight layers (gray curve, experimental) and six and seven layers (black curve, calculated). (Reproduced with permission from J. F. Galisteo-López et al., Opt. Express **15**, 15342, 2007.)

high-energy regime where diffraction by many families of planes takes place and the band structure takes a more complex form. The interest in this spectral range lies in the possibility of coupling to energy bands presenting a low dispersion, with an associated low group velocity, which could be used to enhance light-matter interaction, interesting for processes involving emission of internal light sources[16] or nonlinear processes.[17] To gain insight into the dispersion properties in this spectral regime, measurements were performed for PS artificial opals for reduced frequencies $a/\lambda \sim 1.1$, where the first weakly dispersive energy band marking the onset of out-of-plane diffraction appears (see Figure 3.2.1). As the number of layers increases, we can see how a change in slope develops in the optical phase for a reduced frequency $a/\lambda \sim 1.1$ both in theory and in experiment. This change in slope becomes steeper with the number of layers and eventually undergoes a change of sign to then lose its steepness with further sample thickness (see Figure 3.2.4a and b). If one calculates the difference between the optical phase before and after the change in slope sign, one finds a change of magnitude 2π for both theory and experiment, though such change takes place for a different number of layers (Figure 3.2.4c). Although the meaning of such variations in optical phase is not as immediate as in the low-energy regime, the implications of this behavior in terms of light propagation within the sample will be discussed in the next section.

3.2.3 Group Velocity

The group velocity v_g can be defined as the peak velocity at which an optical pulse traverses a medium and thus is a powerful tool to study the dynamics of light propagation inside that medium. Interest in achieving values for v_g much larger or smaller than c, the speed of light in vacuum, or even negative, has spread over the past few years,[18] and PhCs have played an important role in such a quest. For an infinite ideal PhC, the group velocity at a given frequency is well defined and given by the slope of the energy band to which an incoming beam may be coupled. In the absence of extinction v_g equals the energy velocity v_e, that is, the speed at which electromagnetic energy propagates inside that medium.[19] For an energy gap, where no allowed electromagnetic states are available, neither velocity is defined as no energy can propagate inside the medium. But all these considerations have to be re-examined if one considers a finite medium, as in this case light may traverse a medium even in the presence of an energy gap given a sufficiently thin sample. In this situation $v_g \neq v_e$,[20] and one must be careful when interpreting experimental results. In this section, we will consider how one can extract valuable information from the previous phase measurements regarding group velocities and light propagation inside 3D PhCs in the form of artificial opals.

Once the phase delay has been obtained, as described in the previous section, the effective dispersion relation of the sample is given by $k(\omega) = \phi(\omega)/D$. Then one can retrieve the group velocity associated with light propagation along the (111) direction by just taking the derivative of the dispersion relation. In our case we consider the inverse of the derivative, commonly termed group index:

$$n_g = \frac{c}{v_g} = \frac{c}{D}\frac{d\phi}{d\omega} \tag{3.2.3}$$

Figure 3.2.5 shows n_g derived from measured phase delays as well as from calculated ones for artificial opals made from 705 nm PS spheres. Far from the pseudogap edges, n_g takes on a constant value characteristic of a transparent homogeneous medium and expected from the behavior already observed for n_{eff}. Fabry–Pérot oscillations due to the finite size of the samples, and noisy regions around ~0.7 and 0.9 due to water absorption, are imposed over the background. As the sample thickness increases, we see two peaks of high n_g (i.e., low v_g) develop. These peaks correspond to the bending of the dispersion relation at the pseudogap edges (see Figure 3.2.1). For an ideal infinite crystal, such bending of the energy bands leads to the formation of standing waves for these frequencies. For a real, finite crystal, spectral regions of low v_g, or "slow light," form near the pseudogap edges leading to an enhanced light-matter interaction that in similar samples has been used to improve the efficiency of processes involving light emission, nonlinear processes, or photochemistry (see the review by Galisteo-López[2] for a more complete list).

Finally, for those frequencies contained within the pseudogap, we can see how v_g takes on values well above those in the transparency regions. For samples sufficiently thick, for which Bragg diffraction becomes more efficient, v_g can become $>c$ (i.e., $n_g < 1$), and hence superluminal. Similar values are known to occur also for 1D PhCs[21,22] and such a scenario is known not to be at odds with causality since, as mentioned above, in the presence of extinction v_g is not equal to v_e that remains below c at all moments.[20]

A dramatically different behavior, not found in 1D PhCs, occurs when one considers the high-energy spectral regime discussed in the previous section. Using Equation 3.2.3 one can obtain n_g in this region for samples of different thickness from the data presented in Figure 3.2.4. Experimental results and numerical simulations are shown in Figure 3.2.6. Here we can see how for thin samples n_g takes negative values for frequencies close to the onset of diffraction ($a/\lambda \sim 1.12$). Increasing the number of layers makes

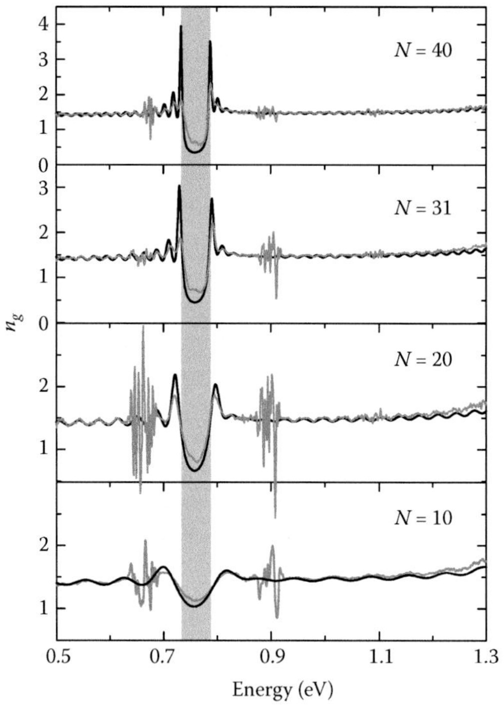

FIGURE 3.2.5
Experimental (dark gray lines) and calculated (black lines) group indices for artificial opals grown from 705 nm polystyrene spheres having different number of layers. The light gray box corresponds to the pseudogap. (Reproduced with permission from J. F. Galisteo-López et al., Phys. Rev. B **73**, 125103, 2006.)

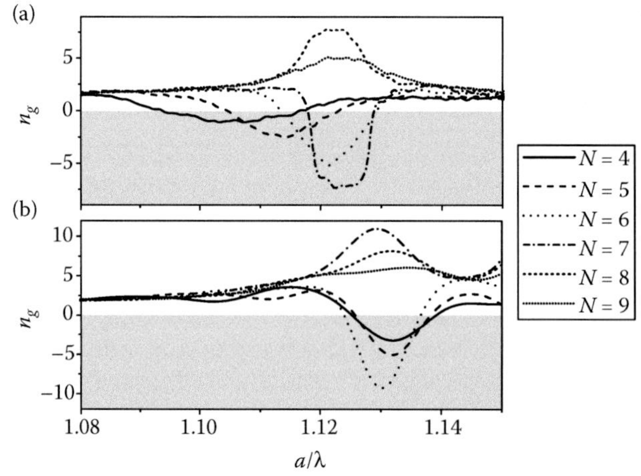

FIGURE 3.2.6
(See color insert.) Experimental (a) and calculated (b) group indices for artificial opals grown from 705 nm polystyrene spheres with different number of layers. The gray box indicates negative values of the group velocity. (Reproduced with permission from J. F. Galisteo-López et al., Opt. Express **15**, 15342, 2007.)

the dip in n_g more pronounced until, for a given thickness (seven layers in the simulations and eight in the experiment), it flips its sign becoming positive and large. These results indicate that in this spectral region finite size effects strongly influence the optical response of the sample and regimes of both slow light and negative n_g can be found. At this point, we would like to mention that the determination of the group velocity from the measured phase remains fully valid even in the presence of absorption. Even in previous studies where anomalous dispersion was associated with extinction (see, e.g., studies by Steinberg et al.[21] and Solli et al.[23]) v_g has proved to maintain its definition.

Here it is interesting to consider the implications of the above results in terms of light propagation inside the samples under consideration. If one takes a sample made from 705 nm PS spheres and having seven layers, we can find an 8 nm spectral region with a nearly constant negative v_g. The shortest Gaussian pulse with such spectral width would have a duration of 290 fs. The time needed for a reference pulse to traverse a distance in vacuum equal to the sample thickness (4 μm) is 13.5 fs. On the other hand, the group delay when traversing the sample is –107 fs, which would result in a pulse advance. Bearing in mind the fact that transmission is 10% in this spectral range, the transmitted beam would remain within a reference pulse that propagated in vacuum. It remains to be further studied whether such advance could be enhanced and what effects it would have in terms of light-matter interaction.

Another configuration of 3D PhCs that has raised a strong interest within the last years is that of an artificial opal with a two-dimensional microcavity in the form of a thin slab contained within the sample and parallel to its surface. For these samples, group velocities extracted from the measured phase can also be obtained and relevant information regarding light transport can be extracted from them.[10] Figure 3.2.7a shows the transmittance measured and calculated using scalar wave approximation[24] for an artificial opal made from 700 nm PS spheres and containing a planar defect in the form of a 230 nm slab of nanocrystalline TiO_2. The spectra show a dip corresponding to the pseudogap inside which a defect state is found through which light can tunnel through the structure (λ_2). The calculated (Figure 3.2.7b) phase now shows a 2π change across the pseudogap as a consequence of an additional resonance introduced in the form of the defect state. If one extracts the group index from the phase (Figure 3.2.7c), besides the two regions of light slowing that occurs at the pseudogap edges, we can see how an additional one occurs for the wavelength associated with the defect state.

To understand the different regimes of light transport taking place in this type of sample, one can calculate the total field intensity at three different wavelengths. For a wavelength λ_3 contained within the pseudogap, the light intensity undergoes an exponential attenuation as it propagates through the sample (Figure 3.2.7d). If one considers the defect state λ_2 (Figure 3.2.7e), the intensity is mainly localized in the spatial region where the planar cavity lies, as it corresponds to a localized state. Finally, for a wavelength near the pseudogap edge λ_1 (Figure 3.2.7f), the intensity is nearly equally distributed throughout the entire sample, a precursor of the infinite sample behavior where a standing wave forms for the edges of the pseudogap. In this way, by a combination of phase-sensitive measurements and numerical calculations of the light intensity within the sample, a thorough characterization of light propagation in this spectral range can be obtained. Such characterization is especially relevant if applications such as lasers[25] or sensors[26] are envisaged where enhanced light-matter interaction takes place.

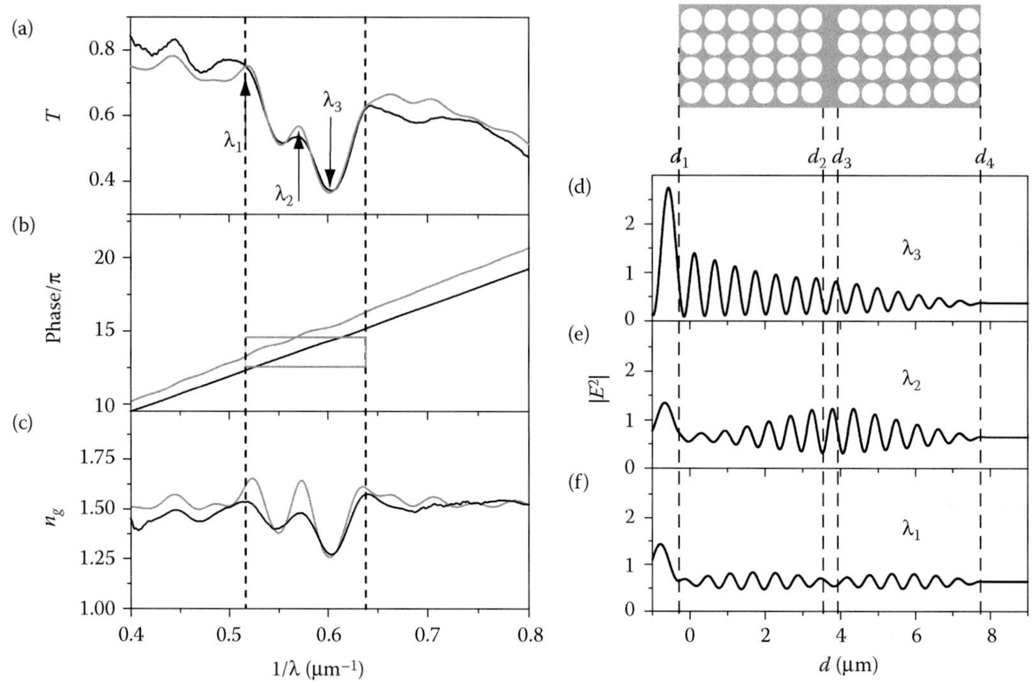

FIGURE 3.2.7
Experimental (black line) and calculated (gray line) transmission (a), optical phase (b) and group index (c) for an artificial opal made from 700 nm polystyrene spheres and containing a 230-nm-thick nanocrystalline TiO_2 slab. (d)–(f) The total field intensity within the sample at three different wavelengths: witin the gap (d), at the defect state (e), and at the band edge (f). (Reproduced with permission from J. F. Galisteo-López et al., Appl. Phys. Lett. **90**, 101113, 2007.)

3.2.4 Scattering Mean Free Path

So far in this chapter we have shown how using phase-sensitive techniques one can retrieve valuable information regarding the dispersive properties of 3D PhCs as well as information on how light propagates through these media in terms of the v_g light can experience. It is also important to explore how such dispersive properties affect light propagation in 3D PhCs with a given amount of disorder. In these samples, reflection and transmission spectroscopy can be combined to unveil unconventional light transport resulting from the interplay between order present in the lattice and disorder arising from structural defects. The dramatic effect of disorder on the light transport properties of PhCs was already the subject of the seminal work of John,[27] where changes in light diffusion and even the possibility of achieving Anderson localization of light for frequencies close to a photonic band gap were pointed out. Pioneering experiments on coherent back-scattering[28] and diffusive light transport[29,30] searched for signatures of Bloch-mode mediated scattering but have merely shown standard light diffusion.[31]

In this section, we explore how the mean free path (ℓ_s) is affected by the dispersion relation of a 3D PhC with a controlled amount of disorder. The expected changes in a system with modified density of states ($N(\omega)$) as compared to the vacuum come from an enhanced light-matter interaction (and hence scattering by defects) when $N(\omega)$ increases at the pseudogap edges and from a suppression of scattering channels within the pseudogap where $N(\omega)$ decreases. In the absence of absorption or out-of-plane scattering, we can assume that scattering losses follow Lambert–Beer's law and extract ℓ_s from a combination of reflection and transmission measurements:

$$T(L) + R(L) = \exp(-L/\ell_s) \tag{3.2.4}$$

The samples under study are polymethil-metacrilate artificial opals grown from 237 nm spheres with a controlled amount of extrinsic disorder in the form of intentionally added vacancies formed by selectively removing fractions of PS spheres added during the growth process. To explore the effect of intentional disorder on the optical response, we study extinction in samples with different amounts of added disorder and an increasing number of layers for a frequency for which no photonic features are present and the sample behaves as a transparent medium ($a/\lambda = 0.52$). Results are shown in Figure 3.2.8, where three distinct regions for light attenuation can be appreciated. In region I (thickness <10 layers), attenuation rapidly increases with the number of layers, probably because of the inferior sample quality observed for thin samples already mentioned in Section 3.2.2. As we enter region II (thickness between 10 and 50 layers), we can see how light attenuation becomes exponential with the number of layers and experimental data can be fitted to Equation 3.2.4 to extract ℓ_s. Finally, in region III (thickness >50 layers), the self-assembly process loses its effectiveness and cracks appear. This behavior, with the three different regions, can be seen not only in the "ideal" samples but also in the ones where different amounts of intentional disorder have been added. We can extract ℓ_s for samples with

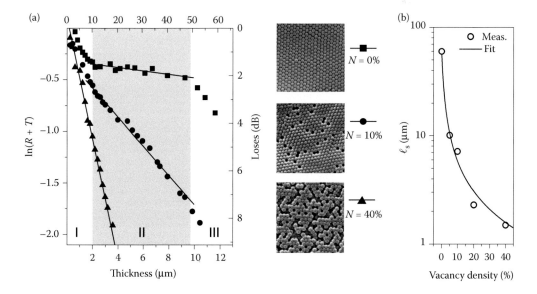

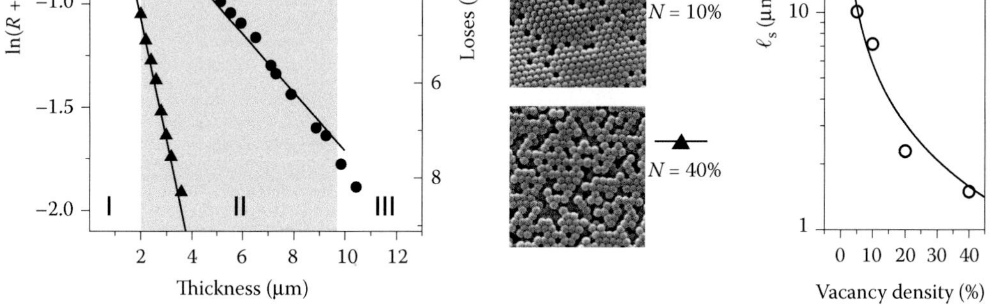

FIGURE 3.2.8
(a) Light attenuation as a function of sample thickness for samples having different amounts of added disorder. Scanning electron microscopy images of the measured samples are shown and the percentage of vacancies is indicated. (b) Evolution of ℓ_s as the amount of intentionally added vacancies is increased. (Reproduced with permission from P. D. García et al., Phys. Rev. B **79**, 241109(R), 2009.)

different amounts of added disorder (see Figure 3.2.8b) and see how ℓ_s increases from 63 μm, a value well above the lattice parameter ($a = 0.33$ μm) and the Bragg attenuation length ($L_B = 3.8 \pm 03$ μm), for the sample without added vacancies to a value comparable to L_B for the sample with 40% vacancies.

From the above measurements, information can also be gathered on how efficiently the added vacancies scatter light within the PhC environment. In random media the mean free path can be shown to be $\ell_s = 1/\rho_s\sigma$, where ρ_s is the scatter number density and σ their scattering cross section.[32] In our case, we can take into account intrinsic disorder already present in the initial samples and the extrinsic contribution in the form of added vacancies. We fitted our data to $\ell_s^{-1} = \rho_s\sigma_s + \rho_0\sigma_0$, where ρ_0 and ρ_s are the density of intrinsic and intentionally added scatters and σ_0 and σ_s are their cross sections, respectively, and fitted the data in Figure 3.2.8b to it, obtaining a cross section for the added vacancies of $\sigma_s = 0.016 \pm 0.002$ μm^2.

Following a similar procedure, one can extract ℓ_s for every frequency in the low-energy regime and obtain its chromatic dispersion. In doing so, the dispersive properties of the sample already discussed earlier in this chapter can be compared. Figure 3.2.9 shows the dispersion of ℓ_s for samples with different amounts of added disorder. For the sample with no added disorder, $a \sim \omega^2$ dependence far from the pseudogap can be observed and attributed to Rayleigh–Gans type of scattering. Values of $\ell_s = 100$ μm in the low-energy regime are obtained, much larger than the values of the transport mean free path (which in general is larger than ℓ_s) measured from samples grown by natural sedimentation[29] or centrifugation[28] pointing to a superior crystalline quality of the present samples.

In the vicinity of the pseudogap, ℓ_s undergoes strong variations that differ from the Rayleigh–Gans shape valid for lower energies. To account for this behavior, one must take into account the fact that the scattering of light by a scatter within the PhC consists of two processes: the incoming beam coupling to the scatter (the strength of which is accounted for by the group index n_g) and the polarized scatter coupling to the radiative electromagnetic field modes (which depends on the local density of states at the scatter position $\rho(\omega, \mathbf{r})$). One must also recall that n_g of the finite crystal is proportional to the spectral function of the system $S(\omega, \mathbf{k})$ that provides the number of states available for the vacuum–crystal

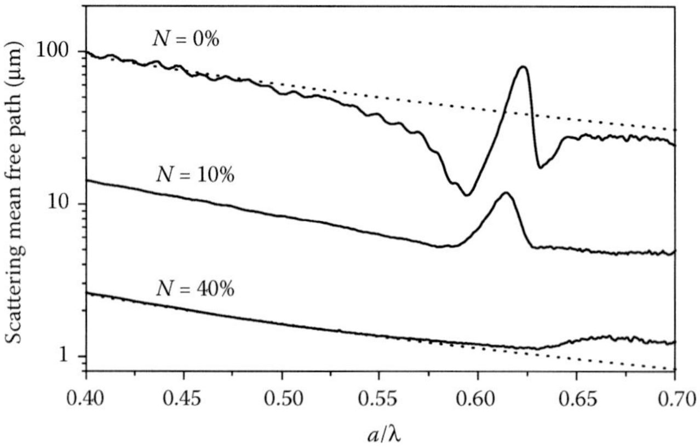

FIGURE 3.2.9
Variation of ℓ_s with light frequency for samples having different amounts of added vacancies. Dotted lines indicate ω^{-2} dependence for frequencies well below the pseudogap. (Reproduced with permission from P. D. García et al., Phys. Rev. B **79**, 241109(R), 2009.)

coupling of a signal for a given direction $\mathbf{k}$. The scattering cross section in the Rayleigh limit can therefore be described as:[33]

$$\sigma_k(\omega) \propto F(\omega)\omega^2 N(\omega) n_g(\omega, \mathbf{k}) \tag{3.2.5}$$

where $F(\omega)$ is a form factor that takes into account corrections beyond Rayleigh scattering and, in our case, is replaced by a Rayleigh–Gans factor $F(\omega) \sim \omega^2$.

The dependence of ℓ_s with n_g, which vanishes in isotropic random media, is of utmost relevance in 3D PhCs, as can be inferred from the results in the previous section and evidenced for the case of 0% vacancies in Figure 3.2.9. At the pseudogap edges, where v_g reaches a minimum, light-matter interaction reaches a maximum and so does the effect of disorder, with ℓ_s dropping a factor of 4 to a value of 11 ± 1 μm for a = 0.59, at the low-energy edge of the pseudogap. As we enter the forbidden spectral range, we see how n_g increases again nearly an order of magnitude with respect to the values observed at its edges. Increasing the amount of intentional defects reduces the effect of n_g, which arises from the order present in the lattice, smoothing the dispersion of ℓ_s. For the sample containing 10% vacancies, we see how the enhancement at the pseudogap edges is the first signature to vanish, as it is extremely sensitive to disorder. Further increasing the vacancy concentration to 40% washes out the effects of $N(\omega)$ and n_g as the sample can be considered a disordered one. The only feature to be discerned in the spectral variation of ℓ_s for this last sample is for higher energies at the spectral position of the first Mie resonance of the individual spheres.

To further evidence the influence of the photonic effects on light transport in this kind of samples, we plotted $N(\omega)^{-1}$ and n_g^{-1} in Figure 3.2.10a. Here it is evident that whereas the total density of states has a slight variation for the spectral range under consideration, it is for the direction-dependent n_g that strong dispersive effects are evident. In Figure 3.2.10b, we plotted an enhancement factor defined as the ratio between ℓ_s and its value far from the pseudogap region: $\ell_s(a/\lambda)/\ell_s(a/\lambda = 0.4)$, which clearly shows that strong fluctuations are present in ℓ_s for those frequencies in that spectral range. Results are plotted for two different samples with diameters of 237 and 600 nm. Although in both cases large changes in the values of ℓ_s are patent, 8- and 20-fold respectively, the larger modulation in the latter case is likely due to a better crystalline quality of the lattice. To explain the large fluctuations in ℓ_s we also plotted $N(\omega)^{-1}n_g^{-1}[\omega, \mathbf{k} = (111)]$ for comparison. Here we can see how it reproduces the spectral shape of $\ell_s(\omega)$ and how, though both quantities $N(\omega)^{-1}$ and $n_g^{-1}[\omega, \mathbf{k} = (111)]$ contribute to the final shape, it is the direction-dependent $n_g^{-1}[\omega, \mathbf{k} = (111)]$ that is the main factor responsible for the change in ℓ_s. Hence, the simple and qualitative model described, despite being heuristic and limited to collinear propagation, accounts well for the spectral shape of the measured $\ell_s(\omega)$. The main discrepancy comes from the asymmetry present in the experimental data. This feature is related to the available states for scattered light to couple to for directions other than the incident one. This set of directions is dictated by the shape of the energy bands in the surroundings of the (111) direction[34] and decreases as one moves away from that direction (see Figure 3.2.1).

An increase of ℓ_s in the band gap and a decrease in the band edge reflect the modified phase space available Δk for light scattering when the photonic modes are concentrated around few k directions or the available scattered states are reduced. This is consistent with John's seminal prediction of a need for a modified Ioffe–Regel criterion[27] for scattering in PhCs to include Δk. Furthermore, here we show that as the phase space is modified, not only the Ioffe–Regel criterion is changed but also ℓ_s is renormalized: light scattering in

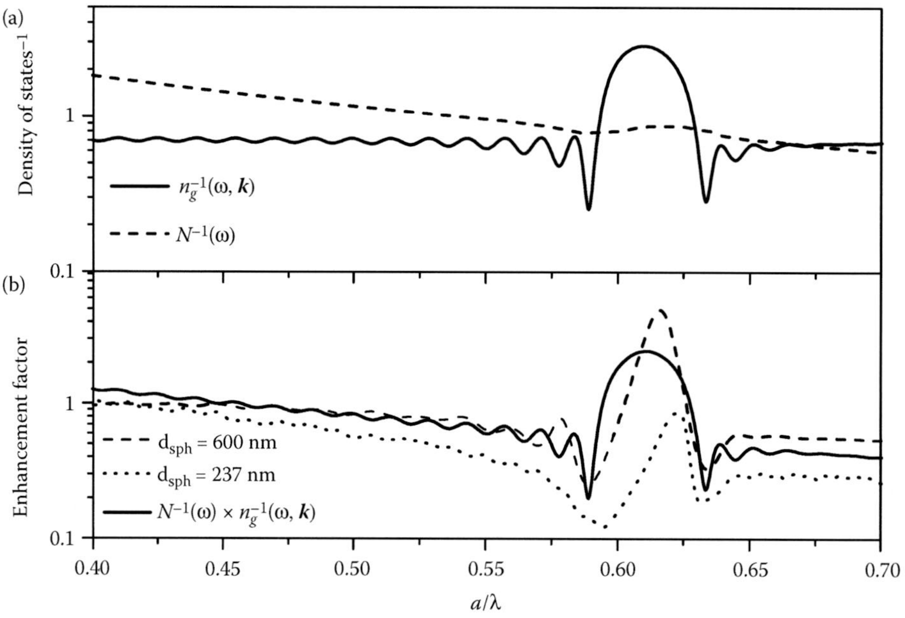

FIGURE 3.2.10
(**See color insert.**) (a) Inverse of the total $N(\omega)$ density of states and group index $n_g[\omega, \boldsymbol{k} = (111)]$ along the incident (ΓL) direction are plotted. (b) Enhancement factor, $\ell_s(a/\lambda)/\ell_s(a/\lambda = 0.4)$, for two opals with no vacancy doping and the quantity $N^{-1}(\omega)/n_g^{-1}[\omega, \boldsymbol{k} = (111)]$. (Reproduced with permission from P. D. García et al., Phys. Rev. B **79**, 241109(R), 2009.)

PhCs is richer than in conventional amorphous media. Complete photonic band-gap materials, such as silicon-inverted opals, amplify the effect presented here and could be proper candidates to observe Anderson localization of light.

3.2.5 Conclusions

In this chapter, we have discussed how one can experimentally retrieve information on light propagation through real 3D PhCs fabricated by self-assembly methods. When dealing with the dispersive properties of these samples, phase-sensitive techniques offer information that conventional reflection and transmission spectroscopy fail to provide. By combining phase-sensitive and conventional transmission spectroscopy, the real and imaginary components of the size-dependent effective refractive index can be obtained. Issues such as spectral regions of anomalous dispersion and size-dependent crystalline quality of the samples have been discussed from the obtained data.

One step further in the experimental study of light propagation in 3D PhCs can be taken by obtaining the group velocity (v_g) of real samples from the measured optical phase delay. Broadband light sources offer the possibility of obtaining v_g over a wide spectral range where different phenomena including slow light, superluminal values of v_g, or even negative ones are observed.

Finally, the possibility of obtaining information regarding light transport in 3D PhCs combining conventional reflection and transmission spectroscopy has been discussed. The chromatic variation of the scattering mean free path (ℓ_s) has been obtained for samples with different amounts of intentional disorder added in the form of vacancies. The study has been performed in a spectral range around the pseudogap and evidence has been found on how light transport is dictated by the interplay between order and disorder. The strong fluctuations in ℓ_s can be accounted for by the total density of states $N(\omega)$ and the direction-dependent spectral function $S(\omega, \mathbf{k})$, the latter being responsible for the main features observed.

References

1. E. Yablonovitch, Phys. Rev. Lett. **58**, 2059 (1987).
2. J. F. Galisteo-López, M. Ibisate, R. Sapienza, L. S. Froufe-Pérez, A. Blanco, and C. López, Adv. Mater. **23**, 30 (2011).
3. Yu. A. Vlasov, M. Deutsch, and D. J. Norris, Appl. Phys. Lett. **76**, 1267 (2000).
4. İ. İ. Tarhan, M. P. Zinkin, and G. H. Watson, Opt. Lett. **20**, 1571 (1995).
5. B. T. Rosner, G. J. Schneider, and G. H. Watson, J. Opt. Soc. Am. B **15**, 2654 (1998).
6. E. Istrate and E. H. Sargent, Appl. Phys. Lett. **86**, 151112 (2005).
7. M. Galli, F. Marabelli, and G. Guizzetti, Appl. Opt. **42**, 3910 (2003).
8. M. Galli, D. Bajoni, F. Marabelli, L. C. Andreani, L. Pavesi, and G. Pucker, Phys. Rev. B **69**, 115107 (2004).
9. J. F. Galisteo-López, M. Galli, L. C. Andreani, and C. López, Appl. Phys. Lett. **90**, 231112 (2007).
10. J. F. Galisteo-López, M. Galli, L. C. Andreani, A. Mihi, R. Pozas, M. Ocaña, and H. Míguez, Appl. Phys. Lett. **90**, 101113 (2007).
11. M. Centini, C. Sibilia, M. Scalora, G. D'Aguanno, M. Bertolotti, M. J. Bloemer, C. M. Bowden, and I. Nefedov, Phys. Rev. E **60**, 4891 (1999).
12. D. M. Whittaker and I. S. Culshaw, Phys. Rev. B **60**, 2610 (1999).
13. E. Pavarini, L. C. Andreani, C. Soci, M. Galli, F. Marabelli, and D. Comoretto, Phys. Rev. B **72**, 045102 (2005); L. C. Andreani, A. Balestreri, J. F. Galisteo-López, M. Galli, M. Patrini, E. Descrovi, A. Chiodoni, F. Giorgis, L. Pallavidino, and F. Geobaldo, *ibid.* **78**, 205304 (2008); J. F. Galisteo-López, M. Galli, M. Patrini, A. Balestreri, L. C. Andreani, and C. López, *ibid.* **73**, 125103 (2006); Opt. Express **15**, 15342 (2007).
14. J. Martorell, R. Vilaseca, and R. Corbalán, Appl. Phys. Lett. **70**, 702 (1997).
15. P. P. Markowicz, H. Tiryaki, H. Pudavar, P. N. Prasad, N. N. Lepeshkin, and R. W. Boyd, Phys. Rev. Lett. **92**, 083903 (2004).
16. M. Scharrer, A. Yamilov, X. Wu, H. Cao, and R. P. H. Chang, Appl. Phys. Lett. **88**, 201103 (2006).
17. M. Botey, M. Maymó, A. Molinos-Gómez, L. Dorado, R. A. Depine, G. Lozano, A. Mihi, H. Míguez, and J. Martorell, Opt. Express **17**, 12210 (2009).
18. R. W. Boyd and D. J. Gauthier, Science **326**, 1074 (2009).
19. A. Yariv and P. Yeh, J. Opt. Soc. Am. **67**, 438 (1977).
20. G. D'Aguanno, M. Centini, M. Scalora, C. Sibilia, M. J. Bloemer, C. M. Bowden, J. W. Haus, and M. Bertolotti, Phys. Rev. E **63**, 036610 (2001).
21. A. M. Steinberg, P. G. Kwiat, and R. Y. Chiao, Phys. Rev. Lett. **71**, 708 (1993).
22. C. Spielmann, R. Szipocs, A. Stingl, and F. Krausz, Phys. Rev. Lett. **73**, 2308 (1994).
23. D. R. Solli, C. F. McCormick, C. Ropers, J. J. Morehead, and R. Y. Chiao, Phys. Rev. Lett. **91**, 143906 (2003).
24. K. W.-K. Shung and Y. C. Tsai, Phys. Rev. B **48**, 11265 (1993).

25. S. Furumi, H. Fudouzi, H. T. Miyazaki, and Y. Sakka, Adv. Mater. **19**, 2067 (2007).
26. F. Fleischhaker, A. C. Arsenault, F. C. Peiris, V. Kitaev, I. Manners, R. Zentel, and G. A. Ozin, Adv. Mater. **18**, 2387 (2006).
27. S. John, Phys. Rev. Lett. **58**, 2486 (1987).
28. A. F. Koenderink, M. Megens, G. van Soest, W. L. Vos, and A. Lagendijk, Phys. Lett. A **268**, 104 (2000).
29. Yu. A. Vlasov, M. A. Kaliteevski, and V. V. Nikolaev, Phys. Rev. B **60**, 1555 (1999).
30. V. N. Astratov, A. M. Adawi, S. Fricker, M. S. Skolnick, D. M. Whittaker, and P. N. Pusey, Phys. Rev. B **66**, 165215 (2002).
31. A. F. Koenderink, A. Lagendijk, and W. L. Vos, Phys. Rev. B **72**, 153102 (2005).
32. E. Akkermans and G. Montambaux, *Mesoscopic Physics of Electrons and Photons* (Cambridge University Press, Cambridge, 2007).
33. P. D. García, R. Sapienza, L. S. Forufe-Pérez, and C. López, Phys. Rev. B **79**, 241109(R) (2009).
34. J. F. Galisteo-López, E. Palacios-Lidón, E. Castillo-Martínez, and C. López, Phys. Rev. B **68**, 115109 (2003).

3.3

Photonic Glasses: Fabrication and Optical Properties

Pedro David Garcia-Fernández
University of Copenhagen

Diederik S. Wiersma
University of Florence
National Institute of Optics

Cefe López
Materials Science Institute of Madrid

CONTENTS

3.3.1 Introduction

Self-assembly techniques have been widely used to grow ordered structures such as opal-based photonic crystals. Such structures consist of periodic arrangements of, usually spherical, building blocks, which, in the ideal case, are identical to each other. In practice, this often means the use of monodisperse spheres. Alternatively, one can also construct disordered configurations of identical building blocks, which by analogy can be termed photonic glasses.[1]

Whereas (ordered) periodic photonic media take advantage of the periodicity in the dielectric constant and the consequent long-range correlation, disordered structures can still strongly affect light transport.[2] Examples of interesting physical (transport) phenomena in random media include coherent backscattering or weak localization of light,[3] random lasing,[4] strong light localization,[5] or long-range intensity correlations.[6] Studies on these phenomena were often performed on materials such as "white paints": random assemblies of irregularly shaped grains such as TiO_2 powder[7] and ZnO powder,[8] for the infrared as GaAs[9] or Ge[10] or random networks of GaP.[11] Materials composed of polydisperse spherical particles have also been available, presented both in solid arrangements such as Al_2O_3,[12] ZnO,[13] and TiO_2,[14] as well as in colloidal suspensions.[15] In all these cases, the particles comprising the system are polydisperse in shape, size, or both.

Fractal media,[16] controlled disordered[17] and quasiperiodic[18] photonic crystals, or nonisotropic complex fluids such as nematic liquid crystals[19] constitute systems that lie somewhere between ordered and fully disordered. Interesting recent new materials in this respect also include polymer-dispersed liquid crystals,[20] which exhibit hugely anisotropic diffusion of which the anisotropy can be controlled electrically, and Lévy glasses,[21,22] an entirely new form of disordered materials in which light exhibits a Lévy flight and hence superdiffusion.

In photonic glasses, the monodispersity of the building blocks add a new feature, namely that of a resonant response at the wavelengths of the Mie resonances of the spherical building blocks. A very important consequence of these resonances, for instance, is that the transport velocity of light can become strongly reduced[23] at certain wavelengths and that random lasing can be controlled.[4,24] In general, photonic glasses constitute a system in which light diffusion becomes strongly dispersive. In this chapter, we will describe in detail the fabrication and optical properties of these fascinating optical materials.

Using the same building block, a dielectric sphere with a size comparable to the wavelength of light, two extreme systems can be built: a perfectly ordered arrangement (photonic crystal) and a completely random arrangement of spheres (photonic glass), as shown in Figure 3.3.1a. The most important property of both solid systems is the monodispersity of their building blocks, in this case polymer microspheres (with diameters from 200 nm to several microns and with a polydispersity less than 2%). The disordered new

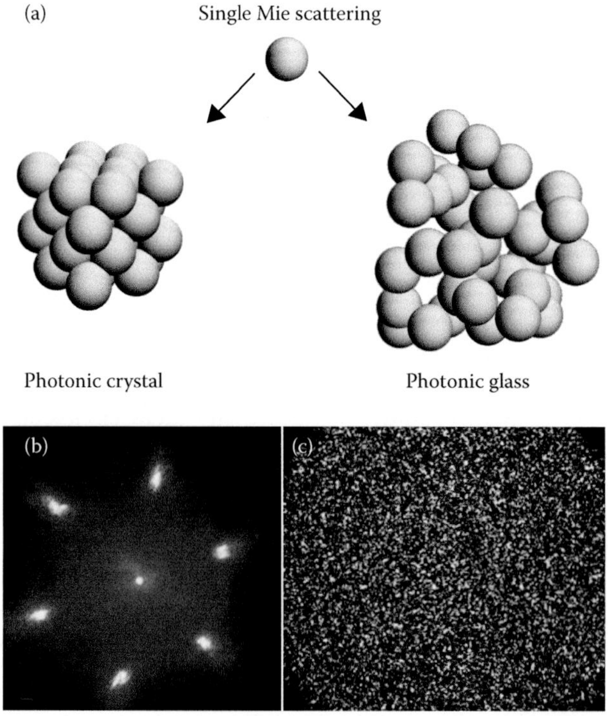

FIGURE 3.3.1
(See color insert.) (a) Ordered and disordered packing of a single-scattering particle, a dielectric sphere in this case. (b) and (c) Light speckle pattern arising from an ordered face-centered cubic arrangement of dielectric microspheres and from a random arrangement of the same spheres, respectively, and collected on a far-field screen. (Adapted from P. D. García, R. Sapienza, and C. López, Adv. Mater. **22**, 12, 2010.)

compact and solid system is just an example of an old very-well-known material[25] (in this case, polymer microspheres) arranged in a different manner. It offers the possibility to probe a range of interesting phenomena affected by the monodispersity of the spheres. Figure 3.3.1a shows how the topological difference between a photonic glass and a photonic crystal leads to very different optical properties. A simple light transmission experiment can give a diffraction pattern for a photonic crystal (Figure 3.3.1b) or a speckle pattern, a granular distribution of light intensity, for a photonic glass (Figure 3.3.1c).

3.3.2 Photonic Glasses: Colloidal Instability and Preparation Method

The key property necessary to achieve colloidal photonic crystals is the colloidal solution stability. A colloid is a two-phase (at least) system in which one substance (the internal phase) is divided into minute particles (colloidal particles) with dimensions between 1 nm and several microns dispersed throughout a second substance (the external phase). Size is not the most important property of colloids; the overwhelmingly important property of colloids is their very large surface area. To some degree, they are all surface and their properties are those of their surfaces. The large area emphasizes surface effects relative to volume effects, giving colloids properties different from those of bulk matter. Hereafter, we will only deal with sols composed of polymer spheres [polymethil-metacrilate (PMMA) or polystyrene (PS)] dispersed in a liquid (water).

Several methods have been tested to obtain disordered, solid, and random arrangements of dielectric microspheres. The first unsuccessful attempts that focused on mechanical perturbation of the colloidal suspension were tried as, for example, the forced sedimentation of microspheres by the action of high-speed centrifugation. By placing a colloidal suspension of microspheres (PS or PMMA) in a high-speed centrifuge (6000 rpm), the sedimentation of the spheres is forced by the artificially increased gravity (~1200 *g*, where *g* is the gravitational acceleration on the earth's surface). By doing so, a solid white sediment is obtained after the evaporation of the solvent. On a first examination, no colors are found in the reflectance from the bottom (system substrate) surface of the sediment but, on the contrary, the top (system air) surface shows visible iridescent due to Bragg reflections. A deeper inspection using scanning electron microscopy (SEM) reveals a very high degree of ordering in the bulk material, corresponding to the top surface and a random distribution of spheres corresponding to the bottom surface. This is an example of how ordered surfaces may hide very disordered bulk systems: a high reflectance from the surface of a sample is not at all a proof of the existence of an underlying ordered structure. Reflectance measurements should be taken very carefully as characterization method. Transmittance measurements, on the contrary, may provide, in combination with reflectance ones, enough information from the bulk of the system.

A mechanical action over the colloidal suspension is not enough to force a random distribution of spheres. A second different failed strategy followed in this wrong way. Opal-based photonic crystals are usually grown by vertical deposition.[26] In this technique, a clean microscope slide is placed in a vial containing a colloidal suspension of spheres. By the action of water evaporation, the meniscus formed between the suspension and the substrate allows the spheres to self-assemble. The formation of the meniscus is, therefore, crucial in this technique and it is only possible if the substrate is hydrophilic. An initial hydrophobic micro slide can be hydrophilized with a light chemical etching of its surface.

Without this etching process, the formation of the meniscus is prevented in an otherwise hydrophobic substrate. This was tried in order to obtain a disordered arrangement of spheres, a photonic glass. Once again, it failed and a very-low-quality photonic crystal was obtained, still showing Bragg iridescences from the surface. An important concept follows from this: bad photonic crystals are also bad photonic glasses.

The reason why these previous attempts failed is because the ordering of the spheres in a face-centered cubic structure takes place during water evaporation. Some examples of not completely disordered systems can be found in the literature,[27] where traces of order reveal the invalidity of a mechanical method to obtain random arrangements of spheres. The water evaporation process, and in particular surface tensions, is what forces the spheres to self-assemble in an ordered fashion and it is actually this process that should be prevented. In order to do that, a more sophisticated method is needed. It is necessary to force the flocculation of the spheres and to induce the formation of random clusters to prevent the spheres from ordering. We pursued this strategy with the colloidal stability as the central argument. It is related to the surface charge of the particles, which ensures the suspension stability preventing them from sticking.[28] If two particles collide, they will stick together and make a bigger particle or, eventually, a cluster. Clusters get larger than the critical size to be suspended and to settle.

In most lyophobic colloids, the particles are electrically charged with the same sign, and this keeps them apart, as they repel one another. As lyophobic sols are stabilized by electric charge, adding extra charge (electrolytes) generally destroys the sol. For example, when rivers reach the sea with their loads of colloidal sediment, the ions in sea water coagulate the sol and the load is deposited forming the delta.

The modeling of the sphere–sphere potential can be performed on the basis of Derjaguin–Landau–Verwey–Overbeek repulsion,[28] whose detailed formulation can be found elsewhere.[29] Figure 3.3.2 shows the total interaction, $U(r)$, between two particles as a function of the electrolyte concentration. As a matter of fact, there is a high probability of flocculation as a result of the collision between two particles if the potential barrier $U(r_M) \leq 10k_BT$, where r_M is the reduced particle–particle distance at which $U(r)$ has the maximum value. This fact is pointed out in Figure 3.3.2b, where energies below these values are shaded in gray. For electrolyte concentrations for which $U(r_M) \leq 10k_BT$, the instability of the colloidal suspension is ensured and colloidal flocculation will take place. In this particular case, for an electrolyte concentration of $\rho_e = 1 \times 10^{-2}$ M, the potential barrier is $U(r_M) \sim 6k_BT$ and the Brownian energy is enough to force the colloidal flocculation.

PS spheres used to grow photonic glasses are synthesized by the Goodwin method,[30] which gives rise to a negative surface charge of the particles. PMMA spheres were synthesized following a similar method.[31] In both cases, positive electrolytes are needed to screen the negative surface charge of the colloids (which otherwise prevent the natural agglomeration) and can be obtained by adding salts (which dissociate producing positive and negative ions in dissolution) or acids (which also dissociate producing protons and negative radicals). The charge, Z, of the ions or the number of protons dissociated from an acid is an important parameter. The concentration of salt or acid needed to provoke the colloidal flocculation is inversely proportional to their charge. For example, half the concentration of a salt such as $CaCl_2$ (which dissociates producing Ca^{2+} ions) is needed rather than a salt such as NaCl (which dissociates producing Na^+ ions) to provoke the colloidal flocculation. The attenuation of the repulsive potential gives rise to a net attractive potential between spheres. In this case, the number of effective collisions between spheres increases and clusters are formed by flocculation in the suspension. When the size of the clusters is larger than the critical size they settle. The formation of clusters inhibits the self-assembling

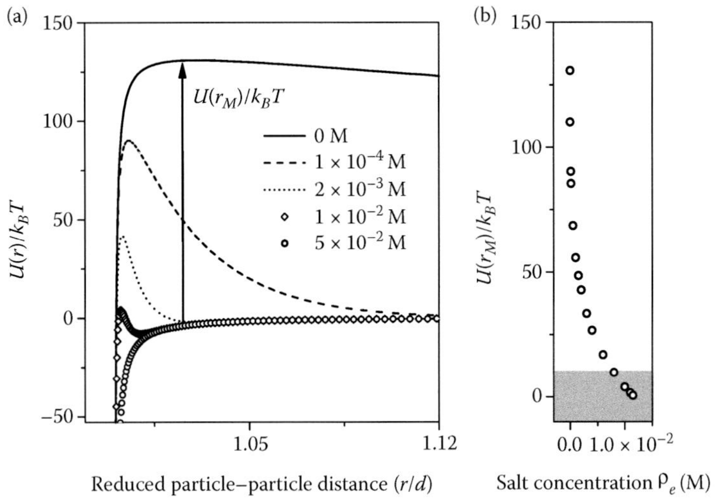

FIGURE 3.3.2
(a) Plot of the total interaction potential between two particles relative to the thermal energy as a function of the concentration of electrolytes in the suspension according to the proposed solution. (b) The magnitude $U(r_M)$ represents the potential barrier that prevents the colloids from flocculating. This barrier decreases when the electrolyte concentration increases. Energies bellow $10k_BT$ are shaded in gray. If the potential barrier is lower than this energy, the colloids flocculate. In this particular case, this happens for $\rho_e \geq 7.5 \times 10^{-3}$ M. The specific parameters used for this particular exemplification are a particle diameter $d = 1220$ nm with a surface potential $\psi = -30$ mV (measured with a standard electrophoretic mobility experiment) and the electrolytes (Ca^{2+} from $CaCl_2$) added to this suspension have $Z = 2$ and their concentration varies from $\rho_e = 0$ M to $\rho_e = 5 \times 10^{-2}$ M. (Adapted from P. D. García, R. Sapienza, and C. López, Adv. Mater. **22**, 12, 2010.)

process that takes place during the liquid evaporation process, and, consequently, a random distribution of disordered clusters is obtained. To control the thickness and the area of the system, a methacrylate cylinder of height h and diameter d is glued with flexible and impermeable gum to a clean hydrophilic glass microscope slide (see Figure 3.3.3a). In a typical procedure, a total volume $V_T = 3$ ml of an aqueous suspension of PS spheres with a diameter of 1220 nm and $CaCl_2$ is prepared as follows: a volume $V_s = 2.5$ ml of a colloidal suspension of PS spheres with a concentration $\rho_s = 20$ g/l (2% weight) is added to a volume $V_e = 60$ μl of $CaCl_2$ with a concentration $\rho_e^i = 0.5$M To reach the total volume, $V_w = 0.44$ ml of deionized water is added. The final concentration of electrolytes in the final suspension is $[Ca^{2+}] = \rho_e^f = 0.01$M. This aqueous suspension is shaken under ultrasound for 5 min to force the flocculation of spheres. The cylinder of diameter $d_c = 2$ cm is then filled with the suspension that is let to evaporate in an oven with constant temperature, typically at $T = 45$°C. The sample is left inside the oven for an adequately long time to allow the total evaporation of the liquid. When the liquid is completely evaporated, the cylinder is removed from the substrate, the photonic glass remaining attached to it (see Figure 3.4.3b and SEM images from the system in Figure 3.3.3). To avoid possible cracking (see Figure 3.4.3c) or peeling of the sample from the substrate, it is important that the cylinder is not touching the substrate, leaving a small air chamber between the cylinder, the substrate, and the gum.

As a matter of fact, a small fraction of colloidal suspension fills this air chamber by capillarity and the sample is more strongly attached to the substrate (see Figure 3.4.3b). Proceeding this way, no cracking appears when removing the methacrylate cylinder from

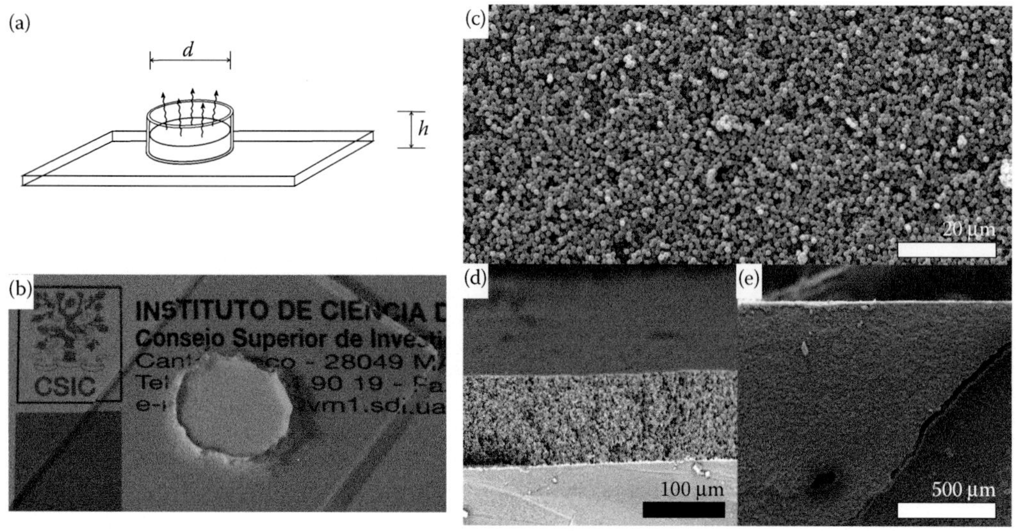

FIGURE 3.3.3
(**See color insert.**) (a) Schematic of the photonic glass growth method. A methacrylate cylinder of high $h \sim 1$ cm is fixed with impermeable gum to a clean, hydrophilic glass substrate. It is then filled with a charged colloidal suspension previously prepared and shaken under ultrasound. Then it is placed in an oven under constant temperature (~45°C) to force the evaporation of the liquid phase. Finally, the methacrylate cylinder is removed from the glass substrate. (b) A photonic glass grown on a substrate. The sample shows a high degree of planarity, apart from the irregularity of the edge. Scanning electron microscopy images from different parts of a photonic glass made with polystyrene spheres with $d = 1220$ nm. (c) Surface of the sample. The random arrangement of the spheres is revealed here. (d) and (e) Cleaved edges of the sample. In both cases, the images reveal the planarity of the surface over millimeters. The crack visible in (e) is produced when removing the methacrylate cylinder from the glass substrate. (Adapted from P. D. García, R. Sapienza, and C. López, Adv. Mater. **22**, 12, 2010.)

the substrate. The typical thickness obtained in this case is $t = 31 \pm 5$ μm constant over centimeters (see Figure 3.3.3b and c). The thickness of the photonic glasses is measured using a low-magnification objective and measuring the focusing on the substrate and on the glass surface and subtracting both measures. If the salt concentration is increased, the repulsive barrier is lowered and, consequently, the size of the clusters formed by flocculation is increased. If the salt concentration increases, the number of effective collisions required to attach two particles decreases. The size of the clusters has an important effect on the filling fraction, f, of the system and may strongly affect the physical magnitudes that describe light transport.

3.3.3 Static Optical Properties: Resonant Mean Free Path

A remarkable property of photonic glasses is the resonant behavior in the diffuse light when propagating through them.[32,33] In multiply-scattering dilute media, light transport can often be described as a diffusion process. In that case, interference effects can be obviated in a first approximation. Diffusion processes are ubiquitous in nature: from molecules in a pressure gradient to (matter, electromagnetic, sound, seismic) waves in random systems. Light entering a disordered media is scattered (elastically) numerous times and when it emerges from the material still conserves the color of the incident light, whatever the detected incident direction. As ambient light covers the full visible spectrum, the diffusive medium appears white. White materials as white paint owe their appearance to multiple light scattering.

Photonic crystals instead exhibit color variations that are not related to absorption bands such as in standard pigments or to Bragg scattering such as in photonic crystals. The optical properties of photonic glasses lie in the light scattering response of its composing dielectric spheres.

A single dielectric microsphere with a size comparable to the wavelength of light (a Mie sphere) can sustain electromagnetic resonances.[34] When all the spheres are identical (within ~2% in our case), these modes occur all at the same frequency and thus are not washed out. In a macroscopic photonic glass, the diffuse-like transport of light is therefore strongly affected by the Mie modes. It is possible to observe resonances in the light transport. These electromagnetic modes are excited when the electromagnetic field wavelength is comparable to the optical diameter of the spheres. These resonances have been probed by static and dynamic experiments as an extended optical characterization of photonic glasses. A detailed optical study of the resonant behavior of diffuse light transport through such a system has also been recently provided.[33] By means of independent static and dynamic measurements, resonances have been shown in the transport mean free path, diffusion constant, and also energy velocity of light.

Figure 3.3.4 shows a direct measurement of the total diffuse light transmission through different photonic glass slabs (thickness $L \sim 100\ \mu m$) on white-light illumination (500 to 920 nm). To be able to compare the optical response of different sphere sizes (790, 930, 1000, and 1220 nm), the measurements are plotted as a function of rescaled energy (dn/λ), where n is the refractive index. Oscillations in transmission and spectral dependence are due to the existence of modes for the electromagnetic field in the spheres. The spectral positions of these Mie modes depend exclusively on sphere diameter, d, and on its refractive index, n. Figure 3.3.4a shows a clear and simple evidence of the resonant behavior of light transport in a broad energy interval. To remark on this fact and also to clarify the conditions under which the modes can be collectively excited, the resonant behavior of these four different sphere sizes has been compared with two different no-resonant dielectric random systems. In Figure 3.3.4b, the total transmission through two reference samples is plotted, which, for two different reasons, do not exhibit resonant behavior. As pointed out through Figure 3.3.4b, there is no trace of resonances in the transport of light for these two reference systems. The first one (dashed curve) is a photonic glass composed of PS spheres with a diameter of 200 nm. The small size of the spheres compared to the light wavelength illumination (dn/λ) ~ 0.4 whose modes have energies in other range. Therefore, those spheres behave, on this particular light energy illumination, as point-like scatterers giving rise to Rayleigh scattering (where scatterer structure can be neglected) instead of Mie scattering (where resonances can be sustained). Resonances are expected in other energy ranges (in the ultraviolet) for this particular system. The second one (solid curve) is composed of TiO_2 nonspherical powders with a polydispersity of about 36% (see inset of Figure 3.3.4b) and a mean diameter of about 850 nm. A different situation arises with the nonresonant light transport through TiO_2 powder. In this case, TiO_2 particles are large enough ($dn/\lambda \sim 2.5$) to sustain Mie modes in this wavelength interval. However, Figure 3.3.4b shows no trace of oscillations in the light transport because resonances are smoothed out by polydispersity and the arbitrary nonspherical shape of the scatterers. Mie modes are defined by the morphology of the scatterer. When the scatterers are nonspherical and also polydisperse, as in the TiO_2 case, each building block sustains resonances for different wavelengths that smoothes out the collective response, giving rise to an overall nonresonant behavior.

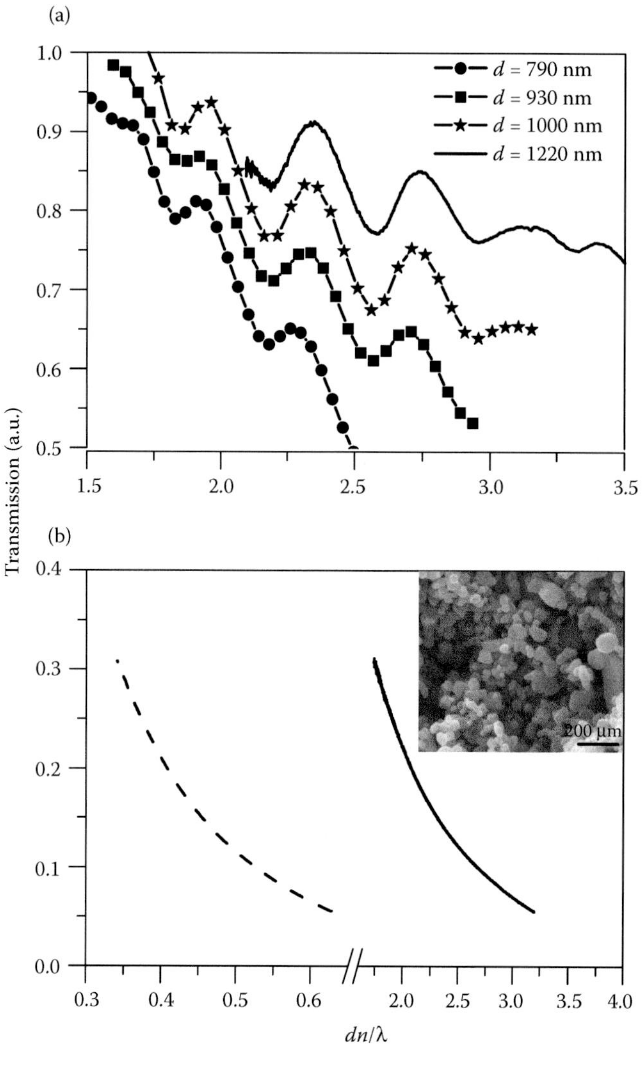

FIGURE 3.3.4
(a) Normalized total transmission of white light through photonic glasses as a function of the reduced parameter dn/λ. Samples are composed of spheres with four different diameters and thickness is about 100 μm in all cases. (b) Total transmission of white light through two different reference samples as a function of the reduced parameter dn/λ. The dashed line represents total transmission through photonic glasses made of polystyrene spheres of $d = 200$ nm. The solid line represents total transmission through a powder made of polydisperse TiO_2 of averaged $d = 850$ nm (scanning electron microscopy image inset of the sample where the scale bar represents 200 nm). Both present no-resonant light transmission. (Adapted from P. D. García, R. Sapienza, and C. López, Adv. Mater. **22**, 12, 2010.)

As in electronic transport and in the absence of absorption, doubling the thickness of the (optical) conductor halves the transmission. This is the optical equivalent of Ohm's law. The diffusion of light in a disordered dielectric slab and, in particular, a photonic glass leads to the photonic Ohm's law. The total light transmission through a photonic glass slab is directly proportional to the transport mean free path, ℓ_t, and inversely proportional to the slab thickness, $T(\lambda) \sim \ell_t/L$. Therefore, with static measurements of the total light

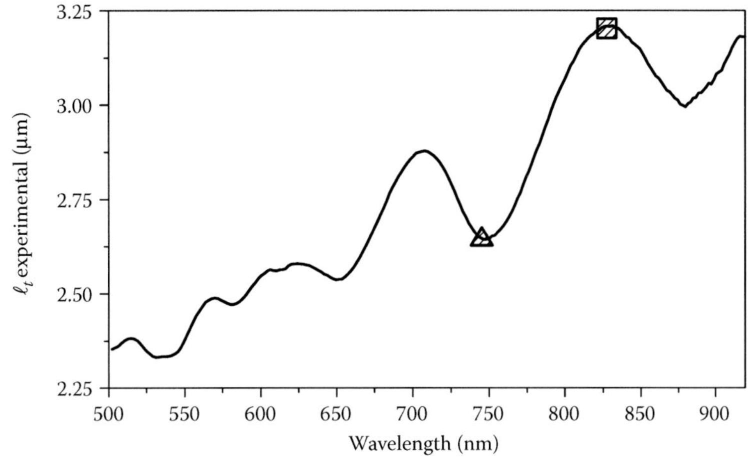

FIGURE 3.3.5
Experimental transport mean free path in a photonic glass made of polystyrene spheres of $d = 1220$ nm. Two particular wavelengths are marked with a triangle ($\lambda_1 = 744$ nm) and a square ($\lambda_2 = 828$ nm). (Adapted from P. D. García et al., Phys. Rev. A **78**, 023823, 2008.)

transmission through a slab with known thickness, it is possible to obtain the absolute value of the transport mean free path, ℓ_t.[35] We can extract the transport mean free path of a photonic glass just by measuring T^{-1} (L, λ) as a function of the slab thickness from a set of samples composed of spheres with the same diameter and with different thicknesses. Figure 3.3.5 shows this for an extended wavelength range for photonic glasses composed of spheres with diameter $d = 1220$ nm. It reveals a clear resonant behavior of $\ell_t(\lambda)$. In the curve, a triangle and a square mark the spectral positions of a minimum and a maximum of $\ell_t(\lambda)$. The scattering cross section is enhanced when a Mie mode is excited in a sphere. The scattering is more efficient at those wavelengths and, therefore, the transport mean free path becomes minimum.

3.3.4 Dynamic Optical Properties: Resonant Diffusion Constant

Like the mean free path, the diffusion constant too is resonant in these structures. It is possible to probe its resonances with a time-resolved measurement with the help of a streak camera that permits one to monitor the time evolution of a very short laser pulse in its propagation through the glass. In particular, we performed the experiment with 2 ps pulses provided by a Ti:Al_2O_3 laser tunable within 700–920 nm. An example of a time-resolved transmission measurement is plotted in Figure 3.3.6 for two different wavelengths ($\lambda_1 = 744$ nm and $\lambda_2 = 828$ nm), which correspond to the minimum and the maximum of a Mie resonance analyzed in the previous section. The different slope of $T(t, \lambda)$ at long times (the time decay) at these two different wavelengths accounts for the presence of a Mie resonance. The thickness of the photonic glasses used in these time-resolved experiments is about 1 mm. Figure 3.3.7 evidences the resonant behavior of $D(\lambda)$. Its value has been obtained by fitting the experimental time profile of $T(t, \lambda)$ with the dynamical solution of

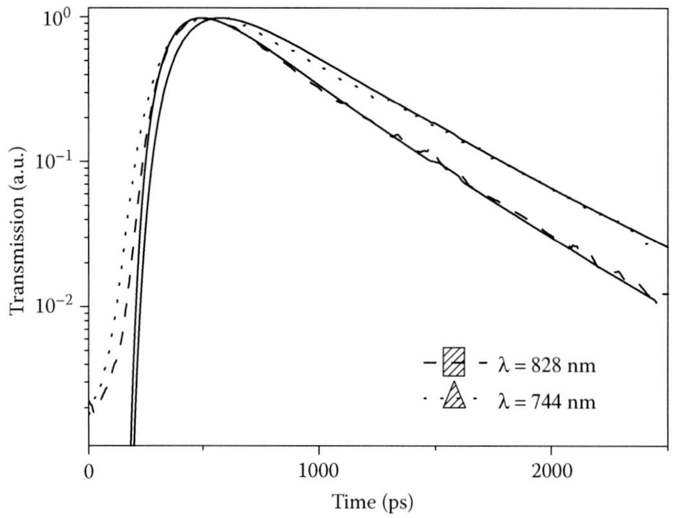

FIGURE 3.3.6
Time-resolved transmission measurements at $\lambda_1 = 744$ nm and $\lambda_2 = 828$ nm, where the different slope of $T(t)$ gives rise to a minimum, $D(744) = 166$ m^2/s, and maximum, $D(828) = 205$ m^2/s, in the diffusion constant. In this case $L = 1120$ μm. (Straight lines represent the fit of the experimental measurements with the diffusion equation.) (Adapted from P. D. García et al., Phys. Rev. A **78**, 023823, 2008.)

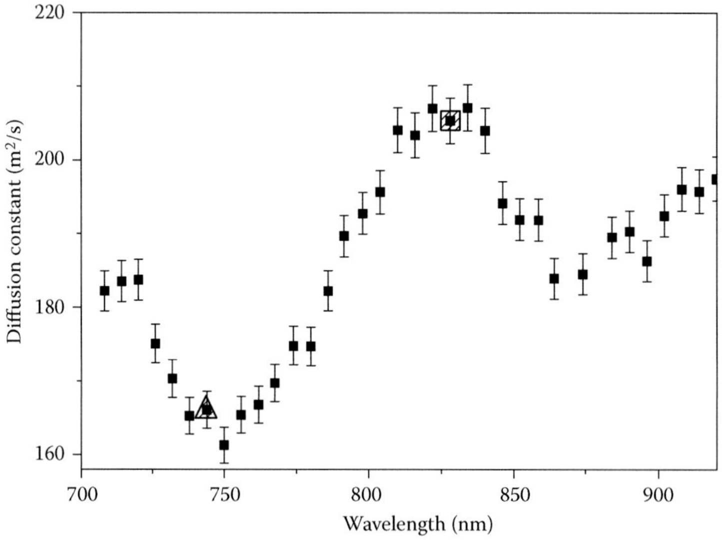

FIGURE 3.3.7
Experimental diffusion constant in a photonic glass made of polystyrene spheres with $d = 1220$ nm obtained by fitting experimental $T(t)$ curves. Two particular wavelengths are marked with a triangle ($\lambda_1 = 744$ nm) and a square ($\lambda_2 = 828$ nm). (Adapted from P. D. García et al., Phys. Rev. A **78**, 023823, 2008.)

the diffusion equation.[36] The solution to the diffusion equation, with appropriate boundary conditions, fits well the experimental data.

We can conclude that the resonant behavior of $D(\lambda)$ is due to Mie modes and presents maxima and minima at the same spectral positions as in the static measurements of $\ell_t(\lambda)$.

3.3.5 Resonant Energy Velocity

The appropriate velocity that describes diffusive light transport in photonic glasses is the energy velocity. A discussion about the possible resonant behavior of this magnitude can be found in the literature.[22,37,38] Figure 3.3.8 represents the energy velocity, v_e, obtained from independent measurements of $\ell_t(\lambda)$ (static) and $D(\lambda)$ (dynamic) and with the help of the expression:

$$D = \frac{1}{3} v_e \ell_t$$

The experimental value is below the average value of the group velocity obtained from the sample average refractive index, $v_g = c/\langle n \rangle \sim 0.77c$, while showing clear resonances. Qualitatively, the minima of $\ell_t(\lambda)$, for which the scattering strength is maximum, correspond to wavelengths that excite a Mie mode, and therefore experience a longer dwell time. This turns into a minimum transport velocity (Figure 3.3.8). For comparison, we show at the bottom of Figure 3.3.8 the lack of velocity resonances of a TiO_2 powder sample measured in the same experimental conditions.

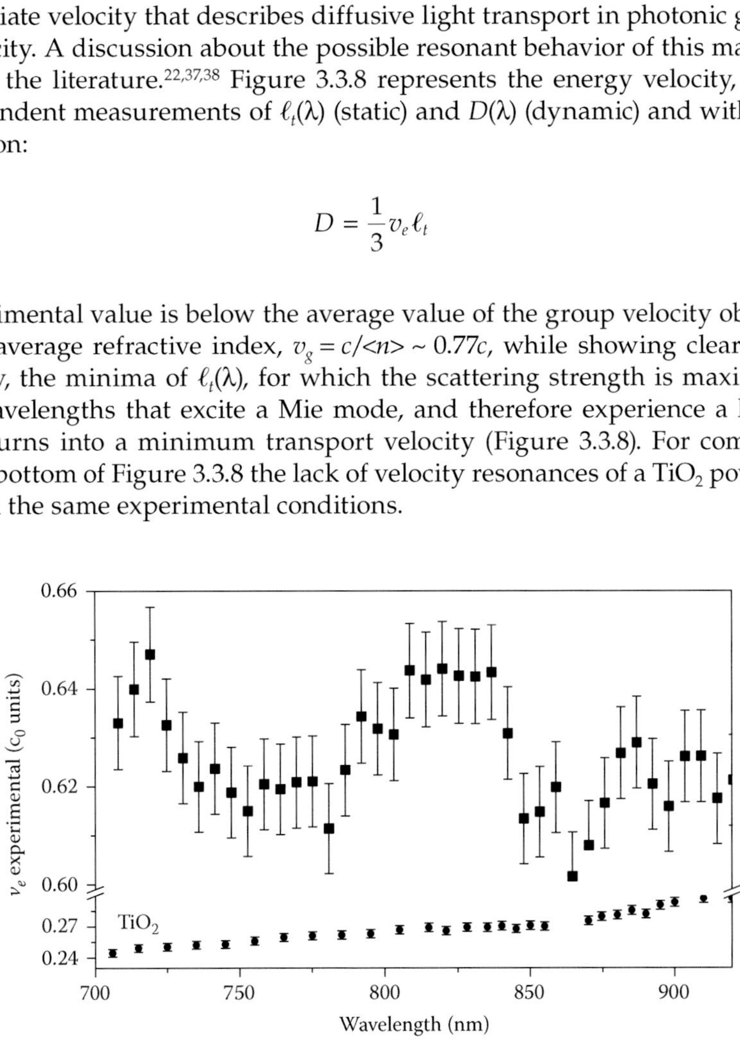

FIGURE 3.3.8
Experimental values of the energy velocity as a function of the wavelength for a photonic glass made of polystyrene sphere with diameter $d = 1220$ nm and for TiO_2 powder. For the photonic glass, a full oscillation of around 5% amplitude is shown. The averaged medium group velocity is $0.77c$. For the TiO_2 powder, flat velocity dispersion is shown, obtained with the same procedure (note the different scales on the y-axis). (Adapted from P. D. García et al., Phys. Rev. A **78**, 023823, 2008.)

Recently, Störzer et al.[39] have reported a wavelength-dependent diffusion constant in 15–20% polydisperse, irregularly shaped titania powders that, in the light of our findings, is unlikely to be the result of Mie resonances. The importance of the morphology of the sphere is evident, as a broad size distribution or a random shape of the dielectric resonator is expected to wash out the resonant properties.[40] One step further in the experimental study of light propagation in 3D photonic crystals can be taken by obtaining the group velocity (v_g) of real samples from the measured optical phase delay (see Chapter 3.2). Broadband light sources offer the possibility of obtaining v_g over a wide spectral range where different phenomena including slow light, superluminal values of v_g, or even negative ones are observed.

Figure 3.3.9 shows the comparison between the experimental and theoretical plots of $\ell_t(\lambda)$ obtained by analytically solving the Mie modes in a single dielectric sphere. This theory can be rigorously applied to the single scattering regime or in the case of diluted systems and independent scattering events.[41] The partial disagreement between experimental and theoretical data comes from the fact that the photonic glass cannot be considered diluted with a filling fraction of 0.55. When the density of scatterers is very low, a multiple scattering theory with a single scatter T-matrix suffices. This holds roughly until the optical (and not physical) sizes of the scatterers start to overlap. Then, the modes start to interact, thus affecting the scattering matrix of the sphere. Finally, it is expected that the resonances will disappear when the scatterers occupy a volume fraction of 100%. The coherent-potential approximation,[38] which is a mean field theory and a first order in scattering density, predicts that the scattering resonances weaken at high *ff* and even disappear for *ff* ~ 50–60%. Nevertheless, at a filling fraction as high as 0.55, it is still possible to experimentally resolve the effect of Mie resonances in the transport mean free path and transport velocity, as we show. It is interesting to look further into the physics involved in the resonant behavior at intermediate filling fractions. When optimizing light diffusion, one often wants to increase the refractive index contrast and to maximize the scatterer

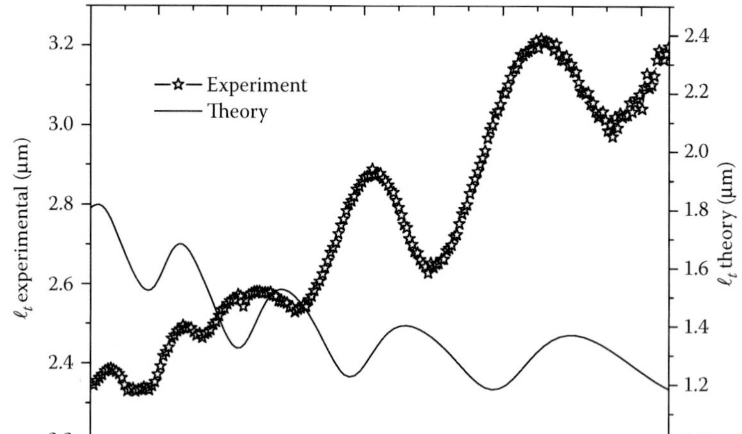

FIGURE 3.3.9
Plot of the transport mean free path: experimental (stars) and theoretical prediction of single-sphere Mie theory for ideal spheres (straight line). (Adapted from P. D. García et al., Phys. Rev. A **78**, 023823, 2008.)

density in order to minimize the scattering mean free path. The price to be paid when increasing the *ff* is to induce both correlations in the scatterer relative positions (in the close-packing limit, only the crystalline face-centered cubic lattice is allowed) and interaction between the Mie modes inside the scatterers.

3.3.6 Conclusions

The realization of photonic glasses, solid random distributions of monodisperse spheres, has been demonstrated to be a new encouraging route in which light diffusion can be controlled. The novel material for photonics is a self-assembled *disordered* material where the building blocks are identical dielectric spheres. By manipulating the colloidal suspensions we can switch the ordering process on and off, giving rise to a self-assembly process that provides very thick and completely random solid samples. These materials combine light dispersion with light diffusion, a mix that is crucial to control the diffuse flow of light in analogy to what photonic crystals do for ballistic light. Photonic glasses exhibit resonances in the diffuse light transport parameters: resonant transport mean free path, diffusion constant, and energy velocity. Active photonic glasses can be used to control random lasing and tune the lasing emission wavelength. The high contrast of dielectric spheres in air secures strong light-matter interaction; the Mie resonances provide spectral selectivity, opening a novel route to active disorder-based photonic devices.[24] Photonic glasses are expected to be important for the field of Anderson localization of light which could be achieved for some resonant wavelength ranges, if the refractive index of the material were increased as, for example, in silicon photonic glasses.[42] Photonic glasses by themselves or integrated with photonic crystals may give rise to new range of applications in future photonic devices.

References

1. P. D. Garcia, R. Sapienza, A. Blanco, C. Lopez, Adv. Mater. **19**, 2597 (2007); P. D. García, R. Sapienza, and C. López, *ibid*. **22**, 12 (2010).
2. P. Sheng, *Introduction to Wave Scattering, Localization, and Mesoscopic Phenomena* (Academic Press, San Diego, 1995).
3. Y. Kuga and A. Ishimaru, J. Opt. Soc. Am. A **8**, 831 (1984); M. P. van Albada and A. Lagendijk, Phys. Rev. Lett. **55**, 2692 (1985); P. E. Wolf and G. Maret, *ibid*. **55**, 2696 (1985).
4. M. A. Noginov, *Solid-State Random Lasers*, (Springer, Berlin, 2005).
5. P. W. Anderson, Phys. Rev. **109**, 1492 (1958); Philos. Mag. B **52**, 505 (1985).
6. A. Z. Genack, N. Garcia, and W. Polkosnik, Phys. Rev. Lett. **65**, 2129 (1990).
7. N. Garcia, A. Z. Genack, Phys. Rev. Lett. **63**, 259 (1989).
8. H. Cao, Y. G. Zhao, H. C. Ong, S. T. Ho, J. Y. Dai, J. Y. Wu, and R. P. H.Chang, Appl. Phys. Lett. **73**, 3656 (1998).
9. D. S. Wiersma, P. Bartolini, A. Lagendijk, and R. Righini, Nature **390**, 671 (1997).
10. J. Gomez Rivas, R. Sprik, L. D. Noordam, C. W. Rella, and A. Lagendijk, Phys. Rev. E **62**, R4540 (2000).
11. P. M. Johnson, B. P. J. Bret, J. G. Rivas, J. J. Kelly, and A. Lagendijk, Phys. Rev. Lett. **89**, 243901 (2002).

12. N. Garcia and A. Z. Genack, Phys. Rev. Lett. **66**, 1850 (1991).
13. X. H. Wu, A. Yamilov, H. Noh, H. Cao, and E. W. Seelig, R. P. H. Chang, J. Opt. Soc. Am. B **21**, 159 (2004).
14. R. H. J. Kop, P. de Vries, R. Sprik, and A. Lagendijk, Phys. Rev. Lett. **79**, 4369 (1997); M. Reufer, L. F. Rojas-Ochoa, S. Eiden, J. J. Saenz, and F. Scheffold, Appl. Phys. Rev. **91**, 171904 (2007).
15. N. M. Lawandy, R. M. Balachandran, A. S. L. Gomes, and E. Sauvain, Nature **368**, 436 (1994); L. F. Rojas-Ochoa, J. M. Mendez-Alcaraz, J. J. Saenz, P. Schurtenberger, and F. Scheffold, Phys. Rev. Lett. **93**, 073903 (2004).
16. H. Takayasu, *Fractals in Physical Science* (Manchester University, Manchester, 1990); M. V. Berry, J. Phys. A: Math. Gen. **12**, 781 (1979); J. Uozumi and T. Asakura, in *Current Trends in Optics*, edited by J. C. Dainty (Academic, London, 1994), p. 83.
17. P. V. Braun, S. A. Pruzinsky, and F. Garcia-Santamaria, Adv. Mater. **18**, 2665 (2006); A. Arsenault, F. Fleischhaker, G. von Freymann, V. Kitaev, H. Miguez, A. Mihi, N. Tetreault, E. Vekris, I. Manners, S. Aitchison, D. Perovic, and G. A. Ozin, *ibid.* **18**, 2779 (2006).
18. A. Ledermann, L. Cademartiri, M. Hermatschweiler, C. Toninelli, G. A. Ozin, D. S. Wiersma, M. Wegener, and G. von Freymann, Nat. Mater. **5**, 942 (2006).
19. D. Langevin and M. A. Bouchiat, J. Phys. (Paris) C **1**, 197 (1975); A. Yu. Val'kov and V. P. Romanov, Zh. Eksp. Teor. Fiz. **83**, 1777 (1982) [Sov. Phys. JETP **56**, 1028 (1982)].
20. S. Gottardo, S. Cavalieri, O. Yaroshchuk, and D. S.Wiersma, Phys. Rev. Lett. **93**, 263901 (2004).
21. P. Barthelemy, J. Bertolotti, and D. S. Wiersma, Nature **453**, 495 (2008).
22. J. Bertolotti, K. Vynck, L. Pattelli, P. Barthelemy, S. Lepri, and D. S. Wiersma, Adv. Funct. Mater. **20**, 965 (2010).
23. M. P. van Albada, B.A. van Tiggelen, A. Lagendijk, and A. Tip, Phys. Rev. Lett. **66**, 3132 (1991).
24. S. Gottardo, Riccardo Sapienza, P. D. Garcia, A. Blanco, D. S. Wiersma, and C. Lopez, Nat. Photonics **2**, 429 (2008).
25. Y. Xia, B. Gates, Y. Yin, and Y. Lu, Adv. Mater. **12**, 693 (2000).
26. P. Jiang, J. F. Bertone, K. S. Hwang, and V. L. Colvin, Chem. Mater. **11**, 2131 (1999).
27. J. Ballato, J. Dimaio, A. James, and E. Gulliver, Appl. Phys. Lett. **75**, 1497 (1999).
28. B. V. Derjaguin, and L. Landau, Acta Physicochim. URSS **14**, 633 (1941); E. J. W. Verwey and J. T. G. Overbeek, *Theory of the Stability of Lyophobic Colloids* (Elsevier, Amsterdam, 1948).
29. K. L. Wu and S. K. Lai, Langmuir **21**, 3238 (2005).
30. J. W. Goodwin, J. Hearn, C. C. Ho, and R. H. Ottewill, Colloid Polym. Sci. **252**, 464 (1974).
31. M. Mueller, R. Zentel, T. Maka, S. G. Romanov, and C. M. Sotomayor Torres, Chem. Mater. **12**, 508 (2000).
32. R. Sapienza, P. D.Garcia, J. Bertolotti, M. D. Martin, A. Blanco, L. Viña, C. López, and D. S. Wiersma, Phys. Rev. Lett. **99**, 233902 (2007).
33. P. D. García, R. Sapienza, J. Bertolotti, M. D. Martín, Á Blanco, A. Altube, L. Viña, D. S. Wiersma, and C. López, Phys. Rev. A **78**, 023823 (2008).
34. G. Mie, Ann. Phys. **25**, 77 (1908).
35. N. Garcia, A. Z. Genack, and A. A. Lisyansky, Phys. Rev. B **46**, 14475 (1992).
36. M. S. Patterson, B. Chance, and B. C. Wilson, Appl. Opt. **28**, 2331 (1989).
37. C. M. Soukoulis, S. Datta, and E. N. Economou, Phys. Rev. B **49**, 3800 (1994).
38. K. Busch and C. M. Soukoulis, Phys. Rev. B **54**, 893 (1996).
39. M. Störzer, C. M. Aegerter, and G. Maret, Phys. Rev. E **73**, 065602(R) (2006).
40. S. C. Hill and R. E. Benner, in *Optical Effects Associated with Small Particles*, edited by P. W. Barber and R. K. Chang (World Scientific, Singapore, 1988); G. H. Watson, P. A Fleury and S. L. McCall, Phys. Rev. Lett. **58**, 945 (1987).
41. G. Labeyrie, E. Vaujour, C. A. Muller, D. Delande, C. Miniatura, D. Wilkowski, and R. Kaiser, Phys. Rev. Lett. **91**, 223904 (2003).
42. M. Ibisate, D. Golmayo, and C. López, Adv. Mater. **21**, 2899 (2009).

3.4

Superdiffusion of Light in Lévy Glasses

Kevin Vynck
University of Florence

Jacopo Bertolotti
University of Florence
University of Twente

Pierre Barthelemy
Delft University of Technology

Diederik S. Wiersma
University of Florence
National Institute of Optics

CONTENTS

3.4.1 Introduction: From Normal to Anomalous Diffusion

Disordered optical materials are extremely common in nature and yet are a source of complex and remarkable phenomena for light. A great variety of transport regimes with distinct properties indeed exist, despite the fact that all of them rely on the same physical ground.[1] The macroscopic properties of a random medium essentially depend on the different length scales involved in the system on the microscopic level. For instance, when both the size of the scatterers and the average distance between them are much smaller than the wavelength, the random medium behaves as a homogeneous effective medium. This approximation is no longer valid when the refractive index of the medium displays fluctuations on length scales comparable with or larger than the wavelength. In such media, light undergoes a series of scattering events that randomizes the direction and phase of propagating waves.

This multiple scattering process can often be described within the *diffusion approximation*. The average intensity $I(\mathbf{r},t)$ in the medium is then given by the diffusion equation

$\partial_t I(\mathbf{r},t) = D\nabla^2 I(\mathbf{r},t)$, where D is the diffusion constant. The diffusive contribution to transport appears at the lowest-order solution for the intensity propagation in the multiple scattering theory that describes the multiple scattering process perturbatively by expanding the field in powers of the scattering potential.[2] Higher orders of the expansion include interference effects that give rise to, for instance, speckle correlations,[3] universal conductance fluctuations,[4] weak localization,[5] and Anderson localization.[6]

An intuitive picture for light diffusion in disordered materials is the Brownian random walk model, in which multiply-scattered light is described as particles performing steps of varying length in random directions. By virtue of the Central Limit Theorem, when the variance of the step-length distribution is finite and the steps are independent, the mean step-length distribution after many steps becomes approximately normally distributed.[7] The process therefore becomes diffusive irrespective of the microscopic transport mechanism. The ubiquity of diffusive processes in nature is inherent to the fact that the normal distribution is the attractor for random variables with finite variance. When, however, the step-length distribution of the random walk has a diverging variance, that is, the probability to perform arbitrary long steps becomes significant, the limit distribution becomes a so-called α-stable Lévy distribution and the transport process no longer obeys standard diffusion dynamics, that is, it becomes *anomalous*.[8,9]

The α-stable Lévy distribution, named after Paul Pierre Lévy for his work on the problem of sums of random variables,[10] constitutes the whole stable distribution family. Its probability density function has no general analytical expression but can be defined through the Fourier transform of its characteristic function. For symmetric and centered distributions, it is given by:[11]

$$f(x) = \frac{1}{2\pi}\int_{-\infty}^{\infty} e^{-|ck|^{\alpha}} e^{-ikx} dk \tag{3.4.1}$$

where $\alpha \in (0,2]$ is the so-called stability index and $c > 0$ is a scale parameter that indicates the width of the distribution. One recognizes easily the normal and the Cauchy distributions for $\alpha = 2$ and $\alpha = 1$, respectively. The importance of the α parameter comes from the generalized form of the Central Limit Theorem,[12] which states that the sum of many random variables, whose distribution asymptotically decays as $|x|^{-(\alpha+1)}$ with $0 < \alpha \leq 2$, converges to a distribution with stability index α. The heavy tail of α-stable Lévy distributions entails that transport following Lévy statistics is dominated by rare but very long steps, and thus appears radically different from that of a Brownian motion, as shown in Figure 3.4.1. The power-law decay of the step-length distribution is such that Lévy-type motions have the property of scale invariance. As a matter of fact, they were first studied by Benoît Mandelbrot in the framework of random walks over fractals.[13] Mandelbrot named this type of motion a *Lévy flight* and used it himself to model the fluctuations of cotton price.[14] Lévy flights are now commonly used in the study of economy indices.[15]

The possibility for random walkers to perform very long steps in Lévy flights yields the remarkable feature that random walkers can spread faster than linearly with time. The spreading of particles is defined by the mean-square displacement $\langle r^2(t)\rangle$, which at long times scales as

$$\langle r^2(t)\rangle \sim Dt^{\gamma} \tag{3.4.2}$$

Whereas $\gamma = 1$ for Brownian motion, it can be shown that $\gamma = 2/\alpha$ for Lévy flights with $0 < \alpha < 2$, yielding $\gamma > 1$.[9] Such transport processes are then said to be *superdiffusive*. In the

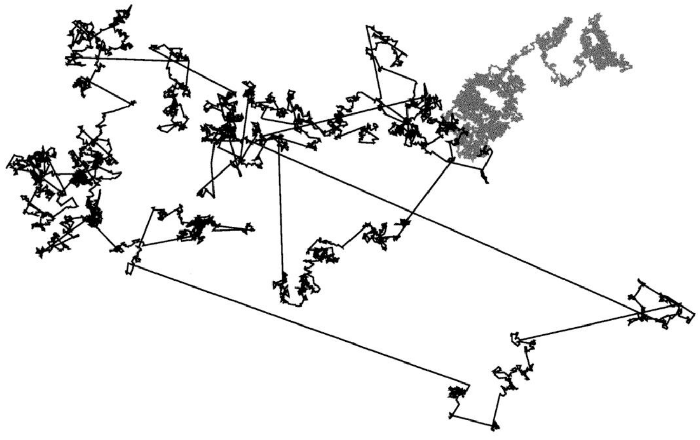

FIGURE 3.4.1
Comparison between a Brownian random walk (gray) and a Lévy random walk (black) with $\alpha = 1.5$ (10,000 steps). Whereas in Brownian motion, each step contributes equally to the average transport properties of the random walker, Lévy motion is dominated by a few very long steps.

study of real systems exhibiting Lévy statistics, a Lévy flight model might however produce unphysical results. Two additional considerations therefore need to be made.

First, each step in a Lévy flight is assumed to be performed in a unit time and thus is independent of the length of the step. A more physical model of Lévy motion is given by *Lévy walks*,[16] which include a spatiotemporal coupling (that is, each step is performed at a finite velocity). In this case, one finds that:[17]

$$\gamma = \begin{cases} 2, & 0 < \alpha < 1 \\ 3 - \alpha, & 1 < \alpha < 2 \end{cases} \tag{3.4.3}$$

Second, one should consider the fact that, in real systems, steps larger than a certain length cannot be made. Such processes are known as *truncated Lévy flights* (or *walks*),[18] as a result of the presence of an upper truncation in the step-length distribution. This truncation is such that the variance of the distribution is finite, which implies that transport eventually becomes diffusive. The convergence toward diffusion can be extremely slow and pure superdiffusive behavior can be observed up to a certain time. The step-length distribution and mean-square displacement of a truncated Lévy walk with $\alpha = 1.5$ and $\ell_m = 300$ is shown in Figure 3.4.2. Transport is superdiffusive with a mean-square displacement growing as $t^{1.5}$, in agreement with Equation 3.4.3, up to the truncation time $t_m = \ell_m/v$, after which it becomes diffusive again.

Superdiffusive processes have been found in a wide range of biological and physical systems in nature,[19] describing for instance human travel,[20] cellular transport,[21] animal foraging patterns,[22] transport in rotating flows,[23] and radiation transport in hot atomic vapors.[24] However, for almost all systems studied so far, control over the degree of superdiffusivity α has been out of reach, thereby making systematic investigations of superdiffusive processes difficult. This chapter is devoted to the experimental, numerical, and theoretical study of disordered optical materials, dubbed Lévy glasses, in which light transport is superdiffusive.[25–28] A recipe for the realization of Lévy glasses with an arbitrary degree of superdiffusivity α is proposed in Section 3.4.2. In Section 3.4.3, attention is given to the optical characterization

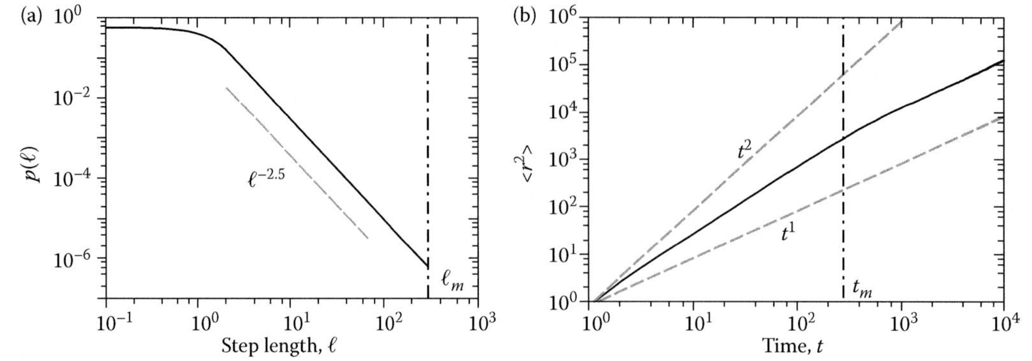

FIGURE 3.4.2
Truncated Lévy walk with $\alpha = 1.5$, a truncation length $\ell_m = 300$ and $v = 1$. (a) Step-length distribution. (b) Mean-square displacement versus time.

of Lévy glasses, in which deviations from diffusive transport are observed. A numerical study of superdiffusive transport in two-dimensional (2D) quenched Lévy glasses by means of Monte Carlo simulations is then presented in Section 3.4.4. Finally, a semianalytical tool based on fractional diffusion that allows one to calculate the optical properties of finite superdiffusive media, including weak interference effects, is introduced in Section 3.4.5.

3.4.2 A Recipe for Superdiffusive Lévy Glasses

The behavior of light in a disordered material is known to depend critically on the scattering particles that compose it and their spatial distribution. A material with a relatively large scattering strength, that is to say spatial fluctuations of the refractive index on the wavelength scale, is required to reach the multiple scattering regime. Powders and etched networks of semiconductors, such as TiO_2, GaP, and Si, have appeared as an ideal choice for studying diffusive processes, as well as weak and strong localization,[29] owing to their large availability and high refractive index. Alternatively, disordered sets of monodisperse spheres exhibiting Mie resonances can offer a fine spectral control over transport. Such materials, dubbed *photonic glasses*,[30] have recently been used for random lasing purposes.[31] In both of the above cases, the scattering elements are distributed quite homogeneously in space, such that the step length of light has a finite variance. The mean step length is then normally distributed with an average value given by the scattering mean free path $\ell_s = (N\sigma_s)^{-1}$, to a first approximation, where N is the density of scatterers and σ_s their scattering cross section, eventually leading to a diffusive type of transport. In order for light to perform a Lévy walk, the step-length distribution $p(\ell)$ should decay as a power law as $p(\ell) \sim \ell^{-(\alpha+1)}$ with $0 < \alpha < 2$. Light should therefore have the possibility to perform steps on length scales of several orders of magnitude, implying that the disordered materials should display large fluctuations of the local density of scatterers. Some of us recently proposed to realize superdiffusive materials by embedding transparent spheres of diameters covering orders of magnitude in a diffusive medium.[25] The configuration of these so-called Lévy glasses is such that light is only scattered in between the spheres and performs long steps when crossing them. An example of a 2D Lévy glass, constructed by a computer

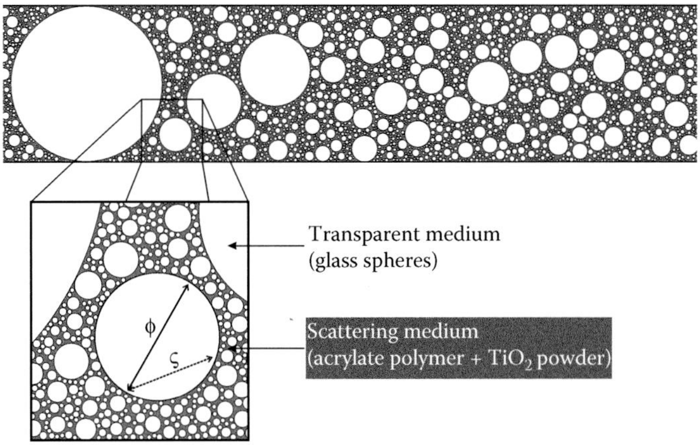

FIGURE 3.4.3
Two-dimensional Lévy glass in a slab geometry, constructed from a set of 20 exponentially spaced disk diameters, spanning over two orders of magnitude. The interstices between the various disks (white regions) are filled by the scattering medium (gray regions). The zoomed view shows the diameter of a disk ϕ and one of its chords ς. (J. Bertolotti et al., Adv. Funct. Mater. **20**, 965, 2010. Copyright Wiley-VCH Verlag GmbH & Co. KgaA. Reproduced with permission.)

program, is shown in Figure 3.4.3. It is particularly interesting to note that Lévy glasses are statistically self-similar over a certain range of length scales. In this sense, transport in Lévy glasses is expected to resemble a random walk over a random fractal.[32]

As the anomalous transport in Lévy glasses is produced by the large-scale fluctuations of the density of scatterers, the step-length distribution $p(\ell)$ in a Lévy glass is essentially determined by the sphere diameter distribution $p(\phi)$. Establishing the relation between these two distributions is therefore the key for designing Lévy glasses with an arbitrary degree of superdiffusivity α. The fact that light should perform steps of length varying by orders of magnitude according to a power law suggests that the sphere diameter distribution should also fall down as a power law. A parameter β, to be chosen during sample preparation, can thus be defined to tune the asymptotic decay of $p(\phi)$. The derivation below, given for d-dimensional space with $d > 1$, is intended to establish a relation between β and α.

The probability $p(\phi)$ for light to enter a sphere of diameter ϕ in a Lévy glass is proportional to the surface of the d-dimensional sphere, indicating how likely it is for light in the diffusive medium to be in the vicinity of the sphere, and to the number of spheres with diameter ϕ present in the sample. Therefore we can write:

$$p(\phi) = A \frac{d\pi^{d/2}(\phi/2)^{d-1}}{\Gamma(d/2+1)} \frac{1}{\phi^{\beta+1}} \tag{3.4.4}$$

where Γ is the gamma function and A is a normalization constant. The actual length of a ballistic step in the sphere, however, is not the sphere diameter ϕ but the length of one of its chords ς. For a d-dimensional sphere of diameter ϕ, the chord-length distribution is given by:[33]

$$p(\varsigma|\phi) = \frac{(d-1)}{\phi^2} \varsigma \left(1 - \frac{\varsigma^2}{\phi^2}\right)^{\frac{d-3}{2}} \quad (0 \le \varsigma \le \phi) \tag{3.4.5}$$

The marginal chord-length distribution $p(\varsigma)$ can be calculated from the knowledge of $p(\phi)$ and $p(\varsigma|\phi)$. For practical reasons, an actual Lévy glass will generally be composed of monodisperse spheres of different diameters, meaning that the distribution function of sphere diameters $p(\phi)$ has to be discretized, taking the form $p(\phi) = \Sigma_i\, p(\phi_i)\delta(\phi - \phi_i)$. This discretization makes important the choice of the distribution sampling, that is, the spacing between successive diameters (e.g., linear, quadratic, exponential), as different spacings are expected to yield different asymptotic behaviors for $p(\varsigma)$. The sampling can be implemented in the continuous limit by using the Riemann–Stieltjes integral form $p(\varsigma) = \int_{\phi_0}^{\infty} p(\varsigma|\phi)p(\phi)dg(\phi)$, where $g(\phi)$, the integrator, maps the diameter spacing on a linear scale, and ϕ_0 is the minimal sphere diameter. In particular, the integrator is given by $g(\phi) = \phi^{1/s}$ for polynomial spacing of power s and $g(\phi) = \log(\phi)$ for exponential spacing. Closed-form solutions can be found in both cases, yielding $p(\varsigma) \sim \varsigma^{-(\beta-(d-2)+1/s+1)}$ and $p(\varsigma) \sim \varsigma^{-(\beta-(d-2)+1)}$, respectively.

We now establish a relation between $p(\varsigma)$ and $p(\ell)$, assuming for simplicity that light crosses exactly one sphere between two scattering events. This assumption is reasonable when the scattering mean free path ℓ_s in the diffusive medium is of the order of the average distance between neighboring spheres. After each scattering event, light travels a distance ε in the diffusive medium, crosses a sphere along one of its chords on a distance ς, and re-enters the diffusive medium (see Figure 3.4.4a). The conditional step-length distribution can then be written as:

$$p(\ell|\varepsilon,\varsigma) = \frac{1}{\ell_s}\left(\Theta[\varepsilon-\ell]e^{-\frac{\ell}{\ell_s}} + \Theta[\ell-(\varepsilon+\varsigma)]e^{-\frac{\ell-\varsigma}{\ell_s}}\right) \qquad (3.4.6)$$

where Θ is the Heaviside step function. Further assuming near close-packing of the spheres, the distance ε is expected to be of the order of the smallest sphere diameter ϕ_0. Thus, for $\varepsilon \ll \varsigma$, one finds that the asymptotic power-law decay of the marginal step-length

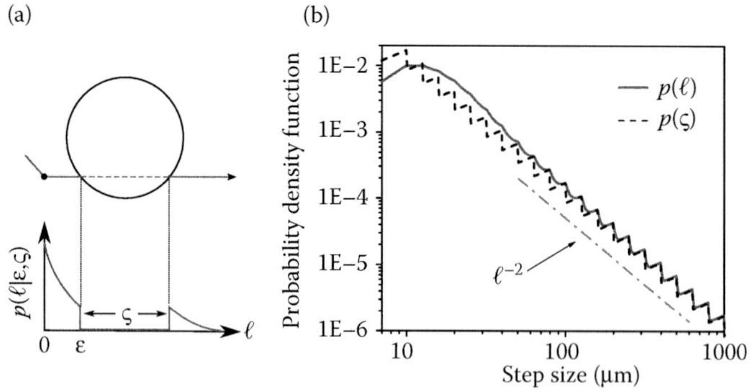

FIGURE 3.4.4
Chord-length model. (a) Derivation of the conditional step-length distribution. Light travels a distance ε in the diffusive medium, travels along a chord of the sphere on a distance ς, and re-enters the diffusive medium. The probability to be scattered in the sphere equals zero. (b) Comparison between $p(\varsigma)$ (black dashed curve) and $p(\ell)$ (gray solid curve) in a three-dimensional Lévy glass with exponentially spaced sphere diameters and $\beta = 2$. The gray dash-dotted curve shows the slope of an exact ℓ^{-2} distribution as a comparison. (Panel (b) (J. Bertolotti et al., Adv. Funct. Mater. **20**, 965, 2010. Copyright Wiley-VCH Verlag GmbH & Co. KgaA. Reproduced with permission.)

distribution $p(\ell) = \int_0^\infty p(\ell|\varsigma)p(\varsigma)d\varsigma$ follows that of $p(\varsigma)$, leading us finally to the following relations between α and β in d-dimensional space:

$$\beta = \begin{cases} \alpha + (d-2) + 1/s & \text{(polynomial spacing)} \\ \alpha + (d-2) & \text{(exponential spacing)} \end{cases} \qquad (3.4.7)$$

In summary, the value of α that characterizes the power-law decay of the step-length distribution $p(\ell)$, and thus the degree of superdiffusivity of the material, can be controlled during sample preparation by an appropriate choice of the number of spheres of each category, parametrized by β. Figure 3.4.4b shows a comparison between $p(\varsigma)$ and $p(\ell)$ for a three-dimensional (3D) Lévy glass with exponentially spaced sphere diameters and $\beta = 2$. The resulting chord-length distribution exhibits a characteristic saw-tooth profile, in accordance with Equation 3.4.5 for $d = 3$. As expected, the step-length distribution tends toward the chord-length distribution for longer steps and decays asymptotically as $p(\ell) \sim \ell^{-2}$, that is, $\alpha = 1$.

Let us now discuss the actual realization of 3D Lévy glasses. In practice, the samples were made by embedding transparent spheres of different diameters in a diffusive medium, composed by scattering particles and a host medium. The materials were chosen such that their absorption was very low in the visible range to avoid any unnecessary difficulty in the interpretation of the results. Glass spheres (highly transparent in the visible) were obtained as near-monodisperse collections with diameters ranging from micrometers to centimeters. Powders of submicrometer TiO_2 particles, commonly used for the optical study of disordered materials as a result of their high refractive index (2.7 at a wavelength of 632 nm), were chosen to constitute the scattering medium. As a host medium, we used an acrylate monomer that polymerizes on ultraviolet light irradiation and that is index-matched with the spheres to avoid reflections at their surface. Polymers have the clear advantage that they suffer a very low shrinkage on solidification and have a very good stability in time. Polymer-based samples are actually expected to maintain their optical properties for at least several months and most likely even years.

Following the chord-length model above, the preparation of the samples with a given value of α requires a definition of the sphere diameter distribution, parametrized by β, using Equation 3.4.7. A relatively large set of spheres with different diameters is needed to construct a chord-length distribution decaying as a power law. In our case, the samples were made using spheres with exponentially spaced diameters ranging over two orders of magnitude (the exponential spacing being the most sensible option when covering a wide range of length scales). The sphere diameter distribution was constructed by weighting each sphere category separately and the ensemble of spheres was then mixed with the TiO_2 powder and the acrylate monomer. Whereas the amount of TiO_2 particles determines the scattering mean free path ℓ_s in the diffusive medium, the amount of acrylate monomer determines the total packing fraction of the glass spheres and thus ε. Under the assumption that $\varepsilon \simeq \phi_0$ for near close-packing, the scattering mean free path was set to be of the order of the smallest sphere diameter in the sample. The mixing process was performed while keeping the sample at 150°C on a hot plate to reduce the monomer viscosity. The resulting paste was then pressed between two glass slides in such a way that the sample thickness would correspond to the largest diameter in the sample. The largest sphere

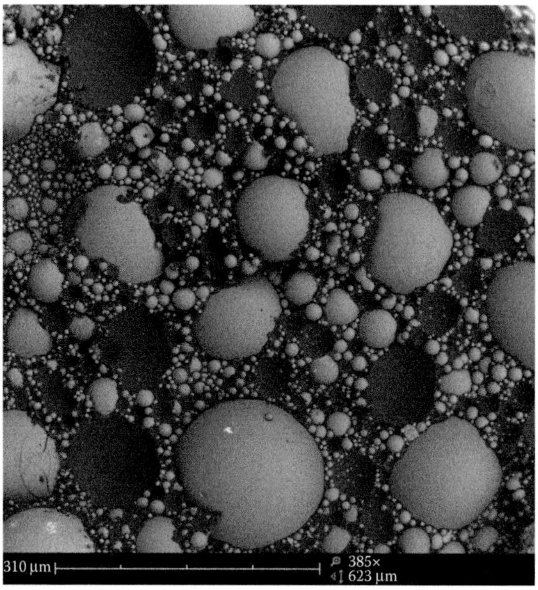

FIGURE 3.4.5
Scanning electron microscopy picture of a three-dimensional Lévy glass with β = 1.0, obtained by splitting the sample into two pieces. The dark and light gray regions correspond to the polymer host medium and to the glass spheres, respectively. (Courtesy of Romolo Savo.)

indeed determines the truncation of the step length on the far end of the distribution and should be kept as large as possible. Finally, the glass spheres and the scattering particles were "frozen" in space by irradiating the sample under ultraviolet light to polymerize the monomer.

Figure 3.4.5 shows an image made by scanning electron microscopy of a 3D Lévy glass (β = 1.0). The self-similarity of the structure can be seen, as well as the near close-packing of the spheres. The optical characterization of such 3D Lévy glasses is presented in the next section.

3.4.3 Optical Characterization of Lévy Glasses

Disordered materials are known to support a variety of different optical phenomena, each of them having very peculiar features. Demonstrating a given physical effect by experiments therefore comes to find clear signatures of it, which in turn may provide information on the microscopic transport mechanism taking place. A measure of the coherent backscattering on a random medium, for instance, is a signature of the weak localization effect and allows one to recover the transport mean free path.[34] Now, considering disordered materials with unusual, or even unknown, optical properties naturally raises the question of how to optically characterize them in a proper way. As Lévy glasses are expected to support anomalous diffusion, looking at deviations in their optical features compared to those of standard diffusive samples may be a sound option. In the following, we propose two possible approaches for the optical characterization of Lévy glasses.

The first one is to study how the total transmission of light through a sample varies as a function of the sample thickness. For materials with diffusive properties, the total transmission is known to follow the optical equivalent of Ohm's law,[1] that is, the total amount of transmitted light T decreases linearly with the slab thickness L. As this expression is derived in the diffusion approximation, one may observe deviations from Ohm's law in the case of Lévy glasses. Owing to the possibility of making very long steps, the total transmission is expected to decrease slower with the thickness than in the diffusive regime. In fact, a scaling reasoning makes it possible to generalize Ohm's law for non-absorbing media with an arbitrary value of α and perfectly absorbing boundaries as:[35,36]

$$T \sim L^{-\alpha/2} \tag{3.4.8}$$

for large L. This relation can be used to show that our samples are truly superdiffusive and retrieve *a posteriori* the value of α. We therefore fabricated a series of Lévy glasses with $\beta = 2$ (i.e., $\alpha = 1$) with different thicknesses in the range of 30–550 μm, the sample thickness always corresponding to the largest sphere of the material and keeping constant the volume ratio of glass spheres, polymer, and TiO_2 powder. Total transmission measurements were performed by using a collimated He–Ne laser ($\lambda = 632.8$ nm) so as to have a 1 mm^2 light spot impinging at normal incidence on the sample. The total transmission was then collected by means of an integrating sphere and a standard lock-in technique was used to reduce the noise. Each measurement was repeated 10 times on different parts of the sample to improve the ensemble average. A fit of the data (shown in Figure 3.4.6) with the generalized Ohm's law yields $\alpha = 0.948 \pm 0.09$, which is in excellent agreement with the expected value $\alpha = 1$. This deviation from the classical Ohm's law clearly shows that the samples support a Lévy-type transport with, in principle, a controllable parameter α. Let us note that for very thick samples ($L \gg 500$ μm), the presence of a small amount of absorption as a result of the TiO_2 powder used as a scattering medium starts to affect the total transmission. Owing to the lack of a more complete theory on Lévy walks, comparing measurements for very thick samples with Equation 3.4.8 is not possible. Also, this

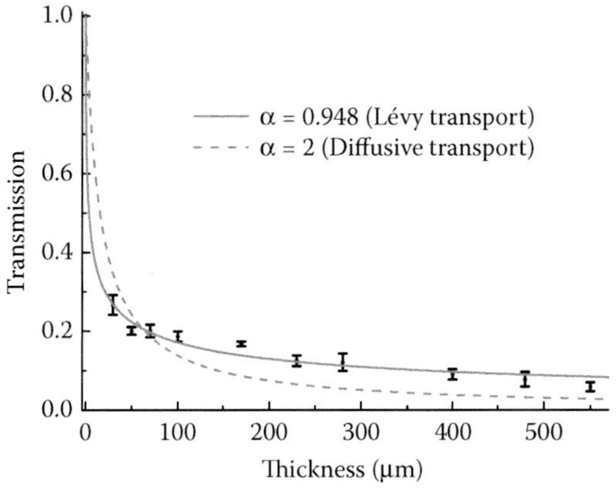

FIGURE 3.4.6
Total transmission measurements as a function of the slab thickness. The error bars represent the standard deviation of measurements repeated on different spots on the sample. The solid line represents the fit with Equation 3.4.8 and shows a good agreement with data, whereas the dashed line represents the fit with the classical Ohm's law. (Reproduced from P. Barthelemy, J. Bertolotti, and D. S. Wiersma, Nature **453**, 495, 2008. With permission.)

approach has the disadvantage that the retrieval of α requires several samples of different thicknesses to be realized.

A second possible approach for the optical characterization of Lévy glasses is to look at the intensity profile at the output of the sample in response to light focused on the input surface. In diffusive materials, this profile is a rounded bell-shaped curve close to a Gaussian.[2] In Lévy glasses, the long tail of the step-length distribution implies that light can cross the sample in just a few steps. These rare but important events are expected to modify significantly the shape of the intensity profile depending on α. Lévy glasses with different values of β were therefore prepared. The intensity profile of each individual sample was measured by illuminating it with a He–Ne laser focused on a 2 μm spot size and the output surface was imaged using a Peltier-cooled CCD camera. The sample was placed between crossed polarizers in order to suppress the ballistic light and thus collect only the scattered component. Interestingly, but as expected, the strong inhomogeneity of Lévy glasses led to important fluctuations in the transmitted profile (see Figure 3.4.7a) as compared to diffusive samples, in which fluctuations are generally limited to speckle patterns that average out rapidly. To perform a good average over disorder realization, the intensity profiles were obtained by performing 3000 measurements on well-separated points at the surface of the sample. The experimental results obtained for different values of α are shown in Figure 3.4.7b (upper panel) and compared with Monte Carlo simulations (lower panel). Both reveal a very marked cusp and slowly decaying tails, which are not present in the diffusive case. The small discrepancy between experimental and numerical results may be attributed to internal reflections at the boundaries of the sample, which were not taken into account in Monte Carlo simulations. Nevertheless, this change of behavior for samples with different values of α suggests that a measure of the output intensity profile may be a reliable technique to determine the degree of superdiffusivity of arbitrary optical materials. In addition, this approach has the advantage, compared to the previous one, that only one sample is needed for the measurements.

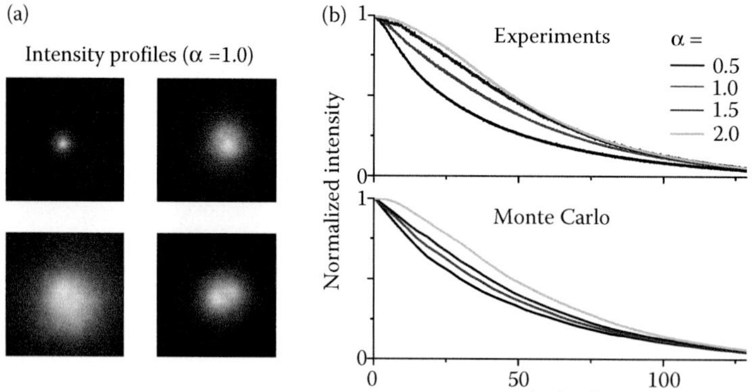

FIGURE 3.4.7
(See color insert.) (a) Spatial distributions of the transmitted intensity for a Lévy glass with β = 2.0, obtained by focusing light on different points of the sample. The transmitted profiles strongly differ from one to the other. (Reproduced from P. Barthelemy, J. Bertolotti, and D. S. Wiersma, Nature **453**, 495, 2008. With permission.) (b) Upper panel: measured intensity profiles for Lévy glasses with different sphere diameter distributions. Lower panel: intensity profiles calculated by Monte Carlo simulations of Lévy walks with the same nominal value of α as in the experiments. Δ*r* represents the distance from the peak of the transmission profile. (J. Bertolotti et al., Adv. Funct. Mater. **20**, 965, 2010. Copyright Wiley-VCH Verlag GmbH & Co. KgaA. Reproduced with permission.)

To summarize, the two optical characterization techniques presented in this section have revealed deviations from standard diffusive transport in the optical features of Lévy glasses, their significance depending on the degree of superdiffusivity α. Although experiments so far have been limited to the steady-state regime, there is a clear interest in studying light superdiffusion through time-resolved experiments. Concurrently, numerical studies provide a good point of comparison for experiments, as well as some greater insight into the physics involved at the microscopic scale. A numerical study of transport in 2D Lévy glasses is presented in the next section.

3.4.4 Transport in 2D Quenched Lévy Glasses

Brownian motion, as discussed above, constitutes an elegant and simple way of modeling light transport in most common disordered materials. The diffusion process stems from the fact that steps between consecutive scattering events are small and independent. As we have seen, the former assumption is not valid in Lévy glasses because of the presence of very large voids. Now, it is also interesting to evaluate the validity of the latter assumption. As a matter of fact, a random walker just crossing a large void has a non-negligible probability to be scattered in its vicinity as a result of the presence of the scattering medium and, therefore, to cross again the very same void. The occurrence of such an event implies that correlations between steps do exist. Disordered media in which step correlations are induced by their geometry are commonly referred to as *quenched* systems, as opposed to *annealed* or memoryless systems. Quenched disorder in one-dimensional (1D) systems has been shown to deeply affect transport,[37,38] suggesting that intricate effects may also take place in two dimensions. In such situations, numerical simulations can provide a greater insight into the underlying physical processes compared to experiments, as they allow one to monitor both microscopic and macroscopic scales simultaneously and, hopefully, to understand how they relate to each other. In this section, we present Monte Carlo simulations of transport in 2D Lévy glasses with particular attention to the role of disorder correlations on superdiffusion.

The 2D Lévy glasses were constructed by placing a large set of nonoverlapping disks of diameter varying by orders of magnitude at random positions in a large square area. The disks were taken from 24 different categories with exponentially spaced diameters ranging from 10 to 14,000 (unit length) and a parameter $\beta = 1$, yielding $\alpha = 1$ (Equation 3.4.7). The sample size was determined from the choice of the disk filling fraction f. In practice, the disks were placed randomly and uniformly in the sample area, from the largest to the smallest, checking after each placement that disks did not overlap each other. This procedure, known as *random sequential addition*,[39] can become problematic for high disk filling fractions. In our case, we used $f = 70\%$, which made the sample realization relatively easy.

Monte Carlo simulations were implemented as follows. The random walks were set to start in the scattering regions only[38] within a square area smaller than the sample area in such a way that the random walkers would probe the most possible local environments, yet without possibly reaching the sample boundaries. The displacement consisted in small steps of length $\delta_l = 1$ at each time step of the simulation and the behavior of each random walker was determined locally by the scattering probability p_s as:

$$p_s = \begin{cases} 0 & \text{in the voids} \\ 1 - e^{-\delta_l/\ell_s} & \text{in the scattering medium} \end{cases} \tag{3.4.9}$$

where ℓ_s is the scattering mean free path. A uniformly distributed random number $0 \leq y < 1$ was generated and isotropic scattering would occur for $y < p_s$. The number of random walkers used to simulate transport in Lévy glasses was set to 10^5, after having checked that convergence was reached.

The step-length distribution $p(\ell)$ in a Lévy glass was evaluated by recording the distance performed between successive scattering events and the scattering mean free path ℓ_s was set to be equal to the minimal disk diameter $\phi_0 = 10$. As shown in Figure 3.4.8a, the step-length distribution is found to decay as a power law with exponent –2.56, corresponding to $\alpha = 1.56$. The agreement with the distribution predicted by the chord-length model is fairly good. The deviations between these two can be explained by the fact that the chord-length model neglects the possibility for random walkers to not only cross more than one disk in only one step but also to perform more than one step in the scattering medium. As the area occupied by the scattering medium is significant ($1 - f = 30\%$), we expect these steps to play a certain role in transport. In addition, the step-length distribution in Monte Carlo simulations is evaluated dynamically, unlike that derived from the chord-length model, and thus possibly includes step correlations. The mean-square displacement $\langle r^2 \rangle$ evaluated from Monte Carlo simulations is shown in Figure 3.4.8b. The superdiffusive behavior is clear and corresponds to a Lévy walk with $\gamma = 1.44$, that is a consistent value $\alpha = 1.56$ according to Equation 3.4.3.

These simulations have therefore confirmed the fact that Lévy glasses do exhibit a superdiffusive behavior but also that the scattering medium can play a significant role in transport. Intuitively, an important parameter here should be the ratio of the average distance separating neighboring disks $\langle \Delta \rangle$ on the scattering mean free path ℓ_s, which expresses how difficult it is for a random walker to escape from its local environment. An average over 100 disks in the 2D Lévy glasses gave $\langle \Delta \rangle = 9.82$, which is close to the minimal disk diameter $\phi_0 = 10$.

To apprehend the specific role of quenched disorder, Monte Carlo simulations were performed on their annealed counterpart. The annealed systems were modeled using the

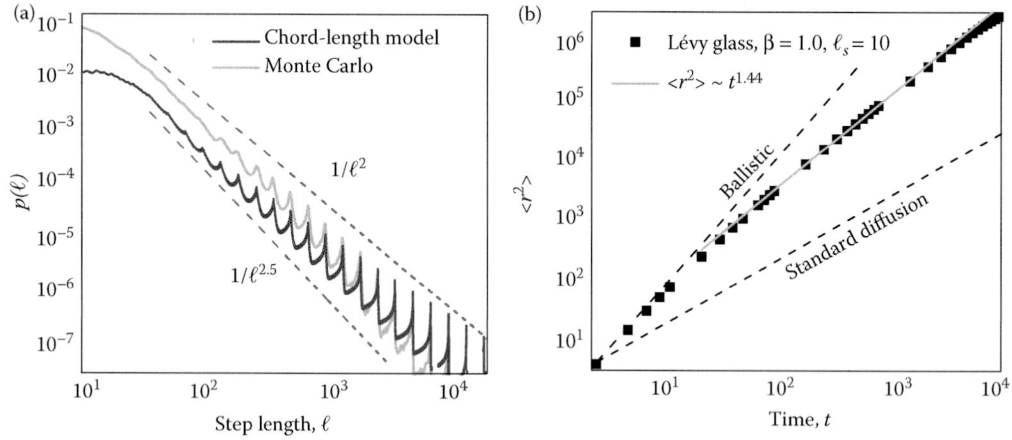

FIGURE 3.4.8
Superdiffusive transport in a two-dimensional Lévy glass ($\ell_s = 10$, $\beta = 1$). (a) Step-length distribution $p(\ell)$ obtained from Monte Carlo simulations (thin light curve) and from the chord-length model (thick dark curve). (b) Mean-square displacement $\langle r^2 \rangle$ calculated from Monte Carlo simulations (black squares) and from the standard Lévy walk model (solid gray line). The mean-square displacement is shown to increase faster than linearly with time, with an exponent $\gamma = 1.44$. (Reproduced from P. Barthelemy et al., Phys. Rev. E **82**, 011101, 2010. With permission.)

step-length distributions $p(\ell)$ retrieved from Monte Carlo simulations in the quenched Lévy glasses and the inversion method was used to generate the steps:[40] the length ℓ at each step was generated according to $\ell = F^{-1}(y)$, where $F(x)$ is the cumulative distribution function $F(x) = \int_0^x p(\ell)d\ell$ and $0 \le y < 1$ a uniformly distributed random number.

In Figure 3.4.9a, we compare the time evolution of the mean-square displacement in the quenched and annealed systems for two different values of ℓ_s. As expected, transport is found to depend strongly on the mean free path ℓ_s in the scattering medium. The difference between the annealed and quenched systems is relatively small for $\ell_s = 10$ and becomes significant for $\ell_s = 2$. This trend can be understood considering that the probability for a random walker to be backscattered into the disk it just left goes with the optical thickness of the scattering medium $\langle\Delta\rangle/\ell_s$. A random walker re-entering a void is likely to perform a step of length comparable to the previous one, thereby reducing its contribution to the mean-square displacement. In this sense, quenched disorder induces some *trapping* effects that force random walkers to stay in the vicinity of a given void for a certain amount of time. Such trapping is shown in Figure 3.4.9b. The random walker for $\langle\Delta\rangle/\ell_s = 5$ has a low

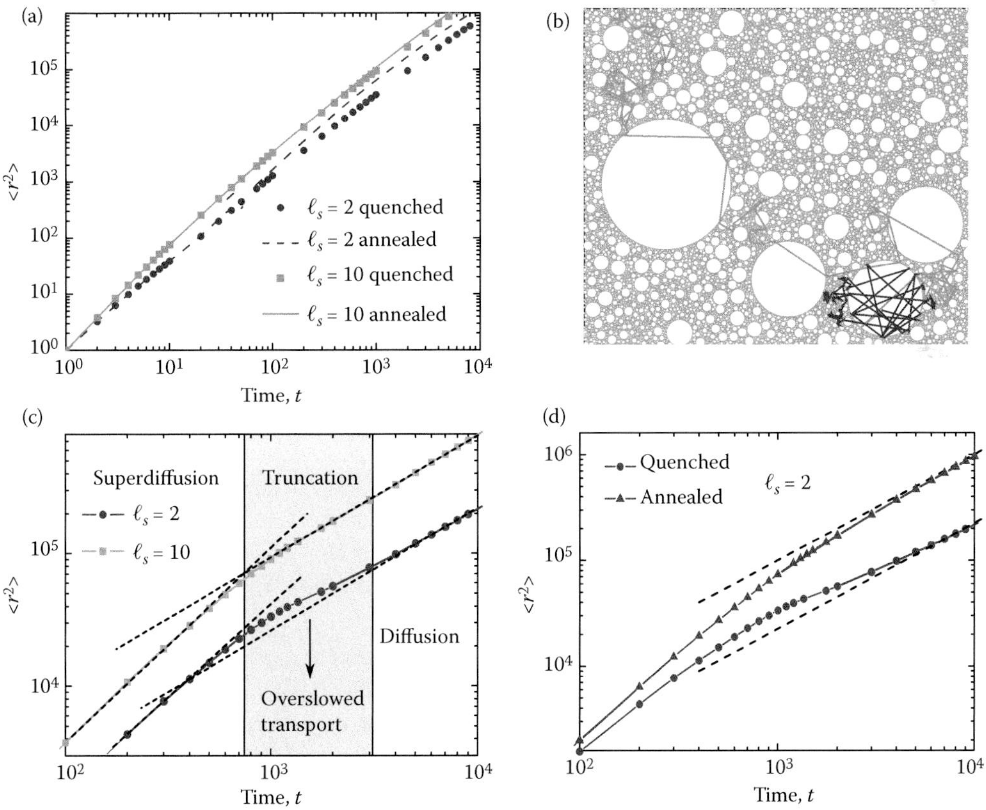

FIGURE 3.4.9
(**See color insert.**) Influence of quenched disorder on transport for large mean free path ($\ell_s = 10 \approx \langle\Delta\rangle$, green) and small mean free path ($\ell_s = 2 < \langle\Delta\rangle$, red). (a) Mean-square displacement versus time in the quenched Lévy glasses (dots/squares) and annealed systems (dashed/solid lines). (b) Typical trajectory of random walkers in two-dimensional (2D) Lévy glasses. (c) Mean-square displacement at the truncation time in 2D quenched Lévy glasses with maximum diameter $\phi_m = 1000$. (d) Comparison of quenched (dots) and annealed (triangles) systems for $\ell_s = 2$. (Reproduced from P. Barthelemy et al., Phys. Rev. E **82**, 011101, 2010. With permission.)

probability to cross the interdisk region without being backscattered, compared to the one for $\langle\Delta\rangle/\ell_s = 1$, which can easily explore the whole sample and is much less sensitive to quenching effects. For very small values of $\langle\Delta\rangle/\ell_s$, the role of quenched disorder becomes negligible and transport can be described by a Lévy walk with a very good approximation.[27]

Up to now, we have considered transport at times smaller than the truncation time of the system, that is the time needed to cross the largest disk of the sample $t_m = \phi_m/v$. Studying how the spread of random walkers evolves at the transition between the superdiffusive and diffusive regimes may however bring valuable information about transport in the system. We therefore set the maximum diameter to $\phi_m = 1000$ to observe this transition at times close to $t = 1000$. As shown in Figure 3.4.9c, the transition for $\ell_s \approx \langle\Delta\rangle$ is smooth and regular. The superdiffusive transport ($\gamma > 1$) gradually slows down to reach a diffusive behavior ($\gamma = 1$). Instead, for small mean free paths ($\ell_s < \langle\Delta\rangle$), a strong overslowed transport ($\gamma < 1$) is observed close to the truncation time. As shown in Figure 3.4.9d, a comparison between the quenched and annealed systems indicates that this transport regime is due to step correlations. Indeed, most of the random walkers that have performed the largest possible step at the truncation time are backscattered into the same void and forced to travel backward, the result being a strongly reduced mean-square displacement over a period of the order of the truncation time. Interestingly, this effect is also accompanied by a reduction of the diffusion constant, which gains importance as $\langle\Delta\rangle/\ell_s$ increases. The ratio between the annealed and quenched diffusion constants is about 5 when $\langle\Delta\rangle/\ell_s = 5$ and about 2 when $\langle\Delta\rangle/\ell_s = 1$. This large reduction of the asymptotic diffusion constant therefore allows one to evaluate the importance of quenched disorder on transport.

In summary, we confirmed by means of extensive Monte Carlo simulations that transport in 2D Lévy glasses is indeed superdiffusive. In addition, we found that quenched disorder can influence superdiffusive transport quite significantly, demonstrating in this way the relevance of microscopic phenomena in the transport properties of Lévy glasses. Recent studies on similar Lévy-type systems corroborate this finding.[41,42] In the next section, we present a semianalytical theoretical approach to describe superdiffusion in finite media, including weak interference effects.

3.4.5 Fractional Diffusion in Finite Media

Superdiffusive processes have been identified in a large and growing number of physical systems, motivating the scientific community to develop solid theoretical models to describe them. The model of Lévy flights has gained in popularity over the years because it gives an intuitive and rather simple picture of superdiffusion through the study of how Lévy random walkers spread in time. The generalized Central Limit Theorem makes it possible to treat all random walk processes for which $p(\ell) \sim \ell^{-(\alpha+1)}$ with $0 < \alpha \leq 2$ macroscopically by a diffusion-like equation generalized to fractional order derivatives:[43,44]

$$\partial_t C(\mathbf{r},t) = D_\alpha \nabla^\alpha C(\mathbf{r},t) \tag{3.4.10}$$

where $C(\mathbf{r},t)$ is the concentration of random walkers at position $\mathbf{r}$ and time t, D_α is a generalization of the diffusion constant, and ∇^α is the symmetric Riesz fractional derivative with respect to spatial coordinates, also called fractional Laplacian, which includes the superdiffusive behavior. Equation 3.4.10 naturally becomes the standard diffusion equation for $\alpha = 2$.

Although the fractional diffusion equation can be solved easily in Fourier space for systems with translational invariance,[9] the introduction of boundaries is known to make the derivation more complicated. In fact, the fractional Laplacian has the property of being spatially nonlocal, meaning that long-range correlations exist, which in turn make the Fourier transform have very limited use. This spatial nonlocality makes the definition of boundary conditions unclear[45] and, thus, the description of quantities observable on real superdiffusive systems particularly difficult.[46] The superdiffusive propagator in noninfinite media is actually known only in a few particular cases.[47]

The problem of nonlocality can now be circumvented by a proper discretization of the fractional Laplacian.[48] When applied to the function C, the fractional Laplacian ∇^α should be replaced by an $M \times M$ matrix that converges to the continuum operator when M goes to infinity. In simpler terms, the continuous time random walk should be replaced by discrete hops on a lattice: at each time interval, the concentration of random walkers is redistributed as $C_j(t_{n+1}) = \Sigma_i \omega_{|i-j|}\, C_j(t_n)$, where $\omega_{|i-j|}$ is the probability to perform a jump from site i to site j, yielding:[49]

$$\frac{C_j(t_{n+1}) - C_j(t_n)}{\tau} = \frac{1}{\tau}\sum_{i=1}^{M}\left(\omega_{|i-j|} - \delta_{i,j}\right)C_j(t_n) \tag{3.4.11}$$

where $\delta_{i,j}$ is the Kronecker delta. Taking the limit $\tau \to 0$, the left- and right-hand sides of this equation become $\partial_t C_j(t)$ and $D_\alpha \nabla^\alpha C_j(t)$, respectively. Writing Ω as the matrix of transition probabilities, we therefore have:

$$D_\alpha \nabla^\alpha = \lim_{\tau \to 0}\frac{1}{\tau}(\Omega - 1) \tag{3.4.12}$$

Any microscopic redistribution process could be used at this point provided that it leads to a superdiffusive process, the exact choice of the transition probability matrix being mainly a matter of convergence. The expression $\omega_{|i-j|} \propto |i-j|^{-(\alpha+1)}$, for instance, is known to suffer from a poor convergence as $\alpha \to 2$.[36] A better convergence has been observed when using a direct discretization of the fractional Laplacian,[50] leading to:

$$\omega_{|i-j|} - \delta_{i,j} = -\frac{\Gamma\left(-\frac{\alpha}{2} + |i-j|\right)\Gamma\left(1 + \frac{\alpha}{2}\right)}{\Gamma\left(1 + \frac{\alpha}{2} + |i-j|\right)\Gamma\left(-\frac{\alpha}{2}\right)} \tag{3.4.13}$$

where Γ is the Euler gamma function.

This approach is particularly suitable to the study of real systems. Setting $\omega_{|i-j|} = 0$ when j is outside a given interval $[0,L]$ is equivalent to the situation in which walkers stepping out of the interval cannot ever re-enter it. Thus, reducing the infinite size matrix Ω to an $M \times M$ matrix comes to defining a finite system with absorbing boundaries.[50] More generally, several features of real systems can be implemented simply by modifying the transition probabilities $\omega_{|i-j|}$. For instance, truncated step-length distributions, which are unavoidable in finite systems, can be defined by setting to zero all matrix elements that admit $|i-j| \geq \ell_m$ and properly renormalizing Ω. Similarly, partially reflecting boundaries can be defined by mapping all matrix elements that correspond to steps overpassing a reflecting boundary to their mirror image.[28]

Now, it is important to note that a physical model for light superdiffusion should rely on Lévy walks rather than Lévy flights as light propagates at a finite velocity. The resulting spatiotemporal coupling[51] makes the analytical description of Lévy walks rather complicated.[52] In the steady-state regime, however, the time required for light to make a step plays no role such that light superdiffusion can be described by Equation 3.4.10. The propagator $f(x,x_0)$ for a continuous point-source at position x_0 in a 1D superdiffusive medium is then given by the following time-independent fractional differential equation:

$$D_\alpha \nabla^\alpha f(x, x_0) = -\delta(x - x_0) \tag{3.4.14}$$

Knowing the eigenvalues λ_i and eigenfunctions Ψ_i of the matrix $(\Omega - 1)$, Green's function f can be determined by eigenfunction expansion, writing it as $f(x) = \sum_i a_i \psi_i$. After substitution into Equation 3.4.14 and having $\psi_j(x)\psi_i(x) = \delta_{i,j}$, we find that $a_i = -\psi_i(x_0)(D_\alpha\lambda_i)^{-1}$, yielding:

$$f(x, x_0) = -\sum_{i=1}^{M} \frac{\psi_i(x)\psi_i(x_0)}{D_\alpha \lambda_i} \tag{3.4.15}$$

The continuum limit is obtained by rescaling the eigenvalues as $\lambda_i \rightarrow \lambda_i(M/L)^\alpha$. The normalized Green's function f, shown in Figure 3.4.10a, is found to depend strongly on the value of α. Whereas the triangular shape characteristic of the diffusive regime is found for $\alpha = 2$, the profile of f becomes more and more cusped as α is decreased.

Studying the optical features of real superdiffusive materials, such as the output intensity profile in Lévy glasses (see Section 3.4.3), now requires one to extend this approach to higher dimensions. Considering the simplest case of a slab geometry oriented such that its interface is normal to the x-axis, the system becomes translationally invariant in both y and z directions. The 3D counterpart of Equation 3.4.14 can then be written in terms of the Fourier transform of $f(\mathbf{r},\mathbf{r}_0)$ in the yz plane as:

$$D_\alpha \left(\nabla_x^\alpha - k_\perp^\alpha\right) f\left(x, x_0, \mathbf{k}_\perp\right) = -\delta(x - x_0) \tag{3.4.16}$$

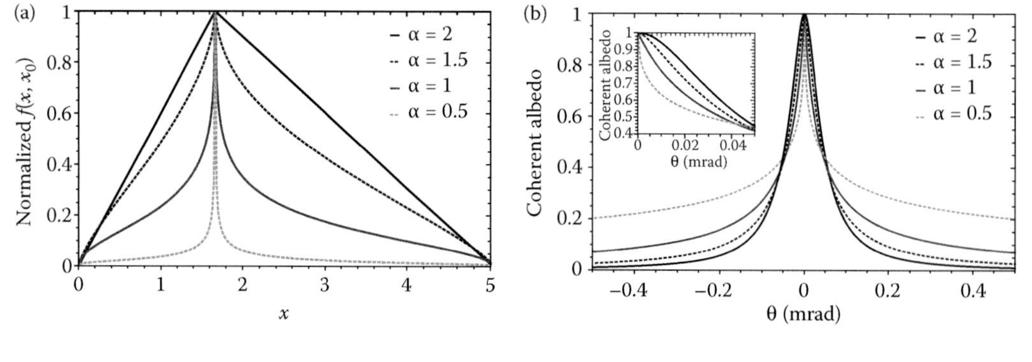

FIGURE 3.4.10
(a) Normalized Green's function $f(x, x_0)$ for a slab with $L = 5$, $x_0 = L/3$, $M = 300$, $D_\alpha = 1$ and absorbing boundaries, for various values of α. (b) Normalized coherent albedo for a slab with $L = 5$, $M = 300$ and $D_\alpha = 1$, for various values of α. (Reproduced from J. Bertolotti, K. Vynck, and D. S. Wiersma, Phys. Rev. Lett. **105**, 163902, 2010. With permission.)

yielding:

$$f(x,x_0,\mathbf{k}_\perp) = -\sum_{i=1}^{M} \frac{\psi_i(x_0)\psi_i(x)}{D_\alpha\left(\lambda_i - k_\perp^\alpha\right)} \tag{3.4.17}$$

where $k_\perp = |\mathbf{k}_\perp|$. The 3D Green's function of the superdiffusive system can be obtained by an inverse Fourier transform of Equation 3.4.17 and the full intensity profile in the medium on integration over a suitable source.

Beyond the derivation of the intensity distribution in a system, Green's function also makes it possible to study fundamental interference effects such as speckles and weak localization.[2] Interferences in the multiple scattering regime are known to play a particularly important role in transport. The weak localization effect, often described as a precursor for Anderson localization, arises from interferences between counterpropagating waves and is expected to hold as long as the system is reciprocal. In light backscattering experiments, these counterpropagating paths give rise to a narrow cone of enhanced albedo known as the coherent backscattering cone. Its peculiar triangular cusp is formed by the longer paths in the system. As in superdiffusive media there is no *a priori* reason for reciprocity to break down, we expect longer paths to contribute more than in standard diffusive media, and thus expect a sharper peak.

From a microscopic point of view, the disorder averaged intensity I observed at a point $\mathbf{R}$ outside a multiple scattering medium can be written as:[2]

$$I(\mathbf{R}) = \frac{4\pi}{c}\int d\mathbf{r}_1\, d\mathbf{r}_2\, d\mathbf{r}_3\, d\mathbf{r}_4\, \Psi(\mathbf{r}_1)\Psi^*(\mathbf{r}_2) f(\mathbf{r}_1,\mathbf{r}_2,\mathbf{r}_3,\mathbf{r}_4) G(\mathbf{r}_3,\mathbf{R}) G^*(\mathbf{R},\mathbf{r}_4) \tag{3.4.18}$$

where ψ is the coherent propagator for the amplitude from outside the sample to the first scattering event, G is the averaged propagator for the amplitude from the last scattering event to the point of observation $\mathbf{R}$, and $f(\mathbf{r}_1,\mathbf{r}_2,\mathbf{r}_3,\mathbf{r}_4)$ is the four-vertex propagator that contains all information about transport. Neglecting recurrent scattering, only the ladder term, which describes the incoherent transport, and the most-crossed term, which corresponds to counterpropagating paths in the system, contribute to f.

The coherent component of the albedo A can be calculated from Equation 3.4.18 by setting $\mathbf{r}_1 = \mathbf{r}_4$ and $\mathbf{r}_2 = \mathbf{r}_3$. Considering a planewave at normal incidence on the slab interface, using the Fraunhofer approximation for Green's functions and assuming that the step distribution follows a power law, we can write:

$$\Psi(\mathbf{r}_1) = \sqrt{\frac{cI_0}{4\pi}}\, x_1^{-(\alpha+1)/2} e^{i\mathbf{k}_\mathrm{i}\cdot\mathbf{r}_1}$$

$$\Psi^*(\mathbf{r}_2) = \sqrt{\frac{cI_0}{4\pi}}\, x_2^{-(\alpha+1)/2} e^{-i\mathbf{k}_\mathrm{i}\cdot\mathbf{r}_2}$$

$$G(\mathbf{r}_2,R) \simeq \left(\frac{x_2}{\cos\theta}\right)^{-(\alpha+1)/2} e^{-i\mathbf{k}_\mathrm{e}\cdot\mathbf{r}_2}\,\frac{e^{ikR}}{4\pi R}$$

$$G^*(R,\mathbf{r}_1) \simeq \left(\frac{x_1}{\cos\theta}\right)^{-(\alpha+1)/2} e^{i\mathbf{k}_\mathrm{e}\cdot\mathbf{r}_1}\,\frac{e^{-ikR}}{4\pi R}$$

where I_0 is the incident intensity, θ is the angle with respect to the normal of the slab, $\mathbf{k}_i$ and $\mathbf{k}_e$ the wavevectors of the incident and emergent planewaves, respectively, and $\mathbf{k}_\perp = (\mathbf{k}_i + \mathbf{k}_e)_\perp$. After substitution in Equation 3.4.18, Fourier transform, and using the discretized approximation in Equation 3.4.17 for the propagator $f(x_1,x_2,\mathbf{k}_\perp)$, we arrive at the following expression for the coherent albedo from a superdiffusive slab:

$$A(\theta) \propto -\sum_{x_1,x_2} \left(\frac{x_1 x_2}{\cos\theta}\right)^{-(\alpha+1)} \sum_{i=1}^{M} \frac{\psi_i(x_1)\psi_i(x_2)}{D_\alpha\left(\lambda_i - |k\sin\theta|^\alpha\right)}. \tag{3.4.19}$$

The normalized profile of the coherent backscattering cone is shown in Figure 3.4.10b for different values of α. As expected, the longer steps, which gain importance with the decrease of α, make the profile of the cone more cusped. It is interesting to note that a similar effect has been predicted for fractal media,[53] considering that Lévy glasses can be assimilated to random fractals on a certain range of scales.

These results suggest that the semianalytical formulation of Green's function presented in this section constitutes a suitable approach to investigate multiple scattering effects in superdiffusive media.

3.4.6 Concluding Remarks

This chapter was concerned with a fundamental transport process known as *superdiffusion*. Lying at the basis of the most varied systems, from animal foraging to the spread of diseases to transport in turbulent flows, superdiffusive processes in general can hardly be studied in a systematic way and are often blurred by additional complex effects.

After introducing superdiffusion through the concepts of Lévy motion, we presented a recipe to fabricate disordered optical materials, dubbed Lévy glasses, in which light transport is superdiffusive. Lévy glasses can be made with ease by embedding transparent spheres with diameters varying by orders of magnitude in a diffusive medium. A chord-length model showed that materials with an arbitrary degree of superdiffusivity α can be made by an appropriate choice of the distribution of sphere diameters. Lévy glasses therefore offer the opportunity to investigate superdiffusive transport processes in a laboratory with light.

Toward this aim, we proposed two techniques to optically characterize Lévy glasses, looking in particular at deviations from a diffusive behavior. The total transmission through Lévy glasses was shown to follow a generalized Ohm's law that depends explicitly on α and the output intensity profile in response to a point source was found to exhibit a sharp cusp on its top, the exact shape of the profile strongly varying with α. The agreement between the experimental and theoretical/numerical results was good, thereby confirming that light in Lévy glasses propagates superdiffusively. Let us note that time-resolved experiments may be a fruitful approach for future investigations as they are likely to give additional information on the superdiffusive process. Eventually, the elaboration of proper optical characterization techniques could provide a way to recover the microscopic transport properties of arbitrary highly inhomogeneous materials.

To gain insight into the intimate relation between the microscopic and macroscopic properties of Lévy glasses, we performed extensive Monte Carlo simulations. We found in particular that the diffusive medium surrounding the voids can play an important role in transport in Lévy glasses. The horizon of possible steps for a random walker depends in a critical way on its exact position in the sample. Monte Carlo simulations showed that quenched disorder could induce some trapping effects that slow down superdiffusion and lead to a transient subdiffusive-like regime at the truncation time of the system. The observation of these effects is expected to motivate further investigations on the theoretical and experimental levels.

In the last part of this chapter, we took a first step toward the theoretical description of light transport in superdiffusive media of finite size. Green's function of the system was derived using a discretized version of the fractional Laplacian, its knowledge giving access to several measurable quantities (e.g., total transmission, output intensity profile). The implementation of complex boundary conditions (e.g., partially reflecting) and a truncation in the step-length distribution is possible, making this approach particularly relevant for the study of real systems. Furthermore, Green's function allows for the study of weak interference effects in the *superdiffusion* approximation. As an example, we calculated the coherent backscattering from superdiffusive media and found that its profile strongly depends on α.

All in all, Lévy glasses provide a great opportunity to study superdiffusive processes in a controlled way. The results presented in this chapter give a flavor of the possible outcomes of such investigations for fundamental and applied sciences. Significant efforts are still required to fully understand superdiffusive transport in Lévy glasses. Yet, by bringing together the fields of anomalous transport and mesoscopic physics, Lévy glasses open the space for exploring novel effects and functionalities, such as random lasing and Anderson localization in superdiffusive media.

Acknowledgments

We would like to acknowledge Stefano Lepri, Roberto Livi, Rafaella Burioni, Alessandro Vezzani, Joseph Klafter, Lorenzo Pattelli, and the entire group Optics of Complex Systems at LENS for fruitful discussions. This work was supported in part by the Italian FIRB project "Anomalous Transport of Light in Complex Systems," the European Network of Excellence "Nanophotonics for Energy Efficiency," and ENI S.p.A Novara.

References

1. P. Sheng, *Introduction to Wave Scattering, Localization and Mesoscopic Phenomena* (Springer, Berlin/Heidelberg, 2006), 2nd ed.
2. E. Akkermans and G. Montambaux, *Mesoscopic Physics of Electrons and Photons* (Cambridge University Press, Cambridge, 2007).
3. S. Feng and P. A. Lee, Science **251**, 633 (1991).
4. F. Scheffold and G. Maret, Phys. Rev. Lett. **81**, 5800 (1998).

5. Y. Kuga and A. Ishimaru, J. Opt. Soc. Am. A **1**, 831 (1984).
6. P. W. Anderson, Phys. Rev. **109**, 1492 (1958).
7. O. Kallenberg, *Foundations of Modern Probability* (Springer-Verlag, New York, 1997), 2nd ed.
8. J. Bouchaud and A.Georges, Phys. Rep. **195**, 127 (1990); R. Metzler and J. Klafter, J. Phys. A: Math. Gen. **37**, R161 (2004); R. Klages, G. Radons, and I. M. Sokolov, *Anomalous Transport* (Wiley-VHC, Weinheim, 2008).
9. R. Metzler and J. Klafter, Phys. Rep. **339**, 1 (2000).
10. P. Lévy, *Théorie de l'addition des Variables Aléatoires* (Gauthier-Villars, Paris, 1937).
11. J. P. Nolan, *Stable Distributions: Models for Heavy Tailed Data* (Birkhauser, Boston, MA, 2011), in progress, Chapter 1 online at academic2.american.edu/~jpnolan.
12. B. V. Gnedenko and A. N. Kolmogorov, *Limit Distributions for Sums of Independent Random Variables* (Addison-Wesley, Boston, MA, 1954).
13. B. Mandelbrot, *The Fractal Geometry of Nature* (V.H. Freeman & Co., San Francisco, 1977).
14. B. Mandelbrot, J. Bus. **36**, 394 (1963).
15. R. N. Mantegna and H.E. Stanley, Nature **376**, 46 (1995); S. Ghashghaie, W. Breymann, J. Peinke, P. Talkner, and Y. Dodge, *ibid.* **381**, 767 (1996).
16. M. Shlesinger, B. West, and J. Klafter, Phys. Rev. Lett. **58**, 1100 (1987).
17. G. Zumofen and J. Klafter, Phys. Rev. E **47**, 851 (1993).
18. R. N. Mantegna and H. E. Stanley, Phys. Rev. Lett. **73**, 2946 (1994).
19. C. Tsallis, S. V. F. Lévy, A. M. C. Souza, and R. Maynard, Phys. Rev. Lett. **75**, 3589 (1995).
20. D. Brockmann, L. Hufnagel, and T. Geisel, Nature **439**, 462 (2006).
21. A. Caspi, R. Granek, and M. Elbaum, Phys. Rev. E **66**(1), 011916 (2002).
22. F. Bartumeus, M. G. E. da Luz, G. M. Viswanathan, and J. Catalan, Ecology **86**, 3078 (2005); G. M. Viswanathan, E. P. Raposo, and M. da Luz, Phys. Life Rev. **5**, 133 (2008).
23. T. H. Solomon, E. R. Weeks, and H. L. Swinney, Phys. Rev. Lett. **71**, 3975 (1993).
24. N. Mercadier, W. Guerin, M. Chrevrollier, and R. Kaiser, Nat. Phys. **5**, 602 (2009).
25. P. Barthelemy, J. Bertolotti, and D. S. Wiersma, Nature **453**, 495 (2008).
26. J. Bertolotti, K. Vynck, L. Pattelli, P. Barthelemy, S. Lepri, and D.S. Wiersma, Adv. Funct. Mater. **20**, 965 (2010).
27. P. Barthelemy, J. Bertolotti, K. Vynck, S. Lepri, and D. S. Wiersma, Phys. Rev. E **82**, 011101 (2010).
28. J. Bertolotti, K. Vynck, and D. S. Wiersma, Phys. Rev. Lett. **105**, 163902 (2010).
29. D. S. Wiersma, P. Bartolini, A. Lagendijk, and R. Righini, Nature **390**, 671 (1997); F. J. Schuurmans, D. Vanmaekelbergh, J. vande Lagemaat, and A. Lagendijk, Science **284**, 141 (1999); M. Störzer, P. Gross, C. M. Aegerter, and G. Maret, Phys. Rev. Lett. **96**, 063904 (2006).
30. P. D. García, R. Sapienza, A. Blanco, and C. López, Adv. Mater. **19**, 2597 (2007).
31. S. Gottardo, R. Sapienza, P. D. García, A. Blanco, D. S. Wiersma, and C. López, Nat. Photon. **2**, 429 (2008).
32. D. ben-Avraham and S. Havlin, *Diffusion and Reactions in Fractals and Disordered Systems* (Cambridge University Press, Cambridge, 2000).
33. A. Mazzolo, J. Math. Phys. **44**, 6195 (2003).
34. M. P. van Albada and A. Lagendijk, Phys. Rev. Lett. **55**, 2692 (1985); P.-E. Wolf and G. Maret, *ibid.* **55**, 2696 (1985).
35. A. Davis and A. Marshak, in *Fractal Frontiers*, edited by M. M. Novak and T. G. Dewey (World Scientific, Singapore, 1997), pp. 63–72.
36. S. Buldyrev, S. Havlin, A. Kazakov, M. da Luz, E. Raposo, H. Stanley, and G.Viswanathan, Phys. Rev. E **64**, 041108 (2001).
37. R. Kutner and P. Maass, J. Phys. A **31**, 2603 (1998); C. W. J. Beenakker, C. W. Groth, and A. R. Akhmerov, Phys. Rev. B **79**, 024204 (2009).
38. R. Burioni, L. Caniparoli, S. Lepri, and A. Vezzani, Phys. Rev. E **81**, 011127 (2010).
39. S. Torquato, *Random Heterogeneous Materials: Microstructure and Macroscopic Properties* (Springer Science + Business Media, Inc., New York, 2001).
40. A. Janicki and A. Weron, *Simulation and Chaotic Behavior of α-Stable Stochastic Processes* (Marcel Dekker, Inc., New York, 1994).

41. P. Buonsante, R. Burioni, and A. Vezzani, Phys. Rev. E **84**, 021105 (2011).
42. C. W. Groth, A. R. Akhmerov, and C. W. J. Beenakker, Phys. Rev. E **85**, 021138 (2012).
43. A. S. Chaves, Phys. Lett. A **239**, 13 (1998).
44. V. V. Yanovsky, A. V. Chechkin, D. Schertzer, and A. V. Tur, Physica A **282**, 13 (2000).
45. A. V. Chechkin, R. Metzler, V. Y. Gonchar, J. Klafter, and L. V. Tanatarov, J. Phys. A **36**, L537 (2003).
46. G. Zumofen and J. Klafter, Phys. Rev. E **51**, 2805 (1995); P. M. Drysdale and P. A. Robinson, *ibid.* **58**, 5382 (1998).
47. N. Krepysheva, L. Di Pietro, and M.-C. Néel, Phys. Rev. E **73**, 021104 (2006).
48. I. Podlubny, A. Chechkin, T. Skovranek, Y. Chen, and B. M. VinagreJara, J. Comput. Phys. **228**, 3137 (2009).
49. R. Gorenflo, G. De Fabritiis, and F. Mainardi, Physica A **269**, 79 (1999).
50. A. Zoia, A. Rosso, and M. Kardar, Phys. Rev. E **76**, 021116 (2007).
51. I. M. Sokolov and R. Metzler, Phys. Rev. E **67**, 010101(R) (2003).
52. M. Meerschaert, D. Benson, H. Scheffler, and P. Becker-Kern, Phys. Rev. E **66**, 060102 (2002).
53. E. Akkermans, P.-E. Wolf, R. Maynard, and G. Maret, J. Phys. (Paris) **49**, 77 (1988).

3.5

Optical Properties of Low-Contrast Opal-Based Photonic Crystals

Alexander A. Kaplyanskii
Ioffe Physical-Technical Institute of the Russian Academy of Sciences

Alexander V. Baryshev
Ioffe Physical-Technical Institute of the Russian Academy of Sciences
Toyohashi University of Technology

Mikhail V. Rybin
Ioffe Physical-Technical Institute of the Russian Academy of Sciences
National Research University of Information Technologies, Mechanics and Optics

Alexander V. Sel'kin
Ioffe Physical-Technical Institute of the Russian Academy of Sciences

Mikhail F. Limonov
Ioffe Physical-Technical Institute of the Russian Academy of Sciences
National Research University of Information Technologies, Mechanics and Optics

CONTENTS

3.5.1 Introduction: Opals—Brilliant Objects for Optical Studies

In 1995, a research team from the Ioffe Physical-Technical Institute at the Russian Academy of Sciences (St. Petersburg) published a paper entitled "Optical spectroscopy of opal matrices with CdS embedded in its pores: quantum confinement and photonic band gap effects."[1] In this pioneer work, it has been demonstrated that synthetic opals are in fact three-dimensional (3D) photonic crystals (PhCs). This work has formed a basis for a series of technological, structural, and optical studies carried out at the Ioffe Physical-Technical Institute in recent years.[2–14] The publication[1] has initiated intense investigation of the photonic properties of opals, opal-like PhCs, and inverted structures in many research centers all over the world.[15–23] In this book, 10 chapters, in part or wholly, are devoted to opal-like PhCs.

In the title of this section, opals are referred to as "brilliant objects for optical studies." The wordplay "brilliant" at once comes to mind because the structural parameters and physical properties of opals make these gemstones not only beautiful decorations but also perfect objects for optical studies. Some opal properties exhibiting the most promise *for the study of disorder effects* in PhCs are as follows.

1. Opals are built up of quasispherical particles of amorphous silica (a-SiO_2), each particle having a rather hard coat and porous nucleus. The amorphous structure of the a-SiO_2 particles leads to two important properties that are the focus of our attention in this chapter. Porosity of the a-SiO_2 particles produces (i) *inhomogeneity of dielectric permittivity* of a single a-SiO_2 particle and (ii) *dissimilarities in size and permittivity* of a-SiO_2 particles in ensemble forming a bulk sample. As a result opals always have, strictly speaking, a disordered structure that can be described as face-centered cubic (FCC) only as a rough approximation. The two structural properties of a-SiO_2 particles govern distinct and very prominent optical effects in opal spectra, among them the photonic band selective immersion and Fano resonance (see Sections 3.5.5 and 3.5.6).[8,10,11,24]
2. The real structure of opals is far short of an ideal FCC lattice not only because of the amorphous structure of the a-SiO_2 particles. It is characterized by a great variety of *defects, including vacancies and dislocations*. In the majority of the samples, the FCC lattice *twinning* initiated by randomly packed (111) layers along the [111] growth axis[13] is observed, without which it is impossible to explain a number of experimental results.[4,6,7,25]
3. Various disordered photonic structures assembled from *essentially different sizes* of a-SiO_2 particles can be fabricated.[26]

Being a model object for an experimental study of disorder in PhCs, opals possess a collection of specific properties:

1. Opals are highly transparent throughout the visible and infrared ranges.
2. The a-SiO_2 particles can be varied in size over a wide range from 0.2 to 1 μm,[27] which permits PhCs to be synthesized that possess photonic stop-bands both in the visible and in the near-infrared ranges.
3. Opals have an overlapping net of air voids in between the a-SiO_2 particles, which allows one to infill the voids with various materials. Of particular interest is infilling opals with liquid dyes to study enhanced stimulated emission and lasing effects.[28] In this connection, synthesis of inverted opal-based structures and fabrication of 3D PhCs with rather high dielectric contrast is an important goal because such structures allow us to obtain a photonic complete band gap.[29,30]
4. The average dielectric constant of opals, as stated in the studies by Rybin et al.[10] and Iler,[31] is about 1.9–2.0. This value is eminently suitable for investigation of the basic phenomena appearing in optical spectra of 3D PhCs under conditions of a

low dielectric contrast. For example, making use of distilled water ($\varepsilon = 1.78$) and propylene glycol ($\varepsilon = 2.05$) taken in an appropriate proportion allows us to infill opals and selectively suppress photonic bands using the method of "immersion spectroscopy" described in this chapter.[8,10,11]

The main goal of this chapter is to describe the results of the studies of opal-based PhCs when the low dielectric contrast $\Delta\varepsilon$ occurs between the *a*-SiO_2 particles and the opal matrix filler. Let us define the dielectric contrast through the relation $\Delta\varepsilon = (\varepsilon_{max} - \varepsilon_{min})/\varepsilon_{av}$, where ε_{max} and ε_{min} are the maximum and minimum values, respectively, of the spatially periodic dielectric function, $\varepsilon(\mathbf{r})$, of a PhC medium, and ε_{av} is its volume averaged value. To clarify what we mean by low dielectric contrast, the $\Delta\varepsilon$ scale for some materials should be mentioned. As the lowest value of the dielectric contrast, one can assign a range of 10^{-6}–10^{-5}, which is characteristic of regular crystals when carrying out experiments on x-ray scattering.[32] In PhCs made of photorefractive chalcogenide glasses $\Delta\varepsilon \sim 0.5$.[33] On the opposite edge of the scale are semiconductor opal-based PhCs, whose dielectric contrast may run as high as ~2. We define low dielectric contrast PhCs as those with $\Delta\varepsilon < 0.1$.

In the low dielectric contrast regime, it is *possible to use the Bragg approximation in description of photonic band structure in a wide range of propagation vectors and energies*, including photonic bands corresponding to high-index (*hkl*) crystallographic planes of the opal FCC lattice. In the case when the band overlapping is negligible, each photonic Bragg band can be marked by the (*hkl*) indices related to a certain plane system. In the framework of the Bragg approximation, we are able to explain the peculiarities of transmission spectra presented in this chapter and the diffraction pattern discussed in Chapter 3.6.

By *photonic stop-bands* we usually mean energy bands with practically *zero light transmission* in certain directions in the crystal lattice. However, such stop-bands may not be observed experimentally because of a low dielectric contrast, finite sample dimensions, structural defects, and other factors. For this reason, depending on the parameters of the observable dips in transmission spectra and for clarity of discussion, two terms will be used in this chapter: *photonic stop-band* (or simply *stop-band*) and *photonic Bragg band* (or simply *Bragg band*).

3.5.2 The *a*-SiO_2 Particles in Opals: Inherent Inhomogeneity and Dissimilarity

Opals are the crystalline-like structures made up of nearly spherical particles of amorphous silica *a*-SiO_2. These particles tend to form by self-assembling a close-packed structure with a symmetry, to a first approximation, of an FCC lattice.[2,9,22]

The bulk opal samples that one usually prepares using the sedimentation technique[34] are then postannealed at a high temperature for sintering. Annealing gives rise to pronounced variations in the permittivity values[35] that can be ascribed to the changes in the inherent structure of the spheres. A complete densification of spheres does not occur even at 600°C, so their refractive index is consistently lower than that of fused quartz.

To study the inherent structure of the individual *a*-SiO_2 particles and the opals they form, scanning electron microscopy (SEM) and transmission electron microscopy (TEM) are utilized. A high resolution TEM image of the spheres reveals their smooth surface and amorphous microstructure, as shown in Figure 3.5.1. Nano-sized pores inside the silica spheres were observed in all samples under investigation,[11] supporting the results of previous

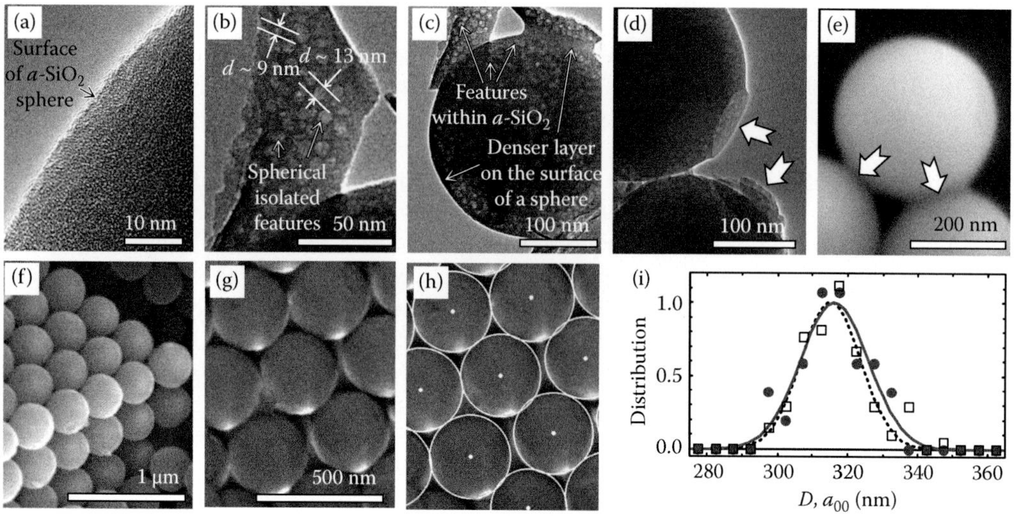

FIGURE 3.5.1
(a)–(e) Transmission electron microscopy images of a-SiO$_2$ particles. (f), (g) Scanning electron microscopy images of opals. (h) Image (g) of the (111) opal plane processed with the approach based on the Hough-type procedure.[37,38] The circles and their centers identify the size and position of a-SiO$_2$ particles. (i) The distribution of the circle diameters D (black circles and solid line) and the distances between circle centers a_{00} (squares and dotted line) obtained by processing a data array of ~180 a-SiO$_2$ particles, whose fragment is shown in (g), (h). The images were obtained by G. Yushin, Georgia Institute of Technology, USA. (Panels (a)–(c), (e), and (f) adapted from M. V. Rybin et al. Phys. Rev. B **77**, 205106 (2008). Panels (g)–(i) adapted from M. V. Rybin et al. Phys. Rev. Lett. **103**, 023901, 2009.)

studies.[36] These pores appear as lighter areas in the micrographs (Figure 3.5.1b and c) as a result of the smaller projected potential and, hence, the weaker scattering of electrons as they travel through the pore-containing silica layer. The sphere-to-sphere sintering zones (see arrows in Figure 3.5.1d and e) as well as the 10–30 nm near-surface layer do not contain pores and are therefore expected to have a higher density with respect to the particle center. This variation in optical density across the spheres indicates that the a-SiO$_2$ particles are *inhomogeneous* and as-produced synthetic opals have a *multicomponent structure* (see Chapter 2.7).

To obtain valuable information of the inherent *dissimilarity* of the a-SiO$_2$ particles in size and *statistical characteristics of the whole ensemble of particles* in the investigated samples, SEM micrographs were analyzed[11,37] using an original procedure based on the Hough transform.[38] From the micrograph shown in Figure 3.5.1h, several important structural parameters of the (111) growth layer were found, such as the average diameter of the a-SiO$_2$ particles, $\bar{D} = 316\,\text{nm}$, and full-width at half-maximum of its normal distribution, $\Delta D = 23$ nm. The latter corresponds to a deviation in the diameter ΔD equal to ~7%. The average center-to-center distance $\bar{a}_{00}$ between particles was found to be 315 nm, whereas $\bar{D} > \bar{a}_{00}$ because of sintering of neighboring a-SiO$_2$ spheres.

Our goal is to provide a comprehensive description of optical spectra of opal-based PhCs possessing low dielectric contrast. Unambiguous interpretation of the effects observed in optical spectra can be obtained only on the bases of extended structural information. As follows from electron microscopy data, the structure of real opals is always imperfect because of the a-SiO$_2$ particles being *dissimilar in size and permittivity* and *inherently inhomogeneous in dielectric properties.*[3,8,10,11,31,37]

Figure 3.5.2 demonstrates the relation between structural peculiarities of opals and spectroscopic effects to be analyzed in this chapter.

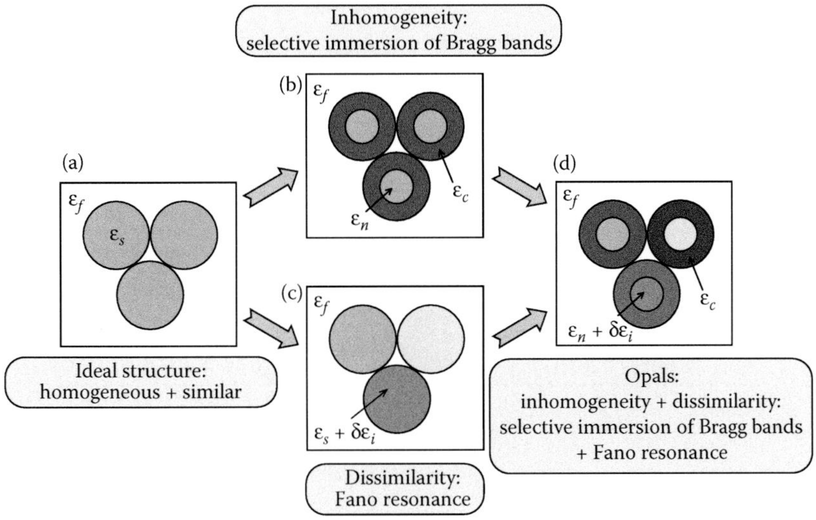

FIGURE 3.5.2
Schematic illustration of ordered and disordered opal-like structures. (a) Ordered two-component structure: *similar homogeneous* spheres. (b) Ordered three-component structure: *similar inhomogeneous* spheres made up of a homogeneous nucleus and a homogeneous coat. (c) Disordered structure made up of *homogeneous* but dielectrically *dissimilar* spheres. (d) Disordered structure approaching most closely the real *a*-SiO_2 opal structure, made up of both *inhomogeneous* and *dissimilar* spheres.

Inhomogeneity of the *a*-SiO_2 particles leads to the concept of multicomponent photonic crystals (MPhCs) that was introduced in Chapter 2.7. The theory presented in this chapter holds: owing to dielectric inhomogeneity of particles, the splitting effect is expected to be observed in immersion dependencies of the different (*hkl*) photonic bands. In other words, different (*hkl*) photonic bands have to disappear in transmission spectra of opals at different values of a filler permittivity ε_f. These effects will be described in Section 3.5.5 and in Chapter 3.6.

Dissimilarity of the *a*-SiO_2 particles brings about another, rather surprising, effect. Because of dissimilarity in permittivity the transmission spectra demonstrate a wide-band background scattering that interferes with narrow Bragg transmission lines. As a result, the interference gives rise to an impressive transformation of a Bragg stop-band to a Bragg pass-band. This phenomenon will be analyzed in Section 3.5.6 in the framework of the Fano-resonance concept[39] as it is discussed in Chapters 2.1 and 4.6.

3.5.3 Mie Scattering in Opals

Before beginning a discussion of the optical properties of low-contrast opal-based PhCs, we outline the main features of light scattering by spherical or nearly spherical particles. During the interaction of light with particles, diffraction, refraction, reflection, and absorption result in light scattering patterns characteristic of the particle size that is conventionally described by a dimensionless *size parameter* D/λ for a light with wavelength λ incident on a spherical particle with diameter D.

If the size of a dielectric spherical scatterer is comparable to the wavelength of incident light, $D/\lambda \sim 1$, the scattering is in the so-called Mie regime. Mie scattering is not strongly

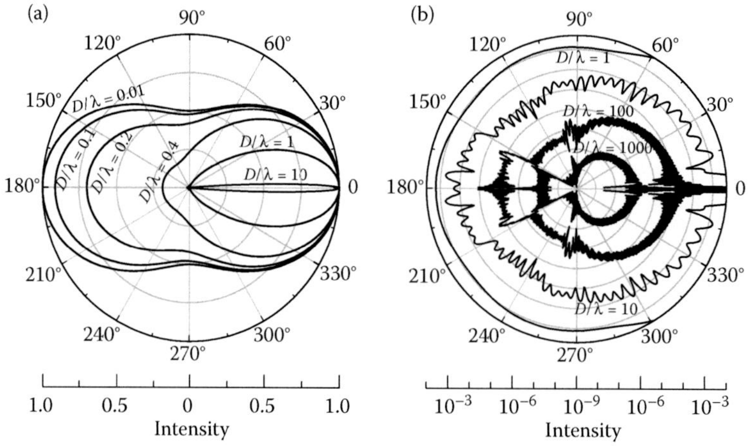

FIGURE 3.5.3
Scattering diagrams illustrating Rayleigh-to-Mie regime transition $D/\lambda = 0.01$–10 (a) and Mie-to-geometrical optics regime transitions $D/\lambda = 1$–1000 (b). Curves are angular dependence of intensity of unpolarized light scattered from a silica ($n = 1.46$) sphere in air.

wavelength-dependent, but exhibits an anisotropic scattering indicatrix with a sharper and more intense forward lobe (see Figure 3.5.3). Such a regime is described by the theory that gives an analytical solution of Maxwell's equations for the elastic scattering of electromagnetic waves from a sphere.[40,41] The Mie solution to the scattering problem is valid for all possible diameter-to-wavelength ratios, although the technique results in a numerical summation of infinite series. In particular, this theory gives solutions for two extreme situations: *Rayleigh regime* at $D/\lambda < 0.1$ and *optical regime* (Fraunhofer diffraction) at $D/\lambda > 100$. As for the Mie regime, it is also of interest because of its own resonances. Mie resonances of a sphere can be considered by analogy with Fabry–Pérot resonances of a plate with parallel surfaces.

The Rayleigh regime is realized if the particle size is considerably smaller than the wavelength of the incident light $D/\lambda < 0.1$ and Mie theory reduces to the Rayleigh approximation. For an unpolarized light source, the intensity of the scattered radiation in the Rayleigh regime is given by

$$I = I_0 \frac{1 + \cos^2\theta}{8R^2}\left(\frac{\pi}{\lambda}\right)^4\left(\frac{\varepsilon - 1}{\varepsilon + 2}\right)^2 D^6, \tag{3.5.1}$$

where R is the distance between the particle and the observer, θ is the scattering angle, and ε is the particle permittivity. Equation 3.5.1 predicts the spectral dependence of intensity to be proportional to λ^{-4}, which governs, in particular, the blue color of the sky. Note that an individual molecule of air does not have a well-defined refractive index and diameter. Instead, the polarizability of a molecule is used in theoretical calculations.

The angular distributions of Rayleigh scattering are governed by the $(1 + \cos^2\theta)$ term and are the same for the forward and backward directions. The intensity of the Rayleigh scattered radiation decreases rapidly as the ratio D/λ decreases. The greater the particle size, the more light is scattered in the forward direction, so we *gradually* proceed from the Rayleigh regime to the Mie one. The Rayleigh scattering approximation becomes invalid when the sphere size exceeds the wavelength of incident radiation by about 10% (Figure 3.5.3a). Note that the value of the size parameter $D/\lambda = 0.1$, which defines the boundary between Rayleigh and Mie regimes, is conditional and depends on the dielectric contrast between the sphere and the surrounding medium.

Another limit corresponds to the case when the particle size is considerably larger than the wavelength of light, $D/\lambda >> 1$. The conventional boundary between optical and Mie regimes is $D/\lambda = 100$. When the wavelength is much smaller than the sphere size, one can consider scattering using the geometrical optics approach (in this approximation, the scattering by spheres is nonselective in λ). One can calculate the angular intensity of scattered light using elementary trigonometry. Geometrical considerations predict that at some angles the intensity becomes infinite. Such angles are referred to as rainbow angles. For example, the geometrical considerations predict two scattering maxima from a water droplet, first at 138° (primary rainbow) and at 129° (secondary rainbow). In case of silica spheres, the rainbow angles have the following values: 100°, 153°, and 180°. Indeed, at $D/\lambda > 100$ these three lobes are formed in intensity angular dependence (Figure 3.5.3b). Note that for the optical regime no analytical solution is similar to Equation 3.5.1.

Regarding opaline structures, let us define the regime of light scattering for specific Bragg bands. If the wavelength λ_{111} of light normally incident on the (111) system of planes in the FCC opal lattice defines the spectral position of the corresponding Bragg band (111), it obeys the relationship $\lambda_{111} = \sqrt{8/3}\varepsilon D \approx 2D$.[9] So we have the estimation $D/\lambda > 0.5$. This means that the (111) Bragg band and a number of high-indexed (*hkl*) Bragg bands corresponding to the shorter wavelengths λ_{hkl} are always in the range of the Mie regime, independent of the size of *a*-SiO_2 particles forming the opal structure. In this case, one can also study Mie resonances that have been probed by static and dynamic experiments as an extended optical characterization of photonic glasses[42] (see Chapter 3.3). When scanning λ to the far infrared or to the far ultraviolet, we satisfy a regime of the Rayleigh scattering ($D/\lambda < 0.1$) in opals or that of geometrical optics ($D/\lambda > 100$), respectively (see Figure 3.5.4).

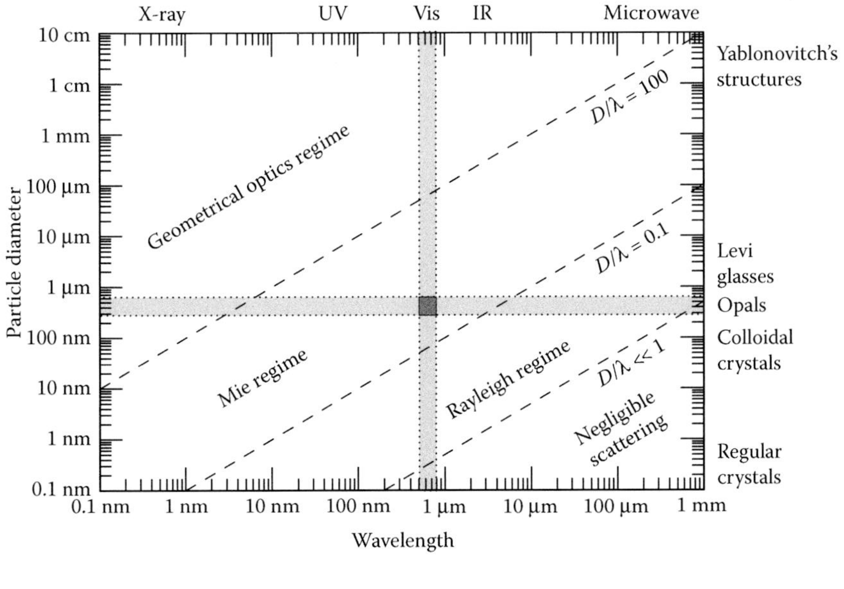

FIGURE 3.5.4
Scattering regimes for various ranges of the size parameter D/λ. The ordinate is the scattering particle diameter D and the abscissa is the wavelength of incident radiation λ. Conditional boundaries between different regimes are shown by dashed lines. Shaded regions correspond to the visible range and typical *a*-SiO_2 particle sizes. Boundaries between regimes correspond to the low dielectric contrast $\Delta\varepsilon < 1$.

Considering PhCs composed of a periodic array of similar spheres, the interference of scattered waves results in transformation of Mie scattering into Bragg scattering and gives rise to formation of the photonic band structure. The underlying Mie scattering is therefore hidden in perfectly ordered PhCs, and, as a result, it has been insufficiently studied and its role is clear only for the case of perfect structures. The resulting Bragg scattering, on the other hand, has been examined, both experimentally and theoretically.[4,7,16,20,21,23,30,43,44] In particular, Bragg diffraction of light from opals is considered in detail in Chapters 3.6 and 3.7.

To deepen our understanding of light scattering in opal-based PhCs, a number of challenging problems can be formulated: What can we expect if a narrow Bragg band interacts with a broad spectrum originating from certain scattering mechanisms such as Mie scattering? Is it possible to observe experimentally the consequences of this interaction or simply Mie scattering? What are the effects of the inherent disorder in the structural components of opal-based PhCs (see Section 3.5.2) beyond the well-known broadening and degradation of stop-bands? We will try to answer these questions in the following sections of this chapter.

3.5.4 Photonic Band-Gap Structure of Low-Contrast Opals: Bragg Approximation

When studying optical properties of low-contrast PhCs, the question arises of whether we can use the extremely suitable and simple Bragg approximation similar to that used for the analysis of x-ray diffraction in regular crystals. X-ray diffraction peaks are attributed to the scattering from certain crystallographic planes prescribed by the Miller indices, whereas the angular dependence of the peak energies is defined by Bragg's formula. A greater dielectric contrast in PhCs leads to a broadening and overlapping of the photonic stop-bands so that the electrodynamic problem for the photonic band structure can be solved only by sophisticated numerical methods.[44,45] In particular, the calculations made with plane-wave expansion[30] and Koringa–Kohn–Rostoker[29] methods show that the FCC structure of inverted opals possesses a complete photonic band gap between the eighth and ninth bands. However, such calculations lack the interpretative clarity and simplicity of the Bragg approach to x-ray data analysis. Note that the dispersion of low-energy stop-bands found from a numerical treatment for high-contrast PhCs[29,30] agrees well with that of Bragg wavelengths for the {111} and {200} low Miller-index planes. Most optical investigations have been focused on the {111}, {200}, and {220} photonic bands.[1,2,5,6,8,16] Higher index {*hkl*} photonic bands have, however, attracted little attention.

Opals are known to be made up of *a*-SiO_2 spheres that are assembled into (111) hexagonal layers superimposed on one another along the [111]-axis to form an FCC lattice or other close-packed structures.[4,7] In this chapter, the [111] growth axis will be denoted as $\Gamma \rightarrow L_g$ to differentiate this direction from the other three equivalent nongrowth directions in the FCC lattice, $\Gamma \rightarrow L$. High symmetry points on the Brillouin zone (BZ) surface that contains the L_g point will also be marked by the subscript g (as illustrated in Figure 3.5.5).

By *photonic band structure* one means a set of the dispersion dependences for (*hkl*) photonic bands expressed in the "energy versus wave vector **k**" coordinates or, what is equivalent, in the "wavelength versus propagation angle" coordinates (see Figure 3.5.5e). For reasonably full information of the photonic band structure to be obtained with an account

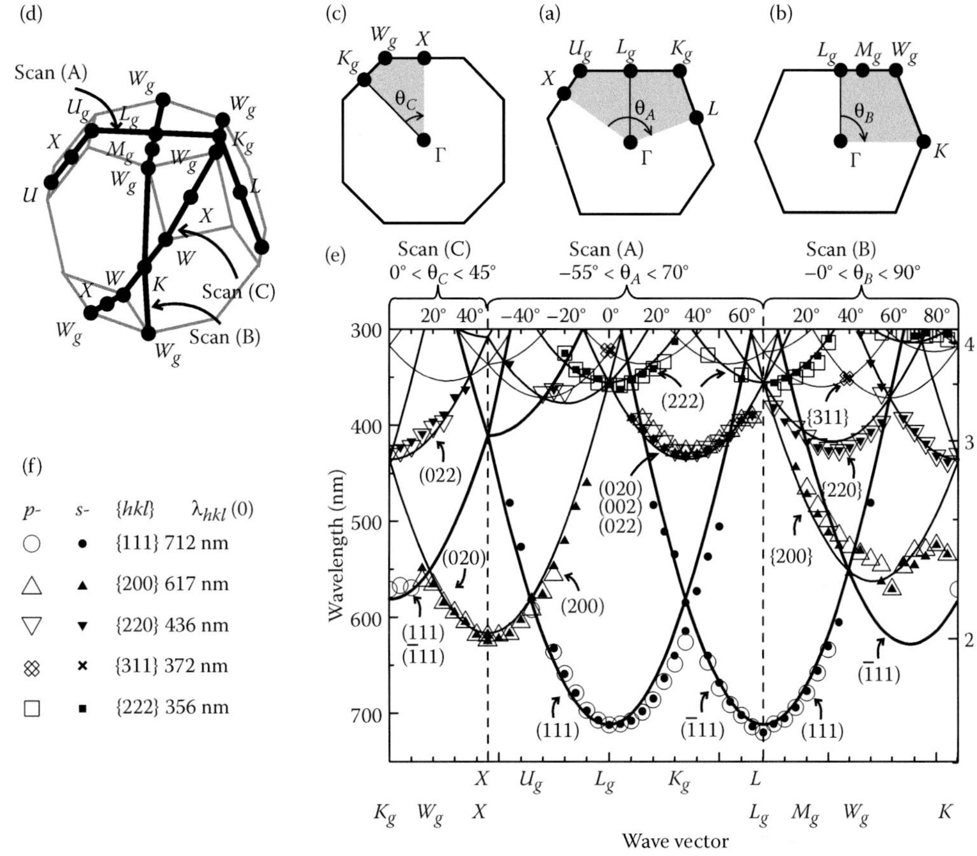

FIGURE 3.5.5
(**See color insert.**) (a)–(c) Three cross sections of the Brillouin zone (BZ) of the face-centered cubic (FCC) lattice made by the scanning planes A (a), B (b), and C (c). (d) The BZ of the FCC lattice showing three scanning paths (A), (B), and (C). (e) Photonic band structure of low-contrast opaline photonic crystals. Solid curves are theoretical dependences of the Bragg wavelength versus angles of light incidence onto the (*hkl*) planes in the FCC lattice; the dependencies are calculated from Equation 3.5.3 for three scanning paths. Symbols are dip positions in the experimental transmission spectra. The symmetry points along the paths are shown in the lower abscissa. (f) Symbols used for denoting the {*hkl*} Bragg bands for *p*- and *s*-polarizations in panel (e) and the theoretical Bragg wavelength $\lambda_{hkl}(\theta)$ at $\theta_{hkl} = 0$ (see Equation 3.5.3). (Panels (a)–(c) and (e) adapted from M. V. Rybin et al., Phys. Rev. B **77**, 205106, 2008. Panels (d) and (f) adapted from M. V. Rybin, K. B. Samusev, and M. F. Limonov, Photonics Nanostruct. Fundam. Appl. **5**, 119, 2007. With permission.)

of the FCC lattice BZ symmetry, it is sufficient to analyze three scans of the wave vector **k** across the BZ (Figure 3.5.5):[9,10]

$$X \to U_g \to L_g \to K_g \to L \quad \text{at scan (A),} \tag{3.5.2a}$$

$$L_g \to W_g \to K \quad \text{at scan (B),} \tag{3.5.2b}$$

$$K_g \to W_g \to X \quad \text{at scan (C).} \tag{3.5.2c}$$

The scan planes (A) $\Gamma L_g K_g L$ and (B) $\Gamma L_g W_g K$ contain the common direction of $\Gamma \rightarrow L_g$ (the [111] growth axis) and are normal to each other, whereas the scan planes (A) $\Gamma L_g K_g L$ and (C) $\Gamma K_g W_g X$ contain the $\Gamma \rightarrow K_g$ common direction and are also normal to each other.

The dependences of the vacuum Bragg wavelengths on the light beam incidence onto the (*hkl*) planes in the FCC lattice were calculated from the formula[46]

$$\lambda_{hkl}(\theta_{hkl}) = 2d_{111}\sqrt{\varepsilon_{av}}\left(\frac{3}{h^2+k^2+l^2}\right)^{1/2}\left|\cos\theta_{hkl}\right|, \tag{3.5.3}$$

where $d_{111} = \sqrt{2/3}D$ is the distance between the adjacent (111) planes, D is the diameter of the a-SiO_2 spheres, $\varepsilon_{av} = 0.74\tilde{\varepsilon}_{SiO_2} + 0.26\varepsilon_f$ is the average dielectric constant, and θ_{hkl} is the angle between the wave vector **k** and reciprocal lattice vector $\mathbf{g}_{hkl}$. Light diffraction from an FCC lattice may be due to only the family of planes with the same Miller indices parity.[46] Therefore, Figure 3.5.5 demonstrates the dispersion dependences of the Bragg wavelengths calculated from formula 3.5.3 for {111}, {200}, {220}, {311}, and {222}. Calculations were performed for $D = 316$ nm and $\varepsilon_{av} = 1.90$ (opals filled with water). In the study by Rybin et al.,[9] the plane families {400}, {331}, and {333} were also examined, but these {*hkl*} families are not discussed in the sections that follow.

Transmission spectra as a function of the wave vector **k** orientation relative to the FCC lattice are shown in Figure 3.5.6. The measurements were made with a 5° step for the *p*- and *s*-linear polarizations, with the *p*-component corresponding to the electric field vector

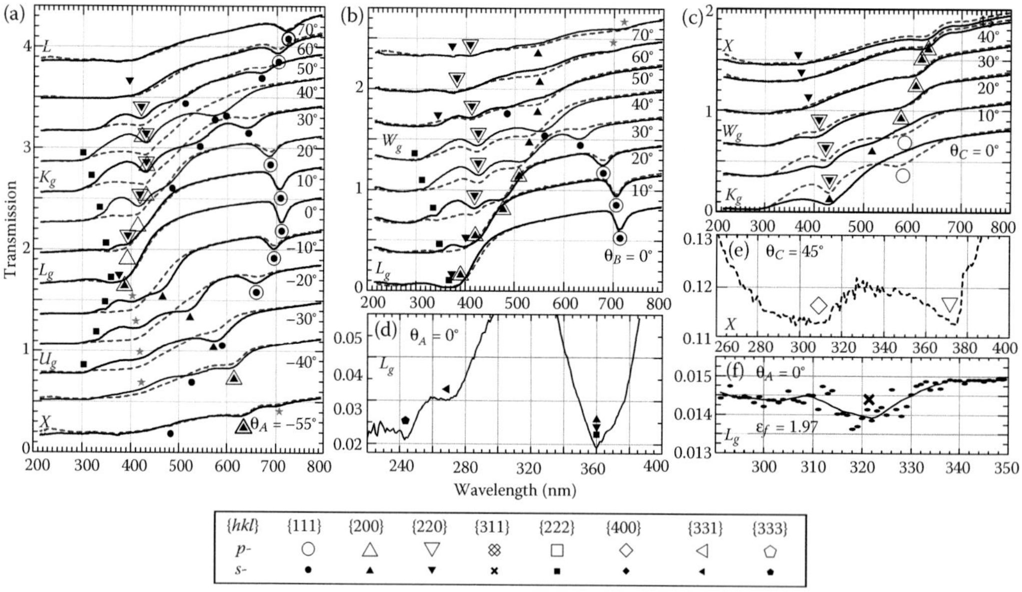

FIGURE 3.5.6
Polarized light transmission spectra of an opal sample in water, obtained in scan A (a), in scan B (b), and in scan C (c) as a function of the scanning angle θ. The spectra are shifted along the axis by a value proportional to the difference between the angles θ in the current and lower spectrum (0.03 per 1°). Solid curve shows *s*-polarization, dashed curve shows *p*-polarization. Bragg bands are denoted as in Figure 3.5.5. Short wavelength region of the *s*-polarized spectrum for opal in water, $\theta_A = 0$ (d), the *p*-polarized spectrum for opal in water $\theta_C = 45°$ (e), and the *s*-polarized spectrum for opal in a propylene glycol–water solution, $\theta_A = 0$ (f). (Reprinted from M. V. Rybin, K. B. Samusev, and M. F. Limonov, Photonics Nanostruct. Fundam. Appl. **5**, 119, 2007. With permission.)

E in the scan plane. The experimental samples were perfect synthetic opals predominantly with the FCC structure.

The superposition of the experimental data with the theoretical dispersion dependences $\lambda_{hkl}(\theta)$ found from formula 3.5.3 is presented in Figure 3.5.5e. This allows a comparison of the majority of the bands that were observed in the transmission spectra with the respective {*hkl*} Bragg band families. Let us consider each family individually.

For each {*hkl*} family, we will give a maximum value $\lambda_{hkl}(\theta)$ that is measured for normal incidence of light onto a corresponding plane (*hkl*), that is, at $\theta_{hkl} = 0$, and the points satisfying this condition on the BZ surface. As an example, the numerical values of $\lambda_{hkl}(\theta)$ are presented for opals with the sphere diameter of $D = 316$ nm filled by water.

- The {111} family, $\lambda_{111}(0) = 2d_{111}\sqrt{\varepsilon_{av}} = 712\,\text{nm}$, at L (L_g) points. An unambiguous assignment of the {111} dips is due to the fact that the quantity $\lambda_{111}(0)$ (the $\Gamma \rightarrow L_g$ direction) refers to a Bragg band with the longest wavelength.
- The {200} family, $\lambda_{200}(0) = \sqrt{3/4}\lambda_{111}(0) = 617\,\text{nm}$, at X points. The spectral range of interest for scan A includes all of the three {200} Bragg bands, namely, (200), (020), and (002). The (200) Bragg band is observed in scan A, with the maximum wavelength at X ($\theta_A = -55°$) point with $\lambda_{200} \approx 625$ nm.
- The {220} family, $\lambda_{220}(0) = \sqrt{3/8}\lambda_{111}(0) = 436\,\text{nm}$, at K (K_g) points. The (022) band energy in scan A is the same as the (020) and (002) band energies in this scan. To distinguish these bands, one should turn to scan B or C, whose planes are normal to the plane of scan A (Figure 3.5.5). In these scanning directions the degeneracy is eliminated, making the {200} and {220} planes identifiable, with the respective wavelengths at X point being quite different: 619 nm for the (020) planes and 370 nm for the (022) ones. The maximum wavelength $\lambda_{220}(0) = 431$ nm is found at K (K_g) points on the BZ surface. Thus, the proper choice of the scanning directions provides an unambiguous differentiation between the {200} and {220} Bragg bands.
- The {311} family, $\lambda_{311}(0) = \sqrt{3/11}\lambda_{111}(0) = 372$ nm, at the <311> directions at $\theta_A \approx -30°$ and $\theta_B \approx -58°$. The spectra of the experimental water-filled opals did not show the presence of the {311} Bragg bands. But these bands are observed when the filler permittivity $\varepsilon_f = 1.97$, which provides slightly higher contrast. There are three dispersion curves corresponding to (311), (131), and (113) Bragg bands intercepting at the $\Gamma \rightarrow L_g$ ($\theta_A = 0$) direction (Figure 3.5.5). Here, a low intensity dip was observed with a center at about 320 nm, practically coinciding with the wavelength of 327 nm calculated from formula 3.5.3. Also, a feature at about 358 nm was observed at the $\Gamma \rightarrow W_g$ ($\theta_B \approx 40°$) direction. For this direction, the calculation reveals an intercepting of (311) and (131) dispersion curves at 356 nm. This allows us to assign these bands (320 and 358 nm) to the {311} Bragg bands.
- The {222} family, $\lambda_{222}(0) = \sqrt{3/12}\lambda_{111}(0) = 1/2\lambda_{111}(0) = 356$ nm, at the L (L_g) points. The (222) Bragg bands of second-order diffraction from the (111) growth planes are observable in scans A and B at wavelengths twice as short as $\lambda_{111}(\theta)$.
- In the study by Rybin et al.,[9] several additional bands were reported as well, which corresponded to the {400} family ($\lambda_{400}(0) = \sqrt{3/16}\lambda_{111}(0) = 1/2\lambda_{200}(0) = 308$ nm, at X points), the {331} family ($\lambda_{331}(0) = \sqrt{3/19}\lambda_{111}(0) = 283\,\text{nm}$, at the <331> directions), and the {333} family ($\lambda_{333}(0) = \sqrt{3/27}\lambda_{111}(0) = 1/3\lambda_{111}(0) = 237\,\text{nm}$, at L (L_g) points) (see Figure 3.5.6).

An interesting effect should be noted that is observed for p-polarized light in the spectra of low-contrast opal-based PhCs. In Chapter 2.7, an expression (2.7.7b) is given for the Bragg band width in p-polarization $\Delta\omega_p = \Delta\omega_s|\cos 2\theta_{hkl}|$, where $\Delta\omega_s$ is the band width in s-polarization of light. Therefore, at $\theta_{hkl} = \pm 45°$ the photonic Bragg band has to disappear in p-polarized light. From the definition of θ_{hkl} angles, by obvious relationships with the scan angles θ_A, θ_B, and θ_C (Figure 3.5.5) we can find simple conditions for the disappearance of any (hkl) Bragg band in p-polarized light. For example, ignoring light refraction on a PhC boundary, we find that the (111) band disappears at $\theta_A = \pm 45°$, the $(\bar{1}11)$ band disappears near $\theta_A \approx 25°$ or $\theta_A \approx 115°$, and the (200) band disappears at $\theta_A \approx -10°$ and $\theta_A \approx -100°$.

Such an effect is actually observed in reflection[17,19,47,48] and transmission experiments.[5,6,9,10,48] From Figures 3.5.5 and 3.5.6, one can see that the Bragg bands (111) and (200) are absent for p-polarized light at $\theta_A = \pm 45°$ when scan A is performed, whereas for the s-polarization they are observed at the same scan.

This kind of polarization dependence was interpreted as being due to an effect similar to the classical Brewster effect in homogeneous media where the reflected component of the incident light vanishes in the p-polarization at the Brewster angle.[5,6,10,47]

Thus, we have described the photonic band-gap structure of low-contrast opals and demonstrated the applicability of the simple Bragg approach to the description of optical spectra in a wide range of wavelengths and wavevectors.

3.5.5 Inhomogeneity of a-SiO_2 Particles: Selective Immersion of Bragg Bands in Opals

In this section, we will describe some of the optical properties of real opals. Let us consider effects due to the inherent *inhomogeneity* of each single a-SiO_2 particle. In doing so we will neglect dissimilarity of the a-SiO_2 particles, that is, opals will be considered as *an ordered FCC structure,* as shown in Figure 3.5.2b. The specific features of the photonic band structure of opals originating from the inhomogeneity of a-SiO_2 particles can be conveniently discussed in the context of MPhCs, which is considered in detail in Chapter 2.7.

In theory,[8,10] properly tuning the permittivity of one of the components in *ordered* MPhCs enables one to achieve selective disappearance of any nonresonant (hkl) photonic Bragg band in the transmission spectra. According to the terminology used in Chapter 2.7, the effect of Bragg band disappearance in the MPhC optical spectra will be called the "immersion," because, experimentally, this effect is observed when opals are immersed in certain liquids. The permittivity value that corresponds to the immersion condition for the (hkl) Bragg band will be denoted by $\varepsilon_f^0(g_{hkl})$.

To detect the selective immersion effects predicted for MPhCs, the immersion behavior of the maximum number of different (hkl) Bragg bands has to be studied and compared to each other. The results described in the above section form a basis for such studies. Indeed, the detailed analysis of the opal transmission spectra allowed us to find a number of different families of {hkl} Bragg bands and to determine an optimum condition for observation, that is, the wavevector **k** orientation in the FCC lattice.

An experimental study of the transmission spectra of opals performed for different filler permittivities ε_f revealed that the dependences of the {111}, {200}, {220}, {311}, and {222} Bragg band parameters on the dielectric contrast are different. The various bands vanish in transmission spectra at notably different values of the filler permittivity $\varepsilon_f^0(g_{hkl})$, as clearly seen in Figures 3.5.7 through 3.5.9.[8,10]

- The {111} family, the immersion regime at $\varepsilon_f^0(g_{111}) = 1.82$. The experimental data indicate that the filler permittivity $\varepsilon_f^0(g_{111}) = 1.82$ determines the immersion regime for both the (111) growth planes and the ($\bar{1}$11) nongrowth planes, and this is valid for the $\Gamma \rightarrow L_g$ and $\Gamma \rightarrow K$ geometries, as well as for the whole range of wavelengths and wavevectors we have studied. Scans A, B, and C were performed for the *s*- and *p*-polarizations, as illustrated in Figure 3.5.8. The results of the spectral analysis showing the normalized square root of the intensity of the (*hkl*) dips as a function of ε_f are presented in Figure 3.5.9a. By the term "dip intensity" we imply intensity of the peak in the extinction spectra $\chi(\lambda) = -\ln T(\lambda)/d$, assuming that the amplitude of the wave, having a frequency within the Bragg band, decreases exponentially with travel distance *d* within a PhC.[14] The dependence for the {111} band family is represented here as that of the (111) photonic band, measured in the $\Gamma \rightarrow L_g$ geometry.
- The {200} family, the immersion regime at $\varepsilon_f^0(g_{200}) = 1.63$. The immersion dependence of the (200) photonic band demonstrates that this band is observed at all values studied, $1.78 \leq \varepsilon_f \leq 2.05$, as shown in Figure 3.5.7d. Keeping in mind the linear dependence for the (111) band and assuming a similar dependence for the

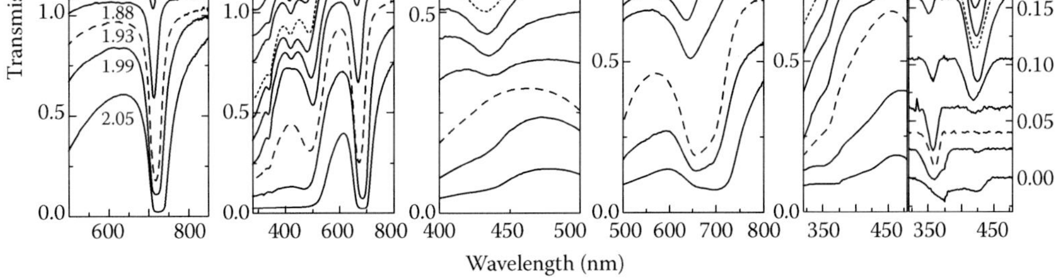

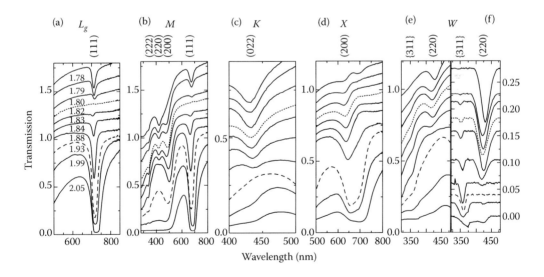

FIGURE 3.5.7
(a)–(e). Transmission spectra evolution in the range of the filler permittivity $1.78 \leq \varepsilon_f \leq 2.05$. The spectral regions and scattering geometries (with special points on the BZ surface) indicate the most typical immersion behavior of the photonic band families {111} (L_g, *M*), {200} (*M*, *X*), {220} (*M*, *K*, *W*), {311} (*W*), {222} (*M*). (f) The processed spectra are shown in panel (e). The filler permittivity values ε_f given in panel (a) also refer to the respective spectra in panels (b)–(f). Dotted lines are for the filler $\varepsilon_f^0(g_{111}) = 1.82$ that immerses the {111} bands, dashed lines are for the filler $\varepsilon_f^0(g_{220}) = 1.93$ that immerses the {220} bands. (Reprinted from M. V. Rybin et al., Phys. Rev. B 77, 205106, 2008.)

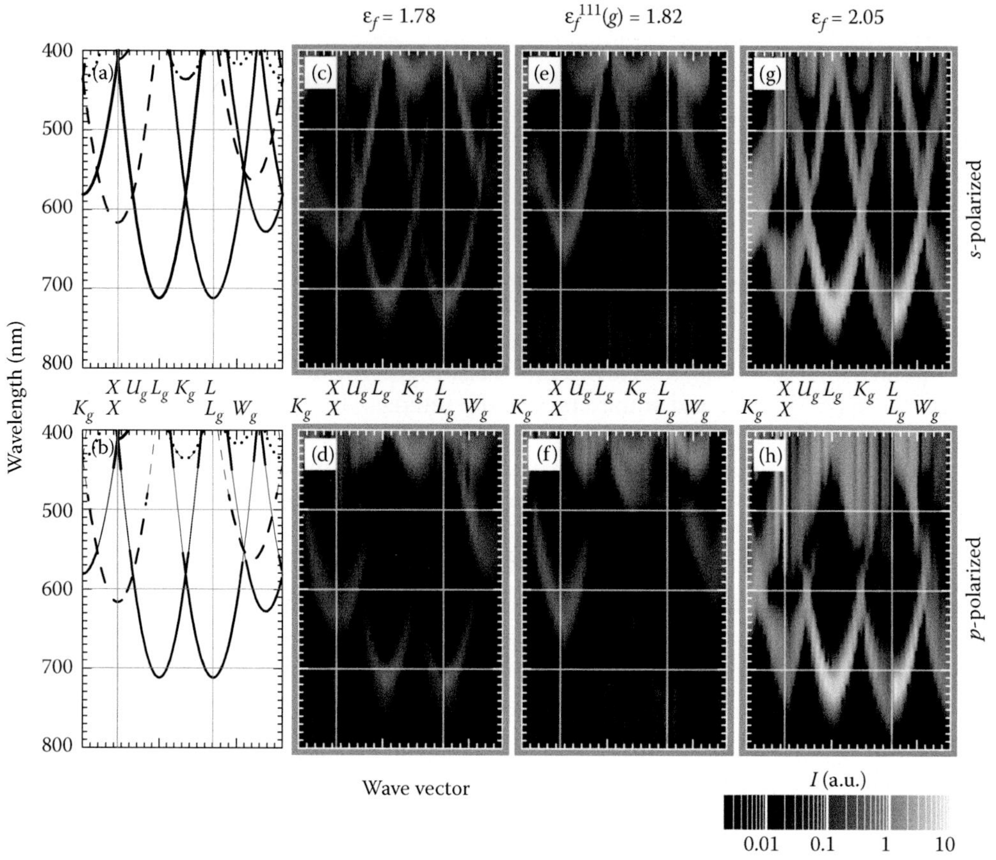

FIGURE 3.5.8
(**See color insert.**) The photonic band structure of opal. (a) Bragg wavelengths $\lambda_{hkl}(\theta)$ for *s*-polarization versus the angle of incidence for diffraction from different {*hkl*} plane families, calculated from Equation 3.5.3 for three scanning paths (A–C) (see Equation 3.5.2). Solid lines represent the {111} family of planes, dashed lines the {200} family of planes, and dotted lines the {220} family of planes. (b) Bragg wavelengths $\lambda_{hkl}(\theta)$ for *p*-polarization. The dispersion regions near the Brewster angle $\cos 2\theta_{hkl} = 0$ (see Equation 2.7.7 in Chapter 2.7) are represented by thin curves. (c)–(h) The photonic band structure of opals for three fillers and two linear polarizations is shown on the same scales of wavelengths and wave vectors and in the same scanning planes as the calculations in panels (a) and (b). The intensity scale of photonic bands is given in the lower right corner. (c), (d) $\varepsilon_f^0(g_{111}) = 1.78$, *s*- and *p*-polarization; (e), (f) $\varepsilon_f = 1.82$, *s*- and *p*-polarization; (g), (h) $\varepsilon_f = 2.05$, *s*- and *p*-polarization. (Adapted from M. V. Rybin et al., Phys. Rev. B **77**, 205106, 2008.)

(200) band, we can find the value of $\varepsilon_f^0(g_{200}) = 1.63$, at which the (200) Bragg band is to be immersed by the filler.

- The {220} family, the immersion regime at $\varepsilon_f^0(g_{220}) = 1.93$. The immersion dependence of the (022) band is shown in Figure 3.5.7c for the $\Gamma \rightarrow K$ scattering geometry and in Figure 3.5.7e and f for the $\Gamma \rightarrow W$ geometry. Unlike the (111) and (200) photonic bands, the (022) band becomes immersed at a much higher filler permittivity $\varepsilon_f^0(g_{220}) = 1.93$.
- The {311} family, the immersion regime at $\varepsilon_f^0(g_{311}) = 1.75$. It is difficult to understand the nature of the immersion dependence for the {311} band from the raw

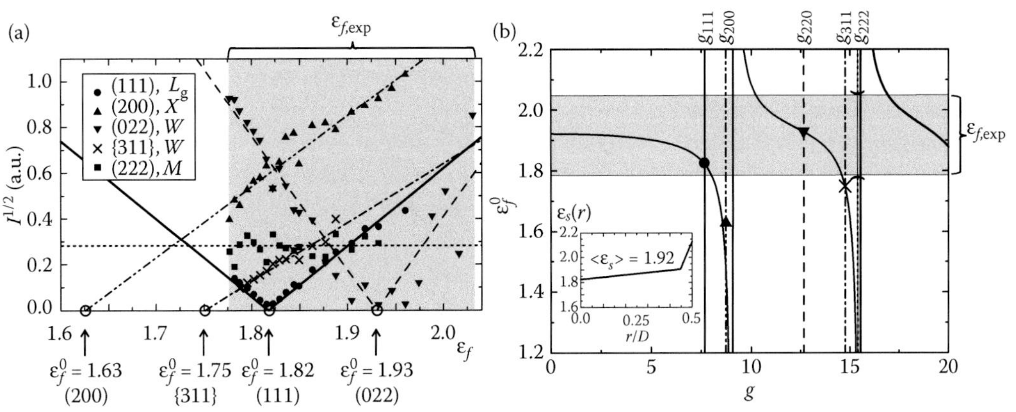

FIGURE 3.5.9
(**See color insert.**) (a) The square root of the intensity of the (*hkl*) Bragg dips as a function of the filler permittivity ε_f. Straight lines represent the fitting of experimental points by a linear function. The gray region is for the experimental range of ε_f. (b) The immersion conditions for an opal structure. The filler permittivity $\varepsilon_f^0(g)$ that corresponds to the immersion condition as a function of the reciprocal lattice vector module $g = |\mathbf{g}|$ calculated from Equation 3.5.4 for the permittivity profile $\varepsilon_s(r)$ simulating the *a*-SiO_2 spheres. The $\varepsilon_s(r)$ profile is given in the inset. The modules of the shortest reciprocal lattice vector g_{hkl} are shown by vertical lines. Symbols represent experimental values of $\varepsilon_f^0(g_{hkl})$ for four investigated (*hkl*) Bragg bands. (Reprinted from M. V. Rybin et al., Phys. Rev. B **77**, 205106, 2008. With permission.)

spectra in Figure 3.5.7e. However, the dependence of its integral intensity can be seen fairly well in Figures 3.5.7f and 3.5.9a. Similar to the (200) Bragg band, the (311) band exists for all of the fillers with a permittivity value of $1.78 \le \varepsilon_f \le 2.05$. Assuming a linear dependence of this band intensity, we find the value of $\varepsilon_f^0(g_{311}) = 1.75$, at which the {311} photonic bands are to be immersed.

- The {222} family, the resonance Bragg bands without the immersion regime. The (222) dip intensity is practically independent of the filler permittivity in the range of $1.78 \le \varepsilon_f \le 2.05$, as seen in Figure 3.5.9a. Recall that the {222} Bragg bands originate from the second-order diffraction in the {111} growth planes, the {111} and {222} families demonstrating absolutely different immersion dependences. Whereas the {111} photonic bands are immersed by a filler with $\varepsilon_f^0(g_{111}) = 1.82$, the {222} bands are not immersed in the ε_f range of interest and have the resonance nature or are similar to resonance bands, which can be immersed only by fillers with very high or very low (negative) permittivities $\varepsilon_f^0(g_{222})$.

In Chapter 2.7, a formula derived for ε_f^0 as a function of g in the case of 3D MPhCs with opaline FCC structure is given here in the following form:

$$\varepsilon_f^0(g) = \frac{\sin(gr_s) - gr_s\cos(gr_s)}{g^2\int_0^{r_s} \frac{1}{\varepsilon_s(r)} r\sin(gr)\,dr}, \tag{3.5.4}$$

As seen in Equation 3.5.4, the *only* fitting parameter is an arbitrary profile of the permittivity of inhomogeneous *a*-SiO_2 spheres, $\varepsilon_s(r)$. The simulation of $\varepsilon_s(r)$ was based on the data

(Figure 3.5.1) indicating that the sphere surface was continuous, and should have a permittivity close to that for bulk silica, $\varepsilon_{SiO_2} = 2.13$. An additional restriction on fitting results was imposed by the magnitude of the average permittivity $\tilde{\varepsilon}_s$ of the *a*-SiO_2 spheres, which, according to the literature data, is 1.9–2.0.[31] When fitting, a model of point-contacting spheres, with $r_s = 0.5$ in the units of lattice constant a_{00}, was considered. Further, Figure 3.5.1 indicates that the dense coat has a small thickness as compared with the sphere radius r_s and rapidly transforms to the porous nucleus. Taking all these factors into account, we approximated the $\varepsilon_s(r)$ function with a piecewise linear function and used nonlinear optimization procedures to calculate $\varepsilon_f^0(g)$, which describes the experimental data well, as seen in Figure 3.5.9b.[10] By averaging $\varepsilon_s(r)$ over the sphere volume, we get $\tilde{\varepsilon}_s = 1.92$, a value that agrees fairly well with $\tilde{\varepsilon}_s = 1.97$ derived from the processing of the reflection spectra for opals.[49] So, all the optical and structural data listed were successfully described in the framework of the MPhC model by the use of only three fitting parameters for $\varepsilon_s(r)$.[10]

In Chapter 2.7, we described a resonant behavior of the $\varepsilon_f^0(g)$ function, which is due to the alternating-sign denominator becoming zero in Equation 3.5.4. The first two resonances of this function governed by the profile of $\varepsilon_s(r)$ lie at $g^{res,1} = 9.07$ and $g^{res,2} = 15.54$. The resonant values of g are close to $g_{200} = 8.89$ and $g_{222} = 15.40$, which follows from the general symmetry property of the FCC lattice, and are therefore model-independent. The found relationships $g_{222} \approx g^{res,2}$ directly explain the photonic properties of opals: the independence of the (222) Bragg band intensity on ε_f indicates clearly its resonant character.

Thus, the *inhomogeneity* of *a*-SiO_2 particles in opals produces effects characteristic of MPhCs so that different nonresonance (*hkl*) Bragg bands disappear at different values of the filler permittivity $\varepsilon_f^0(g_{hkl})$ in the optical spectra of opals. In Chapter 2.7, a theoretical model was presented that enabled a description of the MPhC optical spectra. This model, based on an analysis of the form factor $F(\mathbf{g})$, was developed for the case of an ordered lattice comprising similar spheres (Figure 3.5.2b). For such a lattice, one can introduce the Wigner–Seitz cell and analyze the equation $F(\mathbf{g}) = 0$ from which the immersion conditions are derived. These equations define the conditions for the *complete disappearance of photonic Bragg bands in ordered PhCs*. As already pointed out, however, *a*-SiO_2 particles are *dissimilar* and opals cannot be considered as ordered PhCs. The answer to the question "Do photonic Bragg bands in synthetic opals indeed completely disappear at $\varepsilon_f = \varepsilon_f^0(g_{hkl})$?" will be explored in the following section.

3.5.6 Dissimilarity of *a*-SiO_2 Particles: Fano Effect in Opals

In this section, we will continue the discussion of optical properties of opal-based PhCs. Following the phenomena arising from the *inhomogeneity* of each *a*-SiO_2 particle, we will focus on the optical effects associated with the *dissimilarity* of *a*-SiO_2 particles. As pointed above, the dissimilarity is a consequence of *dispersion in the size and permittivity* of *a*-SiO_2 particles in the ensemble making up real *disordered* opal samples. As seen in Figure 3.5.1, *a*-SiO_2 particles in opals exhibit dispersion in size. Also, the specific structure of *a*-SiO_2 particles, which contain in their internal part randomly distributed spherical pores 7–15 nm in diameter, assumes a dissimilarity of permittivity of *a*-SiO_2 particles. A theoretical calculation of opal transmission spectra reveals that the most remarkable optical effects originate from the *dissimilarity of the permittivities* of a-SiO_2 particles rather than from a dissimilarity of their size.[11] Note that in the case of one-dimensional (1D) PhCs (see Chapter 2.1) the dissimilarity of the permittivity gives rise to the most prominent effects in transmission spectra, among them the Fano resonance between narrow Bragg lines and a disorder-induced Fabry–Pérot background. Related effects are also the subject of this chapter.

The optical effects in the transmission spectra arising from fluctuations in $\tilde{\varepsilon}_s$ become evident when the dielectric contrast $| \varepsilon_f - \varepsilon_f^0(g_{hkl}) |$ is comparable with the fluctuations $\Delta\tilde{\varepsilon}_s$. Therefore, if we want to detect the effects due to the *dissimilarity* of a-SiO_2 particles, we should more carefully investigate the immersion range $\varepsilon_f \rightarrow \varepsilon_f^0(g_{hkl})$, where one could expect the complete disappearance of $\{hkl\}$ Bragg bands.

Figure 3.5.10 shows a sharp (111) Bragg band in the transmission spectrum, and this fact allows us to analyze its shape with particular thoroughness. The results clearly indicate a remarkable transformation of the (111) band when the filler permittivity ε_f is changed (Figure 3.5.11a). At first, the well-known effect of diminishing the Bragg band intensity is observed when the filler permittivity approaches a certain value that was found to be $\varepsilon_f^0(g_{111}) = 1.816$ for the family of {111} Bragg bands.[10] We can clearly observe an asymmetry of the (111) dip that is inverted around a central frequency when the sign of the dielectric contrast $(\varepsilon_f - \varepsilon_f^0(g_{111}))$ is reversed. For $(\varepsilon_f - \varepsilon_f^0(g_{111})) < 0$ (three bottom curves in Figure 3.5.11a), the long-wavelength wing of the (111) band is relatively flat in contrast to a steep short-wavelength wing. For $(\varepsilon_f - \varepsilon_f^0(g_{111})) > 0$ (three curves at the top), the situation reverses: the long-wavelength wing of the (111) dip becomes abrupt as compared to the now relatively flat short-wavelength wing of the band. It should be emphasized that there is *no value of* ε_f *at which the* (111) *Bragg band completely disappears*. Moreover, when the filler permittivity is $\varepsilon_f = \varepsilon_f^0(g_{111})$ (the curve at the middle), a *transmission Bragg rise is observed* instead of the conventional transmission Bragg dip, that is, an enhanced Bragg transmission takes place.[11] The transmission Bragg rise could be observed only in a very narrow range $\varepsilon_f \approx 1.816 \pm 0.003$. This phenomenon was overlooked in many papers[8,10,50] in which

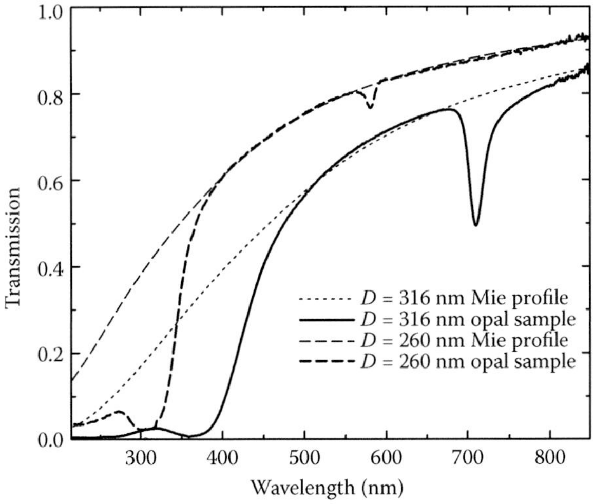

FIGURE 3.5.10
Transmission spectra of two opal samples in the $\Gamma \rightarrow L_g$ geometry and the results of calculations of the transmission spectra of a disordered ensemble of isolated spherical particles with use of the Mie theory. (Reprinted from M. V. Rybin et al., Phys. Rev. Lett. **103**, 023901, 2009. With permission.)

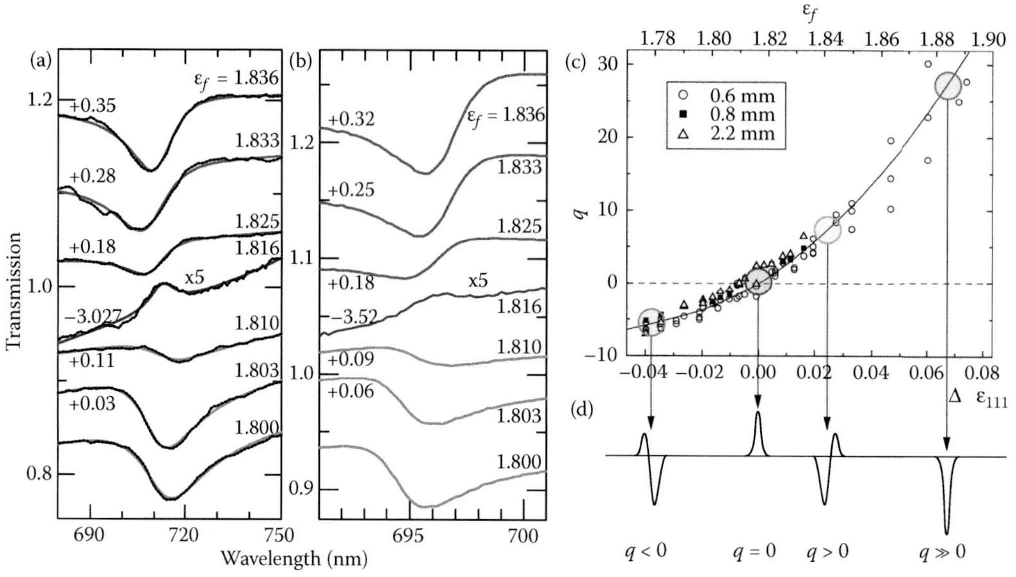

FIGURE 3.5.11
(**See color insert.**) (a) The measured transmission spectra of an opal sample ($D = 316$ nm) as a function of the filler permittivity ε_f in the region of the (111) photonic band (dotted curves). The solid curves are the results of the fitting with a Fano function. (b) The results of calculations of transmission spectra using the "quasi-3D" model of disordered opal structure.[11,24] In (a) and (b) the curves are shifted vertically by the values shown. (c) The Fano asymmetry parameter q as a function of the filler permittivity ε_f, determined by fitting the transmission spectra of the three samples with different thicknesses of 0.6, 0.8, and 2.2 mm. The solid line is a guide for the eyes only. (d) Schematic of the Bragg band shape observed in the transmission spectra on changes in the filler permittivity ε_f, which, in turn depends on Fano parameter q. (Adapted from M. V. Rybin et al., Phys. Rev. Lett. **103**, 023901, 2009; M. V. Rybin et al., Photonics Nanostruct. Fundam. Appl. **8**, 86, 2010. With permission.)

the filler permittivity was changed in a larger step of $\Delta\varepsilon_f \geq 0.01$. For the same reason, the Bragg rise is virtually lacking in Figure 3.5.7a and faintly visible in Figure 3.5.7b.[10]

From a theoretical point of view, light scattering in PhCs is inherently a multilevel process. The first level represents scattering by a single particle with its own scattering properties. At the higher levels, multiple scattering events by similar particles arranged in a periodic structure add up, resulting in the formation of the Bloch waves of a perfect PhC. Disorder, however, induces additional random phase shifts and amplitude variations at the first-level single-scattering event. This, in turn, modifies higher-level multiple-scattering events and causes degradation of bands and Bloch modes, as well as giving rise to the appearance of a broad background. Bloch waves and the background can interfere, giving rise to the unusual transmission pattern observed experimentally (see Figure 3.5.11a).

To describe the experimentally observed phenomenon, a theoretical model should incorporate both scattering mechanisms (Bragg and Mie) in a way that naturally permits their coupling. In the model proposed by Rybin et al.,[11,24] the scattering properties of a disordered array of spherical particles forming a two-dimensional layer corresponding to the (111) plane of the FCC opal structure are calculated at the first stage with the use of the Mie theory.[41] In-layer diffraction and multiple scattering processes are neglected in this model, which restricts the model applicability to the low-contrast regime and implies that we neglect ordering of the particles within the layer. In this approximation, the transmittance and reflectance of the layer are directly related to the forward and backward scattering amplitudes $S(\theta = 0)$ and $S(\theta = 180°)$ of the Mie theory,[41] respectively, through

$$t_n = e^{ikh}\left(1 - \frac{2\pi\eta}{k^2}\tilde{S}_n(\theta = 0)\right), \quad (3.5.5a)$$

$$r_n = e^{ikh}\frac{2\pi\eta}{k^2}\tilde{S}_n(\theta = 180^{\circ}), \quad (3.5.5b)$$

where n is the layer index, k is the modulus of the wavevector in the filler material, h is the layer thickness, η is the density of the particles in the layer, and the tilde above the scattering amplitudes implies intralayer averaging over the size and permittivity of particles. This averaging is conducted for every single layer by numerical integration of the scattering amplitudes around average values $\tilde{\varepsilon}_s$ and $\tilde{r}_s$, which vary from layer to layer according to normal distribution.

At the second stage, the optical properties of the FCC PhC built from a sequence of such (111) layers are determined by the conventional 2×2 transfer matrix technique.[51] The transfer matrix of every layer of the structure is expressed through transmission and reflection coefficients obtained at the previous stage with an account of interlayer disorder; that is, the permittivity of particles and their size also vary from layer to layer. Finally, the calculated transmittance is averaged over a sufficient number of realizations. This model therefore can be referred to as "quasi-3D" as the Bragg diffraction is purely 1D, whereas the disorder-induced scattering is 3D (both intralayer and interlayer disorder are taken into account). In numerical calculations, the particle size distribution was taken to be normal with a full-width at half-maximum of 7% for an average diameter $\bar{D}$ = 316 nm (Figure 3.5.1). The same model was taken for modeling the distribution of permittivity. The only adjustable parameter was an average sphere permittivity. As can be seen from calculated spectra shown in Figure 3.5.11b, all peculiarities of the experimental spectra, such as asymmetry of the Bragg dip, *transformation of the Bragg dip into the symmetric Bragg rise* corresponding to enhanced Bragg transmission, and the overall behavior of the transmission spectrum are revealed by our model.

The effects discussed, as applied to spectra of 3D opal-based PhCs, are in their physical nature the same as those for 1D disordered PhCs described in Chapter 2.1. Here, we keep in mind the interference between narrow Bragg lines and a wide-band Fabry–Pérot background, which is due to permittivity disorder in 1D PhCs. Such a phenomenon was considered in the framework of the Fano resonance concept,[39] and we apply this concept to the case of 3D PhCs. From the general point of view, the Fano resonance between continuum and discrete states is a phenomenon well-known across many different branches of physics[39,52] (see also Chapter 4.6). Figure 3.5.11 demonstrates a beautiful example of a Fano-type resonance in 3D PhCs that originates from interference between a narrow Bragg band and the disorder-induced continuous Mie spectrum.

In transmission spectra of opals one can identify at least two components, more specifically, broad-band background and comparatively narrow Bragg bands, with both spectral components being of the same origin, namely, Mie scattering from isolated a-SiO_2 particles. Figure 3.5.10 displays calculated transmission spectra for a disordered ensemble of isolated spherical particles by the Mie theory.[11] These spectra typically have a smooth shoulder whose position in the spectrum is determined by particle diameter D. In the case where D = 250–300 nm, the shoulder falls within the visible spectral region. The calculated transmission spectra fit nicely onto the background component of the opal transmission spectra. If a narrow Bragg band interacts with the continuum spectrum through an interference effect constructively or destructively, we can expect a Fano-type interaction. To verify the possibility of describing transmission spectra of opals with the Fano formula,[39] experimental results obtained for samples of different thickness, 0.6, 0.8, and 2.2 mm, were fitted using the following relation:

$$I(\Omega) = \frac{1}{1+q^2}\frac{(\Omega+q)^2}{\Omega^2+1} + \left(C_0 + C_1\Omega + C_2\Omega^2\right), \tag{3.5.6}$$

where $\Omega = (\omega - \omega_B)/(\gamma_B/2)$, ω_B is the frequency, γ_B is the width of the Bragg band, and q is the Fano asymmetry parameter. The first term in Equation 3.5.6 describes the Bragg band, and the Mie-type background spectrum of the opal in the region of the (111) band is approximated by the second term in the form $(C_0 + C_1\Omega + C_2\Omega^2)$. As follows from Equation 3.5.6, at intermediate values $-\infty < q < 0$ and $0 < q < +\infty$, the Bragg band's profile is asymmetric, the asymmetry changing to its mirror image with respect to frequency ω_B as parameter q reverses its sign. For $q \to \pm\infty$, the band has the symmetrical Lorentzian shape $F(\Omega) = 1/(\Omega^2 + 1)$. As $|q| \to 0$, the line asymmetry continues to grow, until at $|q| \approx 0$ the line flips over on the intensity scale. At $q = 0$, the narrow band transforms to a symmetric band $F(\Omega) = \Omega^2/(\Omega^2 + 1)$ inverted with respect to the band observed at all values of q outside the region with $|q| \approx 0$. Thus, we observe altogether four line shapes that are displayed schematically in Figure 3.5.11d.

In fact, all the four shapes of the (111) Bragg band that can be described by the Fano formula have been observed experimentally. With the use of Equation 3.5.6, it is possible to describe well the spectra shown in Figure 3.5.11a. The fitting procedure revealed that the Fano asymmetry parameter q varies continuously from negative to positive values with the filler permittivity ε_f increasing within the range covered. Significantly, the results obtained by fitting the spectra measured for three opal samples are in excellent agreement, as illustrated in Figure 3.5.11d.

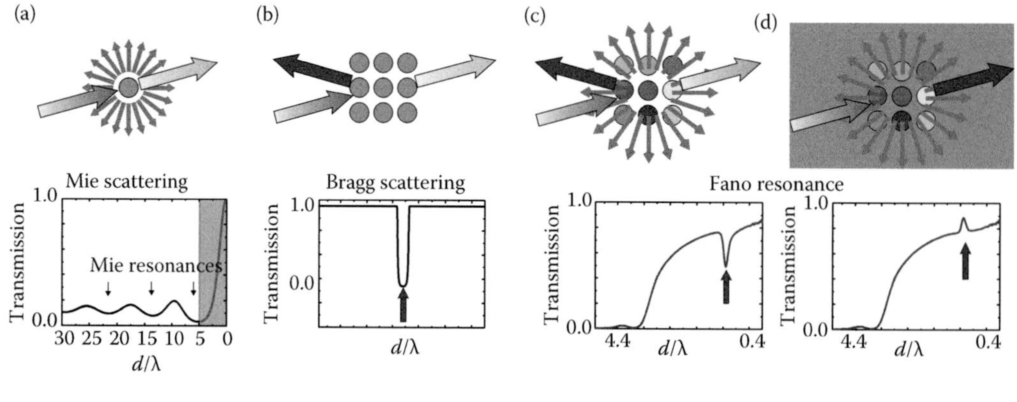

FIGURE 3.5.12
Schematic describing the mechanisms of the transmission spectra formation in opal-based photonic crystals (PhCs). (a) Light scattering from a sphere (Mie scattering). (b) Light scattering in the ordered PhCs (Bragg scattering). (c), (d) Light scattering in the PhCs assembled from dissimilar spheres with dispersed permittivity values: Fano resonance between Bragg scattering and disorder-induced Mie scattering.

To sum up, the *dissimilarity* in the permittivity of *a*-SiO_2 particles gives rise to the appearance of an additional scattering, which accounts for the background component in opal transmission spectra. It is the interference of this background Mie scattering with Bragg scattering that creates the asymmetry and reversal of the Bragg band. This behavior of the spectra finds adequate description in the framework of the Fano effect, as illustrated in Figure 3.5.12.

3.5.7 Conclusions: Transmission Spectra of Low-Contrast Opal-Based PhCs

We have now elucidated a scenario of light scattering in low-contrast opal-based PhCs (Figure 3.5.13) that are always imperfect owing to the presence of inherent inhomogeneity and dissimilarity of *a*-SiO_2 particles. Shown in the upper row (a–c) are FCC-like structures

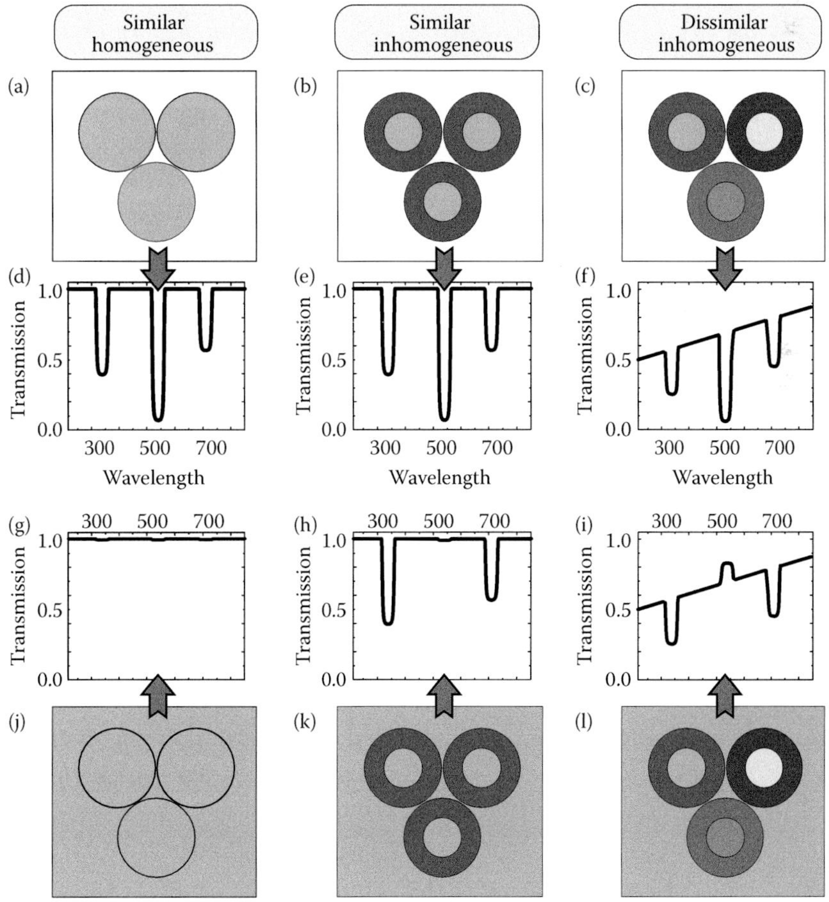

FIGURE 3.5.13
Schematic image of the structure of opal-like two-component photonic crystals (PhCs) and multicomponent PhCs at $\varepsilon_f \neq \varepsilon_f^0(g_{hkl})$ (a)–(c) and at $\varepsilon_f = \varepsilon_f^0(g_{hkl})$ (j)–(l). Schematic representation of transmission spectra of PhCs at $\varepsilon_f \neq \varepsilon_f^0(g_{hkl})$ (d)–(f) and at $\varepsilon_f = \varepsilon_f^0(g_{hkl})$ (g)–(i) corresponding to the above structures.

of three types, which consist of (a) homogeneous similar spheres, (b) inhomogeneous similar spheres, and (c) *inhomogeneous dissimilar spheres, as in the case of real opals.* The homogeneous filler in the space between the spheres has a variable permittivity $\varepsilon_f \neq \varepsilon_f^0(g_{hkl})$ The bottom row (j–l) depicts the same structures with a filler corresponding to the immersion condition $\varepsilon_f = \varepsilon_f^0(g_{hkl})$. The row (d–f) illustrates schematically the transmission spectra corresponding to the permittivity $\varepsilon_f \neq \varepsilon_f^0(g_{hkl})$, and the row (g–i) is for the immersion condition $\varepsilon_f = \varepsilon_f^0(g_{hkl})$.

The left column displays the structure and spectra of an ordered two-component PhC. All Bragg bands disappear simultaneously in the spectra at $\varepsilon_f = \varepsilon_s$, which makes the structure completely optically transparent. The transmission is 100% throughout the spectral interval covered. Column 2 corresponds to an ordered three-component PhC as a particular case of an ordered MPhC. This structure cannot be made completely transparent by varying the permittivity of one of the components only. Using the above theoretical model, however, one can select such a value $\varepsilon_f = \varepsilon_f^0(g_{hkl})$ at which a predetermined (*hkl*) Bragg band will disappear. Significantly, there are "resonant" Bragg bands that do not disappear under variation of ε_f within experimentally accessible limits. The (222) Bragg band in synthetic opals may serve as an illustration of such a resonant (or near-resonant) Bragg band. Outside the Bragg bands, the transmission of an ordered MPhC is 100% throughout the spectral range of interest (if one disregards surface effects and light interference at the sample boundaries).

The third column exemplifies disordered MPhCs consisting of inhomogeneous dissimilar spheres, that is, the structure that can be to a greater extent assigned to real opals. The disorder breaks a condition of ideal Bragg scattering, which determines all properties of perfect structures, and gives rise to an additional scattering component. This disorder-induced extra scattering produces background radiation and is responsible for the formation of a continuum spectrum with its specific characteristics determined by the light scattering from elements constituting PhCs. In particular, in disordered 1D PhCs composed of dielectric slabs, it would be extra Fabry–Pérot scattering owing to fluctuations in slab permittivity (see Chapter 2.1), whereas in 3D PhCs formed from spherical particles, this mechanism is Mie scattering owing to fluctuations in the permittivity of spheres. That is, it is the disorder-induced Mie scattering that is responsible for and determines the character of the continuum spectrum of opals. The latter is characterized by the decrease in the transmission observed in the short-wavelength part of the spectra $\lambda < \lambda_{111}/2$ for various PhCs formed from spherical particles.[3,8,10,18,23] An increase in the scattering intensity responsible for this feature is well known from the Mie theory.[41] The continuum disorder-induced Mie spectrum interferes with narrow Bragg bands, and, according to calculations, this effect is related to inhomogeneity of the a-SiO_2 particles with respect to permittivity ε_s. A very good fitting of the experimental spectra by calculation using Equation 3.5.6 confirms that Fano-type interference between the continuum disorder-induced Mie spectrum and narrow Bragg band takes place. As one crosses over from the $q < 0$ range to $q > 0$, the Bragg band undergoes mirror inversion on *the wavelength scale,* with the asymmetry of the Bragg band becoming mirror-inverted. In the same region, at $q \sim 0$, the Bragg band flips over on *the intensity scale,* which means essentially that the photonic *stop-band* transforms into a photonic *pass-band.*

Thus, low-contrast opal-based PhCs reveal a rich pattern of remarkable optical effects that should be assigned primarily to different kinds of inherent disorder in the structural elements making up a PhC. It is worth noting that the spectroscopic effects of interest governed by disorder are also observed when studying diffraction of light and small-angle x-ray diffraction from opals (see Chapters 3.6 and 3.7). Moreover, a novel class of disordered optical materials called "photonic glasses" was created on the basis of opal-like structures.[42] Nontrivial optical properties of photonic glasses are described in Chapter 3.3.

Acknowledgments

This work was done through collaborations with M. Inoue, A. B. Khanikaev, A. V. Moroz, A. K. Samusev, K. B. Samusev, and G. Yushin. The main experimental results presented in this chapter were obtained by studying synthetic opals grown by M. I. Samoylovich (Moscow). The work is supported by Russian Foundation for Basic Research (Grants Nos. 10-02-01094 and 11-02-00865) and OFN RAN 2011.

References

1. V. N. Astratov, V. N. Bogomolov, A. A. Kaplyanskii, A. V. Prokofiev, L. A. Samoilovich, S. M. Samoilovich, and Yu. A. Vlasov, Nuovo Cimento D **17**, 1349 (1995).
2. Yu. A. Vlasov, V. N. Astratov, O. Z. Karimov, A. A. Kaplyanskii, V. N. Bogomolov, and A. V. Prokofiev, Phys. Rev. B **55**, R13357 (1997).
3. Yu. A. Vlasov, V. N. Astratov, A. V. Baryshev, A. A. Kaplyanskii, O. Z. Karimov, and M. F. Limonov, Phys. Rev. E **61**, 5784 (2000).
4. A. V. Baryshev, A. A. Kaplyanskii, V. A. Kosobukin, K. B. Samusev, D. E. Usvyat, and M. F. Limonov, Phys. Rev. B **70**, 113104 (2004).
5. A. V. Baryshev, A. B. Khanikaev, H. Uchida, M. Inoue, and M. F. Limonov, Phys. Rev. B **73**, 033103 (2006).
6. M. V. Rybin, A. V. Baryshev, M. Inoue, A. A. Kaplyanskii, V. A. Kosobukin, M. F. Limonov, A. K. Samusev, and A. V. Sel'kin, Photonics Nanostruct. Fundam. Appl. **4**, 146 (2006).
7. A. V. Baryshev, V. A. Kosobukin, K. B. Samusev, D. E. Usvyat, and M. F. Limonov, Phys. Rev. B **73**, 205118 (2006).
8. A. V. Baryshev, A. B. Khanikaev, M. Inoue, P. B. Lim, A. V. Sel'kin, G. Yushin, and M. F. Limonov, Phys. Rev. Lett. **99**, 063906 (2007).
9. M. V. Rybin, K. B. Samusev, and M. F. Limonov, Photonics Nanostruct. Fundam. Appl. **5**, 119 (2007).
10. M. V. Rybin, A. V. Baryshev, A. B. Khanikaev, M. Inoue, K. B. Samusev, A. V. Sel'kin, G. Yushin, and M. F. Limonov, Phys. Rev. B **77**, 205106 (2008).
11. M. V. Rybin, A. B. Khanikaev, M. Inoue, K. B. Samusev, M. Steel, G. Yushin, and M. F. Limonov, Phys. Rev. Lett. **103**, 023901 (2009).
12. A. K. Samusev, K. B. Samusev, I. S. Sinev, M. V. Rybin, and M. F. Limonov, Phys. Solid State **53**, 1415 (2011); D. A. Mazurenko, R. Kerst, J. I. Dijkhuis, A. V. Akimov, V. G. Golubev, D. A. Kurdyukov, A. B. Pevtsov, and A. V. Sel'kin, Phys. Rev. Lett. **91**, 213903 (2003); A. V. Akimov, Y. Tanaka, A. B. Pevtsov, S. F. Kaplan, V. G. Golubev, S. Tamura, D. R. Yakovlev, and M. Bayer, *ibid.* **101**, 033902 (2008); G. M. Gajiev, V. G. Golubev, D. A. Kurdyukov, A. V. Medvedev, A. B. Pevtsov, A. V. Sel'kin, and V. V. Travnikov, Phys. Rev. B **72**, 205115 (2005); A. B. Pevtsov, D. A. Kurdyukov, V. G. Golubev, A. V. Akimov, A. A. Meluchev, A. V. Sel'kin, A. A. Kaplyanskii, D. R. Yakovlev, and M. Bayer, *ibid.* **75**, 153101 (2007); S. F. Kaplan, N. F. Kartenko, D. A. Kurdyukov, A. V. Medvedev, A. G. Badalyan, and V. G. Golubev, Photonics Nanostruct. Fundam. Appl. **5**, 37

(2007); V. N. Bogomolov, S. V. Gaponenko, I. N. Germanenko, A. M. Kapitonov, E. P. Petrov, N. V. Gaponenko, A. V. Prokofiev, A. N. Ponyavina, N. I. Silvanovich, and S. M. Samoilovich, Phys. Rev. E **55**, 7619 (1997); V. V. Pavlov, P. A. Usachev, R. V. Pisarev, D. A. Kurdyukov, S. F. Kaplan, A. V. Kimel, A. Kirilyuk, and T. Rasing, Appl. Phys. Lett. **93**, 072502 (2008).
13. L. M. Sorokin, V. N. Bogomolov, J. L. Hutchison, D. A. Kurdyukov, A. V. Chernyaev, and T. N. Zaslavskaya, Nanostruct. Mater. **12**, 1081 (1999).
14. Yu. A. Vlasov, M. I. Kaliteevski, and V. V. Nikolaev, Phys. Rev. B **60**, 1555 (1999).
15. Yu. A. Vlasov, X. Z. Bo, J. C. Sturm, and D. J. Norris, Nature **414**, 289 (2001); Z.-Y. Li and Z.-Q. Zhang, Phys. Rev. B **62**, 1516 (2000); J. F. Galisteo-López, E. Palacios-Lidón, E. Castillo-Martínez, and C. López, *ibid.* **68**, 115109 (2003); A. F. Koenderink, A. Lagendijk, and W. L. Vos, *ibid.* **72**, 153102 (2005); S. G. Romanov and C. M. S. Torres, Phys. Rev. E **69**, 046611 (2004); R. Rengarajan, D. Mittleman, C. Rich, and V. Colvin, *ibid.* **71**, 16615 (2005); M. S. Thijssen, R. Sprik, J. E. G. J. Wijnhoven, M. Megens, T. Narayanan, A. Lagendijk, and W. L. Vos, Phys. Rev. Lett. **83**, 2730 (1999); V. Yannopapas, N. Stefanou, and A. Modinos, *ibid.* **86**, 4811 (2001); W. L. Vos and H. M. van Driel, Phys. Lett. A **272**, 101 (2000); H. Takeda and K. Yoshino, Appl. Phys. Lett. **80**, 4495 (2002); A. A. Fedyanin, O. A. Aktsipetrov, D. A. Kurdyukov, V. G. Golubev, and M. Inoue, *ibid.* **87**, 151111 (2005); P. D. García, J. F. Galisteo-López, and C. López, *ibid.* **87**, 201109 (2005); A. F. Koenderink and W. L. Vos, J. Opt. Soc. Am. B **22**, 1075 (2005); L. M. Goldenberg, J. Wagner, J. Stumpe, B. R. Paulke, and E. Gornitz, Physica E **17**, 433 (2003).
16. S. G. Romanov, T. Maka, C. M. S. Torres, M. Muller, R. Zentel, D. Cassagne, J. Manzanares-Martinez, and C. Jouanin, Phys. Rev. E **63**, 056603 (2001).
17. J. F. Galisteo-López, F. López-Tejeira, S. Rubio, C. López, and J. Sánchez-Dehesa, Appl. Phys. Lett. **82**, 4068 (2003).
18. J. F. Galisteo-López and C. López, Phys. Rev. B **70**, 035108 (2004).
19. H. M. van Driel and W. L. Vos, Phys. Rev. B **62**, 9872 (2000).
20. E. Pavarini, L. C. Andreani, C. Soci, M. Galli, F. Marabelli, and D. Comoretto, Phys. Rev. B **72**, 045102 (2005).
21. A. Balestreri, L. C. Andreani, and M. Agio, Phys. Rev. E **74**, 036603 (2006).
22. H. Míguez, C. López, F. Meseguer, A. Blanco, L. Vázquez, R. Mayoral, M. Ocana, V. Fornés, and A. Mifsud, Appl. Phys. Lett. **71**, 1148 (1997).
23. F. García-Santamaría, J. F. Galisteo-López, P. V. Braun, and C. López, Phys. Rev. B **71**, 195112 (2005).
24. M. V. Rybin, A. B. Khanikaev, M. Inoue, A. K. Samusev, M. J. Steel, G. Yushin, and M. F. Limonov, Photonics Nanostruct. Fundam. Appl. **8**, 86 (2010).
25. A. V. Baryshev, A. A. Kaplyanskii, V. A. Kosobukin, M. F. Limonov, K. B. Samusev, and D. E. Usvyat, Phys. Solid State **45**, 459 (2003); J.-M. Meijer, V. W. A. de Villeneuve, and A. V. Petukhov, Langmuir **23**, 3554 (2007); A. V. Petukhov, I. P. Dolbnya, D. G. A. L. Aarts, G. J. Vroege, and H. N. W. Lekkerkerker, Phys. Rev. Lett. **90**, 028304 (2003); X. Checoury, S. Enoch, C. Lopez, and A. Blanco, Appl. Phys. Lett. **90**, 161131 (2007).
26. E. Palacios-Lidón, B. H. Juárez, E. Castillo-Martínez, and C. López, J. Appl. Phys. **97**, 63502 (2005).
27. W. Stöber, A. Fink, and E. Bohn, J. Colloid Interface Sci. **26**, 62 (1968).
28. H. P. Schriemer, H. M. van Driel, A. F. Koenderink, and W. L. Vos, Phys. Rev. A **63**, 011801 (2000).
29. A. Moroz and C. Sommers, J. Phys. Condens. Matter **11**, 997 (1999).
30. K. Busch and S. John, Phys. Rev. E **58**, 3896 (1998).
31. R. K. Iler, *The Chemistry of Silica* (John Wiley & Sons, New York, 1979).
32. A. Guinier, in *Crystals, Imperfect Crystals, and Amorphous Bodies* (W.H. Freeman & Co., San Francisco, 1963).
33. B. J. Eggleton, B. Luther-Davies, and K. Richardson, Nat. Photonics **5**, 141 (2011).
34. P. J. Darragh and J. L. Perdrix, J. Gemmol. **14**, 215 (1975).
35. A. A. Chabanov, Y. Jun, and D. J. Norris, Appl. Phys. Lett. **84**, 3573 (2004).
36. V. N. Bogomolov, L. M. Sorokin, D. A. Kurdyukov, T. M. Pavlova, and J. L. Hutchison, Phys. Solid State **39**, 1869 (1997).
37. K. B. Samusev, G. Yushin, M. V. Rybin, and M. F. Limonov, Phys. Solid State **50**, 1280 (2008).

38. R. C. Gonzalez and R. E. Woods, *Digital Image Processing* (Addison-Wesley Publishing Company, New York, 1992).
39. U. Fano, Phys. Rev. **124**, 1866 (1961).
40. M. Born and E. Wolf, *Principles of Optics* (Cambridge University Press, Cambridge, 2006), 7th ed.
41. C. F. Bohren and D. R. Huffman, *Absorption and Scattering of Light by Small Particles* (Wiley-VCH, Weinheim, 1998).
42. P. D. García, R. Sapienza, and C. López, Adv. Mater. **22**, 12 (2010).
43. J. V. Sanders, Acta Crystallogr. Sect. A **24**, 427 (1968); J. F. Galisteo-López, F. García-Santamaría, D. Golmayo, B. H. Juárez, C. López, and E. Palacios-Lidón, Photonics Nanostruct. Fundam. Appl. **2**, 117 (2004); H. S. Sözüer, J. W. Haus, and R. Inguva, Phys. Rev. B **45**, 13962 (1992); Modinos, N. Stefanou, and V. Yannopapas, Opt. Express **8**, 197 (2001).
44. A. Moroz, Phys. Rev. B **51**, 2068 (1995).
45. K. Sakoda, *Optical Properties of Photonic Crystals* (Springer, Berlin/Heidelberg, 2004), 2nd ed.K. Kunz and R. Luebbers, *The Finite Difference Time Domain Method for Electromagnetics* (CRC Press, Boca Raton, 1993); K. M. Ho, C. T. Chan, and C. M. Soukoulis, Phys. Rev. Lett. **65**, 3152 (1990); J. B. Pendry and A. MacKinnon, *ibid.* **69**, 2772 (1992); X. Wang, X.-G. Zhang, Q. Yu, and B. N. Harmon, Phys. Rev. B **47**, 4161 (1993).
46. C. Kittel, *Introduction to Solid State Physics* (John Wiley & Sons, Inc., New York, 1986), 6th ed.
47. A. G. Bazhenova, A. V. Sel'kin, A. Y. Men'shikova, and N. N. Shevchenko, Phys. Solid State **49**, 2109 (2007).
48. A. V. Baryshev, A. B. Khanikaev, R. Fujikawa, H. Uchida, and M. Inoue, Phys. Rev. B **76**, 014305 (2007).
49. O. A. Kavtreva, A. V. Ankudinov, A. G. Bazhenova, Y. A. Kumzerov, M. F. Limonov, K. B. Samusev, and A. V. Sel'kin, Phys. Solid State **49**, 708 (2007).
50. M. N. Shkunov, M. C. DeLong, M. E. Raikh, Z. V. Vardeny, A. A. Zakhidov, and R. H. Baughman, Synth. Met. **116**, 485 (2001).
51. P. Yeh, A. Yariv, and C.-S. Hong, J. Opt. Soc. Am. **67**, 423 (1977).
52. F. Cerdeira, T. A. Fjeldly, and M. Cardona, Phys. Rev. B **8**, 4734 (1973); B. Friedl, C. Thomsen, and M. Cardona, Phys. Rev. Lett. **65**, 915 (1990); M. F. Limonov, A. I. Rykov, S. Tajima, and A. Yamanaka, *ibid.* **80**, 825 (1998); J. J. Hopfield, P. J. Dean, and D. G. Thomas, Phys. Rev. **158**, 748 (1967); V. Madhavan, W. Chen, T. Jamneala, M. F. Crommie, and N. S. Wingreen, Science **280**, 567 (1998); N. M. Kabachnik and I. P. Sazhina, J. Phys. B **9**, 1681 (1976); J.-P. Connerade and A. M. Lane, Rep. Prog. Phys. **51**, 1439 (1988); A. E. Miroshnichenko, S. Flach, and Y. S. Kivshar, Rev. Mod. Phys. **82**, 2257 (2010); B. Luk'yanchuk, N. I. Zheludev, S. A. Maier, N. J. Halas, P. Nordlander, H. Giessen, and C. T. Chong, Nat. Mater. **9**, 707 (2010).

3.6

Light and Small-Angle X-Ray Diffraction from Opal-Like Structures: Transition from Two- to Three-Dimensional Regimes and Effects of Disorder

Anton K. Samusev, Kirill B. Samusev, Ivan S. Sinev, Mikhail V. Rybin, and Mikhail F. Limonov
Ioffe Physical-Technical Institute of the Russian Academy of Sciences
National Research University of Information Technologies, Mechanics and Optics

Natalia A. Grigoryeva
St. Petersburg State University

Sergey V. Grigoriev
Konstantinov St. Petersburg Nuclear Physics Institute

Andrei V. Petukhov
Utrecht University

CONTENTS

3.6.1 Introduction

In this chapter, we present an overview of the diffraction of visible light and x-rays in opal-like structures. Special attention will be paid to the transformation of diffraction patterns on building up the opal structure from a two-dimensional (2D) single layer of silica particles *a*-SiO_2 toward a bulk three-dimensional (3D) opal structure as well as to the effects of disorder such as stacking faults. The consideration here will be limited to structures with low permittivity contrast $\Delta\varepsilon$. For small-angle x-ray diffraction, this condition is automatically fulfilled for all materials: $\Delta\varepsilon \sim 10^{-5}$–$10^{-6}$. For optical studies, such a situation can be realized for different opal-based structures by filling the voids of the structure with a medium having nearly matching ε. An enhancement of $\Delta\varepsilon$ can lead to broadening of the Bragg reflections and a modification of the relative intensity of different Bragg reflections as well as to significant extinction of the incident wave (see Chapter 3.7), which are not considered here. A different approach considering the full set of Maxwell equations to calculate the electromagnetic eigenmodes and photonic band structure is unavoidable for high $\Delta\varepsilon$, especially if one moves toward the full band-gap materials. This approach is considered in Chapter 3.7 in detail, whereas in this chapter the case of high dielectric contrast is left out.

Opal-like structures are photonic crystals (PhC) that possess stop-bands in the visible range owing to the typical size of the constitutive particles of some hundreds of nanometers. This provides a unique chance to study photonic properties not only by traditional methods such as registering transmission or reflection with a spectrometer but also by directly observing diffraction patterns on a screen disposed behind or around the sample.

For periodic structures made of identical spherically symmetric building blocks, the diffraction intensity $I(\mathbf{q})$ can be written as a product of the squared modulus of form factor $|F(q)|^2$, which describes scattering on a single particle, and the squared modulus of scattering structure factor $|S(\mathbf{q})|^2$, which describes the interference between waves scattered by different particles: $I(\mathbf{q}) \propto |S(\mathbf{q})|^2 \cdot |F(q)|^2$. Here the scattering vector $\mathbf{q} = \mathbf{k}_s - \mathbf{k}_i$ is the difference between the wavevectors $\mathbf{k}_s$ and $\mathbf{k}_i$ of the scattered (s) and incident (i) waves, respectively. Note that for the structures under consideration both $S(\mathbf{q})$ and $F(q)$ magnitudes are real-valued; therefore, the modulus signs are omitted in further text for simplicity. The structure factor $S(\mathbf{q})$ possesses maxima when the condition for constructive interference $\mathbf{q}\mathbf{r}_{ij} = 2\pi n$ is fulfilled for all pairs of particles, where $\mathbf{r}_{ij}$ is the radius vector connecting i-th and j-th particles and n is an integer.

As the transition from 2D to 3D diffraction regimes is of a particular interest here, let us consider the two limiting cases. A 2D periodic structure (a single monolayer of colloidal particles in our case) can be generated using two primitive vectors $\mathbf{a}_1$ and $\mathbf{a}_2$. The condition of constructive interference is fulfilled if the projection $\mathbf{q}_{\|}$ of the scattering vector on the 2D structure obeys

$$\mathbf{q}_{\|} = \mathbf{g}_{hk} \equiv h\mathbf{b}_1 + k\mathbf{b}_2 \tag{3.6.1}$$

where $\mathbf{g}_{hk}$ is the 2D reciprocal lattice vector, and $\mathbf{b}_1$ and $\mathbf{b}_2$ are primitive vectors of the 2D reciprocal lattice, which are given by

$$\mathbf{b}_1 = 2\pi \frac{\mathbf{a}_2 \times \mathbf{n}}{\mathbf{a}_1 \cdot (\mathbf{a}_2 \times \mathbf{n})}, \quad \mathbf{b}_2 = 2\pi \frac{\mathbf{n} \times \mathbf{a}_1}{\mathbf{a}_1 \cdot (\mathbf{a}_2 \times \mathbf{n})} \tag{3.6.2}$$

with $\mathbf{n}$ denoting the layer normal.

For a 3D structure, which can be generated by three primitive lattice vectors $\mathbf{a}_1$, $\mathbf{a}_2$, and $\mathbf{a}_3$, the condition of constructive interference leads to a set of Laue equations:[1]

$$\mathbf{q} \cdot \mathbf{a}_1 = 2\pi h, \quad \mathbf{q} \cdot \mathbf{a}_2 = 2\pi k, \quad \mathbf{q} \cdot \mathbf{a}_3 = 2\pi l, \tag{3.6.3}$$

where h, k, and l are integers. The solution of 3.6.3 can be written in the form of the Bragg's law

$$\mathbf{q} = \mathbf{g}_{hkl} \equiv h\mathbf{b}_1 + k\mathbf{b}_2 + l\mathbf{b}_3, \tag{3.6.4}$$

where $\mathbf{g}_{hkl}$ is a 3D reciprocal lattice vector normal to (*hkl*) lattice planes (see Figure 3.6.1a) and $\mathbf{b}_1$, $\mathbf{b}_2$, and $\mathbf{b}_3$ are primitive vectors of the 3D reciprocal lattice, which are given by[1]:

$$\mathbf{b}_1 = 2\pi\frac{\mathbf{a}_2 \times \mathbf{a}_3}{\mathbf{a}_1 \cdot (\mathbf{a}_2 \times \mathbf{a}_3)}, \quad \mathbf{b}_2 = 2\pi\frac{\mathbf{a}_3 \times \mathbf{a}_1}{\mathbf{a}_1 \cdot (\mathbf{a}_2 \times \mathbf{a}_3)}, \quad \mathbf{b}_3 = 2\pi\frac{\mathbf{a}_1 \times \mathbf{a}_2}{\mathbf{a}_1 \cdot (\mathbf{a}_2 \times \mathbf{a}_3)}. \tag{3.6.5}$$

As we consider only elastic scattering ($|\mathbf{k}_s| = |\mathbf{k}_i|$), Equation 3.6.4 implies that in ordered 3D crystals Bragg diffraction follows the law of *specular reflection* from (*hkl*) crystallographic planes with surface normal parallel to $\mathbf{g}_{hkl}$. Moreover, the end of scattering vector $\mathbf{q}$ must lie on the so-called Ewald sphere (a sphere of radius $2\pi/\lambda$ in reciprocal space, where λ is the radiation wavelength).[2] Diffraction is observed only when the Ewald sphere crosses at least two reciprocal lattice nodes.

Let us first consider the diffraction of visible light on an opal lattice. The lattice spacing is determined by the diameter of close-packed spheres $D \sim 10^2$–10^3 nm, which is of the order of the typical visible light wavelength $\lambda \sim (4 \div 7)10^2$ nm. Diffraction is then observed at large diffraction angles between $\mathbf{k}_s$ and $\mathbf{k}_i$ (Figure 3.6.1c). Moreover, optical Bragg diffraction can only be observed for low Miller index planes. For example, for a face-centered cubic (FCC) lattice, the interplanar spacing is given by $d_{hkl} = D/\sqrt{(h^2 + k^2 + l^2)/2}$. The

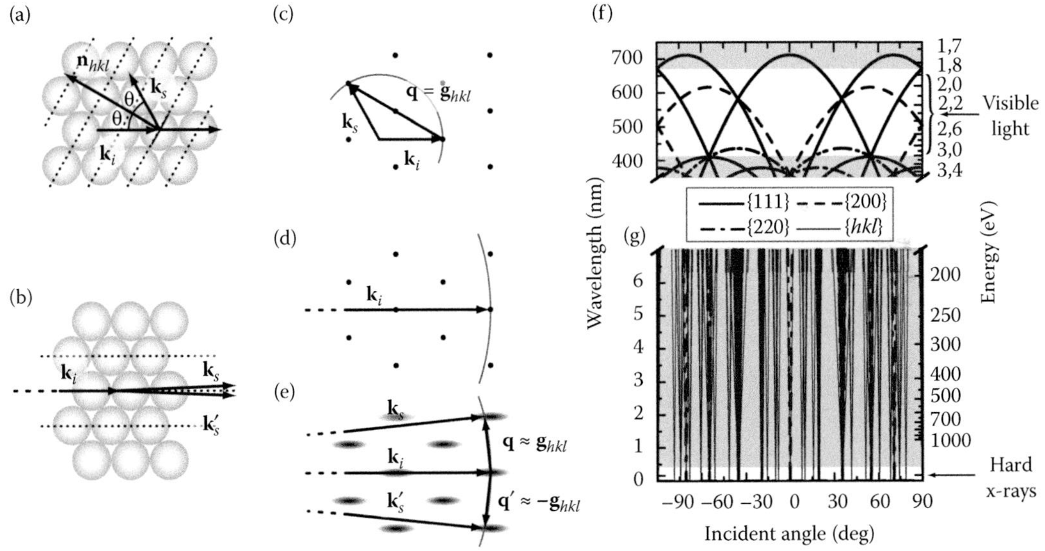

FIGURE 3.6.1
(a) Illustration of the light Bragg diffraction on a system of (*hkl*) planes in real space. (b) Illustration of the small-angle x-ray diffraction on a system of (*hkl*) planes in real space. Owing to the large difference between the structure period and the wavelength the diffracting incident beam propagates almost parallel to the (*hkl*) planes. (c) The Ewald sphere construction in reciprocal space for the Bragg diffraction of light. (d) The Ewald sphere construction in reciprocal space for incident x-ray wave propagating along the (*hkl*) planes of an ideal infinite opal-like structure ($|\mathbf{k}_i| \gg |\mathbf{g}_{hkl}|$). (e) The same as in (d) but including broadening of the reciprocal lattice nodes for a realistic structure. Broadening of the reciprocal lattice nodes and the curvature of the Ewald sphere are exaggerated for clarity. (f) Photonic band structure of low-contrast opal-like photonic crystals for the near-infrared, visible, and near-ultraviolet range of the spectrum, that is, the theoretical dependences of the Bragg wavelength versus angles of light incidence onto the {*hkl*} planes in the face-centered cubic lattice (see details in Figure 3.5.5 of Chapter 3.5). (g) Extrapolation of the photonic band-gap structure into the x-ray range of the spectrum.

diffraction can only be observed on sets of crystal planes with low Miller indices such as {111}, {200}, and {220} as for higher Miller indices the factor $\sqrt{(h^2+k^2+l^2)/2}$ makes d_{hkl} smaller than half of the visible light wavelength in the medium. This result is also reflected in Figure 3.6.1f, where the dependence of the Bragg wavelength on the angle of incidence is shown. This scheme can be also presented as the dependence of the energy of photonic stop-bands on the diffraction wavevector **q**. Such a representation known as the photonic band structure is discussed in detail in Chapter 3.5.

To get access to Bragg reflections with high Miller indices, one can use radiation with a shorter wavelength λ. In the ultraviolet part of the spectrum the Bragg diffraction is precluded for many materials, which become opaque as a result of the presence of numerous electronic resonances. One has to go all the way to hard x-rays, where the photon energy ($\hbar\omega > 10$ keV) is significantly higher than the energy of electronic transitions. Here the transparency increases sufficiently for all materials to allow Bragg diffraction experiments again. The change of the photon energy by almost four orders of magnitude between visible light and x-rays implies several important consequences. For x-rays the permittivity of materials differs from 1 by as little as $|1-\varepsilon| \sim 10^{-5}$–$10^{-6}$. This means that the contrast is intrinsically very weak and the Born approximation is fulfilled with very high accuracy. Diffraction on sets of crystal planes with high Miller (*hkl*) indices can now be accessed as a result of the short wavelength of x-rays, $\lambda \sim 0.1$ nm. On the other hand, Bragg angles for opal-like structures become very small, in the range of 0.1–10 μrad as illustrated in Figure 3.6.1b. Measuring diffraction experimentally at such small angles is a challenge and one usually needs to use synchrotron x-rays. In addition, the radius of the Ewald sphere $2\pi/\lambda$ is very large for x-rays in comparison to the scale of interest. As discussed in more detail in Section 3.6.5, for most cases of practical interest the curvature of the Ewald sphere is negligible in comparison, for example, with the broadening of the crystal reflections due to the finite size effects as schematically sketched in Figure 3.6.1e. As a result, many Bragg reflections can be observed simultaneously in a single diffraction pattern. It is also reflected in Figure 3.6.1g, where the photonic band structure is extrapolated into the x-ray spectrum. One can see that here many bands converge at a single value of the incident angle.

Bragg diffraction of light was studied experimentally and theoretically on different opal-like structures, including bulk samples of synthetic opals[3–5] and opal films.[6–11] These works involved studies of both specular reflection spectra and diffraction patterns, which were observed on screens, captured and processed using various software. In particular, for an opal film in case of normal beam incidence, either three or six diffraction reflections are observed depending on the sample quality, thickness, the "matrix-filler" contrast of the permittivity and its other parameters. The main features of the **k**-resolved spectra of 3D PhCs can be successfully analyzed in terms of single or multiple Bragg diffraction. The possibility of the visual study of the formation and alteration of photonic band structure during the diffraction experiments was demonstrated by Baryshev et al.[4] by comparing diffraction patterns and synthetic opals' transmission spectra. The theory of 3D Bragg polarized light diffraction from opals based on Green's electrodynamic functions method is presented in the study by Baryshev et al.[5] This theory takes into account the effects of random packing of growth (111) layers in opal structures.

In general, self-assembled opal-like crystals contain disorder of various types. The stacking disorder is one of the most common and intensively studied types of disorder. Colloidal opal-like crystals are composed of close-packed layers, which can form different structures depending on their stacking sequence: FCC structure (...ABCABC...), hexagonal close-packed (HCP) structure (...ABABAB...), as well as random hexagonal close-packed (RHCP) structure, in which close-packed layers are arranged in a random fashion.[1] The colloidal crystals synthesized by the sedimentation method (bulk opals) often demonstrate RHCP structure. The colloidal crystals grown by vertical deposition (film opals) are composed of large-scale FCC fragments separated by stacking faults or HCP fragments.[12]

The optical diffraction method is distinguished by its simplicity and informativity. Wide capabilities of optical diffraction for studying different types of structure disorder in opal-like PhCs is demonstrated in Figure 3.6.2 where four diffraction patterns are shown. They are obtained during the studies of opals with different degrees of lattice disorder—from a virtually "non-twinned" sample to a fully disordered sample.

In this chapter, we present a detailed study of 2D diffraction of light from thin opal films (Section 3.6.2) and the transition from 2D to 3D diffraction (Section 3.6.3). Then we discuss the selective switching of (*hkl*) reflections in diffraction patterns of thick opal films depending on the value of permittivity of infiltration material (filler) ε_f (Section 3.6.4). The effects of the sample thickness in x-ray diffraction are discussed in Section 3.6.5. The order and disorder of opal-like structures of bulk and film samples will be considered using the example of microradian x-ray diffraction (Section 3.6.6). The advantage of this approach stems from the intrinsically low contrast for x-rays, which leads to clear single scattering data. Moreover, x-rays can be applied to characterize opaque inverse opals as well as the samples with a periodicity less than 300 nm, whose structure cannot be investigated by conventional optical methods. An example of how x-ray diffraction can serve the technology of opal growth will be given in Section 3.6.7. In Section 3.6.8, the principles of 3D reciprocal space reconstruction of large-scale opal-like structures will be presented on the basis of microradian x-ray diffraction patterns. The chapter concludes with a discussion of the complementarity of visible light diffraction and microradian x-ray diffraction for investigation of photonic opal-like structures (Section 3.6.9).

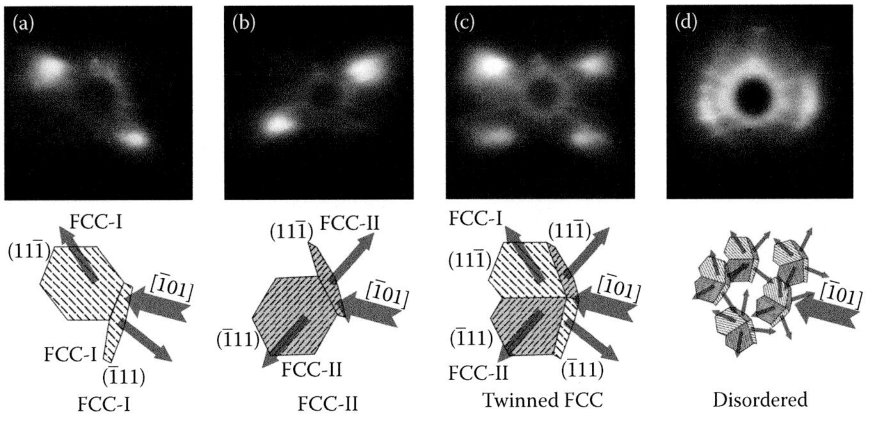

FIGURE 3.6.2
Top row: diffraction patterns from various opal samples in the $\mathbf{k}_i \parallel [\bar{1}01]$ incidence geometry: (a), (b) are the best-ordered monodomains of the face-centered cubic (FCC)-I (...ABCABC...) and FCC-II (...ACBACB...) lattices; (c) is a slightly disordered region of FCC twins; (d) is a strongly disordered region made up of numerous microcrystallites with their growth axes oriented randomly relative to one another. Bottom row: a schematic arrangement of the crystal planes involved in diffraction, together with the incident and diffracted beams. (Reprinted from M. V. Rybin et al., Photonics Nanostruct.: Fundam. Appl. **4**, 146, 2006, Copyright 2006, with permission from Elsevier.)

3.6.2 Two-Dimensional Optical Diffraction from Thin Opal Films

In this section, we focus primarily on the experimental and theoretical study of 2D diffraction from thin opal films. We should emphasize two key features of the diffraction studies reported in the literature.[9,10] First of all, a novel technique of recording the diffraction patterns is proposed. The commonly used scheme[5,7,13] is illustrated in Figure 3.6.3a–c. It involves investigating diffraction patterns displayed on a *flat screen* positioned before or after the sample. Among the disadvantages of this scheme are pattern distortions due to the nonsphericity of the screen and, more importantly, the limitation of the spatial angle where the diffraction pattern is registered. It should be mentioned that in most cases one

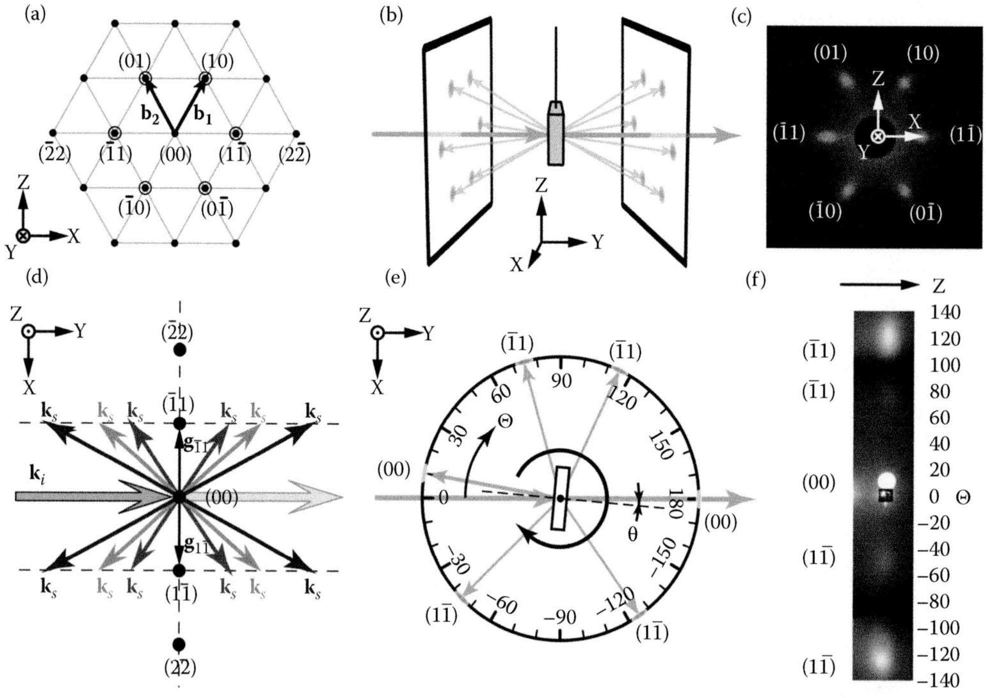

FIGURE 3.6.3
(See color insert.) (a) The reciprocal lattice of a two-dimensional (2D) hexagonal layer placed in the XZ plane normally to the incident beam. The lattice consists of an array of nodes, the position of each determined by the 2D vector $\mathbf{g}_{hk} = h\mathbf{b}_1 + k\mathbf{b}_1$, where $\mathbf{b}_1$ and $\mathbf{b}_2$ stand for the basis vectors of the reciprocal lattice, and h and k are integers. The nodes defining the first order of diffraction for the beam incident along the film normal (along the Y axis) are marked with circles. (b) A schematic illustration of the experimental setup designed to observe transmitted or reflected light on a flat screen. (c) The experimental pattern of monochromatic ($\lambda = 532$ nm) light diffraction from a thin opal film observed on a flat screen positioned behind the sample. Six observed reflections are marked with the respective pairs of 2D diffraction indices (hk). (d) A schematic illustration of 2D diffraction conditions in the XY plane $\mathbf{q}_{\parallel} = \mathbf{g}_{\bar{1}1}$ and $\mathbf{q}_{\parallel} = \mathbf{g}_{1\bar{1}}$ with the incident white light beam propagating along the Y axis and the hexagonal layer positioned in the XZ plane. (e) A sketch of the experimental setup designed to observe light diffraction in the XY plane on a cylindrical screen. Six diffracted beams are shown. Three of them are scattered forward with respect to the incident beam (i.e., the Y axis), whereas the other three are scattered backward. The angle of incidence θ and the angle of scattered light registration Θ are indicated. Oblique incidence is exemplified here: $\theta = 5°$. (f) A surface development of the white light diffraction pattern from opal film observed on the cylindrical screen in the XY plane at $140° \leq \Theta \leq 140°$. (Adapted from A. K. Samusev et al., Phys. Solid State **53**(5), 1056, 2011. With permission.)

does not necessarily need the information about light diffraction in full solid angle (4π). The method considered in the study by Samusev et al.[9] is suggested as a tool to investigate the diffraction in the horizontal plane containing the incident beam. Diffraction patterns are displayed on a *cylindrical* screen with a sample fixed in its center using a vertical holder as shown in Figure 3.6.3e. In this approach, diffraction reflections are displayed on the cylinder surface in the full angle (2π) in the XY plane without any distortion. Therefore, at the near-normal incidence one observes five diffraction reflections lying in the horizontal plane on the cylindrical screen, as can be seen in Figure 3.6.3e. Note that the sixth diffraction spot (00) at $\Theta = 180°$ is out of the experimentally investigated range of $-140° \leq \Theta \leq 140°$ (Figure 3.6.3f). Rotating the sample holder allows us to change the angle of incidence θ within the range $-90° \leq \theta \leq 90°$. The diffraction pattern on the narrow cylindrical screen is photographed from three different fixed points. The full diffraction pattern (Figure 3.6.3f) is formed by processing the obtained images onto a flat stripe line. Note that in the diffraction literature angle θ is usually used for the Bragg angle. Here, however, θ denotes sample orientation to comply with notations used in the majority of photonics literature and the notation used in other chapters of this book.

The second key feature of these studies is the experimental data presentation.[9,10] Most commonly, 2D diffraction patterns are presented as recorded on a flat screen. Here an alternative approach is used: a collection of a large number of the cylindrical screen color photographs are presented as a function of the angle of light incidence θ and the angle of diffracted light registration Θ. The diffraction patterns are collected for the angles of incidence within the range $-90° \leq \theta \leq 90°$ with the angular step size of $\Delta\theta = 2.5°$. Hereby, 72 narrow "strip patterns," similar to those presented in Figures 3.6.3f and 3.6.4a, are obtained and joined together into a single image similar to the one presented in Figure 3.6.4b.

The advantage of this presentation is that one can easily *distinguish 2D and 3D diffraction,* along with singling out a variety of effects induced by the disorder in the sample. Indeed, in the case of 2D diffraction, Equation 3.6.1 produces a complex dependence between Θ and θ angles for the allowed, that is, experimentally observed, reflections. At the same time, the case of 3D Bragg diffraction reduces to the specular reflection of the beam from a set of (*hkl*) crystallographic planes, and is thus described by a simple equation $\Theta = 2(\theta - \theta_{hkl})$ as discussed in more detail in the next section. Therefore, *in the* (θ,Θ) *coordinates the allowed Bragg reflections will be represented by parallel straight lines,* and any deviation or broadening of the lines will be associated with some kind of sample disorder.

Let us now consider light diffraction from thin opal films. It is worth recalling that a commonly observed diffraction pattern with the incident beam perpendicular[6–10] to the sample surface ($\theta = 0$) has the C_6 symmetry and consists of six intensive reflections symmetrical with respect to the incident beam (see Figure 3.6.3c). Note that six diffraction reflections in the given geometry can be observed in two cases. The first one is first-order 2D light diffraction from the hexagonal layer with the C_6 symmetry; the second is 3D Bragg diffraction from the twinned FCC lattice[6] oriented so that the [111] axis coincides with the incident beam direction. Each FCC twin has the C_3 symmetry in the direction [111] and defines a triplet of diffraction reflections. As the twins in the FCC lattice turn into each other after a rotation by 60° about the [111] axis, the result will bear the C_6 symmetry and will look like the image shown in Figure 3.6.3c.

To distinguish 2D and 3D diffraction cases, one should investigate light scattering patterns for arbitrary incident angles ($-90° \leq \theta \leq 90°$) and present the result in the (θ,Θ) coordinates. Let us use the scheme with a cylindrical screen described above and register the reflections formed as a result of 2D diffraction in the XY plane (see Figure 3.6.3d–f). In these experiments, the investigated film should be oriented so that two out of six

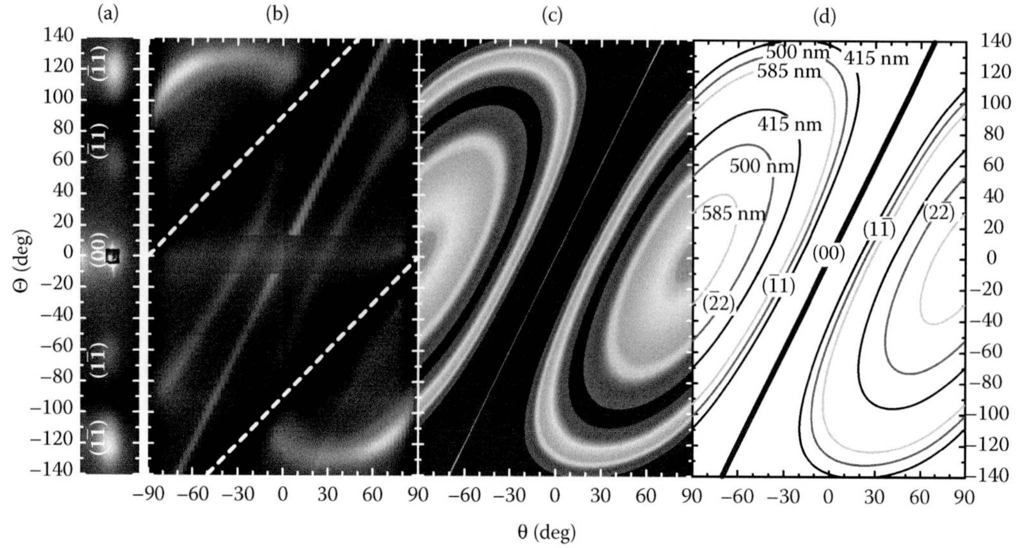

FIGURE 3.6.4
(See color insert.) (a) The surface development of a white light diffraction pattern recorded on a cylindrical screen at the normal incidence ($\theta = 0$) on a six-layer-thick opal film consisting of a-SiO_2 particles with diameter 720 nm. (b) The combined experimental pattern of light diffraction in the XY plane (see Figure 3.6.3e) for the angle of light incidence varying within the range $-90° \leq \theta \leq 90°$. Dashed lines correspond to the angles $\Theta = \theta \pm 90°$. (c) The numerical simulation of the two-dimensional (2D) diffraction pattern in the Born approximation. The square of the scattering structure factor $S^2(\mathbf{q})$ was calculated using Equation 3.6.6. (d) The calculation of positions of the zeroth-, first-, and second-order 2D diffraction reflections obtained from Equation 3.6.7 for three different wavelengths: $\lambda = 415$, 500, and 585 nm. (Adapted from A. K. Samusev et al., Phys. Solid State **53**(5), 1056, 2011. With permission.)

reflections observed on the flat screen are situated strictly along the X axis. On the cylindrical screen, one can then observe the reflections that correspond to the zeroth order of 2D diffraction (diffraction condition being $\mathbf{q}_{\|} = 0$), the first order (diffraction conditions being $\mathbf{q}_{\|} = \mathbf{g}_{\bar{1}1}$ and $\mathbf{q}_{\|} = \mathbf{g}_{1\bar{1}}$), the second order ($\mathbf{q}_{\|} = \mathbf{g}_{\bar{2}2}$ and $\mathbf{q}_{\|} = \mathbf{g}_{2\bar{2}}$), and so on. Note that according to the 2D diffraction condition each reciprocal lattice node $\mathbf{g}_{hk}$ gives rise to a pair of diffraction reflections, one corresponding to forward scattering with respect to the incident beam $\mathbf{k}_i$, the other to backscattering (see Figure 3.6.3d). The possibility of the experimental observation of a certain reflection in the visible spectral range is determined by a-SiO_2 particle diameter and the order of diffraction.

The experimental results presented in Figure 3.6.4b are obtained using a six-layer-thick opal film consisting of a-SiO_2 particles with diameter $D = 720$ nm.[9] With the incident beam normal to the sample surface ($\theta = 0$), diffraction reflections, which are specified by the zeroth and the first-order diffraction, are observed on the cylindrical screen. Note that the intensity of the "forward-scattered" ($|\Theta| > 90°$) reflections is higher than the intensity of the "backscattered" ($|\Theta| < 90°$) ones. This can be associated with the nature of Mie scattering on a spherical particle, which has an asymmetric radiation pattern with the forward scattering being dominant (see Figure 3.5.3 in Chapter 3.5 and the study by Bohren and Huffman[14]).

Also, in Figure 3.6.4b one can distinguish three types of diffraction features: (i) a straight line of white light corresponding to the scattering angles $\Theta = 2\theta$; (ii) two intense color curves (ovals) formed by diffracted white light with ruptures in the vicinity of the angles $\Theta = \theta \pm 90°$, marked with dashed lines; and (iii) less intense diffraction features situated within the intensive color curves, which are also interrupted at the angles close to $\Theta = \theta \pm 90°$. From Figure 3.6.4b it follows at once that light diffraction from the six-layer-thick opal film differs strongly from the 3D Bragg diffraction considered in more detail in Section 3.6.4. For experimental data interpretation we will use the results of two calculations—numerical and analytical,[9] illustrated in Figure 3.6.4c and d, which are described below.

For diffraction pattern calculation in the case of low-contrast PhCs (which include synthetic opals), it is usually sufficient to use the Born approximation.[15] Main features of light diffraction are then determined by the scattering structure factor and the scattering form factor. The calculations show that for interpretation of the experimental data (Figure 3.6.4b) it is sufficient to consider only the scattering structure factor $S(\mathbf{q})$. In the Born approximation, its square is given by the following expression:[16]

$$S^2(\mathbf{q}) = \frac{1}{NLM} \cdot \frac{\sin^2(N\mathbf{q}\mathbf{a}_1/2)}{\sin^2(\mathbf{q}\mathbf{a}_1/2)} \cdot \frac{\sin^2(L\mathbf{q}\mathbf{a}_2/2)}{\sin^2(\mathbf{q}\mathbf{a}_2/2)} \cdot \frac{\sin^2(M\mathbf{q}\mathbf{a}_3/2)}{\sin^2(\mathbf{q}\mathbf{a}_3/2)}, \tag{3.6.6}$$

where $\mathbf{q} = \mathbf{k}_s - \mathbf{k}_i$ is the scattering vector, and $\mathbf{a}_1$, $\mathbf{a}_2$, and $\mathbf{a}_3$ are the translation vectors of FCC opal lattice primitive cells. Equation 3.6.6 allows one to calculate $S(\mathbf{q})$ for different angles of scattering and sample orientation determined by the $\mathbf{k}_s$ and $\mathbf{k}_i$ vector directions. The size of the sample is the calculation variable and is determined by the number of scatterers N, L, M along the directions of $\mathbf{a}_1$, $\mathbf{a}_2$, and $\mathbf{a}_3$ vectors, respectively. For the calculations, we used the values $D \equiv |\mathbf{a}_i| = 720$ nm, $N = 1$, $L = M = 1500$, which correspond to the linear dimensions of the sample being about a millimeter and the sample thickness being one layer of a-SiO_2 particles. The results of the numerical calculation are in excellent agreement with the experimental results, as seen in Figure 3.6.4b and c. However, this calculation cannot interpret all the features observed in the diffraction patterns.

Equation 3.6.1 describing diffraction from a 2D periodic system of scatterers allows us to perform a detailed analysis of experimental data. Let us place the centers of spheres at the nodes of the hexagonal Bravias lattice with a nearest-neighbor distance D. One can then obtain a simple formula for the diffracted light wavelength:

$$\lambda_{hk} = \frac{D\sqrt{\varepsilon_{av}}}{2}\left(\frac{3}{h^2 + hk + k^2}\right)^{1/2} \left|\sin(\Theta - \theta) - \sin\theta\right|, \tag{3.6.7}$$

where θ and $(\Theta - \theta)$ are the angles of incidence and scattering counted from normal to scattering plane, h and k are integers, ε_{av} stands for the average permittivity, and D is the a-SiO_2 particle diameter, which in the first approximation can be equated to the modulus of translation vectors $|\mathbf{a}_i|$.

Expression 3.6.7 implies that the zeroth-order diffraction ($h = k = 0$) can be observed only when $\sin(\Theta - \theta) = \sin\theta$, that is, $\Theta = 2\theta$ or $\Theta = 180°$. The first condition corresponds to specular reflection of white light from film surface, whereas the second corresponds to directly transmitted beam (two (00) beams in Figure 3.6.3e). The first-order diffraction meets the condition $(h^2 + hk + k^2)^{1/2} = 1$, which is satisfied for the pairs of (h,k) indices being equal to $(\bar{1}1)$, $(1\bar{1})$, (1,0), (0,1), $(0\bar{1})$, or $(\bar{1}0)$, whereas the investigated diffraction in the XY plane corresponds to $(\bar{1}1)$ and $(1\bar{1})$ pairs of indices. The second-order diffraction meets the condition $(h^2 + hk + k^2)^{1/2} = 2$, whereas the diffraction in the XY plane is defined by $(\bar{2}2)$ and $(2\bar{2})$ pairs of indices. Higher orders of diffraction can be treated similarly. The experimental data can be interpreted as solutions $\Theta = f(\theta,\lambda)$ of Equation 3.6.7 for the first three orders (0,0), $(\bar{1}1)$, $(1\bar{1})$, $(\bar{2}2)$, $(2\bar{2})$ of diffraction as illustrated in Figure 3.6.4d for monochromatic light with wavelengths $\lambda = 415$, 500, and 585 nm.

Comparing the results presented in Figure 3.6.4b and d, one can easily interpret all observed features. The zeroth-order diffraction corresponds to the specular reflection from the hexagonal plane on all wavelengths and is observed experimentally as a bright white stripe at $\Theta = 2\theta$, which splits Figure 3.6.4b into two areas. These areas correspond to light diffraction of $(1\bar{1})$ and $(2\bar{2})$ orders, that is, "to the right" with respect to $\mathbf{k}_i$ and of $(\bar{1}1)$ and $(\bar{2}2)$ orders, that is, "to the left" (see Figure 3.6.4d). Theoretical dependences $\Theta = f(\theta,\lambda)$ for a nonzero order of diffraction represent ovals, with the oval corresponding to a higher order of diffraction lying within the oval of the lower order for the given wavelength. Note that on the experimental pattern diffraction is not observed close to the angle values $\Theta = \theta \pm 90°$, which are marked with dashed lines in Figure 3.6.4b. This diffraction geometry implies propagation of diffracted light inside the film and exit of light from the film edge that significantly decreases the intensity of the registered light. Therefore, we can conclude that *light diffraction from a six-layer-thick opal film has a vivid 2D nature.*

3.6.3 Two-Dimensional to Three-Dimensional Transition in Optical Diffraction

The alteration of physical properties with the transition from 2D to 3D objects creates tremendous interest for researchers, and optical diffraction is no exception.[17] When considering 2D diffraction patterns, represented in (θ, Θ) coordinates as in Figure 3.6.4, a challenging question arises: What is the mechanism of transformation of the continuous ovals into a set of straight parallel lines expected in the case of Bragg diffraction from a low-contrast 3D photonic structure?

To answer this question one can use numerical modeling of the diffraction patterns computing the square of scattering structure factor $S^2(\mathbf{q})$ using Equation 3.6.6 as a function of the film thickness. Thickness is determined by the varying number of layers N whereas the other geometrical parameters are fixed: $L = M = 3000$ and $D = 330$ nm, which corresponds to the linear dimensions of the sample of about a millimeter. As is well known, most opal samples have twinned FCC structure. To describe the diffraction patterns observed in the experiment, the computation is performed for the twinned structure assuming that both twins (FCC-I and FCC-II) contribute equally to the diffraction patterns.[10]

Light scattering from a single hexagonal layer formed by close-packed a-SiO_2 spheres corresponds to the 2D diffraction case ($N = 1$) (Figure 3.6.5a). The calculated diffraction pattern presented in the (θ, Θ) coordinates and in the range $-90° \leq \theta \leq 90°$ of angles consists of two interrupted ovals and a straight line $\Theta = 2\theta$ separating them that corresponds to the specular reflection from the hexagonal layer. A presentation of the same result in the range of angles $0° \leq \theta \leq 180°$ would merge the semiovals into one solid oval that represents 2D diffraction from a single hexagonal layer for a given wavelength range. In this calculation, the diameter of a-SiO_2 particles was assumed to be equal to 330 nm, which corresponds to the diameter of the particles of the sample studied by Samusev et al.[10] Note that the 2D diffraction pattern shown in Figure 3.6.5a differs significantly from the one given in the previous section for the sample with the diameter of particles of 720 nm (Figure 3.6.4). The

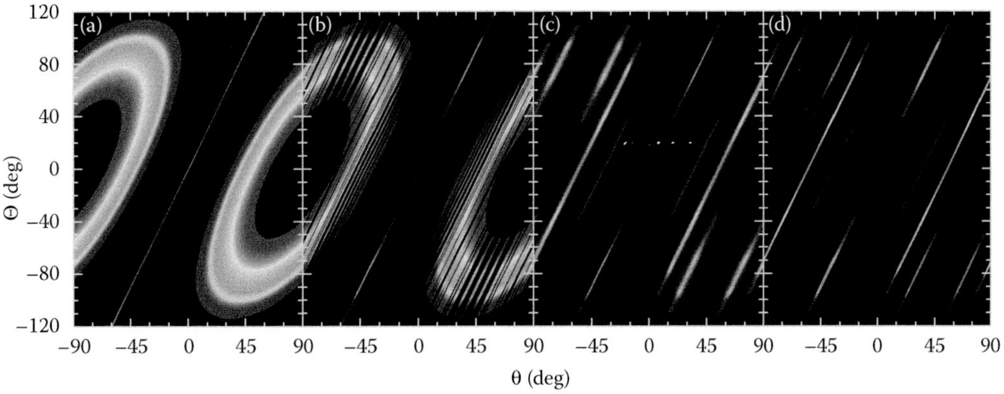

FIGURE 3.6.5
(**See color insert.**) Intensity of the visible (400 ÷ 700 nm) light diffraction from the twinned face-centered cubic opal structure with different number of layers: $N = 1$ (a), $N = 20$ (b), $N = 50$ (c), and $N = 3000$ (d). θ is the angle of incidence and Θ is the angle of scattered light registration. The calculations of the square of scattering structure factor $S^2(\mathbf{q})$ were performed for the structures with a-SiO_2 particle diameter of 330 nm. The opal filler is water ($\varepsilon_{H_2O} = 1.78$). (From A. K. Samusev et al., Phys. Solid State **53**(7), 1415, 2011. With permission.)

most important difference is the lack of the inner diffraction oval in Figure 3.6.5a, which falls into the ultraviolet spectral range and is therefore not shown in Figure 3.6.5.

Increasing the sample thickness leads to a transition from 2D to 3D diffraction regimes. A 2D structure is constructed with only two primitive vectors so that the diffraction condition should satisfy only two Laue equations (3.6.3) out of three. In contrast, for a 3D structure all three equations are to be used, thus implying additional limitations to the diffraction conditions and reducing the number of solutions.

The transformation of the diffraction patterns in the (θ, Θ) coordinates during the transition from 2D to 3D is presented in Figure 3.6.5. With increasing number of layers, certain regions of both the semiovals and the diagonal fade out, unlike the initial 2D diffraction pattern. First, the closed semiovals break up into bands, then the bands narrow down, and, finally, in the case of the bulk sample consisting of $3000 \times 3000 \times 3000$ unit cells, these regions turn into a set of straight segments parallel to the $\Theta = 2\theta$ line (see Figure 3.6.5d). Note the transformation of the line $\Theta = 2\theta$ itself. It changes from a strip of reflected white light into colored segments formed by diffracted light. As a result, the expected Bragg diffraction pattern is indeed reached. A detailed interpretation of the 3D diffraction pattern represented in the (θ, Θ) coordinates is given in the following section.

3.6.4 Three-Dimensional Optical Diffraction and Selective Switching of Diffraction Reflections

In Chapter 2.7, a calculation of photonic stop-band width in low-contrast multicomponent photonic crystals (MPhCs) was presented. MPhCs are structures composed of three or more homogeneous or inhomogeneous components.[10,18] The calculation results indicate that different $\{hkl\}$ stop-bands can be selectively "switched off" in MPhCs while varying the permittivity of one of the components. In Chapter 3.5, it was demonstrated that opal-based PhCs belong to the class of MPhCs. The multicomponent nature of the opal structure is created by the fact that the constitutive a-SiO_2 particles consist of smaller amorphous silica dioxide grains and possess an inhomogeneous inner structure: the porous nucleus covered by a denser thin coat.[18,19] The opal transmission spectra studies have established that the parameters of photonic stop-bands corresponding to the different $\{hkl\}$ families of crystal planes have different dependences on the permittivity of the filler ε_f.[18] For the bulk opal sample, the stop-bands of the {111} family significantly weaken when the filler permittivity approaches the value of $\varepsilon_f^0(\mathbf{g}_{111}) \approx 1.82$, whereas the intensities of the dips corresponding to the {220} family stop-bands significantly decrease when ε_f approaches $\varepsilon_f^0(\mathbf{g}_{220}) \approx 1.92$. Note that at the critical points $\varepsilon_f = \varepsilon_f^0(\mathbf{g}_{hkl})$ the stop-bands in opal spectra do not vanish, but undergo a complex transformation leading to the formation of *pass*-bands. This effect was discussed in the context of the Fano resonance model[19,20] and is described in Chapter 3.5.

In low-contrast PhCs, stop-bands in transmission spectra and the reflections in the diffraction patterns both originate from the Bragg light reflection from certain $\{hkl\}$ crystallographic plane sets. Taking into account the effects discovered in opal transmission spectra, one can expect some nontrivial transformation of the Bragg $\{hkl\}$ reflections in the diffraction patterns, depending on the dielectric contrast. Such effects are indeed experimentally observed in light diffraction from bulk opal samples while varying ε_f.[10] The experimental setup illustrated in Figure 3.6.3e is used. An opal sample is fixed on a holder in the center of a cylindrical flask serving as a screen for observing and capturing the diffraction patterns. The experiments are performed using the immersion spectroscopy method, which implies consecutive acquisition of the diffraction patterns as a function of the permittivity of a liquid filling the opal matrix.[18] The cylindrical flask was filled with the liquid serving as the opal filler as well as the ambient medium. During the experiment distilled water ($\varepsilon_{H_2O} = 1.78$), glycerine ($\varepsilon_{gl} = 2.16$), and their mixtures were used. This allowed us to vary the filler permittivity within the range $1.78 \le \varepsilon_f \le 2.16$.

The results of the experimental studies of white light diffraction from bulk opal samples as a function of the filler permittivity ε_f are presented in Figure 3.6.6. We are primarily concerned with the behavior of the Bragg reflections corresponding to the different (hkl) plane sets. For the experimental data analysis, one must associate the reflections shown in Figure 3.6.6 in the (θ, Θ) coordinates with specific (hkl) plane sets of the FCC opal lattice. For this purpose, the Bragg wavelength λ_{hkl} of the light beam reflected from the (hkl) plane set was calculated as a function of the incident angle θ, which is counted from the normal to (111) growth plane set. The expression for λ_{hkl}, which is also used in Chapter 3.5 (see Equation 3.5.3), is given by

$$\lambda_{hkl}(\theta) = 2d_{111}\sqrt{\varepsilon_{av}}\left(\frac{3}{h^2+k^2+l^2}\right)^{1/2} |\cos(\theta - \theta_{hkl})|, \tag{3.6.8}$$

where $d_{111} = \sqrt{(2/3)}D$ stands for the distance between neighboring (111) planes, $\varepsilon_{av} = 0.74\varepsilon_{SiO_2} + 0.26\varepsilon_f$ is the average permittivity of the opal FCC structure, and θ_{hkl} is the angle between the [111] and $[hkl]$ directions. The calculation is performed for {111}, {200}, and {220} plane families of the twinned FCC lattice (see Figure 3.6.6a). Bragg wavelengths $\lambda_{hkl}(\theta)$ corresponding to higher-index $\{hkl\}$ plane sets for the sample formed by 330 nm a-SiO_2 particles fall within the ultraviolet spectral range ($\lambda_{hkl}(\theta) < 400$ nm) that was not studied experimentally by Samusev et al.[10] The calculation of the dispersion curves $\lambda_{hkl}(\theta)$ was performed for the incident wavevector $\mathbf{k}_i$ lying in the scanning plane $X \to L_g \to K_g \to L$

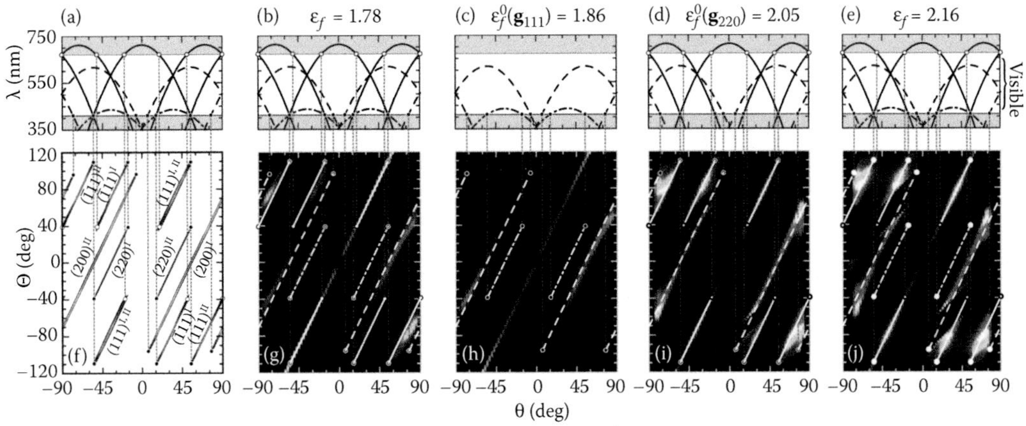

FIGURE 3.6.6
(See color insert.) (a)–(e) The dispersion relations $\lambda_{hkl}(\theta)$ of the Bragg wavelengths for the twinned face-centered cubic (FCC) opal lattice formed by a-SiO_2 particles with D = 330 nm. $\lambda_{\{111\}}(\theta)$ are indicated by solid curves, $\lambda_{\{200\}}(\theta)$ by dashed curves, and $\lambda_{\{220\}}(\theta)$ by dot-dash curves (see Equation 3.6.8). Panel (c) misses the dispersion curves $\lambda_{\{111\}}(\theta)$ and panel (d) misses the dispersion curves $\lambda_{\{220\}}(\theta)$. The visible spectral range of electromagnetic waves is separated out by two horizontal lines; the ultraviolet and infrared spectral ranges are darkened. (f) The calculated intensity of visible (400 ÷ 700 nm) light diffraction from the twinned FCC opal structure consisting of 3000 × 3000 × 3000 scatterers immersed in water. (g)–(j) The experimental patterns of light diffraction from the synthetic opal sample immersed in fillers with dielectric permittivity ε_f = 1.78 (g), 1.86 (h), 2.05 (i), and 2.16 (j). θ is the angle of incidence counted from growth (111) plane set of opal FCC lattice and Θ is the angle of scattered light registration. (From A. K. Samusev et al., Phys. Solid State **53**(7), 1415, 2011. With permission.)

of FCC-I Brillouin zone, as this case was realized in the experiment. In the Brillouin zone of FCC-II opal twin, this plane coincides with the $L \to K_g \to L_g \to X$ scanning plane. The representation of the Brillouin zone of the FCC lattice and its cross sections can be found in Chapter 3.5 and in the study by Samusev et al.[10]

The angle of the specular Bragg reflection from different (hkl) plane sets is given by

$$\Theta = 2(\theta - \theta_{hkl}). \quad (3.6.9)$$

The angles of interest θ_{hkl} between [111] and [hkl] directions are $\theta_{\bar{1}11} = 70.5°$, $\theta_{200} = -54.7°$, and $\theta_{022} = 35.3°$ for the FCC-I lattice and $\theta_{\bar{1}11} = -70.5°$, $\theta_{200} = 54.7°$, and $\theta_{022} = -35.3°$ for the FCC-II lattice. Therefore, Bragg diffraction reflections associated with different (hkl) plane sets are expected to appear in the (θ, Θ) diagram as a set of parallel straight lines shifted by θ_{hkl} along θ axis with respect to the Θ = 2θ diagonal corresponding to the (111) reflection common for both FCC-I and FCC-II twins ($\theta_{111} = 0$).

By comparing the upper panels (a–e) in Figure 3.6.6, which contain the Bragg wavelength dispersion relations $\lambda_{hkl}(\theta)$ for {111}, {200}, and {220} families, with the lower panels (f–j), one can interpret all the observed diffraction reflections. Note that in the lower panels theoretical (Figure 3.6.6f) and experimental (Figure 3.6.6g–j) data are presented only for the visible range, which form segments rather than continuous straight lines passing through the entire image. The interruptions correspond to diffraction in either the ultraviolet or infrared range.

The effect of the selective weakening of the reflections corresponding to {111} and {220} families of planes appears to be the most important experimental result. Let us first consider this effect for (022) reflections, which are observed as blue stripes at $-50° < \theta < -15°$ (FCC-I), $15° < \theta < 50°$ (FCC-II), and $\Theta = 2(\theta \pm 35.3°)$ ("+" for FCC-I, "−" for FCC-II) angles within the range of registration angles $-40° < \Theta < 40°$. Note that outside of this range (022) reflections are not visible as $\lambda_{022}(\theta)$ falls within the ultraviolet range, as clearly seen in Figure 3.6.6a–c and e. From the dispersion curves $\lambda_{hkl}(\theta)$ represented in Figure 3.6.6d ($\varepsilon_f = 2.05$), both branches of the (022) plane set are eliminated. This simplifies singling out the diffraction associated with the other {*hkl*} families. The intensity of (022) reflections are seen to significantly depend on ε_f. The lowest intensity of (022) reflections corresponds to the value of $\varepsilon_f^0(\mathbf{g}_{220}) = 2.05$, whereas the reflections of {111} and {200} families are still clearly observed on a diffraction pattern with the same ε_f (Figure 3.6.6i).

Let us now discuss the immersion dependence of {111} family reflections, marked with solid lines in Figure 3.6.6. The {111} family of planes is represented by reflections from (111) and ($\bar{1}11$) crystal planes. The position of the (111) reflection meets the requirement $\Theta = 2\theta$ and therefore coincides with the specular reflection of incident white light. Nevertheless, the (111) reflection is experimentally observed within the range of angles $-55° < \theta < -20°$ and $20° < \theta < 55°$, while its color changes through the entire visible range from blue to red in full accordance with the computational dependence shown in Figure 3.6.6f. Within the range of angles $-20° < \theta < 20°$, (111) reflection is not registered as $\lambda_{111}(\theta)$ falls within the infrared range, whereas within the range of angles $\theta < -55°$ and $\theta > 55°$, $\lambda_{111}(\theta)$ falls within the ultraviolet range so this reflection is also not visible.

The intensity of all reflections of the {111} family exhibits pronounced dependence on the dielectric contrast with a minimum at $\varepsilon_f^0(\mathbf{g}_{111}) = 1.86$. From the dispersion curves $\lambda_{hkl}(\theta)$ represented in Figure 3.6.6, all branches of the {111} family are eliminated. This simplifies singling out the diffraction associated with the other {*hkl*} families, specifically with {200} (these reflections are of virtually all colors of the visible range—from blue to orange) and with {220} (only blue reflections).

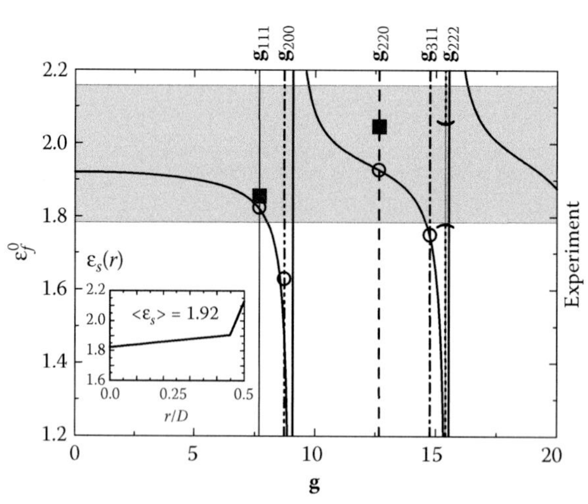

FIGURE 3.6.7
The filler permittivity $\varepsilon_f^0(\mathbf{g})$ that corresponds to the immersion condition for an opal structure as a function of the reciprocal lattice vector modulus **g** calculated from Equation 3.5.4 for the permittivity profile $\varepsilon_s(r)$ simulating the *a*-SiO_2 spheres. The $\varepsilon_s(r)$ profile is given in the inset. The moduli of the shortest reciprocal lattice vectors $\mathbf{g}_{hkl}$ are shown by vertical lines. Open circles represent values of $\varepsilon_f^0(\mathbf{g}_{hkl})$ obtained in the transmission experiments (Figure 3.5.9 in Chapter 3.5); black squares represent the diffraction data. The gray region is for the experimental range of ε_f.

The fitting of the $\varepsilon_f^0(g)$ function to experimental data is discussed in detail in Chapter 3.5. The result is presented in Figure 3.6.7 where both transmission and diffraction experimental data are displayed. It should be mentioned that the values of ε_f^0 obtained in diffraction experiments on opal films differ from the values derived from the transmission spectroscopy of bulk opal samples (Chapter 3.5) because of a technological difference in the preparation of a-SiO_2 particles forming opal structures for transmission and diffraction experiments.

Therefore, one can conclude a possibility of selective control over the intensity of Bragg diffraction reflections in low-contrast MPhCs. It is important to note that the phenomenon of selective stop-band switching discussed in Chapter 3.5 and the effect of selective attenuation of $\{hkl\}$ diffraction reflections are complementary.

3.6.5 Sample Thickness Dependence in Small-Angle X-Ray Diffraction

X-ray diffraction is described by the same laws as light diffraction. However, about four orders of magnitude difference in the wavelength results in a number of essential differences in the appearance of the diffraction phenomenon. To illustrate the differences, let us consider the effects of the sample thickness in x-ray diffraction. Figure 3.6.8a–c illustrates the modification of the reciprocal lattice of the opal-like structure. If only a single close-packed layer is deposited on the substrate, the reciprocal lattice consists of scattering rods normal to the crystal plane XZ (Figure 3.6.8a). In this case, the scattering structure factor $S(\mathbf{q})$ does not change along the rods and the decay of the scattering intensity is solely determined by the particle form factor $F(q)$. On increase of the number of layers, the interference between their contributions will lead to appearance of additional structure in $S(\mathbf{q})$ along these Bragg scattering rods, which eventually will split into separate reflections, which are broadened along the layer normal Y (Figure 3.6.8b). With further increase of the number of layers, these reflections will transform into delta-function-like Bragg reflections characteristic for an ideal large 3D crystal (Figure 3.6.8c).

The appearance of the diffraction pattern depends on the sample orientation, which is illustrated in Figure 3.6.8d and e in the (θ, Θ) coordinates introduced above. Here only monochromatic x-rays with a wavelength of $\lambda = 0.1$ nm are considered. The crystal is assumed to consist of N hexagonal close-packed layers of touching spheres with the diameter $D = 400$ nm. One can see that x-ray diffraction is very much concentrated around the forward direction, that is, around $\Theta = 180°$. For a single ($N = 1$) crystal plane (Figure 3.6.8d), observation angle Θ slowly varies on sample rotation by angle θ. For a crystal consisting of $N = 15$ layers, the interference between them splits the monotonic lines in Figure 3.6.8d into segments, where most of the diffraction intensity is concentrated (Figure 3.6.8e).

Note an essential difference between the intensity distributions for visible light diffraction in Figures 3.6.3 through 3.6.6, and that for x-ray diffraction in Figure 3.6.8d and e. For x-rays, the diffraction "traces" appear to be nearly symmetric with respect to $\Theta = 180°$. The diffraction peaks at $\Theta > 180°$ and $\Theta < 180°$, which correspond to $\mathbf{q} = \mathbf{g}_{hkl}$ and $\mathbf{q} = -\mathbf{g}_{hkl}$, can be simultaneously observed at a single θ value, even in the case of a multilayer crystal (Figure 3.6.8e). Moreover, multiple diffraction orders, $2\mathbf{g}_{hkl}$, $3\mathbf{g}_{hkl}$, and so on, can be simultaneously detected. Such observations are impossible with visible light. The reason for these effects is that for x-rays the radius $2\pi/\lambda$ of the Ewald sphere is very large in comparison to the scale of interest. As a result, for a crystal with $N = 15$ layers the Ewald sphere can be simply approximated by a plane. However, on further increase of the crystal thickness and

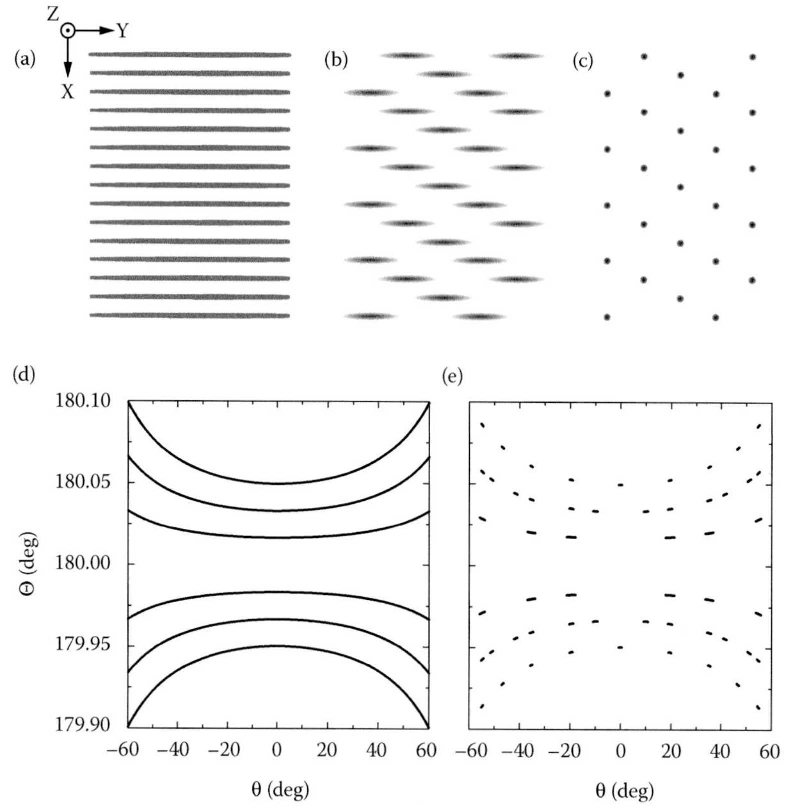

FIGURE 3.6.8
Sketch of the reciprocal lattice of a single monolayer of microspheres (a), a thin multilayer crystal (b), and a thick crystal (c). The film is situated in the XZ plane, and the number of layers in (a)–(c) grows along the Y axis. (d), (e) Calculated positions of hard x-ray ($\lambda = 0.1$ nm) diffraction reflections in the (θ, Θ) coordinates for a monolayer film sample (d) and twinned face-centered cubic crystal with $N = 15$ layers. The calculations of the square of scattering structure factor $S^2(\mathbf{q})$ were performed for the structures with a-SiO_2 particle diameter of 400 nm. The length of the segments in panel (e) is selected to display the full-width at half-maximum of every maximum in $S^2(\mathbf{q})$. It is assumed in panel (e) that the crystal consists of two twin domains in equal proportion, whereas in (b) and (c) only a single domain with unique stacking direction is illustrated.

further sharpening of the crystal reflections, effects of the curvature of the Ewald sphere should show up.[21]

Let us now estimate the sample thickness at which the curvature effects can be observed and the possible effect of the curvature of the Ewald sphere in x-ray diffraction. Let us assume that the x-rays are incident *exactly* along the (*hkl*) crystallographic plane set as sketched in Figure 3.6.1b and d. From a simple geometrical consideration one can see that the *hkl* site of the reciprocal lattice of an FCC crystal will deviate from the Ewald sphere by $\Delta = \lambda g_{hkl}^2/(4\pi) = \pi\lambda(h^2 + k^2 + l^2)/(2D^2)$. Because of this deviation the Ewald sphere should miss the reciprocal lattice node and no Bragg diffraction should be observed in a *perfect infinite* opal-like crystal. In practice, however, the nodes of the reciprocal lattice of a real opal-like crystal are usually broadened because of finite thickness of the sample, presence of disorder, and so on.

Just to give a few typical numbers, let us consider diffraction of x-rays with $\lambda = 0.1$ nm on an FCC crystal of spheres with diameter $D = 400$ nm. For the (220) reflections the deviation of the Ewald spheres from a plane is $\Delta = 8 \cdot 10^{-6}$ nm^{-1}. For a crystalline film of a finite thickness L the broadening of the reciprocal lattice nodes is of the order of $2\pi/L$ (full-width at half-maximum). To observe the effect of the curvature of the Ewald sphere, one requires $2\pi/L < \Delta$, or $L > 800$ μm, that is, 2000 times the sphere diameter!

In practice, the thickness of opal crystals is usually far too small to observe the effects of the curvature of the Ewald sphere. The diffraction pattern is then equal to the squared modulus of the Fourier transform of the 2D projection of the electron density in the sample on a plane normal to the x-ray beam.

3.6.6 Order and Disorder from Microradian X-Ray Diffraction

Here one can find examples of the characterization of structure and disorder in PhCs using small-angle x-ray diffraction with microradian resolution (microradian x-ray diffraction). For more detailed interpretation of the microradian diffraction patterns and the analysis of structural order and disorder of opals, we refer the readers to our previously published articles.[21–26]

To characterize the 3D crystal structure one usually needs to measure the microradian x-ray diffraction patterns with different sample orientations. Some typical examples of diffraction patterns are presented in Figure 3.6.9a–c for film opals recorded at different angles of rotation θ. As shown in Section 3.6.5, the Ewald sphere curvature is negligible, so many subsequent diffraction orders can be observed simultaneously. The patterns are measured at normal incidence $\theta = 0°$ (a) and after sample rotation by $\theta = -35°$ (b) and $\theta = 55°$ (c). Figure 3.6.9d–f presents the expected Bragg reflections for an FCC crystal with corresponding orientations. One can see that most of the Bragg peaks in Figure 3.6.9a–c correspond to an FCC structure with the size of the conventional cubic unit cell of $a = \sqrt{2}D = d_{hkl}\sqrt{h^2 + k^2 + l^2}$, where $d_{hkl} = 2\pi/g_{hkl}$ is the interplane distance corresponding to an (hkl) Bragg reflection.

The diffraction peak intensities are proportional to the product of the squared structure factor $S^2(\mathbf{q})$—as a result of scattering from the structure—and the squared form factor $F^2(q)$—as a result of scattering from a particle that is the base element of this structure. Figure 3.6.10 shows the squared form factor of the homogeneous spherical particles described by $F^2(q) = [\sin(qR) - qR\cos(qR)]^2/(qR)^6$, where R is the radius of a colloidal particle. The strength of the form factor for a number of Bragg reflections is illustrated by points at the corresponding q-values. This allows us to clarify not only the positions of the Bragg peaks but also their relative intensities. It is clear from Figure 3.6.10 that the form factor for the (111) reflection is slightly higher than that for the (220) reflection, whereas both are two orders of magnitude higher than that for the (200) reflection. Furthermore, the low visibility of the (422) and (222) in Figure 3.6.9a and b, respectively, can be explained by their proximity to the form factor minima (Figure 3.6.10). Such theoretical calculations are in good agreement with the results obtained from the experiment (Figure 3.6.9a–c).

In addition, in Figure 3.6.9a–c there are a number of additional peaks that do not correspond to an ideal FCC structure. These features can be caused by the stacking faults parallel to the substrate as well as by finite crystal thickness. The more striking are the rods of diffuse scattering seen in Figure 3.6.9b. These Bragg scattering rods are parallel to the $\langle 111 \rangle$ crystallographic directions and indicate the presence of planar defects parallel to two types of HCP layers that make an angle of 70.5° with the substrate.

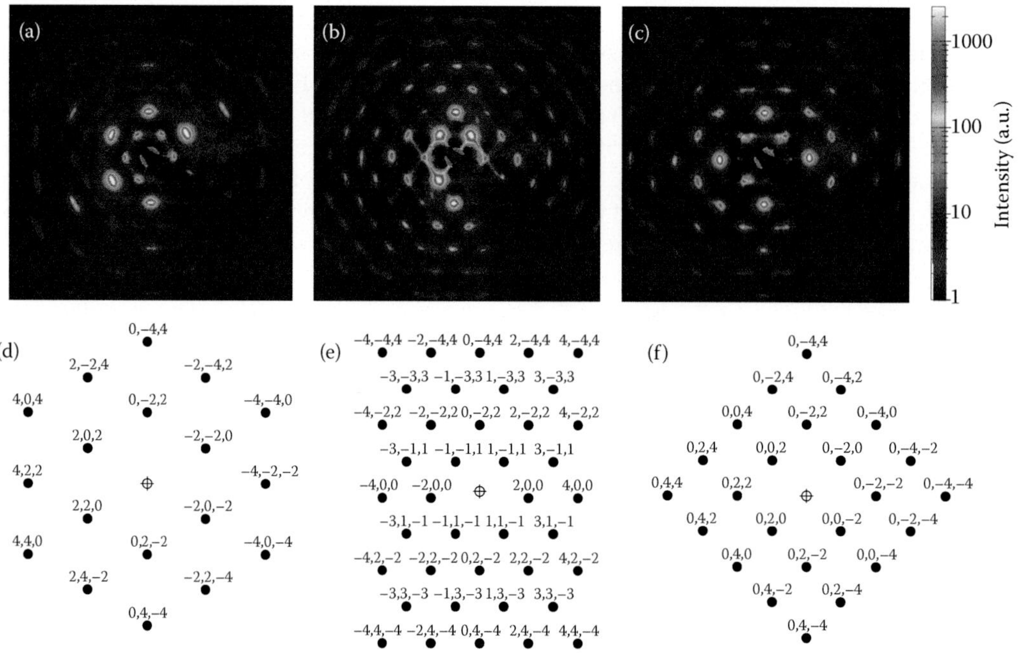

FIGURE 3.6.9
(**See color insert.**) Microradian x-ray diffraction patterns of opal-like photonic films of polystyrene colloidal spheres. The film is about 20 layers thick. The patterns are measured at normal incidence $\theta = 0°$ (a) and after sample rotation by $\theta = -35°$ (b) and 55° (c). The sketches (d–f) below each pattern (a–c) present the expected Bragg reflections for a face-centered cubic crystal with corresponding orientations.

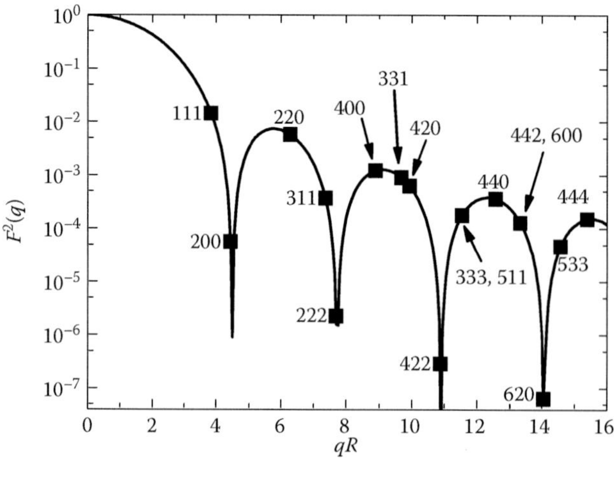

FIGURE 3.6.10
Theoretical squared scattering form factor $F^2(q)$ of a single uniform sphere (solid line). The symbols display the squared form factor values for different (*hkl*) reflections of a face-centered cubic structure of touching uniform spheres.

Figure 3.6.11 presents the intensity distribution along the rods between the (111) and (200) reflections in Figure 3.6.9b. The intensity distribution along the Bragg scattering rods is usually described by Wilson's theory,[2] which assumes that the probability α of finding a close-packed FCC layer is independent of the stacking environment of the neighboring layers. In this model, the fraction of FCC-type layers in a crystal is labeled α. Figure 3.6.11 shows the measured intensity profile along a first-order Bragg rod. It also shows the profile along a Bragg rod, calculated within Wilson's theory for a crystal with dominant FCC stacking ($\alpha = 0.9$). One can see that the theory predicts two times more maxima than experimentally observed. The reason for this discrepancy is related to the fact that in the Wilson model a single HCP layer changes the direction of the FCC growth from the ...ABCABC...- to ...ACBACB...-type stacking, giving rise to an additional set of Bragg reflections along the Bragg rod (dashed line in Figure 3.6.11). If stacking disorder is caused by pairs of HCP layers and only one growth direction dominates, then Wilson's theory should be modified by adapting Paterson's model[27] to fit the data correctly (solid line in Figure 3.6.9). In this model, each subsequent layer is stacked with a probability β in a particular direction. With the probability $(1 - \beta)$, however, a fault is created by selecting alternative stacking direction. For β close to 1 the original stacking direction will most probably be restored in further layers, leaving behind a stacking fault, for example, in the sequence ...ABC**A**CABC... the central two layers are in an HCP environment with a deformation stacking fault between them. Such a "double"-stacking fault is equivalent to sliding one part of a perfect FCC crystal (...ABCA | in the example above) relative to the other part along the stacking fault (| BCAB... becomes then | CABC...). One can see that the intensity profile $I(\mathbf{q}) \propto S^2(\mathbf{q}) \cdot F^2(q)$ along the first-order

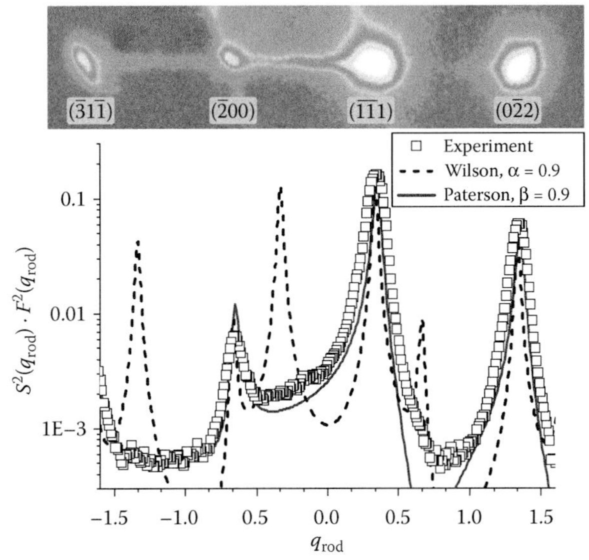

FIGURE 3.6.11
Intensity profile along one of the Bragg scattering rods in Figure 3.6.9b. The corresponding piece of the diffraction pattern is reproduced above the graph. The coordinate q_{rod} along the Bragg rod is normalized to the length of $\mathbf{q}_{111}$ wavevector corresponding to the spacing $d_{111} = a/\sqrt{3}$ between close packed layers. $q_{rod} = 0$ corresponds to the point closest to the origin of the reciprocal space. The dashed and solid lines display results of model calculations using the Wilson theory and Paterson model, respectively.

Bragg rod, calculated with the modified theory[27] for $\beta = 0.9$ (Figure 3.6.11), closely follows the experimental result, confirming the presence of 10% double-stacking faults in at least two out of four $\langle 111 \rangle$ directions in this crystal with a dominating FCC structure.

3.6.7 How X-Ray Diffraction Can Serve the Technology of Opal Growth

Here we present an example of a study where x-ray diffraction is used to properly adjust the fabrication process of opal films. A direct current electric field is applied normal to the substrates during vertical deposition of colloidal crystals. The external electric field is shown to be an excellent tool to control the crystallization process of the opal-like structures. It can act as the driving force, which moves colloidal particles to the substrate or induces dipole–dipole interparticle interactions, and therefore can improve the crystal quality and even change the crystal structure.[28]

Colloidal crystals made of polystyrene microspheres were grown by the vertical deposition technique in the presence of an external electric field $U = -1.5 \div +1.5$ V.[24] Negatively charged polystyrene spheres were deposited onto indium tin oxide glass vertically aligned electrodes (substrates) in a cylindrical glass cell (5 cm diameter) using a Solartron 1287 potentiostat. The distance between the electrodes was 3 cm. The electrodes were fixed precisely parallel to each other. X-ray studies were performed using a microradian x-ray diffraction setup.[26] To achieve angular resolution needed to study these large-period structures, the beam was focused at the detector by a set of compound refractive lenses.[29] The colloidal films were first mounted perpendicular to the x-ray beam. Samples were then rotated around the vertical axis (the $[10\bar{1}]$ axis of the crystal) within the range $-75° \le \theta \le 75°$, and the diffraction patterns were recorded at each degree of rotation.

Figure 3.6.12 shows typical examples of microradian diffraction patterns measured at $\theta = 0°$ (Figure 3.6.12a, d, and g) and at $\theta = -35°$ (Figure 3.6.12b, e, and h) for colloidal crystals synthesized at $U = -1.5$, 0, and +1.5 V. In the diffraction patterns one can clearly identify a large number of Bragg reflections, which can be assigned to the reciprocal lattice of an ideal FCC crystal structure with lattice constant $a = 750$ nm. Corresponding indices are shown in Figure 3.6.12. In addition to the assigned Bragg peaks, the diffraction patterns also show features that cannot be assigned to the FCC structure. They can be truly Bragg reflections (e.g., corresponding to an HCP structure coexisting with FCC), or can be sections of diffuse objects in the reciprocal space. The latter can be related to the finite film thickness and/or to the presence of stacking faults along the [111] direction. Complete information on this type of disorder could be inferred from the distribution of diffracted intensity in the 3D reciprocal space (see Section 3.6.8).

Figure 3.6.12c, f, and i show the intensities of (220), $(1\bar{1}\bar{1})$, $(1\bar{1}1)$, and (202) reflections observed in Figure 3.6.12a and b, d and e, as well as g, and h, respectively, as functions of the sample rotation angle θ. We note that in contrast to the example considered in the previous section, where we dealt with a single crystal with unique stacking direction, here one observes the coexistence of twin domains with different stacking directions. The relative amounts of ...ABCABC... and ...ACBACB... stacking sequences can be estimated by comparing the integral intensities of the corresponding peaks: (111)-type at $\theta = -19.5°$ (...ABCABC...) and $\theta = +19.5°$ (...ACBACB...); (202)-type at $\theta = -54.7°$ (...ABCABC...) and $\theta = +54.7°$ (...ACBACB...).

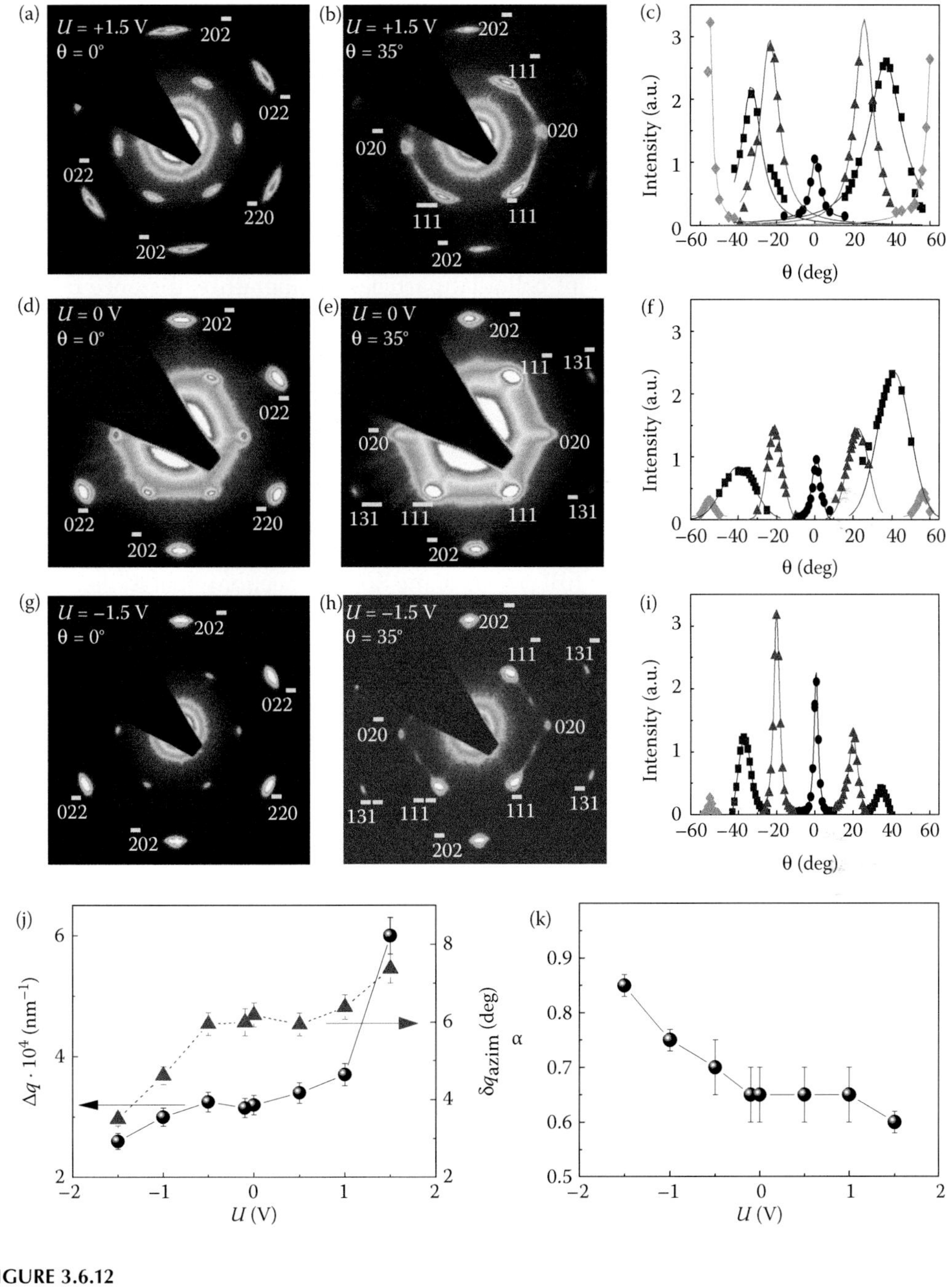

FIGURE 3.6.12
(See color insert.) The diffraction patterns obtained at rotation angles $\theta = 0°$ (a, d, g) and 35° (b, e, h) for samples synthesized at a few selected voltage values $U = -1.5$ V (g, h), 0 V (d, e), and +1.5 V (a, b). Panels (c), (f), and (i) present corresponding rocking curves: the θ dependence of the intensity for (220) (circles), $(1\bar{1}\bar{1})$ (triangles), $(1\bar{1}1)$ (squares), and (202) (rhombs) Bragg reflections. Panel (j) displays the longitudinal (Δq) and transversal (δq_{azim}) width of $(1\bar{1}1)$ reflection (measured at $\theta = 19°$) as a function of the potential applied on colloidal film synthesis. The dependence of stacking probability α of a colloidal film on applied voltage U is given in panel (k). Negative (positive) values of U correspond to cathode (anode) polarizations.

The variation of the crystal quality can also be derived from the width of the diffraction reflections. The results obtained from the patterns measured at $\theta = 0°$ are summarized in Figure 3.6.12j. The full-width at half-maximum of the diffraction maxima in the azimuthal (δq_{azim}) and radial (Δq) directions characterizes the mosaicity of the colloidal films and the average crystallite size (Λ), respectively. One can see that the mosaicity δq_{azim} of the colloidal crystals decreases from 8° to 3° when the applied voltage changes from +1.5 to −1.5 V. Higher values of applied potential lead to significant disorientation of the domains in colloidal films grown at anode polarizations. One can see in Figure 3.6.12c, f, i, and j that the optimum value of an applied voltage U for the conditions we have used (given concentration of the suspension, pH, charge of colloidal particles, temperature, distance between electrodes, etc.) is around −1.5 V.

3.6.8 Three-Dimensional Reciprocal Space Reconstruction

The collected 2D images shown in Figure 3.6.9 or Figure 3.6.12 can be compiled to reconstruct a full 3D reciprocal space of the sample. This process, called 3D reciprocal space reconstruction, allows one to reveal full details of the Bragg reflections, that is, their broadening in different directions, their integrated intensities, and so on.

An example of a 3D map in reciprocal space is shown in Figure 3.6.13a for the colloidal crystal formed on a cathode at $U = 1.5$ V, described in Section 3.6.7. One clearly sees the presence of the extended rods of diffuse scattering and of localized reflections with well-defined round shapes. This reciprocal lattice is typical for a close-packed structure with stacking faults.

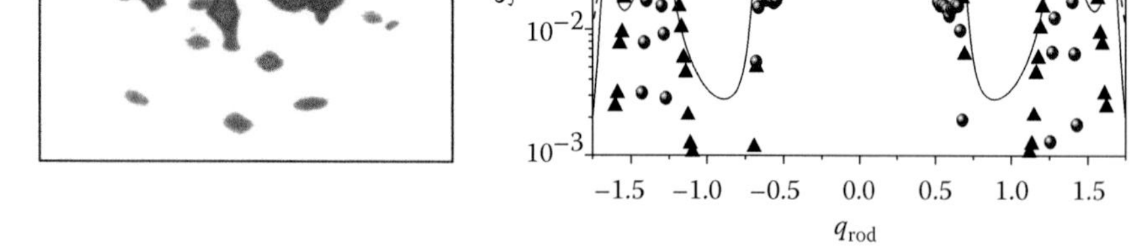

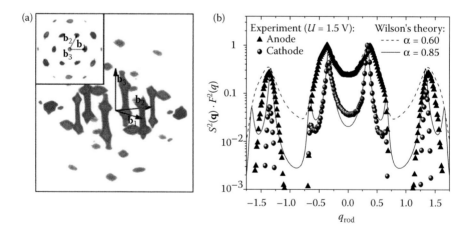

FIGURE 3.6.13
(a) Three-dimensional reconstruction of the reciprocal space for the colloidal crystal obtained on the cathode at $U = 1.5$ V. Only the reflections within $q < 0.03$ nm^{-1} are shown. Hexagonal basis is depicted by the vectors $\mathbf{b}_1$, $\mathbf{b}_2$, and $\mathbf{b}_3$. The inset illustrates a view on the reciprocal space along $\mathbf{b}_3$ (i.e., at $\theta = 0°$). (b) Normalized intensity variation along Bragg rods for two samples prepared on electrodes with two opposite polarizations at ±1.5 V. Intensity profile calculated within Wilson's theory for $\alpha = 0.85$ and 0.60 are shown by solid and dashed lines, respectively. (Reprinted with permission from K. S. Napolskii et al., Langmuir **26**, 2346, 2010. Copyright 2010 American Chemical Society.)

To reconstruct the total reciprocal space, a program for the 3D reconstruction of an object from a set of its 2D cross sections was developed in the MathCad worksheet.[23] The Ewald sphere in the reconstruction region can be considered as planar for the case of small-angle scattering. The set of the initial data consists of (i) diffraction maxima located at the nodes of the reciprocal lattice, (ii) diffuse rods as a result of the stacking faults, and (iii) small-angle diffuse background with a maximum at $q = 0$, which is determined by the scattering of x-rays on the imperfections of the sample structures and on the elements of the optical scheme. As the intensity of the diffuse rods is low, the correct subtraction of the background is an important task determining the possibility of the correct quantitative reconstruction of the reciprocal space pattern.[23]

We analyze the distribution of the diffracted intensity along the Bragg rod with the help of Wilson's theory.[2] Figure 3.6.13b shows the normalized intensity profiles along a Bragg rod extracted from 3D reconstruction for colloidal crystals grown at $U = 1.5$ V on the cathode and the anode. These profiles are compared to the predictions of Wilson's theory with different values of probability α of finding an FCC sequence of stacked layers. The calculated intensity profiles $I(\mathbf{q}) \propto S^2(\mathbf{q}) \cdot F^2(q)$ are shown by lines in Figure 3.6.13b. One can see that while for the crystal grown on the cathode the stacking is mostly FCC ($\alpha \approx 0.85$), a nearly RHCP structure is found on the anode ($\alpha \approx 0.6$). The dependence of parameter α on the applied potential is shown in Figure 3.6.12k.

Thus, microradian x-ray diffraction provides extremely valuable information on the real structure of mesoscopic materials, which cannot be easily obtained by other analytical approaches. Our results based on the analyses of diffraction patterns, 3D reconstructions of the reciprocal space, and Wilson's theory unambiguously show that application of a negative potential can significantly improve the quality of opal-like photonic structures. The suggested synthetic approach is a highly reproducible way for the formation of large-scale high-quality dry colloidal crystals with controllable thickness on conducting substrates.

3.6.9 Conclusions

To summarize, in this chapter diffraction of light and x-rays on opal-like structures is considered. New methodological approaches to collection, processing, and interpretation of experimental data are presented.

In particular, a novel representation of the light diffraction data in the "incident angle – registration angle" (θ, Θ) coordinates is shown to be an effective tool of data analysis. This representation allows one to easily distinguish the reflections originating from 2D diffraction from the ones governed by 3D Bragg diffraction. In addition, structural disorder becomes apparent in the (θ, Θ) representation. It is also demonstrated that the immersion spectroscopy method can be used to selectively switch diffraction reflections. This phenomenon is caused by inhomogeneity of the a-SiO_2 particles that form synthetic opals.

Furthermore, it is demonstrated that microradian x-ray diffraction is a powerful technique that is able to reveal the dominating structure and the presence of disorder in opal-like structures. Short acquisition times, modern 2D detectors, and progress in computing techniques make 3D reconstructions of reciprocal space routinely available. This method provides extremely valuable information on the real structure of mesoscopic materials that cannot be easily obtained by other analytical approaches.

Acknowledgments

The authors are grateful to E. Yu. Trofimova, D. A. Kurdyukov, and V. G. Golubev for providing us with opal films and for numerous fruitful discussions. We thank A. A Kaplyanskii, A. B. Pevtsov, and Yu. E. Kitaev for useful discussions of the work.

The authors express their sincere thanks to students, PhD students, and scientists of the Faculty of Material Sciences of Moscow State University (Russia), to K. S. Napolskii, N. A. Sapoletova, A. A. Eliseev, A. S. Sinitskii, and V. V. Abramova for fruitful cooperation and useful discussions; PhD students of the Physical Faculty of St. Petersburg State University (Russia), to A. A. Mistonov, and to research fellows of the Petersburg Nuclear Physics Institute (Russia) and A. V. Chumakova for their help in conduction of the experiments and preliminary treatment of the data. The synchrotron experiments were performed with active participation of Dmytro Byelov, Jan Hilhorst, Anatoly Snigirev, Wim Bouwman, Irina Snigireva, and Kristina Kvashnina.

The work is supported in part by the RFBR (Project Nos. 10-02-00634 and 10-02-01094), the program of the Russian Government "Development of the scientific potential of the High School" (Nos. 2.1.1/4661, 2.2.2/8155) the RF program (Nos. 02.513.11.3186, 02.513.11.3120, 02.513.11.3352, 02.740.11.5186), and the DAAD programs "Mikhail Lomonosov" and "Leonhard Euler" 2008/2009/2010. We thank the Netherlands Organization for Scientific Research (NWO) for granting us the beam time, and the personnel of the beamline BM-26 of the European Synchrotron Radiation Facility (Grenoble, France) for their excellent support.

References

1. C. Kittel, *Introduction to Solid State Physics* (John Wiley & Sons, Inc. New York, 1986), 6th ed.
2. A. J. C. Wilson, *X-ray Optics* (Methuen & Co. Ltd., London, 1949).
3. A. V. Baryshev, A. A. Kaplyanskii, V. A. Kosobukin, M. F. Limonov, K. B. Samusev, and D. E. Usvyat, Phys. Solid State **45**, 459 (2003); M. V. Rybin, A. V. Baryshev, M. Inoue, A. A. Kaplyanskii, V. A. Kosobukin, M. F. Limonov, A. K. Samusev, and A. V. Sel'kin, Photonics Nanostruct. Fundam. Appl. **4**, 146 (2006).
4. A. V. Baryshev, A. A. Kaplyanskii, V. A. Kosobukin, K. B. Samusev, D. E. Usvyat, and M. F. Limonov, Phys. Rev. B **70**, 113104 (2004).
5. A. V. Baryshev, V. A. Kosobukin, K. B. Samusev, D. E. Usvyat, and M. F. Limonov, Phys. Rev. B **73**, 205118 (2006).
6. R. M. Amos, J. G. Rarity, P. R. Tapster, T. J. Shepherd, and S. C. Kitson, Phys. Rev. E **61**, 2929 (2000).
7. F. García-Santamaría, J. F. Galisteo-López, P. V. Braun, and C. López, Phys. Rev. B **71**, 195112 (2005).
8. L. M. Goldenberg, J. Wagner, J. Stumpe, B. R. Paulke, and E. Gornitz, Physica E **17**, 433 (2003); K. Wostyn, Y. Zhao, B. Yee, K. Clays, A. Persoons, G. Shaetzen, and L. Hellemans, J. Chem. Phys. **118**, 10752 (2003).
9. A. K. Samusev, K. B. Samusev, M. V. Rybin, M. F. Limonov, E. Y. Trofimova, D. A. Kurdyukov, and V. G. Golubev, Phys. Solid State **53**, 1056 (2011).
10. A. K. Samusev, K. B. Samusev, I. S. Sinev, M. V. Rybin, and M. F. Limonov, Phys. Solid State **53**, 1415 (2011).
11. Y. A. Vlasov, X. Z. Bo, J. C. Sturm, and D. J. Norris, Nature **414**, 289 (2001); L. A. Dorado, R. A. Depine, D. Schinca, G. Lozano, and H. Míguez, Phys. Rev. B **78**, 075102 (2008); G. Lozano, J. E.

Mazzaferri, L. A. Dorado, S. Ledesma, R. A. Depine, and H. Míguez, J. Opt. Soc. Am. B **27**, 1394 (2010); S. A. Asher, J. M. Weissman, A. Tikhonov, R. D. Coalson, and R. Kesavamoorthy, Phys. Rev. E **69**, 066619 (2004).
12. P. N. Pusey, W. van Megen, P. Bartlett, B. J. Ackerson, J. G. Rarity, and S. M. Underwood, Phys. Rev. Lett. **63**, 2753 (1989); Y. A. Vlasov, V. N. Astratov, A. V. Baryshev, A. A. Kaplyanskii, O. Z. Karimov, and M. F. Limonov, Phys. Rev. E **61**, 5784 (2000).
13. A. V. Baryshev, A. B. Khanikaev, H. Uchida, M. Inoue, and M. F. Limonov, Phys. Rev. B **73**, 033103 (2006).
14. C. F. Bohren and D. R. Huffman, *Absorption and Scattering of Light by Small Particles* (Wiley-VCH, Weinheim, 1998).
15. J. M. Ziman, Models of disorder: *The Theoretical Physics of Homogeneously Disordered Systems* (Cambridge University Press, Cambridge, 1979).
16. A. Guinier, X-ray diffraction. In *Crystals, Imperfect Crystals, and Amorphous Bodies* (W.H. Freeman and Co, San Francisco, 1963).
17. L. A. Dorado, R. A. Depine, G. Lozano, and H. Míguez, Opt. Express **15**, 17754 (2007).
18. A. V. Baryshev, A. B. Khanikaev, M. Inoue, P. B. Lim, A. V. Sel'kin, G. Yushin, and M. F. Limonov, Phys. Rev. Lett. **99**, 063906 (2007); M. V. Rybin, A. V. Baryshev, A. B. Khanikaev, M. Inoue, K. B. Samusev, A. V. Sel'kin, G. Yushin, and M. F. Limonov, Phys. Rev. B **77**, 205106 (2008).
19. M. V. Rybin, A. B. Khanikaev, M. Inoue, K. B. Samusev, M. Steel, G. Yushin, and M. F. Limonov, Phys. Rev. Lett. **103**, 023901 (2009).
20. M. V. Rybin, A. B. Khanikaev, M. Inoue, A. K. Samusev, M. J. Steel, G. Yushin, and M. F. Limonov, Photonics Nanostruct. Fundam. Appl. **8**, 86 (2010).
21. A. V. Petukhov, D. G. A. L. Aarts, I. P. Dolbnya, E. H. A. de Hoog, K. Kassapidou, G. J. Vroege, W. Bras, and H. N. W. Lekkerkerker, Phys. Rev. Lett. **88**, 208301 (2002); A. V. Petukhov, I. P. Dolbnya, D. G. A. L. Aarts, G. J. Vroege, and H. N. W. Lekkerkerker, *ibid.* **90**, 028304 (2003).
22. V. Abramova, A. Sinitskii, N. Grigor'eva, S. Grigor'ev, D. Belov, A. Petukhov, A. Mistonov, A. Vasil'eva, and Y. Tret'yakov, J. Exp. Theor. Phys. **109**, 29 (2009); A. G. F. De Beer and A. V. Petukhov, J. Appl. Crystallogr. **40**, 144 (2007); I. P. Dolbnya, A. V. Petukhov, D. G. A. L. Aarts, G. J. Vroege, and H. N. W. Lekkerkerker, EPL **72**, 962 (2005); R. P. A. Dullens and A. V. Petukhov, *ibid.* **77**, 58003 (2007); S. V. Grigoriev et al. Phys. Rev. B **79**, 045123 (2009); J.-M. Meijer, V. W. A. de Villeneuve, and A. V. Petukhov, Langmuir **23**, 3554 (2007); J. Hilhorst et al. *ibid.* **25**, 10408 (2009); A. V. Vasilieva et al. J. Phys. Conf. Ser. **247**, 012029 (2010).
23. A. Eliseev et al. JETP Lett. **90**, 272 (2009).
24. K. S. Napolskii et al. Langmuir **26**, 2346 (2010).
25. K. S. Napolskii, A. Sinitskii, S. V. Grigoriev, N. A. Grigorieva, H. Eckerlebe, A. A. Eliseev, A. V. Lukashin, and Y. D. Tretyakov, Physica B **397**, 23 (2007); Sinitskii, A., Abramova, V., Grigorieva, N., Grigoriev, S., Snigirev, A., Byelov, D. V., and Petukhov, A. V., EPL **89**, 14002 (2010).
26. A. V. Petukhov, J. H. J. Thijssen, D. C. 't Hart, A. Imhof, A. van Blaaderen, I. P. Dolbnya, A. Snigirev, A. Moussaïd, and I. Snigireva, J. Appl. Crystallogr. **39**, 137 (2006); J. H. J. Thijssen, A. V. Petukhov, D. C. Hart, A. Imhof, C. H. M. van der Werf, R. E. I. Schropp, and A. van Blaaderen, Adv. Mater. **18**, 1662 (2006).
27. M. S. Paterson, J. Appl. Phys. **23**, 805 (1952).
28. A. I. Plekhanov, D. V. Kalinin, and V. V. Serdobintseva, Nanotech. Russ. **1–2**, 245 (2006); W. D. Ristenpart, I. A. Aksay, and D. A. Saville, Phys. Rev. Lett. **90**, 128303 (2003); A. Yethiraj, A. Wouterse, B. Groh, and A. van Blaaderen, *ibid.* **92**, 058301 (2004); A. L. Rogach, N. A. Kotov, D. S. Koktysh, J. W. Ostrander, and G. A. Ragoisha, Chem. Mater. **12**, 2721 (2000); H. J. Schöpe, J. Phys. Condens. Matter **15**, L533 (2003); M. Trau, D. A. Saville, and I. A. Aksay, Langmuir **13**, 6375 (1997).
29. A. Snigirev, V. Kohn, I. Snigireva, and B. Lengeler, Nature **384**, 49 (1996).

3.7

Interplay of Order and Disorder in the High-Energy Optical Response of Three-Dimensional Photonic Crystals

Gabriel Lozano and Hernán Míguez
Materials Science Institute of Seville

Luis A. Dorado and Ricardo A. Depine
University of Buenos Aires

CONTENTS

3.7.1 Introduction

Opal-like structures, consisting of lattices of dielectric spheres, are the most commonly studied example of three-dimensional (3D) photonic crystals (PCs).[1] Since they were proposed as new materials to mold the flow of light, they have become an important area of research because of their technological potential and fundamental interest.[2] Among all fabrication techniques developed up to date to prepare opaline PCs, those based on evaporation-induced self-assembly (EISA) are some of the most frequently used and thoroughly analyzed.[3] The advent and subsequent improvement of fabrication techniques that take advantage of self-organizing properties of dielectric spheres in the micrometer scale have permitted to obtain solid colloidal crystals that exhibit PC properties,[4,5] whose optical response has been studied in depth in the low-energy range, where the lattice parameter is smaller than the incident wavelength.[6]

In line with this, the so-called high-energy range of artificial opals, where the lattice constant is on the order of or greater than the wavelength of light, presents a number of applications whose exploration has just started.[7] It is well known that the presence of imperfections strongly affects the optical properties of 3D periodic structures.[8] However, recent advances in the comprehension of the growth dynamics of these structures opens the door to rational design of experiments aiming at fabricating lattices in which the density of intrinsic defects is minimized.[9,10] As such imperfections have a dramatic effect on scattered light of wavelength smaller than the lattice constant,[11] the evaluation of the experimental optical response at those energy ranges, based on the comparison to rigorous calculations performed using the vector Korringa–Kohn–Rostoker (KKR) approach (as it has been discussed in depth in Chapter 2.3), is identified as a useful tool to understand many reported optical properties in this spectral range that remain unsatisfactorily explained[12–15] and the most sensitive guide to accurately evaluate the progress toward the actual realization of defect-free colloidal crystals.[3]

In particular, the understanding of the complex optical properties observed in the high energy range has been lively debated during the last years.[11,13–18] We have recently provided a detailed analysis and rationalization of a number of reported observations related to artificial opal films whose cause remained unclear, that is, the physical origin of all the spectral features observed in the specular reflectance and forward transmittance spectra, the fine crystalline structure, the near-field pattern intensity, as well as of the intensity variations of the nonspecularly diffracted beams.[19–24] In this chapter, we will review the optical phenomena that occur when light wavelength is on the order of or shorter than the lattice parameter of the periodic dielectrics, a rich and exciting electromagnetic range in which there exists a strong interplay between light propagation, optical diffraction, sample size, and imperfections. We will analyze the origin of anomalous specular reflectance and ballistic transmittance, diffusely scattered light, and optical diffraction and the influence of imperfections in such phenomena. Emphasis will be put on the analysis of the sensitivity of the optical response to small variations of the structural parameters.

Therefore, the chapter is organized as follows. In Section 3.7.2, we present a brief description of the technique used to fabricate the artificial opals whose optical properties will be analyzed in this chapter. In Section 3.7.3, an overview of the vector KKR method used to calculate the optical response of artificial opals in the high-energy range is presented. Far-field results obtained are addressed in Section 3.7.4. In particular, we analyze the interplay between crystal size and disorder effects in the high-energy range, a dramatic effect being found on the reflectance spectra of thicker crystals. We also present a complete description of nonspecular diffracted beams. The optical patterns obtained from artificial opals when analyzed using scanning near-field optical microscopy (SNOM) are shown in Section 3.7.5. Finally, in Section 3.7.6 we emphasize and discuss some of the more interesting results addressed here.

3.7.2 Sample Characteristics and Experimental Setups

Artificial opal films whose optical properties are presented in this chapter were grown using a variation of the vertical deposition method developed by Jiang and coworkers.[4]

These photonic materials were formed as a result of the evaporation of the liquid phase of a colloidal dispersion at the contact line with the suspension meniscus. Under some specific conditions, the spheres in the film tend to self-organize and form a film that exhibit PC properties. Further details can be found, for example, in the study by Lozano and Míguez.[9] Figure 3.7.1a shows a scanning electron microscopy (SEM) image of the outer (111) plane of such a structure, where the spheres ordered in a triangular lattice can be observed. The diameter of the scatters was selected to be around 750 nm in order to place the higher frequency bands of the resulting photonic structure in the visible part of the spectrum.

Some of the optical results presented in this chapter were obtained by means of specular reflectance spectroscopy. In order to do so, we used a Fourier transform infrared spectrophotometer (Bruker IFS-66) attached to a microscope. A 4× objective with a numerical aperture of 0.1 (light cone angle of 5.7°) was used to irradiate the lattices and collect the reflected light at quasinormal incidence with respect to its surface. This equipment allowed us to obtain absolute values for the specular reflectance at normal incidence, the comparison between measured and calculated spectra being straightforward.

The measurements of the efficiencies of the nonspecular diffracted beams presented in Section 3.7.4 were determined as described in the study by Dorado et al.[23] The experimental setup used for doing so is depicted in Figure 3.7.2a. In this scheme, all relevant geometrical parameters are introduced. The relative orientation of the PC film with respect to the source and the screen or the detector is described using a Cartesian reference system as shown in Figure 3.7.2a.

Details about the technique followed to carry out the analysis of artificial opals by SNOM can be found in the study by Barrio et al.[22] In this case, we used the so-called collection-mode configuration, where the frontal surface of the sample is illuminated directly by a planar wave whereas the light transmitted by the sample is collected using a near-field probe. As this is the illumination configuration used to perform standard far optical field measurements, this setup could allow observing the fine features of the transmitted optical field before reconstruction takes place away from the sample to yield the well-known homogeneous distribution of the far field, as observed in the calculations shown in Figure 3.7.2b.

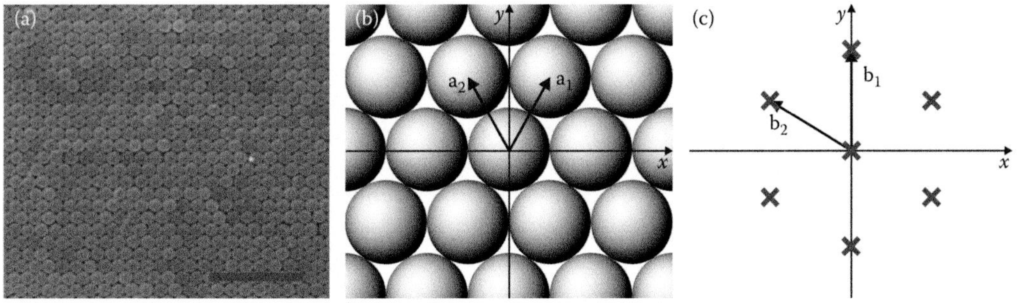

FIGURE 3.7.1
(a) Scanning electron microscopy image of the surface of an artificial opal film made of 750 nm polystyrene spheres. Scale bar is 5 μm. (b)–(c) Diagrams showing the two primitive vectors used to describe the (b) real and (c) reciprocal lattices. (Adapted from G. Lozano et al., Opt. Soc. Am. B **27**, 1394, 2010.)

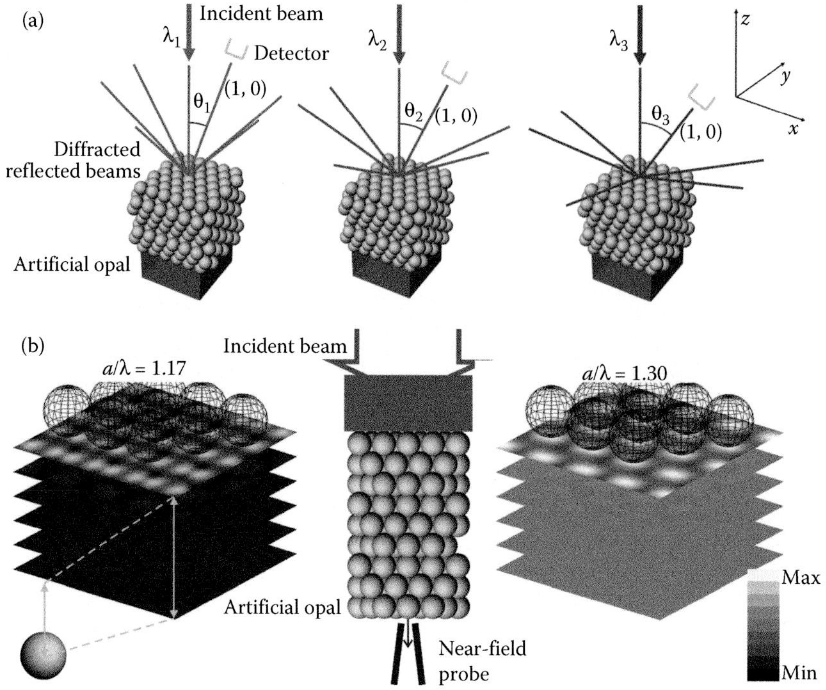

FIGURE 3.7.2
(See color insert.) (a) Scheme of the experimental setup used to measure the spectral dependence of the intensity of the reflected diffracted beams. Note that $\lambda_1 < \lambda_2 < \lambda_3$. (b) Schematic drawings of the setup used to measure the near-field intensity pattern in the collection mode. Calculations showing the evolution of the transmitted near-field intensity profile before reconstruction of the far field takes place for two values of reduced frequency for a 10-layer-thick artificial opal made of spheres of dielectric constant $\varepsilon_s = 2.5 + 0.05i$. Profiles are calculated each $0.3d$ (d being the diameter of the sphere) between 0, that is, the plane tangent to the poles of the spheres, and $1.5d$. Color scale bar is also indicated.

3.7.3 Determination of Simulation Parameters

The observation of fine optical features at $a/\lambda > 1$ (a being the lattice constant of the conventional cubic cell) was first possible when the EISA method on vertical substrates was used to form crystals of dielectric particles of diameters larger than 350 nm.[12,16] At those spectral ranges, a complex band structure describes the photon modes within a sphere face-centered cubic (FCC) lattice, as illustrated in Figure 3.7.3a. Above that energy threshold the reflectance and transmittance spectra of colloidal crystals show a fine peak structure that does not correspond to the presence of any photonic stop-bands or blind modes,[25] as demonstrated. Typical specular reflectance and forward transmittance spectra of a colloidal crystal film are plotted as black lines in Figure 3.7.3b and c, respectively.

However, until very recently, the lack of adequate theoretical models to describe the optical properties at $\lambda < a$ made it impossible to establish a rigorous comparison between experimental measurements shown in Figure 3.7.3b and c and the response predicted for perfectly ordered structures. This prevented a reliable assessment of the quality of a lattice based on such higher-energy response. Nevertheless, the KKR approach described in Chapter 2.3 allowed us to calculate the optical reflectance and transmittance spectra of a

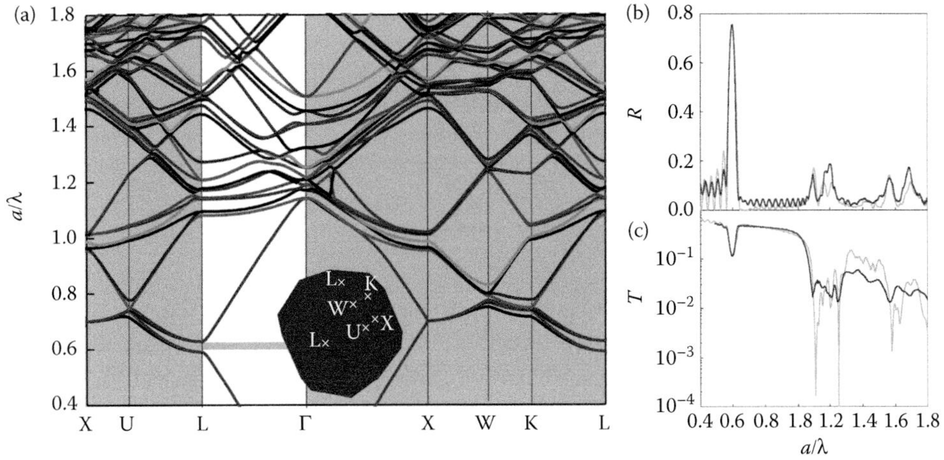

FIGURE 3.7.3
(a) Calculated photonic band structure for a face-centered cubic structure consisting of dielectric polystyrene spheres ($\varepsilon_s = 2.5$) in an air background. (Adapted from G. Lozano et al., Langmuir **25**, 12860, 2009.) (b) Specular reflectance and (c) forward transmittance spectra measured (black lines) and calculated (gray lines) from an artificial opal film. (Adapted from G. Lozano et al. J. Mater. Chem. **19**, 185, 2009.)

perfect lattice of finite size. Remarkably, a strongly fluctuating reflectance and transmittance is expected for such ideal crystals (see Figure 2.3.1 in Chapter 2.3), a large number of peaks and dips being expected for $\lambda < a$. We have already proved that these features can be rigorously related to multipolar resonances of the sphere ensemble excited by an external incident plane wave.[19] Using these ideal structures as a starting point, we have studied the effect of imperfections on the optical properties by introducing a complex dielectric permittivity whose imaginary part ε_i was gradually increased.

As opal films are PC slabs made of sphere layers piled up in the [111] direction, the spheres in each layer are ordered in a triangular lattice. If these layers are parallel to the *xy*-plane, a set of two-dimensional (2D) primitive lattice vectors can be

$$\begin{aligned} \vec{a}_1 &= \frac{d}{2}\left(\hat{x} + \sqrt{3}\hat{y}\right) \\ \vec{a}_2 &= \frac{d}{2}\left(-\hat{x} + \sqrt{3}\hat{y}\right) \end{aligned} \tag{3.7.1}$$

where d is the distance between spheres in the same layer (see Figure 3.7.1b). A set of primitive reciprocal-lattice vectors can be chosen as

$$\begin{aligned} \vec{b}_1 &= \frac{4\pi\sqrt{3}}{3d}\hat{y} \\ \vec{b}_2 &= \frac{2\pi}{d}\left(-\hat{x} + \frac{\sqrt{3}}{3}\hat{y}\right) \end{aligned} \tag{3.7.2}$$

(see Figure 3.7.1c). Thus, any reciprocal-lattice vector can be written as

$$\vec{g} = p\vec{b}_1 + q\vec{b}_2 \tag{3.7.3}$$

where (p, q) is a pair of integers. As it is customary to use the lattice constant a of a classical cubic cell, we have $a = d\sqrt{2}$, so the photon energy is expressed in reduced units a/λ, where λ is the wavelength of the incident light.

The wave vector of a diffracted beam emerging from the slab can be written as

$$\vec{K}_{\vec{g}}^{\pm} = \vec{g} + \vec{k}_{i\|} \pm \sqrt{k^2 - \left|\vec{g} + \vec{k}_{i\|}\right|^2}\,\hat{z} \tag{3.7.4}$$

where the ± sign corresponds to a transmitted (reflected) beam, $\vec{k}_{i\|}$ is the component of the incident wave vector parallel to the surface of the slab (xy-plane), $k = 2\pi n_d/\lambda$, and n_d is the refractive index of the diffraction medium. Each diffracted beam corresponds to a propagating wave if the z-component of $\vec{k}_{\vec{g}}^{\pm}$ is purely real, so a diffraction channel $\vec{g} = p\vec{b}_1 + q\vec{b}_2$ is open when $\left|\vec{g} + \vec{k}_{i\|}\right| < k$ and we have a diffraction cutoff when $\left|\vec{g} + \vec{k}_{i\|}\right| = k$. If $R_{p,q}$ and $T_{p,q}$ are the reflectance and transmittance coefficients of diffraction channel (p, q), respectively, the total reflectance and transmittance will be $R = \sum_{(p,q)} R_{pq}$ and $T = \sum_{(p,q)} T_{p,q}$, where the sums include the open channel (0, 0). The coefficients $R_{p,q}$ and $T_{p,q}$ will also be referred to as efficiencies of channel (p, q) in reflection and transmission, respectively. In the absence of energy losses, conservation of energy implies that $R + T = 1$. As disorder removes energy from the coherently scattered beams, it works as a sort of loss mechanism and we have $R + T < 1$ in actual experiments.

All the experimental results presented in this chapter were analyzed using the vector KKR approach,[11,26] which has been discussed in depth in Chapter 2.3. From the optimum fitting of the measured spectra, in particular the ones presented in Figure 3.7.3b and c, the crystal thickness and the ε_i value were extracted[27] An excellent agreement between the simulated and the experimental spectra measured was found, as shown in Figure 3.7.3b for specular reflectance and in Figure 3.7.3c for forward transmittance, an indication of the validity of this theoretical approximation. In order to fit the optical response in the energy range under analysis ($0.4 < a/\lambda < 2$), it is necessary to consider that the interparticle distance of actual colloidal crystals coincides with the expected diameter for spheres belonging to the same close-packed (111) plane but differs significantly in directions oblique to the [111] one. This leap has something to do with the fact that the spheres flatten along directions oblique to the normal to the substrate they are deposited on.[21]

3.7.4 Effect of the Interplay between Order and Disorder in the Far-Field Optical Response

The first and by far the most extensively used approach to probe the optical properties of PCs has been transmission or reflection experiments in which both source and detector are in far field.[16,17,20,28–31] In this section, we present an overview of some of the most outstanding results that we have found in the analysis of the role that imperfections play in the far-field optical response of 3D periodic structures.

All the optical features occurring at the high-energy range were correctly identified as a sign of the improved order achievable with the EISA technique, as in lower structural quality lattices they were barely detectable. The effect of imperfections is actually expected to be much more dramatic when wavelengths become on the order of or shorter than the typical size of such defects. Results presented in Section 3.7.3 demonstrate that such an effect must be considered in the calculations in order to accurately reproduce the experimental optical response that real structures exhibit.

Therefore, the extraordinary sensitivity of the short wavelength optical response to subtle structural modifications has been used to distinguish between the different variants of the close-packed sphere structure, namely, cubic, hexagonal, or random, in very thin opal films.[30,31] In Figure 3.7.4a, we present the specular reflectance spectra taken from a four-layer-thick opal, measured by Checoury et al.,[31] as a function of a/λ for two different stacking patterns. In the low-energy range ($a/\lambda < 1$), the two spectra are quite similar. In particular, a minimum of transmission always occurs near $a/\lambda = 0.6$. It corresponds to the frequency location of the first pseudogap in the [111] direction in the case of an FCC structure, as indicated in Figure 3.7.3a. These results were expected since for $a/\lambda < 1$ the transmission and reflection spectra at normal incidence are roughly determined by the first Fourier component of the dielectric map along the direction of light incidence and this Fourier component is the same for both structures. As a consequence, in this frequency range, the opals roughly behave as a Bragg mirror with period $d\sqrt{2/3}$. On the contrary, transmission and reflection spectra strongly differ in the high-frequency region ($a/\lambda > 1$), which explains the different colors that these samples exhibit when they are observed under an optical microscope.[31] Indeed, in this high-energy range, the transmission and reflection spectra will depend on the Fourier components of the dielectric map in all space directions and these components strongly depend on the stacking pattern. For this reason, reflection spectra clearly show two different tendencies for $a/\lambda > 1$, as shown in Figure 3.7.4a. To analyze the effect of this sort of disorder in the stacking, we fitted these experimental data using the vector KKR approach. Figure 3.7.4b shows the calculated results for an FCC structure with a repeating ABC pattern (black lines in Figure 3.7.4) or a hexagonal close-packed structure with a repeating AB pattern (gray lines in Figure 3.7.4). By considering imperfections in the model, the overall characteristics of the simulated spectra are in excellent agreement with the measured ones, as it can be observed by comparing Figure 3.7.4a and b. As real opals present imperfections that cause light to be scattered diffusely, it is absolutely necessary to introduce some disorder in the model to reproduce the spectral features experimentally observed.

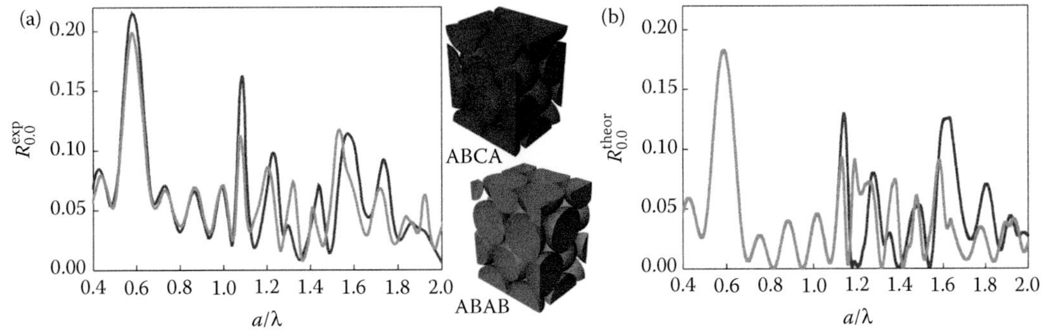

FIGURE 3.7.4
(a) Experimental and (b) calculated specular reflection spectra of a four-layer opal arranged in a face-centered cubic structure, that is, with a stacking of the form ABCA (black line), and in a hexagonal close-packed structure, that is, with a stacking of the form AB (gray line). (Experimental data have been extracted from the study by Checoury et al.,[31] whereas fittings were performed using the Korringa–Kohn–Rostoker approach described in Chapter 2.3.)

3.7.4.1 Interplay between Crystal Size and Disorder Effects

Herein we present the most relevant results of the analysis of the behavior of the optical spectra features in the high-energy range as the width of the crystal slab is gradually increased. We compare the experimental and theoretical evolutions of the specular reflectance spectra as we increase the crystal thickness from 1 to 18 sphere layers. We will focus on the analysis of specular reflectance results, as they can be readily attained by microspectroscopy from single domains of the crystalline film using a low numerical aperture objective, ensuring the thickness uniformity of the tested spot and the normal incidence of the beam. This is a much more complicated task to be performed in transmission mode, as the incident beam is typically focused with a larger numerical aperture objective.

In Figure 3.7.5, the measured and calculated specular reflectance (or $R_{0,0}$) can be seen as a function of a/λ for different crystal slab widths. Very good agreement between the

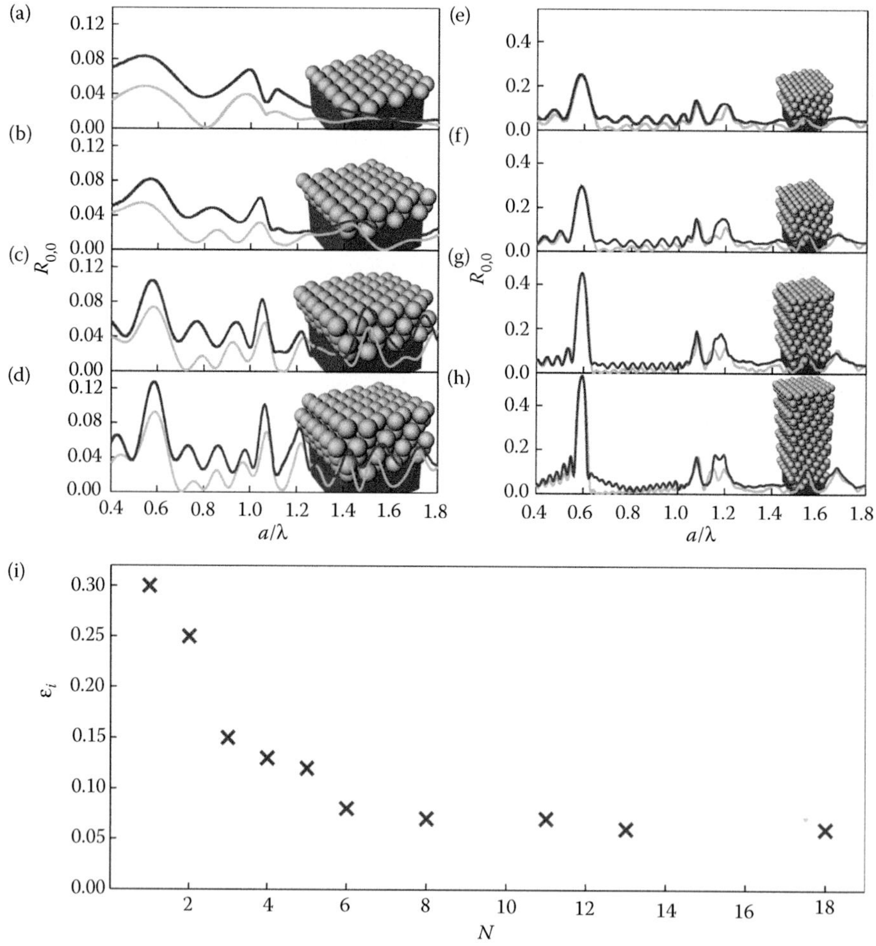

FIGURE 3.7.5
(a)–(h) Measured (black lines) and calculated (gray lines) specular reflectance spectra for artificial opals made of (a) 1, (b) 2, (c) 3, (d) 4, (e) 6, (f) 8, (g) 13, and (h) 18 planes of spheres of dielectric constant $\varepsilon_s = 2.5 + i\varepsilon_i$ in air, with $\varepsilon_i = 0.30$ (a), $\varepsilon_i = 0.25$ (b), $\varepsilon_i = 0.15$ (c), $\varepsilon_i = 0.13$ (d), $\varepsilon_i = 0.08$ (e), $\varepsilon_i = 0.07$ (f), $\varepsilon_i = 0.06$ (g), and $\varepsilon_i = 0.06$ (h). Imaginary part of the dielectric permitivity of the spheres (ε_i) as a function of the number of close-packed layers forming the artificial opal (N). (Adapted from L. A. Dorado et al., Phys. Rev. B **76**, 245103, 2007.)

experimental and theoretical reflectance spectra has been obtained by introducing extinction in the theoretical model, as we have demonstrated for much thicker crystals. Figure 3.7.5a–d shows the experimental and theoretical specular reflectivity spectra for artificial opals made of a few layers (N being the number of layers). The value of ε_i in the calculated spectra was chosen to obtain the best fit. Note that all oscillations present a similar height with respect to the background in both the experimental (black lines) and the simulated (gray lines) spectra, which was actually the criterion followed to choose the best fit. Differences in the baseline are due to the fact that only one interface of the glass substrate is considered for the calculations to avoid the presence of very-short-frequency lobes in the spectra. These arise from the interference between beams reflected at the upper and lower faces of the thick substrate (1 mm) under consideration.

In Figure 3.7.5a, two main peaks for the monolayer ($N = 1$) can be seen: one near $a/\lambda = 0.55$ and the other at $a/\lambda = 1.0$. These peaks move gradually toward higher energies as the slab width is increased, as seen in Figure 3.7.5a–d. Also, secondary peaks related to the Fabry–Pérot oscillations appear between these main peaks for slabs of $N = 2$, 3, and 4 layers. As all our measurements were taken far from the edges of the crystal, in which different types of stacking sequences have been reported,[31] the fairness of the fittings we attain considering only the *ABCABC* . . . sequence indicates that an FCC lattice is built as more monolayers are piled up.

In Figure 3.7.5e–h, the specular reflectance spectra for artificial opals made of $N = 6$, 8, 13, and 18 layers can be appreciated. The main peak in the low-energy band tends to its final position, which is at $a/\lambda = 0.61$, as can be predicted from the photonic band structure of the sphere FCC lattice under consideration, because there is a pseudo-band gap in the ΓL direction centered at this particular energy, as shown in Figure 3.7.3a. The first peak in the high-energy region has moved to $a/\lambda = 1.07$, and a second double degenerated peak near $a/\lambda = 1.2$ is present. It is interesting to note that this second peak becomes double degenerated in slabs with at least six layers. In other words, we must complete at least two *ABC* sequences in the slab in order to obtain the peak at $a/\lambda = 1.2$. Figure 3.7.5 also shows that this peak is difficult to see in the measured data for $N = 6$, 8, and 13 layers, but Figure 3.7.5h for $N = 18$ layers confirms its existence. There is a second peak structure in the range of $1.5 < a/\lambda < 1.8$, which is composed of two peaks of similar amplitudes. They can be observed in Figure 3.7.5c–h, so the number of layers needed must be $N \geq 3$, which corresponds to one or more complete sequences of *ABC*.

In the experimental spectra of Figure 3.7.5, we can appreciate a gradual increase of the reflectance amplitudes as the number of layers is increased. Also, the imaginary part ε_i of the dielectric constant of the spheres decreases as the number of layers N is increased in the calculated spectra. This tendency is plotted in Figure 3.7.5i, where ε_i can be seen as a function of N. Hence, there exists a clear correlation between the extinction introduced in the model to fit the experimental spectra in the high-energy region and the width of the slab, which indicates that diffusely scattered light increases for thinner crystals. In the case of crystals of a few layers, besides the effect of disorder, interactions with the glass substrate produce an additional attenuation of the reflectivity values. Therefore, the increase in ε_i as the crystal width is decreased means that more diffusely scattered light in this energy range is produced, the combination of disorder and glass–crystal interaction being responsible of this effect. Nevertheless, for crystals with six or more layers, ε_i becomes practically a constant, as Figure 3.7.5i shows, which means that any surface effect and glass–crystal interaction are negligible, the value of ε_i being a measure of the degree of disorder in these cases. Thus, we found a method for quantifying disorder in a real crystal because we can use a single parameter such as ε_i, which is determined by fitting the theoretical model to the experimental data, as a measure of the structural quality of the artificial opal. A similar method has already been proposed for the same type of PC films but based on the fitting of the low-energy optical response of the lattices.[29] Interestingly, the same trend of the extinction needed to fit the experimental curve versus the number of close-packed layers in the film is found.

We have found complex spectra in the high-frequency region even for the initial monolayer structure,[32] this peak structure being gradually modified as more layers are piled up. This allowed us to identify which peaks are reminiscent of the optical reflectance features of a single close-packed layer and which are the result of building up a 3D periodic structure. At the same time, we found a clear correlation between the crystal size and the amount of extinction needed to attain the best fit in the high-energy range. The interplay of crystal size and disorder effects on the high-energy optical response is thoroughly analyzed, finding a much more dramatic effect of the former on the reflectance spectra of larger crystals, which indicates that wider crystal slabs produce less diffusely scattered light and provides a method for quantifying disorder in these cases. It can be unambiguously concluded that the effect of imperfections was that of smoothing the strong resonances observed in the ideal structure, only a few robust ones surviving in the spectra of the highest-quality structures available up to date. Furthermore, similar resonant modulation and smoothing effects were measured and fitted for the beams nonspecularly diffracted by the 3D structure, as we will present in Section 3.7.4.2.

3.7.4.2 Analysis of the Intensity of Beams Diffracted by Artificial Opal Films

Diffraction is an interesting and barely explored phenomenon occurring in the high-energy range that has its origin in the interaction between a plane wave and a periodic surface, which results in the opening of diffraction channels. Diffracted modes are propagating waves emerging from the colloidal crystal surface when the incident wavelength is shorter than an energy threshold.

The components of the incident and the diffracted wave vector contained in the plane xy are related by the following expression:

$$\vec{k}_{\hat{g}\|} = \vec{g} + \vec{k}_{i\|} \tag{3.7.5}$$

For a triangular lattice, Equation 3.7.5 can be rewritten as

$$\begin{aligned} \frac{2\pi n_d}{\lambda}\sin\theta_{p,q}\cos\phi_{p,q} &= -q\frac{2\pi}{d} + k_{i\|x} \\ \frac{2\pi n_d}{\lambda}\sin\theta_{p,q}\sin\phi_{p,q} &= \frac{2\pi}{\sqrt{3}d}(2p+q) + k_{i\|y} \end{aligned} \tag{3.7.6}$$

where $\theta_{p,q}$ are $\phi_{p,q}$, respectively, the diffracted zenital and azimutal angles. Thus, angle $\theta_{p,q}$ that forms the diffracted wave vector with the z-axis direction can be expressed as

$$\frac{2\pi n_d}{\lambda}\sin\theta_{p,q} = \left[\left(-q\frac{2\pi}{d} + k_{i\|x}\right)^2 + \left(\frac{2\pi}{\sqrt{3}d}(2p+q) + k_{i\|y}\right)^2\right]^{1/2} \tag{3.7.7}$$

in normal incidence, when an incoming light beam travels in the z-axis direction ($k_{i\|x} = k_{i\|y} = 0$), Equation 3.7.7 reduces to

$$\frac{n_d}{\lambda}\sin\theta_{p,q} = \frac{1}{d}\sqrt{q^2 + \frac{(2p+q)^2}{3}} \tag{3.7.8}$$

As described, each diffracted mode corresponds to a propagating wave if the z-component of $\vec{K}_g^{\pm}$ is purely real. Then, diffraction channels are open when $g < k$ and we have a diffraction cutoff when $g = k$:

$$\frac{a}{\lambda} = \frac{\sqrt{2}}{n_d}\sqrt{q^2 + \frac{(2p+q)^2}{3}} \tag{3.7.9}$$

In air, ($n_d = 1$) six diffraction channels—corresponding to $(p, q) = (1, 0)$, $(1, -1)$, $(0, -1)$, $(-1, 0)$, $(-1, 1)$, $(0, 1)$—open when $a/\lambda \approx 1.63$. The wave vector of these diffracted beams are located along a cone and angle $\theta_{p,q}$ between $\vec{K}_g^{\pm}$ and the z-axis is given by

$$\sin\theta_{p,q} = \frac{2\lambda}{\sqrt{3}d} \tag{3.7.10}$$

which gives angle $\theta_{p,q}$ of the first six diffracted beams as a function of the wavelength. The specularly reflected and forwardly transmitted beams correspond to channel (0, 0), which are always open channels. This situation is illustrated in Figure 3.7.6, where each reflected and transmitted beam is labeled according to the pair of integers (p, q). The fact that the PC is supported on a glass substrate (refractive index $n = 1.51$) introduces a first diffraction cutoff for transmitted beams at $a/\lambda = 1.08$. Although those beams are internally reflected at the glass–air surface and therefore do not propagate in air, they can be experimentally observed by frustrating the total reflection.[14]

In Figure 3.7.6a and b, photographs of the diffraction spots can be observed on a screen for two different wavelengths namely, $\lambda = 484$ nm and $\lambda = 565$ nm, respectively. These spots are produced by the six reflected beams that arise above the diffraction cutoff. The specularly reflected beam (0, 0) passes through a hole perforated in the screen to allow the laser beam reaching the crystal slab. Figure 3.7.6c shows the transmitted beams diffracted from the slab and projected on a screen also parallel to the xy-plane, as indicated in Figure 3.7.6g. In this case, the wavelength is $\lambda = 539$ nm and the spot produced by the forwardly transmitted beam (0, 0) can be seen at the center of the picture. In these photographs, a uniform background due to diffuse light scattering can be observed. The detectors measuring the intensities of the diffracted spots do not capture this energy and then it can be considered as a kind of loss, which is simulated by adding extinction to the dielectric constant of the spheres in the lattice in our theoretical model. The existence of layers having a triangular lattice parallel to the xy-plane and piled up in the sequence *ABCABC . . .* along the z-axis implies a symmetry relation for the efficiencies of diffracted beams when the incident light is polarized. In all the experiments, the incident wave is linearly polarized along the x-axis, as the incident electric-field vector $\vec{E}_i$ indicates in Figure 3.7.6g and h. Under this condition, we have $R_{1,0}$, $R_{1,-1} = R_{0,1}$, $R_{0,-1} = R_{-1,1}$, and $R_{-1,0}$; thus, there are essentially four different diffraction spots. A close look at Figure 3.7.6a–c allows us to observe this symmetry between spots.

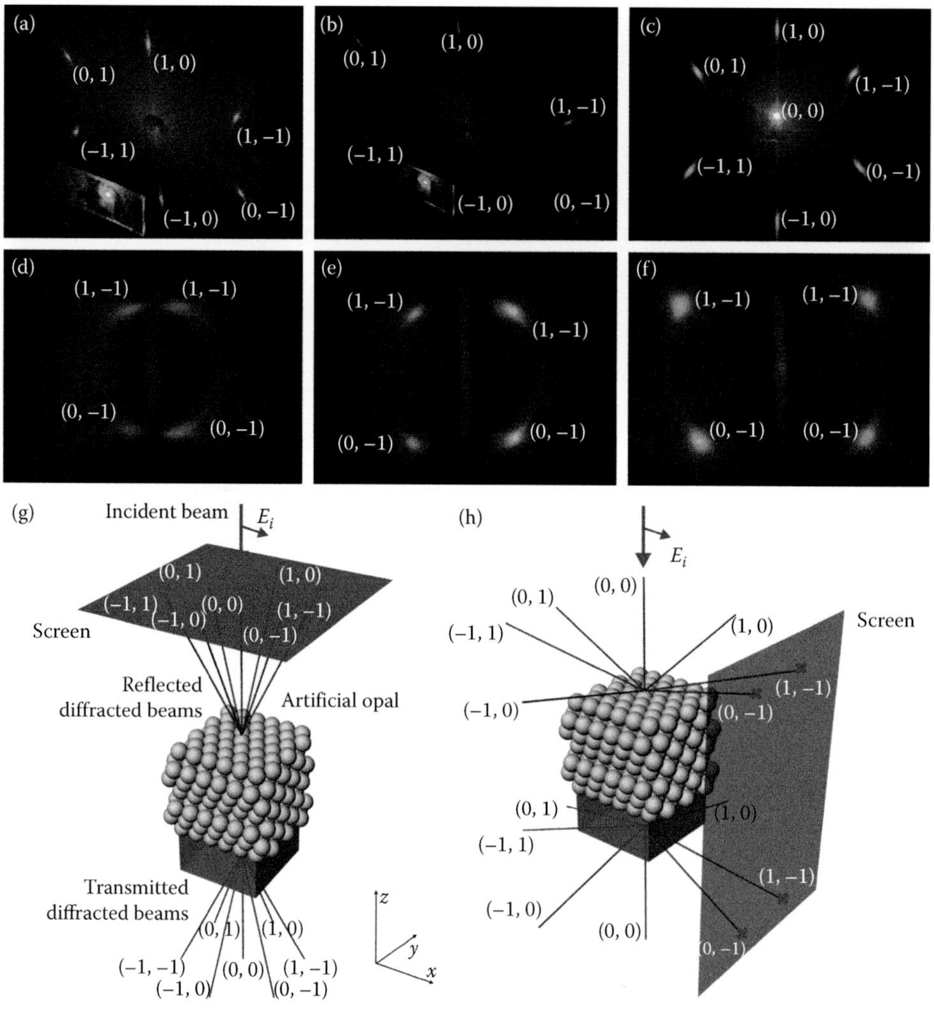

FIGURE 3.7.6
(See color insert.) (a), (b) Diffraction patterns of reflected beams projected on a screen parallel to the xy-plane for: (a) $\lambda = 484$ nm and (b) $\lambda = 565$ nm. (c) Diffraction pattern of transmitted beams projected on a screen parallel to the xy-plane for $\lambda = 539$ nm. (d)–(f) Diffraction patterns of reflected and transmitted beams projected on a screen parallel to the yz-plane for: (d) $\lambda = 622$ nm, (e) $\lambda = 593$ nm, and (f) $\lambda = 512$ nm. Sketches depicting the reflected and transmitted diffracted beams projected on a screen parallel to the (g) xy-plane and (h) yz-plane. (Adapted from L. A. Dorado et al., Phys. Rev. B **78**, 075102, 2008.)

Figure 3.7.6d–f show photographs of diffraction spots obtained by using a screen parallel to the yz-plane for decreasing values of wavelength, namely, (d) $\lambda = 622$ nm, (e) $\lambda = 593$ nm, and (f) $\lambda = 512$ nm. The width (1 mm) of the substrate onto which the sample (around 5 μm wide) is deposited can be seen at the center of each picture and spots of reflected (transmitted) beams (1, –1) and (0, –1) are on the right (left) of the sample. As we have mentioned, diffraction spots of beams (1, –1) and (0, –1) have the same efficiencies as beams (0, 1) and (–1, 1), respectively. In Figure 3.7.6d–f we can see that the spots move away from the sample as the wavelength is decreased, following the change in the direction of the diffracted wave vectors $\vec{K}_g^{\pm}$ predicted by Equation 3.7.10. Substituting the experimental value attained for the sphere diameter $d = 752$ nm in Equation 3.7.8 yields a diffraction cutoff wavelength $\lambda_c = 651$ nm. It is remarkable that these measurements served as a starting point for the optical analysis of the fine crystalline structure of artificial opal films. Further details can be found in the study by Lozano et al.[21] Figure 3.7.6d corresponds to a wavelength ($\lambda = 622$ nm) close to the onset of diffraction spots, and Equation 3.7.10 gives θ near 90°, which is actually what we see in this figure. Thus, looking at the diffracted spots onto a screen perpendicular to the crystal provides a simple way of confirming experimentally the directional properties of the wave vectors of diffracted beams predicted by Equation 3.7.10.

The calculated reflection efficiencies of diffracted channels are plotted in Figure 3.7.7a as functions of the wavelength without extinction ($\varepsilon_i = 0$). The vertical dashed line indicates the diffraction cutoff wavelength, λ_c. Note that diffraction efficiencies of nonspecular beams are all zero for wavelengths greater than λ_c. Although the curves present rapid fluctuations, two main peaks can be appreciated for all the efficiencies, one close to $\lambda = 530$ nm and another near $\lambda = 475$ nm, particularly evident for the beam (1, 0). Angle θ that forms the diffracted beams, given by Equation 3.7.10, is indicated in the upper horizontal scale, where we can see that the main peak near $\lambda = 530$ nm corresponds to $\theta \sim 55°$. As each sphere in the lattice can be considered as a superposition of electric and magnetic dipoles, quadrupoles, octupoles, and so on, these efficiency peaks originate from resonances inside the PC due to the interactions between multipoles. In other words, there is a natural resonance mechanism inside the crystal that is excited for only certain wavelengths. These results are consistent with previous observations and calculations that demonstrated that the origin of the optical response of the reflected and transmitted (0, 0) beams, observed for $a > \lambda$ in 3D PCs, lies on multipolar resonances of the sphere ensemble.[19] Hence, it is not surprising that this mechanism also strongly affects the diffraction phenomena occurring in these lattices. To take into account the effect of disorder, the optical spectra of diffracted beams have been recalculated adding an imaginary part $\varepsilon_i = 0.04$

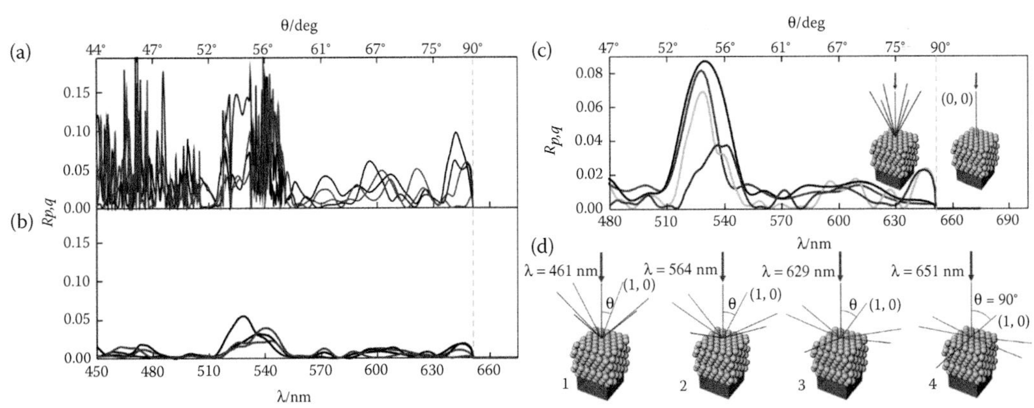

FIGURE 3.7.7
(See color insert.) (a), (b) Calculated efficiencies of nonspecular diffracted reflected beams for a perfectly ordered ($\varepsilon_i = 0$) artificial opal (a) and the same photonic structure after introducing extinction ($\varepsilon_i = 0.04$). (c) Measured and calculated reflection efficiencies of diffracted modes (1, 0) (red and light gray, respectively) and (1, –1) (blue and dark gray line, respectively). Vertical dashed line indicates the onset of diffraction for the reflected beams diffracted by an artificial opal made of spheres of 750 nm in diameter. (d) Scheme showing the evolution of the angle (θ) that forms the wave vector of the diffracted beams ($\vec{K}_g^{\pm}$) with the z-axis for different values of the incident wavelength: that is, $\theta = 45°$ for $\lambda = 461$ nm (1), $\theta = 60°$ for $\lambda = 564$ nm (2), $\theta = 75°$ for $\lambda = 629$ nm (3), and $\theta = 90°$ for $\lambda = \lambda_{\chi} = 651$ nm (4). (Adapted from L. A. Dorado et al., Phys. Rev. B **78**, 075102, 2008.)

to the dielectric constant of the spheres, which is a value that has provided good fitting with measurements of specular reflectance ($R_{0,0}$) and forward transmittance ($T_{0,0}$). The results are shown in Figure 3.7.7b, where a decrease in the intensities of the reflected beams is obtained as well as a smoothing of the curves compared with Figure 3.7.7a. Calculations have been performed for a glass-supported slab as real colloidal crystals are usually deposited on a glass substrate (refractive index = 1.51). Nevertheless, we must point out that the reflected beam efficiencies of the glass-supported slab are basically the same as the ones obtained for the self-standing slab when extinction is introduced in the model. Furthermore, only the first six layers are involved in the optical response of the slab because waves have a finite penetration depth into the slab.[19,20]

Figure 3.7.7c shows the measured (colored lines) and calculated (gray lines) efficiencies of the reflected beam (1, 0) and (1, –1), where a good correlation between theory and experiment is obtained. In this case, a shift of the peaks toward higher wavelengths (red shift) can be appreciated. To fit the main peak positions of the experimental curves we have used $\varepsilon_s = 2.45 + 0.03i$. These results show that, although many fine spectral features are smoothed out as a result of disorder effects, some resonances are strong enough to survive the extinction process and can clearly be appreciated in a real experiment. At the same time, they confirm that extinction due to imperfections plays a crucial role in the optical response of colloidal crystal lattices and strongly determines the result of actual measurements.[33]

It should be mentioned that in Chapter 3.6 an alternative description of the optical diffraction by artificial opals and opal films is presented, in which the effect of disorder is not considered. In that case, the analysis of the experimental data is performed in the frame of a solid-state physics approach rather than the electromagnetic one followed in this chapter, in which the interaction of a plane wave with a finite size opal film is accounted for. The reader is encouraged to compare both approximations to the same problem, as they are the ones more commonly used today in the field.[23,34]

3.7.5 Effect of the Interplay between Order and Disorder in the Near-Field Optical Response

A complete description of the optical response of artificial opals in the high-energy range should take into account not only far-field properties but also near-field properties. Analysis of far-field techniques is adequate for reflectivity studies or stop-gap determinations,[13,23] whereas near-field-based techniques allows one to extract information on the photon local density of states of these sort of lattices.[35]

The development of SNOM has recently opened new possibilities to examine the electric field pattern of one- and two-dimensional metallic structures,[36] but also of PCs in the near-field range.[37–39] Despite being widely studied, results for artificial opals applying the SNOM technique are very limited at present. Flück et al.[38] have reported transmission measurements on polystyrene opals made of 40 monolayers at wavelengths on the order of the lattice parameter in the so-called transmission mode (also known as illumination mode). In this configuration, the PC is illuminated through an optical fiber with a probe tip aperture of 100 nm, which remains at a constant distance of <20 nm from the crystal frontal surface. The transmitted light intensity is collected directly beneath the PC by a photodiode detector in contact with the sample. The SNOM images revealed a modulation of the integrated light intensity at the rear surface with the same lattice constant as the frontal illuminated surface. Larger transmission was observed when the tip was positioned at the top of the spheres. More recently, Bittkau et al.[39] reported patterns for eight monolayer thick polymethyl methacrylate opal films grown from spheres with a radius of 180 nm using the conventional transmission mode. In this case, it was observed that the light transmission intensity map was strongly dependent on intrinsic structural defects along the crystal.

In this section, we report SNOM measurements in polystyrene artificial opal thin films made of spheres with a diameter of 0.75 μm using a SNOM collection mode. In this working configuration, a quasiplanar wave is used to illuminate the sample whereas the transmitted light is detected by an optical probe fiber tip located at few nanometers from the spheres in the rear surface. Details about the sample preparation and the setup used to carry out these measurements as well as the results obtained working in the illumination configuration can be found in the study by Barrio et al.[22] Very briefly, in this other configuration, the observed transmitted intensity pattern reveals that light is launched more efficiently on top of the spheres rather than between the spheres. These results fairly reproduce and confirm the conclusions reported by Flück et al.,[38] who studied the coupling mechanism of light coming from a point-like light source to a colloidal crystal.

In the SNOM collection mode, a quasiplanar wave is used to illuminate the frontal surface of the colloidal crystal whereas the transmitted light is detected by an optical probe located at a few nanometers from the spheres in the rear surface. Notice that this is the same illumination configuration used when standard far optical field measurements, such as reflectance or transmittance, are performed. Therefore, this setup could allow one to observe the fine features of the transmitted optical field before reconstruction takes place away from the sample to yield the well-known homogeneous distribution of the far field, as shown in Figure 3.7.2b.

In Figure 3.7.8, we present the transfer of light transmitted through the sample and collected by a near-field fiber probe. Figure 3.7.8a shows the topographical information obtained simultaneously with the near-field pattern presented in Figure 3.7.8b. The comparison of these two images allows us to analyze the spatial distribution of the transmitted near field when a planar wave impinges on the artificial opal film. As expected, the topographic image presented in Figure 3.7.8a reveals the well-known triangular arrangement of the spheres corresponding to the outer surface of the PC, as shown in Figure 3.7.1a, which corresponds to the (111) plane of a slightly distorted FCC structure.[21] Figure 3.7.8c and e show, respectively, the 2D self-correlation (SC) and the fast Fourier transform (FFT) analysis of the topographic image. In the FFT representation, $\vec{k} = 0$ and the first-order reciprocal lattice is observed. The sharpness of these spots indicates that the lattice is well defined and the SC image supports these assumptions. It is important to mention here that the crystal surface shows imperfections such as dislocations, vacancies, or cracks. This way, the SC is used as a tool for finding a repeating pattern. In our case, this function allows one to reveal a periodic signal, related to the ordered structure, which lay hidden under the unavoidable noise caused by the presence of intrinsic defects. However, despite these anomalies, the triangular arrangement mentioned above is preserved. As a working protocol, zones away from leaps between terraces of different thickness[15] or domain boundaries were always chosen. Interestingly, the 2D FFT and the SC of the optical image, as shown in Figure 3.7.8d and f, respectively, exhibit a diffraction pattern similar to the one displayed in Figure 3.7.8e. This fact indicates that the light-intensity distribution throughout the area analyzed is clearly related to the periodic structure. As far as we know, this is the first time a periodic profile of the near-field pattern is observed from an artificial opal, by using the same illumination configuration employed when standard far optical field measurements are performed.

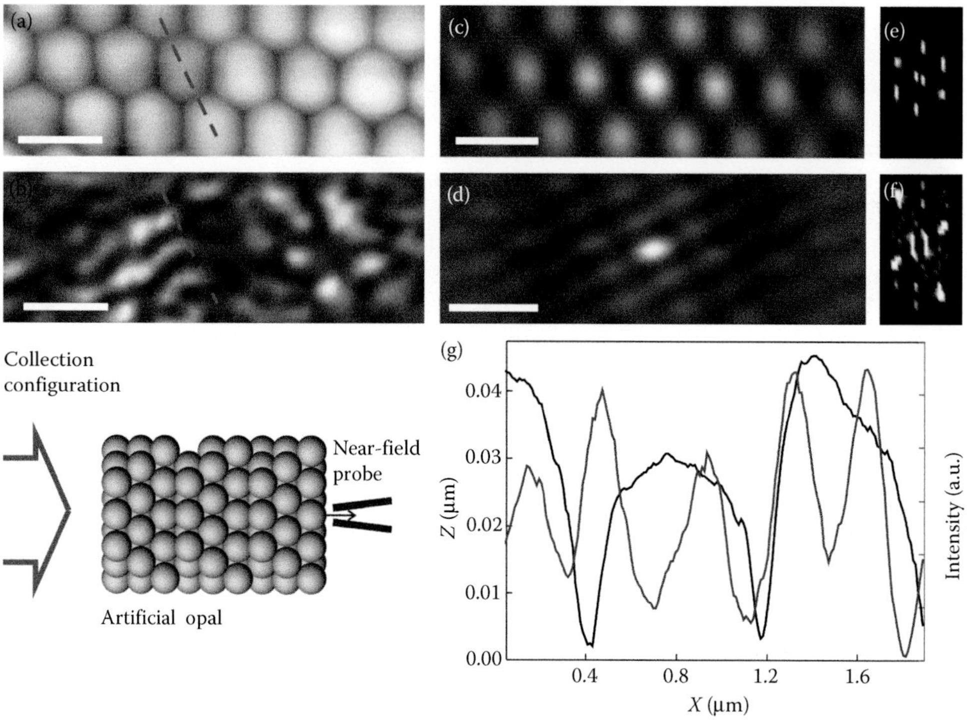

FIGURE 3.7.8
(a) Two-dimensional (2D) topographic image of the outer surface of the artificial opal. (b) 2D light intensity distribution using the scanning near-field optical microscopy collection mode. Self-correlation analysis of the topographic (c) and the optical (d) images presented in (a) and (b), respectively. Scale bars are 1 μm. 2D fast Fourier transform of the topographic (e) and the optical (f) images presented in (a) and (b), respectively. (g) Topographic (black) and optical (gray) profiles obtained along the line traced in (a) and (b), respectively. A scheme of the near-field setup used for the analysis is also shown. (Adapted from J. Barrio et al., J. Appl. Phys. **109**, 083514, 2011.)

Figure 3.7.8g shows the topographic and optical profiles recorded along the line traced in Figure 3.7.8a and b. The periodicity of the topographic pattern corresponds to the arrangement of the polystyrene spheres, which is around 0.75 μm. The FFT and SC analysis of the optical profile reveals a clear periodic modulation of the light intensity, consisting of two maxima and one minimum centered at the position in which each sphere is placed. The same profile structure has been found along different rows of spheres presenting the same relative orientation with respect to the incident electric field. However, the relative peak intensity varies sensibly along the sample that is probably related to the presence of intrinsic defects. This issue makes difficult the direct observation of a well-defined periodic profile when large areas of the crystal are scanned. The observation of the effect of rotating the polarization of the incident wave on the near-field pattern further confirms that the lattice periodicity modulates the near-field intensity profile. Although a detailed description of the polarization dependence of the SNOM patterns is beyond the scope of this study, we could confirm that the periodic light intensity pattern measured rotates with the polarization of the incident wave, as displayed in Figure 3.7.9.

Interpretation of near-field data is complicated, and the support of a simulation model is generally helpful to contrast the experimental data. To provide some insight into the

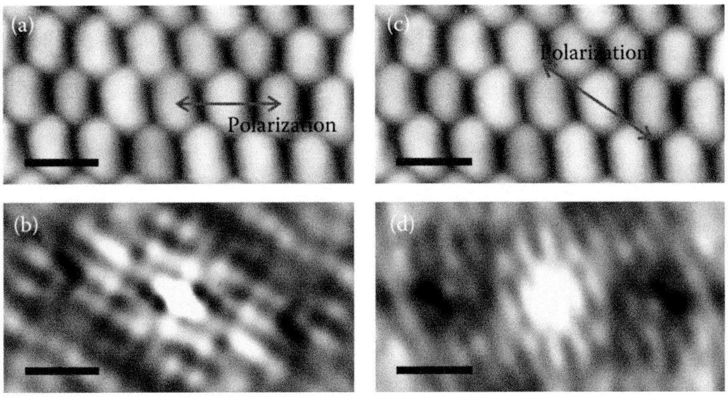

FIGURE 3.7.9
(a), (b) Topographic images taken from a selected region of the rear surface of an artificial opal. (c), (d) Scanning near-field optical microscopy images taken when the incident electric field is linearly polarized along a row of spheres (c) and after rotating the polarization direction 30° (d). The polarization directions are indicated with arrows in the topographic pictures. Scale bars are 1 μm. (Reprinted with permission from J. Barrio et al., J. Appl. Phys. **109**, 083514, 2011. Copyright 2011, American Institute of Physics.)

physical origin of the periodic light-collected intensity pattern and the corresponding polarization dependence, we use the vector KKR-based method. The physical mechanisms that described the fine features of the transmitted optical are based on collective properties of the lattice dynamics, each spectral feature having its origin in the resonance of a specific mode inside the crystal slab, which resonates as a whole entity owing to electromagnetic links between the lattice components.[19] We have already shown that all optical spectral features in the high-energy region of PCs arise from electromagnetic resonances within the ordered array, modified by the interplay between these resonances with the opening of diffraction channels, the presence of imperfections, and finite size effects. Most importantly, the periodic modulation of the intensity of the near-field pattern herein reported constitutes the first experimental evidence of the direct observation of such resonances. In Figure 3.7.10a, we show the calculated transmittance spectrum of a 10-layer artificial opal film in reduced units of a/λ. In all our experiments $\lambda = 532$. This value corresponds to the high-energy range ($a/\lambda = 1.98$), it is well above the predicted cutoff at $a/\lambda = 1.63$ for diffracted beams in air and is indicated by a vertical dashed line in Figure 3.7.10a. In Figure 3.7.10b–d, we show the spatial dependence of the calculated transmitted electric field ($\vec{E}_t$) when an incident plane wave propagates through a 10-layer-thick glass-supported colloidal crystal. In Figure 3.7.10b–d, we plot the $|\vec{E}_t(x,y,z)|$ calculated in the plane tangent to the poles of the spheres that form the outer compact plane in the 10-layer-thick stack for three different values of angle ϕ that forms the incident electric field with the x-axis in our reference system, specifically $\phi = 0°$ in (b), $\phi = 60°$ in (c), and $\phi = 90°$ in (d). Our results indicate on the one hand that the light-intensity distribution throughout this outer plane is related to the periodic structure and on the other hand that the periodic light-intensity pattern rotates with the polarization of the incident wave, which is in excellent agreement with our measurements. As long as the physical origin of the optical response of an artificial opal lies on multipolar resonances of the sphere ensemble, the coupling of light from or to an artificial opal depends on the relative launching position with respect to the crystal lattice and the collective properties of the lattice dynamics. This way, the observation of the periodic profile of the near-field intensity herein reported

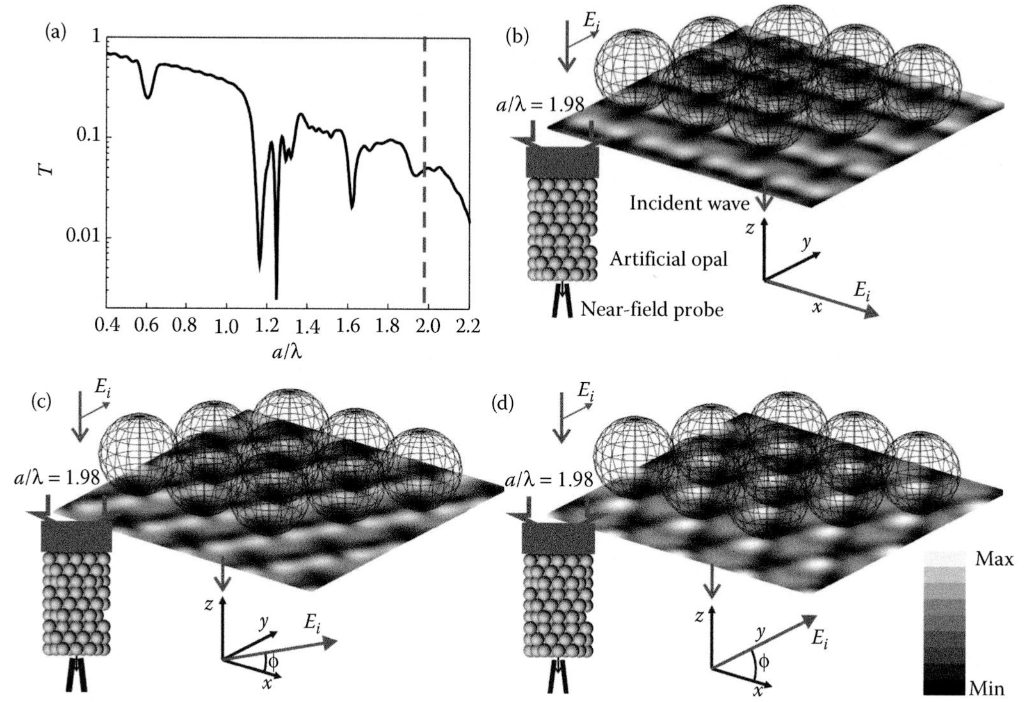

FIGURE 3.7.10
(**See color insert.**) (a) Calculated forward transmittance spectrum for a 10-layer-thick glass-supported colloidal crystal made of spheres of dielectric constant $\varepsilon_s = 2.5 + 0.05i$. Green vertical dashed line indicates the experimental incident wavelength value ($\lambda = 532$). (b)–(d) Spatial distribution of the intensity of the transmitted electric field calculated in the plane tangent to the poles of the spheres that form the outer compact plane in the 10-layer stack. The angle ϕ that forms the incident electric field with respect to the normal of the outer surface of the colloidal crystal (*z*-axis in our reference system) is $\phi = 0°$ in (b), $\phi = 60°$ in (c), and $\phi = 90°$ in (d). Color scales are also indicated. (Adapted from J. Barrio et al., J. Appl. Phys. **109**, 083514, 2011.)

represents the first experimental and theoretical confirmation of the resonances that result in the optical response of an artificial opal.

3.7.6 Conclusions

We have realized an integral approach to the optical response of 3D PCs in the higher-order band energy range. We have demonstrated that extinction owing to intrinsic defects determines the shape of the experimental spectra that have been so far reported. Calculations performed considering extinction show a clear correlation between the behavior of the imaginary part of the wave vector and the optical features observed in that range. By analyzing the gradual modification of that response from one to several monolayers, we could identify the peaks that are reminiscent of the optical reflectance features of a single close-packed layer and those that are the result of building up a 3D periodicity. Furthermore, we have found a clear correlation between the extinction introduced in our theoretical model and the width of the real artificial opal film, which indicates that thicker crystal slabs are actually better ordered than thinner ones and confirming that the extinction parameter introduced provides a method for quantifying disorder. We have also studied the optical response of nonspecular beams diffracted from perfectly ordered 3D PCs, in good agreement with the behavior previously observed for specularly reflected and forwardly transmitted beams. Diffracted beams have been recognized and labeled according to their associated reciprocal-lattice vectors, and the observed spectral dependence of their efficiency has been accurately modeled using a KKR approach. Also in this case, adding extinction to the theoretical model has allowed us to reproduce accurately the experimental diffracted intensity versus angle/wavelength curves. These results revealed that some spectral features in diffracted beams arise from strong resonances that are robust against disorder effects. Finally, by analyzing the near-field optical response that these structures exhibit, we have found that the profile of the near-field intensity is intrinsically related to the periodicity of the structure. The periodic modulation of the intensity of the near-field pattern herein reported constitutes the first experimental evidence of the observation of such resonances.

Acknowledgments

We would like to thank D. Shinca for his valuable help to characterize the spectral dependence of the intensity of the beams diffracted by artificial opals. This work has been partially supported by the Spanish Ministry of Science and Innovation under Grant MAT2007-02166 and Consolider HOPE CSD2007-00007, Junta de Andalucía under Grant FQM3579, Consejo Nacional de Investigaciones Científicas y Técnicas (CONICET), and Agencia Nacional de Promoción Científica y Tecnológica (PICT-11–1785).

References

1. E. Yablonovitch, Phys. Rev. Lett. **58**, 2059 (1987); S. John, *ibid*. **58**, 2486 (1987); C. López, J. Opt. A **8**, R1, (2006); F. Marlow, Muldarisnur, P. Sharifi, R. Brinkmann, and C. Mendive, Angew. Chem. **48**, 6212 (2009); J. F. Galisteo-López, M. Ibisate, R. Sapienza, L. S. Froufe-Pérez, Á. Blanco, and C. López, Adv. Mater. **23**, 30 (2011).
2. A. Blanco, E. Chomski, S. Grabtchak, M. Ibisate, S. John, S. W. Leonard, C. López et al., Nature **405**, 437 (2000).
3. G. Lozano, L. A. Dorado, R. A. Depine, and H. Míguez, J. Mater. Chem. **19**, 185 (2009).
4. P. Jiang, J. F. Bertone, K. S. Hwang, and V. L. Colvin, Chem. Mater. **11**, 2132 (1999).
5. S. Wong, V. Kitaev, and G.A. Ozin, J. Am. Chem. Soc. **125**, 15589 (2003).
6. H. Míguez, C. López, F. Meseguer, Á. Blanco, L. Vázquez, R. Mayoral, M. Ocaña, V. Fornés and A. Mifsud, Appl. Phys. Lett. **71**, (1997); M. Ishii, M. Harada, A. Tsukigase, and H. Nakamura, J. Opt. A **9**, 372 (2007).
7. H. Kosaka, T. Kawashima, A. Tomita, M. Notomi, T. Tamamura, T. Sato, and S. Kawakami, Phys. Rev. B **58**, R10096 (1998); A. Martínez, H. Míguez, A. Griol, and J. Martí, *ibid*. **69**, 165119 (2004); M. Botey, M. Maymó, A. Molinos-Gómez, L. A. Dorado, R. A. Depine, G. Lozano, A. Mihi, H. Míguez, and J. Martorell, Opt. Express **17**, 12210 (2009); M. López-García, J. F. Galisteo-López, A. Blanco, J. Sánchez-Marcos, C. López, and A. García-Martín, Small **6**, 1757

(2010); R. Sapienza, M. Leonetti, L. S. Froufe-Pérez, J. F. Galisteo-López, C. Conti, and C. López, Phys. Rev. A **83**, 023801 (2011).
8. E. Palacios-Lidón, B. H. Juárez, E. Castillo-Martínez, and C. López, J. Appl. Phys. **97**, 063502 (2005).
9. G. Lozano and H. Míguez, Langmuir **23**, 9933 (2007).
10. G. Lozano and H. Míguez, Appl. Phys. Lett. **92**, 091904 (2008).
11. L. A. Dorado, R. A. Depine, and H. Míguez, Phys. Rev. B **75**, 241101(R) (2007).
12. K. Wostyn, Y. Zhao, B. Yee, K. Clays, A. Persoons, G. de Schaetzen, and L. Hellemans, J. Chem. Phys. **118**, 10752 (2003).
13. J. F. Galisteo-López and C. López, Phys. Rev. B **70**, 035108 (2004).
14. F. García-Santamaría, J. F. Galisteo López, P. V Braun, and C. López, Phys. Rev. B **71**, 195112 (2005).
15. A. Balestreri, L. Andreani, and M. Agio, Phys. Rev. E **74**, 036603 (2006).
16. H. Míguez, V. Kitaev, and G. A. Ozin, Appl. Phys. Lett. **84**, 1239 (2004).
17. A. V. Baryshev, V. A. Kosobukin, K. B. Samusev, D. E. Usvyat, and M. F. Limonov, Phys. Rev. B **73**, 205118 (2006).
18. S. Schutzmann, I. Venditti, P. Prosposito, M. Casalboni, and M. V. Russo, Opt. Express **16**, 897 (2008); L. A. Dorado, R. A. Depine, and H. Míguez, Phys. Rev. A **78**, 037801 (2008); R. V. Nair and R. Vijaya, *ibid.* **78**, 037802 (2008); S. G. Romanov, M. Bardosova, I. M. Povey, M. E. Pemble, and C. M. Sotomayor Torres, Appl. Phys. Lett. **92**, 191106 (2008).
19. L. A. Dorado, G. Lozano, H. Míguez, and R. A. Depine, Opt. Express **15**, 17754 (2007).
20. L. A. Dorado, G. Lozano, H. Míguez, and R. A. Depine, Phys. Rev. B **76**, 245103 (2007).
21. G. Lozano, L. A. Dorado, D. Shinca, R. A. Depine, and H. Míguez, Langmuir **25**, 12860 (2009).
22. J. Barrio, G. Lozano, J. Lamela, G. Lifante, L. A. Dorado, R. A. Depine, F. Jaque, and H. Míguez, J. Appl. Phys. **109**, 083514 (2011).
23. L. A. Dorado, G. Lozano, D. Shinca, R. A. Depine, and H. Míguez, Phys. Rev. B **78**, 075102 (2008).
24. G. Lozano, J. E. Mazzaferri, L. A. Dorado, S. Ledesma, R. A. Depine, and H. Míguez, J. Opt. Soc. Am. B **27**, 1394 (2010).
25. F. López-Tejeira, T. Ochiai, K. Sakoda, and J. Sánchez-Dehesa, Phys. Rev. B **65**, 195110 (2002).
26. N. Stefanou, V. Yannopapas, and A. Modinos, Comput. Phys. Commun. **113**, 49 (1998); **132**, 189 (2000).
27. The imaginary part of the dielectric permittivity is not considered as a constant quantity during calculations. We used a smaller value of the ε_i parameter in the high-energy spectral range that represents the different effects of imperfections on the different wavelength ranges.
28. A. Mihi, H. Míguez, I. Rodríguez, S. Rubio, and F. Meseguer, Phys. Rev. B **71**, 125131 (2005); E. Pavarini, L. C. Andreani, C. Soci, M. Galli, F. Marabelli, and D. Comoretto, *ibid.* **72**, 045102 (2005); A. V. Baryshev, A. B. Khanikaev, H. Uchida, M. Inoue, and M. F. Limonov, *ibid.* **73**, 033103 (2006); A. V. Baryshev, A. B. Khanikaev, R. Fujikawa, H. Uchida, and M. Inoue, *ibid.* **76**, 014305 (2007); L. C. Andreani, A. Balestreri, J. F. Galisteo-López, M. Galli, M. Patrini E. Descrovi, A. Chiodoni, F. Giorgis L. Pallavidino, and F. Geobaldo, *ibid.* **78**, 205304 (2008).G. von Freymann, S. John, S. Wong, V. Kitaev, and G. A. Ozin, Appl. Phys. Lett. **86**, 053108 (2005); J. F. Galisteo-López, M. Galli, A. Balestreri, M. Patrini, L. C. Andreani, and C. López, Opt. Express **15**, 15342 (2007).
29. J. F. Galisteo-López, M. Galli, M. Patrini, A. Balestreri, L. C. Andreani, and C. López, Phys. Rev. B **73**, 125103 (2006).
30. E. Vekris, V. Kitaev, D. D. Perovic, J. S. Aitchison, and G. A. Ozin, Adv. Mater. **20**, 1110 (2008).
31. X. Checoury, S. Enoch, C. López, and A. Blanco, Appl. Phys. Lett. **90**, 161131 (2007).
32. S. Yano, Y. Segawa, J. S. Bae, K. Mizuno, S. Yamaguchi, and K. Ohtaka, Phys. Rev. B **66**, 075119 (2002).
33. The angular dependence of the intensity of beams diffracted by artificial opals has also been analyzed in depth. Details of this can be found in the study by Lozano et al.[24]

34. A. Samusev, K. Samusev, M. Rybin, M. Limonov, E. Trofimova, D. Kurdyukov, and V. Golubev, Phys. Solid State **5**, 1056 (2011).
35. K. Sakoda, *Optical Properties of Photonics Crystals.* (Springer, New York, 2001), Vol. 80.
36. C. Ropers, D. J. Park, G. Stibenz, G. Steinmeyer, J. Kim, D. S. Kim, and C. Lienau, Phys. Rev. Lett. **94**, 113901 (2005); T. Rindzevicius, Y. Alaverdyan, B. Sepulveda, T. Pakizeh, M. Käll, R. Hillenbrand, J. Aizpurua, and F. J. García de Abajo, J. Phys. Chem. C **111**, 1207 (2007); J. A. Schuller, E S. Barnard, W. Cai, Y. C. Jun, J. S. Whiteand, and M. L. Brongersma, Nat. Mater. **9**, 193 (2010).
37. P. L. Phillips, J. C. Knight, B. J. Mangan, P. St. J. Russell, M. D. B. Charlton, and G. J. Parker, J. Appl. Phys. **85**, 6337 (1999); M. L. M. Balistreri, H. Gersen, J. P. Korterik, L. Kuipers, and N. F. van Hulst, Science **294**, 1080 (2001); S. I. Bozhevolnyi, V. S. Volkov, T. Søndergaard, A. Boltasseva, P. I. Borel, and M. Kristensen, Phys. Rev. B **66**, 235204 (2002); H. Gersen, J. P. Korterik, N. F. van Hulst, and L. Kuipers, Phys. Rev. E **68**, 026604 (2003); E. Flück, M. Hammer, A. M. Otter, J. P. Korterik, L. Kuipers, and N. F. van Hulst, J. Lightwave Technol. **21**, 1384 (2003); H. Gersen, T. J. Karle, R. J. P. Engelen, W. Bogaerts, J. P. Korterik, N. F. van Hulst, T. F. Krauss, and L. Kuipers, Phys. Rev. Lett. **94**, 073903 (2005); S. Vignolini, M. Burresi, S. Gottardo, L. Kuipers, and D. S. Wiersma, Opt. Lett. **35**, 2001 (2010).
38. E. Flück, N. F. van Hulst, W. L. Vos, and L. Kuipers, Phys. Rev. E **68**, 015601 (2003).
39. K. Bittkau, R. Carius, A. Bielawny, and R. B. Wehrspohn, J. Mater. Sci. **19**, 203 (2008).

3.8

Opal-Based Hypersonic Crystals

Andrey V. Akimov
Ioffe Physical-Technical Institute of the Russian Academy of Sciences
The University of Nottingham

Alexander B. Pevtsov
Ioffe Physical-Technical Institute of the Russian Academy of Sciences

CONTENTS

3.8.1 Introduction

At the end of the twentieth century the start of research activity with photonic crystals (PhCs) required the method of introducing this new and rapidly growing field in a clear way by drawing an analogy with some well-known phenomena from traditional solid-state physics. For semiconductor physics researchers, the simple way to understand the origin of a photonic band gap was to consider the processes lying in the basis of electron spectrum formation in crystalline semiconductors. Scientists working with x-rays may immediately find an analogy between PhCs and x-ray diffraction on a periodic atomic lattice. For those who have studied acoustics, the propagation of sound in artificially fabricated periodic systems mentioning PhCs often reminds of a modern art sculpture by Eusebio Sempere.[1] This famous artwork consists of polished stainless-steel tubes located in periodic order on a base and blocks the propagation of sound waves in a certain frequency band. In the mid 1990s, sonic researchers introduced the term "phononic crystal" (PnC) in analogy with PhCs, which, at that time, already attracted an enormous amount of research groups.

Nowadays, the common way to introduce PnCs is to refer to PhCs. In PhCs the specific optical properties are governed by the spatial periodicity of the dielectric permittivity, ε, whereas PnCs possess specific acoustic properties as a result of the periodicity of the acoustic impedance $Z = s\rho$, where s is the sound velocity and ρ is the density of material.

From the point of basic theory, the difference is in the equations for electromagnetic and sound waves, Maxwell and elastic equations, respectively. Similarly to any wave in a periodic media, the spectrum of elastic waves in a PnC possesses stop bands and, often, depending on the symmetry and the contrast of the acoustic impedances between the elementary blocks, which form a PnC, a full phononic band gap exists.

The number of papers on PnCs is increasing. For those, who would like to start understanding the basics of PnCs on a qualitative level, we would recommend the paper by Liu et al. in *Science*.[2] References on more specific theoretical and experimental aspects studied in early works may be found in the review by Sigalas et al.[3] The technological aspects of microfabrication of PnCs and device applications are described in a review by Olsson and El-Kady.[4]

Several similarities exist in the properties of light and sound in PhCs and PnCs, respectively. The effects of sound localization,[5–7] superlense and superprizm,[8] imaging sound,[9,10] and waveguiding[11] are observed in a number of periodic two-dimensional (2D) and three-dimensional (3D) arrays. Recently, acoustic metamaterials have been introduced.[12] The relationship between phenomena in PhCs and PnCs makes the latter a remarkable object for performing experiments aimed at understanding the basic properties of propagation and localization of waves in periodic media. Often, the goal of the experiments is to study the general phenomena (e.g., localization) due to the periodicity, disorder, or defects in the structure, while the specific origin of the wave (e.g., electromagnetic or elastic) is less important. Acoustic waves of millimeter wavelength correspond to megahertz frequencies and it is easy to fabricate model objects for studying various phenomena typical for periodic arrays. Thus 2D and 3D structures, fabricated from metallic rods and spheres, respectively, show various general phenomena typical to periodic structures.[1,8,9] Recently, by welding millimeter metallic spheres, it was possible precisely to control the effects of disorder and observe Anderson localization of sound.[6,7]

Many works on PnCs have been aimed at developing the principles for practical applications in traditional acoustics and ultrasonics. Most experiments were performed on structures where the period of the acoustic impedance lies in the millimeter to meter range, where the frequencies of the band gaps in these structures correspond to megahertz (ultrasonics) and kilohertz ranges, respectively. Numerous achievements have been reached in this field such as sound filtering, waveguiding, focusing, and others, and PnCs are already used in various applications including imaging and environmental services. However, it is a great challenge to reach gigahertz and terahertz acoustic frequency ranges where structures with a period lying in micrometer and nanometer scales should be used. Such phononic structures are called "hypersonic crystals" (HCs), emphasizing an essential step forward from ultrasonics toward high-frequency elastic limit for solid objects.[13] Achieving the fabrication of HCs with a sufficiently large phononic band gap would give a clue for realizing attractive sound ideas discussed elsewhere.[14] The HCs could be used in communications, microelectromechanical systems and submicrometer imaging similar to their low-frequency analogs.[15] The ideas about integrated gigahertz acousto-optical devices based on periodic structures are considered widely nowadays.[16,17] Another qualitatively new field for application of HCs is the manipulation of heat transport, by using HCs as thermal isolators for phonons with frequencies lying in the phononic band gap. The properties of full phononic band gap in HCs could be used to stop the relaxation for excited electrons on phonons leading to an increase of luminescence quantum efficiency and a decrease of the lasing threshold. For realizing these and other beautiful ideas, the fabrication of high-quality HCs had to be developed and their basic acoustic properties studied.

Similar to PhCs, the PnCs are classified according to their dimension, that is, one-dimensional (1D), 2D, and 3D crystals. High-quality 3D HCs are difficult to fabricate but have significant potential for manipulation with gigahertz and terahertz sound because they may possess a full phononic band gap when the propagation of corresponding sound waves is forbidden in all directions. In theory, full phononic band gaps for various 3D structures have been predicted[18,19] and the effects related to the 3D phononic band gap have been demonstrated experimentally in low-frequency (kilohertz and megahertz) ranges.[2,6,20] In the gigahertz frequency range, the Brillouin scattering experiments with 3D HCs have shown the existence of full phononic band gap in polymer opals.[21]

The technologically most easily accessible objects for studying the properties of 3D HCs are synthetic opals.[21–23] Traditionally, opals are introduced as face-centered cubic (FCC) lattice of close-packed dielectric submicrometer spheres. They have been described in detail as PhCs in a number of chapters in this book, and one will find many similarities in acoustic and optical properties of opals. However, there are three major features that make opals as PnCs essentially different from opals as PhCs:

1. In opals, the contrast of the acoustic impedance between the spheres (Z_s) and material in the voids (Z_v) may be as high as $Z_s/Z_v \sim 10^4$ for bare opals with air in the pores. Such large acoustic contrast easily provides conditions for full phononic band gaps, whereas for FCC dielectric opals the full photonic band gap may only be realized using few available materials whose dielectric contrast is in the order of 10, for example, in so-called inverted silicon opals.[24]
2. In acoustics, there are phonon modes that may be generated at the surface (i.e., phonon surface modes). In homogeneous material, high-frequency surface modes may scatter and easily escape into the bulk. In opals, when the frequency of the surface mode falls into a full phononic band gap, the surface mode will be localized for a long time. Thus, 3D HCs provide the unique possibility to localize gigahertz and sub-terahertz acoustic waves near the surface and realize waveguiding by patterning the surface in a required way.
3. In opal-based HCs, the mechanical contact between spheres plays a crucial role in the formation of phonon dispersion curves and correspondingly the phononic band gap; it is especially important in bare opals constructed from rigid materials such as silica. In these structures, narrow solid bridges originating from the sintering of spheres remain the only paths for high-frequency elastic waves to propagate from one sphere to another. The width distribution of such bridges is usually strongly inhomogeneous, which brings the aspect of disorder in opal-based HCs as one of major importance.

In this chapter, we show the effects that result from using opal structures as the HCs. The chapter describes mainly the experimental studies presented in our recent publications.[23,25] As a preamble, we give a purely descriptive theoretical background in Section 3.8.2. Section 3.8.3 is devoted to the details of preparation of opal films with various sintering and their characterization using a standard optical spectroscopic technique. Section 3.8.4 describes an experimental technique for generation and detection of coherent hypersound in opal films. The experimental results on the studies of hypersound properties in opals with various sintering are presented in Section 3.8.5. Here, the effect of filtering of hypersonic waves generated at the surface of an opal film is demonstrated. In Section 3.8.6, we carry out a qualitative discussion about the role of disorder in HCs. Conclusions that finalize the chapter are given in Section 3.8.7.

3.8.2 Theoretical Background

Calculation of phonon dispersion curves in opals from first principles is a complicated task and lies far beyond the scope of this chapter. Thus, we limit ourselves to general aspects of the theory, and mention in more detail the results that are important for understanding experiments with coherent hypersonic waves in opal films. In Section 3.8.5, these theoretical predictions are used in the interpretation of experimental results. Concerning models, methods, and details of calculations, we suggest that the reader examines original theoretical papers.[18,19,23,26–28]

The main complication of the theoretical description is an inability to consider opal as a structure consisting of a number of rigid spheres connected to each other with identical elastic bonds. Any attempt to do this leads to an obvious result that the spectrum of elastic vibrations in FCC lattice with one type of "atoms" consists only of "acoustic-like" modes and no gaps in the spectrum can be observed.[29] Thus this suggests that opal, being a PnC, does not possess attractive features such as phononic band gap and further consideration becomes irrelevant. In reality this is not true and the confusion comes from using an inadequate model. One may try to reach the aim using another approach where the material that impregnates the voids between the spheres, say air or water, is considered to be a good acoustic media. This task was successfully achieved using the effective medium approximation by Psarobas et al.[19] and clear band gaps have been obtained in the phonon dispersion curves. Apparently, this approximation is good for PnCs working at kilohertz frequencies but should fail for HCs fabricated from submicrometer spheres. There, gigahertz and sub-terahertz elastic waves do not propagate on realistic distance in air or liquid, thus making such an approach invalid.

Another way to understand the underlying physics in real opals fabricated from submicrometer spheres is to start from the vibrational spectrum of a single isolated sphere. The localized modes that form the phonon spectrum in the sphere are well known as Lamb modes, named after Lamb who first calculated these modes in 1881.[30] The frequency of the spheroidal modes may be obtained from the transcendental equation that can be found elsewhere.[26] As an example, the lowest lying vibrational mode in the energy spectrum in sub-micrometer silica sphere has a frequency $\nu_{Lamb} \sim \bar{s}/2D$ GHz, where $\bar{s}$ is the mean sound velocity and D is the diameter of the sphere in opal structures. The next step is to get the spectrum of two elastically bonded spheres. Lamb modes evidently split as a result of elastic coupling between the spheres. On the other hand, new modes appear at the low-frequency side as a result of the vibrations of sphere centers relatively to each other. By increasing the number of spheres bonded to each other, we intuitively get extended optical- and acoustic-like bands and a gap between them could be possible. The optical-like band originates from the Lamb modes of the isolated spheres and the acoustic-like band governs the propagation of sound as if the spheres were infinitely rigid. More specifically, such an approach has been realized in the numerical calculations by Tanaka and Tamura and has resulted in the phonon dispersion curves for 3D opal shown in Figure 3.8.1a for the silica opal with sphere diameter $D = 359$ nm.[23,26] The main result of the calculations in opals is the existence of full phononic band gaps in the vibrational spectrum. The crucial parameter here appears to be the elastic coupling between the spheres that is characterized by the sintering coefficient:

$$\chi = \frac{D}{2a} - 1, \tag{3.8.1}$$

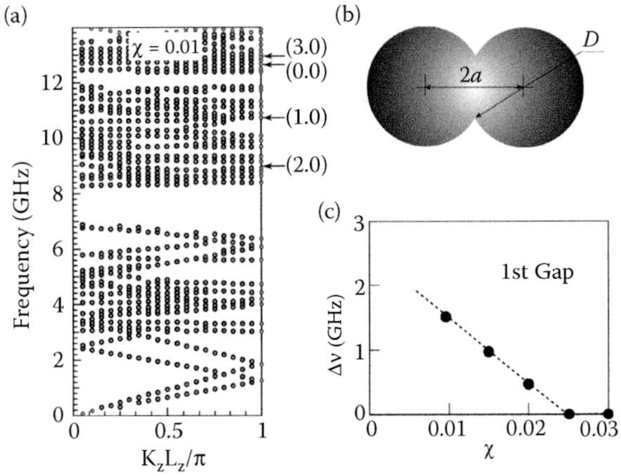

FIGURE 3.8.1
Theoretical calculations of elastic properties of opals. (a) Calculated phononic band structure of the face-centered cubic infinite opal along the $\Gamma - L$ direction;[23,26] the sphere diameter and sintering coefficient are $D = 359$ nm and $\chi = 0.01$, respectively; the arrows labeled by (l, n) are the eigenfrequencies of the low-lying spheroidal vibrational modes in the isolated silica sphere. (b) Diagram illustrating the sintering effect and the origin of the sintering coefficient χ given by Equation 3.8.1; D is initial diameter of opal spheres; $2a$ is the distance between the centers of the neighboring spheres after sintering. (c) Dependence of the width of the first complete phononic band gap on the sintering coefficient χ calculated for silica opal with sphere diameter $D = 359$ nm. (Adapted from A. V. Akimov et al., Phys. Rev. Lett. **101**, 033902, 2008.)

where $2a$ is the distance between the centers of neighboring spheres (see Figure 3.8.1b). It is clear that the increase of sintering eventually will turn opals into a homogeneous bulk silica without any features of HCs. The precise calculations for $\chi \ll 1$ show that in opals the first phononic band gap disappears at $\chi \approx 0.025$ (see Figure 3.8.1c). Another theoretical aspect, which is related to the properties of HCs, is the spectrum of elastic modes at the surface of an opal sample. The spectrum, shown in Figure 3.8.1a, is calculated for an infinite opal structure, but in reality, high-quality opals are fabricated in the form of films and the elastic waves are generated at the surface of the film using a hypersonic transducer. Thus, it is impossible to avoid the generation of surface modes. Moreover, the dream of many researchers in gigahertz acoustics is to exploit elastic surface waves when their spectrum falls into the phononic band gap of the 3D HC. If this happens, then surface waves cannot be scattered to the bulk and it becomes possible to manipulate waveguiding with surface waves. The calculations by Tanaka and Tamura in ideal silica opal films showed existence of the surface mode with a frequency close to the complete phononic band gap.[23] The details of calculations for surface modes in opal-like structures may be found in their works and in other theoretical papers.[27,31]

The disorder plays an important role in the formation of phonon spectrum and in hypersonic properties of opals. Even without any sphere site and diameter disordering, it is impossible in practice to keep the sintering between spheres homogeneous. The aspects of this problem will be discussed later in Section 3.8.6. Theoretically these calculations require certain models, which are not sufficiently developed as yet. It is useful to refer to the recent work by Still et al., where they show experimentally and theoretically that in strongly disordered colloidal structures the band gap may be opened because of hybridization of vibrations of different spheres,[32] that is, to have another nature, than the common

"Bragg" band gap arising because of spatial periodicity of structures. If the "Bragg gap" can be easily destroyed by the disorder, the hybridization gap is robust. This result encourages researchers not to give up in achieving beautiful "sound ideas" using HCs with a certain fraction of disorder.

3.8.3 Fabrication and Characterization of Silica Opals

In this section, we present the specific features of silica opal samples as HCs. These objects are well known as 3D PhCs that possess stop bands in the optical spectra. Thus, light with photon energy in the stop band cannot propagate in the opal sample along certain directions.

The spatial structure of the opals is the FCC lattice formed by close-packed monodisperse spheres of amorphous silica (a-SiO_2). The voids between the opal spheres form a sublattice that can be filled with other materials (up to 26% of the total volume). The opal structure can be controlled by special treatments of the situ-fabricated sample, and one of the methods used for this purpose is high-temperature annealing.

The process of fabrication of the opal films includes three stages. In the first stage, spherical a-SiO_2 particles with a specified diameter (~350 nm) are produced. This is performed by slow alkaline hydrolysis of tetraethoxysilane in an aqueous alcoholic medium (Stöber–Fink–Bohn's method).[33] In the second stage, 3D-ordered opal films with a thickness of 5–15 monolayers are grown from water suspension of a-SiO_2 spheres by a vertical deposition method[34] on 0.5-mm-thick fused silica substrate. The film surface is parallel to the (111) plane of the close-packed a-SiO_2 spheres. The opal films fabricated by this technique have smooth bright surfaces with a small number of microcracks. In the third stage, the grown films are subjected to temperature annealing. The main effect of annealing is the sintering (interpenetration) of the a-SiO_2 spheres. From the scanning electron microscopy (SEM) image (see Figure 3.8.2a) of the annealed opal film, it is seen that

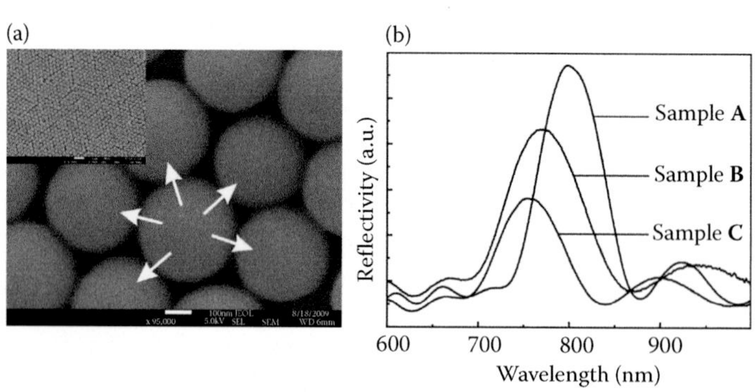

FIGURE 3.8.2
Synthetic silica opals. (a) Scanning electron microscopy image (SEM) of the (111) silica opal surface of sample **B** with sintering coefficient $\chi = 1.6 \times 10^{-2}$. White arrows indicate the places of sintering. Insert is the same SEM image in smaller scale. (b) The optical reflectivity spectra measured at an incidence angle $\theta = 13°$ in samples **A**, **B**, and **C**, which have various sintering coefficients. The high-intensity peaks correspond to the Bragg conditions and respectively to the photonic stop bands. The low-intensity peaks are the result of Fabry–Pérot interference.

neighboring spheres penetrate each other. Thus, a considerable elastic coupling arises between the opal spheres, which may result in the formation of a phononic band structure typical for 3D HCs. For the experiments described in Section 3.8.5, three samples were fabricated and investigated: **A**—as grown, **B**—annealed at 900°C, and **C**—annealed at 1000°C. The sintering coefficient, χ, given by Equation 3.8.1 is directly related to the filling factor, f_0, of silica spheres in opal films. The value of f_0 for moderately sintered spheres ($\chi \leq 2/\sqrt{3} - 1 \approx 0.155$) for the fabricated opal films can be calculated by the following equation:[35]

$$f_0 = \frac{\pi}{3\sqrt{2}} \frac{1 - 3\chi^2(3 - \chi)}{(1 - \chi)^3}. \tag{3.8.2}$$

The restriction imposed on χ obeys the condition for the spatial separation of the interpenetrating regions for various spherical pairs.

The mean value of f_0 for the investigated films can be obtained from the measurements of reflection spectra. Figure 3.8.2b shows reflection spectra measured at the incidence angle, $\theta \approx 13°$, to the normal of the (111) growth planes. The peaks in the spectral range within 750–850 nm correspond to Bragg conditions and arise from the stop bands of the opal-based PhCs. It is seen that the reflection band maximum is shifted toward shorter wavelengths as the annealing temperature is increased. The spectral position (λ_{max}) of the reflection peak obeys the well-known Bragg law:

$$\lambda_{max} = 2d_{111}\sqrt{\varepsilon_0 - \sin^2\theta}, \tag{3.8.3}$$

where d_{111} is the spatial period of the opal structure in the [111] direction, θ is the incidence angle of light onto the film surface, and ε_0 is the average dielectric constant of the opal films and is given by

$$\varepsilon_0 = \varepsilon_a f_0 + \varepsilon_b(1 - f_0), \tag{3.8.4}$$

where ε_a and ε_b are dielectric constants of the a-SiO_2 opal spheres and voids between them, respectively.

The parameters in Equations 3.8.3 and 3.8.4 were calculated by the least-squares method, using the experimental angular dependencies of the Bragg peak spectral positions λ_{max} both for the bare (unfilled) opal samples **A**, **B**, and **C** and for the same samples with the optical oil-filled opal pores. The dielectric constants for air and optical oil are $\varepsilon_b = \varepsilon_{air} = 1$ and $\varepsilon_b = \varepsilon_{oil} = 2.28$, respectively. The obtained fitting parameters ε_0, d_{111}, f_0, ε_a, and χ are presented in Table 3.8.1. It is clear from Table 3.8.1 that the high-temperature annealing of opal films decreases d_{111} and increases the filling factor f_0. As a result, the sintering coefficient of the fabricated opal films rises from 0.46% to 6.1%.

It should be noted that described optical studies have revealed that the value of ε_a is slightly lower than for bulk silica. This has lead to the conclusion that opal spheres consist of nano-sized silica particles with air-filled nanopores between them (see, for instance, the study by Gajiev et al.[36] and references therein). We assumed in our calculations that oil did not penetrate into these pores and that the dielectric constant of the opal spheres, ε_a, remains the same as in the unfilled samples.

TABLE 3.8.1

Calculated Parameters of the Investigated Opal Films

Samples	Filling	ε_0	d_{111} (nm)	f_0	ε_a	χ
A (as grown)	Unfilled	1.74	305	0.75	1.96	0.0046
	Oil-filled	2.03	305			
B (annealing 900°C)	Unfilled	1.76	290	0.77	1.99	0.016
	Oil-filled	2.05	292			
C (annealing 1000°C)	Unfilled	1.95	270	0.85	2.12	0.061
	Oil-filled	2.14	273			

3.8.4 Experimental Methods and Techniques

Experimental studies of elastic vibrations (i.e., phonons) in HCs are performed nowadays using two approaches. The first one is based on Brillouin scattering of light. In homogeneous media, this effect has been used for more than 50 years to obtain the spectrum of the gigahertz acoustic phonons. The momentum and energy conservations for photon–phonon interaction in light scattering events give the shift $\hbar\omega_{\mathbf{q},j}$ (where $\mathbf{q}$ is the phonon wavevector and j is the phonon polarization) of the scattered light spectrum relative to the laser excitation line as a function of propagation direction for the scattered light and sound velocity s_j. In HCs due to phonon folding of dispersion curves, the spectrum of the scattered light may show more peaks than in the homogeneous media, and by studying the dependence of $\hbar\omega_{\mathbf{q},j}$ on $\mathbf{q}$ by varying the angle between the incident and scattered light, it is possible to plot the dispersion curves for gigahertz acoustic waves. The studies on HCs by the Brillouin scattering technique have been performed intensively during the last decade and have provided valuable information about the dispersion curves and phononic band gaps in opals.[21]

The second technique is the extension of traditional acoustic experiments to the gigahertz frequencies measuring the transmission or reflection of coherent sound. Such types of experiments with piezoelectric transducers, which are widely used in the megahertz range, fail for frequencies exceeding several gigahertz owing to technical reasons. The breakthrough in this field was made by Thomsen et al.[37] in 1986 who used short laser pulses to produce hypersonic wave packets in opaque material and detected them optically by the pump–probe technique widely used in ultrafast optics. This started a new field, "picosecond acoustics," which is rapidly growing nowadays. In 1D hypersonic structures (e.g., superlattices and planar microcavities), coherent waves with a frequency as high as 1 THz have been studied.[38] This technique has been used in the gigahertz frequency range in 2D HCs.[10,39]

In this chapter, we describe the experiments performed in opal films using the picosecond acoustic technique.[23,25] The scheme for generation and detection of coherent vibration in opals is shown in Figure 3.8.3. An aluminum film with a thickness of 100 nm was deposited on an opal sample surface and this metal film was excited by laser pulses. As a result the film expands because of the thermoelastic effect, injecting a strain pulse into the opal sample.[40] Thus, the metal film plays the role of the hypersonic transducer for generation of a picosecond strain pulse in the opal film. For excitation, 0.3 ps pulses from a Ti:sapphire laser with a regenerative amplifier (wavelength $\lambda = 800$ nm, repetition rate 250 kHz, and maximum energy per pulse 1 μJ) were used in the experiments described in

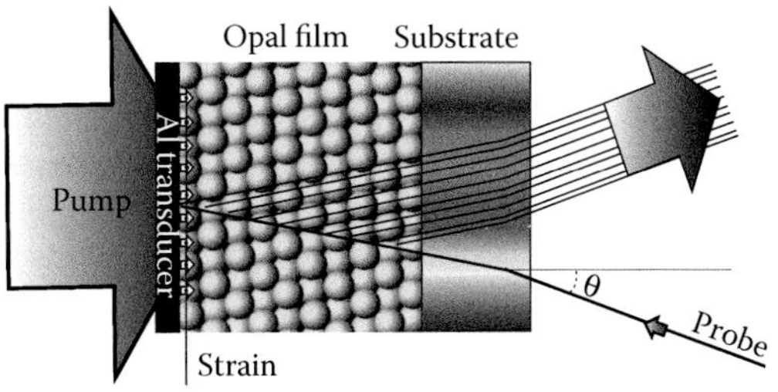

FIGURE 3.8.3
The experimental setup of pump–probe experiments. The pump beam from a femtosecond laser generates picosecond strain pulse in aluminum film that serves as an optoelastic transducer for injection of acoustic wave packet into an opal film. Monitoring of the vibrations in the first opal layer is realized by the probe beam incident on the opposite side of the sample at an angle θ.

Section 3.8.5. The pump beam was sent along a variable delay line and focused (200 μm spot diameter) on the metal film. A strain wave packet generated in this way has frequency components up to 100 GHz, which cover well the frequency range of interest for the opal-based HCs. The amplitude of strain in the nearest to the metal film *a*-SiO_2 spheres is estimated to be $\eta \sim 10^{-3}$.

For detection of the elastic vibrations, the probe pulse was split from the same laser beam and was focused exactly opposite to the pump spot onto the opal film from the silica substrate side with a spot diameter of 20 μm and incidence angle θ. The specularly reflected probe beam was detected by a photodiode. To increase the signal-to-noise ratio, the pump beam was modulated by a mechanical chopper with a frequency of 2 kHz. The intensity of reflected light R was measured as a function of time delay t between pump and probe pulses and as a result the temporal evolution of the reflectivity changes $\Delta R(t)$ was measured.

The contribution from coherent hypersound to the optical probe signal $\Delta R(t)$ is governed by several mechanisms of the interaction of light with coherent elastic vibrations in an opal film. In homogeneous bulk media (without periodic structure), the probe signal is governed by the elasto-optical effect, that is, the strain-induced changes of the refractive index, thus modulating $\Delta R(t)$. In thin films, the signal includes two contributions:[41,42] (i) reflectivity changes governed by the modulation of the thickness of the film owing to the interference of beams reflected from the top and bottom surfaces; and (ii) dynamical interference of the optical beam reflected from propagating acoustic wave in the bulk of the film, known as the coherent Brillouin scattering phenomenon. Contribution (i) gives a signal that is caused only by the displacement of the surfaces in the direction perpendicular to the plane of the film, whereas contribution (ii) is sensitive only to the acoustic waves propagating in the bulk. The essential difference between (i) and (ii) is that (i) gives the real-time evolution of the vibrations at the surface, whereas (ii) is governed by the energy and momentum conservation for light and sound and thus only spectral components that fulfill the equation

$$\nu_j[\text{Hz}] = \frac{2s_j\sqrt{\varepsilon_0 - \sin^2\theta}}{\lambda} \tag{3.8.5}$$

are active in (ii) out of a broad spectrum of generated acoustic frequencies.[41]

In periodic structures such as opal films, the basic mechanisms that govern the probe signal are the same as described above for thin films. Indeed, contribution (i) obviously is identical to the case of a homogeneous film. In the case of Brillouin scattering, several modes may give contributions as a result of folding of the dispersion curves, similar to noncoherent Brillouin scattering of light. An important point in the experimental technique is the sensitivity of probing to the elastic vibrations. It is possible to increase the sensitivity by matching the spectral wing of the optical reflectivity peak (see Figure 3.8.2b) with the probing wavelength. Further discussion of the mechanism of optical probing and sensitivity is given in Section 3.8.5.

The interaction of light with coherent elastic vibrations in opals and other HCs is a very intriguing subject. Indeed, the same periodic structure may possess simultaneously the specific properties as a PhC (e.g., photonic band gap) for light and an HC (e.g., full phononic band gap) for gigahertz sound. Such a combination is one of the elements of revolutionary sound ideas in theoretical works where the photon–phonon interaction is strongly enhanced because of localization of both electromagnetic and elastic waves.[17] In 1D HCs, the combination of phonon and photon microcavities in the same multilayered nanostructure has already proved the benefits of such an approach.[43] Unfortunately, there is no theoretical analysis that provides more detailed information about photon–phonon interaction in 3D "phonotonic" periodic structures.

3.8.5 Detection of Coherent Elastic Vibrations in Opal Film

In this section, we describe the experiments[25] performed by a group of researchers from three institutions: Ioffe Physical-Technical Institute, St. Petersburg, Russia; Dortmund University, Germany; and the University of Nottingham, United Kingdom. These experiments aimed at monitoring the temporal evolution of coherent surface acoustic modes whose frequencies fall in the 3D phononic band gap. The main question, when the experiments were planned, was to understand whether the surface acoustic modes have essentially longer lifetimes than the modes propagating into the bulk of the 3D HC. Coherent broadband acoustic wave packets were generated at the surface of an opal film as described in Section 3.8.4 (see Figure 3.8.3) and the temporal decay of different frequency components was monitored there. In the experiments, the time evolution of the reflectivity $\Delta R(t)$ was measured and then the fast Fourier transform (FFT) of $\Delta R(t)$ was obtained. Modes with frequencies corresponding to theoretically predicted phononic band gaps of the 3D HC (see Figure 3.8.1a) were expected to have a long lifetime owing to localization in the near-surface layer, whereas other elastic modes were expected to escape from the surface to the bulk and quickly lose the coherence because of the inhomogeneities of the elastic bonds between the spheres.

We present results for three opal films (samples **A**, **B**, and **C**), described in detail in Section 3.8.3. These samples were grown under the same conditions but had different annealing temperatures. The difference in the treatment results in these three samples is that they show different sintering coefficients χ (see Table 3.8.1).

Figure 3.8.4 shows the temporal signals $\Delta R(t)$ measured for samples **A**, **B**, and **C** and Figure 3.8.5a presents power FFT of the measured $\Delta R(t)$. The solid lines show the spectra obtained by FFT over a wide time window $t = 0$–1600 ps, from which a striking difference between the samples becomes clear. The spectrum of sample **A** shows a well-isolated peak

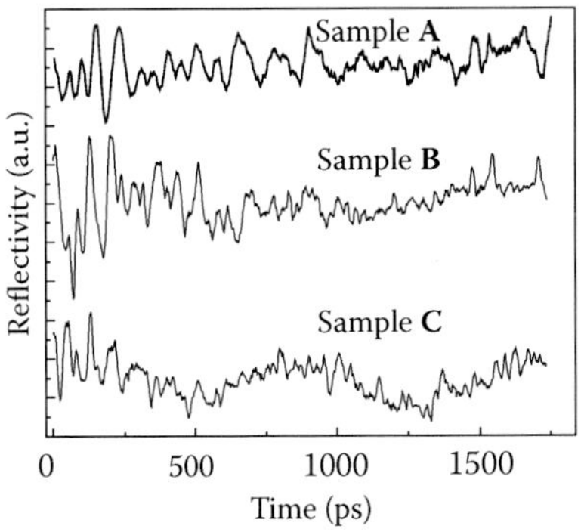

FIGURE 3.8.4
Reflectivity time evolutions $\Delta R(t)$ measured for samples **A**, **B**, and **C**. Sample **A** shows coherent oscillations up to $t = 1$ ns whereas high-frequency (~10 GHz) oscillations in all samples are clearly seen up to $t = 500$ ps (sample **B**) and $t = 200$ ps (sample **C**). Low-frequency (~1 GHz) oscillations correspond to the coherent vibration of the whole opal film and are detected only in sample **C** that has highest sintering coefficient.

centered at $\nu = 7.5$ GHz with an FFT limited spectral width ~1 GHz. This peak is absent in the spectrum of samples **B** and **C**. The other peaks observed in all samples have essentially smaller amplitudes than the $\nu = 7.5$ GHz peak in sample **A**. Dashed curves in Figure 3.8.5a show the FFTs of $\Delta R(t)$ over a shorter time window $t = 0$–600 ps. These spectra have high-amplitude peaks in the frequency range 10–15 GHz as a common feature.

The spectra obtained by FFT for long and short time intervals (solid and dashed lines in Figure 3.8.5a, respectively) provide information about generation and decay of elastic modes in various samples. The short time window spectra show that at early t the modes in the frequency range 10–20 GHz dominate in all three samples. Thus, these modes are efficiently excited by the pump pulse. The absence of these modes in the spectra obtained for long time intervals indicates their relatively fast decay in all samples. The fact that only one mode with $\nu = 7.5$ GHz is still pronounced in the long time interval for sample **A** shows that this mode has a much longer lifetime in this sample than the modes with $\nu > 10$ GHz.

Our qualitative conclusion about the existence of a long-living mode with $\nu < 10$ GHz in sample **A** can also be drawn directly from the temporal evolution of $\Delta R(t)$ (Figure 3.8.4). For $t < 300$ ps, several well-resolved oscillations with a period <100 ps are seen in $\Delta R(t)$ in all samples. At longer $t > 500$ ps, oscillations with a longer period >100 ps are clearly observable for sample **A** only.

To be more specific about the kinetics of the different modes, we show the temporal–spectral evolution of the detected signals in Figure 3.8.5b. The 3D plots in Figure 3.8.5b give the temporal (t) evolutions of the FFT spectra obtained from $\Delta R(t)$ in the time window that starts at t and ends at $t + \Delta t$ ($\Delta t = 660$ ps). A striking difference is observed for samples **A** and **B**. The long-living low-frequency mode at $\nu = 7.5$ GHz is clearly seen for sample **A** (Figure 3.8.5b, upper panel), and dominates in the spectrum for $t > 500$ ps when the modes with $\nu > 10$ GHz have almost disappeared from the spectrum. In contrast, the long-living mode is not observed for sample **B** (see Figure 3.8.5b, bottom panel). The fact that the long-living mode in sample **A** has nonmonotonic behavior with time (the amplitude increases with t at early times and then decreases) can be explained by the beating of several modes with close frequencies, initially generated with opposite phases, which are not resolved in Figure 3.8.5b for the chosen value of Δt. Indeed, there are several modes detected in the frequency region $\nu < 10$ GHz that become resolved in the case of a wider time window

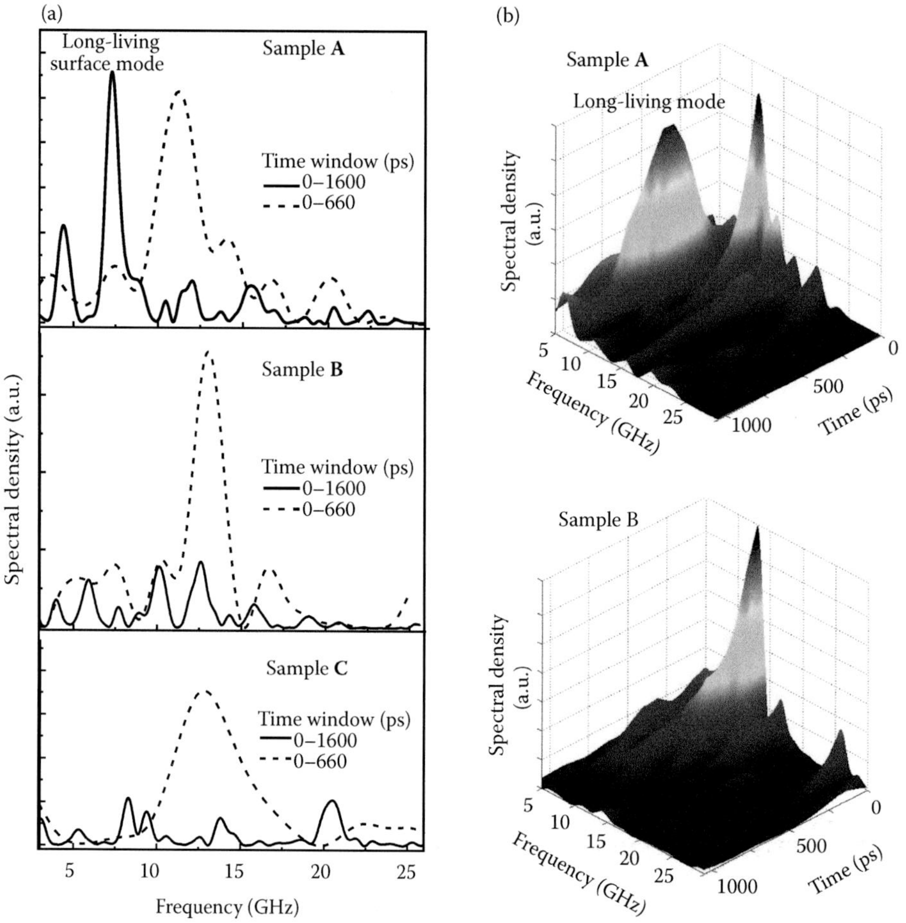

FIGURE 3.8.5
(See color insert.) (a) Power spectra of coherent elastic vibration obtained by fast Fourier transform of the measured reflectivity signals Δt (see Figure 3.8.4) for various opal samples. The solid and dashed curves in (a) give the spectra obtained in long ($\Delta t = 1600$ ps) and short ($\Delta t = 660$ ps) time windows, respectively. (b) Temporal evolutions of the vibrational spectra for samples **A** and **B**; the time scale in (b) corresponds to the start of the time window that was fixed at $\Delta t = 660$ ps. The long-living mode is observed only in sample **A**. (From A. S. Salasyuk et al., Nano Lett. **10**, 1319, 2010. With permission.)

(e.g., the mode at 4.5 GHz—see the solid line in Figure 3.8.5a, sample **A**). Thus, the main experimental result of this work is the observation of an elastic mode in sample **A** with $\nu = 7.5$ GHz, which has much longer lifetime than the vibrations with $\nu > 10$ GHz.

The explanation is based on the existence of elastic modes that are resonant with the complete 3D phononic band gap of the opal HC. We assume these modes to be present at the surface at which the elastic wave packet is initially injected. The theoretically predicted[23,26] band structure of elastic vibrations in ideal silica opal is shown in Figure 3.8.1a. From there we see that in the opal films studied in the present work the first band gap should be centered at $\nu_0 \approx 7.5$ GHz. For $\chi = 4.6 \times 10^{-3}$ (corresponding to sample **A**), the theoretically predicted value for the spectral width $\Delta\nu$ of the band gap is 2 GHz. Theoretically, $\Delta\nu$ decreases rapidly with increasing χ (see Figure 3.8.1c), and at $\chi \geq 2.5 \times 10^{-2}$ the complete phononic band gap vanishes according to the calculations. Thus, if a surface mode at $\nu = \nu_0$ is excited, then it will be localized at the surface only in the samples with sufficiently small χ. In contrast, modes with frequencies above and below the gap correspond to 3D-extended elastic states and therefore are expected to propagate from the surface to the bulk, independently of the χ-value.

These theoretical predictions are in full agreement with the experimental observations if we consider the measured signal $\Delta R(t)$ to be governed mostly by the elastic vibrations at the surface layer where the initial elastic wave packet has been injected from the metal transducer (aluminum film). The experimentally observed long-living mode at $\nu = 7.5$ GHz exactly matches the center frequency of the predicted band gap ($\nu_0 \approx 7.5$ GHz) in sample **A** for which the value of χ is sufficiently small. Elastic modes with $\nu > 10$ GHz, which are excited in the same surface layer, correspond to modes propagating into the bulk and therefore contribute to $\Delta R(t)$ only at early times. For sample **B**, χ is almost four times larger than in sample **A** so that no long-living mode is observed at any ν.

Sample **C** has an even higher sintering parameter $\chi = 6.1 \times 10^{-2}$ and no complete phononic band gap is expected here. Indeed, the spectrum obtained from $\Delta R(t)$ in a long time window (solid line in Figure 3.8.5a, sample **C**) does not show an intensive line around 7.5 GHz. Apparently, the properties of low-frequency modes with $\nu < 3$ GHz (see dispersion curves in Figure 3.8.1a) should be closer to plain silica in samples **A** and **B** because of the higher χ in sample **C**. Thus, in sample **C** it is possible that, during 2 ns measurement period, injected low-frequency phonons fill the opal film and it starts to "breathe" as a whole resulting in slow oscillations with frequency ν_b 1 GHz in $\Delta R(t)$ (see the curve for sample **C** in Figure 3.8.4).

In Section 3.8.4, we mentioned two mechanisms that are responsible for the probe signal $\Delta R(t)$ in opals. Up to now we assumed that $\Delta R(t)$ is governed only by the vibrations in the first opal layer from the side of the aluminum transducer. In principle, both bulk and surface elastic modes may contribute to $\Delta R(t)$ in an ideal opal. The contribution from elastic waves propagating in the bulk of opals (Brillouin contribution) obviously depends on whether these modes keep their coherence. If the coherence is not destroyed, then we can expect the dependence of temporal evolution of $\Delta R(t)$ on θ. Contrarily, θ dependence of $\Delta R(t)$ for coherent vibrations localized near the surface should be absent. Figure 3.8.6 shows the temporal signals $\Delta R(t)$ measured in sample **A** for several incident angles θ. It is seen that the temporal evolution of the signal for various θ is almost identical: the maxima and

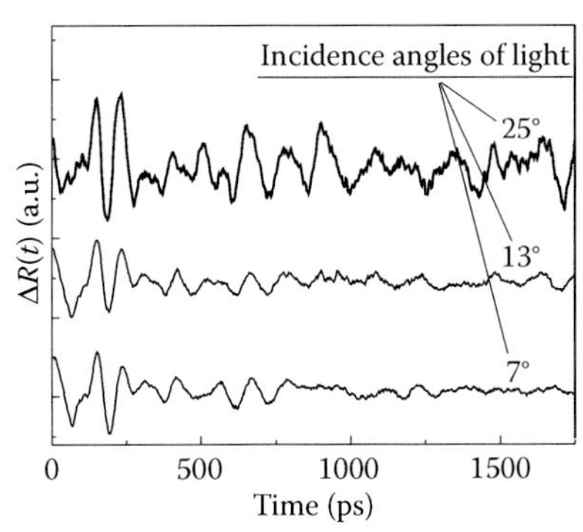

FIGURE 3.8.6
The time evolutions of the reflectivity changes $\Delta R(t)$ for sample **A** measured at various incidence angles θ.

minima in the oscillating signals take place at the same time at $t < 1000$ ps. At longer times, the comparison is not relevant because of higher noise levels for $\theta \approx 7°$ and 13°. The independence of $\Delta R(t)$ on θ points out that the coherent part of $\Delta R(t)$ is governed mostly by vibrations localized at the surface of the opal sample near the metal film where they are initially generated by the pump laser pulse. Actually this is not surprising because it is reasonable to expect that the coherence of the bulk hypersonic waves is already lost at the second layer of the opal film as a result of inhomogeneities of the elastic bonds between the sintered spheres. The role of inhomogeneities and disorder in vibrational dynamics of bonded opal spheres is discussed further in Section 3.8.6.

The results for probing the coherent elastic vibrations at various angles of light incidence (Figure 3.8.6) give an idea about the role of photonic stop band in the detection mechanism. As mentioned earlier in Section 3.8.4, it is possible to increase the sensitivity by matching the probing wavelength ($\lambda = 800$ nm) with the spectral wing of the Bragg reflection peak of the opal film. In sample **A**, the maximum sensitivity to the elastic vibrations is observed at $\theta \approx 25°$. The distinct signal $\Delta R(t)$ is also observed at other θ (see Figure 3.8.6), when the probing wavelength (800 nm) corresponds to the maximum of the spectral line, for example, at $\theta \approx 13°$ as seen in Figure 3.8.2b. At these conditions, the signal is a result of reflectivity amplitude modulation rather than spectral shift. In this case, the detected signal could be due to contributions from the modulation of Fabry–Pérot oscillations that are present in the reflectivity spectra around 800 nm.

3.8.6 The Role of Disorder in Hypersonic Properties of Opals

In Section 3.8.5, we have shown that elastic modes generated at the surface with a frequency $\nu = \nu_0$ have a long coherence lifetime. By making conclusions we were drawing a parallel between the experiment and theory when considering opal as an ideal FCC structure and were referring to the calculated dispersion curves shown in Figure 3.8.1a. Actually, the existence of disorder and several types of defects may cause serious doubts in such an approach. In this section, we discuss the role of disorder and defects in relation to the obtained experimental results. Nowadays this subject is little studied for PnCs and thus we limit the discussion to a qualitative level.

A number of chapters in this book provide valuable information about the role of disorder in opals. In optics the main sources of disorder are the dispersion of sphere size and its shape, stacking faults, inhomogeneous filling factor of pores, and cracks. The disorder smears the Bragg stop bands and decreases both the Bragg reflection peaks and transmission dips. If the disorder is too strong, then the periodicity is destroyed and, as a result, stop bands disappear completely. For further details, we refer the reader to Chapters 2.1 through 2.7 of this book. From this point of view, the opal samples described in Section 3.8.3 may be considered as high-quality PhCs with slight disorder.

The inhomogeneities of sintering are not usually considered in opal-based PhCs, but it becomes extremely important when considering opals as HCs. Let us look at the SEM pattern of the opal sample (Figure 3.8.2a) where the inhomogeneities of sintering are clearly seen in the image. A strong deviation of sintering from bond to bond results from a dispersion of sphere diameter and the annealing condition of opal films. It is clear that a slight change in D results in strong deviations of elastic bonding between the spheres up to breaking a number of bonds completely. This way of thinking brings one to the conclusion that the coherence of the elastic waves is lost as soon as the elastic vibrations, initially generated in spheres at the surface of the opal film, are transferred to the neighboring spheres. Thus, the propagation of hypersound from sphere to sphere turns into a diffusive regime. Such strong disorder should evidently kill the interference of elastic waves on the periodic structure and related Bragg phononic band gap should disappear.

However, why does the experiment show the distinct features of a band gap despite the strong disorder of sintering? To answer this intriguing question let us turn to the processes that lie at the basis of the phononic band-gap formation. As important specific feature of HCs is that there may be two ways to form band gaps. One is governed by the periodicity of the structure, and we call these band gaps "Bragg gaps" because the elastic waves with a frequency in a gap satisfy the Bragg conditions on a periodic structure. These band gaps are known to be quite sensitive to disorder as in the case of PhCs. However, for phonons in opal-like structures there exists another type of band gap that originates from the elastic properties of the coupled spheres and is not strongly sensitive to the periodicity of the structure. These band gaps were first mentioned by Turner et al.[44] who studied sound propagation through coupled spheres of aluminum beads. The gaps appear between the optical- and acoustic-like elastic modes similar to those well known in crystals.[29] The acoustic-like modes in an opal film are formed as a result of the elastic coupling between the absolutely rigid spheres, whereas optical-like modes are governed by the interaction between the Lamb vibrations of the coupled spheres. The maximum frequency of acoustic-like modes increases with bond stiffness, which in turn increases with the sintering coefficient χ. The optical-like modes due to coupling between the spheres transform from narrow Lamb modes into bands. It is clear that there will be a gap between acoustic-like and optical-like modes at small sintering. This gap will decrease when the sintering becomes stronger, and when χ exceeds a certain value the gap disappears and acoustic- and optical-like modes overlap.

The above qualitative arguments for the existence of the phononic band gaps show that there is no need for periodicity and thus a gap may exist in disordered opals. In the experiments by Turner et al.,[44] there was site disorder while the sintering was homogeneous. The same conditions have been realized in recent works where Anderson localization for sound was observed.[6,7]

The disorder causes strong scattering of the elastic waves propagating (i.e., for modes with frequencies not in the gap) in opals. This should lead to a loss of coherence in the next silica sphere layer from the surface, and thus silica opal is not a good material for hypersonic bulk transport. The experimental data described in Section 3.8.5 have shown that indeed there is no coherent signal from the modes propagating the bulk of opals. However, there is a remarkable feature of opals as HCs, as this has the ability to localize modes at the first silica sphere layer and the disorder does not prevent this localization. The question arises whether these localized modes can propagate coherently in the plane of the surface or whether they are localized in the areas limited to several spheres. We think that in the structures used in the described experiments the latter scenario is more likely because the disorder of sintering will break coherence for the 2D hypersonic waves propagating at the surface. Moreover, the theoretically predicted frequency of the surface propagating wave[23] has a value ≈8.7 GHz that is higher than the center of the gap $\nu_0 = 7.5$ GHz. From this respect, in the experiments described in this chapter, the disorder plays a favorable role providing a broad spectrum of surface vibrations that fall into a gap. Actually, the coherence time of these modes is governed by the dephasing of modes at the surface that fall into the gap and thus are defined by the inverse width of the gap ~1 GHz (Figure 3.8.1a and c).

3.8.7 Conclusions and Prospects

Silica opals show HC's properties typical for coherent elastic waves with frequency ~10 GHz. The phonon spectrum in opals with small sintering between the spheres possesses a full phononic band gap despite strong inhomogeneity of the sintering. Opals with full phononic band gap may work as a filter to select long-living elastic vibrations at the surface of the sample. These vibrations have a frequency and lifetime that match the central frequency of the full phononic band gap and its width, respectively.

The described experiments were carried out in 2009–2010 and more experimental studies may be performed in the near future. It is a challenge to control the sintering to be homogeneous but if it is achieved, a breakthrough for the fabrication of hypersonic integrated circuits at the surface of 3D HCs will be possible. The combination of HCs with piezoelectric materials (e.g., piezoelectric polymers) can provide a link with electromagnetic waves, and, in this respect, a combination of opal-based HCs with metamaterials becomes realistic.

Further work with opals and their applications as HCs may go in three directions: the first includes efforts to control the sintering on the level of a single sphere. One of the ways to do this is to fill the voids with some soft material that will keep a high contrast of acoustic impedances in HCs, but will serve as a good acoustic "conductor" between the spheres. The second direction could be aimed at performing various acousto-optical and acousto-microwave experiments that could lead to the developing of high-frequency optical modulators and microwave devices. The third is the development of theoretical models for the HCs with disorder.

Acknowledgments

We greatly acknowledge all our colleagues who have fabricated and characterized opal samples: Dr. S. F. Kaplan, Dr. S. A. Grudinkin, Dr. A. V. Nashchekin, and PhD student S. A. Yakovlev (Ioffe Physical-Technical Institute, St. Petersburg, Russia). The described ultrafast experiments were carried out at the University of Dortmund, Germany, by PhD students A. V. Salasyuk, M. Bombeck, and C. Bruggeman. The work was done in the frame of several projects lead by: Professors A. A. Kaplyanskii, V. G. Golubev and Dr. A. V. Scherbakov (Ioffe Physical-Technical Institute, St. Petersburg, Russia); Professors M. Bayer, D. R. Yakovlev, and Dr. T. Berstermann (University of Dortmund, Germany). The experiments and related publications would not be successful without theoretical support provided by Professors Y. Tanaka and S. Tamura from Hokkaido University, Japan. We acknowledge Professor A. J. Kent for his interest in this activity and E. S. K. Young for the help in preparing the manuscript.

References

1. R. Martinez-Sala, J. Sancho, J. V. Sánchez, V. Gómez, J. Llinares, and F. Meseguer, Nature **378**, 241 (1995).
2. Z. Liu, X. Zhang, Y. Mao, Y. Y. Zhu, Z. Yang, C. T. Chan, and P. Sheng, Science **289**, 1734 (2000).

3. M. Sigalas, M. S. Kushwaha, E. N. Economou, M. Kafesaki, I. E. Psarobas, and W. Steurer, Z. Kristallogr. **220**, 765 (2005).
4. R. H. Olsson III and I. El-Kady, Meas. Sci. Technol. **20**, 012002 (2009).
5. M. Torres, F. R. Monterode Espinosa, D. Garcia-Pablos, and N. Garcia, Phys. Rev. Lett. **82**, 3054 (1999).
6. S. Faez, A. Strybulevych, J. H. Page, A. Lagendijk, and B. A. van Tiggelen, Phys. Rev. Lett. **103**, 155703 (2009).
7. H. Hu, A. Strybulevych, J. H. Page, S. E. Skipetrov, and B. A. van Tiggelen, Nat. Phys. **4**, 945 (2008).
8. J. H. Page, A. Sukhovich, S. Yang, M. L. Cowan, F. Van DerBiest, A. Tourin, M. Fink, Z. Liu, C. T. Chan, and P. Sheng, Phys. Status Solidi B **241**, 3454 (2004); A. Sukhovich, L. Jing, and J. H. Page, Phys. Rev. B **77**, 014301 (2008); S. Yang, J. H. Page, Z. Liu, M. L. Cowan, C. T. Chan, and P. Sheng, Phys. Rev. Lett. **93**, 024301 (2004).
9. A. Sukhovich, B. Merheb, K. Muralidharan, J. O. Vasseur, Y. Pennec, P. A. Deymier, and J. H. Page, Phys. Rev. Lett. **102**, 154301 (2009).
10. R. E. Vines, J. P. Wolfe, and A. V. Every, Phys. Rev. B **60**, 11871 (1999).
11. A. Khelif, A. Choujaa, S. Benchabane, B. Djafari-Rouhani, and V. Laude, App. Phys. Lett. **84**, 4400 (2004).
12. N. Fang, D. Xi, J. Xu, M. Ambati, W. Srituravanich, C. Sun, and X. Zhang, Nat. Mater. **5**, 452 (2006); S. Zhang, C. Xia, and N. Fang, Phys. Rev. Lett. **106**, 024301 (2011).
13. T. Gorishnyy, C. K. Ullal, M. Maldovan, G. Fytas, and E. L. Thomas, Phys. Rev. Lett. **94**, 115501 (2005).
14. T. Gorishnyy, M. Maldovan, C. Ullal, and E. L. Thomas, Phys. World **18**, 24 (2005).
15. K.-Y. Hashimoto, *Surface Acoustic Wave Devices in Telecommunications: Modelling and Simulation* (Springer, Berlin, 2000); O. Holmgren, J. V. Knuuttila, T. Makkonen, K. Kokkonen, V. P. Plessky, W. Steichen, M. Solal, and M. M. Salomaa, Appl. Phys. Lett. **86**, 024101 (2005); T.-T. Wu, L.-C. Wu, and Z.-G. Huang, J. Appl. Phys. **97**, 094916 (2005).
16. M. M. deLimaJr and P. V. Santos, Rep. Prog. Phys. **68**, 1639 (2005); E. J. Reed, M. Soljačić, and J. D. Joannopoulos, Phys. Rev. Lett. **90**, 203904 (2003).
17. M. Maldovan and E. L. Thomas, Appl. Phys. Lett. **88**, 251907 (2006).
18. M. S. Kushwaha, P. Halevi, L. Dobrzynski, and B. Djafari-Rouhani, Phys. Rev. Lett. **71**, 2022 (1993); M. Kafesaki and E. N. Economou, Phys. Rev. B **60**, 11993 (1999); I. E. Psarobas, A. Modinos, R. Sainidou, and N. Stefanou, *ibid.* **65**, 064307 (2002).
19. I. E. Psarobas, N. Stefanou, and A. Modinos, Phys. Rev. B **62**, 278 (2000).
20. S. Yang, J. H. Page, Z. Liu, M. L. Cowan, C. T. Chan, and P. Sheng, Phys. Rev. Lett. **88**, 104301 (2002).
21. W. Cheng, J. Wang, U. Jonas, G. Fytas, and N. Stefanou, Nat. Mater. **5**, 830 (2006).
22. D. A. Mazurenko, X. Shan, J. C. P. Stiefelhagen, C. M. Graf, A. van Blaaderen, and J. I. Dijkhuis, Phys. Rev. B **75**, 161102 (2007); C. Mechri, P. Ruello, D. Mounier, J. M. Breteau, I. Povey, M. Pemble, S. G. Romanov, and V. Gusev, J. Phys. Conf. Ser. **92**, 012030 (2007); X. Shan, X. Zhang, D. A. Mazurenko, A. van Blaaderen, J. I. Dijkhuis, F. Hudert, and T. Dekorsy, *ibid.* **92**, 012035 (2007).
23. A. V. Akimov, Y. Tanaka, A. B. Pevtsov, S. F. Kaplan, V. G. Golubev, S. Tamura, D. R. Yakovlev, and M. Bayer, Phys. Rev. Lett. **101**, 033902 (2008).
24. K. Busch and S. John, Phys. Rev. E **58**, 3896 (1998); Y. A. Vlasov, X.-Z. Bo, J. C. Sturm, and D. J. Norris, Nature **414**, 289 (2001).
25. A. S. Salasyuk, A. V. Scherbakov, D. R. Yakovlev, A. V. Akimov, A. A. Kaplyanskii, S. F. Kaplan, S. A. Grudinkin et al., Nano Lett. **10**, 1319 (2010).
26. Y. Tanaka, S. Tamura, A. V. Akimov, A. B. Pevtsov, S. F. Kaplan, A. A. Dukin, V. G. Golubev, D. R. Yakovlev, and M. Bayer, J. Phys. Conf. Ser. **92**, 012107 (2007).
27. Y. Tanaka, Y. Tomoyasu, and S.-i. Tamura, Phys. Rev. B **62**, 7387 (2000).
28. Z. Liu, C. T. Chan, P. Sheng, A. L. Goertzen, and J. H. Page, Phys. Rev. B **62**, 2446 (2000).
29. C. Kittel, *Introduction to Solid State Physics* (Wiley, New York, 2004), 8th ed.
30. H. Lamb, Proc. London Math. Soc. **s1–13**, 189 (1881).

31. Y. Tanaka and S.-i. Tamura, Phys. Rev. B **58**, 7958 (1998); R. Sainidou, B. Djafari-Rouhani, and J. O. Vasseur, *ibid.* **77**, 094304 (2008); Y. Tanaka, T. Yano, and S.-i. Tamura, Wave Motion **44**, 501 (2007).
32. T. Still, W. Cheng, M. Retsch, R. Sainidou, J. Wang, U. Jonas, N. Stefanou, and G. Fytas, Phys. Rev. Lett. **100**, 194301 (2008).
33. W. Stöber, A. Fink, and E. Bohn, J. Colloid Interface Sci. **26**, 62 (1968).
34. P. Jiang, J. F. Bertone, K. S. Hwang, and V. L. Colvin, Chem. Mater. **11**, 2132 (1999).
35. G. M. Gajiev, V. G. Golubev, D. A. Kurdyukov, A. V. Medvedev, A. B. Pevtsov, A. V. Sel'kin, and V. V. Travnikov, Phys. Rev. B **72**, 205115 (2005).
36. G. M. Gajiev, D. A. Kurdyukov, and V. V. Travnikov, Nanotechnology **17**, 5349 (2006).
37. C. Thomsen, H. T. Grahn, H. J. Maris, and J. Tauc, Phys. Rev. B **34**, 4129 (1986).
38. A. Yamamoto, T. Mishina, Y. Masumoto, and M. Nakayama, Phys. Rev. Lett. **73**, 740 (1994); A. Bartels, T. Dekorsy, H. Kurz, and K. Köhler, *ibid.* **82**, 1044 (1999); C.-K. Sun, J.-C. Liang, and X.-Y. Yu, *ibid.* **84**, 179 (2000); M. F. PascualWinter, G. Rozas, A. Fainstein, B. Jusserand, B. Perrin, A. Huynh, P. O. Vaccaro, and S. Saravanan, *ibid.* **98**, 265501 (2007); R. P. Beardsley, A. V. Akimov, M. Henini, and A. J. Kent, *ibid.* **104**, 085501 (2010); W. Chen, Y. Lu, H. J. Maris, and G. Xiao, Phys. Rev. B **50**, 14506 (1994); A. Huynh, B. Perrin, N. D. Lanzillotti-Kimura, B. Jusserand, A. Fainstein, and A. Lemaître, *ibid.* **78**, 233302 (2008).
39. D. M. Profunser, O. B. Wright, and O. Matsuda, Phys. Rev. Lett. **97**, 055502 (2006).
40. O. B. Wright, Phys. Rev. B **49**, 9985 (1994); G. Tas and H. J. Maris, *ibid.* **49**, 15046 (1994).
41. H.-N. Lin, R. J. Stoner, H. J. Maris, and J. Tauc, J. Appl. Phys. **69**, 3816 (1991).
42. O. B. Wright, J. Appl. Phys. **71**, 1617 (1992).
43. M. Trigo, A. Bruchhausen, A. Fainstein, B. Jusserand, and V. Thierry-Mieg, Phys. Rev. Lett. **89**, 227402 (2002).
44. J. A. Turner, M. E. Chambers, and R. L. Weaver, Acta Acust. United Ac. **84**, 628 (1998).

Section 4

Toward Applications

4.1

Strong Light Confinement by Perturbed Photonic Crystals and Photonic Amorphous Structures

Keiichi Edagawa
The University of Tokyo

Masaya Notomi
NTT Basic Research Laboratories

CONTENTS

4.1.1 Introduction

In 1987, the pioneering works by Yablonovitch[1] and John[2] suggested that a three-dimensional (3D) photonic band-gap (PBG), which is analogous to an electronic band-gap in semiconductor materials, can be realized in artificial periodic dielectric structures, namely, photonic crystals.[3] The PBG was predicted to enable 3D confinement of light in a wavelength-sized volume without accompanying substantial loss, which is difficult to achieve in other media. As tight light confinement can enhance various light-matter interactions and can also realize very efficient nanophotonic devices/circuits, there have been extensive studies during the last two decades.[4] Although the 3D PBG was indeed realized in several structures, the expected strong light confinement has not been achieved yet.[5] This is mainly because 3D PBG can exist only in a very limited class of photonic crystals (most of them are categorized into "diamond" structures[6,7]), which are rather complex structures and thus difficult to fabricate with sufficient accuracy by available methods.

However, strong light confinement was recently achieved in somewhat different structures. Those are based on two-dimensional (2D) photonic crystals that have only a partial PBG. A typical 2D photonic crystal is shown in Figure 4.1.1, which can be fabricated by well-matured lithography and etching techniques. It was proposed and demonstrated that it is possible to greatly suppress the radiation loss originating from its incomplete PBG by choosing appropriate designs. In particular, a photonic crystal with

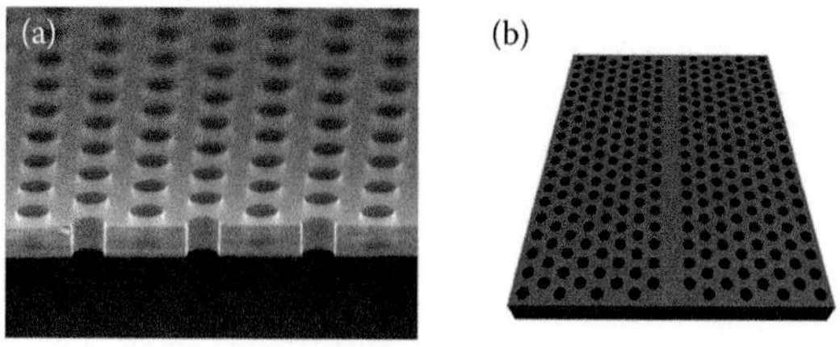

FIGURE 4.1.1
Scanning electron micrograph of a Si two-dimensional (2D) photonic crystal (a) and a schematic of a line defect in a 2D photonic crystal (b).

local structural modulation can exhibit unprecedented high Q (>million) with a small mode volume (~$(\lambda/n)^3$). The first half of this chapter is devoted to this type of perturbed photonic crystals, and will show how strong light confinement can be achieved in them.

The latter half will describe another completely different approach to realize 3D light confinement. Recently, it was demonstrated that a 3D photonic amorphous structure can form a sizable complete 3D PBG, despite absence of the lattice periodicity.[8,9] This amorphous structure consists of a random network of dielectric rods with a diamond-like local tetrahedral configuration and is therefore named "photonic amorphous diamond" (PAD). As it is an amorphous structure, we may have different approaches to fabricate it, and it may be easier than the fabrication of conventional 3D photonic crystals. We will discuss the 3D PBG formation in PAD and the possibility of light confinement by PAD.

In this chapter, we aim to show how light can be strongly confined by playing with order and disorder, which is something beyond the original proposal of the PBG, and may shed another light on a future form of nanophotonic components.

4.1.2 Light Confinement by Perturbed Photonic Crystals

In this section, we examine 2D photonic crystals with local modulation for achieving strong light confinement. A typical structure is shown in Figure 4.1.2a, which is based on a simple line defect in a 2D photonic crystal slab (Figure 4.1.1b), but the center part of the line defect (colored holes in Figure 4.1.2a) is locally modulated. We will show that this structure realizes an ultrahigh-Q nanocavity. In fact, this modulation is a kind of point defect in the photonic crystal, and it is well known that introduction of a point defect in the PBG produces a cavity mode. In this sense, there seems nothing special about the cavity formation in this structure. However, what is remarkable for this structure is the fact that it leads to extraordinary high Q (theoretically, the vertical radiation loss can be practically eliminated), and that such a confinement can be achieved by very small modulation. We examine it in the following.

The cavity shown in Figure 4.1.2a is proposed by one of the present authors and called a width-modulated line-defect cavity.[10,11] This cavity is based on a simple straight line defect in a hexagonal air-hole 2D photonic crystal with the lattice constant of typically 420 nm, and the colored holes are shifted slightly outwards by 9 (red), 6 (green), and 3 nm (blue), respectively. Numerical simulations show that it has a well-confined cavity mode as seen in the right panel of Figure 4.1.2a. The theoretical cavity Q calculated by the 3D

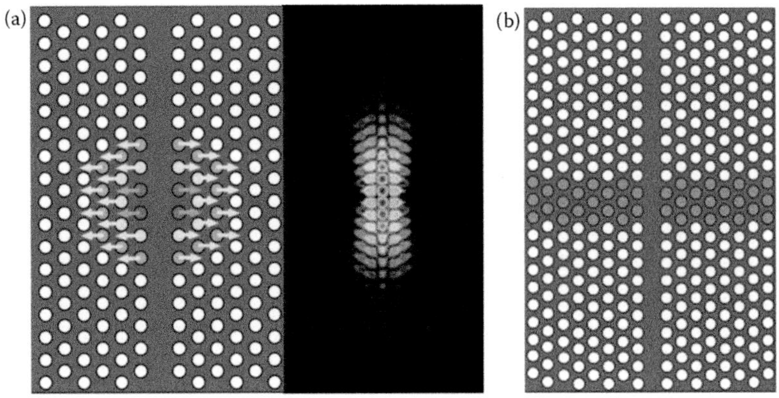

FIGURE 4.1.2
(**See color insert.**) Modulated mode-gap cavities in two-dimensional photonic crystals. (a) Width-modulated line-defect cavity. Schematic and field-intensity profile. (b) Double heterostructure cavity.

finite-difference time-domain (FDTD) method is extremely high ($Q = 2 \times 10^8$) and the estimated effective cavity mode volume,[12] V, is still close to being wavelength-cubic ($V = 1.6(\lambda/n)^3$). Such an ultrahigh Q indicates that the vertical radiation loss is virtually inhibited despite the absence of a PBG in that direction.

The structural design shown in Figure 4.1.2a is one example of *modulated mode-gap cavities* that are composed of locally modulated mode-gap waveguides, which generally exhibit ultrahigh Q. A variety of modulated mode-gap cavities have been proposed by several authors[10,13,14] (another example is shown in Figure 4.1.2b). The local modulation can be realized by modulating the effective width of the line defect shown in Figure 4.1.2a,[10] by changing the lattice constant of the crystal lattice shown in Figure 4.1.2b,[13] and also by modulating the hole size.[14] Here we describe the mechanism of this ultrahigh Q intuitively. Figure 4.1.3a shows the dispersion curve of the base straight line-defect. It has a guided mode outside the radiation cone of the cladding. In other words, none of the k vectors in the guided mode can couple with the radiation modes, which means that there is theoretically no radiation loss.

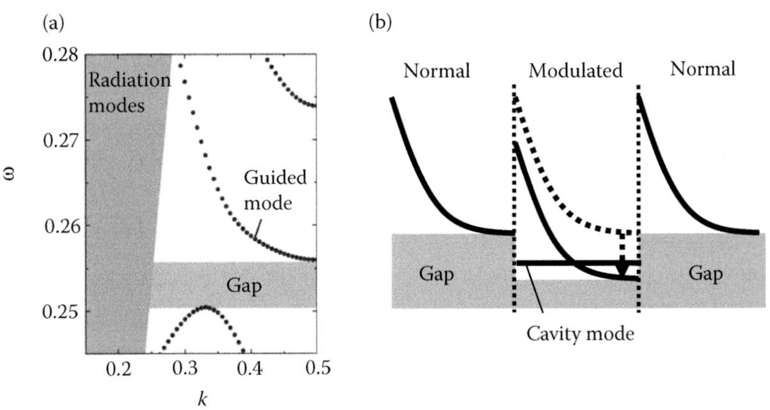

FIGURE 4.1.3
Confinement mechanism of the mode-gap modulated cavity. (a) Dispersion of the base line defect. (b) Schematic for the local modulation.

Generally, guided modes in photonic crystals have mode gaps in their dispersion, which is also the case for this simple line defect (called a W1 waveguide[15]). If we introduce a certain local structural modulation in a line defect (schematically shown in Figure 4.1.3b), the mode-gap frequency becomes spatially modulated. As a result, a cavity mode can be created. The key issue is how this modulation changes the loss-free k distribution of the base line defect. As is the case in Figure 4.1.2, an extremely small structural modulation is sufficient for creating light confinement for this mechanism. Hence, if we carefully design the modulation so that it is spatially smooth, we can keep most of the k distribution still outside the radiation cone of the cladding, which eventually leads to an ultrahigh Q for this 2D PBG cavity even without a PBG in the last dimension.

This 3D confinement mechanism is nothing but a combination of 2D PBG and one-dimensional (1D) total internal reflection, but the above numerical result demonstrates that this design strategy works very well. A Kyoto University group first demonstrated an ultra-high Q (~10^5) in double-heterostructure cavities[13] in which the lattice constant of a photonic crystal was locally altered (later they achieved a Q over 2×10^6 for a similar cavity design).[16] We also achieved an ultrahigh Q in the width-modulated line-defect cavity[10] as shown in Figure 4.1.4a.[11] Figure 4.1.4b shows a transmission spectrum that exhibits an ultra-sharp resonance with a spectral width of 0.87 pm, corresponding to a loaded Q of 1.8×10^6. We also measured the time-resolved intensity decay from the cavity (ring-down measurement)

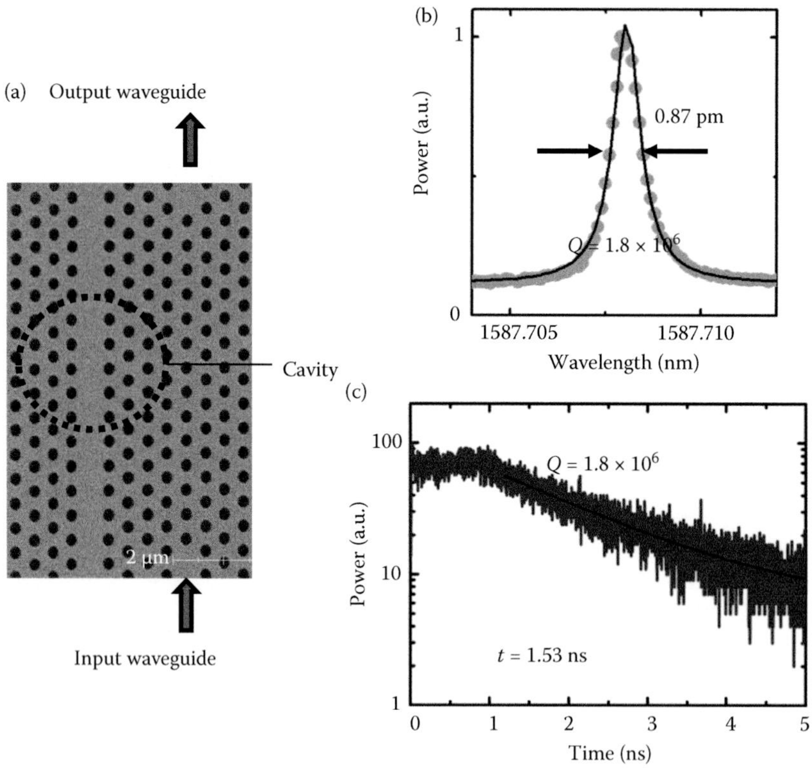

FIGURE 4.1.4
Cavity Q measurements of the width-modulated cavity shown in Figure 4.1.2a. (a) Scanning electron micrograph of the sample. (b) Transmission spectrum showing the resonance width. (c) Ring-down measurement showing the photon lifetime.

as shown in Figure 4.1.4c and found that the photon lifetime was 1.53 ns, corresponding exactly to the same Q as that in Figure 4.1.4b. The unloaded Q estimated from Figure 4.1.4b and c is 2×10^6. These results show that it is now possible to store photons for over 1 ns in a wavelength-sized volume, which had been impossible to achieve with other methods.

These Qs are orders of magnitudes smaller than the theoretical values and currently limited by finite structural imperfections in the fabricated cavities,[17] but they are still much higher than other types of miniaturized cavities such as a microdisc cavity or a micropost cavity. Besides the modulated mode-gap design, several other designs have been proposed and demonstrated for high Q and small cavities based on 2D photonic crystals.[18] They also achieved superior performance to conventional cavities in terms of Q/V. However, the modulated mode-gap cavity exhibits the highest Q, and the highest Q/V ratio. In addition to their high Q, it should be noted that the confinement mechanism itself is quite unique. Contrary to our original naive expectation for 3D PBG cavities, light confinement is achieved by modulating the mode gap in a 2D periodic structure having only a partial PBG, where *very small structural modulation produces ultrastrong light confinement*. This latter feature indicates that we can manipulate strong light confinement by using a small perturbation, which leads to very interesting ways of manipulating light as we will see later. The present confinement mechanism is rather general, and can therefore be applied or extended to other systems. Some examples are provided in the next section.

4.1.3 Light Confinement by Index Modulation of Photonic Crystals

The cavity designs described in the previous section are based on a local structural modulation of a line defect, but a similar confinement mechanism will work if we introduce a local refractive-index modulation.[19,20] The idea is schematically illustrated in Figure 4.1.5a. We confirmed this speculation with numerical simulations,[20] and found that a very tiny index modulation of $\Delta n/n = +0.3\%$ with a Gaussian distribution whose full-width at half-maximum (FWHM) is 2.5 μm leads to an ultrahigh-Q cavity with Q of 4.8×10^9 and V of $2.1(\lambda/n)^3$.

As far as we know, this is the highest theoretical Q obtained for any 2D photonic crystal. Figure 4.1.5b shows the calculated Q and V at various index modulation strengths.

These findings indicate that we can produce an ultrahigh-Q cavity mode by dynamic index modulation possibly induced by optical Kerr nonlinearity or some other dynamic index-modulation techniques. This means that we can freely generate and eliminate a high-Q nanocavity mode on demand. We have shown that if we utilize fast Kerr nonlinearity for this system, we can pin (in other words, freeze) a light pulse running via the waveguide by shining another light from the top.[20] Such a "stopping light" process will be important for future optical or quantum information processing. It is worth noting that these interesting features are derived as a direct consequence of the fact that one can achieve very strong confinement by very small modulation in this system as pointed out before. In addition, we have recently demonstrated short pulse generation from an ultrahigh-Q Si width-modulated line-defect cavity as a result of temporal index modulation caused by optical pumping.[21] The experimental setup and the result are shown in Figure 4.1.5c and d. A short pulse is released from an ultrahigh-Q nanocavity capturing an optical pulse by dynamic refractive-index tuning. For this process, the confinement strength was greatly altered by a very small index modulation ($\Delta n/n < 0.1\%$).

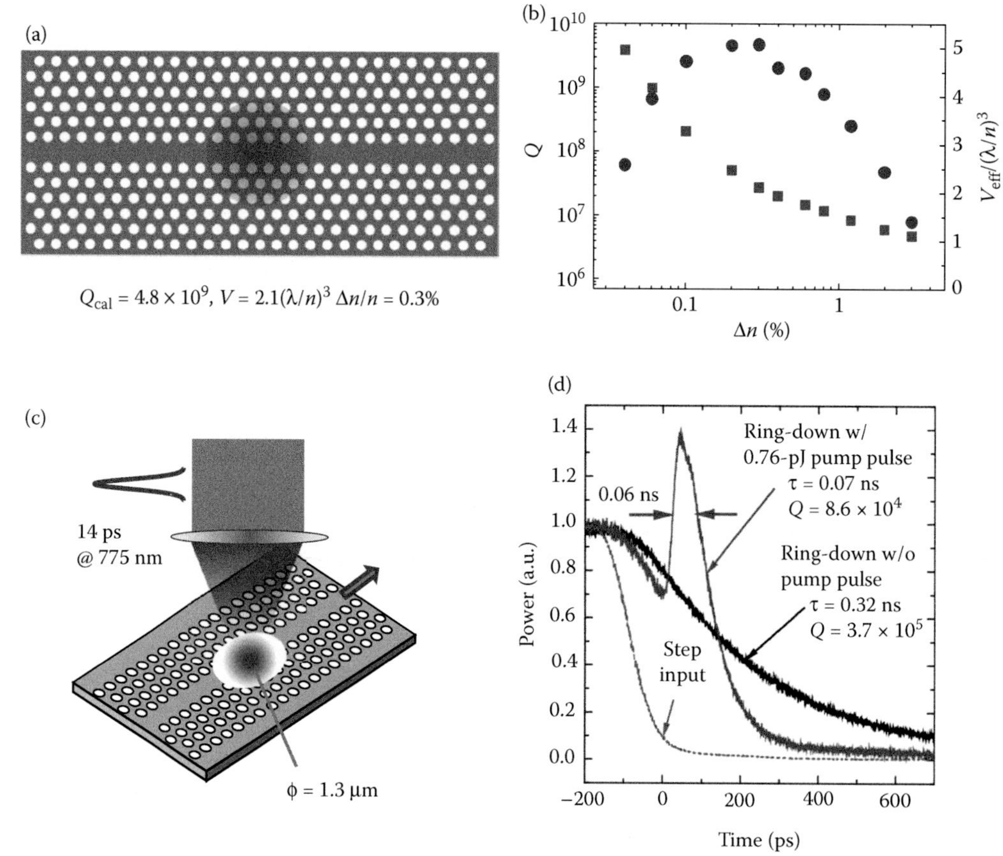

FIGURE 4.1.5
(**See color insert.**) Light confinement by index modulation. (a) Ultrahigh-Q cavity formed by local refractive index modulation. (b) Calculated Q and mode volume for index-modulated cavities. (c) Experimental setup for dynamic Q tuning for a Si width-modulated cavity. (d) Experimental observation of Q tuning. A short pulse is generated in the output waveguide because the coupling Q is dynamically lowered by optical pumping. (Panels (a) and (b) adapted from M. Notomi and H. Taniyama, Opt. Express **16**, 18657, 2008. Panels (c) and (d) adapted from T. Tanabe et al., Phys. Rev. Lett. **102**, 043907, 2009.)

4.1.4 Some Other Variations of Perturbed Photonic Crystal Cavities

Before moving on to the next topic, we add two recent examples of variations of modulated mode-gap cavities. The first one is shown in Figure 4.1.6a through c, which is merely a 1D photonic crystal but we applied essentially the same strategy of making a high-Q cavity.[22] That is, we introduce a certain local modulation for the width or hole size of a periodic waveguide having a mode gap. Surprisingly, such modulated 1D photonic crystal cavities can have ultrahigh Q (~10^8) comparable to that for 2D modulated-mode-gap cavities with an even smaller mode volume. Furthermore, it is possible to design ultrahigh-Q nanocavities even on an oxide cladding (such as SiO_2),[23] although it is well known that ultrahigh-Q design is very difficult for 2D photonic crystals with an oxide cladding. So far, the highest measured Q is around 700,000.[23,24]

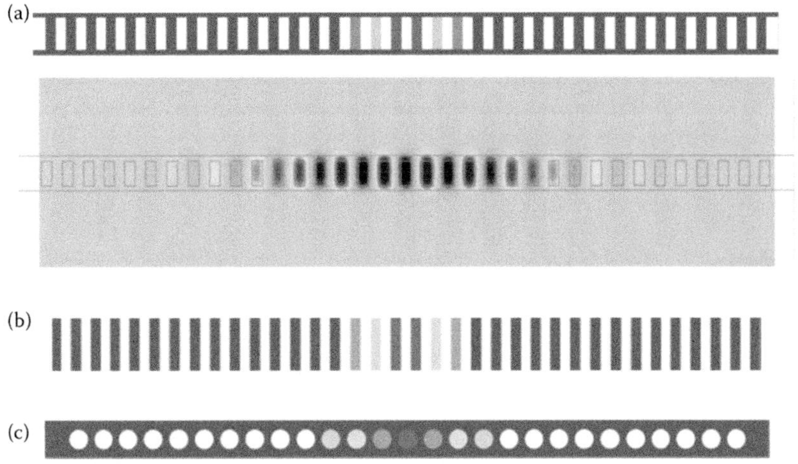

FIGURE 4.1.6
One-dimensional modulated mode-gap cavities. (a) Ladder cavity; $Q = 2 \times 10^8$, $V = 1.4(\lambda/n)^3$. (b) Box cavity; $Q = 6 \times 10^7$, $V = 2.1(\lambda/n)^3$. (Adapted from M. Notomi, E. Kuramochi, and H. Taniyama, Opt. Express **16**, 11095, 2008.) (c) Circular hole ladder cavity; $Q = 3 \times 10^9$, $V = 0.7(\lambda/n)^3$. (Adapted from E. Kuramochi et al., Opt. Express **18**, 15859, 2010.)

Yet another example is shown in Figure 4.1.7, which does not have any modulation in the geometrical structure of the photonic crystal, but we replace the central part of the host material (InP) with another material having slightly larger refractive index (InGaAsP quantum wells) with gain. In the same scenario as that in the previous section, one can expect high-Q cavity formation as a result of the local index modulation. This particular cavity design is especially useful for making a laser, because of the following two reasons. First, carriers and photons are strongly confined in the same area. Second, the heating

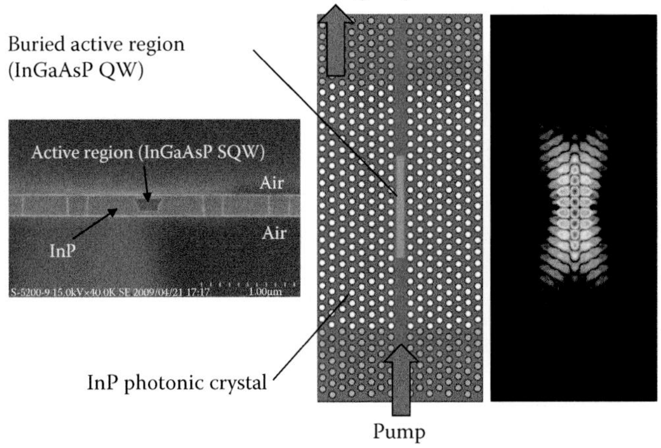

FIGURE 4.1.7
(**See color insert.**) Index-modulated cavity for a photonic crystal buried heterostructure laser. (a) Scanning electron micrograph. (b) Schematic and calculated field-intensity profile.

effect is greatly suppressed as the thermal conductivity of InP is 20 times higher than that of InGaAsP. Recently, we demonstrated room-temperature continuous-wave operation for this laser by optical pumping with the record-low threshold power of 1.5 μW.[25] Moreover, we achieved 20 Gbps modulation with the energy cost of 8.8 fJ/bit, which is two orders of magnitude smaller than previous lasers.[26]

4.1.5 Structure of PAD

We now move on to the second topic, where we discuss the possibility of light confinement by photonic amorphous structures. In this section, we describe the PAD structure, which has recently been found to form a sizable 3D PBG despite the complete absence of lattice periodicity.[8,9]

The PAD structure is based on the continuous random network (CRN) model,[27–29] which has been developed to represent an atomic arrangement of amorphous Si or Ge. An example of a CRN structure is shown in Figure 4.1.8a, which should be compared with the atomic structure of crystalline Si or Ge, that is, the crystalline diamond structure in Figure 4.1.8b. The requirement of the CRN model is simply that it should be a random network in which all the junction points are fourfold. The quality of the model can be evaluated by the amount of local strain, as measured by the degree of deviations from the regular-tetrahedral configuration; the "ideal" CRN structure should be characterized by the lowest spread in the bond-length and bond-angle distributions around those values in the crystalline diamond structure.

In 1985, Wooten et al.[28] proposed an efficient algorithm to generate CRN structures of high quality, which has been widely used since then. This WWW algorithm starts from the crystalline diamond structure (Figure 4.1.8b). First, we randomly pick up a local structure shown in Figure 4.1.9a, and transpose the bonds as in Figure 4.1.9b. This bond transposition preserves the fourfold connectivity of the network. After the bond transposition,

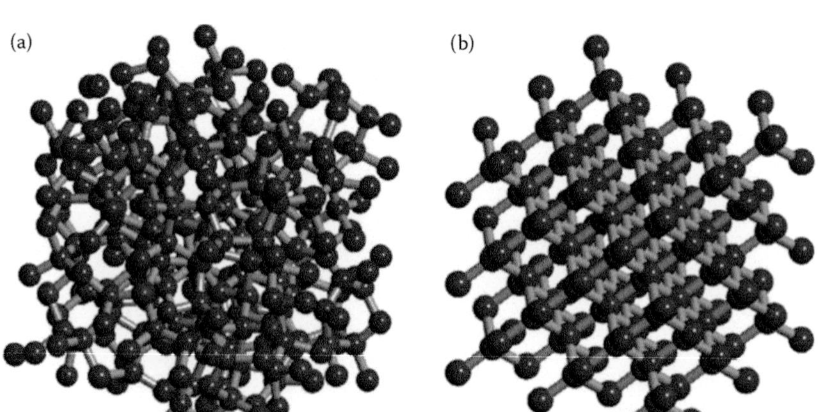

FIGURE 4.1.8
The continuous random network structure constructed by Barkema and Mousseau[29] (a) and the crystalline diamond structure (b).

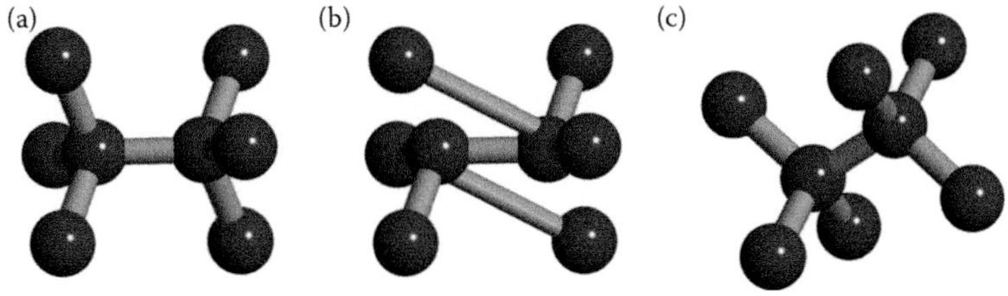

FIGURE 4.1.9
Schematics of the elemental process to generate continuous random network structures in the WWW algorithm.

the structure is relaxed with an interatomic potential such as the Keating potential[30] that favors the local regular-tetrahedral configuration for the five atoms connected with bonds (one at the center and the other four at the vertices of the tetrahedron); the relaxed configuration is schematically shown in Figure 4.1.9c. This relaxation reduces the distortion of the two tetrahedrons in the figure but increases the distortion of the tetrahedrons around them. The elemental process of Figure 4.1.9a–c is sequentially repeated for randomly chosen local structures. This results in a CRN structure but it is usually distorted rather severely. To obtain a CRN structure of high quality, Wooten et al.[28] have proposed to introduce the Metropolis acceptance probability:

$$P = \begin{cases} 1 & (E_b > E_a) \\ \exp[(E_b - E_a)/k_BT] & (E_b < E_a) \end{cases} \tag{4.1.1}$$

where E_b and E_a are the total energies of the system before and after the elementary process of Figure 4.1.9a–c, and k_BT has the usual meaning. First, we adopt an adequately high temperature at which almost all the trials are accepted. This leads to a sufficiently randomized CRN structure with a considerably high energy. After randomization, we anneal the structure at a relatively low temperature, where the trials to decrease the energy are mainly accepted, leading to a CRN structure with less distortion. We may continue the annealing at an even lower temperature if necessary. By choosing an optimal thermal history, we could obtain an ideal CRN structure with a minimum distortion.

In our study, we used the CRN structure constructed and provided by Barkema and Mousseau,[29] where an improved algorithm is adopted. Here, we start from a random configuration, instead of the crystalline diamond structure. This guarantees that the constructed structure is not contaminated by any memory of the initial crystalline structure. Therefore, this CRN structure cannot be regarded as a disordered diamond crystal; the network topology is completely different from that in the diamond crystal, and no trace of lattice periodicity should get into it. This structure consists of a periodic arrangement of a cubic supercell with the size $(11.5d)^3$ (where d is the average bond length). The supercell contains 1000 atoms. The CRN structure shown in Figure 4.1.8a corresponds to a part of this structure. Indeed, we can see no trace of periodic order of the diamond lattice whereas we notice a definite local tetrahedral order. To see this more quantitatively, we calculated the radial distribution function (RDF) and the bond-angle distribution function (BADF) for the CRN structure. These functions have already been calculated by Barkema and Mousseau.[29] Figure 4.1.10a shows the RDF ($f(r)$), where $f(r)$ indicates the average atomic density in the spherical shell $(r, r + dr)$ around an arbitrarily selected atom, where the density is

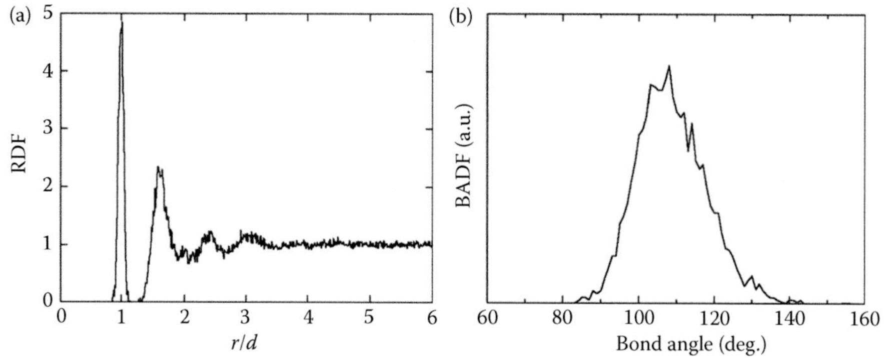

FIGURE 4.1.10
Radial distribution function (a) and bond-angle distribution function (b) for the continuous random network structure.

normalized to the average atomic density; when the structure has no order, $f(r)$ should be $f(r) = 1 =$ constant. We notice in Figure 4.1.10a that the first peak at $r = d$ is very sharp; the FWHM value of the peak is around $0.1d$. This indicates a quite limited spread in the bond-length distribution. On the other hand, BADF in Figure 4.1.10b also shows a considerably small spread around the ideal value of 109°; the FWHM value of the peak is around 20°. From these two facts, we can conclude that this CRN structure is of high quality, that is, it has a good local regular-tetrahedral order. In the RDF in Figure 4.1.10a, the second peak at $r \approx 1.7d$ is also relatively sharp but the subsequent peaks at $r \approx 2.4d$ and $3d$ are very broad; they are marginally observed. Then, no peaks can be detected in the range $r > 3.5d$. This indicates that this CRN structure has no order in the range beyond $r \approx 3.5d$. It should be noted that the range $r > 3.5d$ corresponds to $r > 1.5a$, where $a = 4d/\sqrt{3}$ denotes the lattice constant of the crystalline diamond. This verifies the fact that this CRN structure has no trace of the diamond-lattice periodicity.

We modeled the PAD structure by connecting the tetrahedral bond in the CRN structure with dielectric rods in air background.[8,9] For comparison, we also constructed a structure by connecting the tetrahedral bonds in the crystalline diamond structure with dielectric rods, which we hereafter call a photonic crystalline diamond (PCD) structure. This PCD structure is known to be the best 3D PBG structure;[7,31] it exhibits the largest 3D PBG among all the photonic crystals studied thus far. We may regard PAD and PCD as the photonic versions of amorphous and crystalline Si, respectively, in electronic systems.

4.1.6 PBG Formation in PAD: FDTD Calculations

In this section, we present FDTD calculations of the photonic eigenmodes and their frequency distributions in PAD.[8,9] We adopt here an "order-N" method originally developed by Chan et al.,[32] which is suitable for the systems requiring a large supercell.[33,34] We have also made calculations for PCD for comparison. As described in the preceeding section, our PAD structure consists of a periodically arranged cubic supercell of size $(11.5d)^3$. For PCD, we used a crystalline diamond structure with a fictitious supercell of $(5a)^3 = (11.5d)^3$.

Using this fictitious supercell structure for PCD enables us to compare the results obtained under exactly the same conditions for PAD and PCD. We assumed that the refractive index of rods was $n = 3.6$ (Si), and the absorption was neglected, that is, $\kappa = 0$ (κ: extinction coefficient). The rod radius was assumed to be $r = 0.26d$. Then, the volume fraction of the rod in the structure is 22%.

In this method, we calculate the time evolutions of the magnetic and electric fields, $\mathbf{H}(\mathbf{r}, t)$ and $\mathbf{E}(\mathbf{r}, t)$, by an FDTD method to find the steady states, that is, eigenmodes, in the system. Here, the initial fields must have nonzero projections onto all the eigenmodes in the frequency range in which we search. The eigenmodes in our structures should satisfy the boundary conditions of the Bloch theorem:

$$\mathbf{H}(\mathbf{r} + \mathbf{a}_i, t) = e^{i\mathbf{k}\cdot\mathbf{a}_i}\mathbf{H}(\mathbf{r}, t) \quad \text{and} \quad \mathbf{E}(\mathbf{r} + \mathbf{a}_i, t) = e^{i\mathbf{k}\cdot\mathbf{a}_i}\mathbf{E}(\mathbf{r}, t) \tag{4.1.2}$$

where $\mathbf{a}_i$ ($i = 1, 2$, and 3) represents the translational vectors of the cubic supercell and $\mathbf{k}$ is the Bloch wave vector within the first Brillouin zone. Considering these facts and the requirement of transverse wave for $\mathbf{H}$, we selected the initial fields:[33]

$$\mathbf{H}_{\mathbf{k}}(\mathbf{r}, 0) = \sum_{|\mathbf{G}|<G_{\max}} \mathbf{H}_{\mathbf{G}} e^{i(\mathbf{k}+\mathbf{G})\cdot\mathbf{r}+i\phi_{\mathbf{G}}} \quad \text{and} \quad \mathbf{E}_{\mathbf{k}}(\mathbf{r}, 0) = \mathbf{0} \tag{4.1.3}$$

where $\{\mathbf{G}\}$ is the reciprocal-lattice vectors corresponding to the direct lattice spanned by $\mathbf{a}_i$ ($i = 1, 2$, and 3) and $\phi_{\mathbf{G}}$ is a random phase. $\mathbf{H}_{\mathbf{G}}$ is a unit vector, where the direction of $\mathbf{H}_{\mathbf{G}}$ is randomly chosen within the plane perpendicular to ($\mathbf{k} + \mathbf{G}$). We adopted a sufficiently large value for $G_{\max}$ to cover the frequency range of interest. The time evolutions of the fields $\mathbf{H}_{\mathbf{k}}(\mathbf{r}, t)$ and $\mathbf{E}_{\mathbf{k}}(\mathbf{r}, t)$ were calculated by an FDTD method under the boundary conditions of Equation 4.1.2, where the supercell was discretized into $141 \times 141 \times 141$ meshes and we typically used 2^{14} time steps with each step of about 0.05 d/c (where c is the speed of light). The calculated time evolutions $\mathbf{H}_{\mathbf{k}}(\mathbf{r}, t)$ and $\mathbf{E}_{\mathbf{k}}(\mathbf{r}, t)$ were Fourier-transformed into frequency domain to obtain $\tilde{\mathbf{H}}_{\mathbf{k}}(\mathbf{r}, \omega)$ and $\tilde{\mathbf{E}}_{\mathbf{k}}(\mathbf{r}, \omega)$, where we used a Blackman function as the window function. These functions give the eigenmode profiles at a given frequency ω. The frequency distribution of the eigenmodes was evaluated by calculating the function:

$$I_{\mathbf{k}}(\omega) = \sum_i |\tilde{\mathbf{E}}_{\mathbf{k}}(\mathbf{r}_i, \omega)|^2 \tag{4.1.4}$$

for 42^3 evenly distributed points $\{\mathbf{r}_i\}$ in the supercell. This function should consist of peaks at the eigenmode frequencies at $\mathbf{k}$ in the first Brillouin zone. In principle, we need to calculate $I_{\mathbf{k}}(\omega)$ for all the $\mathbf{k}$ vectors in the zone to pick up all the eigenfrequencies. However, we found that only the $I_{\mathbf{k}}(\omega)$ for $\mathbf{k} = \mathbf{0}$ (Γ-point sampling) is necessary to locate the gap because the supercell is large enough, as shown below. For PCD, the photonic band structure was calculated also by a plane-wave expansion method using "BandSOLVE" (RSoft Design Group, Inc., USA).

Figure 4.1.11a and b shows the spectral intensities ($I_{\mathbf{k}=0}(\omega)$ in Equation 4.1.4) calculated for PAD and PCD, respectively. First, the spectral intensity for PCD in Figure 4.1.11b should be compared with the photonic band structure calculated by a plane-wave method in Figure 4.1.12a. In the calculation of the spectrum in Figure 4.1.11b, we used a fictitious

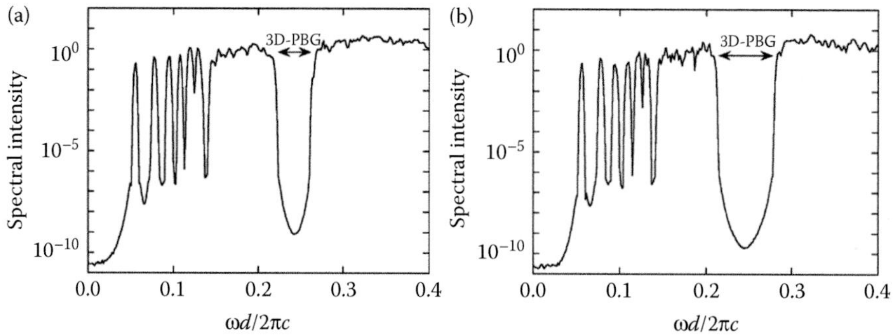

FIGURE 4.1.11
Spectral intensities calculated for photonic amorphous diamond (a) and photonic crystalline diamond (b). They are roughly normalized to the values at the band-edges.

supercell of $(5a)^3$. This is 4×5^3 times larger than the volume of the primitive unit cell of the face-centered cubic diamond lattice. This fact indicates that the volume of the first Brillouin zone is reduced by the factor $(4 \times 5^3)^{-1}$. Then, we can redraw the band structure in the reduced zone, as schematically shown in Figure 4.1.12b. Here, each branch in the band structure in the original zone is broken up into 4×5^3 pieces, which have fallen into the reduced zone. As a matter of course, this is merely a different drawing of physically the same band structure.

The spectrum in Figure 4.1.11b should consist of the peaks at the eigenfrequencies at Γ ($\mathbf{k} = \mathbf{0}$) in the reduced zone shown in Figure 4.1.12b. We should note here that the eigenfrequencies at Γ in the reduced zone correspond to those at 4×5^3 **k** points in the original zone; the larger the supercell the more eigenfrequencies we should pick up. In Figure 4.1.11b, we notice well-isolated peaks in the low frequency region below $\omega d/2\pi c \approx 0.15$. In contrast, above $\omega d/2\pi c \approx 0.15$, the peaks are so dense that the spectrum appears to be a continuous curve. This is because in a higher-frequency region more branches fall into the reduced zone. Owing to such a continuous feature in the spectrum above $\omega d/2\pi c \approx 0.15$, we can locate unambiguously the gap at $\omega d/2\pi c \approx 0.24$. The gap width to the mid-gap frequency ratio ($\Delta\omega/\omega_c$) is about 27%. This gap frequency range agrees perfectly with the frequency range of the 3D PBG shown in the band structures in Figure 4.1.12a and b, indicating the validity of the present calculations.

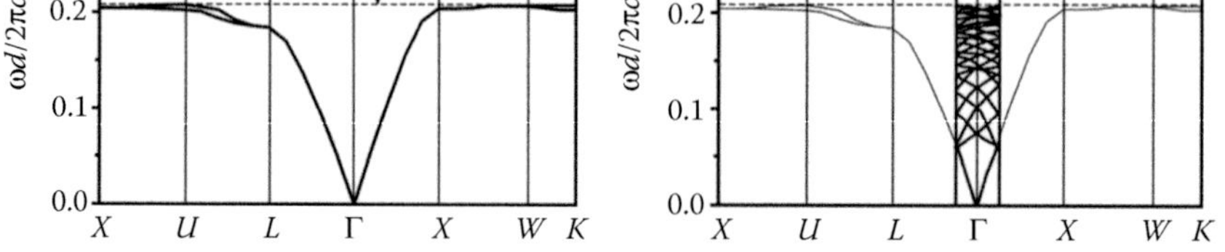

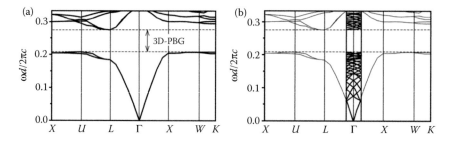

FIGURE 4.1.12
(a) Photonic band structure of photonic crystalline diamond calculated by a plane-wave expansion method. (b) A schematic drawing of the same band structure using the reduced zone.

The spectrum for PAD in Figure 4.1.11a shows similar features to that for PCD in Figure 4.1.11b; we have isolated peaks in the frequency range below $\omega d/2\pi c \approx 0.15$ and a continuous curve above $\omega d/2\pi c \approx 0.15$. A clear spectrum gap is seen at $\omega d/2\pi c \approx 0.24$ with $\Delta\omega/\omega_c \approx 18\%$. The position of the gap is approximately the same as that for PCD, indicating that the supercell periodicity is not relevant to the PBG formation in PAD. The gap depth is also nearly the same as that in PCD. Here, the depth of the gap is approximately 10^{-10}, which is the resolution limit in the present calculations. This fact indicates that the gap in PAD is as clean as that in PCD, with no trace of localized-state formations in the gap. Such a clean gap is critically important in realizing strong light confinement.

In conventional photonic crystals, light confinement can be achieved by introducing defects that disrupt lattice periodicity. Then, what kind of defects should be introduced in our PAD to confine light? The structure of PAD is characterized by the network that consists exclusively of fourfold connections. Thus, we can introduce defects that disrupt this fourfold connectivity of the network. For example, if we remove a rod, we can create locally two threefold junctions, as shown in Figure 4.1.13a. We calculated the eigenmodes for the

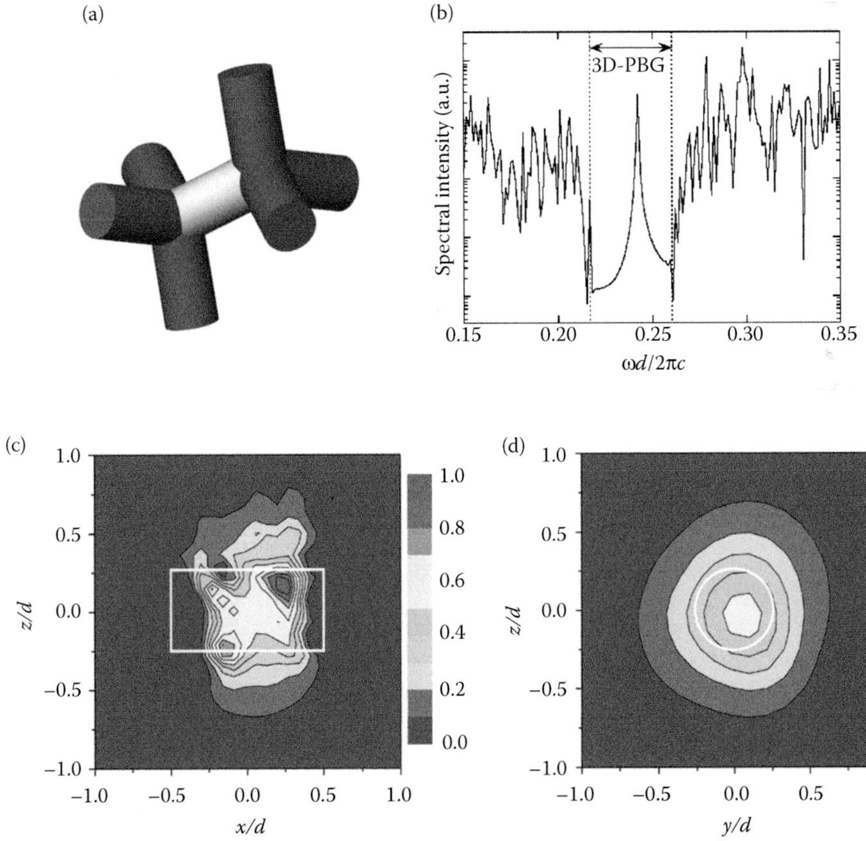

FIGURE 4.1.13
(See color insert.) (a) A schematic illustration of the defect introduced in a photonic amorphous diamond by removing a rod. Here, the light gray rod is removed, creating two threefold junctions. (b) The spectral intensity at the center of the removed rod. (c) and (d) The defect mode profiles on two mutually orthogonal planes, where the position of the missing rod is indicated by a white rectangle or circle. (Panels (b)–(d) adapted from K. Edagawa, S. Kanoko, and M. Notomi, Phys. Rev. Lett. **100**, 013901, 2008.)

3D PAD having such a defect, and found a localized mode with a frequency within the PBG. In Figure 4.1.13b, the spectral intensity at the center of the removed rod is shown, where we can see a sharp defect peak at $\omega d/2\pi c \approx 0.242$ in the 3D PBG. The mode profiles $|\tilde{\mathbf{E}}_{\mathbf{k}}(\mathbf{r},\omega)|^2$ corresponding to this peak are shown in Figure 4.1.13c and d, where the position of the removed rod is indicated by a white circle or rectangle. We can see that this mode is well confined to the defect position. We found that the width of the defect peak in Figure 4.1.13b is determined by the frequency resolution limit in the present calculation, indicating that the cavity Q-value cannot be evaluated properly.

In a previous study,[8] we investigated by an FDTD calculation the attenuation behavior of an evanescent wave with the mid-gap frequency when it enters the structure from the surface. Both PAD and PCD showed exponential decay, that is, $I \propto \exp(-r/r_c)$ (where r_c is the attenuation length) with nearly the same r_c ($\approx 0.6d$). This fact suggests that the Q values of defect modes in PAD can be as high as those in PCD and that PAD can confine light as strongly as PCD.

4.1.7 PBG Formation in PAD: Microwave Experiments

The subject of this section is an experimental substantiation of the 3D PBG formation in PAD. We fabricated PAD and PCD structures in a microwave regime and conducted microwave transmission experiments.[9]

Figure 4.1.14a and b shows examples of the fabricated PAD and PCD structures, respectively, where we used a selective laser sintering method.[35] Mixtures of nylon (PA12) powders (typically 50 μm in diameter) and TiO_2 powders (75–100 μm in diameter) were used for fabrication. TiO_2 powders were added to increase the refractive index n of the rods. The rod had a porosity of approximately 40%. The fabricated structures were soaked in water for 24 h, dried for 2 h, and cooled to 268 K. This resulted in the formation of ice in the pores of the rods, leading to a further increase in the refractive index n. Ice is suitable for increasing n without increasing the extinction coefficient κ. The final values of n and κ were approximately 3.0 and 0.03, respectively, for 30 GHz microwaves ($\lambda = 10$ mm in

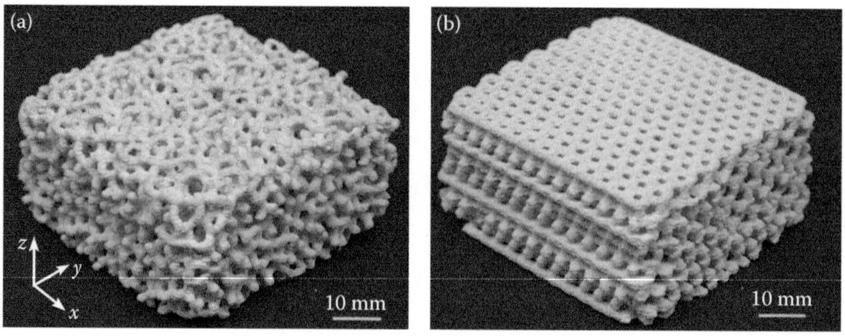

FIGURE 4.1.14
Fabricated photonic structures: (a) photonic amorphous diamond and (b) photonic crystalline diamond. (Adapted from S. Imagawa et al., Phys. Rev. B **82**, 115116, 2010.)

air). The resulting (1/*e*) absorption length is 330 mm; this is so long that the absorption is negligible in our samples. Each rod is $d \approx 3$ mm in length and $r \approx 0.78$ mm ($0.26d$) in radius. The size of the fabricated structures is approximately $x \times y \times z = 70 \times 70 \times 35$ mm^3 ($23.3 \times 23.3 \times 11.7d^3$). Microwave signals were incident on the x–y plane, along the z-axis. Three PAD samples with different orientations were fabricated. These were cut out of the infinite periodic supercell structure. The z directions of the three samples are along [100], [110], and [111] of the cubic supercell, respectively. For PCD, two samples with different orientations were fabricated—samples with the z direction along [100] and [111] of the diamond crystal.

For microwave transmission measurements, we used a free-space method using a vector network analyzer (HP Model 8722D, Agilent Technologies, USA) in the frequency range 18–35 GHz in the configuration schematically shown in Figure 4.1.15a. Conical horns with Teflon microwave lenses were positioned 163 mm in front of and behind the sample. Linearly polarized microwave radiation was emitted from one horn (transmitter), and the parallel- and cross-polarized components of the transmitted field were detected using the other horn (receiver) by changing the relative orientations of the mode selectors attached to the two horns. The transmittances T_p and T_c of the parallel- and cross-polarized components, respectively, were obtained by normalizing the measured intensities to the intensity of the parallel component measured before mounting the sample. The background level was evaluated to be 3×10^{-6}. A total of 10–20 measurements were performed by changing the x–y positions of the sample to obtain the averaged spectra.

Figures 4.1.15b and c present the spectra of T_p and T_c for PAD, and Figure 4.1.15d and e present T_p and T_c for PCD, respectively. The blue and red spectra for PCD in Figure 4.1.15d and e are for the microwave incident directions along [100] and [111], respectively. These spectra exhibit large decreases reaching 10^{-4}–10^{-6} in the transmittance. The dip positions in the spectra for the two incidences are slightly different from each other, indicating an anisotropic gap formation. These are expected to correspond to the bandgaps in the ΓX and ΓL directions, respectively, in the photonic band structure of PCD. These bandgap positions deduced by a plane-wave expansion method are indicated by blue and red bars in Figure 4.1.15d and e, respectively, which indeed agree quite well with the observed gap positions.

In the spectra of T_p and T_c for PAD in Figure 4.1.15b and c, respectively, the six curves are for the three samples with different orientations, and for two mutually orthogonal directions of the incident microwave polarizations for each sample. The black bars indicate the frequency range of the 3D PBG deduced by an FDTD calculation. First, in the spectra of T_c in Figure 4.1.15c, all the six spectra exhibit clear dips and the frequency ranges of the dips coincide well. This is in contrast to the case of PCD, in which the two spectra for the different incident directions show the dips in different frequency ranges. In addition, the frequency ranges of the dips in Figure 4.1.15c agree well with that of the calculated 3D PBG. These facts indicate that the measured dips in T_c in Figure 4.1.15c represent an isotropic 3D PBG in PAD. On the other hand, we notice a peculiar feature in the spectra of T_p in Figure 4.1.15b: it begins decreasing at $\omega d/2\pi c \approx 0.20$, which is far from the band edge, rapidly declining to 10^{-3} at the band-edge frequency of $\omega d/2\pi c \approx 0.26$. Above the gap, the transmission remains at 10^{-3}–10^{-2}. This peculiar feature has been successfully explained by the existence of a 3D PBG and the strong light diffusion outside it, as described in detail in a previous paper.[9] Thus, from this fact, along with the result of T_c, we finally conclude the experimental substantiation of the formation of an isotropic 3D PBG in PAD.

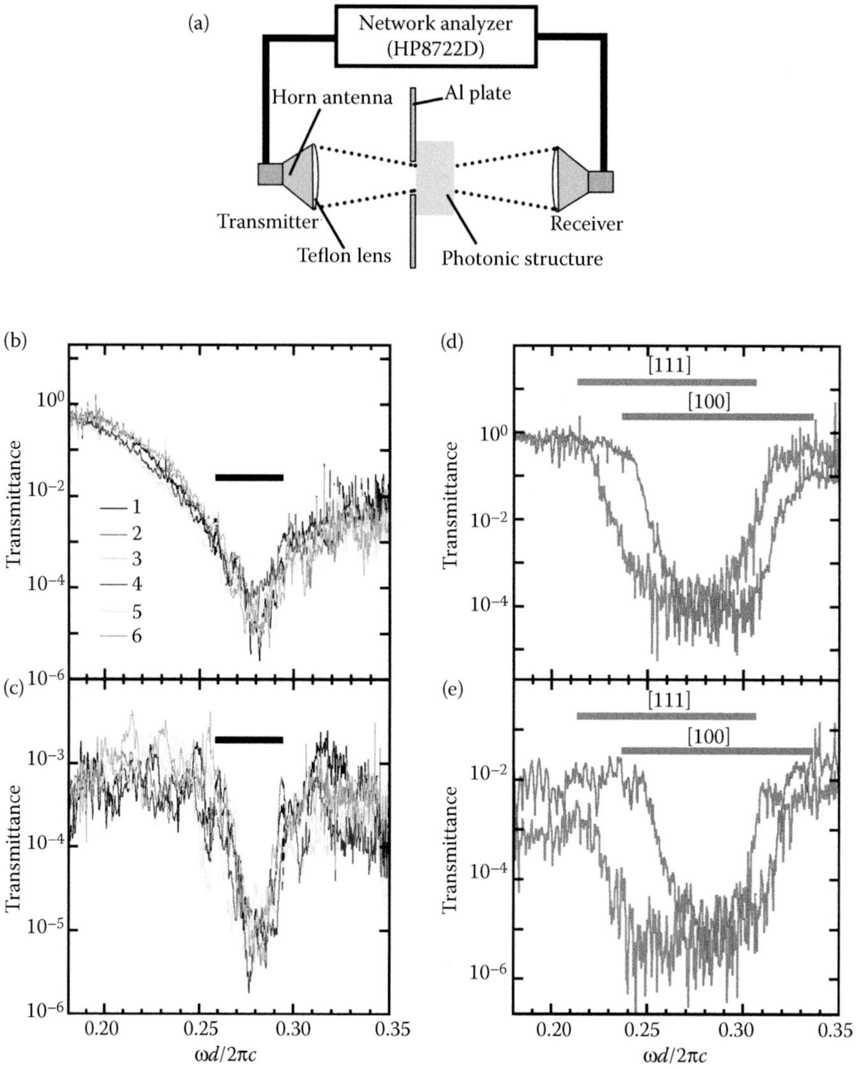

FIGURE 4.1.15

(**See color insert.**) (a) A schematic illustration of the experimental setting for microwave transmission measurements. (b)–(e) Measured transmission spectra: (b) T_p for PAD, (c) T_c for PAD, (d) T_p for PCD, and (e) T_c for PCD. The six curves in (b) and (c) are for three samples with different orientations, and for two mutually orthogonal directions of the incident microwave polarizations for each sample. The blue and red spectra in (d) and (e) are for the microwave incident directions along [100] and [111] of the diamond lattice, respectively. The black bar in (b) and (c) indicates the frequency range of the three-dimensional photonic band-gap in PAD deduced by a finite-difference time-domain calculation. The blue and red bars in (d) and (e) indicate the gap positions of PCD in the Γ*X* and Γ*L* directions, respectively, deduced by a photonic band-structure calculation using a plane-wave expansion method. Note that these gap frequency ranges are different from those in Figures 4.1.11 and 4.1.12 because of the different refractive index.PAD, photonic amorphous diamond; PCD, photonic crystalline diamond. (Panels (b)–(e) adapted from S. Imagawa et al., Phys. Rev. B **82**, 115116, 2010.)

4.1.8 Concluding Remarks

In this article, we reviewed our recent studies on perturbed photonic crystals and photonic amorphous structures, focusing on the possibility of strong light confinement. In the first part, we showed that ultrahigh-Q nanocavities can be formed by spatially modulating 2D PBG materials. Weak perturbation of 2D photonic crystals induces unprecedented strong light confinement with extremely small leakage into the direction perpendicular to the PBG plane. This configuration is favorable for various planar technologies used in electronic and light-wave circuits. We have shown that this concept can be applied for 1D photonic crystals and also for index-modulated structures. The fact that very small perturbation can lead to ultrastrong light confinement is especially important because we can manipulate the light confinement by weakly perturbing the photonic crystal via, for example, weak optical nonlinearity of materials. This feature may open up a novel form of photonic devices that can dynamically control light confinement.

In the second part, we reviewed our numerical and experimental studies on the 3D PBG formation in PAD, after describing its structural characteristics. The formation of a 3D PBG in PAD has proved unambiguously that the lattice periodicity is not essential to the realization of a 3D PBG, contrary to the common belief. We demonstrated that the PAD can confine light strongly at a defect. The 3D PBG in PAD is isotropic, regardless of the wavevector orientation and polarization direction, which, in principle, cannot be realized in conventional photonic crystals. The isotropic nature of the 3D PBG may be advantageous in designing an optical circuit in PAD; it can provide a large geometrical flexibility in the orientations of waveguides and other optic devices. Besides the formation of a 3D PBG described in this article, the PAD has exhibited unique features related to light diffusion and localization outside the PBG, which has been discussed in detail in a previous paper.[9] These findings should provide new insights into PBG formation mechanism, light diffusion and localization, and open new possibilities in the application of photonic materials.

Three-dimensional light confinement by 3D PBGs in photonic crystals was proposed more than two decades ago, but the present two examples show that light confinement can now be established (and also manipulated) by slightly different and more sophisticated manners. We assume that concepts behind these two examples are general and can be applied to other cases. We also believe that there will be a wide variety of photonic structures (beyond simple 3D PBG materials and other than the present two examples) that can realize 3D light confinement.

References

1. E. Yablonovitch, Phys. Rev. Lett., **58**, 2059 (1987).
2. S. John, Phys. Rev. Lett., **58**, 2486 (1987).
3. J. Joannopoulos, S. G. Johnson, R. Meade, and J. Winn, *Photonic Crystals: Molding the Flow of Light* (Princeton University Press, New Jersey, 2007), 2nd ed.
4. M. Notomi, Rep. Prog. Phys. **73**, 096501, (2010).
5. E. Yablonovitch, T. J. Gmitter, and K. M. Leung, Phys. Rev. Lett. **67**, 2295 (1991); K. M. Ho, C. T. Chan, C. M. Soukoulis, R. Biswas, and M. Sigalas, Solid State Commun. **89**, 413 (1994); S. Y. Lin, J. G. Fleming, D. L. Hetherington, B. K. Smith, R. Biswas, K. M. Ho, M. M. Sigalas,

W. Zubrzycki, S. R. Kurtz, and J. Bur, Nature **394**, 251 (1998); S. Noda, K. Tomoda, N. Yamamoto, and A. Chutinan, Science **289**, 604 (2000).
6. K. M. Ho, C. T. Chan, and C. M. Soukoulis, Phys. Rev. Lett. **65**, 3152 (1990).
7. M. Maldovan and E. L. Thomas, Nat. Mater. **3**, 593 (2004).
8. K. Edagawa, S. Kanoko, and M. Notomi, Phys. Rev. Lett. **100**, 013901 (2008).
9. S. Imagawa, K. Edagawa, K. Morita, T. Niino, Y. Kagawa, and M. Notomi, Phys. Rev. B **82**, 115116 (2010).
10. E. Kuramochi, M. Notomi, S. Mitsugi, A. Shinya, T. Tanabe, and T. Watanabe, Appl. Phys. Lett. **88**, 041112 (2006).
11. T. Tanabe, M. Notomi, E. Kuramochi, A. Shinya, and H. Taniyama, Nat. Photonics **1**, 49 (2007); T. Tanabe, M. Notomi, E. Kuramochi, and H. Taniyama, Opt. Express **15**, 7826 (2007).
12. The definition of the mode volume is given by $V_{eff} = \int \varepsilon(\mathbf{r})\ |\mathbf{E}(\mathbf{r})|^2\ d^3\mathbf{r}/\varepsilon(\mathbf{r}_{max})\ |\mathbf{E}\ (\mathbf{r}_{max})|^2$. Further discussions about the definition of the mode volume are given by Notomi.[4]
13. B. S. Song, S. Noda, T. Asano, and Y. Akahane, Nat. Mater. **4**, 207 (2005).
14. R. Herrmann, T. Sunner, T. Hein, A. Loffler, M. Kamp, and A. Forchel, Opt. Lett. **31**, 1229 (2006).
15. M. Notomi, K. Yamada, A. Shinya, J. Takahashi, C. Takahashi, and I. Yokohama, Phys. Rev. Lett. **87**, 253902 (2001).
16. Y. Takahashi, H. Hagino, Y. Tanaka, B. S. Song, T. Asano, and S. Noda, Opt. Express **15**, 17206 (2007).
17. M. Notomi, T. Tanabe, A. Shinya, E. Kuramochi, H. Taniyama, S. Mitsugi, and M. Morita, Opt. Express **15**, 17458 (2007).
18. T. Tanabe, A. Shinya, E. Kuramochi, S. Kondo, H. Taniyama, and M. Notomi, Appl. Phys. Lett. **91**, 021110 (2007); Z. Y. Zhang and M. Qiu, Opt. Express **12**, 3988 (2004).
19. S. Tomljenovic-Hanic, M. J. Steel, C. M. de Sterke, and D. J. Moss, Opt. Lett. **32**, 542 (2007).
20. M. Notomi and H. Taniyama, Opt. Express **16**, 18657 (2008).
21. T. Tanabe, M. Notomi, H. Taniyama, and E. Kuramochi, Phys. Rev. Lett. **102**, 043907 (2009).
22. M. Notomi, E. Kuramochi, and H. Taniyama, Opt. Express **16**, 11095 (2008).
23. E. Kuramochi, H. Taniyama, T. Tanabe, K. Kawasaki, Y. G. Roh, and M. Notomi, Opt. Express **18**, 15859 (2010).
24. P. B. Deotare, M. W. McCutcheon, I. W. Frank, M. Khan, and M. Loncar, Appl. Phys. Lett. **94**, 121106 (2009).
25. S. Matsuo, A. Shinya, T. Kakitsuka, K. Nozaki, T. Segawa, T. Sato, Y. Kawaguchi, and M. Notomi, Nat. Photonics **4**, 648 (2010).
26. S. Matsuo, A. Shinya, C.-H. Chen, K. Nozaki, T. Sato, Y. Kawaguchi, H. Taniyama, and M. Notomi, Opt. Express **19**, 2242 (2011).
27. D. E. Polk, J. Non-Cryst. Solids **5**, 365 (1971).
28. F. Wooten, K. Winer, and D. Weaire, Phys. Rev. Lett. **54**, 1392 (1985).
29. G. T. Barkema and N. Mousseau, Phys. Rev. B **62**, 4985 (2000).
30. P. N. Keating, Phys. Rev. **145**, 637 (1966).
31. C. T. Chan, S. Datta, K. M. Ho, and C. M. Soukoulis, Phys. Rev. B **50**, 1988 (1994).
32. C. T. Chan, Q. L. Yu, and K. M. Ho, Phys. Rev. B **51**, 16635 (1995).
33. E. Lidorikis, M. M. Sigalas, E. N. Economou, and C. M. Soukoulis, Phys. Rev. B **61**, 13458 (2000).
34. Y. S. Chan, C. T. Chan, and Z. Y. Liu, Phys. Rev. Lett. **80**, 956 (1998).
35. S. Kumar, J. Minerals Metals Materials, **55**, 43 (2003).

4.2

Nonlinear Optics in Silicon Photonic Crystal Nanocavities

Lucio Claudio Andreani, Paolo Andrich, Matteo Galli, and Dario Gerace
University of Pavia

Liam O'Faolain and Thomas F. Krauss
University of St. Andrews

CONTENTS

4.2.1 Introduction

Optical microcavities and nanocavities,[1] in which light is confined on a spatial scale of the order of its wavelength, have undergone substantial progress in the last few years.[2,3] Advances in design and nanofabrication have allowed the realization of structures with a high quality (Q) factor and/or low mode volume (V) that are suited for a variety of applications. In this chapter, we shall be concerned with nanocavities realized in photonic crystal (PhC) slabs, or waveguide embedded PhCs: such *PhC slab nanocavities* are characterized by unprecedented values of the ratio Q/V, especially when fabricated in silicon membranes, taking advantage of superior material properties and highly developed nanofabrication techniques.

Although nanocavities with a high Q/V ratio are believed to be ideal for enhancing radiation–matter interaction and nonlinear optical properties, silicon is usually considered to be a poor material for exploiting nonlinearities such as second-harmonic generation (SHG).[4,5] Indeed, in the centrosymmetric diamond lattice the second-order nonlinear susceptibility $\chi^{(2)}$ vanishes in the dipole approximation, leaving only weak electric quadrupole and magnetic dipole contributions, as well as surface-related contributions that add to the bulk quadrupole.[6,7] Third-harmonic generation (THG) is allowed by lattice symmetry, but being a higher-order contribution it is usually small and difficult to observe. Also, for Si-based devices optically pumped in the telecommunication bands at $\lambda \sim 1.3$–1.55 μm, both SHG and THG frequencies fall into the visible spectral region, well above the Si absorption edge at $\lambda = 1.1$ μm, and are therefore strongly absorbed. Moreover, in the same wavelength range the phenomenon of two-photon absorption (TPA) leads to generation of free carriers that produce further absorption and contribute to reducing the harmonic generation efficiency.

In this chapter we shall show that nonlinear optical properties can be substantially enhanced in silicon PhC nanocavities, up to a point where they become macroscopically observable even when using a continuous-wave (CW) diode laser as a pump beam. The main breakthrough that leads to the observation of these effects is the realization of PhC slab nanocavities that are optimized for highly efficient coupling to the far field in the direction normal to the slab plane. Such geometry under optimized coupling conditions allows the increase of the pump electromagnetic field in the cavity region to maximum level, whereas the generated harmonic fields are also observed in directions out of the slab plane and are therefore weakly affected by absorption. Moreover, the high surface-to-volume ratio given by the background PhC lattice reduces the free-carrier recombination lifetime to a few 100 ps, thereby contributing to the reduction of undesirable effects such as free-carrier absorption (FCA).

Several kinds of PhC slab nanocavities have been thoroughly investigated, including H1, Ln, heterostructure, and width-modulated line-defect cavities (see the review by Notomi[3]). All of these cavities are based on a two-dimensional (2D) triangular lattice on a high index (usually silicon) membrane. The H1 cavity is obtained by removing one hole, whereas Ln cavities consist of n missing holes along the Γ–K direction. The Q factor can be strongly enhanced by optimization of the local geometry, that is, by shifting and/or shrinking the holes nearest to the cavity: it has been shown for L3 nanocavities that local geometry optimization leads to "gentle confinement" and to the removal of the radiative components of the k-space distribution of ground cavity mode.[8] Heterostructure and width-modulated cavities are both based on line-defect waveguides of the W1 type, that is, a line of missing holes along the Γ–K direction. Heterostructure cavities are obtained by varying the period along the waveguide axis,[9] whereas width-modulated coupling relies on a variation of the waveguide (channel) width in the perpendicular Γ–M direction.[10] Among these structures, heterostructure and width-modulated cavities have the highest Q factors (higher than 10^6 and even reaching $2–4 \times 10^6$; see the studies by Kuramochi et al.[11] and Takahashi et al.[12]), whereas H1 cavities have the lowest mode volumes. Ln cavities with geometry optimization can have Q factors higher than 10^5 (we have demonstrated $Q = 1.1$, 2, and 4×10^5 for L3, L5, and L7, respectively[13]) and mode volume down to $0.6(\lambda/n)^3$. In other words, Ln cavities represent a good solution for optimizing the ratio Q/V. Additionally, they exhibit very good yields (98%+) and repeatibility. This makes them an excellent platform for the study of many phenomena. However, Ln cavities usually have a poor optical coupling to an incoming beam in the far field; this is also implicit in the "gentle confinement" approach, which is designed to remove radiative components from the k-space distribution of cavity modes. To study harmonic generation processes with out-of-plane coupling (which is the optimal configuration to reduce material absorption), special designs have to be used to increase far-field optical coupling. Far-field optimized Ln cavities have been studied by Tran et al.[14] and Toishi et al.[15]

In Section 4.2.2, we shall discuss our approach for systematic optimization of Ln cavities with the goal of increasing far-field coupling to an incoming beam. We shall also present characterization measurements of the fabricated cavities by means of a resonant (cross-polarized) light scattering technique,[16,17] which is especially suited to measure the Q factor of isolated (non waveguide-coupled) cavities in an out-of-plane configuration.

The generation of harmonic beams by nonlinear optical effects involves the conversion of n photons at a fundamental frequency ω into one photon at frequency $n\omega$ and is described by the nth-order nonlinear susceptibility $\chi^{(n)}(\omega)$. The harmonic intensity for the nth-order process scales as $(I_{\omega})^n$, where I_{ω} is the pump field intensity. Resonant cavities (or, more generally, high Q resonances) can be used to increase the pump intensity by a factor Q/V and, therefore, to enhance the harmonic intensity by $(Q/V)^n$. Thus, the use of microcavities and nanocavities with high Q factors and low mode volumes can be expected to yield a substantial increase of the nonlinear conversion efficiency. Indeed, this approach has proven to be very effective and to result in low-power CW harmonic generation processes, such as SHG in III–V-based PhC cavities[18,19] and THG in silica microtoroids.[20] In silicon-based systems, however, enhanced harmonic generation has only been observed under pulsed excitation with high peak power ($P_{peak} > 10$ W).[21,22] Enhancement of THG with a pump beam at $\lambda = 1.55$ μm, resulting in emission of green light, has also been achieved in silicon PhCs by exploiting slow-light waveguides with modified geometry to increase the mode group index.[23]

In Section 4.2.3, we shall show that the use of far-field optimized PhC nanocavities leads to a substantial enhancement of harmonic generation processes in a silicon chip and to simultaneous generation of second- and third-harmonics in the visible region by using a low-power CW laser in the telecom band as a pump source. Theoretical analysis allows one to attribute the observed SHG and THG processes to surface and bulk nonlinear susceptibilities, respectively. The results lead to unprecedented normalized conversion efficiencies and show the great potential of PhC nanocavities for nanoscale nonlinear optics.

In addition to harmonic generation, another important class of nonlinear phenomena in silicon is related to the active control of light by light, that is, *all-optical switching*. Such effects are also enhanced by high Q nanocavities, as demonstrated in pioneering works using TPA-induced nonlinearities. All-optical bistable switching was demonstrated using L3 or L4 nanocavities coupled to W1 waveguides using either thermo-optic[24] or free-carrier nonlinearities,[25] as well as in heterostructure cavities.[26] The switching time related to carrier-induced nonlinearity is faster than in other silicon photonic microdevices like ring resonators[27] and it can be made shorter than 50 ps using ion-implanted samples to reduce the free-carrier lifetime.[28] All-optical bistable switching is the basis for realization of logic ports, bit memories, and logic operation on a chip using infrared light at telecommunication wavelengths.[3] The enhancement of TPA in silicon is another important application for high Q/V nanocavities. As the important telecom wavelengths correspond to frequencies falling below the absorption edge of silicon, the key function of photodetection is normally absent—a serious problem for silicon-based photonic-integrated circuits. However, the relatively high TPA coefficient[29] of silicon provides a potential solution. With the enhancement in confined light intensity provided by a suitable nanocavity design, this absorption may reach significant levels. By integrating a *pin* diode with a width-modulated PhC cavity, Tanabe et al.[30] have recently been able to measure the TPA induced photocurrent. They show that the nanocavity enhances the signal by five orders of magnitude, bringing it to a level where there is useful signal at high speeds (0.1 Gbit/s) even for input powers of just a few tens of microwatts, thus making this very promising as a pure silicon photodetector.

All-optical switching and logic port operation based on free-carrier nonlinearities can also be realized in out-of-plane coupling configurations, in which the pump beam is focused onto the cavity region.[31] A natural step is then to combine harmonic generation with free-carrier-induced nonlinearities in order to shift the bistable switching behavior to the visible spectral region. In Section 4.2.4, we shall study optical bistability in out-of-plane coupling configurations and focus especially on the interplay with harmonic generation nonlinearities, leading to the new effect of *bistability in harmonic beams*. This phenomenon is very promising in view of achieving active control of a visible beam by acting on a CW pump beam at telecom wavelengths, with interesting prospects for applications to photodetectors and to other kinds of integrated optical devices.

4.2.2 Far-Field Optimized Nanocavities

Increasing the vertical coupling efficiency of PhC cavities is essential to many proposed applications, such as nanolasers or generic light-emitting devices. In the specific case of applications in nonlinear optics, it is crucial to efficiently couple light into the resonators in order to increase the field intensity, thus enhancing nonlinear effects at the nanoscale. A few works have shown possible strategies to optimize the far-field coupling for such cavity modes. Some early attempts have tried to achieve a high Q cavity design by reducing the out-coupling efficiency and simultaneously manipulating the far-field profile.[32] Later, far-field optimization of hexapole modes in H1 cavities has been obtained[33] by properly placing a distributed-Bragg reflector below the membrane to achieve constructive interference of the vertically emitted beam. More recently, Römer and Witzigmann[34] presented a systematic numerical study of simultaneous Q factor and far-field optimization, mainly for the H1-type cavity. Optimization of H1-type cavity far-field has been also addressed by Larque et al.[35]

However, H1-type cavities intrinsically suffer from a relatively limited maximum achievable Q factor. Working on Ln-type cavity modes, which have larger theoretical Q, a grating approach has been proposed and verified to concentrate light emission around the vertical direction, thus enhancing the out-coupling efficiency.[14,15] The idea is to consider that Fourier components lying outside the light cone can be folded back to $k = 0$, that is, around the normal direction to the sample surface, by superimposing a lattice with twice the periodicity of the underlying PhC structure. This way, leakage will be mainly determined by the harmonic components oscillating with a wavevector $k \sim \pi/a$ of the original lattice, which in the Brillouin zone of the modified lattice with period $2a$ are folded exactly around $k = 0$. For an L3-type cavity, the schematic cavity design is illustrated in Figure 4.2.1a. One can start from an L3 cavity optimized to have a large Q factor according to the principle of gentle confinement,[8,36] with nearby holes along the Γ–K direction (in the ideal hexagonal lattice) that have been shifted and shrunk by amounts $\Delta x/a = 0.16$ and $\Delta r'/a = (r' - r)/a = -0.06$, respectively. In Figure 4.2.1a, we identify from holes with modified radius r'' the ones that can be used to superimpose a second lattice with periodicity $2a$. Although cavity Q factors are reduced by the effects of fabrication disorder,[37] the procedure of far-field optimization is anticipated to be very robust against disorder.[14]

PhC cavities are fabricated on a standard SOITEC silicon-on-insulator wafer, with the buried oxide layer underneath being selectively under-etched to leave the PhC section as a suspended silicon membrane (see the study by O'Faolain et al.[38] for details). The lattice constant for all devices was 420 nm, with a nominal hole radius $r/a \sim 0.28$, and the dimensionless parameters, $\Delta x/a$ and Δ'/a, were also held constant. The radii of modified holes have been reduced/increased in steps of 3 nm, that is, from $\Delta r'' = -21$ nm to $\Delta r'' = +21$ nm. The exposure conditions were carefully chosen to allow such precise increments in hole radii. A scanning electron microscopy image of a typical device is shown in Figure 4.2.1b, where the modified holes for far-field optimization can be clearly identified. As L3 cavities

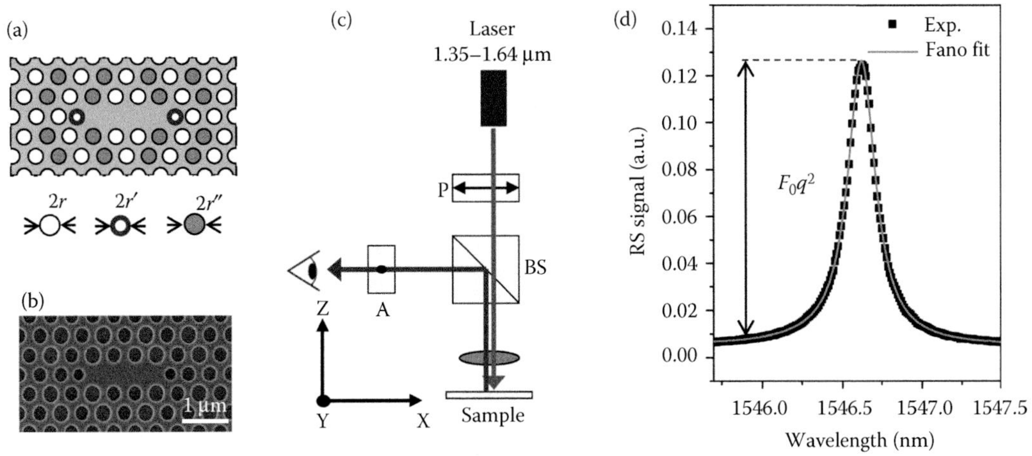

FIGURE 4.2.1
(a) Schematic of an L3-type far-field optimized cavity. Holes with bold edge are shrunk and shifted to optimize the Q factor. Shaded holes are modified to increase the vertical out-coupling. (b) A scanning electron microscopy image of a fabricated device on a silicon membrane. Holes corresponding to the filled circles in (a) are enlarged by $\Delta r'' = 18$ nm in this image. The lattice constant is $a = 420$ nm. (c) Scheme of a resonant scattering setup. (d) A typical resonant scattering spectrum on an L3 cavity with $\Delta r'' = 18$ nm.

have maximum theoretical Q factors around 1.5×10^5,[36,39] L5- and L7-type cavities, which nominally have larger Q factors, have also been designed and fabricated. For such cavity types, similar far-field optimization principles hold, promising useful coupling efficiencies at possibly higher Q factors, as detailed in the study by Portalupi et al.[13]

A systematic optical characterization of the devices can be performed by resonant scattering (RS) from the sample surface.[16] The experimental setup is illustrated in Figure 4.2.1c and detailed in the study by Galli et al.[17] It consists of measuring reflectance at normal incidence from the cavity in a crossed-polarization geometry defined by a polarizer (P) and an analyzer (A). The cavity must be oriented at 45° with respect to both the polarizer and the analyzer in order to achieve simultaneous coupling of incoming and outgoing polarizations with the fundamental cavity mode, therefore maximizing the resonant signal over the background. Asymmetric Fano lineshapes are in general observed and can be fitted with the function

$$F(\omega) = A_0 + F_0 \frac{[q + 2(\omega - \omega_0)/\Gamma]^2}{1 + [2(\omega - \omega_0)/\Gamma]^2}, \tag{4.2.1}$$

where q is the Fano parameter that determines the asymmetry of the lineshape and A_0 and F_0 are constant factors. The quality factor is determined as $Q = \omega_0/\Gamma$. Notice that for $q \gg (\omega - \omega_0)/\Gamma$ the Fano lineshape reduces to a symmetric Lorentzian. In this case, the quantity $F_0 q^2$ represents the intensity of the RS signal at resonance with the cavity mode. A close to Lorentzian spectrum for an L3 cavity with $\Delta r'' = 0$ is shown in Figure 4.2.1d together with the Fano lineshape fit. The Q factors extracted from the RS measurements are usually compared to guided-mode expansion[40] calculations, showing a reproducible and very good agreement in all the parameters range.[13] It is interesting to point out that the RS technique can be used to simultaneously give a qualitative estimation of the vertical coupling efficiency of our devices. In fact, the quantity $F_0 q^2$ in Equation 4.2.1 is proportional to the light intensity that has been coupled to the cavity mode and reflected back to the detector in crossed polarizations.[15] To normalize this quantity, we determine the intensity I of the incident light by replacing the sample with a nearly ideal dielectric mirror and measuring the reflected intensity, under the same focusing conditions but with parallel polarizations. Thus, we define the RS efficiency as F_0q^2/I. The latter quantity is taken as a measure of cross-polarized scattering due to resonant coupling with the cavity mode. A few representative RS spectra are shown in Figure 4.2.2a for the L3 cavities with different hole modifications. The RS efficiencies, $\eta_{RS} = F_0q^2/I$, are reported in Figure 4.2.2b together with the corresponding Q factors. The minimum RS efficiency occurs in correspondence with the maximum Q factor, as expected; in the experiment, it is the cavity with $\Delta r'' = -3$ nm that simultaneously displays the largest Q factor and the smallest RS efficiency. Here, $\Delta r'' = r'' - r > 0$ for larger modified holes, whereas $\Delta r'' < 0$ for smaller holes. The measured out-coupling efficiency generally increases at the expense of the Q factor. However, a compromise can be easily found from such kind of systematic characterization, in which a reasonably large Q is accompanied by a significant fraction of the incoming (or outgoing) light being coupled to the cavity mode.

To give a qualitative interpretation of the latter results, the measured devices can be modeled by three-dimensional (3D) finite-difference time-domain (FDTD) simulations. The cavity mode excitation is simulated with an internal dipole source, the near-field intensity is recorded at the sample surface on a 2D monitor much larger than the cavity area, and a standard near-to-far-field projection[33] is applied to get the fields on a hemispherical surface at a large distance from the cavity. A few normalized far-field patterns are shown in Figure 4.2.3, clearly displaying the qualitative evolution as a function of $\Delta r''$. In particular, the far-field intensity becomes more and more Gaussian and centered around the normal direction as $|\Delta r''|$ is increased, at the expense of the cavity Q factor. For values of $|\Delta r''| < 6$ nm, the far-field profile is strongly mismatched with respect to the collection optics, which causes the experimental RS efficiency to be lower as shown in Figure 4.2.2b. For $|\Delta r''| > 6$ nm, the far-field profile is well matched with a collection objective. The

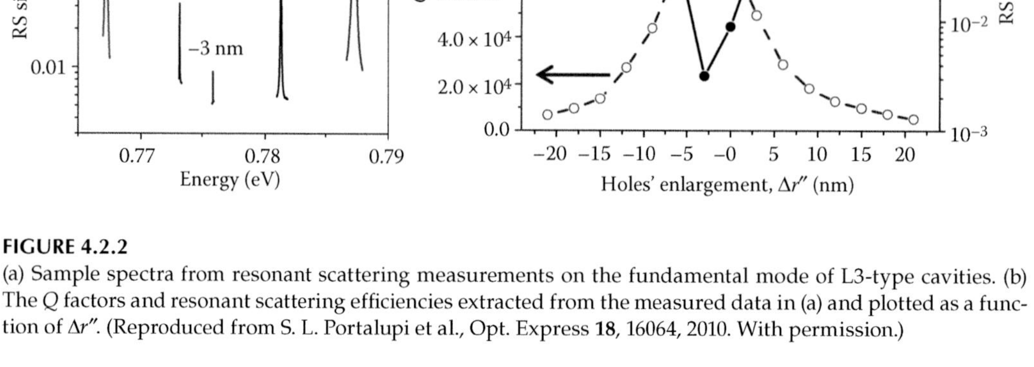

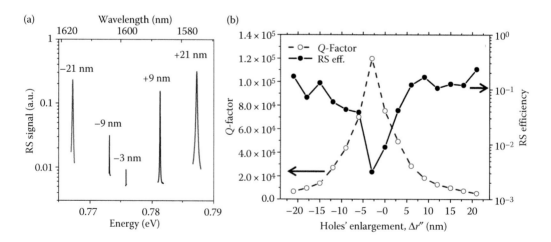

FIGURE 4.2.2
(a) Sample spectra from resonant scattering measurements on the fundamental mode of L3-type cavities. (b) The Q factors and resonant scattering efficiencies extracted from the measured data in (a) and plotted as a function of $\Delta r''$. (Reproduced from S. L. Portalupi et al., Opt. Express **18**, 16064, 2010. With permission.)

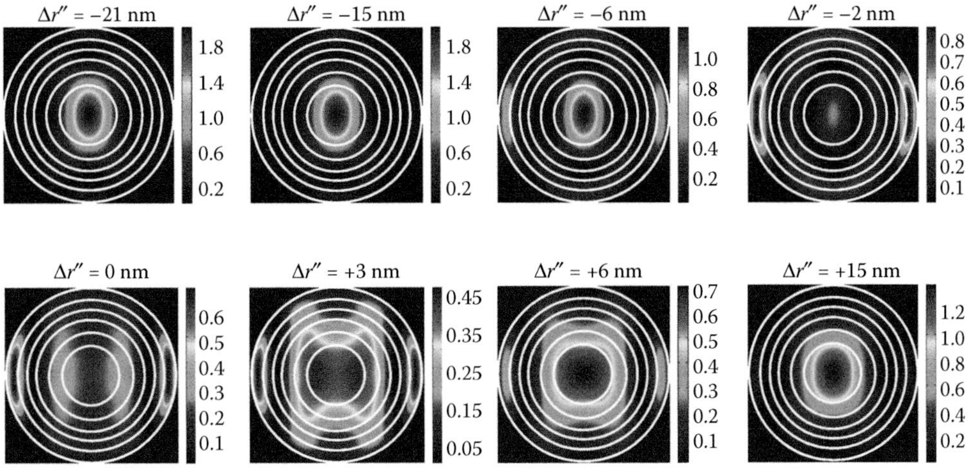

FIGURE 4.2.3
(**See color insert.**) A selection of calculated far-field patterns (electric field intensity profile, $|\mathbf{E}|^2$) by three-dimensional finite-difference time-domain simulations, corresponding to the measured devices whose results are shown in Figure 4.2.2. Field intensities are normalized to the total emitted power in the vertical half-space. Concentric circles correspond to $\theta = 20°, 30°, 40°, 50°, 60°$, and $90°$ from the inner to the outer one, respectively. (Reproduced from S. L. Portalupi et al., Opt. Express **18**, 16064, 2010. With permission.)

optimal condition with good coupling efficiency and relatively high Q (about 30,000) is obtained for $\Delta r'' = 6$ nm. Performances can be improved by changing cavity design to allow for a larger intrinsic Q factor, for example, with an L7 cavity design.[13]

To summarize this section, we have shown how to design, fabricate, and characterize a series of silicon cavities with modified geometry to improve coupling of cavity modes in the far field with an incoming/outgoing beam at telecom wavelengths. A systematic investigation of, for example, the L3 cavity geometries allows us to quantify the Q factors and the out-coupling efficiency. These results demonstrate that far-field optimized PhC cavities can have simultaneously high coupling efficiency and quality factors, and can be important for enhancing nonlinear effects in the nanoscale.

4.2.3 CW Harmonic Generation

As discussed in the previous section, the use of far-field optimized PhC nanocavities enables a very efficient coupling of an infrared laser beam to the cavity mode. Thus, a very efficient harmonic generation can be achieved by resonantly pumping the PhC nanocavities with a low-power CW excitation,[41] as reported in Figure 4.2.4a and b. Here, a 1 mW (~0.15 mW coupled power) infrared laser tuned to resonance with the fundamental mode at $\omega = \omega_c$ is used as a pump beam, leading to the simultaneous observation of a bright visible emission at $\omega = 3\omega_c$ (THG) and a less intense one at $\omega = 2\omega_c$ (SHG). The harmonic emission wavelength can be easily tuned across the blue/green visible window (450–525 nm) for THG or in the red window (675–790 nm) for SHG by choosing far-field optimized PhC cavities with fundamental resonant wavelengths λ_c in the range 1350–1580 nm, that is, in the main telecom bands. In particular, the bright THG light emission can be recorded by a simple snapshot using a commercial CCD camera, as shown in insets of Figure 4.2.4a and b. On the other hand, the spectrally resolved CW SHG and THG of the devices were probed using the same experimental setup used for RS characterization (see Figure 4.2.1c), in which the crossed polarizers have been removed and the beam-splitter replaced by a dichroic mirror. The low-power pump beam from the 1350–1640 nm tuneable laser was tuned to resonance with the cavity mode and focused on the sample from the top by means of a 50× microscope objective with NA = 0.8. Visible second-harmonic (SH) and third-harmonic (TH) emissions were collected through the same objective and then dispersed by a 500 mm focal length monochromator equipped with a liquid-nitrogen-cooled Si CCD.

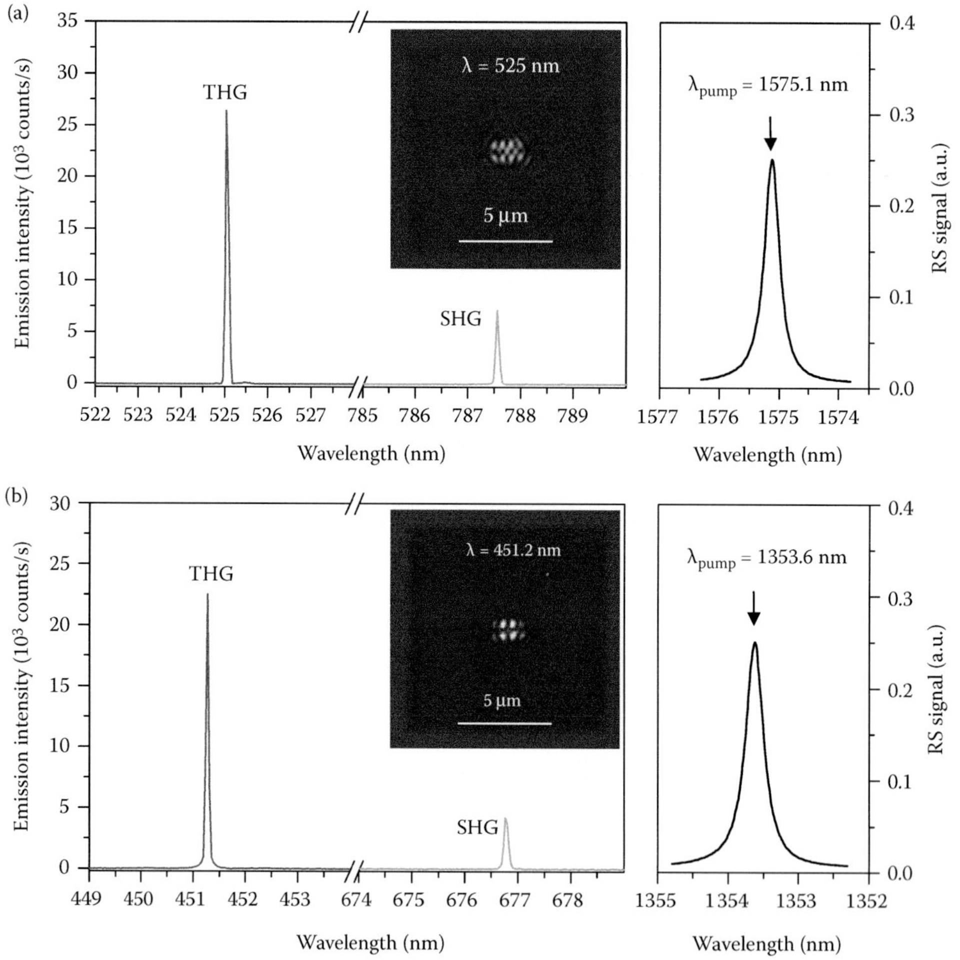

FIGURE 4.2.4
(**See color insert.**) (a) Resonant scattering spectrum (right) of a PhC nanocavity with fundamental mode at $\lambda_c = 1575.1$ nm (pump wavelength). SHG and THG emission spectra (left) at deep-red and green wavelengths $\lambda_{SH} = 787.5$ nm and $\lambda_{TH} = 525$ nm, respectively. (b) Resonant scattering spectrum (right) of a PhC nanocavity with fundamental mode at $\lambda_c = 1353.6$ nm (pump wavelength). SHG and THG emission spectra (left) at red and blue wavelengths $\lambda_{SH} = 676.8$ nm and $\lambda_{TH} = 451.2$ nm, respectively. Insets show THG images taken with a commercial CCD camera.

To confirm that light emission shown in Figure 4.2.4a and b is indeed due to harmonic generation processes, the scaling behavior of SHG and THG as a function of pump power is reported in Figure 4.2.5a. For low-power excitation, that is, for coupled powers <100 μW, both the SHG and THG emissions closely follow the predicted scaling behavior. However, at higher coupled power a slight saturation sets in as a result of FCA induced by TPA, which come into play on increasing the electromagnetic energy stored inside the cavity. As a result, a reduction in the conversion efficiency is observed because of the combined action of thermo-optic effect and degradation of the cavity Q factor. The first effect causes thermal heating of the PhC cavity and induces a red-shift of the cavity mode.[24,26] Then, the detuning of the resonance wavelength with respect to the pump beam leads to a reduced coupled power and a decreased harmonic generation efficiency. On the other hand, the significant FCA at the pump frequency degrades the Q factor of the cavity and, once again, reduces the harmonic generation efficiency. Moreover, strong FCA also occurs at the SH and TH frequencies, further reducing the conversion efficiency. The FCA and TPA nonlinearities are responsible for the bistable behavior that will be discussed in Section 4.2.4. A further confirmation of the scaling behavior is obtained by measuring the spectral lineshapes across the SH and TH frequencies. This is shown in Figure 4.2.5b, which reports the measured SHG and THG intensities recorded as the pump laser is scanned across the cavity resonance. The RS spectrum of the cavity mode is also reported with a best-fit to a Lorentzian lineshape. We see that the SHG and THG spectra are very well described by the squared and cubed Lorentzian lineshapes used to fit the fundamental resonance, respectively. These observations represent definite proof that the observed light emission should be interpreted as cavity-enhanced nonlinear frequency conversion by SHG and THG.

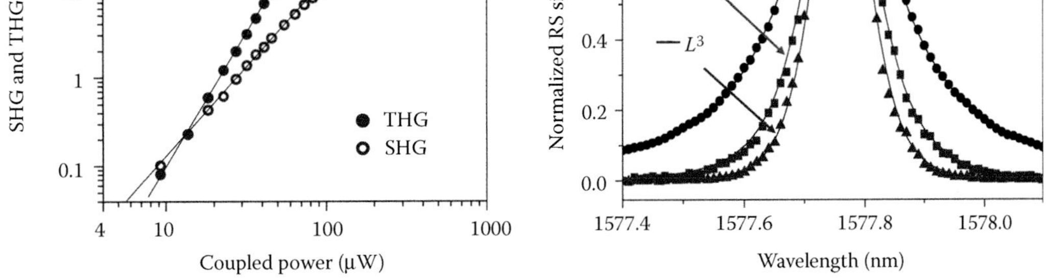

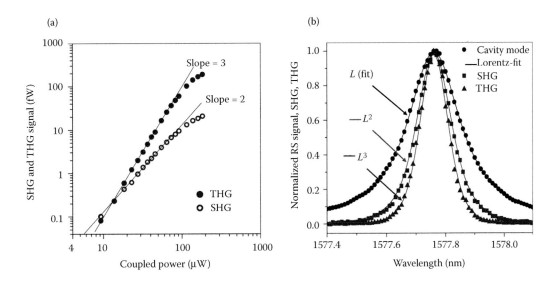

FIGURE 4.2.5
(a) SHG (open circles) and THG (full circles) emission intensity versus coupled pump power for a PhC nanocavity with $\Delta r'' = +21$ nm ($\eta_{RS} = 0.2$ and $Q = 5.2 \times 10^3$), at fixed pump wavelength. (b) Resonant scattering spectrum for the same PhC nanocavity (dots), and best-fit to a Lorentzian lineshape (line, L). SHG and THG spectra recorded while scanning the pump laser across the cavity resonance (squares and triangles, respectively). Lines interpolating the SHG and THG data are the squared (L^2) and cubed (L^3) Lorentzians, respectively. (Reproduced from M. Galli et al., Opt. Express **18**, 26613, 2010. With permission.)

Another important scaling for cavity-enhanced harmonic generation is the dependence of SHG and THG emission intensity on the Q factor of the PhC nanocavity fundamental mode. In fact, the harmonic conversion efficiency is expected to scale as $(\eta_{cav}Q/V)^n$, where η_{cav} is the in-coupling efficiency of the pump beam to the PhC nanocavity that is assumed to be equal to η_{RS} (as reported in Section 4.2.2), and n is the order of the generated harmonic. Therefore, assuming that the mode volume of each PhC cavity remains almost unchanged, one expects the harmonic generation signal to scale as $(Q\eta_{cav})^2$ for the SH process and as $(Q\eta_{cav})^3$ for the TH one. Then, very clear Q^2 and Q^3 scaling behaviors for the SHG and THG processes are obtained by normalizing the measured signals from each cavity by the experimental η_{cav}^2 and η_{cav}^3, respectively, as shown explicitly in Figure 4.2.6.

The nonlinear conversion efficiency for SHG and THG processes is usually defined as $\eta_{NL} = P_{out}/P_{in}$. Accordingly, an estimation of the maximum η_{NL} observed in these devices can be evaluated from the plots of Figure 4.2.4a. From these data we estimate $\eta_{SH} \sim 1 \times 10^{-10}$ for the SHG process and $\eta_{TH} \sim 2 \times 10^{-9}$ for the THG one. However, as we are dealing with CW conversion efficiencies, a more appropriate comparison with literature results is obtained by evaluating the normalized conversion efficiencies as $\rho_{SH} = P_{out}/P_{in}^2$ for the SHG process and $\rho_{TH} = P_{out}/P_{in}^3$ for the THG one. These quantities are independent of the experimental conditions and give a measure of the intrinsic harmonic generation efficiency of the device. This yields $\rho_{SH} \sim 3 \times 10^{-6}/W$ and $\rho_{TH} \sim 1/W^2$, respectively. For comparison, the normalized SH conversion efficiency for a focusing area of ~1 μm² is of the order of $10^{-14}/W$ in bulk silicon[21] and it is enhanced by no more than two to three orders of magnitude in microcavities and 2D PhCs.[22] For THG, the highest reported normalized conversion efficiency is $10^{-10}/W^2$.[23] This clearly demonstrates the crucial role of silicon PhC

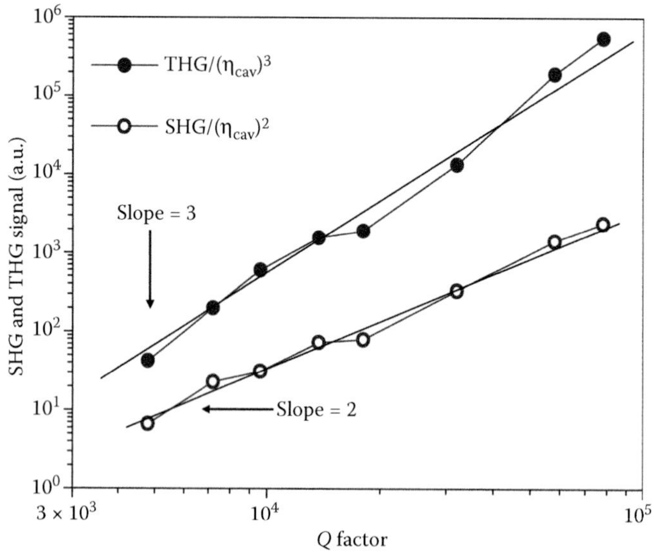

FIGURE 4.2.6
Normalized SHG and THG emission intensity (see text) as a function of the cavity Q factor showing clear Q^2 and Q^3 scaling, respectively. Each SHG and THG data corresponds to measurements on different photonic crystal nanocavities with increasing Q factors. These devices correspond to those characterized in Figure 4.2.2b for $0 \leq \Delta r'' \leq 21$ nm. (Reproduced from M. Galli et al., Opt. Express **18**, 26613, 2010. With permission.)

nanocavities to enhance optical nonlinearities and to reduce pump powers by several orders of magnitude.

Another important aspect concerns the microscopic origin of the harmonic generation process in silicon. Owing to its centrosymmetric nature, bulk crystalline silicon has a third-order nonlinear susceptibility tensor but no bulk electric dipole second-order susceptibility. However, SHG is possible in silicon as a surface contribution, which adds to the weak bulk quadrupole and magnetic dipole contributions.[4–6] The bulk contribution is measurable but small[7,21] and it will be neglected in the following.

The spatial regions within the PhC cavity giving rise to these two contributions can be experimentally identified by taking the optical images of the emitting cavity selectively filtered at the THG and SHG wavelengths, as shown in Figure 4.2.7a and c, respectively. Optical images are obtained by using a 50× microscope objective with NA = 0.8 in combination with f = 400 mm tube lens, to realize a 100× magnified image of the PhC cavity and emission pattern on the liquid-nitrogen-cooled Si CCD. These two different contributions can be modeled with 3D FDTD simulations of the electric field generated from suitably determined nonlinear polarizations determined by the cavity field profile, following a procedure similar to the one already used by McCutcheon et al.[18] The calculated near-field profiles of the electric field at $\omega = \omega_c$, decomposed in the two orthogonal components E_x and E_y, are shown in Figure 4.2.7b and d. One can immediately see that the SHG image is characterized by two bright lobes separated by a dark region just at the cavity location, indicating that only the E_x component is contributing to SHG emission and confirming its surface-related nature. On the other hand, the THG image shows up all of the intensity maxima corresponding to the sum of both $|E_x|^2$ and $|E_y|^2$ contributions to the electric field.

As a last result, we shall now discuss the emission properties of the generated harmonic from the PhC nanocavities, that is, the polar emission profiles of the emitted light in the far-field. These can be experimentally measured using an optical Fourier-imaging technique that consists of creating an image of the back-focus of the objective lens used to collect the harmonic light directly on the Si CCD. Such images are shown in Figure 4.2.8a and b. We notice that the two Fourier images differ considerably from each other, reflecting the different near-field distribution of the field components giving rise to the SHG and THG signals. To interpret the experimental images, the SH and TH far-fields are calculated by a 3D FDTD modeling, as shown in Figure 4.2.8c and d. These simulations agree fairly well with the measured patterns. In particular, the four-lobe TH and the double-lobe SH

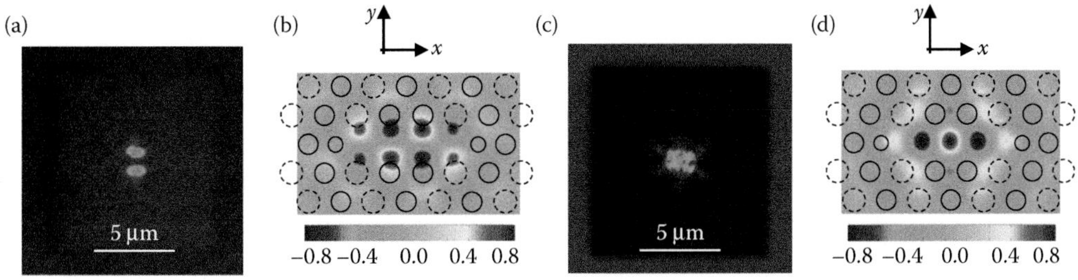

FIGURE 4.2.7
(**See color insert.**) Spectrally filtered optical image of (a) second-harmonic generation and (c) third-harmonic generation emission taken with a high-sensitivity Si CCD. Calculated electric field inside the photonic crystal cavity at the resonance frequency for (b) E_x and (d) E_y components, respectively. (Reproduced from M. Galli et al., Opt. Express **18**, 26613, 2010. With permission.)

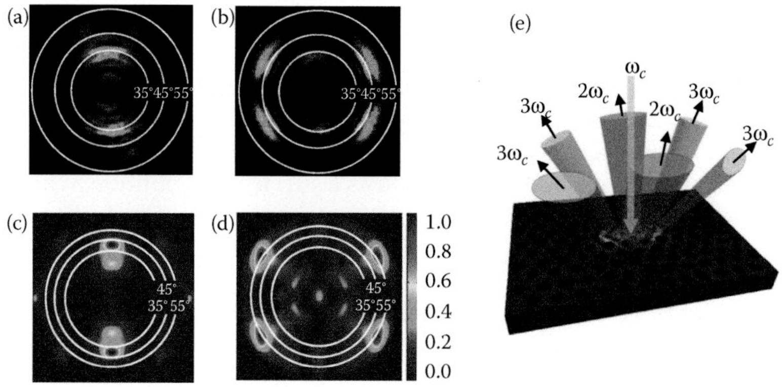

FIGURE 4.2.8
(See color insert.) Experimental Fourier images showing the polar far-field emission profile of (a) second-harmonic generation and (b) third-harmonic generation light. The corresponding far-field intensity patterns in (c) and (d) are calculated by three-dimensional finite-difference time-domain for (a) second-harmonic and (b) third-harmonic frequencies, respectively. (e) An artist's impression of the SHG and THG emission from the photonic crystal nanocavity. (Reproduced from M. Galli et al., Opt. Express **18**, 26613, 2010. With permission.)

far-field emissions are nicely reproduced within the model. The TH nonlinear polarization is calculated as:[7]

$$P_{x,y}(3\omega_c) = 3\chi^{(3)}_{1212} \mid \mathbf{E}(\omega_c) \mid^2 E_{x,y}(\omega_c), \tag{4.2.2}$$

where $\chi^{(3)}_{1212}$ is the third-order nonlinear susceptibility. The small nonspherical terms are neglected in Equation 4.2.2. The nonlinear TH polarization components depend on the field profiles shown in Figure 4.2.7b and d. We can assume a discrete set of dipoles oscillating at frequency $3\omega_c$ to be placed in the relative maxima of $\mathbf{P}(3\omega_c)$: five y-polarized dipoles are placed along the cavity axis, whereas eight x-polarized dipoles are placed off-axis close to the PhC holes around the cavity. The amplitudes of the dipoles are proportional to the local values of the nonlinear polarization and a phase of π is given to the dipoles placed in the relative maxima with opposite signs. The generated near-field is recorded just above the PhC cavity surface and then propagated with a near-to-far field transformation, as for the calculations shown in Figure 4.2.3.

The SHG is modeled in a similar fashion, but the nonlinear interaction in the centrosymmetric silicon material is now dominated by surface effects. The second-order nonlinear polarization is determined by the surface contribution at the silicon-hole boundaries. The field maxima in Figure 4.2.7b occur at the positions of the eight holes close to the cavity and the electric field at the maxima is along the x direction. Modeling the second-order surface nonlinear tensor in the approximation of an isotropic vertical sidewall surface,[4] the nonlinear polarization is found to be along the y direction and it is given by (notice that the normal to the surface is along y):

$$P_y(2\omega_c) = \chi^{(2)}_{311} E_x^2(\omega_c). \tag{4.2.3}$$

Considering the second-order nonlinear tensor components for the microscopic symmetry of the surface,[7] for a cavity axis along the [110] direction of the cubic silicon lattice, yields the same result as the electric field is dominated by the x component.

From such a detailed analysis of the harmonic generation and its propagation to the far field, we can conclude that THG is a bulk effect, as opposed to SHG that originates from surface contributions at the vertical hole sidewalls.

4.2.4 Bistability Effects

The concept of optical bistability is of great interest in the field of integrated photonics, potentially providing a means to implement all-optical transistors, switches, logical gates, and memories. For a system to display optical bistability, two features are required: nonlinearity and feedback.[42] In the case of a PhC cavity, the medium provides the nonlinearity whereas the variation in the optical properties of the medium itself due to the stored energy provides the feedback control.[24,36,43] As a consequence of these features, the output signal is a strong nonlinear function of the input signal, and may even display a hysteresis loop. Furthermore, the RS spectra of the cavities are distinctly altered. By measuring the modification of the optical response as a function of the input power, we were able to appreciate once more the nonlinear efficiency of these systems and their usefulness as nanodevices with advanced functionalities for integrated optical circuits.

At first, the effects on the cavities spectra are studied using the RS setup previously described. The results for one of the samples are reported in Figure 4.2.9a. When P_{in} is 4 μW, the measured lineshape is nearly symmetric, as expected. As the input power is

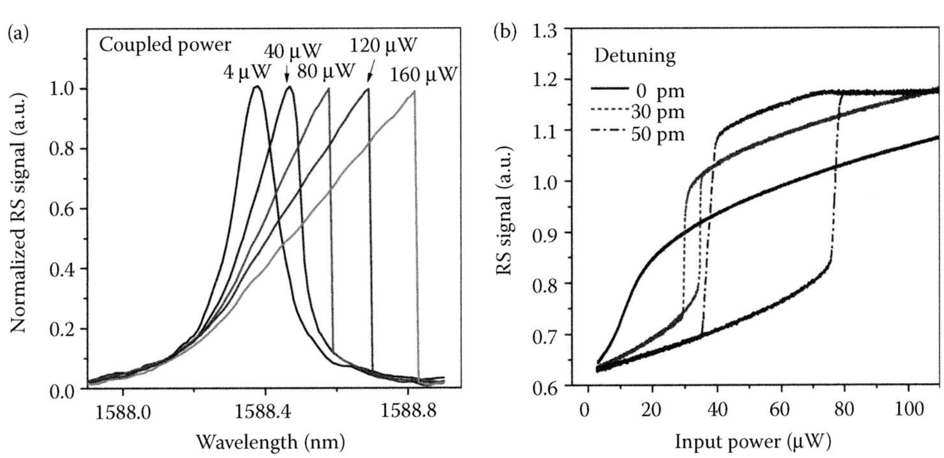

FIGURE 4.2.9
(a) Normalized resonant scattering spectra of a photonic crystal cavity mode with $Q \cong 15{,}000$ measured for increasing coupled optical power of the probing laser. (b) Input/output (P_{out} vs. P_{in}) curves showing the typical hysteresis loop for different wavelength detunings, $\Delta\lambda$, with respect to the natural cavity resonance.

increased, the cavity resonance starts to red-shift as a result of the thermo-optic effect in silicon and the measured spectrum becomes asymmetric. The sharp drop in the output power on the longer wavelength side, which is visible at powers above 80 μW, is direct evidence of the bistable behavior of the system. As a further proof of optical bistability, the hysteresis loop in the characteristic P_{out} versus P_{in} was measured. To this purpose, an electro-optical modulator was introduced to continuously modulate the input power, in order to collect and reproduce the characteristic P_{out} versus P_{in} behavior. The resulting curves for several values of the initial pump-cavity detuning $\Delta\lambda$ are reported in Figure 4.2.9b. In this case, as P_{in} is increased, the nonlinear shift of the cavity resonance modifies the power coupled to the cavity mode. The combined effect of these interrelated mechanisms leads to the opening of a hysteresis loop provided that the initial pump-cavity detuning is above a threshold value determined by the quality factor of the cavity.[44] We notice that a well-defined hysteresis loop is clearly measured for input power as low as 30 μW, and this clearly demonstrates the nonlinear efficiency of silicon PhC nanocavities and their potential to act as ultra-low-power micrometric-sized logic elements in the telecom wavelength range.

Besides the bistable behavior of the fundamental resonant mode in PhC nanocavities, the very efficient harmonic generation of these devices, as demonstrated in the previous sections, can be exploited to provide optical bistability also at the TH frequency, as shown in Figure 4.2.10. Here the visible (green) THG signal from the same PhC nanocavity is recorded while scanning the 1.588 μm pump laser across the cavity resonance. The optical spectra of the radiated THG light, shown in Figure 4.2.10a, displays a characteristic asymmetric lineshape, together with a sharp drop at the longer wavelength side, thus demonstrating a bistable behavior at the THG frequency. As before, to further confirm the THG bistability, we report in Figure 4.2.10b the opening of a hysteresis loop in the curve P_{out} versus P_{in} as obtained by monitoring the THG light emission as a function of an

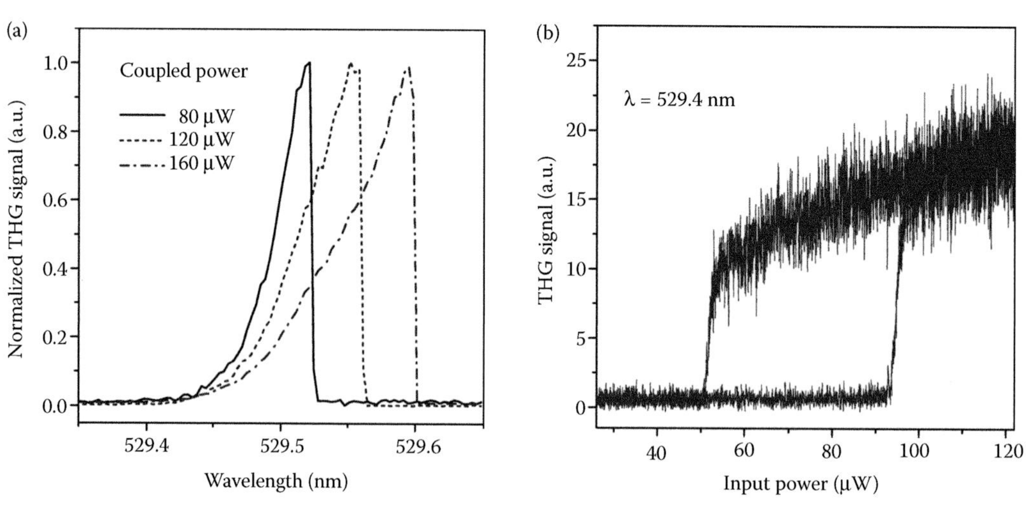

FIGURE 4.2.10
(a) Normalized THG emission for the same device as in Figure 4.2.9 measured for increasing coupled optical power of the probing laser. (b) Input/output curves showing the opening of an hysteresis loop in the THG output for fixed input wavelength detuned by $\Delta\lambda = 50$ pm with respect to the natural cavity resonance.

intensity-modulated infrared pump. This result extends the capability of PhC nanocavities to perform logical operations to wavelengths in the visible region of the spectrum, while still using very low input powers for the infrared pump beam.

4.2.5 Conclusions

In this chapter, we have discussed the application of PhC slab nanocavities to nonlinear frequency conversion in silicon. Although integrated devices based on PhC waveguides and nanocavities rely on optical propagation and processes in a waveguide, we have chosen to work in an out-of-plane configuration, which allows one to reduce TPA for the pump beam and absorption losses for the harmonic beams. The well-known problem of inefficient coupling of high Q nanocavity modes to an external beam is solved by working with Ln nanocavities that are optimized for coupling to the far field, while maintaining a reasonably high Q factor. We can use a similar experimental configuration for cavity mode characterization, which is done by cross-polarized light scattering from the surface without the need of using coupling waveguides, and for harmonic generation and switching experiments.

The results show that strong SHG and THG in a silicon system can be obtained by using a low-power CW laser in the telecommunication range as a pump beam. The conversion efficiencies normalized to the incident power are enhanced by several orders of magnitude as compared to previous experiments and hold promise of further enhancement by using nanocavities with an intrinsically higher Q factor. Detailed analysis demonstrates that SHG arises from surface nonlinear dipoles at the vertical hole sidewalls surrounding the cavity, whereas THG stems from a bulk effect. Moreover, mapping of the cavity mode is achieved through the harmonic beams, especially THG whose intensity reflects the mode profile in the Si PhC cavity. The all-optical switching and bistability effects that have been previously demonstrated at the pump frequency are now translated to the visible region through the process of frequency conversion, giving rise to the phenomenon of TH bistability. This new effect may be used for nonlinear switching or modulation of a visible beam using a weak infrared beam as a pump. In general, we believe that there is ample room both for improving the present results and for unraveling novel physical effects related to harmonic generation, frequency mixing, and parametric processes, with bright perspectives for nanoscale nonlinear optics using Si PhC cavities.

Acknowledgments

We gratefully acknowledge Simone Luca Portalupi and Roberto Lo Savio (University of Pavia) for their essential contributions to optical measurements, Daniele Bajoni and Marco Liscidini (University of Pavia) for many useful discussions and suggestions, and Giorgio Guizzetti (University of Pavia) for his continuous support and interest. We are deeply indebted to Christopher Reardon and Karl Welna (University of St. Andrews) for their help with device fabrication.

References

1. K. J. Vahala, Nature **424**, 839 (2003).
2. S. Noda, M. Fujita, and T. Asano, Nat. Photonics **1**, 449 (2007); J. D. Joannopoulos, S. G. Johnson, J. N. Winn, and R. D. Meade, *Photonic Crystals: Molding the Flow of Light* (Princeton University Press, Princeton, 2008).
3. M. Notomi, Rep. Progr. Phys. **73**, 096501 (2010).
4. Y. R. Shen, *The Principles of Nonlinear Optics* (Wiley, New York, 1984).
5. R. Boyd, *Nonlinear Optics* (Academic Press, California, 1992).
6. N. Bloembergen, R. K. Chang, S. S. Jha, and C. H. Lee, Phys. Rev. **174**, 813 (1968); J. A. Litwin, J. E. Sipe, and H. M. van Driel, Phys. Rev. B **31**, 5543 (1985); P. Guyot-Sionnest, W. Chen, and Y. R. Shen, *ibid*. **33**, 8254 (1986).
7. J. E. Sipe, D. J. Moss, and H. M. van Driel, Phys. Rev. B **35**, 1129 (1987).
8. Y. Akahane, T. Asano, B. S. Song, and S. Noda, Nature **425**, 944 (2003).
9. B. S. Song, S. Noda, T. Asano, and Y. Akahane, Nat. Mater. **4**, 207 (2005).
10. M. Notomi, A. Shinya, A. Mitsugi, S. Kuramochi, and H.Y. Ryu, Opt. Express **12**, 551 (2004).
11. E. Kuramochi, M. Notomi, S. Mitsugi, A. Shinya, T. Tanabe, and T. Watanabe, Appl. Phys. Lett. **88**, 041112 (2006).
12. Y. Takahashi, H. Hagino, Y. Tanaka, B. S. Song, T. Asano, and S. Noda, Opt. Express **15**, 17206 (2007).
13. S. L. Portalupi, M. Galli, C. Reardon, T. F. Krauss, L. O'Faolain, L. C. Andreani, and D. Gerace, Opt. Express **18**, 16064 (2010).
14. N.-V.-Q. Tran, S. Combrié, and A. De Rossi, Phys. Rev. B **79**, 041101(R) (2009).
15. M. Toishi, D. Englund, A. Faraon, and J. Vučković, Opt. Express **17**, 14618 (2009).
16. M. McCutcheon, G. W. Rieger, I. W. Cheung, J. F. Young, D. Dalacu, S. Frédéric, P. J. Poole, G. C. Aers, and R. Williams, Appl. Phys. Lett. **87**, 221110 (2009).
17. M. Galli, S. L. Portalupi, M. Belotti, L. C. Andreani, L. O'Faolain, and T. F. Krauss, Appl. Phys. Lett. **94**, 071101 (2009).
18. M. W. McCutcheon, J. F. Young, G. W. Rieger, D. Dalacu, S. Frédérick, P. J. Poole, and R. L. Williams, Phys. Rev. B **76**, 245104 (2007).
19. S. Combrié, A. De Rossi, Q. V. Tran, and H. Benisty, Opt. Lett. **33**, 1908 (2008); K. Rivoire, Z. Lin, F. Hatami, W. T. Masselink, and J. Vučković, Opt. Express **17**, 22609 (2009).
20. T. Carmon and K. J. Vahala, Nat. Phys. **3**, 430 (2007).
21. M. Falasconi, L. C. Andreani, A. M. Malvezzi, M. Patrini, V. Mulloni, and L. Pavesi, Surf. Sci. **481**, 105 (2001).
22. T. V. Dolgova, A. I. Maidykovski, M. G. Martemyanov, A. A. Fedyanin, O. A. Aktsipetrov, G. Marowsky, V. A. Yakovlev, and G. Mattei, Appl. Phys. Lett. **81**, 2725 (2002); P. P. Markowicz, H. Tiryaki, H. Pudavar, P. N. Prasad, N. N. Lepeshkin, and R. W. Boyd, Phys. Rev. Lett. **92**, 083903 (2004); M. G. Martemyanov, E. M. Kim, T. V. Dolgova, A. A. Fedyanin, and O. A. Aktsipetrov, Phys. Rev. B **70**, 073311 (2004); C. Comaschi, G. Vecchi, A. M. Malvezzi, M. Patrini, G. Guizzetti, M. Liscidini, L. C. Andreani, D. Peyrade, and Y. Chen, Appl. Phys. B **81**, 305 (2005).
23. B. Corcoran, C. Monat, C. Grillet, D. J. Moss, B. J. Eggleton, T. P. White, L. O'Faolain, and T. F. Krauss, Nat. Photonics **3**, 206 (2009).
24. M. Notomi, A. Shinya, S. Mitsugi, G. Kira, E. Kuramochi, and T. Tanabe, Opt. Express **13**, 2678 (2005).
25. T. Tanabe, M. Notomi, S. Mitsugi, A. Shinya, and E. Kuramochi, Appl. Phys. Lett. **87**, 151112 (2005).
26. T. Uesugi, B. S. Song, T. Asano, and S. Noda, Opt. Express **14**, 377 (2006).
27. V. R. Almeida, C. A. Barrios, R. R. Panepucci, and M. Lipson, Nature **431**, 1081 (2004).
28. T. Tanabe, K. Nishiguchi, A. Shinya, E. Kuramochi, H. Inokawa, M. Notomi, K. Yamada et al., Appl. Phys. Lett. **90**, 031115 (2007).
29. A. D. Bristow, N. Rotenberg, and H. M. van Driel, Appl. Phys. Lett. **90**, 191104 (2007).

30. T. Tanabe, H. Sumikura, H. Taniyama, A. Shinya, and M. Notomi, Appl. Phys. Lett. **96**, 101103 (2010).
31. M. Belotti, J.F. Galisteo-López, S. De Angelis, M. Galli, I. Maksymov, L. C. Andreani, D. Peyrade, and Y. Chen, Opt. Express **16**, 11624 (2008); M. Belotti, M. Galli, D. Gerace, L.C. Andreani, G. Guizzetti, A. R. Md Zain, N. P. Johnson, M. Sorel, and R. M. De La Rue, *ibid.* **18**, 1450 (2010).
32. S. G. Johnson, S. Fan, A. Mekis, and J. D. Joannopoulos, Appl. Phys. Lett. **78**, 3388 (2001).
33. S.-H. Kim, S.-K. Kim, and Y.-H. Lee, Phys. Rev. B **73**, 235117 (2006).
34. F. Römer and B. Witzigmann, J. Opt. Soc. Am. B **25**, 31 (2008).
35. M. Larque, T. Karle, I. Robert-Philipp, and A. Beveratos, New J. Phys. **11**, 033022 (2009).
36. L. C. Andreani, D. Gerace, and M. Agio, Photonics Nanostruct. Fundam. Appl. **2**, 103 (2004).
37. D. Gerace and L. C. Andreani, Photonics Nanostruct. Fundam. Appl. **3**, 120 (2005).
38. L. O'Faolain, X. Yuan, D. McIntyre, S. Thoms, H. Chong, R. M. De La Rue, and T. F. Krauss, Electron. Lett. **42**, 1454 (2006).
39. C. Sauvan, P. Lalanne, and J.-P. Hugonin, Phys. Rev. B **71**, 165118 (2005).
40. L. C. Andreani and D. Gerace, Phys. Rev. B **73**, 235114 (2006).
41. M. Galli, D. Gerace, K. Welna, T. F. Krauss, L. O'Faolain, G. Guizzetti, and L. C. Andreani, Opt. Express **18**, 26613 (2010).
42. H. Gibbs, *Optical Bistability: Controlling Light with Light* (Academic Press, Orlando, 1985).
43. L. D. Haret, T. Tanabe, E. Kuramochi, and M. Notomi Opt. Express **17**, 21108 (2009).
44. J. Bravo-Abad, A. Rodriguez, P. Bermel, S. G. Johnson, J. D. Joannopoulos, and M. Soljacic, Opt. Express **15**, 16161 (2007).

4.3

Random Lasing Highlighted by π-Conjugated Polymer Films

Randy C. Polson and Z. Valy Vardeny
University of Utah

CONTENTS

4.3.1 Introduction

The first functioning laser used an optically pumped solid-state medium, which was a ruby crystal excited by a flashlamp, created by Maiman in 1960.[1] Theoretical work had been done in previous years,[2] and four years after the demonstration of the laser, the 1964 Nobel Prize in Physics was awarded to Charles H. Townes, Nicolay G. Basov, and Aleksandr M. Prokhorov for "fundamental work in the field of quantum electronics which has led to the construction of oscillators and amplifiers based on the maser-laser principle." Following that breakthrough organic-dye-based lasers were demonstrated in 1966.[3] Discussion of the possibility of random lasing were initiated soon after.[4]

Random lasers are broadly defined as systems that produce laser emission but that do not have an obvious engineered cavity. The development of random lasers followed a similar path to conventionally engineered lasers. First, an optically pumped solid-state medium was shown to have many properties of laser emission.[5] Theoretical work predicted the actual observation, by decades in this case.[4] Soon after solid-state random lasing was demonstrated; random lasing was also shown in optically pumped dye and scatterers.[6]

There are now numerous systems in which random lasing has been demonstrated, all have a gain medium and scatterers. There are two broad types of systems; one has the same materials for gain media and scatterers, whereas the other has separate materials for gain media and scatterers. Some examples of gain and scatterers as the same material include ground Nd:YAG particles,[5] clusters of zinc oxide nanoparticles,[7] and films of π-conjugated polymers.[8] Some examples of gain and extrinsic scatterers include mixtures of laser dye and titanium dioxide (TiO_2) nanoparticles,[6] dye in photonic fibers,[9] dye and liquid crystals,[10] dye and glass slides,[11] dye and colon tissue,[12] and dye in bones.[13]

Light emission from an active medium depends on the excitation intensity. For example, in polymer films at low excitation intensity the emission spectrum is dominated by a broad photoluminescence (PL) band of about 70 nm. At higher intensities, there can be a population inversion and amplified spontaneous emission (ASE) that drastically reduces the emission spectral width to about 7 nm. PL and ASE are properties of the gain medium. If there exists some optical feedback, then the emission spectrum also develops very narrow emission lines on the ASE background, with a width of less than 1 nm. In engineered cavities, the feedback is typically from highly reflective end mirrors. In random lasers, feedback is from either waveguiding effects or multiple scattering.

4.3.2 Lasing in Polymer Films

To fully explore random lasing, we choose one of the well-studied systems and investigate multiple properties of the emission. The specific system we choose is films of a π-conjugated polymer; a typical random lasing spectrum is shown in Figure 4.3.1. This semiconducting polymer has a backbone that is composed of π-conjugated carbon atoms, and organic side chains to increase its solubility, chemically known as poly(2,5-dihexyloxy-p-phenylenevinylene) or DHO-PPV. It is seen that the lasing emission spectrum contains many peaks riding on top of a broader ASE background. The inset of the figure shows the PL spectrum at low excitation intensity, and the narrower region of ASE within the PL spectrum. The PL band is about 70 nm wide, the ASE is about 10 nm wide, and the random lasing peaks are

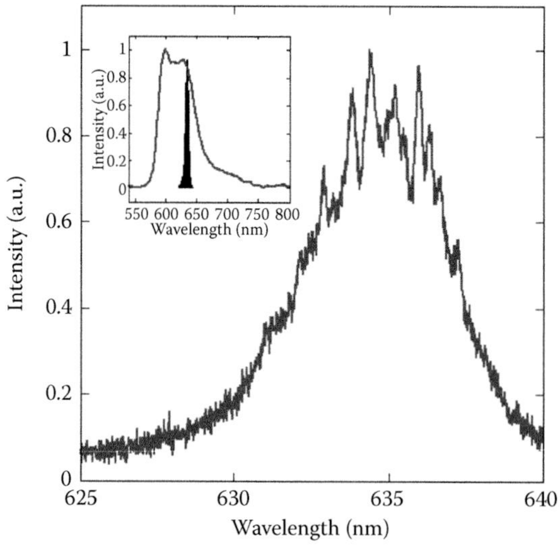

FIGURE 4.3.1
Random lasing spectrum of poly(2,5-dioctyloxy-p-phenylenevinylene) or DOO-PPV polymer film. Inset: Photoluminescence spectrum and region of spectral narrowing.

about 1 nm wide. On the scale of the PL spectrum in the inset, the random lasing peaks are too narrow to be seen.

In random lasing systems, the specific configuration of excited gain medium and scatterers determines the emission spectrum. Changing the configuration leads to a change in the emission.

For liquid systems, the emission spectrum drastically changes with the diffusion of the particles, and therefore steady-state lasing is on the order of milliseconds with dye and suspended particles. This time dependence will be exploited later for averaging purposes. Changing the configuration of scatterers for a solid sample is only possible by exciting a physically different region of the sample that is readily achieved by moving the excitation area on the sample. Each excitation area produces a unique random lasing emission spectrum. However, as the medium is solid, the observed spectrum remains the same for repeated measurements on the same illuminated area of the sample.

Figure 4.3.2 shows multiple excitations of the same excited area for the polymer film; the emission spectrum is nearly identical for all excitations. The small spectral variations are likely related to small intensity fluctuations of the excitation laser. At different locations, the collection configuration of gain and scatterers is different; therefore, the emission spectrum seems to be different. To reiterate, each location has a repeatable spectrum, but different locations have different spectra. The emission spectrum in Figure 4.3.1 is actually a single emission spectrum from Figure 4.3.2b. This property will be used to perform random lasing spectrum averaging.

A π-conjugated polymer is a long chain of smaller repeat units. Common polymers such as nylon or polyethylene have a backbone chain that is composed of carbon atoms joined to other carbon atoms having a single bond; therefore, all outer electrons of the carbon atoms participate in the bonding. The backbone of π-conjugated polymers, in contrast, have carbon atoms each with three planar σ bonding and perpendicular π electrons that may bond to form alternating single and double bonds. Thus, the π bonds overlap and this allows for delocalization of the π electrons, high electrical mobility, and electrical conductivity. The π-conjugated polymers were initially developed a few decades ago because the electrical conductivity nears the very best metallic conductors such as

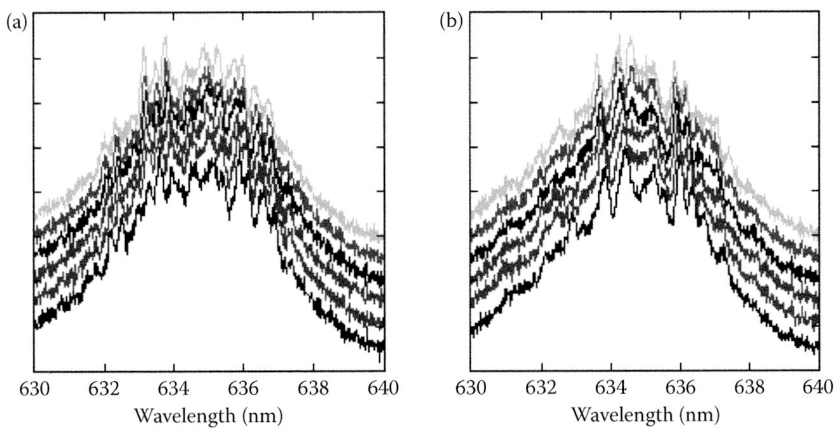

FIGURE 4.3.2
Multiple exposures of same excitation location for poly(2,5-dioctyloxy-p-phenylenevinylene) or DOO-PPV polymer film. (a) and (b) show two different excitation spots with similar but not identical emission spectra.

copper.[14] The 2000 Nobel Prize in Chemistry was awarded to Profs. Alan Heeger, Alan G. MacDiarmid, and Hideki Shirakawa "for the discovery and development of conductive polymers." Besides conductivity, another useful property of many π-conjugated polymers is that they are luminescent. Other structures that have an alternating collection of single and double bonds are the high-gain organic dye molecules, namely rhodamine 6G for example.

An excited electron on a π-conjugated polymer may further develop in a variety of decay channels. Optical studies that follow the ultrafast photoexcitation decay have shown numerous species of excited electrons that vary from singlet and triplet excitons to polaron pairs, polarons, and bipolarons. Time-resolved studies using pump-probe spectroscopy has shown stimulated emission, and thus the primary intrachain photoexcitations are singlet excitons. The main relaxation channel of the initial photoexcitation is a 200 fs decay into a longer lived excited state, identified as intrachain singlet excitons. Time-resolved dynamics showed that conducting polymers may exhibit gain even as films.[15] The stimulated emission dynamics are faster than ASE or PL.[16] Dramatic spectral narrowing in π-conjugated films has been reported in the 1990s in many polymer films and solutions.[17] Stimulated emission is on the path to laser emission. The next necessary step is to have a high enough gain for amplification. The first demonstration of lasing in a π-conjugated polymer was with an optically pumped poly[2-methoxy-5-(2′-ethylhexyloxy)-p-phenylene vinylene] or MEH-PPV solution.[18] The sides of a cuvette provided enough reflection to provide feedback that is necessary to produce laser emission.

The gain of the π-conjugated polymers is quite high, and the polymer films cast from solutions do not quench the emission or optical gain. Polymer films were shown to have laser emission in 1996.[8,19] An obvious advantage of polymer films is that they can emit light when electrically excited; many light-emitting diodes have been fabricated from polymer films,[20] and this application has initially pushed the field of polymer lasers. However, for laser emission it is necessary to achieve a high enough population inversion to produce amplified stimulated emission. The polarons that are the dominant charge excitations in π-conjugated polymers unfortunately have absorption bands close to the emission spectrum, and this coincidence diminishes the optical cross-section for emission, thereby decreasing the optical gain. So the prospect of electrically pumped polymer lasers is very much in doubt.

4.3.3 Random Lasing in Polymer Films

Random lasing in π-conjugated polymer films was first seen in a film of poly(2,5-dioctyloxy-p-phenylenevinylene) or DOO-PPV.[21] The film thickness is a critical parameter to observe laser emission. The thickness needs to be large enough to support a guided wave, that is, it has to be roughly a wavelength thick. If it is too thin, the emission escapes from the film and amplification does not occur. Alternatively, if the thickness is too large then there is not enough confinement to obtain ASE.

The most convincing proof of lasing is temporal and spatial coherence. The temporal coherence of random lasing can be measured by photon counting techniques. The arrival of photons from a small wavelength range and short time range may be used to reveal time coherence of the emitted light.[22] Coherent and incoherent light have different distributions of arrival times, and by fitting the observed arrival times the coherence of light may be determined. Coherence of random lasing in zinc oxide has been confirmed,[23] and soon after also in films of DOO-PPV.[24] Thus, the very narrow emission peaks in the lasing spectra of polymer films have been shown to be coherent.[24]

To a first approximation, all random lasing spectra have the same form (compare Figure 4.3.2a and b). There are numerous (>20) narrow lines within the gain profile and different configurations of gain and scatterers produce different emission spectra. As each position on a solid medium has a slightly different emission spectrum, a comprehensive analysis of different spectra is difficult. Simply averaging the emission spectra only leads to a spectrum that equals the gain spectrum; thus, further analysis requires another tool. One such tool is the Fourier transform of the emission spectrum.[25]

Fourier transform is often used in looking at a time varying signal, a sound waveform for example. It can reveal which frequencies comprise the waveform. Fourier transform is also commonly encountered in quantum mechanics relating the uncertainty of a particle's position and momentum. In optics, Fourier transform infrared spectroscopy records an interferrogram and subsequently the entire spectrum is recovered with a Fourier transform. We apply the Fourier transform technique to random lasing emission spectra to get a deeper look at the underlying mechanism of random lasing.

For a linear engineered cavity with two end mirrors and a gain medium, the output spectrum is described by a Fabry–Pérot line separation of $\Delta\lambda = \lambda^2/2nL$, where λ is the emission wavelength, n is the index of refraction, and L is the separation of the end mirrors. If the emission spectrum has regularly spaced emission lines, then the Fourier transform will have regularly spaced peaks, with a peak separation of nL/π.[26] Anticipating later results, a circular cavity has a round trip of πD, where D is the cavity diameter. In the Fourier transform or a circular cavity, nL is replaced by $\pi D/2$. Before attempting to analyze random lasing emission spectra, it is useful to analyze a well-defined, engineered resonator cavity. One characteristic of a laser resonator is the quality factor, or Q factor. This describes how much light escapes the cavity and can be defined as the inverse relative linewidth, $\lambda/\Delta\lambda$. For high Q resonators, little light escapes in each round trip and the output linewidths are narrow. Once the cavity enters into the lasing regime, then the laser emission linewidth is determined by several factors and not just by the cavity Q factor. A simple and high Q resonance cavity is a glass fiber coated with a thin (1 μm) film of the π-conjugated polymer.[27] This is easy to fabricate and provides a high confinement resonator for whispering gallery-type feedback.[28] A whispering gallery is a circular structure where one whisperer can be heard by an observer across the other side, and was originally used to describe sound waves in the top of St. Paul's cathedral in London. The sound waves are reflected and confined by the wall, and can thus propagate to the other side of the dome close to the circumference of the "cavity." For the polymer and glass microring structure, the polymer air interface plays the role of the wall. A ray of light is totally internally reflected if the angle of light hits the interface at an angle greater than the critical angle determined by Snell's law ($\sin(\theta_c) = n_1/n_2$). For an index of refraction of ~1.7 for the polymer in the visible spectral range, the critical angle is ~36°.

Figure 4.3.3 is the emission spectrum from a fabricated circular cavity known as a microring laser. The microring structure is optically excited and the emission is collected and recorded. The emission spectrum consists of a series of very narrow peaks that are observed when the polymer gain medium is optically excited above the lasing threshold. An analysis of circular resonance cavities leads to Bessel functions, designated by J_m, where m is the integer index. For the electric field distribution of a microring, a complete description uses Bessel functions for the interior of the microdisk, and Hankel functions for the exterior, with appropriate boundary conditions at the polymer glass and polymer air interfaces. In the microring emission spectrum of Figure 4.3.3, the resonant lines are very narrow. This

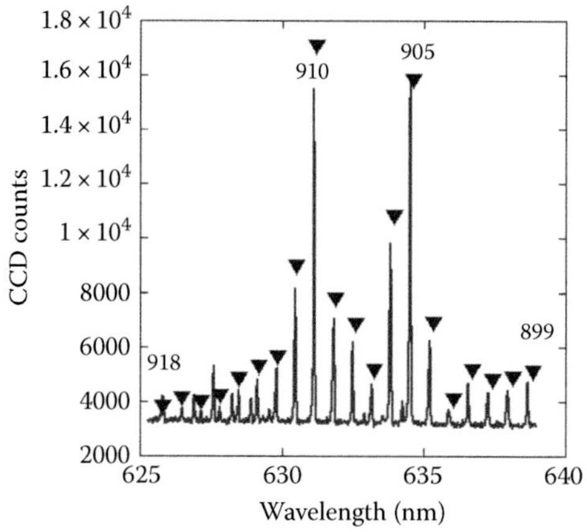

FIGURE 4.3.3
Spectrum of polymer microring laser. Numbers indicate integer Bessel functions.

implies that the Q factor is high and little light escapes. This high-Q structure allows for a simplifying approximation to be made, that the argument of the Bessel function is zero at the air polymer boundary. This condition can be written as $J_m(\pi nD/\lambda) = Z_m$, where Z_m is the first zero of J_m. Zeros of Bessel functions are numbers that a numerical software can easily calculate. The condition can be inverted to calculate a given wavelength knowing nD and the zeros of Bessel functions, namely $(\pi nD/Z_m) = \lambda_c$. Consequently for a given value of nD, a series of emission peaks can be calculated within a gain spectrum.

Figure 4.3.4 is the Fourier transform of the microring spectrum shown in Figure 4.3.3. There is a series of equally spaced diminishing peaks that occur at an interval of nD/π. The numeric value for nD is 186.47. The highest peak in the emission spectrum occurs at 631.21 nm. Looking through a table of first zeros of higher-order Bessel functions gives the first zero of Bessel function 910 to be 928.09. Putting the nD value together with the Bessel function zero gives a predicted wavelength of 631.20. The zeros of successive Bessel

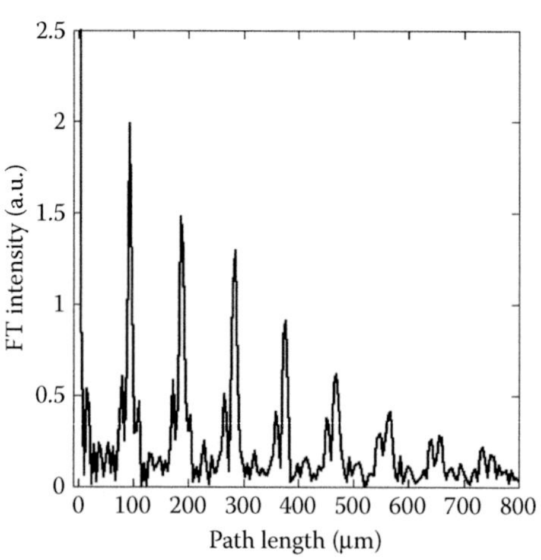

FIGURE 4.3.4
Fourier transform of Figure 4.3.3, microring emission spectrum.

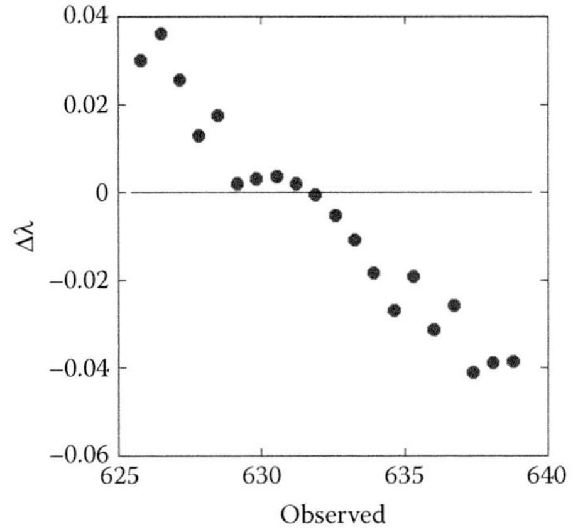

FIGURE 4.3.5
Difference between calculated wavelength and observed wavelength for microring laser.

functions fit the successive observed emission peaks. The entire emission spectrum fits a series of peaks with integer Bessel functions running from 899 for the highest wavelength to 918 for the lowest wavelength in the emission spectrum. We conclude that the emission spectrum of the fabricated circular cavity is completely explained by Bessel functions. Alternatively, fabricated microdisks can be also used as a model engineered cavity to demonstrate Fourier transform analysis.[29] The difference between observed and calculated peaks in the emission spectrum is shown in Figure 4.3.5. When using the same value of nD for all the peaks, the largest difference between observed and calculated peak positions is about 0.04 nm. Looking carefully at Figure 4.3.5, there seems to be a systematic deviation. This can be traced back to the dispersion relation of the polymer index of refraction. The index of refraction n is not constant across the emission spectrum for all lines, but varies slightly with the frequency of the light within the material.[30]

Returning to the random lasing spectrum in Figure 4.3.1a, its Fourier transform is shown in Figure 4.3.6. Understandably, this is not as simple as the Fourier transform of the microring

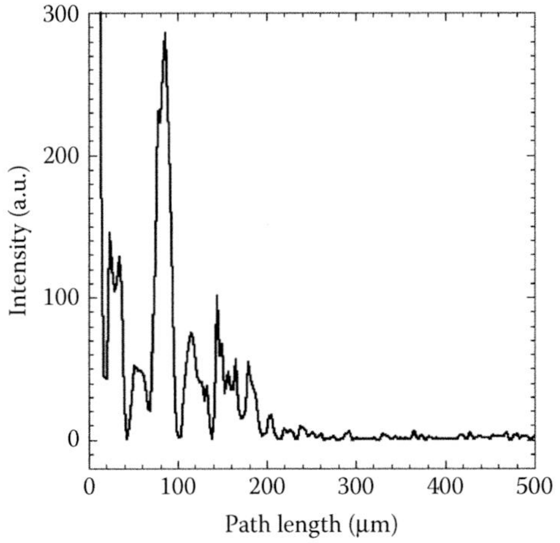

FIGURE 4.3.6
Fourier transform of Figure 4.3.2a.

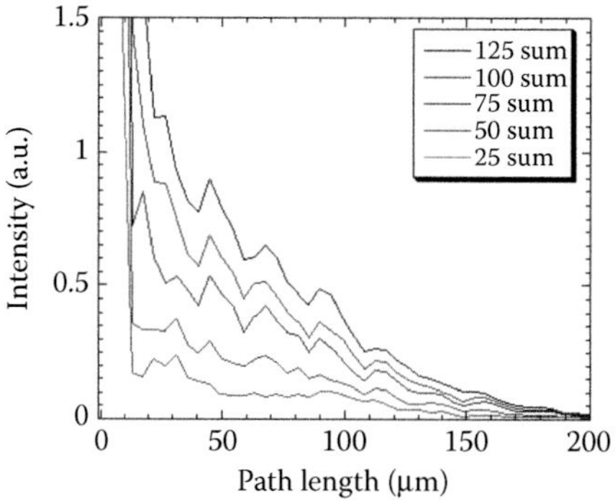

FIGURE 4.3.7
Average Fourier transform of successive numbers of individual locations on poly(2,5-dioctyloxy-p-phenylenevinylene) or DOO-PPV polymer film.

laser emission spectrum. If there is a cavity present that gives the random lasing emission, then neither the spectrum nor the Fourier transform clearly reveals its existence.

When many random lasing spectra are averaged, the resulting composite spectrum is simply the gain spectrum of the material. However, something rather unexpected happens if many Fourier transforms of random lasing emission spectra that come from *different film locations* are averaged together; surprisingly, a more regular structure appears.[24] Figure 4.3.7 shows the results of averaging successive numbers of laser emission Fourier transforms. A more regular structure appears. This can be interpreted by considering that within the polymer film, there is a preferred resonance structure size. To explain this result we have used the idea that random cavities in the polymer films form optical resonators. To achieve lasing, the gain must exceed total losses in the resonator. This is achieved with either high gain or low loss. The cavities that lase first are typically the highest-Q resonators. Longer path lengths are exponentially less likely to occur than short path lengths. However, short path lengths may not have enough gain to overcome loss and thus do not lase easily. At a given excitation intensity, or optical gain value in the polymer film, the minimum path length for lasing is thus fixed. Random resonators that lase first need to be near this minimum path length, as there are cutoffs on each side of this critical length.[31]

4.3.4 Mapping of Random Lasers

To search for resonance cavities within a film, a separate mapping experiment was performed. The goal of spectral mapping is to correlate the physical locations on a film with the specific random lasing spectrum in the collected laser emission spectrum. This is a challenge, as changing the excitation area can alter the random lasing emission spectrum. To achieve the spectral/spatial correlation the polymer film was placed in a vacuum and excited optically. The emission was then collected and collimated with a second lens, and a third lens focused light onto the entrance slit of a spectrometer. The third lens was placed on a linear translation stage. Figure 4.3.8 is the schematic diagram of this experiment configuration. By moving the third lens across the entrance slit of the spectrometer, a complete spectral map of the emission region could be produced and recorded on a CCD array.

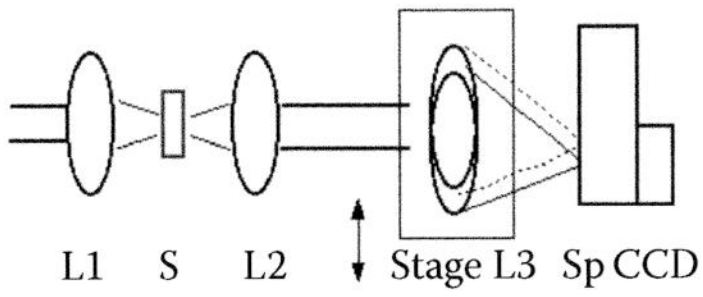

FIGURE 4.3.8
Schematic diagram of emission mapping measurement. (Adapted from R. C. Polson and Z. V. Vardeny, Opt. Lett. **35** (16), 2801, 2010.)

Fixing the excitation spot, film locations and collection optics are necessary to ensure that the same spectrum will be recorded. The output of the CCD is a two-dimensional array of numbers, $F(\lambda, h)$, representing the number of detected photons at a certain wavelength, λ, interval and vertical position, h, of the slit. If the coupling lens is scanned across the entrance slit of the spectrometer in the x-direction, then the emission coming from the sample film can be mapped, with certain amplification according to the lens arrangement. The collected data are a series of two-dimensional images, with lens position the third variable $F(\lambda, h, x)$. The dataset can then be analyzed in several ways. Figure 4.3.9 is a typical random lasing spectrum achieved by integrating $F(\lambda, h, x)$ with respect to x and h, and Figure 4.3.10 is the Fourier transform of the spectrum. As a complete spectrum is obtained at all points, postprocessing image maps can be produced. The main peak in Figure 4.3.9 is used to determine where in the illuminated spot on the sample it originates from. There are two regions on the sample where it originates from, as shown in Figure 4.3.11. Thus, the position of the emission in the film can be mapped and shows roughly circular cavities.[32] We emphasize that the emission line does not come equally from all locations of the film, but instead is concentrated at a few locations. The line emission is concentrated in two regions of the film as clearly visible. Spectral mapping in the polymer film has the additional complication of emission that originates in a localized region of the illuminated spot on the film and then propagates through the gain medium. This can make emission seem to come from everywhere equally.

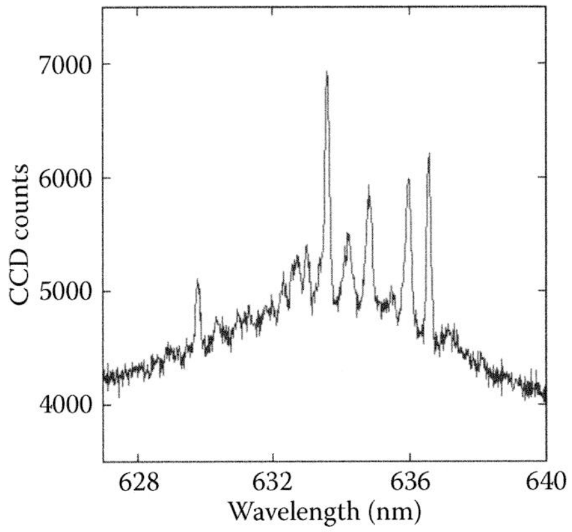

FIGURE 4.3.9
Individual emission spectrum from mapping configuration. (Adapted from R. C. Polson and Z. V. Vardeny, Opt. Lett. **35** (16), 2801, 2010.)

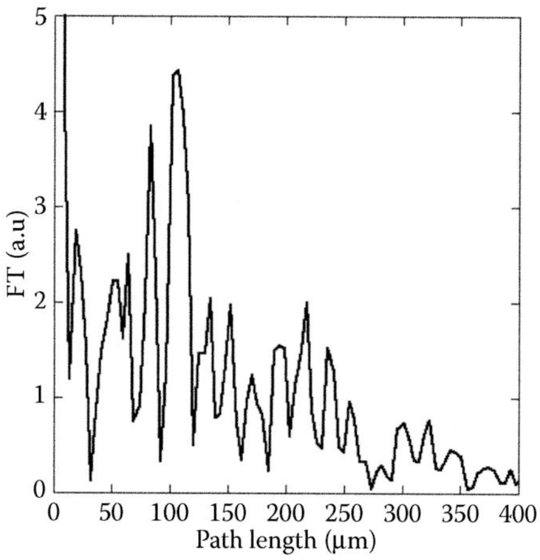

FIGURE 4.3.10
Fourier transform of Figure 4.3.9. (Adapted from R. C. Polson and Z. V. Vardeny, Opt. Lett. **35** (16), 2801, 2010.)

Different regions have different emission characteristics in both spectra and lasing thresholds. To gain insight into both these aspects, the number of positions that lase at a given excitation intensity have been determined. The results are binned and presented in a histogram. A single excitation area can support multiple resonance structures. Multiple excitation intensities at the same location show a changing emission spectrum as new resonators cross the lasing threshold. Figure 4.3.12 demonstrates statistical properties of lasing polymer films. By looking at the appearance of lines in the Fourier transform as well as emission spectra lasing thresholds could be determined. The first threshold is simple to determine; narrow lines do appear in the emission spectrum. Sine one 100 × 100 μm area can support multiple resonance structures, and additional thresholds can be observed at different excitation intensities. We recently showed that the distribution of lasing thresholds follows a general scaling law, which was theoretically predicted.[33]

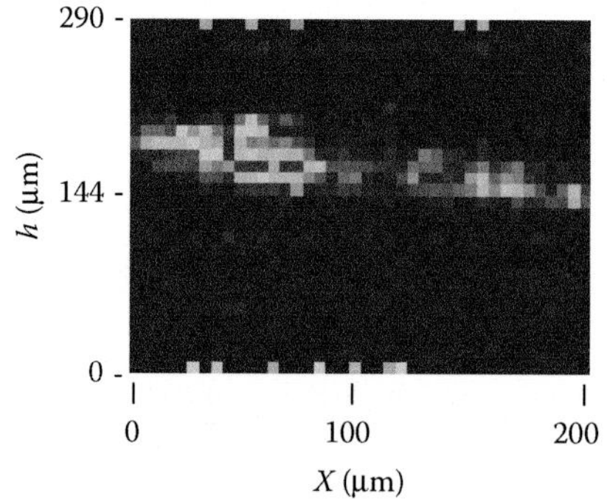

FIGURE 4.3.11
(See color insert.) Emission intensity map from poly(2,5-dioctyloxy-p-phenylenevinylene) or DOO-PPV polymer film. (Adapted from R. C. Polson and Z. V. Vardeny, Opt. Lett. **35** (16), 2801, 2010.)

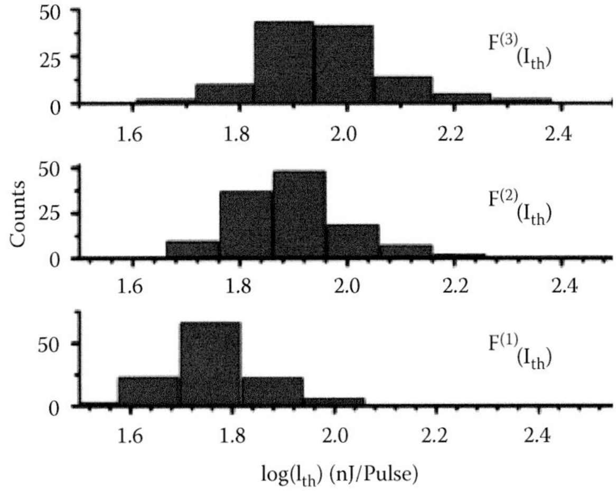

FIGURE 4.3.12
Histogram of excitation intensities needed for polymer film to achieve lasing. (Adapted from A. Tulek, R. C. Polson, and Z. V. Vardeny, Nat. Phys. **6**, 303, 2010.)

4.3.5 Random Laser Statistical Analysis

The scattering length can be altered by the addition of nonluminescent scatterers. TiO_2 balls uniformly scatter visible light. By including such nanoballs into a film, the scattering increases which means that the scattering length decreases. The same threshold measurements as described above can be performed with similar films having different percentages of scatterers. Figure 4.3.13 shows threshold values for scattering length histograms.

The pure polymer films have the lowest threshold for lasing. The average intensity necessary to lase increases with an increasing density of TiO_2 ball scatterers in the film. For a

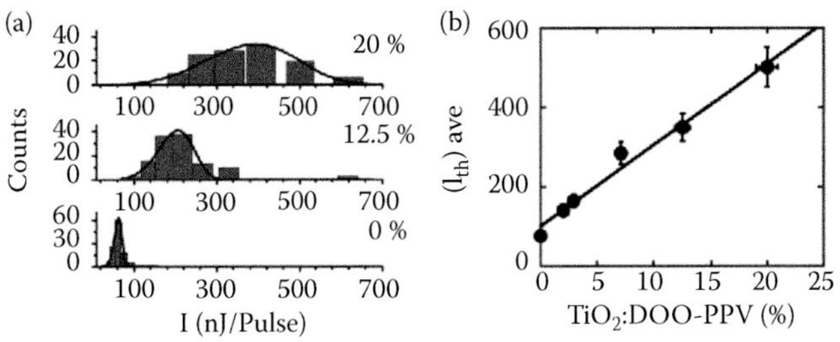

FIGURE 4.3.13
(a) Histogram for lasing threshold for poly(2,5-dioctyloxy-p-phenylenevinylene) or DOO-PPV polymer film with titanium dioxide scatterers. (b) Average threshold for different concentrations of scatterers. (Adapted from A. Tulek, R. C. Polson, and Z. V. Vardeny, Nat. Phys. **6**, 303, 2010.)

given scatterer concentration, the threshold statistics of cavities in film have been measured.[34] There is indeed a minimum excitation intensity needed for lasing, and lasing thresholds are peaked near this measured value.

Most of the random lasing spectra observed to date are excited far above threshold. Therefore the spectra contain many narrow emission lines that complicate the analysis. In contrast, very simple emission spectra may be obtained from polymer films excited near the threshold, where only few resonators may lase. Figure 4.3.14 shows the random laser emission obtained from a film excited just above the lasing threshold. The emission peaks are well separated and still very narrow. We conjecture that excitation of the polymer film just above the lasing threshold excites only the highest-Q resonator. These spectra can be therefore analyzed in exactly the same manner as the emission from fabricated circular microcavities discussed above. The Fourier transform shown in the inset to Figure 4.3.14 is more regular than the Fourier transforms of a film excited at high intensity as in Figure 4.3.6, or that averaged from many locations as in Figure 4.3.7. Therefore, the simpler random laser emission spectrum near the threshold can be actually fit with the same procedure as the microring spectra.

Finding circular structures in a disordered film is somewhat surprising. However, there are several theoretical works directly supporting lasing in circular structures in disordered gain media. One set of works shows explicitly that circular wave-guided structures are the most likely.[33,35] Other theoretical works show localized modes at the boundary of excitation in a mixture of gain and scatterers medium.[36] Coherent backscattering has dealt with scattering from loops,[37] where circular loops are composed of many scattering events.[38] One theoretical work has shown narrow lines from long loops, where each line is from a single long loop.[39]

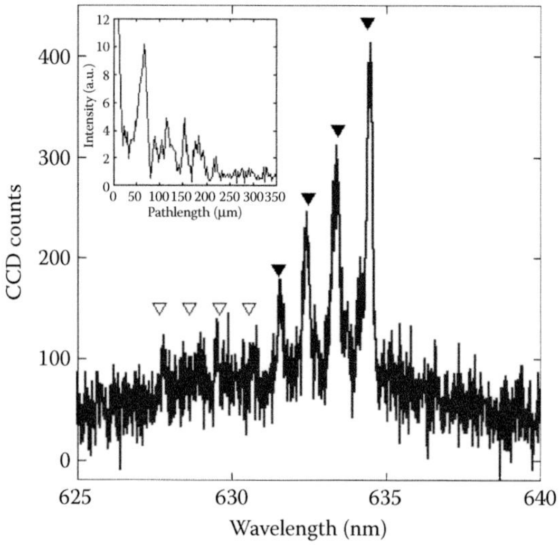

FIGURE 4.3.14
Emission of random lasing film excited just above threshold. Closed symbols represent Bessel functions; open symbols indicate where other predicted peaks would occur. Inset: Fourier transform of emission spectrum. (Adapted from R. C. Polson and Z. V. Vardeny, Synth. Metals, **162**, 276, 2012.)

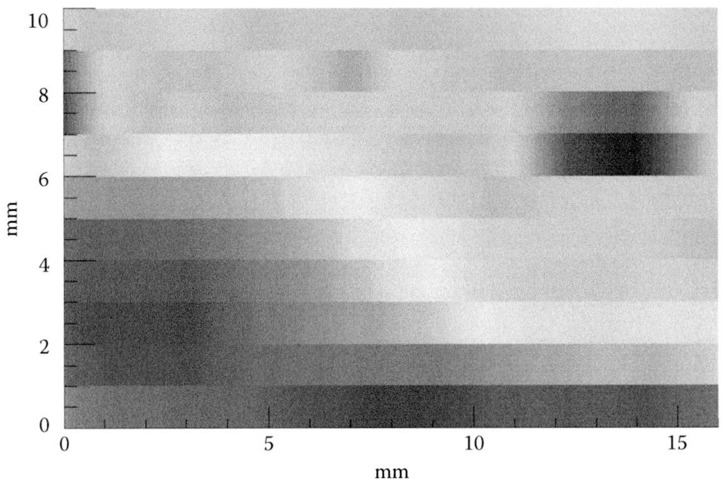

FIGURE 4.3.15
(**See color insert.**) Mapping of relative health of tissue based of averaged Fourier transform of lasing tissue. (From R. C. Polson and Z. V. Vardeny, J. Opt. **12**, 24010, 2010. With permission.)

4.3.6 Random Lasing Applications

Some practical examples of random lasing can be found in medical studies. One study used random lasing in bones.[13] The bone was soaked in a laser dye and random laser emission was studied. Another study looked at random laser emission from tissues.[40] Another medical application has been to distinguish healthy and cancerous tissues on the basis of random lasing emission.[12] When the random laser emission Fourier transforms of healthy tissues are averaged together, the resulting spectrum is similar to that shown in Figure 4.3.7. However, when cancerous tissues are averaged together, then the resulting average has no extra peaks. This difference can further be exploited to generate a computed map based on a local average of laser emission Fourier transforms. Figure 4.3.15 is a false color map generated by looking at the total emission intensity in a local average of the Fourier transform. A higher value means there is more regular structure, and is therefore nominally healthier.[41]

4.3.7 Summary

Random lasing challenges the typical conventions of what elements are required to produce laser emission. In some of the systems one can literally shake the required materials together and produce laser emission; this happens in dye and scatterer suspensions. In one sense laser emission is not terribly surprising; Arthur Schawlow, Nobel Laureate in 1981, once said: "Anything will lase if you hit it hard enough." In random lasing systems, the surprise is the unconventional manner in gain and scattering interaction to obtain coherent, narrow linewidth laser emission.

Acknowledgments

We are grateful to Drs. Sergey Frolov who initiated the organic random laser research at the University of Utah, Mikhail Raikh for suggesting the Fourier transform analysis of random laser emission spectra, Maxim Shkunov who studied organic random lasing in biological tissues, Abdullah Tulek for the random laser statistical studies, and Leonard Wojcik for synthesizing the polymers used in this work. We also acknowledge fruitful discussions with Profs. S. John, N. Lawandy, and B. Shapiro. This work has been supported over the years by grants from NSF Grant No. 08-03325, DOE Grant No. 04-ER46109, and the TCP program at the University of Utah.

References

1. T. H. Maiman, Nature **187**, 493 (1960).
2. N. Basov, B. Vul, and Y. Popov, Sov. Phys. JETP **10**, 416 (1959); A. Prokhorov, Zh. Eksp. Teor. Fiz. **34**, 1658 (1958); A. L. Schawlow and C. H. Townes, Phys. Rev. **112** (6), 1940 (1958).
3. P. P. Sorokin and J. R. Lankard, IBM J.Res.Dev. **10** (2), 162 (1966); F. P. Schafer, W. Schmidt, and J. Volze, Appl. Phys. Lett. **9** (8), 306 (1966).
4. V. Letokhov, Sov. Phys. JETP **26**, 835 (1968).
5. C. Gouedard, D. Husson, C. Sauteret, F. Auzel, and A. Migus, J. Opt. Soc. Am. B. **10** (12), 2358 (1993).
6. N. M. Lawandy, M. R, Balachandran, A. S. L. Gomes, and E. Sauvain, Nature **368**, 436 (1994); W. Sha, C. H. Liu, and R. Alfano, Opt. Lett. **19** (23), 1922 (1994).
7. Y. Sun, J. Ketterson, and G. Wong, Appl. Phys. Lett. **77**, 2322 (2000); S. F. Yu, C. Yuen, S. P. Lau, and H. W. Lee, *ibid*. **84**, 3244 (2004).
8. S. V. Frolov, M. Ozaki, W. Gellermann, Z. V. Vardeny, and K. Yoshino, Jpn. J. Appl. Phys. **35** (10B Part 2), L1371 (1996).
9. C. J. S. de Matos, L. de S. Menezes, A. M. Brito-Silva, M. A. Martinez Gámez, A. S. L. Gomes, and C. B. de Araújo, Phys. Rev. Lett. **99** (15), 153903 (2007).
10. G. Strangi, S. Ferjani, V. Barna, A. De Luca, C. Versace, N. Scaramuzza, and R. Bartolino, Opt. Express **14** (17), 7737 (2006).
11. V. Milner and A. Z. Genack, Phys. Rev. Lett. **94** (7), 073901 (2005).
12. R. C. Polson and Z. V. Vardeny, Appl. Phys. Lett. **85** (7), 1289 (2004).
13. Q. Song, S. Xiao, Z. Xu, J. Liu, X. Sun, V. Drachev, V. M. Shalaev, O. Akkus, and Y. L. Kim, Opt. Lett. **35** (9), 1425 (2010).
14. C. Chiang, C. Fincher Jr, Y. Park, A. Heeger, H. Shirakawa, E. Louis, S. Gau, and A. MacDiarmid, Phys. Rev. Lett. **39** (17), 1098 (1977).
15. M. Yan, L. Rothberg, B. R. Hsieh, and R. R. Alfano, Phys. Rev. B **49** (14), 9419 (1994).
16. C. W. Lee, K. S. Wong, J. D. Huang, S. V. Frolov, and Z. V. Vardeny, Chem. Phys. Lett. **314**, 564 (1999).
17. X. Long, M. Grell, A. Malinowski, D. Bradley, M. Inbasekaran, and E. Woo, Opt. Mater. **9** (1–4), 70 (1998); W. Holzer, A. Penzkofer, S.-H. Gong, A. Bleyer, and D. Bradley, Adv. Mater. **8**, 974 (1996); R. Friend, G. Denton, N. Tessler, and M. A. Stevens, *ibid*. **9**, 547 (1997); S. V. Frolov, A. Fujii, D. Chinn, M. Hirohata, R. Hidayat, M. Taraguchi, T. Masuda, and K. Yoshino, *ibid*. **10** (11), 869 (1998); H.-J. Brouwer, V. V. Krasnikov, A. Hilberer, J. Wildeman, and G. Hadziioannou, Appl. Phys. Lett. **66** (25), 3404 (1995); C. Zenz, W. Graupner, S. Tasch, G. Leising, K. Mullen, and U. Scherf, *ibid*. **71** (18), 2566 (1997); A. Schülzgen, C. Spiegelberg, M. M. Morrell, S. B. Mendes,

B. Kippelen, N. Peyghambarian, M. F. Nabor, E. A. Mash, and P. M. Allemand, *ibid.* **72**, 269 (1998); S. C. Jeoung, Y. H. Kim, D. Kim, J.-Y. Han, M. S. Jang, J.-I. Lee, H.-K. Shim, C. M. Kim, and C. S. Yoon, *ibid.* **74** (2), 212 (1999); T. Virgili, D. G. Lidzey, D. D. C. Bradley, G. Cerullo, S. Stagira, and S. D. Silvestri, *ibid.* **74** (19), 2767 (1999); G. Wegmann, H. Giessen, A. Greiner, and R. F. Mahrt, Phys. Rev. B **57** (8), R4218 (1998); M. Nisoli, S. Stagira, M. Zavelani-Rossi, S. De Silvestri, P. Mataloni, and C. Zenz, *ibid.* **59** (17), 11328 (1999); V. Bulovic, V. G. Kozlov, V. B. Khalfin, and S. R. Forrest, Science **279** (5350), 553 (1998); V. Doan, V. Tran, and B. J. Schwartz, Chem. Phys. Lett. **288**, 576 (1998).

18. D. Moses, Appl. Phys. Lett. **60** (26), 3215 (1992).
19. N. Tessler, G. Denton, and R. Friend, Nature **382** (6593), 695 (1996); F. Hide, M. A. Diaz-Garcia, B. J. Schwartz, M. R. Andersson, Q. Pei, and A. J. Heeger, Science **273** (5283), 1833 (1996).
20. J. Burroughes, D. Bradley, A. Brown, R. Marks, K. Mackay, R. Friend, P. Burns, and A. Holmes, Nature **347** (6293), 539 (1990).
21. S. Frolov, Z. Vardeny, and K. Yoshino, Phys. Rev. B **57** (15), 9141 (1998).
22. R. Loudon, *The Quantum Theory of Light* (Oxford University Press, New York, 1983).
23. H. Cao, Y. Ling, J. Y. Xu, and C. Q. Cao, Phys. Rev. Lett. **86** (20), 4525 (2001).
24. R. C. Polson, M. E. Raikh, and Z. V. Vardeny, IEEE J. Sel. Top. Quantum Electron. **9** (1), 120 (2003).
25. R. C. Polson, J. D. Huang, and Z. V. Vardeny, Synth. Met. **119**, 7 (2001).
26. D. Hofstetter and R. L. Thornton, Appl. Phys. Lett. **72** (4), 404 (1998).
27. S. V. Frolov, M. Shkunov, Z. V. Vardeny, and K. Yoshino, Phys. Rev. B **56** (8), R4363 (1997).
28. L. Rayleigh, Philos. Mag. **20** (120), 1001 (1910).
29. R. C. Polson, Z. V. Vardeny, and D. A. Chinn, Appl. Phys. Lett. **81** (9), 1561 (2002).
30. R. C. Polson, G. Levina, and Z. V. Vardeny, Appl. Phys. Lett. **76** (26), 3858 (2000).
31. R. C. Polson, M. E. Raikh, and Z. V. Vardeny, C. R. Phys. **3** (4), 509 (2002).
32. R. C. Polson and Z. V. Vardeny, Opt. Lett. **35** (16), 2801 (2010).
33. V. M. Apalkov and M. E. Raikh, Phys. Rev. B **71** (5), 054203 (2005).
34. A. Tulek, R. C. Polson, and Z. V. Vardeny, Nat. Phys. **6**, 303 (2010).
35. V. M. Apalkov, M. E. Raikh, and B. shapiro, J. Opt. Soc. Am. B **21**, 132 (2004); V. M. Apalkov, M. E. Raikh, and B. Shapiro, Phys. Rev. Lett. **89** (1), 016802 (2002).
36. H. E. Tureci, L. Ge, S. Rotter, and A. D. Stone, Science **320** (5876), 643 (2008).
37. M. P. van Albada, B. A. van Tiggelen, A. Lagendijk, and A. Tip, Phys. Rev. Lett. **66** (24), 3132 (1991).
38. R. Frank, A. Lubatsch, and J. Kroha, Phys. Rev. B **73** (24), 245107 (2006).
39. S. Mujumdar, M. Ricci, R. Torre, and D. S. Wiersma, Phys. Rev. Lett. **93** (5), 053903 (2004).
40. M. Siddique, L. Yang, Q. Z. Qang, and R. R. Alfano, Opt. Commun. **117**, 475 (1995).
41. R. C. Polson and Z. V. Vardeny, J. Opt. **12**, 24010 (2010).
42. R. C. Polson and Z. V. Vardeny, Synth. Metals, **162**, 276 (2012).

4.4

Self-Optimization of Optical Confinement and Lasing Action in Disordered Photonic Crystals

Alexey Yamilov
Missouri University of Science and Technology

Hui Cao
Yale University

CONTENTS

4.4.1 Introduction

Light scattering is usually regarded detrimental to optical confinement in conventional lasers. In contrast, in random lasers, the confinement is *caused* by disorder-induced scattering. In strongly scattering media, the lasing is defined by the high-quality modes of the passive system. Thus, by incorporating and optimizing a degree of order, one can dramatically reduce the threshold of a random laser to the values comparable to those of photonic crystal (PhC) lasers. Unlike the latter, where the optical cavity has to be carefully designed and impeccably fabricated, in disordered systems the modes originate from the structure imperfections unintentionally introduced during the fabrication process. Optical gain selectively amplifies the high-quality modes of the passive system. Consequently, in PhC slab geometry, for example, the in-plane and out-of-plane leakage rates of the lasing modes can become automatically balanced in the presence of disorder. Such self-optimization of optical confinement makes disordered PhC structures a competitive platform for large-scale low-cost production of microlasers with fabrication requirements much less stringent than those of PhC lasers with designed cavities.

4.4.2 Lasing in Random Media

4.4.2.1 Light Propagation in a Random Medium with Optical Gain

Incorporating optical gain in a random medium adds a new dimension to studies of light transport. In 1968, Letokhov[1] considered the effect of light amplification on the photons propagating through a scattering medium. It was predicted that under some conditions the number of photons will increase exponentially similar to neutron reaction in an atomic bomb. One can understand this incoherent process, called random lasing, within the framework of one of the following two pictures.

Semiclassically, repeated scattering of a spontaneously emitted photon increases its path-length inside the medium. When the system dimensions are increased beyond a certain critical size, every photon, on an average, generates another photon before escaping radiatively. This random walk process can be described by models based on diffusion equations.[1,2]

A modal picture provides another intuitive description of the process of random lasing.[3,4] Spatially and spectrally overlapping leaky modes of the random medium compete for gain. Compensating the radiative losses for a large number of such modes by gain results in a narrowing of the gain spectrum. However, unlike conventional lasers the limiting linewidth is still broad, commonly, in a range of several nanometers, and the emission is not fully coherent.

4.4.2.2 Random Lasing in Strongly Scattering Media

For a little over a decade, there have been many studies on random lasers with *coherent feedback*.[5,6] In a strongly scattering active medium, light may return to a coherence volume it has visited before, and self-interference provides resonant feedback for lasing. With sufficient gain, lasing oscillation might occur at discrete frequencies that are determined by the interference of scattered light.

Similar to the incoherent random lasing introduced above, both semiclassical and modal descriptions can provide valuable insights. Indeed, the photons with semiclassical loop trajectories would have to experience a self-interference effect.[7] The resulting process of wavelength selectivity resembles that in a distributed feedback resonator.

Alternatively, a finite open system of scattering particles can be characterized by a set of quasistationary (leaky) optical modes. When optical gain is introduced to such a system and it is sufficient to compensate the loss in at least one mode, lasing occurs. Thus, the mode with the smallest loss or the highest quality tends to lase first, and its quality factor Q determines the lasing threshold.

Both semiclassical and the highest-Q modal pictures oversimplify the real problem—coherent random lasing and, in particular, its threshold, also depends on many factors.

A realistic estimate should involve a detailed account of the gain material properties and its spatial distribution, the variation of the local density of states in the system and its effect on light-matter interaction, the pumping scheme, the reabsorption of laser light, and so on. Therefore, finding the threshold of random laser theoretically is a complicated problem. Nevertheless, in the important case of a uniform gain and strong scattering, Q can become the determining factor. Studies of the high-Q modes supported by a random medium advanced our understanding of coherent random lasing.[8–14]

4.4.2.3 Absorption-Induced Confinement

Coherent random lasing has been realized also in weakly scattering random media.[15] Tight focusing of pump light was necessary to observe discrete lasing peaks, namely, the pump beam had to be focused to a region of size much smaller than the entire sample. Imaging of laser light on the sample surface revealed that the lasing modes were not extended over the entire random medium but instead were located inside the pumped region with an exponential tail outside it.[16] As the quasimodes of a random system far from the onset of localization are usually extended states, the lasing modes were initially regarded as some types of anomalously localized states, either almost-localized states[17] or prelocalized states.[16,18] Although the anomalously localized states should be rare in the diffusive samples, the experiments always showed lasing modes that are spatially confined in the pumped region independent of where on the sample the pump beam is focused. Moreover, the lasing threshold did not fluctuate much as the pump spot was moved across the random medium.

In the study by Yamilov et al.,[19] it was shown that the above contradiction to the theory of anomalously localized states originates from the assumption that lasing occurs in the quasimodes of the passive random medium. This assumption is not valid when absorption at the emission wavelength is significant outside the pumped volume. The reabsorption of emitted light suppresses the feedback from the unpumped part of the sample and effectively reduces the system size. The lasing modes are dramatically different from the quasimodes of the whole system without gain or absorption. Even if all the quasimodes of a passive diffusive system are extended across the entire sample, the experiments find that the lasing modes are still confined in the pumped volume with only an exponential tail outside it.

The reduction of the effective volume of the system V_{eff} as a result of absorption leads to a decrease of an important parameter that characterizes wave transport—Thouless number $\delta \equiv \delta\nu/\Delta\nu$, where $\delta\nu$ and $\Delta\nu$ are the average mode linewidth and spacing, respectively.[20] In a three-dimensional (3D) diffusive system, $\delta\nu \propto V_{\text{eff}}^{-2/3}$ and $\Delta\nu \propto V_{\text{eff}}^{-1}$, therefore, $\delta \propto V_{\text{eff}}^{1/3}$. The smaller the value of δ, the larger the fluctuation of the decay rates γ of the quasimodes.[18,21] The variance of the decay rates[18] $\sigma_\gamma^2 = \langle\gamma\rangle^2/\delta$, where the average decay rate $\langle\gamma\rangle \sim D/V_{\text{eff}}^{2/3}$. Broadening of the decay rate distribution along with the decrease of the total number of quasimodes within V_{eff} should reduce the number of lasing modes (besides nonlinear mode competition via spatial hole burning, discussed in the study by Tureci et al.[4]), and, therefore, should be responsible for the observation of discrete lasing peaks in the tight focusing experiments. Despite its reduced value, the effective Thouless number is still much larger than unity owing to weak scattering. As a result, the lasing modes are the extended states within the effective volume. Because $\sigma_\gamma/\langle\gamma\rangle \ll 1$, the minimum decay rate is still close to $\langle\gamma\rangle$, leading to high threshold for lasing and its relatively small fluctuations.[9,10,22]

4.4.2.4 From Random Medium to PhCs with Disorder

Random lasing has generated great excitement in the physics community.[6,23] One of the most formidable problems on the way to practical applications is that lasing thresholds are too high because of incomplete confinement of light. To improve the confinement, one approach that has been used with some success is to maximize the scattering strength by using Mie resonances.[24–26]

Another approach can be traced back to the pioneering work by John[27] who proposed to manipulate the effective momentum of light by introducing periodicity into the system. Ioffe-Regel criterion[28] for light localization in the presence of periodicity is replaced by $k_c\ell \sim 1$, where k_c is crystal momentum,[29] which is much smaller than the optical wavevector k near the band edge. This criterion compares the scattering mean free path ℓ to the effective wavelength $\propto k_c^{-1}$; it originates from the semiclassical description of light propagation. In the language of modes, the disorder creates high-Q defects states[10,12,13,30] in the spectral tails that extend into the residual photonic band gap (PBG).[31]

4.4.3 Confinement in Disordered PhCs

4.4.3.1 Disorder in Passive and Active PhC Structures

In passive PhC devices, uncontrollable disorder introduced during the fabrication process has a detrimental effect[32–35] as it contributes to optical losses and limits light propagation length even in the highly precise PhC slab waveguides.[36–39] However, it is not immediately clear how the disorder would affect the performance of an *active* device such as, for example, a PhC slab laser.

Below we show that structural disorder in a PhC laser may not be as detrimental as it is in a passive PhC waveguide. In Section 4.4.3.2, we demonstrate theoretically that weakly disordered PhC structures can support high-Q optical cavities.[10,12–14] Such cavities can be observed experimentally[40] and they can also facilitate lasing action.[14,41–43]

A disordered PhC can be considered as a transition from a perfectly ordered structure to a completely random medium. In Section 4.4.3.2, we systematically investigate this disorder-induced transition in two-dimensional (2D) PhCs. We find that there exists an optimal degree of disorder that leads to the maximum (on average) confinement.[10] In fact, we estimate that Q factor and mode volume of the quasimodes in such spontaneously formed microcavities is comparable to those in carefully designed defect modes in the perfectly ordered PhC. Of course, when making the generalizations one has to take into a consideration the particular type of disorder at hand, for example, correlated or uncorrelated disorder, particle size, and/or position disorder.[44] However, we believe that the main conclusion about the existence of the optimal degree of disorder remains generally valid. This can be understood with a simple argument. The localization length is shortest at the frequency in the middle of the PBG.[27] A disorder affects the PBG by making it spectrally broader and shallower—that is, increasing the localization length at the center of the gap.[45] However, because we are interested in creating an active device—a microlaser—we also have to take into consideration the probability of creating an optical cavity in any given configuration of disorder. When disorder is weak, the local fluctuations of the photonic band edge create so-called tail states[31] that are the photonic equivalent of the Lifshitz states in the condensed matter systems.[46] The density of these states falls off rapidly, exponentially, for the frequencies deep inside the PBG where the localization length is the shortest. This means that, at weak disorder, the tail states cannot take advantage of the maximum confinement. Therefore, the optimum degree of disorder appears as a balance between density of the tail states and their degree of confinement that show opposite trends with an increase of the disorder strength.[10]

The argument outlined above does not account for a possibility of purposefully creating a defect state in the middle of the PBG via certain PhC structure modifications. In Section 4.4.4.3, using a more realistic example of PhC slabs, we show that in the presence of disorder such a defect state may not be superior when other factors such as mode volume and vertical leakage are taken into account. Indeed, we show that because spatially uniform optical gain selectively amplifies the high-Q modes of the passive system, in PhC slabs the in-plane and out-of-plane leakage rates may be automatically balanced in the presence of disorder. The spontaneous optimization of in-plane and out-of-plane confinement of light in a PhC slab can lead to a reduction of the lasing threshold. Our experiments support these conclusions. Furthermore, concentration of the modes in the vicinity of the photonic band edge can enable an efficient extraction of gain. It also offers a possibility to fine-tune the lasing wavelength by, for example, changing the lattice constant (c.f. Section 4.4.4.3).

In the above, we approached the problem of minimization of the threshold of a random laser from the perspective of the *optimal degree of disorder* in a PhC. In fact, in our prior work we followed the opposite approach by searching for the *optimal degree of order* in random medium. This can be accomplished by studying the milestones in the transition from a random to periodic medium. First, we investigated experimentally and theoretically the effect of the zinc oxide (ZnO) scatterer size dispersion.[24] Also, we noted that PBG effects may become already noticeable in the randomly packed monodisperse spheres as a result of a short-range order—formation of the small ordered clusters.[47] This supposition was further confirmed in ZnO nano-structured dielectric films,[48] which exhibited a varying degree of short-range order.[25] ZnO structures with 2D long-range order[14,42,49] and 3D long-range order[43,50] were the third step in our systematic study of lasing in disordered media with variable degree of order. Small feature size, required in order to overlap the PBG with the gain spectrum of ZnO at near-ultraviolet frequency, resulted in relatively large degree of the residual fabrication disorder. Systems with 2D periodicity—PhC slab—gave us particularly many degrees of freedom to perform the combined experimental and theoretical studies of the effect of disorder. In the following, we will concentrate on the last step in our three-step study of the transition from random to ordered media—weakly disordered PhC slabs.

4.4.3.2 Optimal Degree of Disorder

In this section, we use the finite-difference time-domain method[51] to find the highest-quality modes in open passive 2D random systems with various degrees of ordering. We consider a 2D $L \times L$ (up to $9\lambda \times 9\lambda$) PhC made of N ($\propto L^2$) cylinders with diameter $d = 98$ nm and refractive index $n_0 = 2.2$. The cylinders were arranged into a hexagonal lattice with nearest-neighbor distance $a = 140$ nm. In the absence of disorder, the infinite system with these parameters has full band gap in the range [361 nm, 426 nm] for transverse magnetic modes (electric field along the cylinder axis). The disorder in the system is introduced in two ways: by uniformly randomizing the refractive index n of different cylinders in the range $[n_0 - w_n(n_0 - 1), n_0 + w_n(n_0 - 1)]$ and diameter $[d(1 - w_d), d(1 + w_d)]$. Special care should be taken to avoid the uncontrollable disorder due to discretization of the grid. Disorder in the system is characterized with parameter $\delta\varepsilon = \left\langle \int (\varepsilon(\mathbf{r}) - \varepsilon_0(\mathbf{r}))^2 \, d\mathbf{r} \right\rangle^{1/2} / (\int \varepsilon_0^2(\mathbf{r}) \, d\mathbf{r})^{1/2}$, where $\varepsilon_0(\mathbf{r})$ and $\varepsilon(\mathbf{r})$ are the dielectric constant distributions in ordered and disordered samples, respectively; $\langle\ldots\rangle$ stands for the average over different disorder configurations. Here we study the systems with 11 different disorder strengths: 1 to 10 had w_n from 0.1 to 1.0 with the increment 0.1 and $w_d = 0$, the 11th has $w_n = 1.0$, and $w_d = 0.43$. This leads to variations of dielectric constant from weak $\delta\varepsilon = 0.08$ to strong $\delta\varepsilon = 0.95$ disorder. Later we will discuss the effect of this particular choice of the types of disorder. To mimic an open system, a buffer layer of air (150 nm thick) is kept around the sample, followed by uniaxial perfectly matched absorbing layers.[51] To excite the system, we initially launched a short ~10 fs pulse at every grid point. The frequency ω_e of the pulse is chosen to lie at the center of the band gap (391 nm) of the ordered structure. In the frequency domain, the full-width at half-maximum of the excitation pulse is of the order of the band-gap width. Thus, the pulse excites all the modes within the stop band and near the band edges.

Right after the initial pulse, the competition between the modes[52] with different lifetimes leads to the complicated evolution of total electric energy $\mathcal{E}(t) = 1/2\int\varepsilon(\mathbf{r})\mathbf{E}^2(\mathbf{r})d\mathbf{r}$. However, after a sufficient time only the mode with the longest lifetime (highest quality factor) survives. $\mathcal{E}(t)$ followed a monoexponential decay $Re[\exp 2i\omega_m(1 + i/2Q_m)t]$, from which we extract the frequency ω_m and quality factor Q_m of the longest-lived mode in this particular realization of disorder. At the same time, the spatial pattern, $\mathbf{E}(\mathbf{r})$, stabilizes and the mode profile can be seen. Generally, the time needed to reach the monoexponential decay regime varies from about 0.5 ps for the smallest system to 10 ps for the largest. Finally, the Q_m is averaged over 1000 ($N = 75$), or 100 ($N = 137, 188, 261, 368, 449, 608$) disorder realizations.

Figure 4.4.1 shows the dependence of $\langle Q_m\rangle$ normalized by N as a function of the disorderness $\delta\varepsilon$; different curves correspond to different system sizes. This particular normalization makes it easy to see the deviation from diffusion-predicted[1] dependence $\langle Q_m\rangle \propto L^2 \propto N$. One can see that significantly different scalings at different $\delta\varepsilon$ lead to a maximum of $\langle Q_m\rangle$ at the finite disorder strength.[10]

The understanding of this behavior comes from observing the frequencies ω_m of the highest-quality modes in Figure 4.4.2. For small $\delta\varepsilon$ the frequencies are concentrated at lower (long wavelength) band edge, and they (as well as Q_m) are independent of the frequency ω_e of the excitation pulse. The reason for this is the way the disorder was introduced into the system. The long-wavelength modes are mostly concentrated in the dielectric cylinders, which are disordered by the refractive index fluctuations. At $w_n = 0.1$,

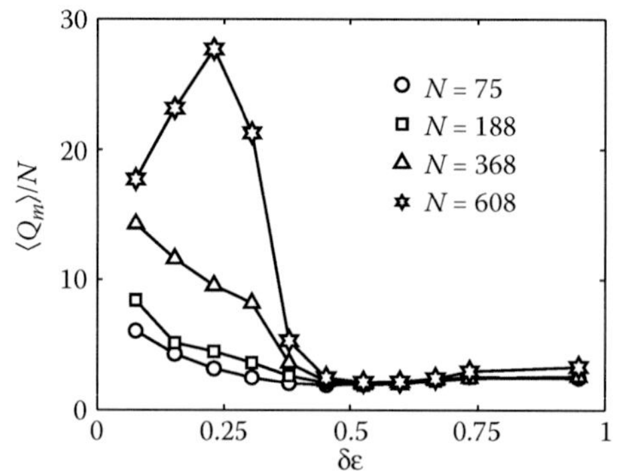

FIGURE 4.4.1
$\langle Q_m\rangle$ normalized by total number of scatterers N ($\propto L^2$) as a function of disorderness $\delta\varepsilon$. The figure shows that with an increase of the system size a maximum of $\langle Q_m\rangle$ develops at a finite value of disorder strength $\delta\varepsilon$. (Reprinted from A. Yamilov and H. Cao Phys. Rev. A **69**, 031803, 2004.)

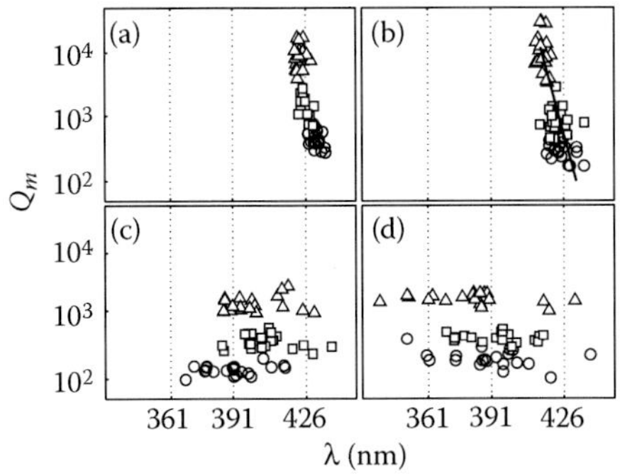

FIGURE 4.4.2
Q_m versus the corresponding mode wavelength for 20 realizations of disorder. Circles, squares, and triangles correspond to N equal to 75, 188, and 608, respectively. Four graphs correspond to different disorder parameters: (a) $w_n = 0.1$ ($\delta\varepsilon \simeq 0.08$), (b) $w_n = 0.2$ ($\delta\varepsilon \simeq 0.15$), (c) $w_n = 0.6$ ($\delta\varepsilon \simeq 0.45$), and (d) $w_n = 1.0$ and $w_d = 0.43$ ($\delta\varepsilon \simeq 0.95$). Concentration of the modes near one of the band edges of the photonic band gap (PBG) (361–426 nm) demonstrates the presence of the residual gap at weak disorder (a, b). Migration of the modes toward the PBG center in (b) with the increased system size causes the decrease of the localization length ξ. (Reprinted from A. Yamilov and H. Cao Phys. Rev. A **69**, 031803, 2004.)

ω_m fell in the immediate vicinity of the band edge (Figure 4.4.2a). Lasing from the band-edge modes is well studied in the case of ordered structures,[63–65] with $Q_m \propto L^3$.[11] The latter indeed gives a good fit to our results.

At the increased disorder, $w_n = 0.2$–0.3, the dependence of $\langle Q_m \rangle$ on the system size L became exponential, as expected for localized modes.[66] In the unit of wavelength the localization length ξ, obtained by fitting, decreased from 1.44λ to 1.27λ as w_n increased from 0.2 to 0.3. Figure 4.4.2b provides an insight into the physics behind the varying ξ. The quality factor can be estimated as $\langle Q_m \rangle \propto \exp[L/2\xi(\omega_m)]$, where $\xi(\omega_m)$ is the "typical" value of the localization length at the frequency ω_m. From Figure 4.4.2b one can see that even for the same disorder strength, the increase in system size leads to the advance of ω_m toward the band-gap center, where ξ is the smallest. This peculiar behavior should lead to superexponential dependence of $\langle Q_m \rangle$ on L even for fixed disorder strength. The frequency migration with the increase of the system size can be explained by the fact that in the small system it is unlikely to find the modes deep into the band gap owing to the low density of states there. An Urbach-like behavior can be expected.[31,67] This is also qualitatively supported by Figure 4.4.2b where the exponential dependence is apparent. Assuming Urbach-like dependence of the density of states, the advancement of ω_m can be estimated from the condition that the total number of defect states (proportional to the number of cylinders N) times the probability of having a state located $\Delta\omega(N)$ away from the band edge, $\exp[-\alpha(\delta\varepsilon)\cdot\Delta\omega(N)]$, is equal to one. Here, $\alpha(\delta\varepsilon)$ is the exponential slope of the density of states that should decrease with the increase of the disorder $\delta\varepsilon$. For small disorder $\alpha^{-1}(\delta\varepsilon) \ll \Delta E_{PBG}$, where ΔE_{PBG} is the width of the PBG. Therefore, for weak disorder (or small system size) the band-edge-type modes have the highest Q. The crossover to the superexponential dependence of $\langle Q_m \rangle$ occurs when the Q of the localized states with the shortest localization length $\xi[\Delta\omega(N)]$ available for this size N exceeds that of the band-edge-type mode: $\exp[N^{1/2}a/\xi(\Delta\omega(N))] \sim N^{3/2}$. Stronger size dependence in the superexponential regime means that the $N^{3/2}$ band-edge-type dependence observed at smaller disorder would eventually switch to the superexponential dependence as N increases. However, the latter can be expected to saturate at larger size *or* disorder when ω_m reaches the band-gap center: $N\cdot\exp[-\alpha(\delta\varepsilon)\Delta E_{PBG}/2] \sim 1$, where the localization length is the smallest. Therefore, we expect the limiting scaling of $\langle Q_m \rangle$ to be exponential: $\langle Q_m \rangle \propto \exp[N^{1/2}a/2\xi(\Delta E_{\mathrm{PBG}}/2)]$.

The sharp drop in $\langle Q_m \rangle$ at $\delta\varepsilon \simeq 0.34$ in Figure 4.4.1 is attributed to the removal of the residual band gap. This can be seen from the loss of the hexagonal symmetry of the observed mode profiles as well as the sensitivity of the modes to the excitation pulse position ω_e. In this regime, ω_m is not associated with the PBG, which does not exist anymore. However, to make a direct comparison with the ordered case, we kept the excitation pulse the same as before. The exact mechanism of the band-gap removal can depend on the disorder.[10,31,33,34,68] In our particular case, we found a simple explanation for the behavior of $\langle Q_m \rangle$ in the way the disorder was introduced. Indeed, the fluctuating index of refraction leads to the fluctuation of the frequency of the Mie resonances of the particles. For box distribution of n, there exists a value of $w_n = 0.6$ when the Mie resonance of some defect cylinders falls into the gap. This value matches the value of disorder parameter $\delta\varepsilon$, where the sharp decrease of $\langle Q_m \rangle$ is observed in Figure 4.4.1. Moreover, Figure 4.4.2c shows that at this crossover disorder, the modes avoid the region of strong single-particle scattering. This is the consequence of the sharp boundary in the distribution of n. It also indicates the presence of the residual band gap, where the ω_m are concentrated.

At $w_n \geq 0.7$ the PBG ceases to exist, and $\langle Q_m \rangle$ acquires the diffusion[1] scaling dependence: $\langle Q_m \rangle \propto L^2 \propto N$ (c.f. Figure 4.4.1). Deviations from this dependence can be seen in the same figure at the largest sizes studied, where $L > \xi_{\mathrm{Anderson}} = 2.54\lambda$ is inferred and the states become localized again as a result of Anderson localization.[66] The exponential dependence of $\langle Q_m \rangle$ on L becomes especially pronounced at the largest disorder studied, where the transition from L^2 to exponential dependence comes at small system sizes. We want to point out that even at such strong disorder, the obtained modes had a collective nature, rather than the single particle's high-order resonances, which are concentrated at higher frequencies. Comparing the localization length of these states to that of band-gap nature we see a difference of a factor of two, which makes the latter preferable (c.f. Figure 4.4.1).

To summarize the results of this section, by varying the strength of disorder, we identified five different scaling regimes of the ensemble-averaged $\langle Q_m \rangle$ with the system size: (a) photonic band edge, L^3, (b) transitional superexponential, (c) band-gap-related exponential, (d) diffusive, L^2, and (e) disorder-induced exponential, due to Anderson localization, regimes. The difference in scaling behavior allows one to draw the following conclusions that provide an intuitive phenomenological picture of the disorder-induced transition: (i) For sufficiently wide band gaps, $\langle Q_m \rangle$ reaches a maximum at some finite strength of disorder; (ii) at this "optimal" degree of disorder, $\langle Q_m \rangle$ is determined by the localization length similar to that of single defect in the ordered structure, leading to a similar quality factor; (iii) with an increase of the system size the optimal disorder strength decreases; (iv) near this optimal disorderness, $\langle Q_m \rangle$ should scale superexponentially with the sample size, owing to the frequency migration of the highest-quality modes toward the band-gap center and the associated decrease of their localization lengths.

4.4.4 Lasing in a Disordered PhC Slab

4.4.4.1 PhC Slab Laser: Design Considerations

A PhC slab utilizes index guiding to confine light to the plane of the slab.[69–71] In-plane confinement is realized either via a defect state located inside a PBG[72–77] or a band-edge state with vanishing group velocity.[64,65,78–81] Over the past decade, tremendous progress has been made in design and fabrication of PhC slab lasers that operate at infrared or near-infrared frequencies.[65,72–77,80,82] To realize a near-ultraviolet PhC slab laser, the feature size has to be reduced roughly by a factor of four[42,49] compared to the infrared PhC. Fabrication of such fine structures inevitably generates random deviations from the perfectly ordered structures.

Design of a PhC slab laser involves determination of structural parameters to ensure spectral overlap between the PBG and the gain spectrum. In addition, lasing mode volume optimization is required owing to a planar geometry of the device. This is because light may escape from the PhC slab vertically through the top/bottom interfaces or laterally via the edge of the periodic pattern into air or unpatterned part of the slab. The vertical leakage rate is characterized by the out-of-plane energy loss per optical cycle $Q_\perp^{-1}$, and the lateral by $Q_\parallel^{-1}$. A defect state spatially localized in the vicinity of an intentionally introduced structural defect typically has large leakage in the vertical direction, that is, $Q_\perp^{-1} \gg Q_\parallel^{-1}$. For a band-edge state, the lateral leakage usually dominates over the vertical one, $Q_\parallel^{-1} \gg Q_\perp^{-1}$. The total loss is described by $Q_{tot}^{-1} = Q_\perp^{-1} + Q_\parallel^{-1}$. Low lasing threshold demands maximization of Q_{tot}, which is hindered by $Q_\perp$ for a defect state and $Q_\parallel$ for a band-edge state. Several designs aim at optimization of PhC slab lasers by balancing $Q_\perp$ and $Q_\parallel$ via "gentle localization,"[75] for example, phase-slip,[74,83] double-heterostructure.[77]

Using disorder-induced defect modes allows one to relax stringent requirements for the residual fabrication disorder. Furthermore, the last step in optimization—mode–volume optimization—can be accomplished spontaneously by the disorder. Therefore, disorder optimization appears as an additional design parameter that can maximize Q_{tot} without the need to repeat previous optimization steps. Below we illustrate the above approach with the combined experimental and theoretical study of ZnO-based disordered PhC slab near-ultraviolet lasers.

4.4.4.2 Structure Parameters and the PBG

A typical PhC slab made of III–V semiconductors is a free-standing membrane structure, whose substrate is selectively etched away so that there is air both above and below the photonic layer. Although such a structure can achieve wider in-plane PBG and better light confinement in the vertical direction, it poorly dissipates heat and it is usually mechanically fragile. Also it is difficult to make a large area fabrication and cannot be applied to on-chip fabrication easily. In our case, the ZnO photonic layer is fabricated on a lattice-matched sapphire substrate. Such a structure is much more robust and easier for large-scale on-chip applications. On the other hand, owing to the lower refractive index contrast between the photonic layer and the substrate, light confinement in the vertical direction is worse than that in the free-standing layer, and the in-plane PBG is also narrower. In this case, the defects introduced during fabrication are supposed to be even more detrimental to PhC slab lasers.

To simulate a PhC slab on a dielectric substrate, we modified[42] the super-cell technique[70] within the plane-wave expansion method for the photonic band-structure calculation.[84] A substrate with high refractive index is expected to significantly mix the polarization of eigenmodes of the PhC slab. However, our calculations demonstrate that although the substrate indeed induces asymmetry of wavefunctions, they still remain strongly transverse magnetic- or transverse electric-polarized for low-order bands.[42] A high filling fraction air-hole-in-ZnO-matrix geometry can possess a complete PBG for transverse electric bands;

meanwhile, ZnO film with c-axis along the growth direction emits mainly into transverse electric-polarized modes. This enabled us to build a near-ultraviolet PhC slab laser.[42] In the photonic band-structure calculation, we find the optimum set of parameters for maximum PBG: $R/a \simeq 0.24$ and $t/a \simeq 1.45$, where R is the hole radius, a is the lattice constant, and t is the slab thickness. These parameters are significantly different from the typical parameters for infrared PhC slabs. This is because of several factors: (i) the presence of the sapphire substrate breaks vertical symmetry of the PhC slab; (ii) the refractive index contrast of ZnO/sapphire is lower than that of InP/air commonly used in an infrared PhC slab; (iii) to preserve guiding in the photonic layer, the filling fraction and the thickness of the ZnO PhC slab need to be significantly increased.

Overlapping the PBG with the emission spectrum of ZnO requires precise control of the designed pattern with $a \simeq 123$ nm, $t \simeq 180$ nm, and $R \simeq 30$ nm. This has been achieved with the focused ion beam (FIB) etching technique.[42] The maximum relative width of PBG that can be achieved via optimization is 5%. This is significantly smaller than what is used in the free-standing membrane in air[65,72,76] or in the case of low refractive index substrate.[71,85] A narrow gap makes it difficult to align an intentionally introduced defect mode inside a PBG. However, there should always be some defect modes with frequencies inside PBGs formed as a result of the disorder introduced unintentionally during the fabrication process.

4.4.4.3 Spontaneous Optimization of In-Plane and Out-of-Plane Confinement

Using the parameters obtained in the previous section we realized the first near-ultraviolet PhC slab laser.[42] ZnO films were grown on sapphire substrates by plasma-enhanced metalorganic chemical vapor deposition.[48] Hexagonal arrays of cylindrical air voids were patterned in the ZnO films by the FIB etching technique. Post-thermal annealing was used to remove the FIB damage. Single-mode lasing at room temperature was realized with optical pumping. The scanning electron micrograph (SEM) of a ZnO PhC slab is shown in Figure 4.4.3. Despite the long-range periodicity, Figure 4.4.3 reveals the deviation of the fabricated pattern from the ideal honeycomb structure. Such "crescent" deviation[35] caused optical scattering on the length scale of a few lattice constants. It was expected to enhance radiative leakage of a PhC slab laser based on either defect state[12] or band-edge mode. Moreover, the propagation loss in a passive PhC slab caused by random[36,38,86] scattering was predicted to increase dramatically near a photonic band edge,[37] where the band-edge-type PhC slab laser operates. Despite these pessimistic expectations based on passive systems, we show that the performance of a PhC slab laser may be less susceptible to the detrimental effects of structural disorder. This is because optical gain predominantly amplifies the mode with the highest quality factor Q_{tot}. For the highest-Q_{tot} mode, the vertical and lateral leakage rates may be automatically balanced in the presence of disorder. This implies that an appropriate amount of structural disorder could lead to spontaneous optimization of in-plane and out-of-plane confinement of light in a PhC slab.

To investigate how the disorder affects the rates of vertical and lateral leakages of light from a PhC slab, we consider a system schematically depicted in Figure 4.4.4a. A dielectric slab of thickness 180 nm and refractive index $n = 2.35$ is sandwiched between air and substrate ($n_{sub} = 1.78$). Within the slab, N infinitely long grooves run parallel to the y-axis. The width of a groove is 22 nm; the lattice constant of the disorderless structure is 100 nm. We consider light propagating in the x-z plane, with the electric field along the y-axis. Such a system is 2D, which allows numerical simulation of large statistical ensembles of random systems. Despite the simplification, the system in Figure 4.4.4a retains the property essential for our study of the PhC slab laser—the possibility of vertical (along z-axis) *and* lateral

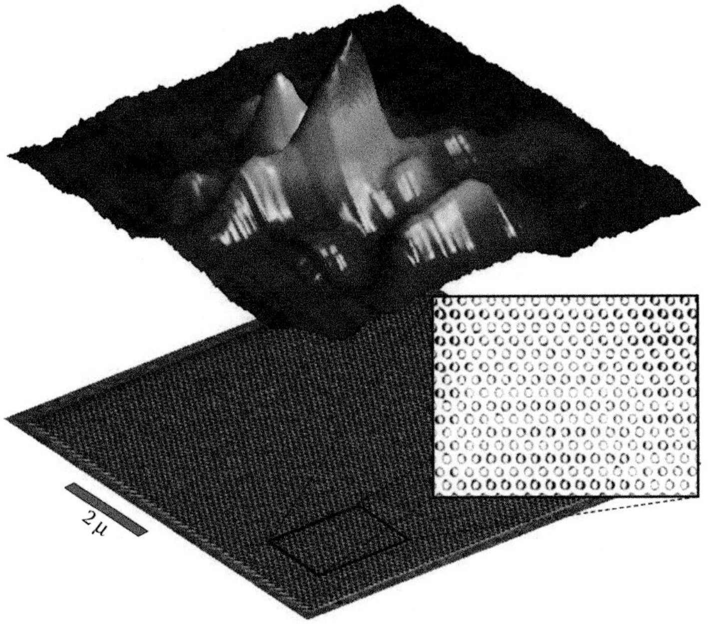

FIGURE 4.4.3
(**See color insert.**) Top-view scanning electron micrograph (SEM) of a ZnO photonic crystal (PhC) slab. Difference between the digitized SEM of a real sample and the perfect honeycomb lattice (blow-out box) reveals the structural disorder. Superimposed is the measured intensity distribution of the lasing mode in a ZnO PhC slab with $a = 115$ nm and $R = 0.25a$. The pattern measures 8 × 8 μm.

(along x-axis) radiative escape. Using the finite-difference time-domain method, we find the mode of the passive system that has the highest Q_{tot}.[10] A Gaussian pulse was launched at all spatial points in the slab and the energy was allowed to leak out radiatively. Simulation area is terminated by a uniaxially perfectly matched absorbing layer that absorbs all outgoing waves. The pulse excites all modes within a 30 nm wavelength range around 400 nm. After the initial multimode decay the field distribution is stabilized and the longest-lived mode can be seen. This is further confirmed by observing a monoexponential decay of the total energy[10,81,87] stored in the system that allows determination of Q_{tot}. By integrating Poynting vector over the corresponding interfaces,[82,87] we obtained the outgoing flux in the vertical and horizontal directions, and $Q_\perp$ and $Q_\parallel$. In our simulation, $Q_{tot}^{-1} = Q_\perp^{-1} + Q_\parallel^{-1}$ relation was satisfied numerically to within 2%.

Fourier transform of the spatial profile of the electric field at the interface between the slab and substrate gives the mode's distribution in $k_\parallel$ (in-plane component of the wavevector) space. In a perfectly periodic structure, the band-edge mode has the highest Q_{tot}. It is spatially extended in x (c.f. Figure 4.4.4b), and thus has a narrow distribution in $k_\parallel$ (c.f. thick dashed curve in Figure 4.4.5a). Next, we intentionally create a defect by increasing the spacing between two neighboring grooves at the center of the pattern to 150 nm. The highest-Q_{tot} mode is localized around this artificial defect with a localization length of 140 nm. Strong localization in x (c.f. Figure 4.4.4c) results in a broad distribution in $k_\parallel$ (c.f. thin dashed curve in Figure 4.4.5a), with the maximum lying closer to the edge of substrate light-cone (c.f. dash-dotted vertical line in Figure 4.4.5a). Its Q_{tot} is limited by $Q_\perp$, which is about three times smaller than the corresponding $Q_\parallel$ in a system of $N = 24$. In contrast, the

(a)

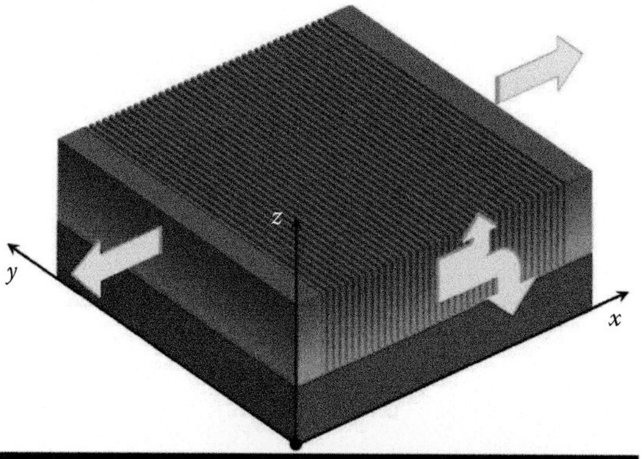

(b)

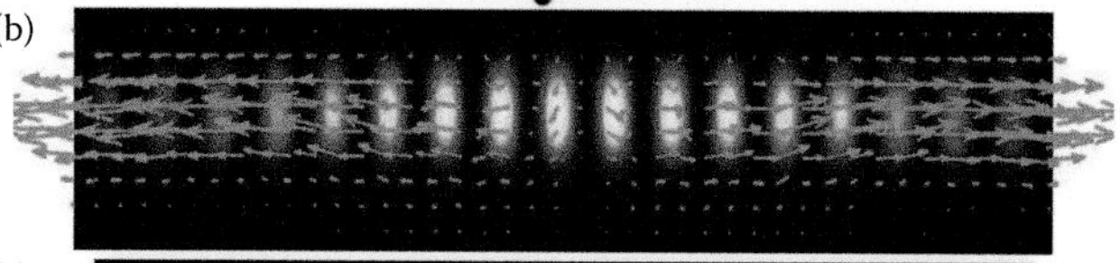

(c)

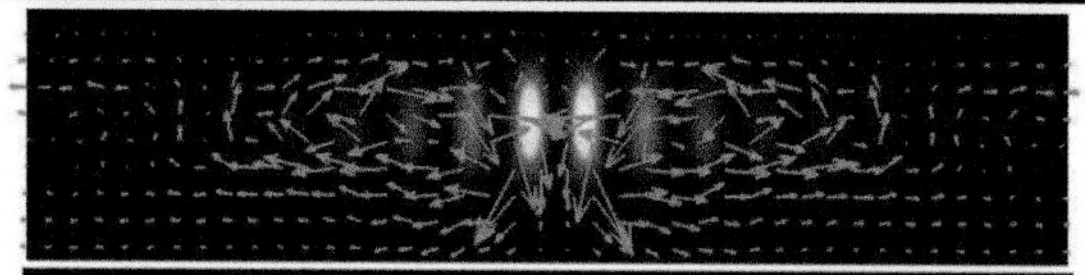

(d)

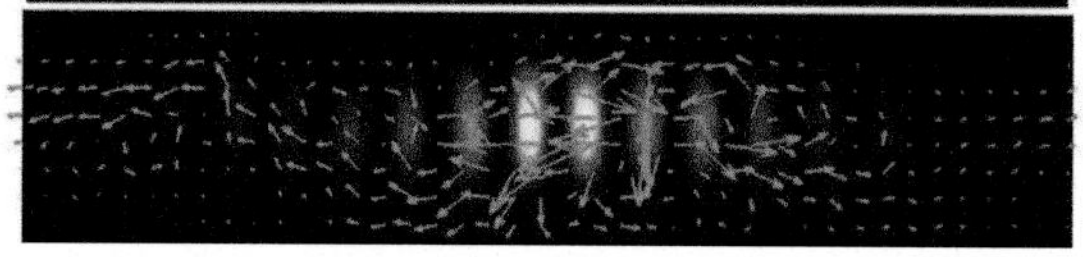

FIGURE 4.4.4
(**See color insert.**) (a) Simplified model of a photonic crystal slab used in numerical simulations. Infinitely long grooves run parallel to the y-axis. Disorder only affects x position of the groves. Although such a system can be modeled in two dimensions—in the xz-plane—it already incorporates the possibilities of vertical and horizontal leakage. Intensity of the mode with the highest Q factor in (a) an ordered system (band-edge mode), (b) an ordered structure with a localized defect, and (c) a disordered structure with a localized defect. Arrows represent the amplitude and direction of the local Poynting vector. The radiative losses are dominated by in-plane leakage in (b) and by out-of-plane leakage in (c), whereas both loss mechanisms are comparable in the disordered system (d).

band-edge mode is concentrated well beyond the light-cone in $k_{\|}$-space; thus its $Q_{\perp}$ is much higher. However, its spatial extension makes the lateral leakage larger; hence its Q_{tot} is limited by $Q_{\|}$.

To simulate the position disorder of air cylinders in real structure (c.f. Figure 4.4.3), random variation of groove position x_n is introduced. We choose Δx_n randomly from a uniform distribution with the standard deviation $\delta x = 5$, 10, and 15 nm, where δx characterizes the "strength" of disorder. As the disorder is introduced, the highest-Q_{tot} state differs from realization to realization, and the correspondent $Q_{\|}$, $Q_{\perp}$ as well as the frequency vary. We study statistical distributions of these parameters and their dependences on disorder strength δx and system size N.

In small systems ($N = 12$ and 24) with an artificial defect and weak disorder ($\delta x = 5$ nm), the highest-Q_{tot} modes always concentrate around the defect at the center of the pattern. These modes become more spatially extended than those without disorder (c.f. Figure 4.4.4d). Therefore, their $k_{\|}$ distribution is narrowed and $k_{\|}$ component within the light-cone is significantly reduced (c.f. Figure 4.4.5a). This reduction leads to a decrease in the

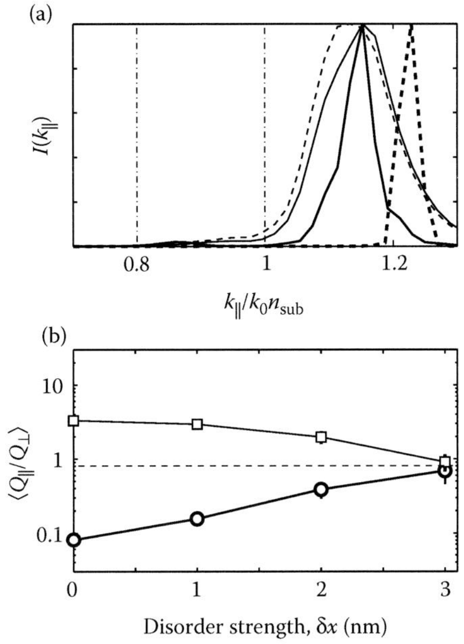

FIGURE 4.4.5
(a) The $k_{\|}$ distributions of the highest-Q_{tot} modes at one pixel beneath the slab/substrate interface. Thin/thick dashed curve represents the mode found in the disorderless system ($N = 24$) with/without an artificial defect. The corresponding solid curves are representative examples of the highest-Q_{tot} modes in these systems with position disorder ($\delta x = 10$ nm). The vertical line marks the substrate light-cone boundary. The inset is a schematic sketch of the simulated structure. (b) Squares/circles represent $\langle Q_{\|}/Q_{\perp}\rangle$, averaged over 300 random realizations of $N = 24$ system with/without the artificial defect, versus disorder strength δx. (Reprinted from A. Yamilov et al., Phys. Rev. Lett. **96**, 083905, 2006.)

vertical leakage and, thus, an increase in $Q_{\perp}$. Meanwhile, $Q_{\|}$ starts increasing as the mode gets less localized in real space. The ensemble-averaged $\langle Q_{\|}/Q_{\perp}\rangle$, shown in Figure 4.4.5b, decreases monotonously to unity with increase in disorder strength. Therefore, *disorder removes the imbalance between vertical and lateral leakages* of a single defect state, making $\langle Q_{\|}\rangle \sim \langle Q_{\perp}\rangle$. As a result, the ensemble-averaged quality factor $\langle Q_{tot}\rangle$ is slightly higher than that without disorder. In a larger system or with stronger disorder, the highest-Q_{tot} mode is no longer pinned at the artificial defect. Instead, it can explore the entire pattern to find the optimum configuration for the best vertical and lateral confinement. This leads to a further increase in $\langle Q_{tot}\rangle$.

With the introduction of disorder, the band-edge mode becomes less extended. As its "tail" moves away from the boundaries of the pattern, the lateral leakage decreases, and thus $Q_{\|}$ increases. Meanwhile, the distribution in $k_{\|}$ space is broadened and shifted closer to the light-cone edge (c.f. Figure 4.4.5a). The increase in vertical leakage results in a decrease in $Q_{\perp}$. The ensemble-averaged $\langle Q_{\|}/Q_{\perp}\rangle$, shown in Figure 4.4.5b, rises continuously to unity with increasing disorder strength. Again, disorder balances the vertical and lateral leakages of the band-edge mode, as it does to the defect state. However, for a band-edge mode the increase in $\langle Q_{\|}\rangle$ is not as large as the decrease in $\langle Q_{\perp}\rangle$; thus $\langle Q_{tot}\rangle$ is slighter lower than that without disorder. Nevertheless, as the pattern size N increases, the total leakage rate decreases monotonically: $\langle Q_{tot}^{-1}\rangle \propto N^{-\alpha}$ (c.f. Figure 4.4.6a). The exponent α decreases from 2.3 at $\delta x = 5$ nm to 1.9 at $\delta x = 15$ nm. Even with the largest disorder

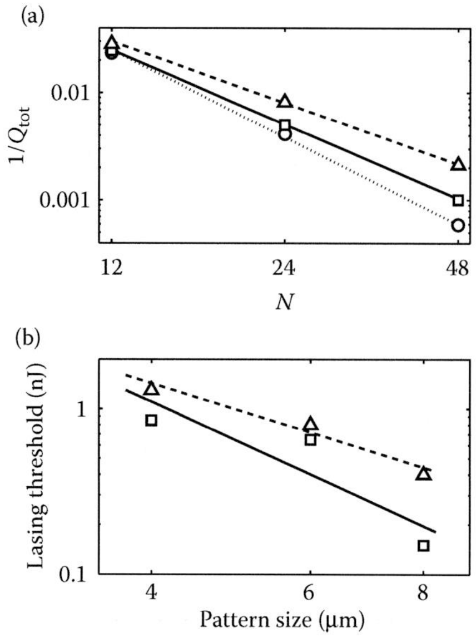

FIGURE 4.4.6
(a) $\langle 1/Q_{tot} \rangle$ for the highest-Q_{tot} modes found numerically in a photonic crystal slab depicted in the inset of Figure 4.4.3a. The average is taken over an ensemble of 300 random realizations. The squares and triangles represent the results for $\delta x = 5$ and 15 nm, respectively. Circles correspond to the disorderless system. Dotted, solid, and dashed lines are $N^{-\alpha}$ fits with $\alpha = 2.7$, 2.3, and 1.9, respectively. (b) Ensemble-averaged incident pump pulse energy at the lasing threshold, measured in the samples of $a = 115$ nm (squares) and $a = 118$ nm (triangles), versus the pattern size. The data are fitted with the power law dependence as in (a) with $\alpha = 2.5$ (solid line) and $\alpha = 1.7$ (dashed line). (Reprinted from A. Yamilov at al., Phys. Rev. Lett. **96**, 083905, 2006.)

we simulated ($\delta x = 15$ nm), no saturation of $\langle Q_{tot}^{-1} \rangle$ with N is observed up to $N = 48$. This behavior differs fundamentally from that of a PhC waveguide, where optical loss increases exponentially with its length. In contrast, a disordered PhC slab laser *benefits* from an increase in the pattern size, simply because a larger system provides a bigger pool of modes from which the highest-Q_{tot} mode can be selected. This effect should be more pronounced in PhC slab microlasers with 2D periodicity (c.f. Figure 4.4.3) as a result of the larger phase space compared to the numerically simulated systems with one-dimensional periodicity.

Experimentally, we fabricated ZnO PhC slabs of dimensions 4 × 4, 6 × 6, and 8 × 8 μm (c.f. Figure 4.4.3). As the complete PBG in a ZnO PhC slab without "undercut" was quite narrow,[49] it was technically challenging to overlap the PBG with the ZnO gain spectrum. By adjusting the magnification of the FIB system, we were able to change the lattice constant a in fine steps of 3 nm over a wide range 100–160 nm. The ratio of the air hole radius R to the lattice constant a was also varied from 0.20 to 0.30. In this way, we could tune the PBG continuously through the ZnO gain spectrum. We also introduced an artificial defect by missing an air hole. Structural analysis as in Figure 4.4.3 gives the average displacement of a hole $\delta r \simeq 0.22R$.

A ZnO PhC slab was optically pumped by the third harmonics of a pulsed Nd:YAG laser ($\lambda = 355$ nm, 10 Hz repetition rate, 20 ps pulse width) at room temperature.[42] In 8 × 8 μm patterns without intentionally introduced structural defect, the ensemble-averaged lasing threshold exhibited a pronounced minimum at $a = 113$–124 nm and $R = 0.25a$

(c.f. Figure 4.4.7a). To understand this phenomenon, we calculated the photonic bands in a ZnO PhC slab using the computational technique described by Yamilov et al.[49] The frequency dependence of the ZnO refractive index was taken into account. In Figure 4.4.7b, the wavelength of the dielectric band edge λ_d for the fundamental PBG of transverse electric modes[42] is plotted against the lattice constant a. The structural parameters were extracted from the SEM of our samples. The ZnO slab thickness $t = 180$ nm, and $R/a = 0.245$. By comparing the lasing wavelength to λ_d in Figure 4.4.7b, we confirmed that the lasing modes were located in the vicinity of the dielectric band edge. This can be explained by two factors: (i) the electric field of the modes near the dielectric band edge is concentrated inside ZnO, and thus experience more gain; (ii) the vanishing group velocity at the band edge enhances light amplification.[64,78,81] The dip in the measured lasing threshold (c.f. Figure 4.4.7a) is attributed to spectral overlap of the dielectric band edge with the ZnO gain spectrum. In Figure 4.4.6b, the measured lasing threshold decreases monotonously with the pattern size for $a = 115$ and 118 nm. These data agree qualitatively with the numerical simulation results shown in Figure 4.4.3a. In all patterns with intentionally missed air holes, the lasing modes were not pinned at the location of the missing hole owing to the existence of better-confined modes away from the defect. This observation is in line with our numerical simulation of large patterns with a single artificial defect.

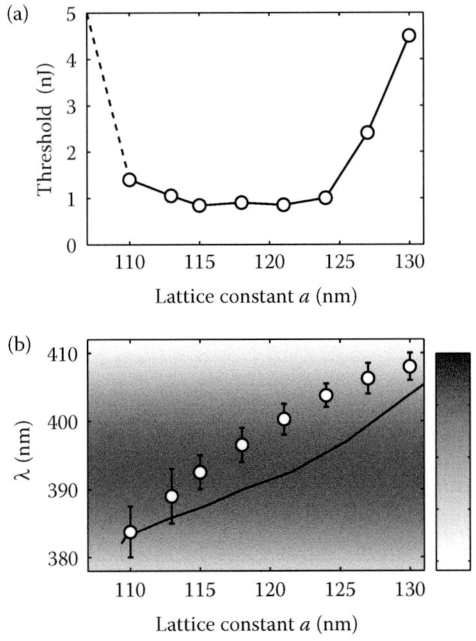

FIGURE 4.4.7
(a) Experimentally measured incident pump pulse energy at the lasing threshold (averaged over 5–10 samples) as a function of lattice constant a. (b) The circles are the measured mean wavelength of lasing modes; the error bar depicts the range of lasing wavelengths. The solid curve represents the wavelength of the calculated dielectric band edge λ_d in a ZnO photonic crystal slab as a function of a. The shade of the background qualitatively describes the position and width of the ZnO gain spectrum. This demonstrates the possibility of tuning lasing frequency of the devices by adjusting its lattice constant. (Reprinted from A. Yamilov et al., Phys. Rev. Lett. **96**, 083905, 2006.)

4.4.5 Summary and Outlook

In summary, with the combined theoretical and experimental studies, we find that the structural disorder may lead to self-optimization of optical confinement in a PhC slab and formation of high-Q_{tot} modes that serve as the lasing modes. In a sufficiently large PhC slab with uncorrelated disorder, a microcavity with balanced $Q_{\perp}$ and $Q_{\parallel}$ can be formed spontaneously without any carefully designed structural defects. In our near-ultraviolet PhC slab laser, random scattering by structural disorder leads to in-plane localization of band-edge modes. The underlying physical mechanism is similar to that of light localization in random media. The reduction of density of states near the photonic band edge enhances the localization effect. The most confined modes are selectively amplified in the presence of optical gain owing to long photon lifetime. We also demonstrate that despite the disorder, the band-edge effect allows one to fine-tune the lasing wavelength from 383 to 407 nm with sample-to-sample fluctuation of about 5 nm in the disordered ZnO PhC slab laser.

Our study questions the conventional wisdom that structural disorder always *degrades* the functionality of optical devices. Instead, we suggest that one needs to determine the type of disorder and understand its effects. Using lasing in disordered PhC slabs as an example, we demonstrate that a disorder can actually *enhance* the functionality of certain applications. Therefore, the strength and properties of the disorder emerge as additional tuning parameters. Although the above conclusions have been drawn on the basis of our analyses of PhC slabs that exhibit 2D periodicity, we believe they are quite general. There have been several reports of lasing in disordered 3D PhCs;[56,88] however, much more work needs to be done to fully understand the intricacies of the interplay between order and disorder in these systems.

Another aspect of our study—the interplay between light transport in disordered systems and amplification—presents a fundamental interest. Indeed, the coherent amplification/absorption nontrivially affects the interferences of multiply-scattered waves and, thus, can promote/suppress localization phenomena. This observation has motivated us to begin systematically exploring an intriguing possibility of localization by gain—an enhancement of the mesoscopic phenomena with an increase of the amplification strength.[89]

In this work, we discussed the disorder-induced transition from a perfectly periodic structure to a completely random medium. Although lasers based on disordered media can exhibit a variety of interesting behaviors, one of their limitations to device applications is the lack of control and reproducibility of the lasing modes. Recently, deterministic aperiodic structures attracted a great deal of attention. They also lie in between the periodic and random structures, ranging from quasicrystals to pseudorandom structures and, hence, spanning the entire spectrum in a hierarchy of complexity. Because of their structural distinction and unusual physical properties, the aperiodic systems have even been called the third form of solid matter.[90] The possibility of engineering lasing modes in deterministic structures with aperiodic ordering appears to be extremely promising.[91]

Acknowledgments

This work was supported by National Science Foundation under Grant Nos. DMR-0093949, ECS-0244457, DMR-0704981, ECCS-0823345, and DMR-0808937.

References

1. V. S. Letokhov, Sov. Phys. JETP **26**, 835 (1968).
2. N. M. Lawandy, R. M. Balachandran, A. S. L. Gomes, and E. Sauvain, Nature **368**, 436 (1994); D. S. Wiersma and A. Lagendijk, Phys. Rev. E **54**, 4256 (1996); R. Frank, A. Lubatsch, and J. Kroha, J. Opt. A **11**, 114012 (2009).
3. L. Florescu and S. John, Phys. Rev. Lett. **93**, 013602 (2004); G. Hackenbroich, J. Phys. A **38**, 10537 (2005); L. Angelani, C. Conti, L. Prignano, G. Ruocco, and F. Zamponi, Phys. Rev. B **76**, 064202 (2007); J. Andreasen et al. Adv. Opt. Photon. **3**, 88 (2011).
4. H. E. Tureci, L. Ge, S. Rotter, and A. D. Stone, Science **320**, 643 (2008).
5. H. Cao, J. Phys. A **38**, 10497 (2005).
6. H. Cao, Opt. Photon. News 16, **24** (2005); D. S. Wiersma, Nat. Phys. **4**, 359 (2008).
7. H. Cao, Y. Zhao, S. T. Ho, E. W. Seelig, Q. H. Wang, and R. P. H. Chang, Phys. Rev. Lett. **82**, 2278 (1999).
8. X. Jiang and C. M. Soukoulis, Phys. Rev. E **65**, 025601 (2002); P. Sebbah and C. Vanneste, Phys. Rev. B **66**, 144202 (2002).
9. M. Patra, Phys. Rev. E **67**, 016603 (2003).
10. A. Yamilov and H. Cao, Phys. Rev. A **69**, 031803 (2004).
11. A. L. Burin, H. Cao, G. C. Schatz, and M. A. Ratner, J. Opt. Soc. Am. B **21**, 121 (2004).
12. A. Rodriguez, M. Ibanescu, J. D. Joannopoulos, S. G. Johnson, Opt. Lett. **30**, 3192 (2005).
13. M. A. Kaliteevski, D. M. Beggs, S. Brand, R. A. Abram, and V. V. Nikolaev, Phys. Rev. E **73**, 056616 (2006).
14. A. Yamilov, X. Wu, X. Liu, R. P. H. Chang, and H. Cao, Phys. Rev. Lett. **96**, 083905 (2006).
15. S. V. Frolov, Z. V. Vardeny, and K. Yoshino, Phys. Rev. B **59**, R5284 (1999); Y. Ling, H. Cao, A. L. Burin, M. A. Ratner, X. Liu, and R. P. H. Chang, Phys. Rev. A **64**, 063808 (2001).
16. H. Cao, Y. Ling, J. Y. Xu, and A. L. Burin, Phys. Rev. E **66**, R025601 (2002).
17. V. M. Apalkov, M. E. Raikh, and B. Shapiro, Phys. Rev. Lett. **89**, 016802 (2002); R. C. Polson, M. E. Raikh, and Z. V. Vardeny, Physica E **13**, 1240 (2002).
18. A. Mirlin, Phys. Rep. **326**, 259 (2000).
19. A. Yamilov, X. Wu, H. Cao, and A. L. Burin, Opt. Lett. **30**, 2430 (2005).
20. D. J. Thouless, Phys. Rev. Lett. **39**, 1167 (1977).
21. A. A. Chabanov, Z. Q. Zhang, and A. Z. Genack, Phys. Rev. Lett. **90**, 203903 (2003).
22. A. L. Burin, H. Cao, and M. A. Ratner, Physica B **338**, 212 (2003).
23. M. A. Noginov, *Solid-State Random Lasers* (Springer Science + Business Media, Inc., New York, 2005).
24. X. H. Wu, A. Yamilov, H. Noh, H. Cao, E. W. Seelig, and R. P. H. Chang, J. Opt. Soc. Am. B **21**, 159 (2004).
25. X. Liu, A. Yamilov, X. Wu, J. Zheng, H. Cao, and R.P.H. Chang, Chem. Mater. **16**, 5414 (2004).
26. S. Gottardo, R. Sapienza, P. D. Garcia, A. Blanco, D. S. Wiersma, and C. Lopez, Nat. Photonics **2**, 429 (2008).
27. S. John, Phys. Rev. Lett. **58**, 2486 (1987).
28. A. F. Ioffe and A. R. Regel, Prog. Semicond. **4**, 237 (1960).
29. S. John, Phys. Rev. Lett. **53**, 2169 (1984).
30. J. Topolancik, B. Ilic, and F. Vollmer, Phys. Rev. Lett. **99**, 253901 (2007).
31. J. M. Frigerio, J. Rivory, and P. Sheng, Opt. Commun. **98**, 231 (1993).
32. S. Fan, P. R. Villeneuve, and J. D. Joannopoulos, J. Appl. Phys. **78**, 1415 (1995); V. N. Astratov, A. M. Adawi, S. Fricker, M. S. Skolnick, D. M. Whittaker, and P. N. Pusey, Phys. Rev. B **66**, 165215 (2002).
33. M. M. Sigalas, C. M. Soukoulis, C. T. Chan, R. Biswas, and K. M. Ho, Phys. Rev. B **59**, 12767 (1999).
34. Z. Y. Li and Z. Q. Zhang, Phys. Rev. B **62**, 1516 (2000).
35. A. F. Koenderink, A. Lagendijk, and W. L. Vos, Phys. Rev. B **72**, 153102 (2005).

36. M. Skorobogatiy, G. Begin, and A. Talneau, Opt. Express **13**, 2487 (2005).
37. S. Hughes, L. Ramunno, J. F. Young, and J. E. Sipe, Phys. Rev. Lett. **94**, 033903 (2005).
38. D. Gerace and L. C. Andreani, Opt. Express **13**, 4939 (2005).
39. R. Ferrini, D. Leuenberger, R. Houdré, H. Benisty, M. Kamp, and A. Forchel, Opt. Lett. **31**, 1426 (2006); L. O'Faolain, T. P. White, D. O'Brien, X. D. Yuan, M. D. Settle, and T. F. Krauss, Opt. Express **15**, 13129 (2007); R. J. P. Engelen, D. Mori, T. Baba, and L. Kuipers, Phys. Rev. Lett. **101**, 103901 (2008).
40. J. Topolancik, B. Ilic, and F. Vollmer, Appl. Phys. Lett. **91**, 201102 (2007).
41. M. N. Shkunov, M. C. DeLong, M. E. Raikh, Z. V. Vardeny, A. A. Zakhidov, and R. H. Baughman, Synth. Met. **116**, 485 (2001); V. Milner and A. Z. Genack, Phys. Rev. Lett. **94**, 073901 (2005).
42. X. Wu, A. Yamilov, X. Liu, S. Li, V. P. Dravid, R. P. H. Chang, and H. Cao, Appl. Phys. Lett. **85**, 3657 (2004).
43. M. Scharrer, H. Noh, X. Wu, M. A. Anderson, A. Yamilov, H. Cao, and R. P. H. Chang, J. Opt. **12**, 024007 (2010).
44. S. F. Liew and H. Cao, J. Opt. **12**, 024011 (2010).
45. C. M. Soukoulis, ed., *Photonic Crystals and Light Localization in the 21st Century* (Kluwer Academic Publishers, Dordrecht, The Netherlands, 2001).
46. I. M. Lifshitz, Sov. Phys. Usp. **7**, 549 (1965).
47. A. Yamilov and H. Cao, Phys. Rev. B **68**, 085111 (2003).
48. X. Liu, W. Wu, H. Cao, and R. P. H. Chang, J. Appl. Phys. **95**, 3141 (2004).
49. A. Yamilov, X. Wu, and H. Cao, J. Appl. Phys. **98**, 103102 (2005).
50. E. W. Seelig, B. Tang, A. Yamilov, H. Cao, and R. P. H. Chang, Mater. Chem. Phys. **80**, 257 (2003); M. Scharrer, X. Wu, A. Yamilov, H. Cao, and R. P. H. Chang, Appl. Phys. Lett. **86**, 151113 (2005); M. Scharrer, A. Yamilov, X. Wu, H. Cao, and R. P. H. Chang, *ibid.* **88**, 201103 (2006).
51. A. Taflove and S. C. Hagness, *Computational Electrodynamics* (Artech House, Boston, 2000), 2nd ed.
52. T. S. Misirpashaev and C. W. J. Beenakker, Phys. Rev. A **57**, 2041 (1998); A. L. Burin, M. A. Ratner, H. Cao, and R. P. H Chang, Phys. Rev. Lett. **87**, 215503 (2001).
63. J. P. Dowling, M. Scalora, M. J. Bloemer, and C. M. Bowden, J. Appl. Phys. **75**, 1896 (1994); V. I. Kopp, B. Fan, H. K. M. Vithana, and A. Z. Genack, Opt. Lett. **23**, 1707 (1998); N. Susa, Jpn. J. Appl. Phys. **40**, 142 (2001).
64. K. Sakoda, K. Ohtaka, and T. Ueta, Opt. Express **4**, 481 (1999).
65. H. Y. Ryu, S. H. Kwon, Y. J. Lee, Y. H. Lee, and J. S. Kim, Appl. Phys. Lett. **80**, 3476 (2002).
66. A. L. Burin, M. A. Ratner, H. Cao, and S. H. Chang, Phys. Rev. Lett. **88**, 093904 (2002).
67. R. C. McPhedran, L. C. Botten, A. A. Asatryan, C. Martijn de Sterke, N. A. Nicorovici, and P. A. Robinson, Aust. J. Phys. **52**, 791 (1999).
68. A. A. Asatryan, P. A. Robinson, L. C. Botten, R. C. McPhedran, N. A. Nicorovici, and C. Martijn de Sterke, Phys. Rev. E **62**, 5711 (2000).
69. P. L. Gourley, J. R. Wendt, G. A. Vawter, T. M. Brennan, and B. E. Hammons, Appl. Phys. Lett. **64**, 687 (1994); R. D. Meade, A. Devenyi, J. D. Joannopoulos, O. L. Alerhand, D. A. Smith, and K. Kash, J. Appl. Phys. **75**, 4753 (1994); T. F. Krauss, R. M. de la Rue, and S. Brand, Nature **383**, 699 (1996).
70. S. G. Johnson, S. Fan, P. R. Villeneuve, J. D. Joannopoulos, and L. A. Kolodzjeski, Phys. Rev. B **60**, 5751 (1999).
71. E. Chow, S. Y. Lin, S. G. Johnson, P.R. Villeneuve, J:D.Joannopoulos, J.R. Wendt, G.A. Vawter, W. Zubrzycki, H. Hou, and A. Alleman, Nature 407, 983 (2000).
72. O. J. Painter, R. K. Lee, A. Scherer, A. Yariv, J. D. O'Brien, P. D. Dapkus, and I. Kim, Science **284**, 1819 (1999).
73. S. Noda, A. Chutinan, and M. Imada, Nature **407**, 608 (2000).
74. M. Lončar, T. Yoshie, A. Scherer, P. Gogna, and Y. Qiu, Appl. Phys. Lett. **81**, 2680 (2002).
75. Y. Akahane, T. Asano, B. S. Song, and S. Noda, Nature **425**, 944 (2003).
76. H. G. Park, S. H. Kim, S. H. Kwon, Y. G. Ju, J. K. Yang, J. H. Baek, S. B. Kim, and Y. H. Lee, Science **305**, 1444 (2004).
77. B. S. Song, S. Noda, T. Asano, and Y. Akahane, Nat. Mater. **4**, 207 (2005).

78. S. Nojima, Jap. J. Appl. Phys. **37** (Part 2), L565 (1998).
79. M. Meier, A. Mekis, A. Dodabalapur, A. Timko, R. E. Slusher, J. D. Joannopoulos, and O. Nalamasu, Appl. Phys. Lett. **74**, 7 (1999); M. Notomi, H. Suzuki, and T. Tamamura, *ibid*. **78**, 1325 (2001); S. Noda, M. Yokoyama, M. Imada, A. Chutinan, and M. Mochizuki, Science **293**, 1123 (2001).
80. M. Imada, S. Noda, A. Chutinan, T. Tokuda, M. Murata, and G. Sasaki, Appl. Phys. Lett. **75**, 316 (1999).
81. L. Florescu, K. Busch, and S. John, J. Opt. Soc. Am. B **19**, 2215 (2002).
82. P. Villeneuve, S. Fan, S. G. Johnson, and J. D. Joannopoulos, IEE Proc.-Optoelectron. 145, **384** (1998); E. Miyai and K. Sakoda, Opt. Lett. **26**, 740 (2001).
83. V. M. Apalkov and M. E. Raikh, Phys. Rev. Lett. **90**, 253901 (2003).
84. S. G. Johnson and J. D. Joannopoulos, Opt. Express **8**, 173 (2001).
85. C. Monat, C. Seassal, X. Letartre et al. Appl. Phys. Lett. **81**, 5102 (2002).
86. J. M. Rico-Garcia, J. M. Lopez-Alonso, and J. Alda, Opt. Express **13**, 3802 (2005).
87. O. Painter, J. Vuckovic, and A. Scherer, J. Opt. Soc. Am. B **16**, 275 (1999).
88. M. N. Shkunov, Z. V. Vardeny, M. C. DeLong, R. C. Polson, A. A. Zakhidov, and R. P. Baughman, Adv. Funct. Mater. **12**, 21 (2002).
89. B. Payne, A. Yamilov, and S. E. Skipetrov, Phys. Rev. B **82**, 024205 (2010); B. Payne, J. Andreasen, H. Cao, and A. Yamilov, *ibid*. **82**, 104204 (2010); B. Payne, H. Cao, and A. Yamilov, Physica B **405**, 3012 (2010); A. Yamilov and B. Payne, J. Mod. Opt. **57**, 1916 (2010).
90. A. N. Poddubny and E. L. Ivchenko, Physica E **42**, 1871 (2010).
91. J.-K. Yang, S. V. Boriskina, H. Noh, M. J. Rooks, G. Solomon, L. Dal Negro, and H. Cao, Appl. Phys. Lett. **97**, 223101 (2010).

4.5

Ultrafast All-Optical Switching in Photonic Crystals

Valery G. Golubev
Ioffe Physical-Technical Institute of the Russian Academy of Sciences

CONTENTS

4.5.1 Introduction

Ultrafast all-optical switching (AOS) is one of the most significant challenges for photonics. Compared with the traditional electro-optical switching, AOS offers the advantage of an ultrafast time response. Ultrafast AOS has received considerable attention because of its promising potential applications in integrated photonic circuits, optical communication, and ultrafast information processing. It would be highly advantageous to have strong and nearly instantaneous interaction of light with light. In principle, this could be achieved by exploiting intrinsic material nonlinearities. Unfortunately, such nonlinearities are quite weak, so it is necessary to look for compromises in interaction time and/or power.

Photonic crystal (PhC) AOS is a promising platform to realize switching on the ultrafast timescale. The principle of the ultrafast PhC AOS is rather simple. The refractive index contrast of a PhC changes dynamically under a high-intensity optical pump pulse, which results in changes in the photonic band gap (PBG) position and width. When the signal light wavelength is at the PBG edge, the PhC transmittivity and reflectivity considerably vary, thus realizing the "on" and "off" AOS states. The advantage offered by strongly photonic crystal switching is that it allows a dynamic control over the density of states (DOS), thus providing switching on or off of light sources in the PBG spectral region.

Three material effects are used to quickly change the refractive index. The first is the Kerr effect, which, in principle, results in modulation rates of 10^{15}–10^{16} Hz. The second is the index change due to a free-carrier generation. The third is the refractive index alteration by a fast phase transition in the material used as a PhC constituent. These approaches are briefly discussed in this chapter.

4.5.2 Physical Principles of PhC AOS

According to the basic electromagnetic theory of PhCs, the PBG strongly depends on the refractive index of each constituent material forming the PhC. Therefore, if a PhC's constituent material exhibits, for example, the Kerr nonlinearity, the refractive index will change dynamically under an applied high-intensity pump beam, which will result in alteration of the PBG. The resulting changes in both the PBG position and width are proportional to the local electromagnetic field intensity. An increase in the refractive index difference widens the PBG, whereas a decrease in the index difference narrows it.

When the signal light is at the PBG edge, its transmission (reflection) energy varies dynamically with respect to the pump beam. The signal light sees either the pass band with a high transmittivity (low reflectivity) under no pump light or the PBG with a low transmittivity (high reflectivity) under a pump light. The contrast in the transmission (reflection) signal intensity will realize the "on" and "off" states of optical switching.

Nonlinear propagation of ultrashort pulses in a one-dimensional (1D) PhC (a distributed Bragg reflector) has been investigated by numerical methods.[1] A multilayer stack of a quarter-wave dielectric material is arranged in such a way that alternating layers have a high index of refraction, n_2, and a low index, n_1, and each layer thickness also alternates and is such that $a = \lambda/4n_1$ and $b = \lambda/4n_2$, where λ is the free-space wavelength. The nonlinear effects cause a dynamical shift in the PBG location and this nonlinear mechanism can induce intensity-dependent pulse transmission and reflection.

To apply this approach to optical switching, multiple beams with slightly differing frequencies are directed on the nonlinear 1D PBG structure described above.[1] It is assumed that I is the intensity of a strong pump capable of altering the index of refraction of the medium and ω_{pumb} is the pump frequency. I_{probe} denotes the probe pulse intensity, such that $I \gg I_{probe}$, the carrier frequency of I_{probe} is ω_{probe}. If ω and ω_{probe} are assumed to be located below the PBG, but ω is relatively far below the PBG, whereas ω_{probe} is near the PBG edge, then both beams will be initially transmitted. As a beam of intensity I alters the index of refraction, its effect is to widen the PBG, and it will do so to such an extent that ω_{probe} will be found inside the PBG. Beam I_{probe} then shuts off as its frequency is now a part of the forbidden frequency range where an extremely high percentage of reflection occurs. This constitutes the operation of the optical switch.[1]

To implement efficient AOS in optical systems, an optimal combination of five major parameters should be achieved. They are a low threshold power of the pump pulse, a fast signal response time, a high switching contrast between the "on" and "off" states, a high spatial homogeneity of the refractive index change, and applicability for integrated optical circuits.

Optical switching is based mainly on the principle of the PBG edge shift. It is highly important that the slope of the PhC transmission (reflection) spectrum be very steep in the PBG edge region. Therefore, the switching device performance will appreciably depend on the PhC sample quality. The higher the PhC perfection, the steeper the PBG edge, and, as a result, a higher switching contrast will be achieved under the same pump power.

It is necessary for the AOS device operation to be ultrafast. If the constituent nonlinear materials have a very fast (almost instantaneous) optical response to the external pump light, the switching response time will be on the order of the pump pulse duration. This means that a state-of-the-art femtosecond laser technique is needed to realize an ultrafast switching speed.

4.5.3 Pump–Probe Setup for Dynamical Study of Ultrafast AOS

To study the switching efficiency, the dynamical response of optical switching should be measured. Therefore, dependences of the probe light reflectance or transmittance changes on the time delay between the pump and probe pulses are measured on the ultrafarst timescale. A state-of-the-art ultrafast time-resolved pump–probe spectroscopic setup for studying switching of various nanophotonic structures has been described in detail.[2]

Two laser pulses (either from the same laser or from two different lasers) are used in the pump–probe experiments. The parameters, such as the wavelength, pulse duration, pulse repetition, and the pump pulse energy, can be tuned according to designs. In the experiments, high pump pulse intensities are often necessary to observe very small variations in probe reflection or transmission signals. This is achieved by using regenerative amplifiers, in which femtosecond pulses from lasers are amplified to pulse energies of up to several millijoules. Unfortunately, the amplification process appreciably reduces the pulse repetition rate, typically from the megahertz to the kilohertz range. The conversion to different wavelengths is performed by using optical parametric amplifiers. The optical delay line is introduced to adjust the temporal relation between the pump and probe laser pulses. The output probe signals are collected by a monochromator and amplified by a computer-controlled electronic system.

The background level in light scattering materials, such as strongly photonic crystals, can be higher than the output probe signals. A versatile measurement scheme that allows compensation for pulse-to-pulse variations in the laser output as well as a subtraction of the pump background from the probe signal has been developed.[2] A 10-fold improvement in the precision of the setup resulting in a measurement accuracy of better than $\Delta R = 0.07\%$ in a 1 s measurement time has been achieved.

4.5.4 Superfast Nonlinear Optical Tuning of Organic PhCs with Kerr Nonlinearity

If the refractive index could be quickly changed, such materials could be utilized for ultrafast AOS in PhCs. One of the main challenges is to develop optically nonlinear PhC materials that are able to efficiently control light diffraction on the ultrafast timescales.[3] This is not an easily realizable task as most conventional nonlinear materials have low optical susceptibilities. Fortunately, there are Kerr nonlinear materials the refractive index of which can be changed under an optical high-intensity pump light. The operating wavelength of optical switching is in the visible spectrum. The aim of this section is to review progress in ultrafast AOS of nonlinear 1D, two-dimensional (2D), and three-dimensional (3D) PhCs in the visible spectral range.

Organic conjugated polymer materials, such as polystyrene, are of major interest because of their relatively low cost, ease of fabrication and integration into devices, a high laser damage threshold, an ultrashort (down to several femtoseconds) nonlinear optical response time, and a high third-order off-resonance nonlinear optical susceptibility ($\chi^{(3)}$~10^{-12} cm^2/W).

A tunable 3D PhC fabricated by the self-assembly formation method of polystyrene spheres with a diameter of 220 nm has been demonstrated.[4] The optical nonlinearity originates from the delocalization of the conjugated p-electron along polymer chains, which leads to a high third-order nonlinear susceptibility and a subpicosecond time response. Transmission changes based on the optical Kerr effect have been measured, and the dynamic PBG shift has been studied. The positive $\chi^{(3)}$ of polystyrene results in the effective refractive index increase with the pump intensity, which provides the PBG shift toward longer wavelengths and the increase of the PBG width with increasing pump intensity. The PBG shifts by about 13 nm when the pump peak intensity is 40.4 GW/cm^2. The time response limited by the pump pulse duration is on the order of several picoseconds.

AOS with a high switching efficiency has been realized within a 2D PhC made of cylindrical air holes in a polystyrene slab.[5] The lattice constant and the air hole radius are 220 and 90 nm, respectively. A transmittance contrast of more than 60% is achieved under the 16.7 GW/cm^2 pump intensity (Figure 4.5.1). The time response of optical switching limited by the experimental time resolution is around 10 ps. The PBG shifts by 10 nm when the peak intensity of the pump light is 16.7 GW/cm^2.

Ultrafast AOS has been demonstrated in high-quality opal-based 3D PhCs prepared by the vertical deposition method using monodisperse polysteryne spheres with a diameter of 240 nm.[6] The PBG shifts by about 10 nm with the excitation of a 27.5 GW/cm^2 pump laser, the transmittance contrast reaches 45%. The achieved response time as short as 120 fs is limited by the experimental time resolution.

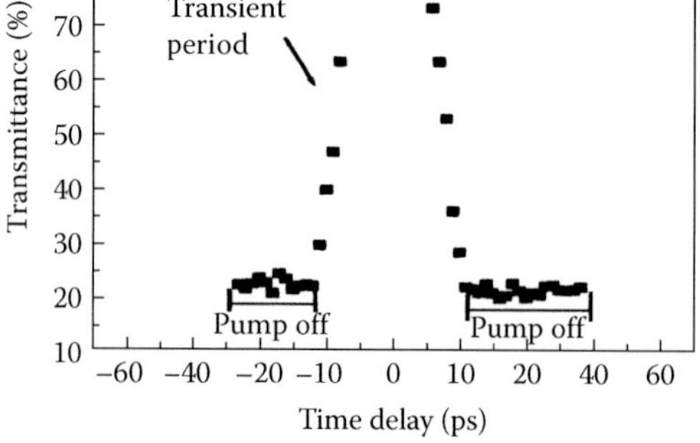

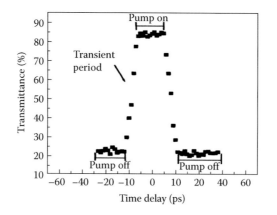

FIGURE 4.5.1
Changes in the probe-light transmittance versus the time delay between the pump and probe pulses. The probe light wavelength is 519 nm and the peak pump light intensity is 16.7 GW/cm^2. The time response varies as follows: first, the time delay changes from 30 ps to 0; then the zero time delay is maintained; finally, the time delay is tuned from 0 to 40 ps. (Reproduced with permission from X. Hu et al., Appl. Phys. Lett. **86**, 121102, 2005.)

Thus, the investigations described above have demonstrated that the shorter the pump pulse duration, the shorter the detected response time of AOS in nonlinear polystyrene-based PhCs. The response time of optical switching can be further shortened to an extremely fast level of 10 fs in a 3D polystyrene opal PhC by implementing an 8 fs pump pulse from a Ti:sapphire pulse laser (a peak pump power of 20.6 GW/cm^2) (Figure 4.5.2).[7] This switching response time approaches the material response limit of polystyrene.

The operating pump intensity of PhC optical switching is typically on the order of several gigawatts per square centimeter owing to relatively low nonlinear optical characteristics of materials. A strategy to achieve ultrafast and low-power AOS of organic PhCs has been reported.[8] Poly[2-methoxy-5-(2-ethylhexyloxy)-1,4-phenylenevinylene] (MEH-PPV), a kind of a nonlinear organic conjugated polymer, is used as a matrix of nonlinear 2D PhCs. MEH-PPV films are fabricated by a slow evaporation method, which leads to poorly conjugated polymer chains. The 2D PhCs consist of square arrays of cylindrical air holes embedded in a MEH-PPV slab. The lattice constant and the air hole diameter are 266 and 213 nm, respectively. Under resonant excitation, MEH-PPV provides a high third-order nonlinear optical susceptibility. A fast switching time of 45.6 ps is maintained by the Forster transfer of excitons from poorly conjugated segments to more conjugated segments. A switching efficiency of 70% is achieved: the transmittance changes from 20% in the PBG to 90% in the pass band at the operating pump energy as low as 514 pJ/cm^2.

A polymer–metal nanoparticle composite material possessing high nonlinear optical coefficients and ultrafast response simultaneously has been reported.[9] The 2D PhCs consist of square arrays of cylindrical air holes embedded in a nano-Ag:MEH-PPV slab. The average diameter of Ag nanoparticles is 15 nm. The lattice constant and the air hole diameter are 260 and 200 nm, respectively. The surface plasmon resonance frequency of Ag nanoparticles drops in the linear absorption band of MEH-PPV. The optical nonlinearity of MEH-PPV and Ag nanoparticles is resonantly enhanced under excitation in the vicinity of the surface plasmon resonance peak. Under resonant excitation of both the polymer matrix and Ag nanoparticles the nonlinear susceptibility reaches 10^{-6} esu. A nonlinear response time of 35 ps is achieved, and an ultrafast AOS with an ultralow pump intensity of 0.2 MW/cm^2 is realized.

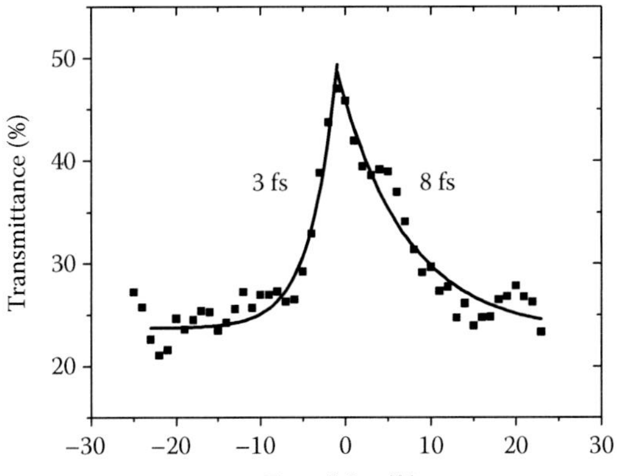

FIGURE 4.5.2
Transmittance change with respect to the time delay between the pump and probe light. The probe light wavelength is 785 nm and the peak pump power is 20.6 GW/cm^2. (Reproduced with permission from Y. Liu et al., Appl. Phys. Lett. **95**, 131116, 2009.)

4.5.5 AOS of Semiconductor PhCs via Free-Carrier Injection

Ultrafast changes in the refractive index can be obtained via resonant processes in which free electrons and holes are created in a semiconductor material. The achievable change in the refractive index due to variations in the carrier concentration within the semiconductor is at least an order of magnitude higher than that resulting from the optical Kerr effect. This process requires a substantially lower pump intensity compared with the optical Kerr effect. Relaxation is limited by the carrier recombination time. Thus, semiconductors have favorable properties for optical switching as excellent constituents for switchable PhCs. Moreover, their high refractive indices are very advantageous for fabrication of strongly photonic crystals.

This section considers a possibility of ultrafast tuning of PhCs with semiconductor constituents. The injection of free carriers in a semiconductor is known to alter both the refractive index and absorption coefficient via the Drude contribution to the refractive index (via the dielectric function). Optically excited electrons and holes induce an ultrafast refractive index change in a semiconductor, thereby causing a shift in the PBG original position and width.

Experimentally, control over the PhC optical properties can be achieved by changing the complex index of refraction ($n' + in''$) of one or both PhC constituent materials by an ultrashort laser pulse. The rise time of the switching will be ultrafast as electrons are excited on the femtosecond timescale. The duration of the state after excitation is determined by the carrier relaxation time. Relaxation times range from 10 ps in amorphous Si to nanoseconds in GaAs and to microseconds in crystalline Si and Ge. Relaxation times are relatively short in structured semiconductor PhCs.[10] As shown in the investigations of the DOS switching in PhCs, extremely large changes in the DOS and ultrafast (100 fs) timescales can be simultaneously achieved.[10]

One of significant challenges is to incorporate switches based on PhCs with a complete PBG into highly integrated photonic circuits. As the complete PBG requires a high refractive-index contrast, Si-based PhC structures are of particular interest.[11] Note that silicon PhC constituent materials will be compatible with other devices on the same silicon chip and have much lower production and operating costs. A number of elements of the Si-based PhC AOS have been proposed. They will be briefly discussed below.

Ultrafast tuning of the PBG edge of a 2D Si/air PhC has been demonstrated for the near infrared.[12] The samples are essentially a triangular lattice of air pores in Si, with a pitch of 500 nm and a pore radius of 206 nm. Changes in the Si refractive index are optically induced by injecting free carriers with 800-nm 300-fs pulses. The rise time of the shift is on the timescale of the laser pulse width (Figure 4.5.3); the recovery time is related to the electron–hole recombination. The PBG edge is observed to shift linearly with pump beam fluence, with a shift in excess of 30 nm for a pump beam fluence of 2 mJ/cm^2.

A strong ultrafast response in the reflectivity of opal–Si composite 3D PhCs has been demonstrated.[13] To prepare opal–Si composites, the voids of silica opal with a SiO_2 ball diameter of 230 nm are filled with Si using thermal decomposition of a 5% SiH_4–Ar gas mixture.[14] Subsequently, the samples are annealed in air at 800°C.

The temporal evolution of a relative transient change in the Bragg reflection $\Delta R(t)/R$ induced by a strong (5 mJ/cm^2) optical pump pulse is shown in Figure 4.5.4. The reflection signal exhibits an abrupt decrease just after the pump pulse arrival ($t = 0$) and partly recovers on a picosecond timescale. Photoinduced changes in the real and imaginary parts of the refractive index suppress the Bragg interference of light inside the PhC and diminish

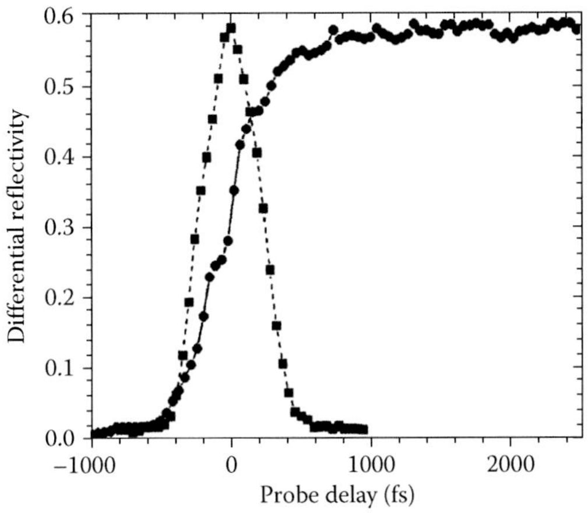

FIGURE 4.5.3
Dependence of differential reflectivity $\Delta R/R$ on probe delay for $\lambda = 1900$ nm and a pump fluence of 1.3 mJ/cm^2 (solid). Also plotted is the pump and probe cross correlation (dashed, arbitrary units). (Reproduced with permission from S. W. Leonard et al., Phys. Rev. B **66**, 161102(R), 2002.)

the Bragg reflection. The changes in the Bragg reflection are the highest at the PBG ($\lambda = 800$ nm) where they reach $\Delta R(t)/R = -46\%$. The inset in Figure 4.5.4 displays the time trace of $\Delta R(t)/R$ measured at $\lambda = 800$ nm at a lower pump power density (70 μJ/cm^2) and higher temporal resolution. Here, the amplitude of the relative change in reflectivity is $\Delta R(t)/R = -1.2 \times 10^{-2}$. The initial peak in $\Delta R(t)/R$ occurs when the pump and probe light pulses overlap in time and can be explained by the instantaneous Kerr effect. The decay at longer delays has a multiexponential shape with the time constants $\tau \sim 0.5$ and ~5 ps. The switching time is less than 30 fs and is determined by the pump pulse duration. The fact that ultrafast switching takes place within 30 fs suggests that the changes in reflectivity are caused by photoexcited carriers in the amorphous-nanocrystalline Si.

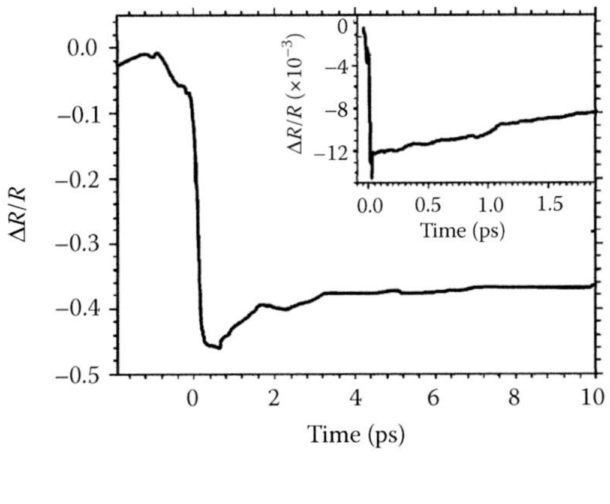

FIGURE 4.5.4
Time-resolved transient differential reflectivity $\Delta R/R$ at 780 nm in the case of a high-power (5 mJ/cm^2) optical pulse excitation. The inset shows $\Delta R/R$ at 800 nm for moderate (70 μJ/cm^2) optical pulse excitation measured with a higher temporal resolution (30 fs). (Reproduced with permission from D. A. Mazurenko et al., Phys. Rev. Lett. **91**, 213903, 2003.)

4.5.6 Homogeneous Switching of Semiconductor PhCs via the Two-Photon Absorption Process

If a PhC made from a semiconductor material is optically excited at the wavelengths longer than the absorption edge of the semiconductor, free carriers are generated homogeneously via the two-photon absorption (TPA) process. The resulting refractive index change modifies the PBG without a significant pump beam absorption that limits the switched material volume and leads to a spatial inhomogeneity in the degree of switching. Owing to TPA, light penetrates at larger distances into the PhC before it is absorbed. Optical pumping of free carriers in semiconductor PhCs via TPA is believed to be an ideal means of controlling the PhC properties in time.[10] The complex index of refraction of excited free carriers in a semiconductor agrees well with the Drude model. Correspondingly, the optical properties of the excited free carriers are determined by the plasma frequency and the electron and hole momentum relaxation times. The optimum excitation maximizes the change in n' and minimizes the increase in n'' to prevent photons from being absorbed.

The nonlinear absorption model, taking into account both linear and TPA processes, has been presented.[15] Optimum pumping conditions have been derived and a homogeneity length scale providing homogeneous switching conditions has been defined from this model. The role of disorder-induced diffusion of the pump beam in PhCs has also been discussed. It has been shown that it is not always sufficient to ensure an optimal homogeneity by simply using TPA instead of linear absorption. The choice of the pumping frequencies that are lower in the TPA regime strongly increases the pumping homogeneity.

The effects of TPA and Kerr nonlinearity on the optical properties of a 1D PhC composed of five periodic pairs of SiO_2 and amorphous Si layers deposited on a glass substrate have been studied.[16] The PBG near 1.5 μm is monitored with a weak probe beam and is modulated by changes in the refractive index caused by a pump pulse at 1.71 μm with a peak intensity of 18 GW/cm^2. As expected, no modulation is observed for negative delays but it is maximal for the time window during which the pump and probe pulses overlap. Modulation of the probe transmission is estimated to be about 400 fs, consistent with the pump and probe pulse cross-correlation measurements. This ultrafast switching process suggests that when the nonlinear index of refraction of Si changes, the PBG shifts slightly and the probe light transmittance changes.

Alteration of the real (Kerr) and imaginary (TPA) parts of the third-order optical nonlinearity has been also used to tune the long (1.6 μm) and short (1.3 μm) wavelength PBG edges in a 2D Si PhC with a triangular lattice arrangement of 560-nm diameter 96-μm-deep air holes with a pitch $a = 700$ nm.[17] The pump and probe intensities are 17.6 and 0.5 GW/cm^2, respectively. The decrease in reflectivity (Figure 4.5.5) is consistent with the PBG edge red-shift due to a positive nondegenerate Kerr index. As shown by pump–probe reflectivity experiments using 130 fs pulses, the short-wavelength PBG edge (1.3 μm) can be red-shifted via the Kerr effect with a pump beam at 2.0 μm. For a pump wavelength of 1.76 μm and probe wavelength of 1.6 μm (the long-wavelength PBG edge), a red-shift also occurs via the Kerr effect. At high pump intensities, generation of free carriers through TPA leads to a blue-shift of the PBG edge via the Drude contribution to the linear dielectric constant.

Ultrafast AOS measurements of Si woodpile PBG crystals have been carried out.[18] The PhCs consist of five layers of stacked polycrystalline Si nanorods that have a refractive index of 3.45 at 1.55 μm. The crystals are spatially homogeneously two-photon excited and probed by measuring reflectivity over an octave in frequency (including the telecommunication range) as a function of time. The complete PBG is shifted to higher frequencies as a

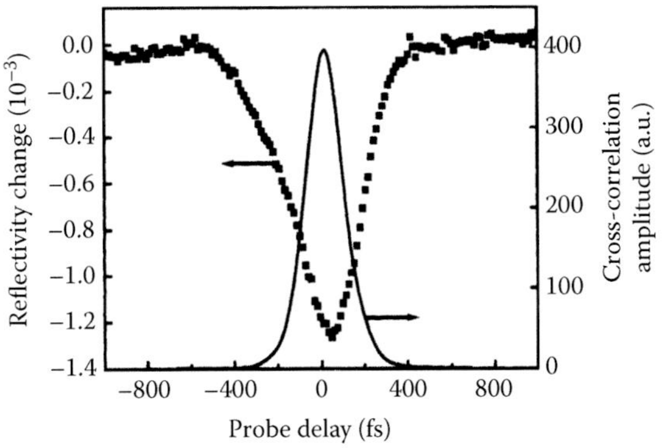

FIGURE 4.5.5
Temporal response of reflectivity change at the 1.3-μm photonic band-gap edge when the photonic crystal is pumped with a 2.0 μm pulse at 17.6 GW/cm^2. Also shown is the cross-correlation trace of the pump and probe pulses. (Reproduced with permission from H. W. Tan et al., Phys. Rev. B **70**, 205110, 2004.)

result of a 300-fs optical excitation of free carriers. The switched state relaxes quickly with a recovery time constant of 18 ps. The induced changes in refractive index are well described by the Drude model with a carrier relaxation time of 10 fs. At 300 fs after excitation the reflectivity displays a decrease $\Delta R/R = -7\%$ at low frequencies (5882 cm^{-1}), whereas at high frequencies (9174 cm^{-1}, blue PBG edge) an increase up to $\Delta R/R = 19\%$ is observed (Figures 4.5.6). This distinct dispersive shape in the differential reflectivity is a clear evidence of a blue-shift of the whole gap.

Femtosecond two-color pump–probe experiments with very high-quality Si inverse opal films have been carried out in a transmission geometry using off-resonant excitation.[19] The Si inverse opal samples are fabricated with a low-pressure self-assembly approach using 860-nm diameter SiO_2 spheres followed by infiltration with Si by the chemical vapor deposition process. Subsequently, the silica template as well as the glass substrate are removed by wet chemical etching, leading to a freestanding Si inverse opal.

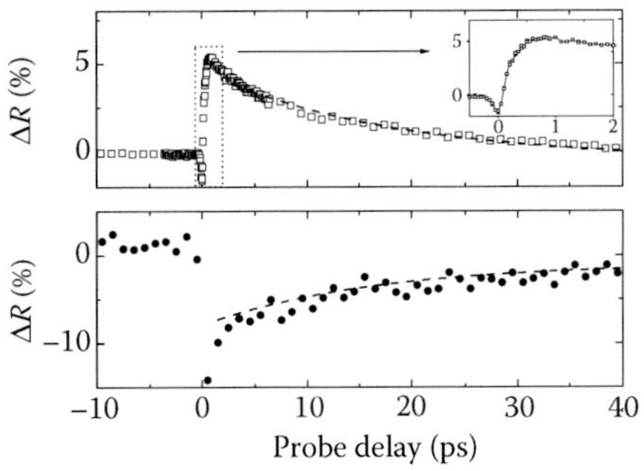

FIGURE 4.5.6
Absolute reflectivity changes versus probe delay at frequency $\omega_{blue} = 9174$ cm^{-1} at the blue photonic band gap (PBG) edge (upper panel) and $\omega_{red} = 5882$ cm^{-1} at the red PBG edge (lower panel). The pump intensity is 16 GW/cm^2. The dashed curves are exponential fits with decay times of 18 ps (upper panel) and 16 ps (lower panel). (Reproduced with permission from T. G. Euser et al., Phys. Rev. B **77**, 115214, 2008.)

The pump and probe have durations of about 130 fs, and excitation is below the electronic Si band gap. The frequency range of the first-order Bragg diffraction has been studied. It is shown that the nonlinear optical response can be modified by sample tempering. Si grown using the chemical vapor deposition process is known to be amorphous; subsequent tempering can bring it into a polycrystalline state. The nonlinear responses are completely different and depend on the electron scattering times. The latter can be tailored by sample tempering causing Drude scattering time to increase drastically from 0.5 to 10 fs and resulting in a strong reduction of the induced absorption. The untempered sample (left column of Figure 4.5.7) has an absorption for all wavelengths and a recovery time of 1 ps. In contrast, the recovery times increase after tempering (right column of Figure 4.5.7) to about 5 ps, and a shift of the PBG is observed. An increased (reduced) transmission occurs on the long-wavelength (short-wavelength) side of the PBG, which results from a reduction of the Si refractive index after excitation with the pump. The transmission changes quadratically with the pump fluence. On the basis of simple calculations using the Drude model, corresponding refractive index changes as large as $\Delta n = -0.5 + i0.07$ are estimated.

Switching of the Si inverse opal, where the complete PBG is predicted, has been achieved by optically exciting free carriers by the TPA process in the frequency range of the second-order Bragg diffraction.[20] The Si inverse opal PhC is made by infiltrating Si in a silica opaline template grown on a Si substrate by a vertical controlled drying method. The template is infiltrated with Si using chemical vapor deposition at 500°C. Subsequently, the sample is annealed for 1 h at 750°C in vacuum. During the annealing process the amorphous Si in the structure crystallizes into poly-Si. Finally, the SiO_2 template is etched away by a buffered hydrofluoric solution, resulting in a high-quality 3D air-sphere crystal supported by a poly-Si backbone. The pump fluence in the experiments is 25 pJ per unit cell area per pulse. A large frequency shift of up to $\Delta\omega/\omega = 1.5\%$ of all spectral features including the peak that corresponds to the PBG is observed with an exponential growth time of 500 fs limited by the pump pulse duration. A corresponding large refractive index change of

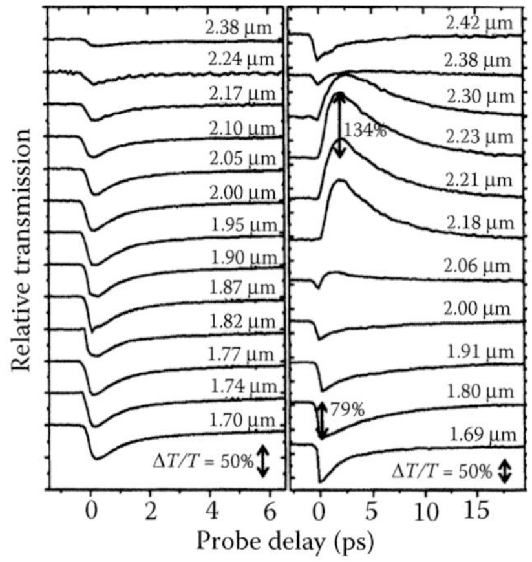

FIGURE 4.5.7
Pump–probe traces for several probe wavelengths (as indicated). Left-hand-side column: as-grown sample (*a*-Si), $\lambda_{pump} = 1.84$ μm, $I_{pump} = 66$ GW/cm²; right-hand-side column: tempered sample (polycrystalline Si), $\lambda_{pump} = 1.73$ μm, $I_{pump} = 90$ GW/cm². All curves in one column are shown on the same scale of transmission relative to the unexcited sample (see arrows). (Reproduced with permission from C. Becker et al., Appl. Phys. Lett. **87**, 091111, 2005.)

$\Delta n'_{Si}/n'_{Si} = 2.0$ %, where n'_{Si} is the refractive index of the Si backbone of the PhC, is obtained. Both the low- and high-frequency edges of the PBGs are shifted. The refractive index change is predicted to strongly modify the DOS inside the crystal. The absorption length is longer than the sample thickness. The observed decay time is 21 ps.

4.5.7 Subpicosecond Switching of the PBG in Opal–Vanadium Dioxide PhC Governed by a Photoinduced Semiconductor–Metal Phase Transition

Another way to switch the spectral PBG position is to use a photoinduced phase transition accompanied by permittivity changes of the constituents forming the PhC. The excitation of the PhC material by intense laser pulses with the photon energy higher than the material fundamental band gap generates hot carriers which, in turn, induce the phase transition directly or indirectly via phonon generation. In both cases, the temporal and spatial evolutions of the photoinduced phase transition during and after the laser pulse are governed by kinetic properties of the photoexcited quasiparticles (electrons, phonons, plasmons, etc.).

The material that demonstrates such an ultrafast photoinduced phase transition and can be used as a basis for a PhC is vanadium dioxide (VO_2). The first-order semiconductor–metal phase transition of VO_2 occurs at $T_c = 68°C$. In the region of red light the transition is accompanied by changes in the real part of the refractive index from $n(VO_2) < 2.9$ in the "cold" semiconductor phase to $n(VO_2) < 2.3$ in the "hot" metallic phase.[21] The phase transition kinetics in VO_2 films has been studied in detail for different timescales.[22] The PBG switching time in the PhC based on VO_2 may be less than 10^{-12} s.

This section presents results on femtosecond PBG shifting in a 3D opal–VO_2 PhCs.[23] The samples are fabricated from a synthetic opal template composed of 240-μm diameter monodispersed (±5%) a-SiO_2 spheres. The opal voids are impregnated with VO_2. Details of the fabrication method and data on the linear optical properties can be found elsewhere.[24]

Time-resolved pump–probe experiments have been carried out by using a pulsed Ti:sapphire laser with a regenerative amplifier (wavelength is 800 nm; pulse duration is 200 fs). A sapphire plate is used to convert the probe into a broadband "white light" pulse. The pulse repetition rate is less than 10 kHz to provide thermal recovery of the sample between pulses. The pump beam density on the sample surface is $W = 10$–25 mJ/cm^2.

The reflectivity spectrum from the (111) facet of the opal–VO_2 composite shows a peak with a maximum at the energy position of the PBG. The inset in Figure 4.5.8 presents two spectra measured under steady-state conditions at $T = 30$ and 90°C corresponding to the semiconductor and metal phases of VO_2, respectively. The high-energy shift of the spectrum and, hence, of the PBG at elevated T is due to changes in the average permittivity of the composite induced by the phase transition. Figure 4.5.8 shows the hysteresis loop of the measured shift $\Delta E(T)$ relative to the energy in the semiconductor phase. A shift up to 90 meV is observed. The hysteresis behavior of $\Delta E(T)$ in the heating and/or cooling cycle is governed by the properties of the phase transition of VO_2 in the opal–VO_2 composite and VO_2-inverted opals (the details can be found elsewhere).[25] It is important to note that 3D film PhCs based on the opal–VO_2 composites have been synthesized; these film PhCs have specified parameters that provide a maximum PBG shift (~170 meV) in the vicinity of ~1.5 μm.[26]

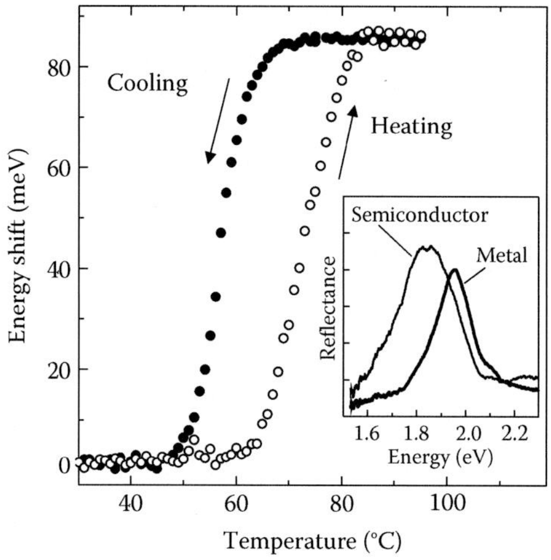

FIGURE 4.5.8
Hysteresis loop in the temperature dependence of the Bragg reflectivity peak energy in the opal-VO_2 composite measured in steady-state experiments. Inset: Reflectivity spectra for the semiconductor $T = 30°C$ and metallic $T = 90°C$ phases. (Reproduced with permission from A.B. Pevtsov et al., Phys. Rev. B **75**, 153101, 2007.)

The spectral position of the reflectivity maximum shifts rapidly toward higher energies in less than a picosecond and then continues to move further on a longer timescale. The inset in Figure 4.5.9 shows the temporal evolution of the spectral shift $\Delta E(t) = E(t) - E_0$ (E_0 is the first moment of the reflectivity in the semiconductor phase) in 125-fs steps. The initial shift $\Delta E_i = 25$ meV occurs almost instantaneously with the laser pump pulse. The shift

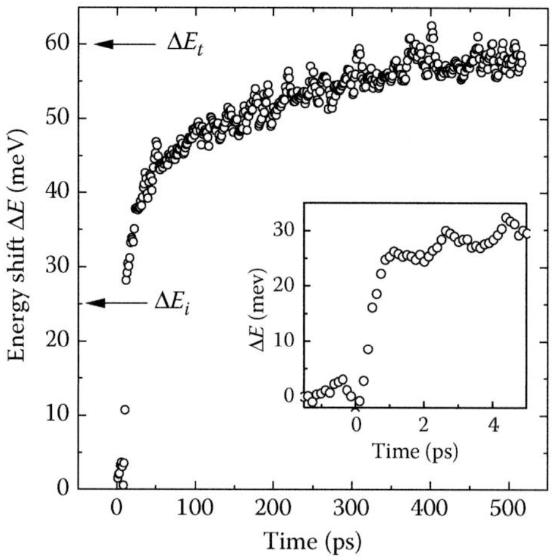

FIGURE 4.5.9
Temporal evolution of the Bragg peak energy in the reflectivity spectrum for a pump excitation density of 20 mJ/cm^2. The horizontal arrows show the values of the ultrafast shift ΔE_i and quasistationary shift ΔE_t. The inset shows the same with a high temporal resolution for the early time range. (Reproduced with permission from A.B. Pevtsov et al., Phys. Rev. B **75**, 153101, 2007.)

shown in the main panel of Figure 4.5.9 is measured using a longer time step. In a subpicosecond, the shift to higher energies goes on and then tends to saturate at a level $\Delta E_t = 60$ meV in a few hundred picoseconds.[27]

Analysis of the experimental results is based on the fact that the photoinduced phase transition of VO_2 in the opal pores starts in a boundary optically excited PhC layer with thickness d_c. The magnitude of d_c is on the order of the penetration depth l_a of the pump light into the opal–VO_2 composite. The ultrafast PBG shift due to the photoinduced phase transition in VO_2 is determined by two quantities: d_c and the fraction α of VO_2 in each pore in the excited boundary layer that has undergone the phase transition. For instance, if the entire VO_2 material in this layer becomes metallic, then $\alpha = 1$. Otherwise $\alpha < 1$, which means that within thickness d_c not all the VO_2 crystallites are metallic. The PBG shift would be equal to the one observed under continuous excitation only if $\alpha = 1$ and d_c was large enough to form the PBG as in a uniform PhC. The kinetics of the PBG shift is governed by the time evolution of d_c and α.

Thus, it has been shown that the ultrafast kinetics of the PBG shift governed by the photoinduced phase transition in a 3D PhC contains two components. The first one is the subpicosecond shift that takes place almost instantaneously with the femtosecond laser pulse. The second component has a transient time of about 100 ps and is governed by a spatial redistribution of the semiconductor and metal phases inside the material volume that undergoes phase transition.

4.5.8 Conclusions

Physical principles of AOS in nonlinear organic and semiconductor PhCs have been discussed, and the experimental progress in exploring ultrafast AOS in PhCs has been described. AOS of PhCs results in ultrafast changes in their reflectivity and transmittivity. Switching of strongly photonic crystals also provides dynamic control over the DOS, which allows switching on or switching off of light sources in the PBG spectral region. It has been demonstrated that ultrafast AOS with a very low operating power is achieved in PhCs at the wavelengths in the visible and telecom ranges. State-of-the-art pump–probe techniques are optimized for spatially homogeneous switching by two-photon excitation of semiconductor PhCs. Thus, PhCs have exciting prospects for fabrication of a new generation of ultrafast all-optical devices.

References

1. M. Scalora, J. P. Dowling, C. M. Bowden, and M. J. Bloemer, Phys. Rev. Lett. **73**, 1368 (1994).
2. T. G. Euser, P. G. Harding, and W. L. Vos, Rev. Sci. Instrum. **80**, 073104 (2009).
3. Y. Liu, F. Qin, F. Zhou, Q.-B. Meng, D.-Z. Zhang, and Z.-Y. Li, Front. Phys. China **5**, 220 (2010).
4. X. Hu, Q. Zhang, Y. Liu, B. Cheng, and D. Zhang, Appl. Phys. Lett. **83**, 2518 (2003).
5. X. Hu, Y. Liu, J. Tian, B. Cheng, and D. Zhang, Appl. Phys. Lett. **86**, 121102 (2005).
6. Y. Liu, X. Hu, D. Zhang, B. Cheng, and D. Zhang, Appl. Phys. Lett. **86**, 151102 (2005).
7. Y. Liu, F. Qin, Z.-Y. Wei, Q.-B. Meng, D.-Z. Zhang, and Z.-Y. Li, Appl. Phys. Lett. **95**, 131116 (2009).

8. P. Jiang, X. Hu, J. Zhang, H. Yang, and Q. Gong, Appl. Phys. B **99**, 187 (2010).
9. X. Hu, P. Jiang, J. Zhang, C. Xin, H. Yang, and Q. Gong, Appl. Phys. Lett. **94**, 031103 (2009).
10. P. M. Johnson, A. F. Koenderink, and W. L. Vos, Phys. Rev. B **66**, 081102 (R) (2002).
11. A. Blanco, E. Chomski, S. Grabtchak, M. Ibisate, S. John, S. W. Leonard, C. Lopez et al., Nature **405**, 437 (2000); Yu. A. Vlasov, X.-Zh. Bo, J. C. Sturm, and D. J. Norris, *ibid.* **414**, 289 (2001).
12. S. W. Leonard, H. M. van Driel, J. Schilling, and R. B. Wehrsophn, Phys. Rev. B **66**, 161102(R) (2002).
13. D. A. Mazurenko, A. V. Akimov, A. B. Pevtsov, D. A. Kurdyukov, V. G. Golubev, and J. I. Dijkhuis, Phys. E **17**, 410 (2003); Lumin. **108**, 163 (2004); D. A. Mazurenko, R. Kerst, J. I. Dijkhuis, A. V. Akimov, V. G. Golubev, D. A. Kurdyukov, A. B. Pevtsov, and A. V. Sel'kin, Phys. Rev. Lett. **91**, 213903 (2003); D. A. Mazurenko, R. Kerst, A. V. Akimov, A. B. Pevtsov, D. A. Kurdyukov, V. G. Golubev, A. V. Sel'kin, and J. I. Dijkhuis, J. Non-Cryst. Solids **338–340**, 215 (2004).
14. V. N. Bogomolov,V. G. Golubev, N. F. Kartenko, D. A. Kurdyukov, A. B. Pevtsov, A. V. Prokofév, V. V .Ratnikov, N. A. Feoktistov, and N. V. Sharenkova, Tech. Phys. Lett. **24**, 326 (1998); V. G. Golubev, J. L. Hutchison, V .A. Kosobukin, D. A. Kurdyukov, A. V. Medvedev, A. B. Pevtsov, J. Sloan, and L. M. Sorokin, J. Non-Cryst. Solids **299–302**, 1062 (2002).
15. T. G. Euser and W. L. Vos, J. Appl. Phys. **97**, 043102 (2005).
16. A. Haché and M. Bourgeois, Appl. Phys. Lett. **77**, 4089 (2000).
17. H. W. Tan, H. M. van Driel, S. L. Schweizer, R. B. Wehrspohn, and U. Gösele, Phys. Rev. B **70**, 205110 (2004).
18. T. G. Euser, A. J. Molenaar, J. G. Fleming, B. Gralak, A. Polman, and W. L. Vos, Phys. Rev. B **77**, 115214 (2008).
19. C. Becker, S. Linden, G. von Freymann, M. Wegener, N. Tétreault, E. Vekris, V. Kitaev, and G. A. Ozin, Appl. Phys. Lett. **87**, 091111 (2005).
20. T. G. Euser, H. Wei, J. Kalkman, Y. Jun, A. Polman, D. J. Norris, and W. L. Vos, J. Appl. Phys. **102**, 053111 (2007).
21. H. W. Verleur, A. S. Barker, and C. N. Berglund, Phys. Rev. **172**, 788 (1968).
22. A. Cavalleri, H. H. W. Chong, S. Fourmaux, T. E. Glover, A. Heimann, J. C. Keeffer, B. S. Mun, H. A. Padmore, and R. W. Schoenlein, Phys. Rev. B **69**, 153106 (2004).
23. D. A. Mazurenko, R. Kerst, J. I. Dijkhuis A. V. Akimov, V. G. Golubev, A. A. Kaplyanskii, D. A. Kurdyukov, and A. B. Pevtsov, Appl. Phys. Lett. **86**, 041114 (2005).
24. V. G. Golubev, V. Yu. Davydov, N. F. Kartenko, D. A. Kurdyukov, A. V. Medvedev, A. B. Pevtsov, A. V. Scherbakov, and E. B. Shadrin, Appl. Phys. Lett. **79**, 2127 (2001).
25. V. G. Golubev, D. A. Kurdyukov, A. B. Pevtsov, A. V. Sel'kin, E. B. Shadrin, A. V. Il'inskii, and R. Boeyink, Semiconductors **36**, 1043 (2002).
26. A. B. Pevtsov, S. A. Grudinkin, A. N. Poddubny, S. F. Kaplan, D. A. Kurdyukov, and V. G. Golubev, Semiconductors **44**, 1537 (2010).
27. A.B. Pevtsov, D. A. Kurduykov, V. G. Golubev, A. V. Akimov, A. M. Meluchev, A. V. Sel'kin, A. A. Kaplyanskii, D. R. Yakovlev, and M. Bayer, Phys. Rev. B **75**, 153101 (2007).

4.6

Resonant Light Scattering in Photonic Devices: Role of Defects

Andrey E. Miroshnichenko and Yuri S. Kivshar
Australian National University

CONTENTS

The smart use of light in science and technology is at the core of an impressive number of high-performance photonic devices ranging from laser chips and optical sensors to all-optical communication systems for high-speed computing and data transfer.[1] Such functionality is achieved because of an extensive use of optical microcavities in various geometries involving coupling of one or several cavities to a waveguide. The optical response of waveguide–cavity systems can be made tunable by adding some cavities with nonlinear response or by using an external control. Microcavities also exhibit high-quality resonances, and they are used for optical modulation and switching. The on/off switching functionality is realized by shifting the resonant frequency either toward or away from the signal frequency.

A powerful principle that could be explored to implement all-optical transistors, switches, and logical gates is based on the concept of optical bistability. The use of photonic crystals enables the system to be of a size of the order of the wavelength of light, consume only a few milliwatts of power, and have a recovery and response time smaller than 1 ps. Several theoretical and experimental studies explored *nonlinear Fano resonances* for the design of optimal bistable switching in nonlinear photonic crystals.[2–5] A photonic crystal provides an optimal control over the input and output, and facilitates further large-scale optical integration.

To understand the key properties of light propagation, we use the powerful method of Green's function approach for Maxwell's equations.[2] It allows us to obtain very accurate results in comparison to the time-consuming direct numerical finite-difference time-domain (FDTD) simulations, even for rather complex geometries of the waveguide–cavity systems. To derive the corresponding equations for scattering and pulse propagation we take into account the explicit temporal dependencies.

4.6.1 Fano Resonances—The Most Promising for Applications

One of the simplest bistable optical devices for applications in future photonic integrated circuits is a nonlinear two-port structure that is connected to other parts of a circuit by one input and one output waveguide. Its scattering properties depend on the light coupled to the input waveguide. One of the realizations of such a device is provided by a waveguide side-coupled to an optical cavity. This system is known to exhibit a Fano resonance with a sharp suppression of the transmission and enhanced reflection.

The physics of Fano resonance is widely known across many different branches of physics.[6] It manifests itself as a sharp asymmetric profile of the transmission or absorption lines, and it is observed in numerous physical systems, including light absorption by atomic systems, Aharonov–Bohm interferometer and quantum dots, resonant light propagation through photonic crystal waveguides, and phonon scattering by time-periodic scattering potentials.[6] From the viewpoint of fundamental physics, Fano resonance may appear in the systems characterized by a certain discrete energy state that interacts with the continuum spectrum through an *interference effect*. Usually, the discrete state is created by a defect that allows one or several additional propagation paths in the wave scattering that interact constructively or destructively. In the transmission line, this interference effect leads to either *perfect transmission* or *perfect reflection*, producing a sharp asymmetric profile.

In a seminal paper,[7] Fano derived a general formula that describes the asymmetric line shape of the transmission or absorption lines:

$$F(\xi) = \frac{(\xi + q)^2}{\xi^2 + 1} \tag{4.6.1}$$

where $\xi = (E - E_F)/(\Gamma/2)$ is the dimensionless energy in the units of the resonance width Γ, q is the asymmetry parameter (Fano factor), and E_F is the resonance energy. In the limit $q \to \infty$, this formula can also describe the so-called Breit–Wigner resonance profile.[6] This quite universal formula is usually used to fit a particular profile observed in experiments and, therefore, to provide proof that the observed phenomenon can be classified as the Fano resonance.

Applications of photonic crystal devices based on Fano resonances are becoming a reality because of recent experimental progress in realizing both linear and nonlinear light transmission in two-dimensional photonic crystal slab structures where a lattice of cylindrical pores is etched into a planar waveguide. In particular, the Noda group realized coupling of a photonic crystal waveguide to a leaky resonator mode consisting of a defect pore of slightly increased radius.[8] Notomi et al.[9] and Barclay et al.[10] observed all-optical

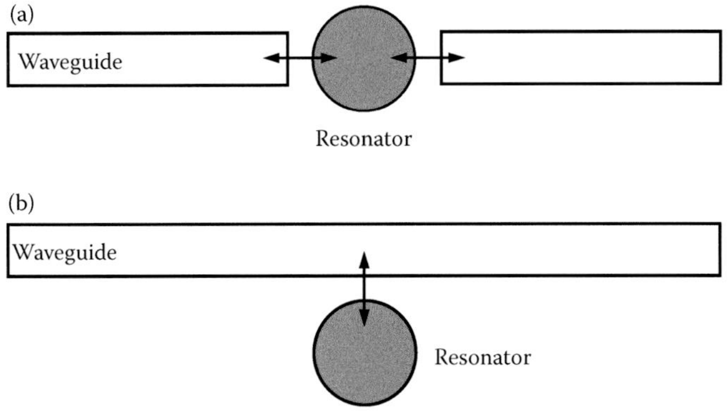

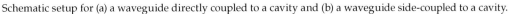

FIGURE 4.6.1
Schematic setup for (a) a waveguide directly coupled to a cavity and (b) a waveguide side-coupled to a cavity.

bistability in directly coupled photonic-crystal waveguide-resonator systems. More recently, Yang et al.[11] demonstrated experimentally Fano-resonance-enhanced bistability in a side-coupled geometry.

The main idea of using the Fano resonance for all-optical switching and bistability is rather simple: one should introduce an element with nonlinear response and achieve nonlinearity-induced shifts of the resonant frequency. Thus, by using *nonlinear Fano resonances* we can achieve bistability in many of the device structures suggested on the photonic crystal platform. For example, for the side-coupled geometry shown in Figure 4.6.1b, one could take advantage of the interference between the propagating wave inside the waveguide and the decaying wave from the cavity, to greatly enhance achievable contrast ratio in the transmission between the two bistable states. This approach was realized by Yanik et al.[4] who demonstrated that such a configuration can generate extremely high contrast between the bistable states in its transmission with low input power.

In the following, we review the basic properties of the simplest waveguide–cavity systems, and discuss several generalizations including all-optical switching structures based on the concepts of Fano resonances.

4.6.2 Green's Function Formalism

Green's function approach[2,12] allows one to obtain very accurate results in comparison to the time-consuming direct numerical FDTD simulations, even for rather complex geometries of the waveguide–cavity systems. To derive the corresponding equations, one takes into account the explicit temporal dependencies, which allow one to study the pulse propagation and scattering.

We consider a photonic crystal created by a periodic square lattice of infinite cylindrical rods parallel to the z-axis. We neglect the material dispersion and assume the dielectric constant $\varepsilon(\vec{r})$ to be periodic in two transverse directions, $\vec{r} = (x, y)$. The evolution of the E-polarized electric field propagating in the (x,y) plane is governed by the scalar wave equation

$$\nabla^2 E_z(\vec{r},t) - \frac{1}{c^2}\partial_t^2[\varepsilon(\vec{r})E_z(\vec{r},t)] = 0, \tag{4.6.2}$$

where $\nabla^2 = \partial_x^2 + \partial_y^2$. We assume that the light field propagating in such structures can be separated into fast and slow components, $E_z(\vec{r},t) = e^{-i\omega t}E(\vec{r},t \mid \omega)$, where $E(\vec{r},t \mid \omega)$ is a slowly varying envelope of the electric field, that is, $\partial_t^2 E(\vec{r},t \mid \omega) \ll \omega\partial_t E(\vec{r},t \mid \omega)$. This allows one to simplify Equation 4.6.2 to the following form:

$$\left[\nabla^2 + \varepsilon(\vec{r})\left(\frac{\omega}{c}\right)^2\right]E(\vec{r},t \mid \omega) \approx -2i\varepsilon(\vec{r})\frac{\omega}{c^2}\partial_t E(\vec{r},t \mid \omega). \tag{4.6.3}$$

Both the straight waveguide and the side-coupled cavity are created by introducing defect rods into a perfect two-dimensional periodic structure. Therefore, the dielectric constant can be represented as a sum of two components, describing the periodic and defect structures $\varepsilon(\vec{r}) = \varepsilon_{\rm pc} + \delta\varepsilon$. We use Green's function of the two-dimensional periodic structure without defects, and rewrite Equation 4.6.3 in the integral form

$$E(\mathbf{x},t \mid \omega) = \int d^2\mathbf{y}G(\mathbf{x},\mathbf{y} \mid \omega)\hat{L}E(\mathbf{y},t \mid \omega), \tag{4.6.4}$$

where we introduce the linear operator

$$\hat{L} = \left(\frac{\omega}{c}\right)^2 \delta\varepsilon(\vec{r}) + 2i\varepsilon(\vec{r})\frac{\omega}{c^2}\partial_t, \tag{4.6.5}$$

and consider the time evolution of the slowly varying envelope as a perturbation to the steady state. The defect rods introduced into the periodic structure can formally be described as follows:

$$\delta\varepsilon(\vec{r}) = \sum_{n,m}\left[\delta\varepsilon_{n,m}^{(0)} + \chi^{(3)}\left|E(\mathbf{x},t \mid \omega)\right|^2\right]\theta(\mathbf{x} - \mathbf{x}_{n,m}), \tag{4.6.6}$$

where we use the θ function to describe the position of a defect rod at site n,m, with $\theta(x) = 1$ for x inside the defect rods, and $\theta(x) = 0$ otherwise; $\delta\varepsilon_{n,m}^{(0)}$ is the variation of the dielectric constant of the defect rod (n,m). Importantly, this approach allows us to incorporate a nonlinear response in a straightforward manner, which is assumed to be of the Kerr type being described by the term $\chi^{(3)}|E|^2$.

Substituting Equation 4.6.6 into the integral Equation 4.6.4 and assuming that the electric field does not change inside the dielectric rods, we can evaluate the integral on the right-hand side of Equation 4.6.4 and derive a set of discrete nonlinear equations

$$i\sigma\partial_t E_{n,m} - E_{n,m} + \sum_{k,l} J_{n-k,m-l}(\omega)\left(\delta\varepsilon_{k,l}^{(0)} + \chi^{(3)}\left|E_{k.l}\right|^2\right)E_{k,l} = 0, \tag{4.6.7}$$

for the amplitudes of the electric field $E_{n,m}(t|\omega) = E(\mathbf{x}_{n,m}, t|\omega)$ calculated at the defect rods. The parameters σ and $J_{k.l}(\omega)$ are determined by using the corresponding integrals of Green's function, where the whole information about the photonic crystal dispersion is now hidden in their specific frequency dependencies, which can be found in studies by Mingaleev et al.[5,13] In this way, Green's function needs to be calculated only once for a given photonic structure, for example, using the approach outlined by Ward and Pendry,[14] and then it can be used to study any photonic circuit in that structure.

4.6.3 Waveguide with a Coupled Defect

We start with the study of the transmission properties of a photonic crystal waveguide with a side-coupled nonlinear defect, as shown in Figure 4.6.2a. It was demonstrated that this structure exhibits bistable transmission even for low input powers.[5,15] For this geometry, the dynamical Equation 4.6.7 can be written as follows:

$$\begin{aligned} -i\sigma\partial_\tau E_n &= \sum_{k=1}^{L} V_{0,k}(\Omega)(E_{n-k} + E_{n+k}) - D_n(\Omega)E_n + \sum_{l=n_1}^{n_2} \delta_{n,l} V_{l-n_1,\alpha} E_\alpha, \\ -i\sigma\partial_\tau E_\alpha &= \kappa_\alpha(\Omega) \mid E_\alpha \mid^2 E_\alpha - D_\alpha(\Omega)E_\alpha + \sum_{l=n_1}^{n_2} V_{\alpha,l-n_1} E_l, \end{aligned} \tag{4.6.8}$$

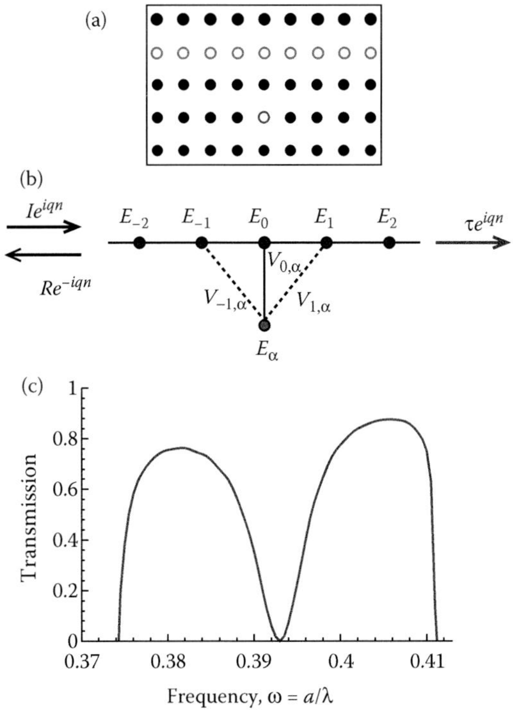

FIGURE 4.6.2
Schematic view of (a) photonic crystal waveguide with an isolated side-coupled cavity and (b) effective discrete system. (c) Typical profile of the corresponding Fano resonance.

where $V_{n,m}(\Omega) = \delta\varepsilon_m J_{n,m}(\Omega)$, $D_n(\Omega) = 1 - \delta\varepsilon_n J_{0,0}(\Omega)$, $D_\alpha(\Omega) = 1 - \delta\varepsilon_\alpha J_{0,0}(\Omega)$, and $\kappa_\alpha(\Omega) = \chi_\alpha^{(3)} J_{0,0}(\Omega)$, and the normalized frequency and time have been introduced as $\Omega = a/\lambda$ and $\tau = 2\pi c/at$, respectively. The side-coupled defect is situated in the middle of the structure near the site $n = 0$, where we take into account the nonlocal coupling to the straight photonic crystal waveguide with $n_1 = -1$ and $n_2 = 1$ (see Figure 4.6.2b).

To study the transmission properties of the system, we impose the scattering boundary conditions

$$E_n(\tau \mid \Omega) = \begin{cases} Ie^{iq(\Omega)n} + R(\tau)e^{-iq(\Omega)n}, & n \ll 0 \\ T(\tau)e^{iq(\Omega)n}, & n \gg 0 \end{cases} \tag{4.6.9}$$

which depend slowly on time.

It is known that in photonic crystal waveguides the effective interaction between defect rods is long-range[2,16] and, thus, in general, we should have $L > 1$. However, the coupling strength decays exponentially with the distance and, as a result, for coupled-resonator optical waveguides the specific discrete array with nearest-neighbor interactions ($L = 1$) gives already an excellent agreement with direct FDTD simulations.[2]

4.6.4 All-Optical Switching and Bistability

The discrete nature of a photonic crystal waveguide suggests that the side-coupled cavity can be attached in two different ways: (i) directly under removed/modified rod, or (ii) between them, in contrast to a continuum system, resulting in different scattering properties. This idea was used to study the resonant transmission of light through a photonic crystal waveguide coupled to a nonlinear cavity, and it was demonstrated how to modify the structure geometry for achieving bistability and all-optical switching at ultra-low powers in the slow-light regime. This can be achieved by placing a side-coupled cavity between two defects of a photonic crystal waveguide assuming that all the defect modes and the cavity mode are having the same symmetry.[5] In this structure, the quality factor grows inversely proportional to the group velocity of light at the resonant frequency and, accordingly, the power threshold required for all-optical switching vanishes as a square of the group velocity (see Figure 4.6.3).

The numerically obtained dependence $Q(v_{gr}) \sim 1/v_{gr}$ is shown in Figure 4.6.3a, and it is in excellent agreement with the theoretical predictions. As the bistability threshold power of the incoming light in waveguide–cavity structures scales as $P_{th} \sim 1/Q^2$,[5] one observes a rapid diminishing of $P_{th} \sim v_{gr}^2$ when the resonance frequency approaches the band edge, as shown in numerical calculations summarized in Figure 4.6.3b and c.

One of the great advantages in using nonlinear photonic crystal cavities is the enhancement of nonlinear optical processes, including nonlinear Fano resonance.[17] Such an enhancement can be very efficient in the regime of the slow-light propagation, which was demonstrated experimentally with the smallest achieved group velocity $c/1000$.[9,18] Because of this success, the interest in slow-light applications based on photonic crystal waveguides is rapidly growing, and posing problems of the design of different types of functional optical devices, which would efficiently operate in the slow-light regime.

Several experimental observations of optical bistability enhanced through Fano interferences have been reported.[11,19] In particular, the use of a high-Q cavity mode ($Q = 30{,}000$) in

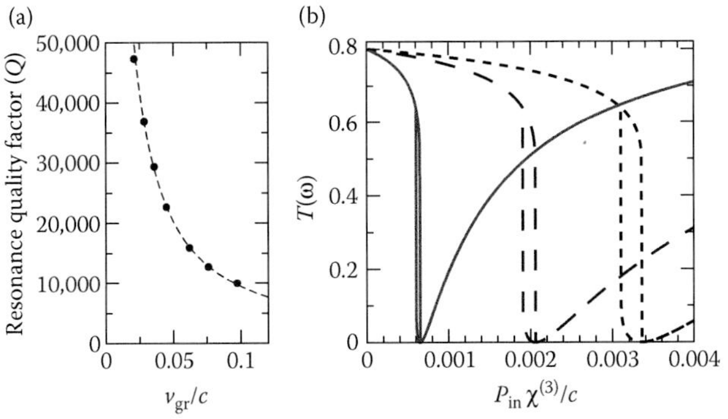

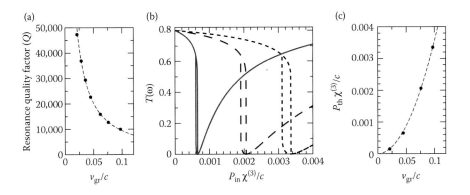

FIGURE 4.6.3
Ultra-low all-optical switching in the slow-light regime. (a) Quality factor Q versus group velocity vg at resonance for the waveguide-cavity structure. (b) Nonlinear bistable transmission at the frequencies with 80% of linear light transmission versus the incoming light power for different values of the rod radius. (c) Switch-off bistability threshold versus the group velocity at resonance. (Adapted from S. F. Mingaleev, A. E. Miroshnichenko, and Y. S. Kivshar, Opt. Express **15**, 12380, 2007.)

a silicon photonic crystal demonstrated Fano-resonance-based bistable states and switching with thresholds of 185 μW and 4.5 fJ internally stored cavity energy that might be useful for scalable optical buffering and logic.[11]

It is important to note that the nonlinear Fano resonance shows dynamical instabilities with plane wave excitations.[20,21] Near the resonance the intensity of the scattered wave starts to grow in time, leading to modulational instability, whereas far from resonance it converges to a steady-state solution (see Figure 4.6.4). However, as demonstrated[20] this

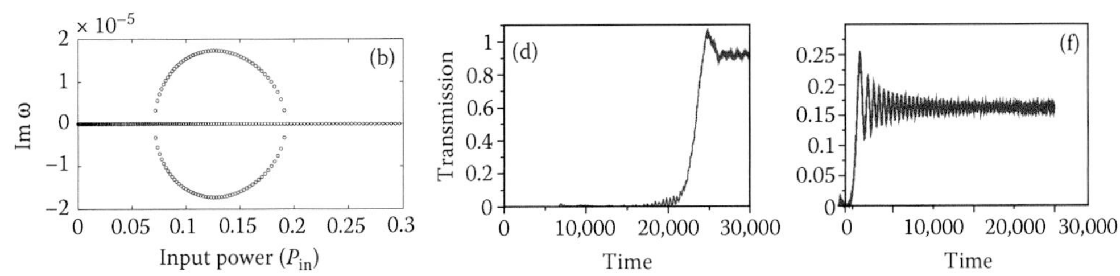

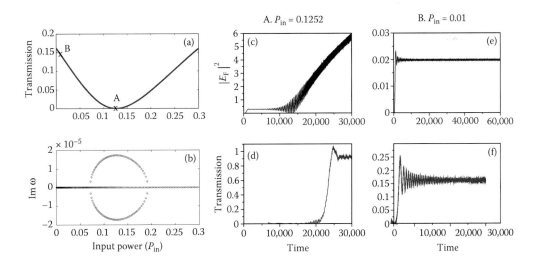

FIGURE 4.6.4
Dynamical instability of the nonlinear Fano resonance. (a) Nonlinear transmission coefficient and (b) imaginary part of eigenvalues of the stability problem versus input power. In the vicinity of the nonlinear Fano resonance, the plane-wave excitation becomes dynamically unstable. Temporal evolution of (c, e) the field inside the side-coupled cavity, and (d, f) the transmission coefficient for two different values of the input power values, indicated in plot (a). Near the resonance the dynamics of the field inside the nonlinear cavity yields a buildup of a modulation instability in time. (Adapted from A. E. Miroshnichenko et al., Phys. Rev. A **79**, 013809, 2009.)

instability can be suppressed for temporal Gaussian pulses excitations, providing an effective method of recovering the bistable transmission.

4.6.5 Overlapping Resonances

A very important effect associated with the Fano resonances in double-resonator photonic structures can be linked to the electromagnetically induced transparency (EIT).[22] Coupled-resonator-induced transparency (CRIT) structures have been introduced in various studies,[23,24] although early work[25] suggested already an idea of macroscopic double-resonator optical system exhibiting the EIT-like effect (see Figure 4.6.5). Recently, the CRIT effect has been observed experimentally in the system of two interacting microresonators (glass spheres of about 400 μm in diameter) with whispering-gallery modes,[26] in a cavity with at least two resonant modes,[27] in integrated photonic chips with two microring resonators,[28] and in planar metamaterial.[29] Providing an efficiently tunable transparency on an optical chip, such CRIT devices are considered as a crucial step toward the development of integrated all-optical chips.[30]

To explain the origin of CRIT resonances, we characterize light transmission by the transmission and reflection coefficients that can be presented in the form

$$|T(\omega)|^2 = \frac{\sigma^2(\omega)}{\sigma^2(\omega)+1}, \quad |R(\omega)|^2 = \frac{1}{\sigma^2(\omega)+1}, \tag{4.6.10}$$

where the detuning function $\sigma(\omega)$ may have quite different types of frequency dependence for different types of waveguide–cavity structures. Zero transmission (total reflection) corresponds to the condition $\sigma(\omega) = 0$, whereas perfect transmission (zero reflection) corresponds to the condition $\sigma(\omega) = \pm\infty$.

For the waveguide–cavity structure shown in Figure 4.6.1b, we obtain[5]

$$\sigma(\omega) \cong \frac{(\omega_\alpha - \omega)}{\gamma_\alpha}, \tag{4.6.11}$$

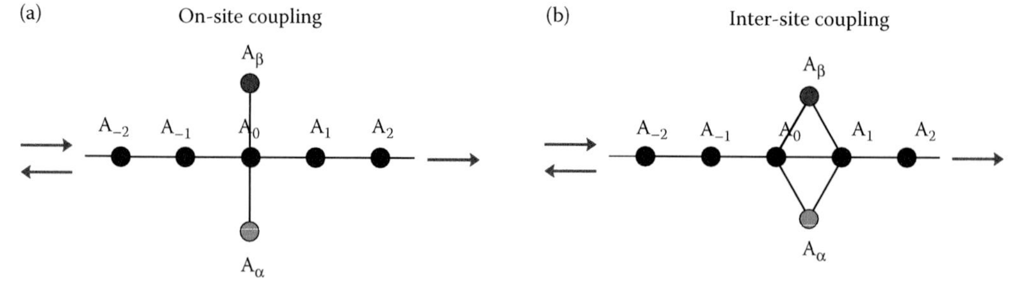

FIGURE 4.6.5
Two types of the geometries of a photonic crystal waveguide side-coupled to two nonlinear optical resonators. Light transmission and bistability are qualitatively different for (a) on-site and (b) inter-site locations of the resonator along the waveguide.

where ω_α is the eigenfrequency of the localized cavity mode of an isolated cavity α. The spectral width γ_α of the resonance is determined by the overlap integral between the cavity mode and the guided mode at the resonant frequency.

To find $\sigma(\omega)$ for the two-cavity structure one can apply a variety of methods, but the simplest approach is based on the transfer-matrix technique.[31] When two cavities are separated by the distance $d = 2\pi m/k(\omega_t)$, where $k(\omega)$ is the waveguide's dispersion relation, m is any integer number, and the frequency ω_T is defined below, and there is no direct coupling between the cavities, we obtain

$$\sigma(\omega) = \frac{(\omega_\alpha - \omega)(\omega_\beta - \omega)}{\Gamma(\omega_t - \omega)} \tag{4.6.12}$$

with the total resonance width $\Gamma = \gamma_\alpha + \gamma_\beta$ and the frequency of perfect transmission $\omega_T = (\gamma_\alpha\omega_\beta + \gamma_\beta\omega_\alpha)(\gamma_\alpha + \gamma_\beta)^{-1}$ lying in between the two cavity frequencies, ω_α and ω_β, of zero transmission.

In the case when the cavities α and β are identical, we obtain a single-cavity resonance and the only effect of using two cavities is the doubling of the spectral width, $\Gamma = 2\gamma_\alpha$, of the resonant reflection line, as illustrated in Figure 4.6.6a. However, introducing even the smallest difference between the two cavities leads to the opening of an extremely

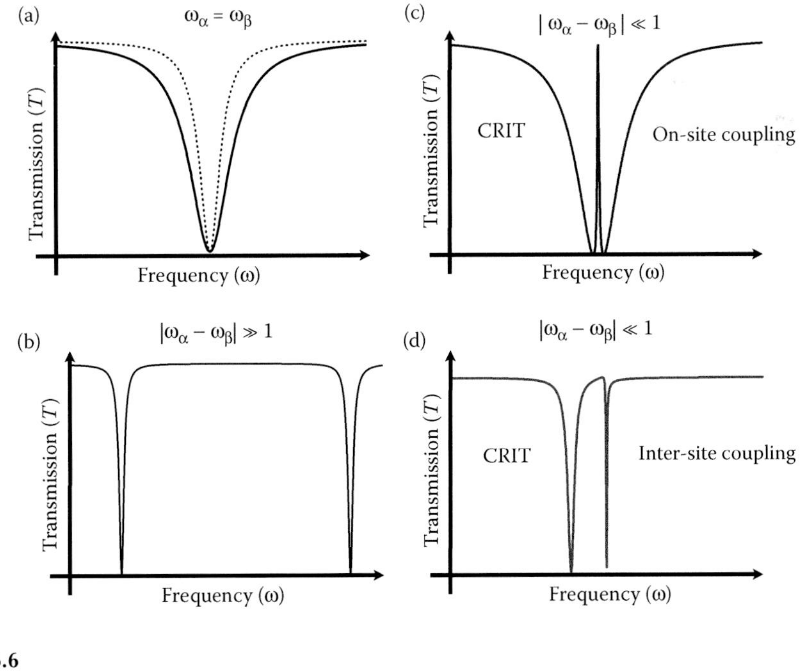

FIGURE 4.6.6
Typical transmission curves for four different cases: (a) two identical side-coupled defects $\omega_\alpha = \omega_\beta$ (solid) (transmission for a single side-coupled cavity is shown by a dashed line); (b) two side-coupled cavities with strongly detuned eigenfrequencies $|\omega_\alpha - \omega_\beta| >> 1$; (c, d) two side-coupled cavities with slightly detuned eigenfrequencies $|\omega_\alpha - \omega_\beta| << 1$ for (c) on-site coupling and (d) inter-site coupling. (Adapted from S. F. Mingaleev, A. E. Miroshnichenko, and Y. S. Kivshar, Opt. Express **16**, 11647, 2008.)

narrow resonant transmission line on the background of this broader reflection line, as illustrated in Figure 4.6.6c. Indeed, for slightly different cavities we may rewrite Equation 4.6.12 in the vicinity of the resonant transmission frequency, $\omega_T = \omega_\alpha + \delta\omega/2$, as $\sigma(\omega) \approx \Gamma_T/(\omega - \omega_T)$, with the line width $\Gamma_T = \delta\omega^2/8\gamma_\alpha$, which can easily be controlled by tuning the frequency difference $\delta\omega$. The quality factor of this transmission line, $Q_T = \omega_T/2\Gamma_T \approx 4\gamma_\alpha\omega_\alpha/\delta\omega^2$, grows indefinitely when $\delta\omega$ vanishes. As mentioned above, this effect is the all-optical analog of the EIT and is now often referred to as the effect of CRIT.[23]

In contrast, the intercoupling between two cavities, as shown in Figure 4.6.5b, manifests itself as a qualitatively new effect of coupled-resonator-induced reflection (CRIR): for small detuning $\delta\omega = \omega_\beta - \omega_\alpha$, one of the resonant reflection frequencies shifts very close to the perfect transmission frequency, ω_T, producing a narrow resonant reflection line, as illustrated in Figure 4.6.6d. The frequency of this line is always close to the frequency ω_α of the cavity mode, whereas its spectral width is determined by the frequency difference $\delta\omega$, growing indefinitely as $\delta\omega$ vanishes.[32]

It should be emphasized that despite such a qualitative difference in their spectral manifestations, both CRIT and CRIR effects have the same physical origin which can be attributed to the Fano–Feshbach resonances[33] that are known to originate from the interaction of two or more resonances (e.g., two Fano resonances) in the overlapping regime, where the spectral widths of resonances are comparable to or larger than the frequency separation between them. In a general situation it leads to a drastic deformation of the transmission spectrum and the formation of additional resonances with sharp peaks. The Fano–Feshbach resonances are associated with a collective response of multiple interacting resonant degrees of freedom, and they have numerous evidences in quantum mechanical systems.[34]

Finally, we discuss the interaction between two Fano resonances[35] that can be used to stop and store light coherently, with an all-optical adiabatic and reversible pulse bandwidth compression process.[36,37] Such a process overcomes the fundamental bandwidth delay constraint in optics and can generate arbitrarily small group velocities for any light pulse with a given bandwidth, without any coherent or resonant light-matter interaction. The mechanism can be realized in a system consisting of a waveguide side-coupled to tunable resonators, which generates a photonic band structure that represents a classical EIT analog.[37,38]

4.6.6 Tuning the Shape of Fano Resonance

Recently, it was experimentally demonstrated that the shape of the Fano resonance in the light scattering by a high-Q planar photonic crystal nanocavity can be controlled by varying the waste of the Gaussian beam.[39] For a tightly focused beam with a spot diameter $d_1 \approx 2\ \mu m$, a strong asymmetric Fano resonance was observed with the asymmetry parameter $q_1 = -0.348$ (see Figure 4.6.7a). On the other hand, for a slightly defocused Gaussian beam with the spot diameter $d_2 \approx 10\ \mu m$, a symmetric Fano resonance was observed with $q_2 = -0.016$ (see Figure 4.6.7b). In this geometry, the light reflected from the nanocavity mimics the scattering through a discrete level, whereas the light reflected from the photonic crystal pattern can be considered as the scattering to the continuum. The interference of these two reflected components leads to the Fano resonance. The

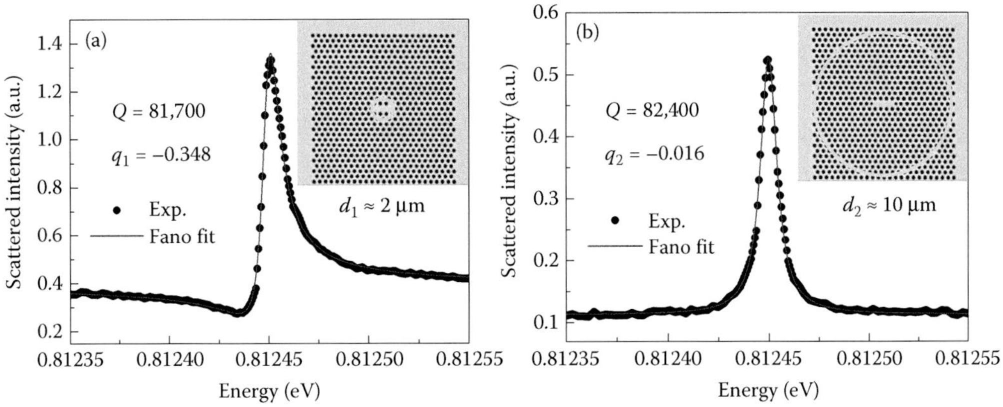

FIGURE 4.6.7
(**See color insert.**) Measured scattering spectra (dots) and fitting by the Fano formula (solid lines) of a photonic crystal nanocavity for two different excitation conditions: (a) a tightly focused and (b) a slightly defocused laser beam of diameters d_1 and d_2, respectively, indicated by circles. Note here, that the actual profiles are inverted ones because of the use of cross-polarized detection. (Adapted from M. Galli et al., Appl. Phys. Lett. **94**, 071101, 2009.)

variation of the Fano profile with the increase in the excitation area can be understood as an enhancement of the scattering to the continuum, leading to the decrease of the asymmetry parameter q. Indeed, the variation of the asymmetry parameter $q_1/q_2 \approx 22$ is proportional to the variation of the excitation areas $(d_2/d_1)^2 \approx 25$. Thus, by changing the excitation conditions it is possible to tune the Fano resonance in the scattering by photonic crystal nanocavity.

4.6.7 Generalizations

The appearance of the sharp asymmetric Fano resonances is not limited to side-coupled geometries only. In the following, we give just a few examples where the approach considered above can be naturally extended for more complex structures.

4.6.7.1 Defects in the Waveguide

The two basic geometries shown in Figures 4.6.1 can be further improved by placing partially reflecting elements into the waveguides.[31,40] These elements allow creating sharp and asymmetric response line shapes. In such systems, the transmission coefficient can vary from 0% to 100% in a frequency range narrower than the full width of the resonance itself.

To illustrate the effect of defects, in Fan's study[31] the response of the structure shown in Figure 4.6.8a was simulated using an FDTD scheme with perfectly matched layer boundary conditions. A pulse is excited by a monopole source at one end of the waveguide. The transmission coefficient is then calculated by Fourier transforming the amplitude of the fields at the other end, and is shown as a solid line in Figure 4.6.8b. In comparison, the transmission spectra for the same structure, but without the two small cylinders in the waveguide, is shown by a dashed line.

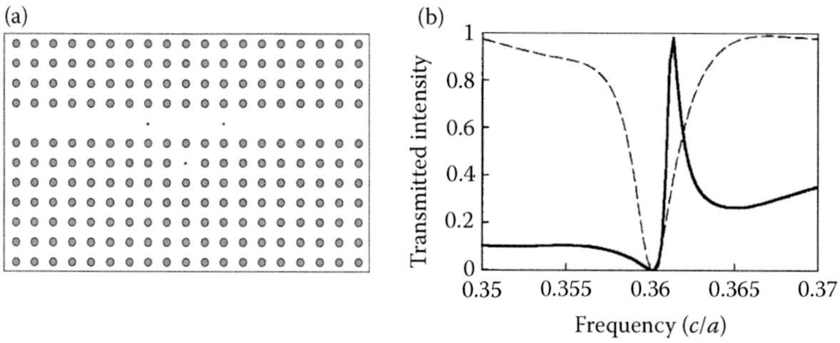

FIGURE 4.6.8
Light propagation in a photonic crystal waveguide with a side-coupled cavity. (a) Photonic crystal waveguide formed by removing a single row of rods. Within the line defect there are two smaller rods. A point defect, created by reducing the radius of a single rod, is placed away from the waveguide. (b) Transmission spectra through the structure (a) with (solid) and without (dashed) the two defects in the waveguide. (Adapted from S. Fan, Appl. Phys. Lett. **80**, 908, 2002.)

Importantly, no detailed tuning of either the resonant frequency or the coupling between the cavity and the waveguide is required to achieve asymmetric line shapes. Also, as the reflectivity of the partially reflecting elements need not be large, the underlying physics here differs from typical coupled-cavity systems, and resembles Fano resonances involving interference between a continuum and a discrete level.

4.6.7.2 Sharp Bends

One of the most fascinating properties of photonic crystals is their ability to guide electromagnetic waves in narrow waveguides created by a sequence of line defects, including light propagation through extremely sharp waveguide bends with nearly perfect power transmission.[41] It is believed that the low-loss transmission through sharp waveguide bends in photonic crystals is one of the most promising approaches to combine several devices inside a compact nanoscale optical chip.

Interestingly, the transmission through sharp bends in photonic crystal waveguides can be reduced to a simple model with Fano resonances, where the waveguide bend hosts a specific localized defect. Miroshnichenko and Kivshar[42] derived effective discrete equations for two types of the waveguide bends in two-dimensional photonic crystals and obtained exact analytical solutions for the resonant transmission and reflection. This approach seems to be useful in getting a deeper insight into the physics of resonant transmission.

4.6.7.3 Add-Drop Filters

Fano resonances can be used for a variety of photonic devices based on resonant tunneling. In particular, if two waveguides interact through a coupling element that supports a localized mode, a channel add-drop filter can be realized via the resonant tunneling between the waveguides.[43–45] The schematic diagram of a generic coupled system of this kind is shown in Figure 4.6.9a. At the Fano resonance, the propagating state excites the resonant modes, which in turn decay into both waveguides. The transmitted signal in the first waveguide is made up of the directly propagating signal and the signal that originates from the second path that visits the coupling region. To achieve complete transfer from one waveguide to another, these two signal components must interfere destructively. The reflected

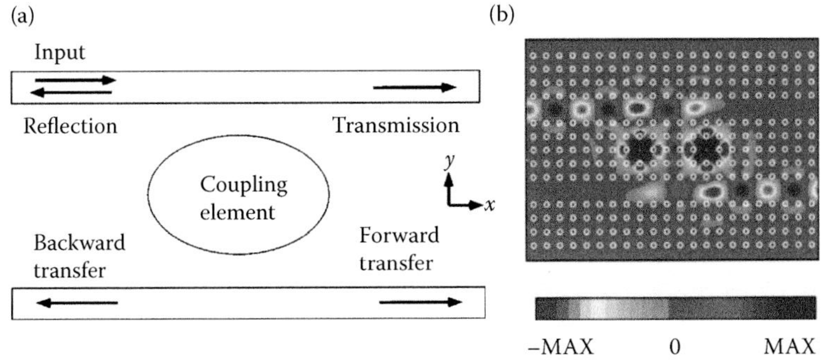

FIGURE 4.6.9
(**See color insert.**) Add-drop filter. (a) Schematic diagram of two waveguides coupled through an element which supports a localized resonant state. (b) Electric field pattern of the photonic crystal at the resonant frequency. The white circles indicate the position of the rods. (Adapted from S. Fan et al., Phys. Rev. Lett. **80**, 960, 1998.)

amplitude, on the other hand, originates entirely from the second path into the coupling region. Hence, at least two states in the coupling region are needed to achieve destructive interference of backscattered waves in the first waveguide. With these conditions satisfied, one may resonantly transfer the excitation from the first into the second waveguide.

This concept was developed in the study by Fan et al.[43] for the propagation of electromagnetic waves in a two-dimensional photonic crystal. To realize this concept, they used two photonic crystal waveguides and two coupled single-mode high-Q cavities. The photonic crystal is made of a square lattice of high-index dielectric rods, and the waveguides are formed by removing two rows of dielectric rods. The cavities are introduced between the waveguides by reducing the radius of two rods. The resonant states have different symmetry. An accidental degeneracy, caused by an exact cancellation between the two coupling mechanisms, is enforced by reducing the dielectric constant of four specific rods in the photonic crystal. The cancellation could equally have been accomplished by reducing the size of the rods instead of their dielectric constant.

This type of four-port photonic crystal system can be used for optical bistability, being particularly suitable for integration with other active devices on a chip.[45] A similar concept can be used for the realization of all-optical switching action in a nonlinear photonic crystal cross-waveguide geometry with instantaneous Kerr nonlinearity. There the transmission of a signal can be reversibly switched on and off by a control input.[46]

4.6.8 Conclusions

In this chapter, we have discussed how to use an effective discrete model to study wave propagation in photonic crystal waveguides with side-coupled cavities. Taking into account the linear coupling between the guided and localized states and applying the time-dependent version of Green's function formalism, we have calculated and characterized analytically the scattering of continuous waves. All examples presented here share the same basic feature—coexistence of resonant and nonresonant paths for propagation of scattering waves. It results in constructive and destructive interference phenomena and asymmetric lineshapes, known as Fano resonances. The resonant reflectivity, which is tunable via nonlinearity, takes the form of a nonlinear Fano resonance because the field at output is composed of linearly transmitted wave and resonantly reflected contribution from the localized cavity. We have demonstrated that in the presence of a nonlinear cavity continuous-wave scattering may show modulational instability near the resonance when the light intensity in the cavity starts growing in time. However, this instability can be largely suppressed by using Gaussian pulses such that the bistable transmission curve can still be recovered in accordance with the analysis of the steady-state transmission. Our analytical results based on an effective discrete model are in excellent agreement with numerical results obtained by direct FDTD simulations. The case of the two nonlinear side-coupled cavities is of special importance, as it demonstrates low-threshold bistable transmission of light owing to the ultra-high Q factor of the asymmetric Fano–Feshbach resonance, not available for any other systems.

Although derived in the context of photonic-crystal-based waveguides and cavities, the approach developed here is valid for other low-loss waveguide-resonator geometries under the condition of weak coupling.

References

1. J.-M. Lourtioz, H. Benisty, V. Berger and J.-M. Gerard, *Photonic Crystals: Towards Nanoscale Photonic Devices* (Springer, London 2008).
2. S. F. Mingaleev and Y. S. Kivshar, J. Opt. Soc. Am. B **19**, 2241 (2002).
3. S. G. Johnson, Y. Fink, M. Soljacic, M. Ibanescu, and J. D. Joannopoulos, Phys. Rev. E **66**, 055601 (2002); A. R. Cowan and J. F. Young, *ibid*. **68**, 046606 (2003); S. F. Mingaleev, A. E. Miroshnichenko, and Y. S. Kivshar, Opt. Express **15**, 12380 (2007); B. Maes, P. Bienstman, and R. Baets, *ibid*. **16**, 3069 (2008).
4. M. F. Yanik, S. Fan, and M. Soljacic, Appl. Phys. Lett. **83**, 2739 (2003).
5. S. F. Mingaleev, A. E. Miroshnichenko, Y. S. Kivshar, and K. Busch, Phys. Rev. E **74**, 046603 (2006).
6. A. E. Miroshnichenko, S. Flach, and Y. S. Kivshar, Rev. Mod. Phys. **82**, 2257 (2010).
7. U. Fano, Phys. Rev. **124**, 1866 (1961).
8. Y. Tanaka, J. Upham, T. Nagashima, T. Sugiya, T. Asano, and S. Noda, Nat. Mater. **6**, 862 (2007).
9. M. Notomi, K. Yamada, A. Shinya, J. Takahashi, C. Takahashi, and I. Yokohama, Phys. Rev. Lett. **87**, 253902 (2001).
10. P. Barclay, K. Srinivasan, and O. Painter, Opt. Express **13**, 801 (2005).
11. X. Yang, C. Husko, C. W. Wong, M. Yu, and D.-L. Kwong, Appl. Phys. Lett. **91**, 051113 (2007).
12. S. F. Mingaleev and Y. S. Kivshar, Opt. Lett. **27**, 231 (2002).
13. S. F. Mingaleev and Y. S. Kivshar, Phys. Rev. Lett. **86**, 5474 (2001).
14. A. J. Ward and J. B. Pendry, Phys. Rev. B **58**, 7252 (1998).
15. A. E. Miroshnichenko, S. F. Mingaleev, S. Flach, and Y. S. Kivshar, Phys. Rev. E **71**, 036626 (2005).
16. S. F. Mingaleev, Y. S. Kivshar, and R. A. Sammut, Phys. Rev. E **62**, 5777 (2000).
17. M. Soljacic and J. D. Joannopoulos, Nat. Mater. **3**, 211 (2004); J. Bravo-Abad, A. Rodriguez, P. Bermel, S. G. Johnson, J. D. Joannopoulos, and M. Soljacic, Opt. Express **15**, 16161 (2007).
18. R. Jacobsen, A. Lavrinenko, L. Frandsen, C. Peucheret, B. Zsigri, G. Moulin, J. Fage-Pedersen, and P. Borel, Opt. Express **13**, 7861 (2005); Y. A. Vlasov, M. O'Boyle, H. F. Hamann, and S. J. McNab, Nature **438**, 65 (2005); H. Gersen, T. J. Karle, R. J. P. Engelen, W. Bogaerts, J. P. Korterik, N. F. van Hulst, T. F. Krauss, and L. Kuipers, Phys. Rev. Lett. **94**, 073903 (2005).

19. E. Weidner, S. Combrié, A. de Rossi, N.-V.-Q. Tran, and S. Cassette. Appl. Phys. Lett. **90**, 101118 (2007).
20. A. E. Miroshnichenko, Y. Kivshar, C. Etrich, T. Pertsch, R. Iliew, and F. Lederer, Phys. Rev. A **79**, 013809 (2009).
21. A. E. Miroshnichenko, Phys. Lett. A **373**, 3586 (2009).
22. M. Fleischhauer, A. Imamoglu, and J. P. Marangos, Rev. Mod. Phys. **77**, 633 (2005).
23. D. D. Smith, H. Chang, K. A. Fuller, A. T. Rosenberger, and R. W. Boyd, Phys. Rev. A **69**, 063804 (2004).
24. L. Maleki, A. B. Matsko, A. A. Savchenkov, and V. S. Ilchenko, Opt. Lett. **29**, 626 (2004); W. Suh, Z. Wang, and S. Fan. IEEE J. Quantum Electron. **40**, 1511 (2004).
25. T. Opatrny and D.-G. Welsch, Phys. Rev. A **64**, 023805 (2001).
26. A. Naweed, G. Farca, S. I. Shopova, and A. T. Rosenberger, Phys. Rev. A **71**, 043804 (2005).
27. J. D. Franson and S. M. Hendrickson, Phys. Rev. A **74**, 053817 (2006).
28. Q. Xu, S. Sandhu, M. L. Povinelli, J. Shakya, S. Fan, and M. Lipson, Phys. Rev. Lett. **96**, 123901 (2006); M. Tomita, K. Totsuka, R. Hanamura, and T. Matsumoto, J. Opt. Soc. Am. B **26**, 813 (2009).
29. V. A. Fedotov, M. Rose, S. L. Prosvirnin, N. Papasimakis, and N. I. Zheludev, Phys. Rev. Lett. **99**, 147401 (2007); N. Papasimakis, V. A. Fedotov, N. I. Zheludev, and S. L. Prosvirnin, *ibid.* **101**, 253903 (2008); N. Papasimakis, Y. H. Fu, V. A. Fedotov, S. L. Prosvirnin, D. P. Tsai, and N. I. Zheludev, Appl. Phys. Lett. **94**, 211902 (2009); N. Papasimakis and N. Zheludev, Opt. Photonics News **20**, 23 (2009).
30. R. W. Boyd and D. J. Gauthier, Nature **441**, 701 (2006).
31. S. Fan, Appl. Phys. Lett. **80**, 908 (2002).
32. S. D. Landobasa, Y. Mario, and M. K. Chin, Opt. Express **14**, 12770 (2006); S. F. Mingaleev, A. E. Miroshnichenko, and Y. S. Kivshar, *ibid.* **16**, 11647 (2008).
33. H. Feshbach, Ann. Phys. (N. Y.) **5**, 357 (1958); F. H. Mies, Phys. Rev. **175**, 164 (1968).
34. A. I. Magunov, I. Rotter, and S. I. Strakhova, Phys. Rev. B **68**, 245305 (2003); M. Raoult and F. H. Mies, Phys. Rev. A **70**, 012710 (2004).
35. K. Hino, Phys. Rev. B **64**, 075318 (2001); A. E. Miroshnichenko, Phys. Rev. E **79**, 026611 (2009).
36. M. F. Yanik and S. Fan, Phys. Rev. Lett. **92**, 083901 (2004).
37. M. F. Yanik, W. Suh, Z. Wang, and S. Fan, Phys. Rev. Lett. **93**, 233903 (2004).
38. B. Maes, P. Bienstman, and R. Baets, J. Opt. Soc. Am. B **22**, 1778 (2005).
39. M. Galli, S. L. Portalupi, M. Belotti, L. C. Andreani, L. O'Faolain, and T. F. Krauss, Appl. Phys. Lett. **94**, 071101 (2009).
40. A. Khelif, B. Djafari-Rouhani, J. O. Vasseur, and P. A. Deymier, Phys. Rev. B **68**, 024302 (2003).
41. A. Mekis, J. C. Chen, I. Kurland, S. Fan, P. R. Villeneuve, and J. D. Joannopoulos, Phys. Rev. Lett. **77**, 3787 (1996); S. Y. Lin, E. Chow, V. Hietala, P. R. Villeneuve, and J. D. Joannopoulos, Science **282**, 274 (1998).
42. A. E. Miroshnichenko and Y. S. Kivshar, Opt. Express **13**, 3969 (2005).
43. S. Fan, P. R. Villeneuve, J. D. Joannopoulos, and H. A. Haus, Phys. Rev. Lett. **80**, 960 (1998).
44. S. Fan, P. R. Villeneuve, J. D. Joannopoulos, M. J. Khan, C. Manolatou, and H. A. Haus, Phys. Rev. B **59**, 15882 (1999).
45. M. Soljacic, C. Luo, J. D. Joannopoulos, and S. Fan, Opt. Lett. **28**, 637 (2003).
46. M. F. Yanik, S. Fan, M. Soljacic, and J. D. Joannopoulos, Opt. Lett. **28**, 2506 (2003).

4.7

Structural Features and Related Optical Responses of Magnetophotonic Crystals

Mitsuteru Inoue
Toyohashi University of Technology

Alexander V. Baryshev
Ioffe Physical-Technical Institute of the Russian Academy of Sciences
Toyohashi University of Technology

Alexander M. Merzlikin
Institute for Theoretical and Applied Electromagnetics

Hironaga Uchida
Tohoku Institute of Technology

Alexander B. Khanikaev
The University of Texas at Austin

CONTENTS

4.7.1 Introduction

Within the last decades the electrodynamics of inhomogeneous media has experienced rapid development. Advances in experimental and applied electrodynamics are connected with the rising potential of modern fabrication techniques, allowing creation of tiny structures with characteristic lengths (periodicity) comparable or even smaller than the wavelength of light. And development of theory is caused by translation (or "mapping") of wave phenomena from quantum theory of solids into electrodynamics. This mapping led to appearance of theories of photonic crystal (PhC),[1] diffusion of light,[2] backscattering,[3] and Anderson localization of light.[4]

Not only the paradigm of PhCs but even the term "crystal" has been transferred into electrodynamics from solid-state physics, where the behavior of electrons in a periodic system is well studied. It is the wave equation describing both the electron wave function and the electromagnetic wave that underlines the common features of a crystal in solid-state physics and electrodynamics. In the one-dimensional (1D) case, Schrödinger's equation

$$-\frac{1}{2m}\frac{\partial^2\Psi}{\partial z^2}+(U(z)-E)\Psi=0 \tag{4.7.1}$$

is identical to the electromagnetic wave equation describing a medium composed of isotropic materials

$$\frac{\partial^2 E}{\partial z^2}+\frac{\omega^2}{c^2}\varepsilon(z)E=0 \tag{4.7.2}$$

In fact, if one makes the following substitutions:

$$-2m(U(z)-E)\to\frac{\omega^2}{c^2}\varepsilon(z)\quad\text{and}\quad\Psi\to E \tag{4.7.3}$$

then Equation 4.7.1 turns into Equation 4.7.2.[5] The one-to-one correspondence between these equations allows us to establish the mapping between solid-state physics and electrodynamics of PhCs. Moreover, this correspondence allows us to transfer the well-known phenomena from physics of conventional solids into electrodynamics of PhCs. Finally, this is the reason why many theories of solid-state physics are applicable to problems in PhCs.

The most striking advantage of PhCs, when compared with natural optical materials, is the existence of photonic band gaps (PBGs), that is, the frequency ranges, where propagation of waves is suppressed by the Bragg diffraction. To illustrate the appearance of a PBG, we consider wave propagation through a 1D PhC [and later through magnetophotonic crystals (MPhCs)]; which is a periodical system of layers with dielectric constants ε_1 and ε_2 and thicknesses d_1 and d_2. The electromagnetic fields at the different surfaces of a single layer are linearly connected owing to linearity of Maxwell equations:

$$\begin{pmatrix}E\\H\end{pmatrix}_{\text{left}}=T\begin{pmatrix}E\\H\end{pmatrix}_{\text{right}} \tag{4.7.4}$$

where T is a transfer matrix. For one homogeneous layer, the T matrix takes the form of

$$T=\begin{pmatrix}\cos\left(\frac{\omega}{c}\sqrt{\varepsilon}d\right) & i\sqrt{\varepsilon}\sin\left(\frac{\omega}{c}\sqrt{\varepsilon}d\right)\\ \frac{i}{\sqrt{\varepsilon}}\sin\left(\frac{\omega}{c}\sqrt{\varepsilon}d\right) & \cos\left(\frac{\omega}{c}\sqrt{\varepsilon}d\right)\end{pmatrix}. \tag{4.7.5}$$

As it follows from Equation 4.7.4, the T matrix of a system of layers is equal to the product of the T matrices of all layers. In an infinite periodical system, Bloch's theorem asserts that the eigensolution is changed only by the phase at propagation through a unit cell consisting of two layers

$$\begin{pmatrix} E \\ H \end{pmatrix}_{\text{left}} = e^{ik_{\text{Bl}}a} \begin{pmatrix} E \\ H \end{pmatrix}_{\text{right}} \tag{4.7.6}$$

If we compare Equations 4.7.6 and 4.7.4, one may find that $e^{ik_{\text{Bl}}a}$ is an eigenvalue of the T matrix of the unit cell, where k_{Bl} is the Bloch wavevector. For a PhC with such a unit cell, we come to a dispersion relation:

$$\cos(k_{\text{Bl}}a) = \cos\left(\frac{\omega}{c}\sqrt{\varepsilon_1}d_1\right)\cos\left(\frac{\omega}{c}\sqrt{\varepsilon_2}d_2\right) - \frac{1}{2}\left(\sqrt{\frac{\varepsilon_1}{\varepsilon_2}} + \sqrt{\frac{\varepsilon_2}{\varepsilon_1}}\right)\sin\left(\frac{\omega}{c}\sqrt{\varepsilon_1}d_1\right)\sin\left(\frac{\omega}{c}\sqrt{\varepsilon_2}d_2\right), \tag{4.7.7}$$

where $a = d_1 + d_2$ is a thickness of the unit cell. For small frequencies, we will get $k_{\text{Bl}} = \omega/c\sqrt{(\varepsilon_1 d_1 + \varepsilon_2 d_2)/a}$. This means that the dispersion relation $k_{\text{Bl}}(\omega)$ is just a linear relation for long waves. At the same time, we have

$$\frac{\omega}{c}\sqrt{\varepsilon_1}d_1 + \frac{\omega}{c}\sqrt{\varepsilon_2}d_2 = \pi \tag{4.7.8}$$

for the Bragg condition, and the right part of Equation 4.7.7 becomes equal to $-1 - \left(\left(\sqrt{\varepsilon_1} - \sqrt{\varepsilon_2}\right)^2 / \left(2\sqrt{\varepsilon_1\varepsilon_2}\right)\right)\sin^2\left(\omega/c\sqrt{\varepsilon_1}d_1\right)$; this negative value is less than –1 and, therefore, k_{Bl} is imaginary. This implies that, firstly, wave propagation is prohibited because of the resonant reflection from an infinite PhC and that a PBG develops for such waves. Secondly, the group velocity tends to zero for modes approaching to edges of the PBG. Figure 4.7.1 illustrates a typical dispersion $k_{\text{Bl}}(\omega)$ for a 1D MPhC. One can see that the dispersion is a line at small frequencies and is flattened at the edges of the PBG.

It should be pointed out that the band edges are distinguished not only by zero group velocities but also by a specific field distribution. It will be shown in Figure 4.7.3 that the electric field of modes corresponding to the low-frequency edge of the PBG is mainly concentrated in the high-permittivity layers. On the contrary, the electric field concentration is greater in the low-permittivity layers for modes from the high-frequency edge of the PBG. This phenomenon can be referred to as the frequency analog of the Borrmann effect,[6] which is known in the x-ray spectroscopy of conventional solids.

In the next sections we will discuss properties of MPhCs, where the translational invariance is broken. For such MPhCs, new solutions of Maxwell's equations can arise within their PBGs. These solutions will be shown to have a resonant behavior, where propagating modes are spatially localized in magneto-optical (MO) constituents of MPhCs; that gives rise to a strong enhancement of their MO response.

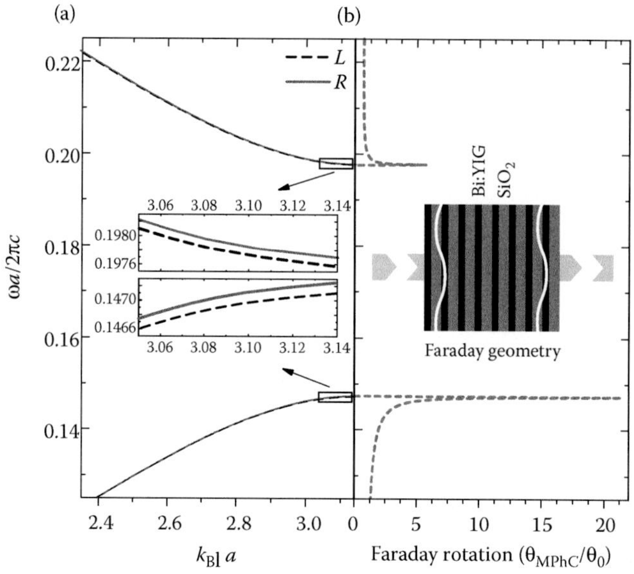

FIGURE 4.7.1
(a) Photonic band structure for circularly polarized waves calculated for normal incidence of light onto a one-dimensional (1D) magnetophotonic crystal (MPhC) composed of SiO_2 layers with the dielectric constant of $\varepsilon_1 = 2.1$ and Bi:YIG layers with the diagonal component of dielectric constant of $\varepsilon_2 = 5.6$ and the off-diagonal component (gyration) of $g = 0.1$; absorption and dichroism in the layers are neglected. (b) Normalized Faraday rotation of a 1D MPhC comprised 20 Bi:YIG/SiO_2 bilayers; here θ_0 is the rotation of a homogeneous Bi:YIG film with a total thickness as that of Bi:YIG in the MPhC.

In spite of the similarity of wave phenomena in different areas of physics, there is a difference between electrodynamics and the quantum theory of solids: the electron wave function is scalar (neglecting effects associated with spin) or two-spinor (taking into account degrees of freedom associated with spin), whereas the electric and magnetic fields are vector quantities. This difference becomes pronounced for MO materials that can be described by a permittivity tensor

$$\hat{\varepsilon} = \begin{pmatrix} \varepsilon & ig & 0 \\ -ig & \varepsilon & 0 \\ 0 & 0 & \varepsilon \end{pmatrix}, \tag{4.7.9}$$

where g describes the MO activity, that is, gyrotropic properties of an MO material.

Application of MO materials as components of MPhCs is of great technological interest, as manipulation of light characteristics (polarization plane and transmittance) is possible at application of an external magnetic field. For example, modes with certain polarizations and the frequencies in the immediate proximity of a PBG edge *may satisfy either passband or band gap condition*, that is, their propagation may be either allowed or suppressed by magnetization. Unfortunately, most optical materials possess very small values of g. That is why it is not possible to directly measure the spectral shift of PBGs. However, one may observe an alteration of PBGs when measuring rotation of the polarization plane—the Faraday effect.[7]

The Faraday effect causes a rotation of the polarization plane of a wave propagating through an MO medium. This rotation is a result of splitting the dispersions for right and left circularly polarized waves:

$$k_{\text{right}}(\omega) = \frac{\omega}{c}\sqrt{\varepsilon - g} \quad \text{and} \quad k_{\text{left}}(\omega) = \frac{\omega}{c}\sqrt{\varepsilon + g}$$

On entering an MO material, a linearly polarized wave splits into cophased left and right circularly polarized waves of the same amplitude. These waves gain a relative phase shift during the propagation along a traveled distance L, and, then, when emerging from the material, they rebuild a linearly polarized wave with the rotated polarization plane. The angle of polarization rotation is calculated by

$$\theta_F = \frac{1}{2}(k_{\text{left}}L - k_{\text{right}}L) = \frac{\omega}{2c}\left(\sqrt{\varepsilon + g} - \sqrt{\varepsilon - g}\right)L \approx \frac{\omega g L}{2c\sqrt{\varepsilon}} \quad (4.7.10)$$

Let us now go back to Figure 4.7.1 and consider the Faraday rotation in the MPhC. The dispersion $k_{Bl}(\omega)$ is close to a line at small frequencies, and the Faraday rotation of the MPhC has the same magnitude as that of an inbuilt homogeneous MO material. However, the dispersion curves $k_{\text{left}}(\omega)$ and $k_{\text{right}}(\omega)$ more and more flatten when approaching the PBG. And, for a wave with a certain frequency of ω^*, this results in the enhancement of the Faraday rotation due to a much larger difference in $|k_{\text{left}}(\omega^*)-k_{\text{right}}(\omega^*)|$.

Another opportunity for enhancing the Faraday rotation is to utilize the concept of the optical resonator based on MPhCs. Resonant propagation of a wave through a resonator may be represented as a *multipass regime in nonreciprocal materials*. This means that the wave bounces inside the resonator, that is, travels across it many times before emerging. Thus, the effective traveled distance L from Equation 4.7.10 can be much greater than the thickness of the built-in MO material. That is why *the Faraday rotation is accumulated owing to nonreciprocity* of the MO material as L, governed by the Q-factor of the resonator, increases.

Initial theoretical studies on MPhCs deal with light propagation in discontinuous magnetic media with a 1D structure,[8] where the Faraday rotation is analyzed using the matrix approach. One-dimensional structures considered in these works are multilayer films composed of bismuth-substituted yttrium iron garnet (Bi:YIG) and SiO_2, which are piled up in an arbitrary sequence. For such multilayers, the enhancement both in transmittance and in the Faraday rotation angle θ_F is shown to originate from the localization of light caused by multiple interference.[9] Theoretical analysis shows that *the angle of Faraday rotation increases as the degree of light localization rises*. Largest enhancement of θ_F can be obtained in 1D MPhCs with a microcavity structure, where a Bi:YIG layer is sandwiched between two Bragg reflectors.[10] Following theoretical predictions, 1D MPhCs were fabricated by different techniques and their experimental evaluation was performed. Below we will discuss experimental MPhCs created by sputtering. Qualitative distinctions among various MPhCs will be reviewed. As for their technical parameters, they can be found in detail in works cited in this chapter or specifically in the study by Inoue et al.[11]

For two-dimensional (2D) [and potentially for three-dimensional (3D)] MPhCs, a regime of multiple Bragg diffraction appears to play a crucial role. This regime takes place when the Bragg resonances for different families of planes of a PhC appear at the spectrally close position or simply overlap. In PBG diagrams, this regime corresponds to the points where two or more bands become degenerate (or nearly degenerate). It appears that the MO activity affects PBGs and optical responses of MPhCs only in the close proximity to such degeneracies.[12] Owing to the MO activity, components of the permittivity tensor "intermix" polarizations corresponding to the degenerate bands; note that this effect is rather marginal for modes that are far from degeneracies of a PBG.[13]

To explain this, a comparison of MPhCs with gyroanisotropic crystals (birefringent media possessing MO activity[7]) is instructive. In the case of gyroanisotropic crystal, birefringence significantly suppresses the effect of magnetization at the condition that the light propagation direction differs from the optical axis of the crystal. This occurs because the eigenmodes of birefringent crystals are nondegenerate linearly polarized ordinary and extraordinary waves. In general, these waves cannot effectively intermix and form the circularly polarized eigenmodes (inherent to conventional MO materials) because of a mismatch in their refractive indexes.[3] For this reason, the eigenmodes of gyroanisotropic crystals are elliptically left- and right-polarized modes propagating with distinct phase velocities. That is why the birefringence prevents nonreciprocal polarization rotation unless the propagation direction is chosen to be collinear with the optical axis. By analogy, we consider an "effectively birefringent" 2D MPhC, where the transverse electric (TE) and transverse magnetic (TM) modes have different effective refractive indexes and are analogous to the ordinary and extraordinary waves of a birefringent crystal (see Figure 4.7.2a).

However, TE and TM modes are flattened and degenerate at the high symmetry points; the modes have the same effective phase refractive index $k_{Bl}c/\omega$. As the birefringence no longer plays any destructive role in MPhCs,[14] the MO activity results in an efficient coupling between the TE and TM eigenmodes and formation of contrariwise rotating elliptically polarized modes. They build a ray with the rotated polarization plane (or the rotated major axis of the polarization ellipse) when emerging from 2D MPhCs.

Despite the similarities between MPhCs and birefringent MO crystals, there are several important differences. First, the dispersion curves are folded owing to the Bragg diffraction in the case of PhCs. This results in the appearance of an infinite number of

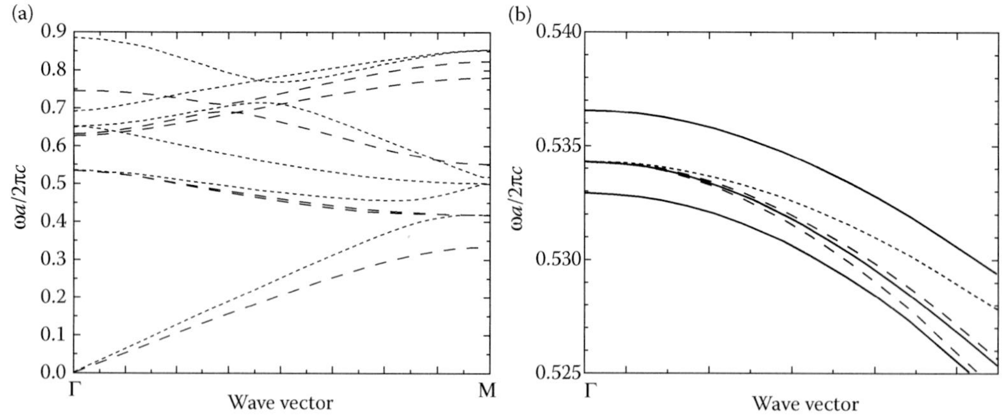

FIGURE 4.7.2
(a) Photonic band structure of a two-dimensional photonic crystal (PhC) with a square lattice of circular dielectric rods: transverse electric (long dash) and transverse magnetic (short dash) eigenmodes of the nonmagnetized structure. (b) Photonic band structure of the magnetophotonic crystal (shown by solid lines) and the nonmagnetized PhC shown in plot (a) in the proximity of the Γ point. Parameters used for calculations are $\varepsilon_{xx} = 6.25$ and $\varepsilon_{xy} = g = 0.06$ for magneto-optical rods made of Ce-substituted bismuth iron garnet in the air; the radius of the rods is $0.4a$ (where a is the lattice constant).

frequency-dependent dispersion branches $\omega(k_{Bl})$ and the formation of PBGs. At high symmetry points, where nonaccidental degeneracies usually take place, $\omega(k_{Bl})$ curves are very flat. We recall again that this reflects the strong Bragg scattering regime realized for the modes with a slow group velocity. The splitting degenerate branches are large, and the difference in effective refractive indexes for eigenmodes exceeds the value for elliptically polarized modes found in "non-PhC-like" MO materials. Second, there exist narrow frequency ranges, where only one of the hybrid modes can propagate. This implies different widths of band gaps for differently polarized modes. Finally, the third difference is that the degeneracy between eigenmodes is not necessarily twofold as in the case of birefringent crystals. This is illustrated in Figure 4.7.2b, where a photonic band structure of a 2D MPhC is threefold degenerate. Generally, all three modes can be coupled giving rise to formation of Bloch waves with a quite complicated polarization state.

4.7.2 MPhCs with the Simplest Design

A realization of 1D MPhCs with the simplest design is shown in Figure 4.7.3. This bounded MPhC is representative of the class discussed in Figure 4.7.1. The multilayer was a $(SiO_2/Bi{:}YIG)^n$ stack of alternating $\lambda/4$-thick SiO_2 (150 nm) and Bi:YIG (100 nm) layers; the index

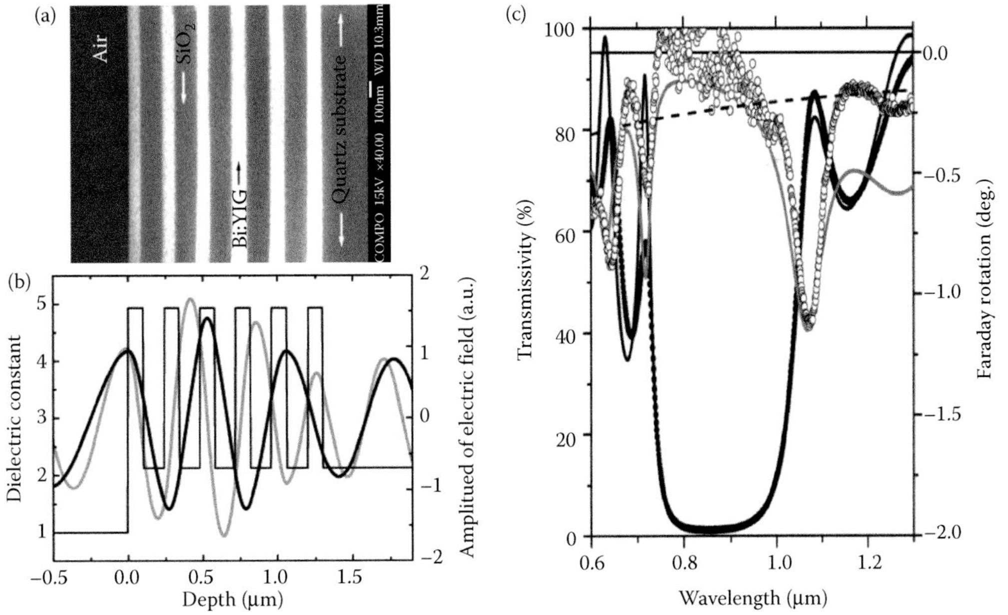

FIGURE 4.7.3
(a) Scanning electron microscope image of a cleaved facet of a one-dimensional (1D) magnetophotonic crystal (MPhC). (b) Sketch of the spatial distribution of dielectric constants and the in-sample distribution of the electric field amplitude for wavelengths corresponding to photonic band-gap edges; gray line is for the short-wavelength edge and black line for the long-wavelength edge. (c) Transmission (solid circles) and Faraday rotation spectra (open circles) of the 1D MPhC at normal incidence and the saturating external magnetic field of 2 kOe. Solid curves show the fitting of the experimental data by the four-by-four matrix approach. Dashed line is for the interference-subtracted spectrum of the Faraday rotation for a Bi:YIG film with the thickness equal to the total thickness of Bi:YIG in the 1D MPhC shown in panel (a).

of $n = 5$ stands for the number of bilayers. A scanning electron microscope (SEM) image of the MPhC shows abrupt interfaces between the layers as well as their reproducible thicknesses. Note that the used materials have low optical losses in the red (and near-infrared) spectral range, and Bi:YIG has a significant Faraday rotation. *There is always a trade-off between the optical and magneto-optical responses for a particular MO material*—called in literature the figure of merit (FOM),[7] as a larger Faraday rotation is accompanied by a rise in absorption. For the MPhC, transmission was strongly suppressed in a range of 730–1020 nm. In this spectral range, θ_F was found to decrease to zero for light escaping from the MPhC. On the contrary, transmittance rose outside the attenuation band, exhibiting interference fringes. The magnitude of θ_F oscillated and had the local maxima at 640 and 720 nm that corresponded to the maxima in the transmission spectrum (see Figure 4.7.3c).

The maximum angle of the Faraday rotation was observed at the long-wavelength edge of the attenuation band, at 1070 nm. The latter was a clear manifestation of the optical Borrmann effect in the artificially stratified media.[6] The electromagnetic field of modes with wavelengths of 720 and 1070 nm is strongly concentrated in the Bi:YIG layers. The electric field augmentation occurred more effectively for waves with the wavelengths corresponding to the long-wavelength edge of the attenuation band (see the maxima of the black curve in Figure 4.7.3b that occur in the garnet layers). We will show in the next sections that the considered $(SiO_2/Bi{:}YIG)^n$ multilayers provide building blocks for 1D MPhCs with various designs.

4.7.3 Light Localization or Defect Modes in Fabry–Pérot 1D MPhCs

One-dimensional MPhCs with a single[10] and dual Fabry-Pérot cavity structure[15] are theoretically shown to be most efficient media to gear up the Faraday rotation of existing MO materials. Single microcavities composed of two dielectric Bragg reflectors and a Bi:YIG defect layer incorporated between them, $(Ta_2O_5/SiO_2)^n/Bi{:}YIG/(SiO_2/Ta_2O_5)^n$, were realized and evaluated in linear and nonlinear optical experiments.[16] One-dimensional MPhCs with $n = 5$ is shown in Figure 4.7.4 together with an in-sample electric field distribution and optical spectra. For transmission and Faraday rotation spectra, formation of sharp peaks at $\lambda = 720$ nm within an attenuation band of 600–850 nm is the result of the well-known resonant light propagation through the microcavity—the light localization inside the Bi:YIG layer. A remarkable feature of such MPhCs is that the *microcavity provides a high transmissivity and a large enhancement* of *Faraday rotation simultaneously*. The resonant transmission of the microcavity-type 1D MPhCs are controllable through the Bi:YIG defect layer; that is, the thickness of Bi:YIG governs spectral positions and the number of localized modes exhibiting enhanced angles of the Faraday rotation.[16]

Generally, transmission peaks located within the PBG of the considered single-cavity MPhCs split on magnetization as a result of the difference in the resonant wavelengths corresponding to the right and left circularly polarized light. This causes an unwanted effect, that is, a reduction of the transmitted signal. That is why magnitudes of transmissivity and the MO response of single-cavity MPhCs is also a fundamental trade-off. To largely increase the responses, 1D MPhCs containing two defects—*dual cavities*—are theoretically analyzed and shown to exhibit a maximal possible transmissivity and Faraday rotation up to 45°.[15] However, a recent experimental realization of the dual-cavity MPhCs showed that even tiny structural defects have an extremely destructive effect, and the responses of the fabricated dual cavities were smaller than predicted ones.[17]

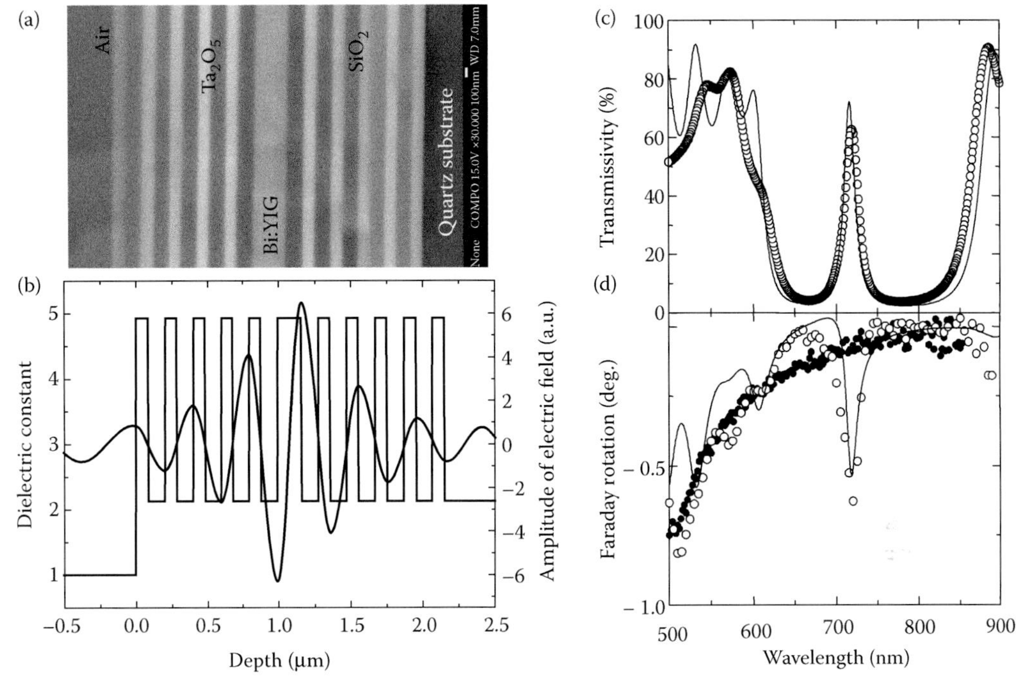

FIGURE 4.7.4
(a) Cross-sectional SEM image of a 1D MPhC—a microcavity with the $(Ta_2O_5/SiO_2)^5$/Bi:YIG/$(SiO_2/Ta_2O_5)^5$ structure. (b) Sketch of the spatial distribution of dielectric constants and the in-sample distribution of the electric field amplitude for a resonant wavelength. (c) Transmission and (d) Faraday rotation spectra of the 1D MPhC.

4.7.4 Optical Tamm States in 1D MPhCs

An intriguing example of 1D MPhC supporting *optical surface states* has been recently demonstrated theoretically[18] and experimentally.[19] It was shown that a boundary between two 1D PhCs causes a localized state to appear, and the frequency of such a state located inside the overlapping PBGs of two PhCs. The eigensolution $f(z)e^{ik_{\mathrm{Bl}}z}$ inside each PhC is the Bloch wave and consists of two factors—the exponential part $e^{ik_{\mathrm{Bl}}z}$, where k_{Bl} is the Bloch wavenumber, and a periodic function $f(z)$. For the band-gap frequencies, the electromagnetic field decreases exponentially from interfaces of a PhC (as the Bloch wavenumber k_{Bl} is an imaginary quantity) and is still modulated by the $f(z)$ function. In the case of two adjoining PhCs, two f functions make it possible to match solutions for two Bloch waves exponentially evanescent from the boundary between the PhCs. This results in a state spatially localized at the interface between two PhCs. This state is intrinsically surface one and is cognate to the Tamm state[14] known in solid-state physics. Another mapping of Tamm's structure onto the electrodynamics would be a system comprising a PhC and a film with the negative permittivity adjoined to the PhC. The existence of optical Tamm states (OTSs) in such a system has been theoretically predicted in various studies.[20] Below we demonstrate OTSs in MPhCs.

Figure 4.7.5a shows one of experimental realizations comprising two adjoining 1D PhCs that were successively sputtered on a quartz substrate. The first nonmagnetic PhC was a dielectric multilayer $Ta_2O_5/(SiO_2/Ta_2O_5)^5$. Then, the magnetic $(Bi{:}YIG/SiO_2)^5$ multilayer

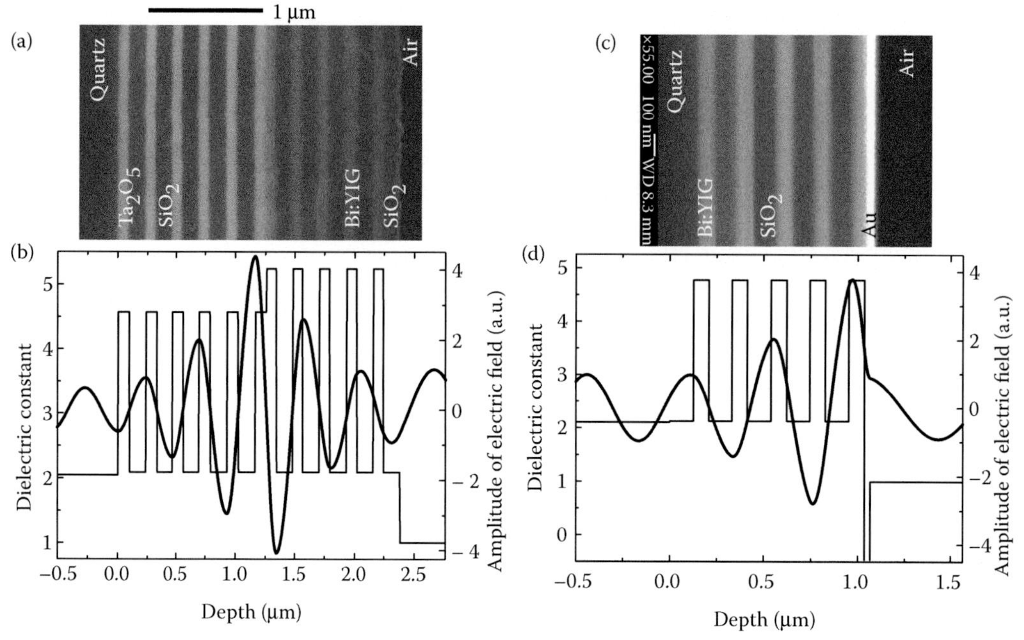

FIGURE 4.7.5
(a) and (c) Scanning electron microscope images of two realizations of optical Tamm structures: a system of two adjoining photonic crystals (magnetic and nonmagnetic slabs) and a multilayer terminated with an Au layer. (b) and (d) Sketches of the spatial distribution of dielectric constants and the in-sample distribution of the electric field amplitude for the resonant wavelength.

was formed so that the resultant MPhC had a structure of quartz substrate/Ta_2O_5/(SiO_2/Ta_2O_5)^5/(Bi:YIG/SiO_2)^5 (Figure 4.7.5a). Another more simple Tamm structure was a quartz substrate/(SiO_2/Bi:YIG)^5/Au multilayer (see Figure 4.7.5c). For resonant wavelengths, the electric field pattern within the MPhCs, together with the dielectric constant profile, is illustrated in Figure 4.7.5b and d. For both realizations, the amplitude is *remarkably high at the interface* between the MO substructure and the adjoined one, and it falls exponentially away from the interface. Such a distribution confirms the formation of the OTS. Also, the designs of the MPhCs were such that all, for the resonant wavelength of OTS, maxima of the field pattern were spatially located inside the Bi:YIG layers.

Figures 4.7.6 and 4.7.7 illustrate measured and calculated responses of the MPhCs shown in Figure 4.7.5. To verify that the peaks associated with OTSs are due to resonant coupling to Bi:YIG, MO spectra of reference samples that were (SiO_2/Bi:YIG)^5 multilayers (discussed in Section 4.7.2) identical to that used in the MPhCs were measured. According to some works,[18,20] if one of the adjoining PhCs is magnetic, an OTS should cause a substantial enhancement of the Faraday rotation.

The Faraday rotation of the magnetic PhC (depicted by the monotonous black solid line in Figure 4.7.6b) followed the ordinary response from the Bi:YIG constituents. As for the combined MPhC, the Faraday rotation was enhanced at the transmission peak of 800 nm; θ_F *was almost one order of magnitude larger* than that of the magnetic PhC. It is worth noting that, qualitatively and in the view point of the measured optical and MO responses, this realization of MPhCs is very close to the single magnetic microcavity (see Section 4.7.3).

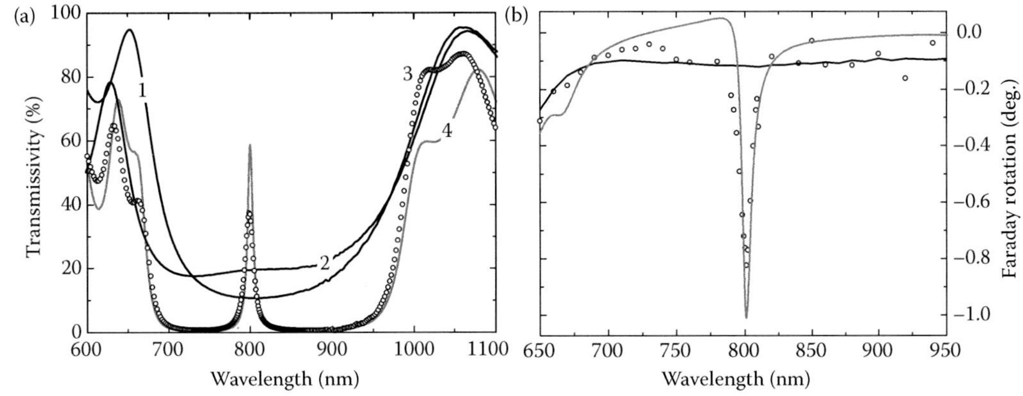

FIGURE 4.7.6
(a) Transmissivity of a nonmagnetic photonic crystal (PhC) and a magnetic PhC are denoted by lines 1 and 2, respectively. Transmissivity of the system of adjoining PhCs [referred in the text as the magnetophotonic crystal (MPhC)] is shown by circles (3); its calculated spectrum is given by curve 4. (b) Angle of Faraday rotation: a reference $(SiO_2/Bi{:}YIG)^5$ multilayer (black solid line), the experimental (circles) and calculated (gray line) spectra of the MPhC.

To finish describing experimental realizations of 1D MPhCs, we now consider an OTS in MPhCs with a structure of $(SiO_2/Bi{:}YIG)^5/Au$, in which the periodicity is terminated by an Au film with negative permittivity. For this MPhC, the OTS with a wavelength of 780 nm appeared inside the attenuation band of 600–850 nm. Intensity of the transmission peak was 25%, and the corresponding enhanced Faraday rotation angle was of –0.4° (see Figure 4.7.7). As for the reference $(SiO_2/Bi{:}YIG)^5$ multilayer, a peak related to the Borrmann-like enhancement is seen at a wavelength of 875 nm. Note again that this Tamm structure represents the direct optical analog of the conventional crystals supporting the electron Tamm states, whereas the system of adjoined PhCs resembles the microcavity-type MPhC discussed in Section 4.7.3.

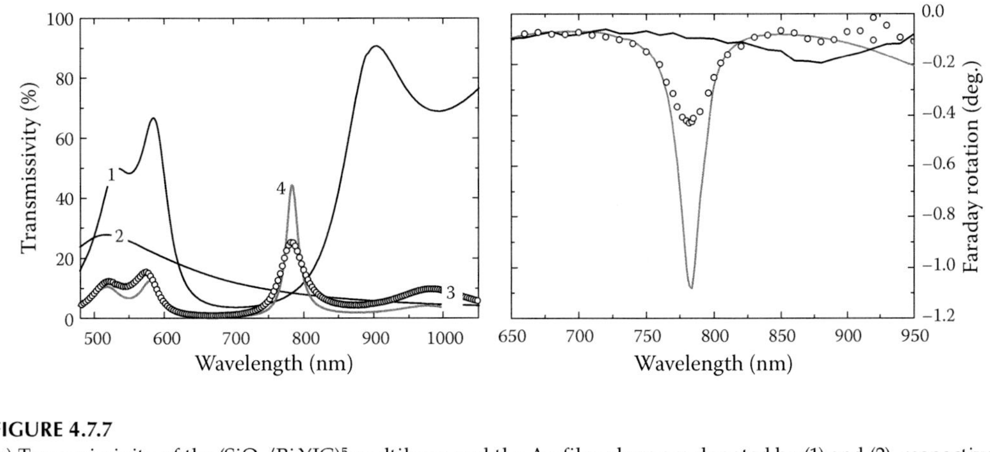

FIGURE 4.7.7
(a) Transmissivity of the $(SiO_2/Bi{:}YIG)^5$ multilayer and the Au film alone are denoted by (1) and (2), respectively. Transmissivity of sample $(SiO_2/Bi{:}YIG)^5/Au$ is shown by circles (3); its calculated spectrum is given by curve 4. (b) Angle of Faraday rotation: the $(SiO_2/Bi{:}YIG)^5$ multilayer (experiment, black solid line) and the $(SiO_2/Bi{:}YIG)^5/Au$–experiment (circles) and theory (gray line).

To underline peculiarities of MPhCs supporting OTSs, one may state that these structures can be employed for localizing light within any active material used as the constitutive layers of PhCs or introduced at the interface between two PhCs. OTSs might provide the additional mechanism to increase the electric field strength in photonic structures called upon to excite long-range surface plasmon-polaritons. Moreover, they can be attractive for sensing applications, since their optical responses may strongly vary with a change of dielectric conditions in the vicinity of open noble metal surfaces.

4.7.5 Multiple Bragg Diffraction in Quasi-2D MPhCs

Up to the present section this chapter dealt with the realizations of 1D MPhCs, although 2D and 3D MPhCs were fabricated by combining different approaches.[21] In fact, the MO responses of such experimental structures are not competitive with those of 1D MPhCs. However, theoretically predicted effects (for instance, magnetic superprism,[12] waveguide circulator[22]) provide room for the next challenges in two and three dimensions. In the search for successful MPhCs with higher dimensionality, MO structures with quasi-2D (Q-2D) structures[23] were fabricated by autocloning.[24] Below we will briefly discuss structural properties and spectra of Q-2D MPhCs.

Two-dimensionality in the Q-2D was set by structured substrates, where 1D arrays of photoresist bars had a lattice constant of 400 nm; the bars had a width of 200 nm (inset in Figure 4.7.8a). Up to seven Bi:YIG/SiO_2 bilayers were stacked; an SEM image in Figure 4.7.8a shows a typical cross-section of the fabricated Q-2D multilayers. (To illustrate the symmetry of typical samples, a reference $(SiO_2/Ta_2O_5)^5$ multilayer is shown.) Symmetry of the photoresist pattern was replicated across and along the MPhC. The cleaved edge reveals Q-2D periodicity, where scattering elements can be denoted by parallelograms (see Figure 4.7.8b). Sputtering was made simultaneously for the Q-2D MPhCs and for a regular, reference 1D MPhC fabricated on the top of a flat substrate (Section 4.7.2). Geometry of the experiment, together with polarization directions for incident light and the orientation of samples, is shown in Figure 4.7.9a. The linearly polarized electric field was set by a polarizer to be of a chosen orientation: the electric field vector was across the photoresist bars for *p*-polarized light ($E_\perp$, $\psi = 0°$—a TM mode discussed in Section 4.7.1), and it was parallel to the bars for *s*-polarized light ($E_\parallel$, $\psi = 90°$—a TE mode).

Transmission spectra and spectra of the Faraday rotation at the normal incidence for the Q-2D $(Bi{:}YIG/SiO_2)^7$ multilayer, and the reference 1D MPhC are shown in Figure 4.7.9. One can see spectrally neighboring, overlapping bands in transmission spectra (Figure 4.7.9b). Light with wavelengths from such an overlap will experience the so-called multiple Bragg diffraction (the effect[25] known from the x-ray spectrography). In the case of MO materials, such diffraction should introduce unusual responses from MPhCs with 2D and 3D structures.

We analyze now the structure and related MO responses in spectra of Q-2D MPhCs. The band in a range of 580–830 nm for a 1D MPhC (thick gray line) can be traced in the spectra of the Q-2D MPhC. Interestingly, the $T_\parallel$- and $T_\perp$-polarized spectra of the sample have a reach structure with a number of resonant peaks and showed large transmission anisotropy through the measured spectral range. The prominent difference in the spectra of the sample was an additional stop band in a range of 535–600 nm. This band with its minimum at $\lambda = 570$ nm is clearly seen in the $T_\perp$ spectrum. For $E_\parallel$-polarized light, the additional band merged with that located in the long-wavelength range. Origin of the additional band is likely due to diffraction from "oblique sine curve piece" (OS) periodical scattering elements

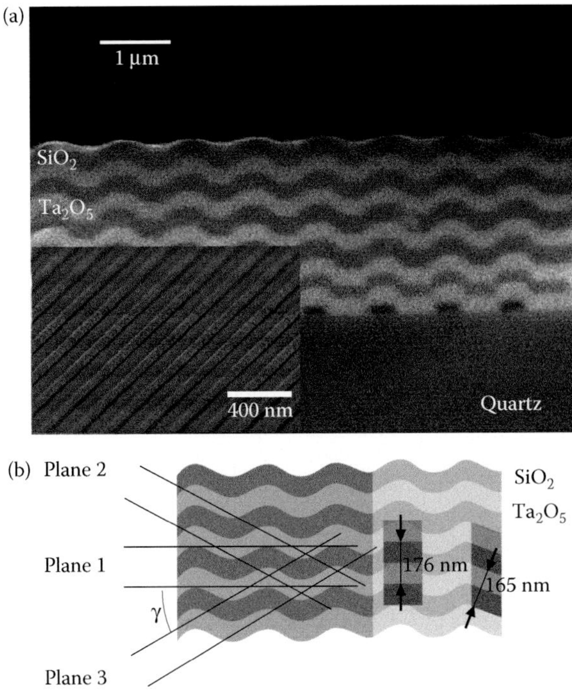

FIGURE 4.7.8
(a) Scanning electron microscope image of a typical quasi-two-dimensional (Q-2D) photonic crystal with a structure of $(SiO_2/Ta_2O_5)^5$ fabricated on the top of the photoresist pattern shown in the inset. (b) Model of the Q-2D structure: diffraction planes, interplanar spacing, and scattering elements are shown. (From S. Baek, A. V. Baryshev, and M. Inoue, Appl. Phys. Lett. **98**, 101111, 2011. With permission.)

of the multilayers, and "inflection sine curve piece" (IS) scattering elements are responsible for the spectral features in long wavelengths. These three sets of effective diffraction planes in the structure can be seen in Figure 4.7.8b. Thus, plane 1 (or IS elements) is responsible for attenuation in the longer range. Planes 2 and 3 (or OS elements) diffract light with shorter wavelengths of $\lambda = 530–600$ nm. Actually, when scanning the angle of incidence α, analysis of spectral features showed that the OS band split into two bands moving apart as α rises (see Figure 4.7.10a and b). Decomposing the spectra into two Gaussians, shifts of minima for the splitting bands can be roughly fitted by the Bragg law, $\lambda = 2dn_{eff}\cos(\gamma - \sin^{-1}(\sin\alpha/n_{eff}))$ with due account of refraction (Figure 4.7.10d). Here an effective refractive index (n_{eff}), an angle (γ) between the Bragg diffraction plane and the sample basal plane, and an interplanar distance (d) were the fitting parameters.

For the reference crystal, the Borrmann-effect-related enhancement of Faraday rotation happened at the band edges ($\lambda = 590$ and 825 nm); line 4 in Figure 4.7.10c shows the enhancement at the short-wavelength edge. Similarly, a significant enhancement of the rotation angle was observed in the $\theta_{||}$ spectrum of the sample for $\lambda = 550$ nm, at the short-wavelength edge of the additional band. If compared to $\theta_{1D\ MPhC}$, one order of magnitude larger rotation angle was observed for light with $\lambda = 610$ nm tunneling through the sample; transmissivity in this range was suppressed. Also, the sign of polarization rotation sharply changes in the range of 590–610 nm.

The observed features are attributed to the multiple Bragg diffraction, where light beams with a fixed wavelength simultaneously experience diffraction events from three

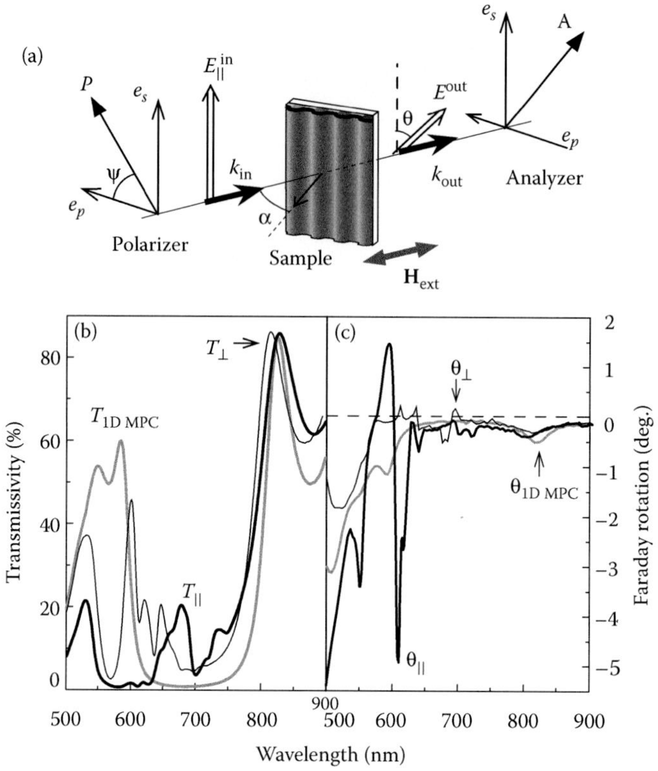

FIGURE 4.7.9
(a) Scheme of experiment. (b) Polarization-resolved transmission and (c) Faraday rotation spectra of a quasi-two-dimensional magnetophotonic crystal (MPhC), the corrugated $(SiO_2/Bi{:}YIG)^7$ multilayer, at normal incidence. Spectrum of a one-dimensioanl MPhC with the same number of $SiO_2/Bi{:}YIG$ bilayers (gray line) is shown for reference. (From S. Baek, A. V. Baryshev, and M. Inoue, Appl. Phys. Lett. **98**, 101111, 2011. With permission.)

diffraction planes. In this regime, light should be maximally trapped inside the sample, and, as expected in accord with nonreciprocity of the Faraday effect, the polarization rotation should be accumulated. However, the $\theta_{\perp}$ spectrum illustrates that the Faraday rotation can be suppressed (see Figure 4.7.10c). Processes contributing to the MO responses and results of diffraction experiments (showing a superprism effect accompanied with the demultiplexing phenomenon) are discussed in more detail in the study by Baek et al.[26]

If the multiple Bragg diffraction is a key factor for altering polarization, detuning the overlap of the adjacent stop bands should bring a notable modification in the Faraday rotation spectra. We prove now this assumption. Figure 4.7.10c shows transmission and Faraday rotation spectra for normal (line 1, in each plot) and oblique (line 2, $\alpha = 2°$ and line 3, $\alpha = 5°$) incidence. Interestingly, a moderate change in α resulted in strong changes of spectra of Faraday rotation: $\theta_{||}$ at $\alpha = 2°$ (line 2) was flipped with respect to $\theta_{||}$ at $\alpha = 0$, and the rotation angle for $\alpha = 5°$ (line 3) did not exceed the rotation angle of the reference 1D MPhC (gray line). Experiments showed that, at other angles of incidence except for $\alpha = 0$ and 2°, Faraday rotation spectra were similar to that given by line 3. With due account taken of the sign of Faraday rotation, the mirror symmetry in the spectra and rotation degradation implies that the MO response was extremely sensitive to the strength of each diffraction channel—one IS and two OS channels.

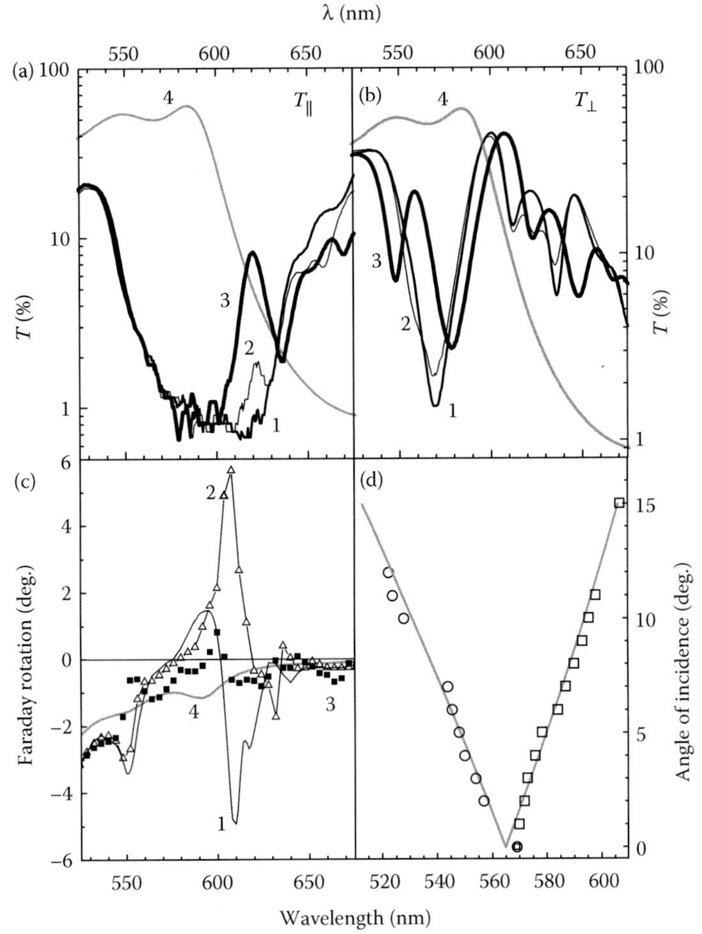

FIGURE 4.7.10
Angle-resolved transmission spectra for $E_{\|}$- and $E_{\perp}$-polarized light are shown in plots (a) and (b), respectively. (c) Angle-resolved Faraday rotation spectra for the $E_{\|}$ polarization. Incident angles were 0° (line 1), 2° (line 2), and 5° (line 3). Lines 4 show spectra for the reference one-dimensional magnetophotonic crystal measured at normal incidence. (d) Angle-resolved split between the bands shown in plot (b). (From S. Baek, A. V. Baryshev, and M. Inoue, Appl. Phys. Lett. **98**, 101111, 2011. With permission.)

Multiple Bragg diffraction resulted in a significant enhancement of the MO response of Q-2D MPhCs. Moreover, the enhancement was accompanied by the change of the polarization rotation direction in a narrow spectral range, where neighboring photonic stop bands overlapped. It is worth mentioning that *the observed effect cannot be found in conventional 1D MPhCs.*

4.7.6 Magneto-Optics of Noble Metal–Garnet Composites

Noble metal–garnet nanocomposites provide an important function—enhancing MO response due to surface plasmon resonances. Previously, MO effects associated with excitation of the bulk plasma and surface plasmons in conventional magnetic materials have been reported. For example, studies[27,28] have established the relationship between the bulk plasma and the Kerr effect in ferromagnetic/noble metal multilayers. More recent works dealt with MO responses associated with the surface plasmon resonance: a theoretical study on nonmagnetic plasmonic particles surrounding a magnetic medium,[29] a study on a plasmonic nano-onion with multicore-shell structure,[30] and a study on the Kerr effect in Au–garnet granular films.[31] Also, an anomalous Faraday effect has been theoretically demonstrated for light in the regime of the extraordinary transmittance in a thick perforated metal film in which holes are filled with an MO material.[32] In this section, we describe recent studies aiming at larger enhancement of MO responses induced by excitation of the localized surface plasmon resonance (LSPR) in thin Bi:YIG films comprising noble metal particles.

Noble metal particles supporting LSPR were fabricated through melting few-nanometer-thick initial metal films, including Au, Ag, and Au–Ag multilayers. A mean diameter and distribution of the particles was defined by the initial film thickness and the heating temperature.[33,34] Usually, the metal particles and Bi:YIG layer were sequentially formed on a quartz substrate as a layered structure. The Bi:YIG deposited by the RF magnetron sputtering method requires annealing in air at 750°C for crystallization.

First composites exhibiting the LSPR-related enhancement of the Faraday rotation were fabricated on quartz substrates and had structures of Au/Bi:YIG/quartz and Bi:YIG/Au/Bi:YIG/quartz in which arrays of Au particles were randomly distributed in Bi:YIG.[35,36] Figure 4.7.11a and b shows the top and oblique views of the Au/Bi:YIG composite and Figure 4.7.11c shows a cross-section of the Bi:YIG/Au/Bi:YIG composite. Au particles were formed by melting a 5-nm-thick Au film at a temperature of 750°C; the choice of temperature is limited by crystallization of Bi:YIG underlayer. The formed Au particles had lateral dimensions in a range of 20–80 nm, and, normally, heights of the particles were 10–40 nm.

Transmission spectra of the Au/Bi:YIG composite for linearly polarized light at normal incidence are shown in Figure 4.7.11d. The absorption band at a wavelength of 620 nm was due to LSPR in the Au particles. Note that the spectral positions of absorption bands in the polarization-resolved spectra ($E_{0°}$ and $E_{90°}$, the polarization plane was set by a polarizer) slightly differed, showing transmission anisotropy of the arrays of Au particles. Furthermore, a rotation of the polarization plane for light propagating through nonmagnetized composites was observed.[36] The transmission anisotropy and the polarization rotation vanish for light with wavelengths beyond the LSPR band (see Figure 4.7.11e). In our case, the polarization rotation is protruded in the regime of LSPR and most likely to originate from the polarization-sensitive structural peculiarities of the composites; for example, an irregular roughness of the surface.

When measuring the MO response of Au/Bi:YIG composites in external magnetic fields, the contribution of the aforementioned anisotropy-related polarization rotation was subtracted. Figure 4.7.11f and g shows transmission and Faraday rotation spectra of Au/Bi:YIG and Bi:YIG/Au/Bi:YIG composites. One can see that the magnitude of the Faraday rotation was enhanced at the wavelengths of the LSPR absorption bands. If one compares the plasmonic composites to reference Bi:YIG films, which did not contain Au particles, the observed enhancement ($\Delta\theta_F = \theta_F^{Au/Bi:YIG} – \theta_F^{Bi:YIG}$) was $\Delta\theta_F = 0.03°$ at 628 nm and $\Delta\theta_F = 0.08°$ at 677 nm for the Au/Bi:YIG and Bi:YIG/Au/Bi:YIG composites, respectively.

Optimal conditions providing larger enhancement of the Faraday rotation by Au/Bi:YIG composites were sought by varying temperatures at which arrays of Au particles were fabricated.[34] Results showed that the enhancement $\Delta\theta_F$ was observed up to a temperature of 1000°C, although MO properties of Bi:YIG degraded at temperatures of more than 700°C. To resolve this problem the fabrication approach was modified, and sputtering of Bi:YIG was processed after fabrication of Au particles on quartz substrates. Figure 4.7.12a–c shows

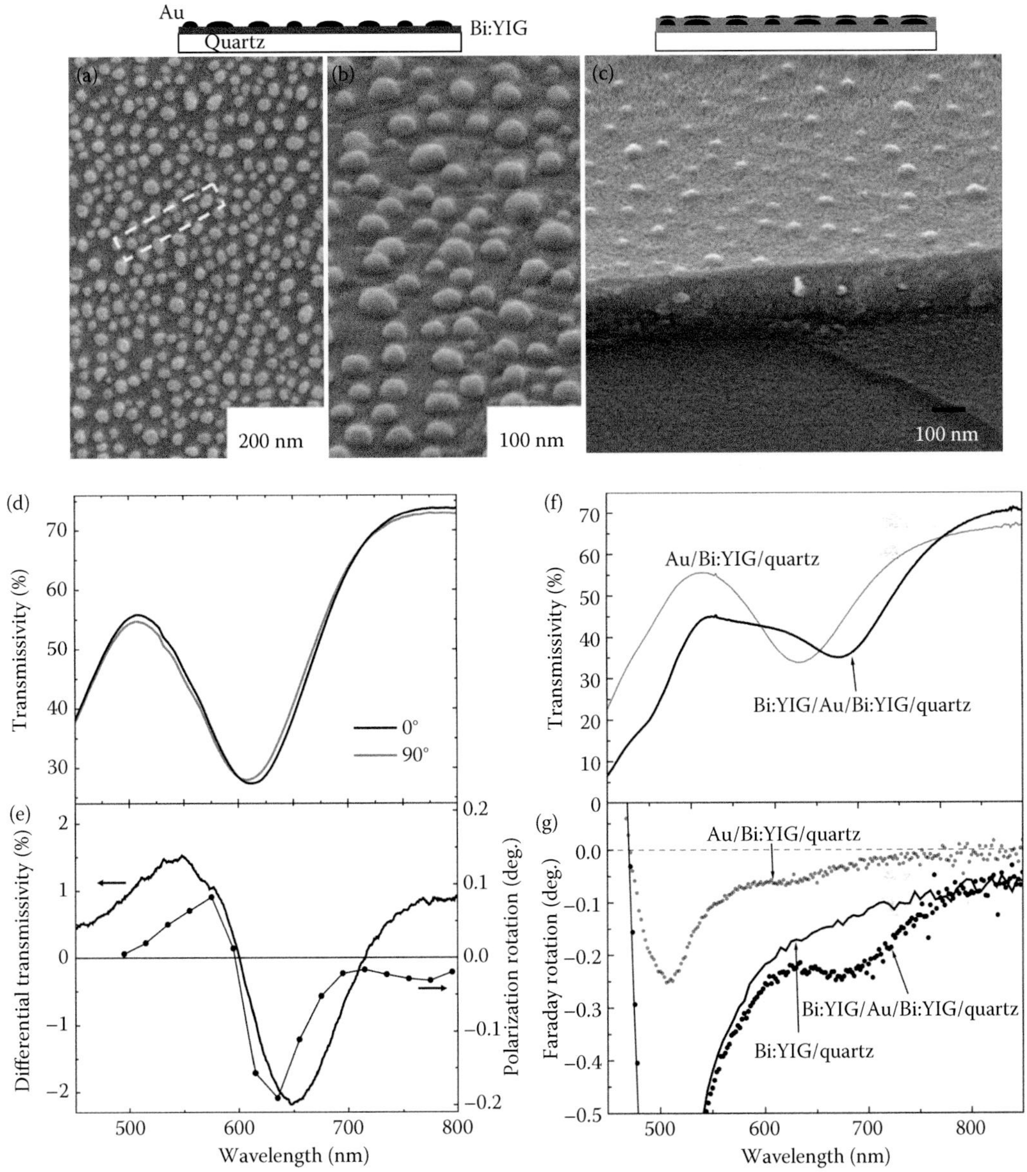

FIGURE 4.7.11
(a) Top and (b) oblique scanning electron microscope images of Au particles on Bi:YIG film (Au/Bi:YIG/quartz). (c) Oblique view of Au particles embedded in Bi:YIG (Bi:YIG/Au/Bi:YIG/quartz). (d) Polarization-resolved transmissivity, (e) differential transmissivity and polarization rotation of the Au/Bi:YIG composite. (f) Transmissivity and (g) Faraday rotation spectra of the Au/Bi:YIG/quartz and the Bi:YIG/Au/Bi:YIG/quartz. Thickness of each Bi:YIG layer was 90 nm, and the reference Bi:YIG film in plot (g) had a thickness of 180 nm.

SEM images of Au particles fabricated at different temperatures from initial 5-nm-thick Au films: the distribution of Au particle size after temperature treatment is shown in Figure 4.7.12d. It is worth noting that Au particles had elongated and distorted shapes when treated at temperatures of less than 700°C. Interestingly, Au particles produced at 1000°C had a spherical shape and a small mean diameter of about 60 nm. At higher

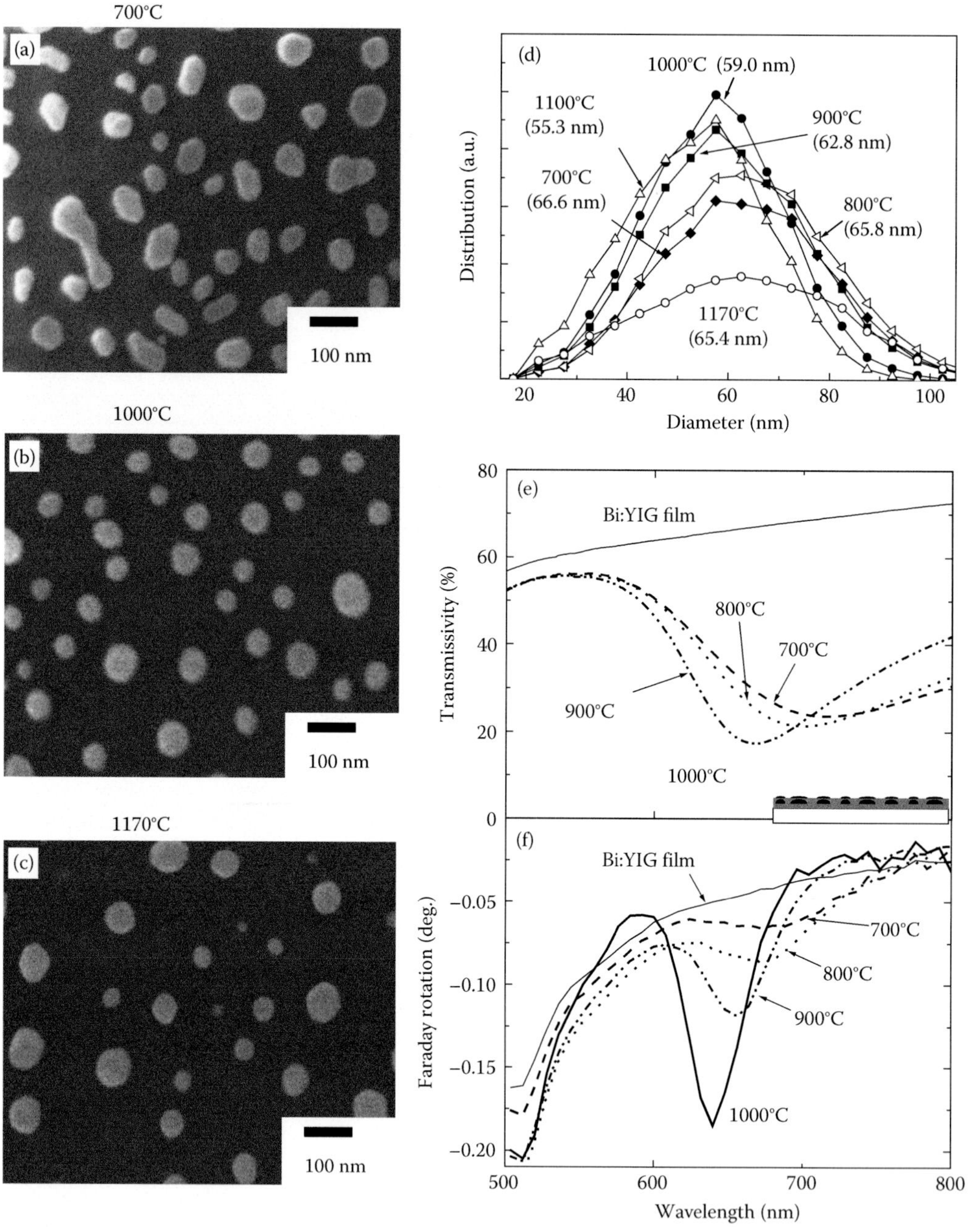

FIGURE 4.7.12
Scanning electron microscope images of Au particles formed by heating at (a) 700°C, (b) 1000°C, and (c) 1170°C. (d) Distributions of Au particle diameters; heating temperatures and mean diameters are denoted. (e) Transmissivity and (f) Faraday rotation spectra of the Bi:YIG/Au composites fabricated on quartz substrates (Bi:YIG/Au/quartz).

temperatures (see Figure 4.7.12c), the number of particles significantly reduced because of Au evaporation. Figure 4.7.12 illustrates these experiments where the magnitude of $\Delta\theta_F$ followed the strength of LSPR governed by the shape and distribution of Au particles.

Optimization of distribution and density of particles was studied by using a repetitive formation method when deposition together with melting Au at 1000°C was repeated several times.[37] A Bi:YIG film with a certain thickness was sputtered onto such arrays of Au particles; the resultant structure of Bi:YIG/Au/quartz was treated at 750°C for garnet crystallization. Results are shown in Figure 4.7.13a through c. The mean diameter of Au particles was 48 nm, and the diameter distribution (full-width at half-maximum) was 24 nm (Figure 4.7.13a) after a one-cycle formation. After repeating the fabrication cycle up to four times, full-width at half-maximum was 51 nm and increased further for the next cycles. Also, the distribution exhibited two peaks, illustrating the presence of majorities with small and large diameters (Figure 4.7.13c).

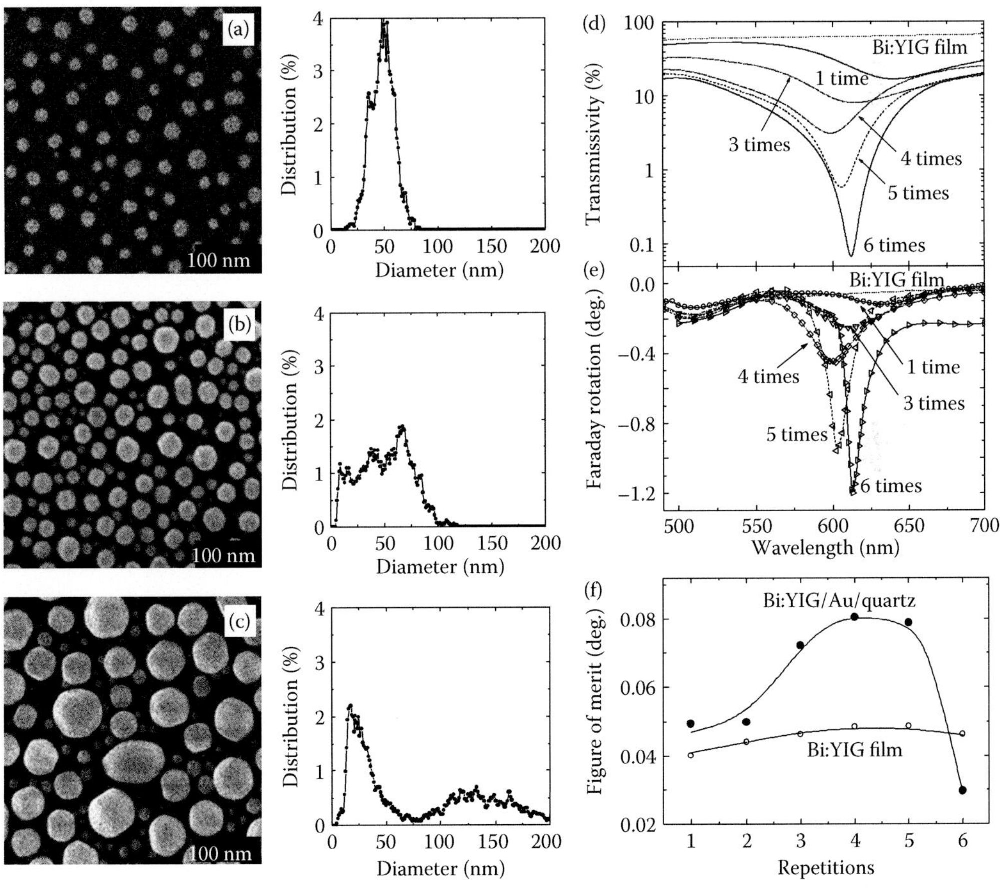

FIGURE 4.7.13
Scanning electron microscope images and diameter distributions of Au particles fabricated by the repetitive formation from 5-nm-thick Au films at 1000°C: (a) one cycle, (b) four cycles, and (c) eight cycles. (d) Transmission, (e) Faraday rotation spectra of the Bi:YIG/Au composites, and (f) their figure of merit $|\theta_F|T^{1/2}$; magnitudes of rotation for reference Bi:YIG films are shown. Thicknesses of sputtered Bi:YIG were 60 nm.

Transmittance of Bi:YIG/Au composites is shown in Figure 4.7.13d. For composites fabricated by several cycles, overall transmittance in the measured spectral range decreased as the volume fraction of Au increased. The Faraday rotation for light with the LSPR wavelengths was as large as −1.2° for the composite formed by six cycles; this was 20 times larger than that of a Bi:YIG film with the same thickness.[38] However, transmittance at LSPR band became as low as 0.06%. Accordingly, it is obvious that the larger Faraday rotation was obtained for the sample with a larger Au extent of the composite's surface, that is, when the openings in Bi:YIG are very small. The FOM (here we use a definition of $|\theta_F|T^{1/2}$) is shown in Figure 4.7.13f. One can see that the maximum FOM of 0.08° was obtained after the four-cycle formation. However, the magnitude of FOM decreased to 0.03° for the six-cycle composite; it was even smaller than that of a reference Bi:YIG film.

For fabricated samples—arrays of Au particles on quartz substrates and Bi:YIG/Au/quartz—the LSPR bands are located at 520 and 620 nm, respectively. To significantly shift spectral positions of LSPRs to short wavelengths, Ag can be used instead of Au.[39] However, Bi:YIG/Ag composites, subjected to heating during the fabrication procedure, did not show any change in θ_F at the wavelength of LSPR band. It is most likely that nonmagnetic Ag_2O-based compound shells separated plasmonic particles from Bi:YIG, degrading the contribution of LSPR to the MO response from Bi:YIG/Ag composites.[37] To reduce chemical reaction at the interface between Ag and Bi:YIG and, moreover, to change the spectral position of LSPR, Ag–Au alloy plasmonic particles and Ag–Au/Bi:YIG composites were studied. The alloy particles were formed on quartz substrates, when heating 5-nm-thick Ag–Au bilayers at 800°C (see Figure 4.7.14a and b).[37] Figure 4.7.14c shows that the spectral position

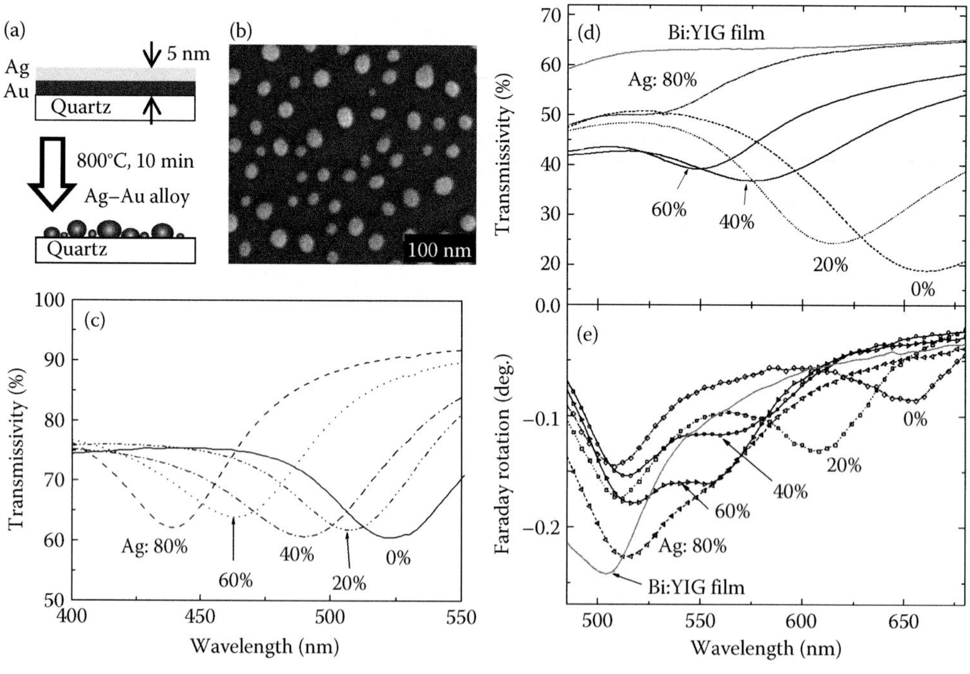

FIGURE 4.7.14
(a) Fabrication process for Ag Au alloy particles. (b) Scanning electron microscope image of the alloy particles with a Ag volume fraction of 60%. (c) Spectral position of the localized surface plasmon resonance band versus Ag volume fraction. (d) Transmission and (e) Faraday rotation spectra of different Bi:YIG/AuAg composites.

of LSPR shifted to short wavelengths with increase in the Ag volume fraction. Figure 4.7.14d and e shows transmission and Faraday rotation spectra of composites with the structure of Bi:YIG/Ag–Au/quartz. With increase in the Ag volume ratio from 0% to 80%, the LSPR band shifted to the short-wavelength region from 660 to 530 nm. All fabricated composites exhibited enhanced Faraday rotation. However, $\Delta\theta_F$ decreased as the volume fraction of Ag rose; the possible reason for this trend was mentioned above.

4.7.7 Summary of Phenomena in MPhCs and MO Plasmonic Structures

	Theory/Numerical Simulation	Experiment/Application
1D MPhC	Enhanced magneto-optics due to localization: Single and dual Fabry-Pérot cavities[8–11,15–17] Borrmann effect[6] Surface Tamm state[18] Nonlinear effects[40] Bragg waveguides[41] Magnetorefractive effect[42] Tunability of PBGs: Helicoidal MPhCs[43] Degenerate bands (intra-Brillouin PBG, Yeh's PBG)[13,44] Unidirectionality and frozen modes in MPhCs[45] Bigyrotropic MPhCs[46]	Enhanced magneto-optics and nonlinear effects[16,47] Single cavity-based spatial light modulators[16,48] Waveguide MPhC switches[49] Refractive index sensing[50]
2D MPhC	Magnetic superprism effect[12] Circulator[22,51]	Enhanced/reversed magneto-optics and demultiplexing with polarization control[23,26]
3D MPhC		Opal-based MPhCs[52] Altered magneto-optics and enhanced nonlinear responses[53]
Magneto-plasmonic structures	Nonreciprocal transmission in noble metal–MO structures[54] and enhanced magneto-optics[28–30,32,55]	LSPR-enhanced MO responses of composites[33–36,38]

4.7.8 Conclusion

Designing a microscopic distribution of fields inside the primitive cell of a PhC allows tuning light interaction with PhC constituents made of active materials and brings a significant change in their optical, magneto-optical, and other responses. In this chapter, we have demonstrated that MPhCs composed of sequences of magnetic and dielectric layers permit the enhancement of responses of known MO materials. Light coupling to MPhCs, where the MO constituents are periodically arranged, built-in as a defect layer into a periodic structure, comprising structures supporting surface states or multiple Bragg diffraction, results in significant enhancement of the polarization rotation. Additionally, we have discussed noble metal/garnet composites exhibiting enhanced MO responses in the regime of plasmon excitation. *Implementation of MPhCs* to optical integrated devises *promises fast on-spin-relaxation*, multimode and multidirectional control of light flow by these miniature magnetic media.

Acknowledgments

We greatly acknowledge all our colleagues who have fabricated and characterized magnetophotonic crystals: D. Kobayashi, Y. Masuda, and Y. Mizutani, and PhD students T. Goto, S. M. Baek, and Dr. M. E. Dokukin. The experiments and related publications would not be successful without activities and discussions provided by Professors O. A. Aktsipetrov, A. B. Granovsky, A. A. Fedyanin, P. B. Lim, A. A. Lisyansky, and A. P. Vinogradov, and Drs. K. Nishimura, R. Fujikawa and A. V. Dorofeenko.

References

1. E. Yablonovitch, Phys. Rev. Lett. **58**, 2059 (1987); S. John, *ibid.* **58**, 2486 (1987); K. Sakoda, *Optical Properties of Photonic Crystals* (Springer, Berlin, 2001); J. D. Joannopoulos, R. Meade, and J. Winn, *Photonic Crystals* (Princeton University Press, New Jersey, 1995).
2. A. D. Greenwood, Proc. Phys. Soc. 71, 585 (1958); J. C. Ward, Phys. Rev. **78**, 182 (1950); M. P. van Albada, B. A. van Tiggelen, B. A. Langedijk, and A. Tip, Phys. Rev. Lett. **66**, 3132 (1991).
3. D. A. de Wolf, IEEE Trans. Antennas Propag. **19**, 254 (1971); L. Tsang and A. Ishimaru, JOSA A **1**, 836 (1984); M. P. van Albada and B. A. Langedijk, Phys. Rev. Lett. **55**, 2692 (1985).
4. P. W. Anderson, Phys. Rev. **109**, 1492 (1958); P. Sheng, B. White, Z. Q. Zhang, and G. Papanicolau, Phys. Rev. B **34**, 4757 (1986); J. E. Sipe, P. Sheng, B. White, and M. H. Cohen, Phys. Rev. Lett. **60**, 108 (1988).
5. P. Sheng, *Introduction to Wave Scattering, Localization, and Mesoscopic Phenomena* (Academic, London, 1995).
6. G. Borrmann, Phys. Z. **42**, 157 (1941); A. P. Vinogradov, Yu. E. Lozovik, A. M. Merzlikin, A. V. Dorofeenko, I. Vitebskiy, A. Figotin, A. B. Granovsky, and A. A. Lisyansky, Phys. Rev. B **80**, 24.7106 (2009); A. B. Khanikaev, A. B. Baryshev, P. B. Lim, H. Uchida, M. Inoue, A. G. Zhdanov, A. A. Fedyanin, A. I. Maydykovskiy, and O. A. Aktsipetrov, *ibid.* **78**, 193102 (2008).
7. A. K. Zvezdin and V. A. Kotov, *Modern Magnetooptics and Magnetooptical Materials* (Taylor and Francis, New York, 1997).
8. M. Inoue, T. Yamamoto, K. Isamoto, and T. Fujii, J. Appl. Phys. **79**, 5988 (1996); M. Inoue, and T. Fujii, *ibid.* **81**, 5659 (1997); M. Inoue, T. Fujii, K. I. Arai, and M. Abe, J. Magn. Soc. Jpn. **22**, 141 (1998).
9. M. P. van Albada and A. Lagendijk, Phys. Rev. Lett. **55**, 2692 (1985); P. E. Wolf and G. Maret, *ibid.* **55**, 2696 (1985).
10. M. Inoue, K. I. Arai, T. Fujii, and M. Abe, J. Appl. Phys. **83**, 6768 (1998); S. Kahl and A. M. Grishin, Appl. Phys. Lett. **84**, 1438 (2004).
11. M. Inoue, A. Khanikaev, and A. Baryshev, in *Nanoscale Magnetic Materials and Applications*, edited by J. P. Liu, E. Fullerton, O. Gutfleisch, and D. J. Sellmyer (Springer, New York, 2009), Chapter 21.
12. A. Khanikaev, A. Baryshev, M. Inoue, A. Granovsky, and A. Vinogradov, Phys. Rev. B **72**, 04.7123 (2005); A. M. Merzlikin, A. P. Vinogradov, M. Inoue, and A. B. Granovsky, Phys. Rev. E **72**, 046603 (2005).
13. A. B. Khanikaev, M. Inoue, and A. B. Granovsky, J. Magn. Magn. Mater. **300**, 104 (2006); A. M. Merzlikin, A. P. Vinogradov, M. Inoue, A. B. Khanikaev, and A. B. Granovsky, *ibid.* **300**, 108 (2006).
14. I. Y. Tamm, Phys. Z. Sowjetunion **1**, 733 (1932).
15. M. J. Steel, M. Levy, and R. M. Osgood, IEEE Photonics Technol. Lett. **12**, 1171 (2000).

16. M. Inoue, R. Fujikawa, A. Baryshev, A. Khanikaev, P. B. Lim, H. Uchida, O. Aktsipetrov, A. Fedyanin, T. Murzina, and A. Granovsky, J. Phys. D **39**, R151 (2006).
17. T. Goto, A. V. Baryshev, K. Tobinaga, and M. Inoue, J. Appl. Phys. **107**, 09A946 (2010).
18. F. Villa and J. A. Gaspar-Armenta, Opt. Commun. **223**, 109 (2003); A. Kavokin, I. Shelykh, and G. Malpuech, Appl. Phys. Lett. **87**, 261105 (2005); F. Villa and J. A. Gaspar-Armenta, Opt. Express **12**, 2338 (2004); A. P. Vinogradov, A. V. Dorofeenko, S. G. Erokhin, M. Inoue, A. A. Lisyansky, A. M. Merzlikin, and A. B. Granovsky, Phys. Rev. B **74**, 045128 (2006); N. Malkova and C. Z. Ning, *ibid.* **73**, 113113 (2006); **76**, 045305 (2007); A. M. Merzlikin, A. P. Vinogradov, A. V. Dorofeenko, M. Inoue, M. Levy, and A. B. Granovsky, Physica B **394**, 277 (2007);
19. M. Kaliteevskii, I. Iorsh, S. Brand, R. A. Abram, J. M. Chamberlain, A. V. Kavokin, and I. A. Shelykh, Phys. Rev. B **76**, 165415 (2007); T. Goto, A. V. Baryshev, M. Inoue, A. V. Dorofeenko, A. M. Merzlikin, A. P. Vinogradov, A. A. Lisyansky, A. B. Granovsky, *ibid.* **79**, 125103 (2009); T. Goto, A. V. Dorofeenko, A. M. Merzlikin, A. V. Baryshev, A. P. Vinogradov, M. Inoue, A. A. Lisyansky, A. B. Granovsky, Phys. Rev. Lett. **101**, 113902 (2008).
20. M. E. Sasin, R. P. Seisyan, M. A. Kaliteevski, S. Brand, R. A. Abram, J. M. Chamberlain, A. Yu. Egorov, A. P. Vasil'ev, V. S. Mikhrin, and A. V. Kavokin, Appl. Phys. Lett. **92**, 251112 (2008); C. R. Rosberg, D. N. Neshev, Y. V. Kartashov, R. A. Vicencio, W. Krolikowski, M. I. Molina, A. Mitchell, V. A. Vysloukh, L. Torner, and Yu. S. Kivshar, Opt. Photonics News **17**, 29 (2006).
21. A. V. Baryshev, T. Kodama, K. Nishimura, H. Uchida, M. Inoue, Trans. Magn. Soc. Jpn. **4**, 290 (2004); J. Appl. Phys. **95**, 7336 (2004); IEEE Trans. Magn. **40**, 2829 (2004); K. Nishimura, T. Kodama, A. Baryshev, H. Uchida, and M. Inoue, J. Appl. Phys. **95**, 6633 (2004); R. Fujikawa, A. V. Baryshev, A. B. Khanikaev, H. Uchida, P. B. Lim, M. Inoue, IEEE Trans. Magn. **42**, 307 (2006).
22. Z. Wan and S. Fan, Appl. Phys. B **81**, 369 (2005).
23. M. E. Dokukin, A. V. Baryshev, A. B. Khanikaev, M. Inoue, Opt. Express **17**, 9063 (2009); S. Baek, M. Dokukin, K. Yayoi, A. Baryshev, M. Inoue, J. Appl. Phys. **107**, 09A923 (2010).
24. S. Kawakami, T. Kawashima, and T. Sato, Appl. Phys. Lett. **74**, 463 (1999).
25. M. Renninger, Z. Phys. **106**, 141 (1937).
26. S. M. Baek, A. V. Baryshev, and M. Inoue, J. Appl. Phys. **109**, 07B701 (2011); Appl. Phys. Lett. **98**, 101111 (2011).
27. T. Katayama, H. Awano, and Y. Nishihara, J. Phys. Soc. Jpn. **55**, 2539 (1986).
28. H. Feil and C. Haas, Phys. Rev. Lett. **58**, 65 (1987).
29. V. A. Kosobukin, Surf. Sci. **406**, 32 (1998).
30. M. Abe and T. Suwa, Phys. Rev. B **70**, 24.7103 (2004).
31. S. Tomita, T. Kato, S. Tsunashima, S. Iwata, M. Fujii, and S. Hayashi, Phys. Rev. Lett. **96**, 167402 (2006).
32. A. B Khanikaev, A. A. Fedyanin, A. V. Baryshev, M. Inoue, and A. B. Granovsky, Opt. Express **15**, 6612 (2007).
33. H. Uchida, H. Fumoto, A. Baryshev, J. Kim, and M. Inoue, IEE J. Trans. Electr. Electron. Eng. **3**, 660 (2008).
34. H. Uchida, Y. Masuda, R. Fujikawa, A. V. Baryshev, and M. Inoue, J. Magn. Magn. Mater. **321**, 843 (2009).
35. T. Okubo, R. Fujikawa, Y. Masuda, A. Fedyanin, A. Baryshev, A. Khanikaev, K. H. Shin, H. Uchida, P. B. Lim, and M. Inoue, Tech. Rep. Inst. Electro. Eng. Jpn. Mag. **06-167** 1 (2006), in Japanese.
36. R. Fujikawa, A. V. Baryshev, J. Kim, H. Uchida, and M. Inoue, J. Appl. Phys. **103**, 07D301 (2008).
37. Y. Mizutani, H. Uchida, Y. Masuda, A. V. Baryshev, and M. Inoue, J. Magn. Soc. Jpn. **33**, 481 (2009).
38. H. Uchida, Y. Mizutani, Y. Nakai, A. A. Fedyanin, and M. Inoue, J. Phys. D **44**, 064014 (2011).
39. S. Link, Z. L. Wang, and M. A. El-Sayed, J. Phys. Chem. B **103**, 4.729 (1999).
40. I. L. Lyubchanskii, N. N. Dadoenkova, M. I. Lyubchanskii, E. A. Shapovalov, and Th. Rasing, J. Phys. D **36**, R277 (2003).

41. M. Levy and P. Kumar, Opt. Lett. **4.7**, 3147 (2010).
42. J. V. Boriskina, S. G. Erokhin, A. B. Granovsky, A. P. Vinogradov, and M. Inoue, Phys. Solid State **48**, 4, 717 (2006).
43. F. Wang and A. Lakhtakia, Appl. Phys. Lett. **92**, 011115 (2008); Phys. Rev. B **79**, 193102 (2009).
44. Hai-Xia Da, Zi-Qiang Huang, Z.Y. Li, Opt. Lett. **34**, 1693 (2009); A. M. Merzlikin, A. P. Vinogradov, M. Inoue, A. B. Khanikaev, and A. B. Granovsky, J. Magn. Magn. Mater. **300**, 108 (2006); A. M. Merzlikin, M. Levy, A. A. Jalali, A. P. Vinogradov, Phys. Rev. B **79** 195103 (2009); M. Levy, A. A. Jalali, J. Opt. Soc. Am. B **24**, 1603 (2007); A. A. Jalali, M. Levy, J. Opt. Soc. Am. B **25**, 119 (2008); A. M. Merzlikin, A. P. Vinogradov, A. V. Dorofeenko, M. Inoue, M. Levy, A. B. Granovsky, Physica B **394**, 277 (2007).
45. A. Figotin and I. Vitebskiy, Phys. Rev. E **74**, 066613 (2006); A. Figotin and I. Vitebsky, J. Magn. Magn. Mater. **300**, 117 (2006); J. Ballato, A. Ballato, A. Figotin, and I. Vitebskiy, Phys. Rev. E **71**, 036612 (2005); A. B. Khanikaev, A. V. Baryshev, M. Inoue, and Yu. S. Kivshar, Appl. Phys. Lett. **95**, 011101 (2009).
46. Yu. S. Dadoenkova, I. L. Lyubchanski, Y. P. Lee, and Th. Rasing, Low Temp. Phys. **36**, 538 (2010); I. L. Lyubchanskii, N. N. Dadoenkova, M. I. Lyubchanskii, E. A. Shapovalov, A. Lakhtakia, and Th. Rasing, Appl. Phys. Lett. **85**, 5932 (2004); I. L. Lyubchanskii, N. N. Dadoenkova, M. I. Lyubchanskii, E. A. Shapovalov, and Th. Rasing, J. Phys. D **36**, R277 (2003).
47. T. V. Murzina, R. V. Kapra, T. V. Dolgova, A. A. Fedyanin, O. A. Aktsipetrov, K. Nishimura, H. Uchida, and M. Inoue, Phys. Rev. B **70**, 012407 (2004).
48. S. Mito, H. Takagi, P. B. Lim, A. V. Baryshev, and M. Inoue J. Appl. Phys. **109**, 07E313 (2011); M. Inoue, A. V. Baryshev, A. B. Khanikaev, M. E. Dokukin, K. H. Jung, J. Heo, H. Takagi, H. Uchida, P. B. Lim, J. Kim, IEICE Trans. Electron. **E91–C(10)**, 1630 (2008).
49. Z. Wu, Miguel Levy, V. J. Fratello, and A. M. Merzlikin, Appl. Phys. Lett. **96**, 051125 (2010).
50. N. K. Dissanayake, M. Levy, A. A. Jalali, and V. J. Fratello, Appl. Phys. Lett **96**, 181105 (2010).
51. K. Yayoi, K. Tobinaga, Y. Kaneko, A. V. Baryshev, and M. Inoue, J. Appl. Phys. **109**, 07B750 (2011).
52. A. V. Baryshev, T. Kodama, K. Nishimura, H. Uchida, M. Inoue, IEEE Trans. Magn. **40**, 2829 (2004); A. V. Baryshev, T. Kodama, K. Nishimura, H. Uchida, M. Inoue, J. Appl. Phys. **95**, 7336 (2004).
53. T. V. Murzina, E. M. Kim, R. V. Kapra, I. V. Moshnina, O. A. Aktsipetrov, D. A. Kurdyukov, S. F. Kaplan, V. G. Golubev, M. A. Bader, and G. Marowsky, Appl. Phys. Lett. **88**, 022501 (2006); K. Napolskii, N. Sapoletova, A. Eliseev, G. Tsirlina, A. Rubacheva, E. Gan'shina, M. Kuznetsov, M. Ivanov, V. Valdner, E. Mishina, A. van Etteger, and Th. Rasing, J. Magn. Magn. Mater. **321**, 7, 833 (2009).
54. A. B. Khanikaev, S. H. Mousavi, G. Shvets, and Yu. S. Kivshar, Phys. Rev. Lett. **105**, 126804 (2010).
55. A. Battula, S. Chen, Y. Lu, R. J. Knize, and K. Reinhardt, Opt. Lett. **32**, 2692 (2007); V. I. Belotelov, D. A. Bykov, L. L. Doskolovich, A. N. Kalish, V. A. Kotov, and A. K. Zvezdin, *ibid*. **34**, 398 (2009); V. I. Belotelov, L. L. Doskolovich, and A. K. Zvezdin, Phys. Rev. Lett. **98**, 077401 (2007); V. I. Belotelov, D. A. Bykov, L. L. Doskolovich, A. N. Kalish, and A. K. Zvezdin, J. Opt. Soc. Am. B **26**, 1594 (2009).

4.8

Inhomogeneous Hybrid Metal–Dielectric Plasmonic–Photonic Crystals

Sergei G. Romanov
University of Erlangen-Nuremberg
Ioffe Physical-Technical Institute of the Russian Academy of Sciences

CONTENTS

4.8.1 Overview

The concept of photonic crystals (PhCs) was put forward in the 1970s and 1980s with the aim to enable control on light emission, propagation, and absorption by purposive structuring of materials.[1] This is achieved through interference of the incident and scattered electromagnetic waves in the materials possessing periodic or quasiperiodic variation of the dielectric permittivity in one or more dimensions. The collection of photon eigenmodes in PhCs is called the photonic band gap (PBG) structure. Since scattering becomes especially efficient if the size of any obstacle approaches the wavelength, λ, the unit cell size in such architectures is inevitably in match to the wavelength, $a \sim \lambda$. Originally, PhCs were designed assuming frequency-independent permittivities of their frameworks. By now, the variability of such PhCs is practically exhausted. That is why in the race for novel functionalities one has to look into physical effects beyond the interference paradigm. The energy exchange between electromagnetic waves and electronic resonances in the PhC material can be considered as one possible option.

The obvious examples of functional PhCs with energy conversion are photoluminescent crystals,[2] but corresponding changes of the dielectric permittivity are minute owing to strong energy dissipation. Alternatively, the powerful pumping can be applied for generating free carriers, heat, phase transition, or composition change in the body of PhC, thus leading to changes of its dielectric permittivity and corresponding alteration of its PBG.[3]

Another approach requires PhC components with electronic excitations of large oscillator strength. The conditions of strong coupling between Bragg resonances and electronic excitations must be fulfilled for creating photon–exciton polaritons.[4] Polaritonic PhCs showing splitting of the dispersion of diffraction resonances are known as the resonant ones.[5] Examples are the Bragg superlattices of semiconductor quantum wells[6] or lattices of holes that are infiltrated with layered semiconductors with strong exciton resonances.[7] Such PhCs process the light flow in a different way compared to a nonresonant PhC of the same topology and refractive index contrast.

Alternative polaritonic PhCs are the periodic metal–dielectric (MD) structures in which electronic resonances are the localized or propagating plasmons that are supported by a metal. For example, the metal diffraction gratings demonstrate Wood anomalies occurring as a result of light coupling to surface plasmon polaritons (SPPs).[8] Apparently, MD PhCs represent more efficient energy transducers, as they can use the light scattering rather than absorption. Because the resonant scattering modifies the permittivity of the PhC, this process allows one to shape the spectral composition of propagating light.

Currently, investigations of MD PhCs focus on such effects as the extraordinary optical transmission (EOT) through subwavelength-size holes in otherwise opaque metal films,[9] guided plasmon–polaritons in lattices of nanostructures on the surface of dielectric waveguides, the formation of an energy band structure for SPPs (plasmonic crystals (PlCs)),[10] and the plasmonic-assisted light absorption in regularly structured metal films.[11] The latest trend in optical MD architectures is toward metamaterials—periodic arrays of MD nanostructures that are designed to operate like homogeneous media for the processed radiation, $a \leq 0.1\lambda$, that is, well beyond the diffractive regime. The ultimate goal of metamaterials is to achieve a negative index of refraction[12] in order to realize novel functionalities ranging from super-resolution imaging to optical illusions.[13] In particular, metamaterials with a spatially engineered refractive index profile enabled the transformation optics,[14] which is aimed at manipulating the beam trajectory in a manner similar to gravity.

It is highly desirable to integrate the MD structures with 3D PhCs in order to exercise the high degree of light control. The colloidal self- and force-assembling is arguably the most frequently used technique for 3D crystal construction.[15] Colloidal crystals offer a convenient template for exploring PhC properties.[16] Originally, the use of metals in PhCs was motivated by the high index contrast between the dielectric and the metal. The omnidirectional PBG was theoretically predicted for lattices of highly reflective metal nanospheres;[17] however, their synthesis remains a challenge. The alternative approach assumes preparation of metal nanoshells on dielectric cores.[18] Unfortunately, the interesting optical properties of nanoshell crystals appear on top of a strong absorption occurring as a result of inhomogeneities of thin metal coatings.[19] Also assembling crystals from dielectric spheres with small metal cores[20] did not bring changes to optical spectra contrary to theoretically expected omnidirectional PBG.[21] The radical change of the metal topology by making the inverted metallic opal[22] does not resolve the absorption problem as well.

On the contrary, the 2D colloidal-based structures consisted of crystal-ordered monolayers (MLs) of metal nanoshells and their metal replicas[23] demonstrate the optical response dominated by plasmon resonances. However, the reported structures were examined only in reflected light owing to their negligible transparency. Thus, in lattices possessing the high volume fraction of metal the optical response is mostly formed in the first layer owing to high metal reflectivity. Hence, there is little sense in development of 3D architectures with high metal content. In contrast, in MD crystals of lower metal fraction, such as ensembles of metal nanoshells, light is rapidly absorbed. To break the circle and realize low-dissipative plasmonic components, a thin but continuous metal film can be integrated in a colloidal crystal. Thus, we arrive at the idea of hybrid MD architectures, the properties of which are reviewed in this chapter.

Hybrid MD architectures have the following properties (Figure 4.8.1): (i) The absorption is small but reflectance is high; that is, the metal film reduces the light admission/loss in/out the PhC interior. (ii) Plasmon resonances mediate the metal transparency thus changing the spectrum and directionality of propagating light.[24] (iii) Plasmon resonances depend on the topology of a metal component. (iv) Synergy of plasmonic and diffraction resonances leads to polaritonic optical response. (v) Localized plasmon resonances can be excited. (vi) Diffraction resonances are preserved.

One can distinguish 2D and 3D PhCs, which are represented by an ordered ML of spheres and a stack of such MLs, respectively (Figure 4.8.1). Next, either a flat metal film can be placed in the near-field zone of a PhC or a periodically profiled metal film can be attached to a PhC surface. Different combinations of these basic configurations could be also imagined. In the 2D slab hybrid PhC, the geometry favors mixing of Bloch and SPP resonances. In contrast, in the 3D case a strong spatial inhomogeneity of electromagnetic field distribution can be expected, because the influence of the plasmonic component decays quickly away from the metal.

Figure 4.8.2 illustrates the effect of attached metal films on the optical response of PhCs. As mentioned, the reduction of hybrid PhCs transmission compared to their parent dielectric templates takes place as a result of high reflectivity of metal films, whereas the actual light absorption in 30–50-nm-thick Au and Ag films is within 2%–5% in the near-infrared and increases toward shorter wavelengths. It is clearly seen that changes are considerable and the way of changing depends on the metal film topology and the PhC dimensionality. In what follows the discussion will be focused on different realizations of hybrid crystals.

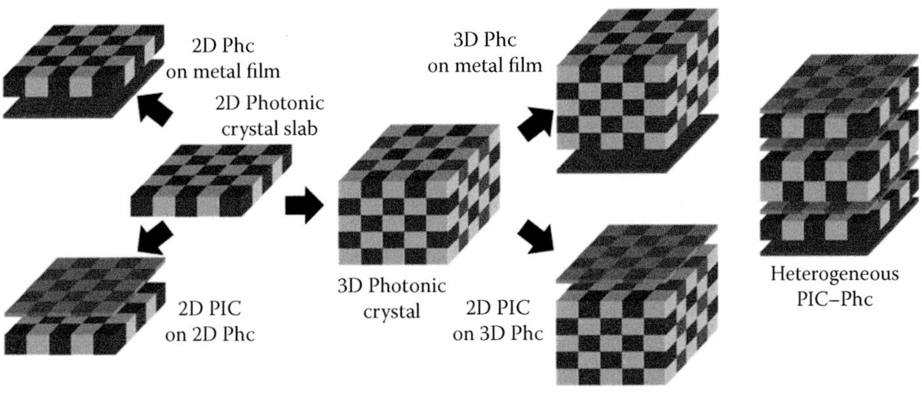

FIGURE 4.8.1
Hybrid architectures based on two-dimensional (2D) slab photonic crystals (PhCs) represented by monolayers of spheres and three-dimensional (3D) PhCs represented by opal films in contact either with flat or corrugated metal films. The latter are defined as 2D plasmonic crystals. Heterostructured hybrid architectures are combinations of PhCs and metal films.

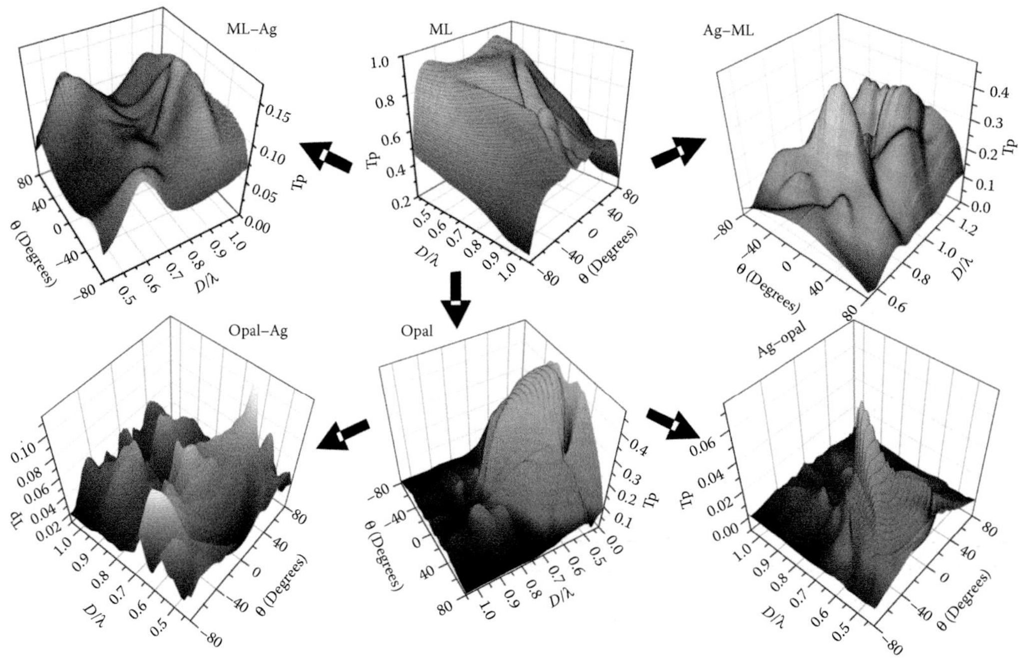

FIGURE 4.8.2
Engineered transmission spectra of hybrid crystals. ML–Ag: monolayer of spheres on Ag film. Ag–ML: Ag film on monolayer. Opal–Ag: opal film on Ag film. Ag–opal: Ag film on opal film. All samples are prepared on glass substrates. Spectra were acquired by measuring the zero-order transmission at different angles of light incidence, θ, with respect to the film normal. Illumination and detection was performed in the *p*-polarized light. The frequency scale is given in dimensionless units, D/λ, where D is the sphere diameter. $D = 431$ nm for all samples except Ag–ML one that is based on 526 nm spheres.

4.8.2 ML of Spheres on Metal Film: Bringing SPPs to Work

Two-dimensional hybrid crystals prepared by self-assembling of ML of spheres on top of a metal sheet (Figure 4.8.3) are known since the early 1990s.[25] This architecture is the analogy to the dielectric diffraction grating assembled on the metal surface, which is used as a coupling element for excitation of SPPs.

A number of sharp resonances with quality factors ranging from 80 to 150 is observed in the reflectance spectra of ML samples deposited on the thick opaque metal films (Figure 4.8.3). These resonances are associated with eigenmodes, which are either index-guided modes of the colloidal crystal (resonances b and d) or plasmon modes that are localized mostly at the interface with the metal (modes a, c, and e). Then the latter are the pure SPPs and the hybridized index-guided and SPP modes. Highly reflecting metal films enable a good mode confinement. This hybrid architecture was used to develop enhanced light emitters and advanced sensors.[25]

In the case of semitransparent thin metal films, the confinement of guided modes is weaker and the spectral features become smoother. Then, one has to measure transmission spectra to account for all losses experienced by the incident light (inset in Figure 4.8.3a). In the first approximation, ML can be represented as a planar waveguide consisting of a stack of three layers of homogeneous dielectric films and possessing the same effective

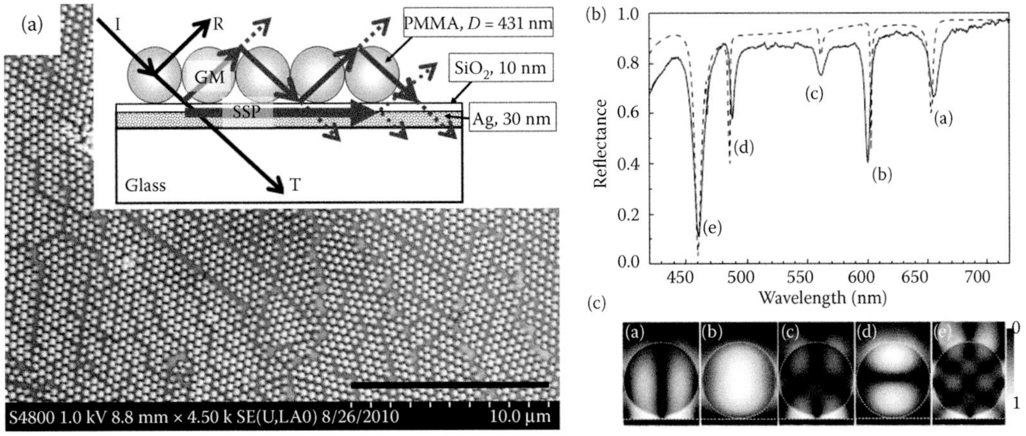

FIGURE 4.8.3
(a) Scanning electron microscope image of the Ag–monolayer (ML) sample. The scale bar is 10 μm. Poly(methyl methacrylate) spheres of $D = 431$ nm (3.5% standard deviation) were used. The ML coverage is about 90% of the substrate area. The 10–15-nm-thick SiO_2 layer ensures the hydrophilicity of the substrate. Inset shows the distribution of the incident light (I) intensity between the reflected (R) and transmitted (T) beams, Fabry–Pérot quasi-bound modes, index-guided modes (GM) light, and surface plasmon polaritons. Dashed lines show losses of guided modes. (b) Measured (solid line) and simulated (dashed) reflection spectra of the hexagonal array of 500 nm polystyrene spheres on a 200-nm-thick Ag film at normal incidence. (c) The calculated distribution of the electric field modulus at corresponding resonances. Horizontal dashed line represents the Ag surface. (Graphics (b) and (c) reprinted with permission from L. Shi et al., Phys. Lett. A, **374**, 1059, 2010.)

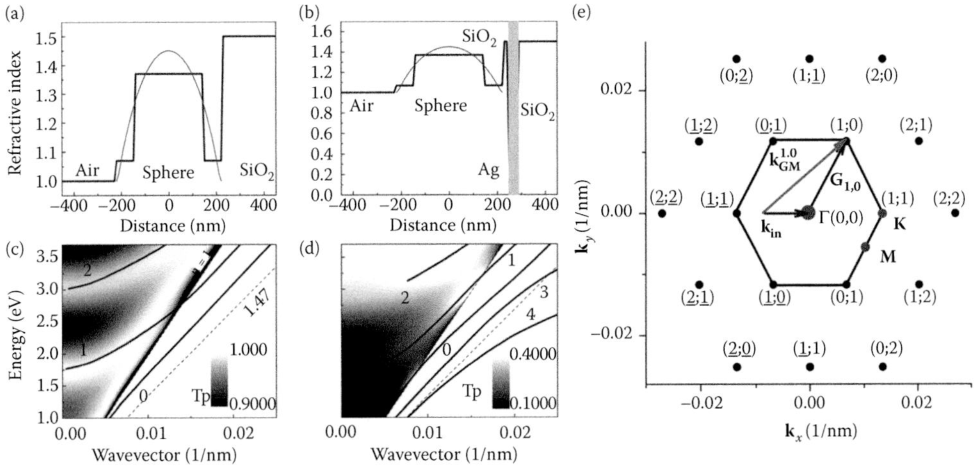

FIGURE 4.8.4
Refractive index profiles in effective three-layer-structures, which approximate a monolayer (ML) of spheres without (a) and with (b) metal film. The complex index of the Ag film is shown by a gray box. (c) and (d) Calculated transmission spectra of layered structures in *p*-polarized light. Dotted lines are light lines of air and the glass substrate. Full lines show the dispersion of the eigenmodes of these structures. Modes 3 and 4 in panel (d) are surface plasmon polariton modes located at the ML- and glass-facing interfaces of a metal film. (e) The reciprocal lattice of two-dimensional hexagonal lattice of spheres. The sketch shows the construction of $\mathbf{k}_{GM}^{1,0}$ diffraction order due to diffraction of the lateral projection of the incident beam $\mathbf{k}_x^{in}$ at the lattice vector $\mathbf{G}_{1,0}$.

index of refraction (Figure 4.8.4a). The calculated transmission reveals the quasi-bound modes that are guided along this three-layer planar structure (Figure 4.8.4c), which are known as Fabry–Pérot oscillations. Adding a metal film (Figure 4.8.4b) leads to the enhancement of quasi-bound mode magnitudes and the inclusion of SPP modes in the set of eigenmodes of the three-layer–Ag hybrid (Figure 4.8.4d). For calculations, the refractive index of Ag was adopted from the study by Jonson and Christy.[26]

The optical properties of 2D crystals of spheres are well documented.[27] The transmission minima demonstrated by self-assembled MLs and MLs prepared by the Langmuir–Blodgett method[28] agree to literature data despite disorder (Figure 4.8.3a). The spectra of the ML–Ag sample (Figure 4.8.5a and b) can be interpreted in terms of a waveguiding model.[29] First, the ML is considered as the planar waveguide (Figure 4.8.4c and d). The diffraction at an ML grating provides coupling of the incident light to the modes with wavevectors outside the light cone (Figure 4.8.4e).[30] The diffraction order is excited if one of the reciprocal lattice vectors of the ML grating matches the propagation constant of the respective guided mode. These modes that are folded back into the light cone bring the fine structure to optical spectra. Their exact association to the diffraction orders can be attained using the momentum conservation law.

It is worth listing the observations against the model predictions: (a) The magnitude of Fabry–Pérot oscillations is enhanced and they dominate the spectra. (b) The diffractively coupled guided modes can be observed in reflectance (Figure 4.8.5a), but they almost vanish in transmission (Figure 4.8.5b). (c) The correlation between eigenmodes in reflectance and transmission spectra is preserved.

The polarization anisotropy spectra of the ML–Ag hybrid unambiguously demonstrate the diffractive coupling of incident light to guided modes (Figure 4.8.5c). The SPP bands are seen owing to the fact that no SPPs can be induced by *s*-polarized incident light in the flat metal film. The SPP mode 4, which is guided at the glass–metal interface, is not observed owing to light attenuation in the metal film.

Summarizing, in the studied ML–Ag hybrid architecture the contribution of diffractively coupled orders of guided modes is weak, mostly, because of the lattice disorder as well as

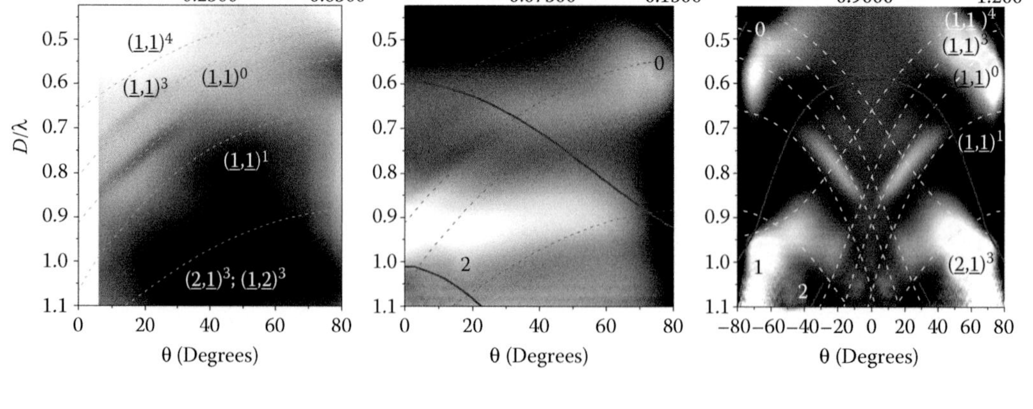

FIGURE 4.8.5
Experimental (a) reflectance and (b) transmission patterns of the ML–Ag on a glass substrate. $D = 431$ nm. Solid lines show the Fabry–Pérot modes extracted from Figure 4.8.4d. Dashed lines show the dispersion of diffracted guided modes labeled according to Figure 4.8.4e. (c) Polarization anisotropy of the transmission $(Tp_{ML\text{-}Ag}/Tp_{Ag})/(Ts_{Ml\text{-}Ag}/Ts_{Ag})$ of ML–Ag hybrid.

the weak confinement provided by a semitransparent metal film. The optical tuneability of such hybrids was demonstrated by changing the topology of a colloidal crystal.[31] Overall, the mixture of strong Fabry–Pérot resonances accompanied by weak diffractively coupled index-guided and SPP modes determines the optical response of this hybrid architecture.

4.8.3 Synergetic Response of Plasmon and Bragg Modes in the Metal-Coated ML

All in all the influence of SPPs on optical properties of ML-on-metal hybrids is rather weak. To increase the coupling strength between the photonic and plasmonic components, (i) the metal film should possess its own resonances, (ii) the metal and PhCs should interpenetrate each other, and (iii) the modes of PlC and PhC should overlap each other. This strategy is based on the concept of templated ordered arrays of nanostructures in porous dielectric matrices that can be traced back to the 1970s.[32] An implementation of this approach is the deposition of a metal film on an ML of spheres to form Au-capped spheres (Figure 4.8.6).[33]

Recently, metal-on-sphere architectures were used to study the EOT in corrugated metal films.[34] However, the reports on angle and polarization dependence of transmitted light in this architecture differ considerably from each other and no conclusive decision was derived concerning its SPP or localized plasmon nature.[35] Actually, the light transit through corrugated metal film differs from that in planar patterned ones[36] because EOT can be realized in the absence of through holes.[37] Moreover, numerical simulations demonstrated that light confinement in Bloch modes of PhC promotes EOT effect due to higher field strength.[38] A comparison of spectra of bare ML and Ag–ML samples demonstrates that relatively shallow transmission minimum occurring because of an excitation of guided modes is replaced by the massive transmission peak (Figure 4.8.7a).

Figure 4.8.7b shows the transformation of the transmission spectrum along the increase of the thickness, t, of the metal cap. As soon as individual Au caps become connected at $t \geq 25$ nm, a sharp drop of transmission and EOT peak take place, whereas the index-guided modes dominate the spectra of bare ML and Au–ML with coating thickness $t < 10$ nm.

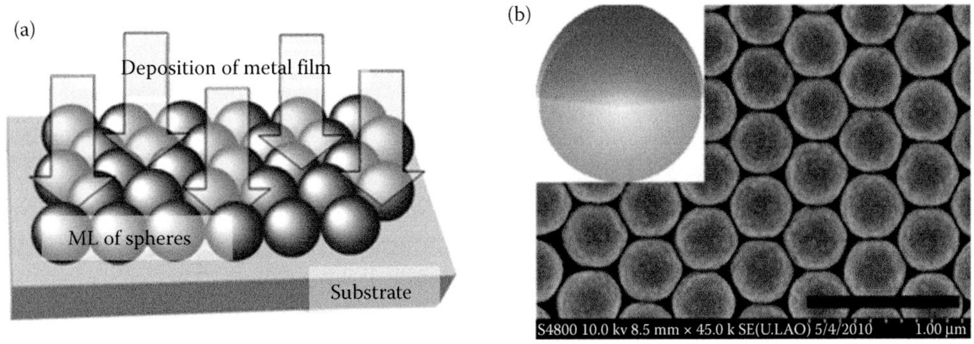

FIGURE 4.8.6
(a) Preparation of a corrugated metal film on a monolayer of spheres. (b) Top view of the Ag-coated poly(methyl methacrylate) spheres of $D = 431$ nm. Scale bar is 1 μm. Inset shows a metal-capped sphere.

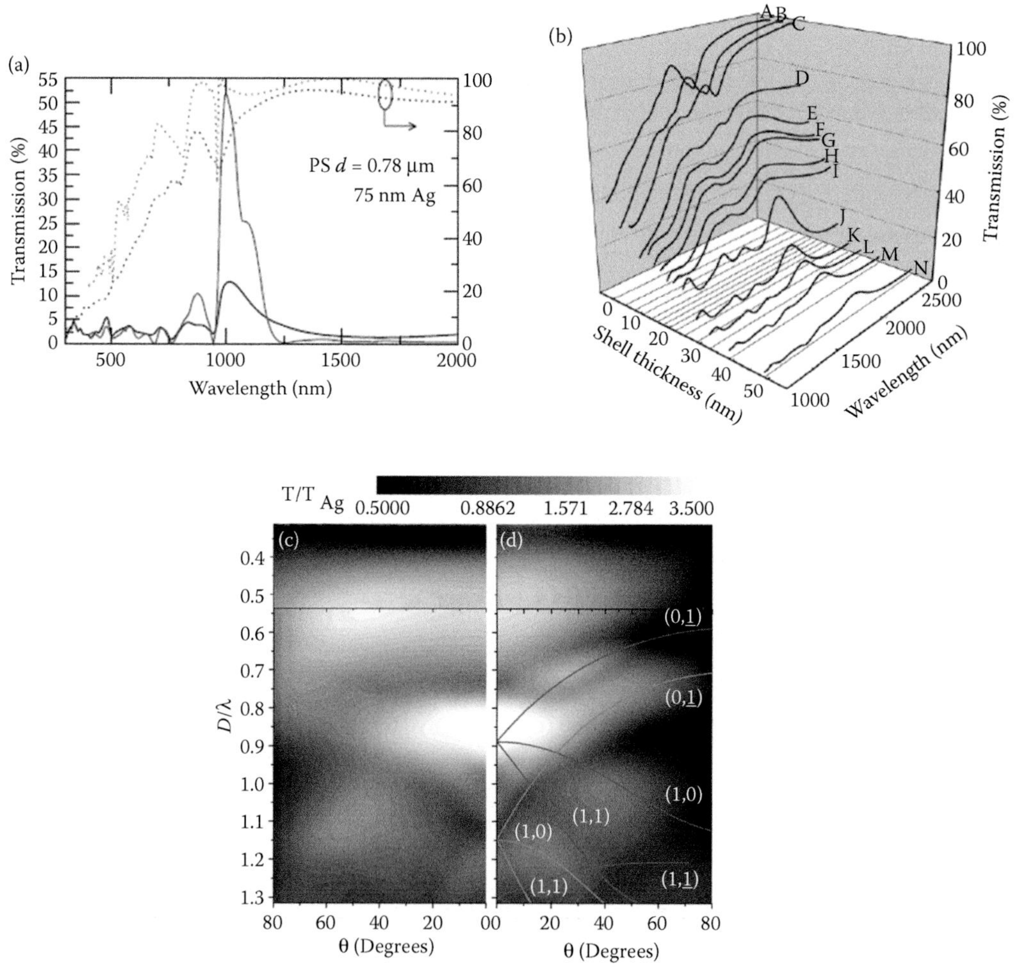

FIGURE 4.8.7
(a) Measured (dark) and modeled (light) zero-order transmission spectra. Bare ML—dotted curves (right y-scale), and with 75 nm Ag–ML—full curves (left y-scale). $D = 0.78$ μm. (Reprinted with permission from L. Landstrom, et al., Opt. Express, **17**, 761, 2009.) The main peak (or dip) scales with D. (b) Transmission spectra of MLs of silica spheres ($D = 1580$ nm) along the increase of the Au coating thickness from 0 (curve A) up to 48 nm (curve N). (Reprinted with permission from P. Zhan et al., Adv. Mater., 18, 1612, 2006.) (c) and (d) Transmission spectra of ML–Ag normalized to the transmission of a 30-nm-thick planar Ag film in s- and p-polarized light, respectively. Lines in (d) show the dispersion of diffractively coupled surface plasmon polariton modes in a corrugated metal film. $D = 526$ nm.

Owing to the preparation method the metal film is forced to adopt the corrugation of the same periodicity as the sphere lattice. Correspondingly, one can expect mixed excitations for these two 2D electromagnetic crystals, and, hence, properties of these hybrids cannot be described by the linear superposition of their component properties.

The transmission of the Ag–ML sample (Figure 4.8.7) remarkably differs from that of the ML–Ag hybrid (Figure 4.8.5): (a) In the PBG spectral range of the 2D slab PhC ($D/\lambda \approx 0.9$), the overall transmission of the Ag–ML is much higher than that of the ML–Ag for the same nominal Ag film thickness. (b) SPP bands are seen in both polarizations (Figure 4.8.7c and d).[36] (c) The elaborated profile of transmission spectrum (Figure 4.8.2, Ag–ML) is the evidence of resonant light transfer.

The coating profile, which reaches $t \approx 30$ nm thickness on top of the spheres, possesses a corrugation depth of up to 50%; hence, the SPPs here are the short-range excitations, because the long-range SPPs can only be realized if the metal film corrugation does not exceed 10% of the period.[39] It is important to emphasize here that the lifetime of SPPs in a strongly corrugated film is much smaller than their characteristic absorption time.

With an increase in the angle of incidence some bands of enhanced transmission follow precisely the dispersion of SPP modes in *p*-polarized light that are calculated for the flat periodically perforated metal film (Figures 4.8.7 and 4.8.8). The EOT peak occurs at the overlap between index-guided modes and SPPs at $D/\lambda \approx 0.85$. Apparently, the transmission minimum occurring in the ML sample as a result of excitation of guided modes (Figure 4.8.2, ML) shares the same spectral range with the SPPs because of the same lattice periodicity. As a result, the transmission peak in the spectrum of Ag–ML substitutes the transmission minimum of the bare ML. Second-order diffraction is another source of this peak, which also requires long-range lattice ordering.

Another resonance at $D/\lambda \approx 0.45 - 0.55$ (Figure 4.8.7c and d), in the Ag–ML hybrid refers to localized plasmons excited in individual semishells,[40] which are of the same nature as nanoshell plasmons.[41] By joining these semishells in a dense ensemble, a discrimination of transverse (dipole) and longitudinal resonances takes place.[42] In ML-based samples with PBG in the visible, the semishell resonance can reveal itself in the near-infrared.

Angular resolved reflectance spectra of the bare ML (Figure 4.8.8a and b) show broad shallow Fabry–Pérot oscillations, which overlap narrow resonances for the diffraction orders of guided modes and the minimum at $\theta \approx 57°$ for *p*-polarized light at the Brewster angle. One can trace these Fabry–Pérot modes also in the Ag–ML as observed from its uncoated ML side (Figure 4.8.8c and d). However, the set of diffractively coupled resonance modes changes as compared to that of bare ML. Satisfactory matching is observed between dips in the pattern and the dispersion of SPP modes on the air and ML sides of the Ag film under *p*-polarized light illumination. Note that almost equal intensity of SPP resonances localized at the air and ML sides of the Ag film points to the strong light coupling to these excitations. Different modes are excited by *s*-polarized light with less steep dispersion.[34] Yet another reflectance pattern is obtained when the hybrid is observed from the Ag film side (Figure 4.8.8e and f). The Fabry–Pérot modes are retained in these patterns,

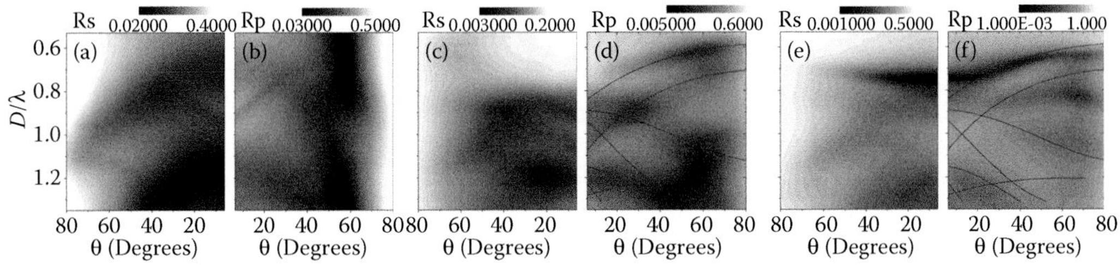

FIGURE 4.8.8
Reflectance spectra in *s*- and *p*-polarized light, as indicated at scale bars. (a), (b) Bare monolayer. (c), (d) Ag–ML from the ML side. (e), (f) Ag–ML from the metal film side. Lines in (d) and (f) mark the surface plasmon polariton dispersions.

but the minimum for SPP at the Ag–ML interface dominates reflectance as the consequence of Bloch mode-enhanced light coupling.[38]

Concluding this section, we can affirm that the transmission and reflectance bands dominating spectra of metal–ML hybrids are connected to SPP and localized plasmon excitations in the corrugated metal film. Apparently, the metal-on-ML architecture represents the extreme case of mode mixing in the plasmonic–photonic crystals.

4.8.4 Opal on Flat Metal Film: Toward Cavity Resonance

The next configuration of hybrid PhCs is obtained by substituting 2D with 3D PhC, that is, the ML of spheres with the opal crystal.[43] Currently, the best crystallinity can be achieved using opals in their thin film form (Figure 4.8.9a).[44] To prepare opal-on-metal hybrids, opal films were self-assembled on metal-coated glass substrates supplied with a 10-nm-thick protective SiO_2 layer. The lattice in thin film opals is slightly stretched compared to the symmetry of the face-centered cubic lattice, if they are prepared by pulling the substrate out of suspension during crystallization.[45] Nevertheless, the Brillouin zone of the face-centered cubic lattice (Figure 4.8.9c) can be used for assigning the diffraction resonances.

Light diffraction in 3D lattice (Figure 4.8.9b) is the source of complex transmission spectra of the opal film (Figure 4.8.2, opal). The low-frequency transmission minimum is caused by diffraction at (111) planes that are parallel to the substrate (Figure 4.8.9a). At higher frequencies many diffraction bands share the same spectral range and cross each other. Growing the opal on a metal film alters considerably the transmission of the opal (Figure 4.8.2, opal–Ag). As expected, the overall transmission level is strongly reduced as a result of the high metal reflectivity, but most diffraction resonances survive in the spectra of opal–Ag hybrids (Figure 4.8.10a).

As the diffraction grating in the close vicinity to the metal film is the same in both opal–Ag and ML–Ag hybrids, one can expect similar involvement of SPP modes in optical spectra. Taking into account weaknesses of these excitations (see Figure 4.8.5b) compared to

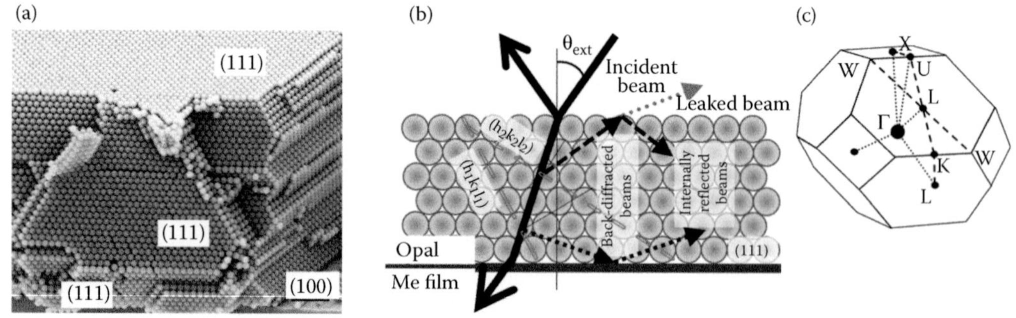

FIGURE 4.8.9
(a) Scanning electron microscope image of a cleaved opal film. Three planes of the {111} family and small fractions of (100) facets are exposed. (Image courtesy of W. Khunsin.) (b) Light propagation in the opal-on-metal hybrid. (c) Brillouin zone of the face-centered cubic lattice. The angle-resolved transmission and reflectance spectra are measured along XULKL line.

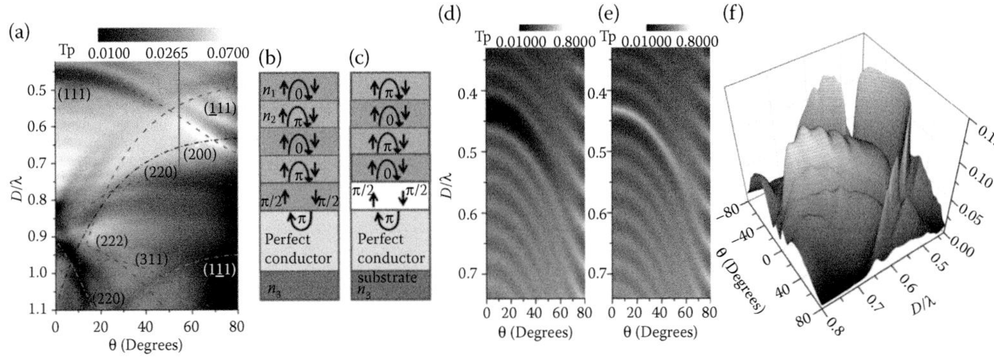

FIGURE 4.8.10
(a) Transmission spectra of the opal–Ag hybrid ($D = 431$ nm, 30 nm Ag film) in p-polarized light. Dashed and dash-dotted lines show the dispersion of Bragg resonances labeled with Miller indices of corresponding planes. The vertical line shows the Brewster angle. (b) Opal–metal hybrid represented as one-dimensional $\lambda/4n_i$ stack of dielectric layers ($n_i = \{n_1, n_2\}$) on a conducting substrate. Refractive indices range as $n_1 < n_2 < n_3$. (c) The same as (b), but with the $\lambda/4n_c$ cavity, where n_c is the index of the cavity dielectric. The phase shift along the light path is indicated in fractions of π. (d) and (e) Calculated transmission patterns of 10 period Bragg mirrors in p-polarized light with the refractive index profile adjusted to that along [111] axis of the opal assembled from 431 nm poly(methyl methacrylate) spheres and occurring in contact with (d) a 30-nm-thick Ag film on a glass substrate and (e) a quarter-wavelength SiO_2 cavity on a 30-nm-thick Ag film. (f) Experimental transmission spectra of the opal–Au hybrid with a 280-nm-thick SiO_2 cavity in s-polarized light.

diffraction resonances in the opal film (Figure 4.8.2, opal), there is no chance to resolve SPP-related features at $D/a > 0.6$.

The multilayer model can be applied to describe optical properties of this hybrid. A stack of (111) planes can be represented by a Bragg mirror consisting of $\lambda/4n_i$ layers with alternating refractive indices (Figure 4.8.10b). The high-index area corresponds to a central part of ML, whereas the low-index part is attributed to ML interfaces. The optical cavity between the Bragg mirror and the metal film is not the resonant one, because conditions of constructive interference, namely, the accumulation of 2π phase shift along one round trip, are not fulfilled (Figure 4.8.10b). This is confirmed by calculations of transmission spectra (Figure 4.8.10d). Instructive, for p-polarized light the reflection of the Bragg mirror changes phase at the Brewster angle $\theta_B \approx 53°$, at which each Fabry–Pérot maximum is replaced by a minimum (Figure 4.8.2, opal–Ag, and Figure 4.8.10a).

The important possibility offered by the opal–metal hybrid is the microcavity formation, if the cavity size allows a round trip with 2π phase shift (Figure 4.8.10c).[46] Experimentally, such a cavity was realized by adding a spacer of $\lambda/4n_c$ thickness. Then, the sharp transmission band appears in the middle of (111) transmission minimum, as it observed both in calculated (Figure 4.8.10e) and experimental (Figure 4.8.10f) spectra.

It is worth mentioning that the strength of cavity resonance between the Bragg and metal mirrors considerably exceeds that of a resonance mode introduced by a planar defect in the (111) bandgap.[47] In the latter, the transmission band comprises at best 50% of the minimum depth, whereas the cavity mode height exceeds this level by more that 10 times (Figure 4.8.10f).

Summarizing, the significance of the opal-on-metal architecture is in offering the resonance structures that can be easily combined with light-emitting or -absorbing materials.

4.8.5 Metal on Opal: Multiple-Resonance Architecture

Until recently, the surfaces of 3D PhCs were not a focus of interest[48] because they cannot, in general, support surface modes. In contrast, the metal-on-opal architectures offer a possibility of a surface control of PhC properties over a broad spectral range.[49] If the metal caps are electrically connected throughout the opal surface, the 2D corrugation lattice is formed (Figure 4.8.11a) and conditions for the SPPs coupling are fulfilled similarly to the metal-on-ML case. In metal-on-opal hybrids the light diffraction cannot be destroyed, but the metal film controls the amount of light that reaches the opal. Moreover, the interaction of 2D plasmonic and 3D photonic Bloch modes is achieved only in the metal film vicinity, thus accomplishing the inhomogeneous PhC configuration.

The transmission spectrum of the Ag–opal hybrid (Figure 4.8.2, Ag–opal) is clearly dominated by the EOT peak that is accompanied at oblique light incidence by SPP transmission pass bands and diffraction-related transmission minima. Fabry–Pérot oscillations overlapping the complex background add completeness to the beauty of this engineered spectrum. To rectify the plasmonic contribution, transmission spectra of the Ag–opal hybrid were normalized to transmission spectra of a bare opal film and that of a flat Ag film. The reduced transmission pattern (Figure 4.8.11d) shows that: (i) the (111) transmission minimum is preserved and becomes relatively deeper owing to a longer optical path length inside the opal film the latter being a result of higher reflectance of the opal–metal interface; (ii) the $(\bar{1}11)$, (200), and (220) transmission minima are converted into transmission bands;[49] (iii) SPPs modes are excited for any polarization of incident light, because lattice diffracted light runs away from the plane of incidence (Figure 4.8.12a and b); (iv) the transmission is enhanced at localized plasmon resonance on metallic semishells at $D/\lambda < 0.7$; (v) EOT peak occurs at $D/\lambda \approx 0.77$; and (vi) the minimum-to-maximum conversion applies to the low-angle fine structure of the transmission pattern at $0.8 \div 0.95D/\lambda$.

Diffraction orders of a 3D crystal that reach the corrugated metal film (Figure 4.8.12a and b) are further expanded in a number of SPP modes (Figure 4.8.4e). Hence, many beams are converted in SPP modes for each single angle of illumination of a metal–opal hybrid.

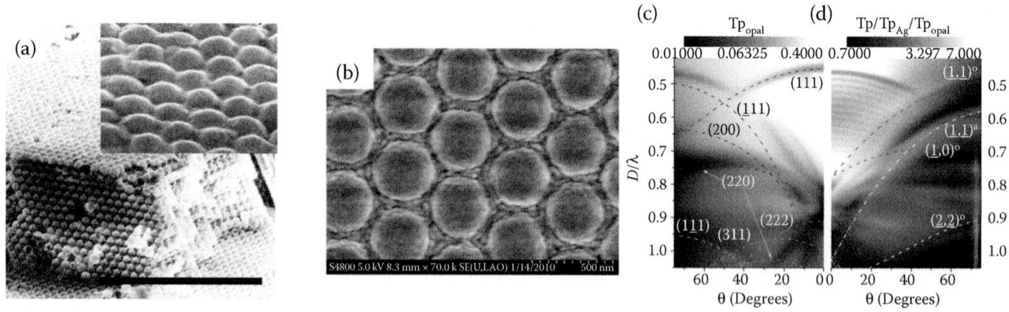

FIGURE 4.8.11
Scanning electron microscope (SEM) image of Au–opal hybrid consisting of a 50-nm-thick Au film deposited on opal film assembled from 560 nm poly(methyl methacrylate) spheres. Scale bar is 10 μm. The inset shows the metal on the opal surface. (Image courtesy B. Ding.) (b) SEM image of thick-film-coated opal. (c) Transmission pattern of the bare opal in p-polarized light. Dashed lines show the dispersion of diffraction resonances. (d) Pattern of reduced transmission of the Ag–opal. Lines show the surface plasmon polariton mode dispersions. $D = 431$ nm, nominally 50-nm-thick Ag coating.

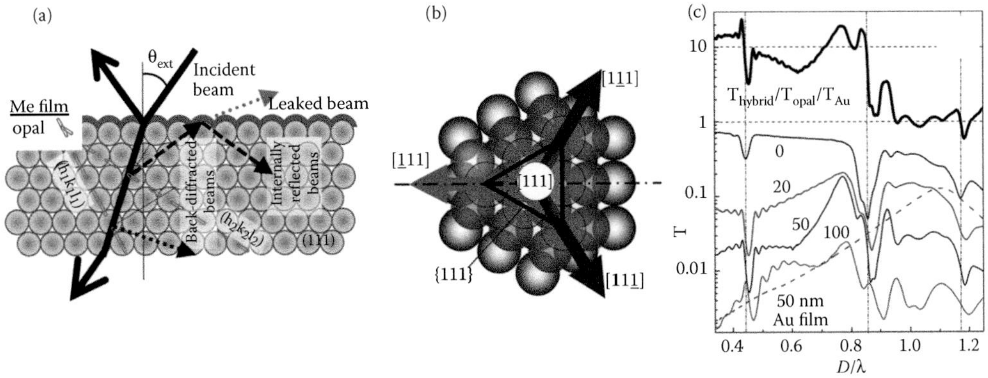

FIGURE 4.8.12
(a) Light propagation in a metal–opal sample. Note, that no additional periodicity is induced by projecting the three-dimensional lattice to its surface. (b) Escape of the {111} diffracted light from the plane of incidence. (c) Transmission spectra of Au–opal samples at $\theta = 0°$. The numbers at the curves show the nominal thickness of the Au films in nanometers. The dashed line shows the transmission of 50-nm-thick Au film. The thick curve is the reduced transmission of the Au–opal with 50-nm-thick coating after normalization to spectra of its respective opal and metal components.

A patchy transmission pattern (Figure 4.8.11d) corresponds to changing of the diffraction pattern with the incidence angle.

The transmission changes along the metal film thickness increase (Figure 4.8.12c) shows (i) a finer structure coming up, (ii) the progressive "blue" shift of diffraction minima, and (iii) nonmonotonous changes of the width and strength of EOT peak. The provisional explanations of these observations include a dependence of the phase shift of reflected light on the metal film thickness,[50] the formation of the Fano-resonance-type line shape,[51] and the cavity resonance between the (111) PBG and the corrugated metal mirror. The sequence of the $\lambda/4$ layers for the metal-on-opal architecture (Figure 4.8.10b) allows 2π phase shift for the light round trip, because metal here penetrates the former low n_1 index layer. If a cavity resonance is developed, it pushes the diffraction resonance toward higher frequencies. An additional source of spectral modification is the localized semishell plasmon.

Another important conclusion can be drawn from the reduced transmission spectrum (Figure 4.8.12c), which discriminates the plasmonic contribution. Remarkably, this ratio achieves a factor of 10 owing to the plasmon involvement at long wavelengths, but it drops down to unity, when SPPs become suppressed at shorter wavelengths by interband transitions in a gold film.[52]

The reflectance of the metal–opal hybrid as seen from the opal side appears similar to that of the bare opal with a dominating peak of the (111) diffraction resonance (Figure 4.8.13).[53] Additionally, the traces of the high-index plane resonances become clearly visible, although they appear as minima, because the diffraction losses apply to bouncing beams. The SPP modes of the PlC cannot be identified. On the contrary, the reflectance spectra obtained from the metal side of this hybrid show only SPPs (Figure 4.8.13c and d). The most pronounced minimum is caused by the (1,1) SPP mode, which is localized at the air–Au interface. Remarkably, SPP features obey other diffraction orders under *s*-polarized light and rapidly disappear at oblique incidence.

Summarizing, the optical response of metal-on-opal hybrids combines the contributions from all resonance phenomena displayed by the other three hybrids.

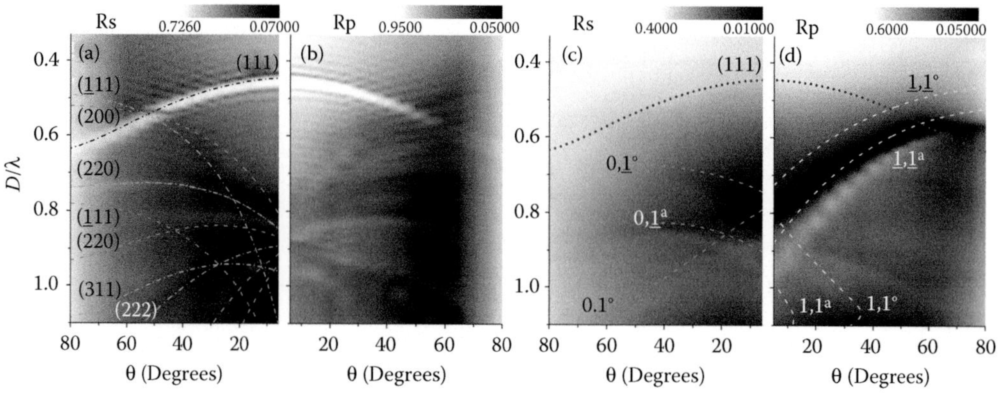

FIGURE 4.8.13
Reflectance patterns of the Au–opal hybrid in *s*- and *p*-polarized light obtained from (a, b) the opal and (c, d) metal film sides of this hybrid. The opal film consists of 560 nm poly(methyl methacrylate) spheres and the 50-nm-thick Au film. Dashed lines in panel (a) show the dispersion of diffraction resonances in the face-centered cubic lattice for LKL and LUX scans together, because the mirror-like reflectance of the metal film reduces the transmission anisotropy of the opal lattice. The dispersion of surface plasmon polariton modes is shown in panels (c) and (d).

4.8.6 Hybrid Heterostructures

The sequential approach to assembling of hybrid crystals allows designing complex heterogeneous architectures (Figure 4.8.1).[54] Such an approach opens new possibilities for engineering the optical response of hybrid crystals, although at the expense of lower transmission. One prominent example of heterostructuring is the opal–metal–opal architecture, in which the metal film acts as a planar defect in depth of the opal film (Figure 4.8.14a). The corrugated metal film in this case is immersed in the structured electromagnetic vacuum and practically isolated from the mode continua of the air and substrate.

In a heterocrystal the coupling–decoupling of SPPs at opposite sides of the metal film is almost equalized, but the lattices of surrounding opal films may not be in registry with each other. Interestingly, the transmission attenuation in (111) resonance depends on the

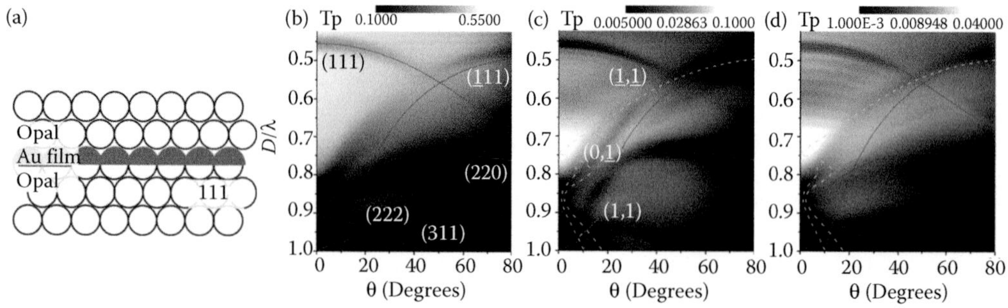

FIGURE 4.8.14
(a) Schematics of the metal film defect embedded in an opal film. (b)–(d) *p*-Polarized transmission spectra of the opal, Au–opal, and opal–Au–opal, respectively. Solid lines represent dispersions of diffraction resonances and dotted lines represent dispersions of surface plasmon polariton resonances at the Au–opal interface. No plasmons exist at $D/a > 0.75$ owing to interband transitions in Au.

sample composition. For the 13 ± 1 (111) planes comprising the first opal film, the attenuation achieves 1.6 times in a bare opal, 5.5 in Au–opal, and only 4 in opal–Au–opal despite the second film added to the latter structure.[55] The dramatic reduction of the width of the dispersive transmission minimum along $(\bar{1},\bar{1})$ resonance is tempting to assign to the resonant coupling of SPPs at similar interfaces (Figure 4.8.14c and d), but it can be also shadowed by the $(\bar{1}11)$ resonance.

The hybrid approach was also applied to enhance the light scattering at the interface between two different PhCs.[56] Using the Langmuir–Blodgett crystal as the cover for the Au–opal hybrid, more than 90% of the incident light intensity was scattered off the zero diffraction order over extremely broad spectral range in omnidirectional and polarization-independent manner.[57] Such extraordinary performance was achieved by exploiting synergy of diffraction and plasmonic mechanisms of light diversion.

Summarizing, hybrid heterocrystals that can be assembled from hybrid blocks as per Figure 4.8.1 bring about new functionalities unreachable by any single hybrid PhC and provide complex wave propagation in a superstructure.

4.8.7 Conclusions and Outlook

Opal-based PhCs are frequently used as the test platform for investigation of fundamental phenomena applicable to other PhC platforms and the discussed hybrid strategy is not an exempt from this rule. In the last 15 years, the work on colloidal PhCs passed several phases. It was started with studies of bare opals and opals infiltrated with different "guest" materials that represent "weak" PhCs with directional bandgaps. Then, new properties have been achieved with the introduction of inverted opals with strong light-to-structure interaction. New dimension was given to colloidal PhCs by invention of hetero-opals, the light propagation in which is sequentially controlled by different PBGs and, additionally, by interfaces between PhCs. The hybrid MD opal architectures can be considered as a new generic class of PhCs, the specific functionality of which is based on the plasmon-assisted resonant light transfer. The novelty comes from the close encounter of the metal and PhC components. Overall, five major resonance mechanisms can be used to engineer the optical properties of MD hybrid crystals. Moreover, these resonances can be tuned almost independently by changing the topology of hybrids. These peculiar properties of hybrid PhCs make them attractive for developing efficient light sources, absorbers, sensors, and nonlinear optical materials.

References

1. V. P. Bykov, Sov. Phys. JETP **35**, 269 (1972); R. Zengerle, J. Mod. Opt. **34**, 1589 (1987); E. Yablonovitch, Phys. Rev. Lett. **58**, 2059 (1987); S. John, *ibid*. **58**, 2486 (1987).
2. S. G. Romanov and U. Peschel, in *Nanocrystals*, edited by Y. Masuda (SCIYO, Rijeka, 2010), Chapter 4.
3. P. M. Johnson, A. F. Koenderink, and W. L. Vos, Phys. Rev. B **66**, 081102 (2002); A. B. Pevtsov, D. A. Kurdyukov, V. G. Golubev, A. V. Akimov, A. M. Meluchev, A. A. Kaplyanskii, A. V. Sel'kin, D. R. Yakovlev, M. Bayer, *ibid*. **75**, 153101 (2007); M. T. Tinker and J-B. Lee, Opt. Express **13**, 7174

(2005); W. Khunsin, S. G. Romanov, C. M. Sotomayor Torres, M. Scharer, L. Aagesen, and R. P. H. Chang, Opt. Lett. **33**, 461 (2008).
4. O. Toader and S. John, Phys. Rev. E **70**, 46605 (2004).
5. E. L. Ivchenko and A. N. Poddubnyi, Phys. Solid State **48**, 581 (2006).
6. E. L. Ivchenko, A. N Nezvizhevskii, and S. Jorda, Phys. Solid State **36**, 1156 (1994); V. P. Kochereshko, G. R. Pozina, E. L. Ivchenko, D. R. Yakovlev, A. Waag, W. Ossau, G. Landwehr, R. Hellmann, and E. O. Göbel, Superlatt. Microstruct. **15**, 471 (1994).
7. T. Fujita, Y. Sato, T. Kuitani, and T. Ishihara, Phys. Rev. B **57** 12428 (1998); A. L. Yablonskii, E. A. Muljarov, N. A. Gippius, S. G. Tikhodeev, T. Fujita, and T. Ishihara, J. Phys. Soc. Jpn. **70**, 1137 (2001); R. Shimada, A. L. Yablonskii, S. G. Tikhodeev, and T. Ishihara, IEEE J. Quantum Electron. **38**, 872 (2002).
8. R. W. Wood, Philos. Mag. **4**, 396 (1902); U. Fano, J. Opt. Soc. Am. **31**, 213 (1941).
9. T. W. Ebbesen, H. J. Lezec, H. F. Ghaemi, T. Thio, and P. A. Wolff, Nature **391**, 667 (1998).
10. W. L. Barnes, S. C. Kitson, T. W. Preist, and J. R. Sambles, J. Opt. Soc. Am. A **14**, 1654 (1997); M. Kretschmann and A. A. Maradudin, Phys. Rev. B **66**, 245408 (2002).
11. T. V. Teperik and V. V. Popov, F. J. García de Abajo, Phys. Rev. B **71**, 085408 (2005); T. V. Teperik, F. J. García de Abajo, A. G. Borisov, M. Abdelsalam, P. N. Bartlett, Y. Sugawara, and J. J. Baumberg, Nat. Photonics **2**, 299 (2008).
12. V. G. Veselago, Sov. Phys. Usp. **10**, 509 (1968).
13. H. Chen, C. T. Chan, and P. Sheng, Nat. Mater. **9**, 387 (2010).
14. U. Leonhardt, Science **312**, 1777 (2006).
15. M. Bardosova, M. E. Pemble, I. M. Povey, and R. H. Tredgold, Adv. Mater. **22**, 3104 (2010); J. F. Galisteo-López, M. Ibisate, R. Sapienza, L. S. Froufe-Pérez, Á. Blanco, and C. López, *ibid*. **23**, 30 (2011).
16. V. N. Astratov, V. N. Bogomolov, A. A. Kaplyanskii, A. V. Prokofiev, L. A. Samoilovich, S. M. Samoilovich, Yu. A. Vlasov, Il Nuovo Cimento **17D**, 1349 (1995); S. G. Romanov, A. V. Fokin, V. V. Tretiakov, V. Y. Butko, V. I. Alperovich, N. P. Johnson, and C. M. Sotomayor Torres, J. Cryst. Growth **159**, 857 (1996).
17. Z. Wang, C. T. Chan, W. Zhang, N. Ming, and P. Sheng, Phys. Rev. B, **64**, 113108 (2001).
18. Z. Liang, A. S. Susha, and F. Caruso, Adv. Mater. **14**, 1160 (2002); C. Graf and A. van Blaaderen, Langmuir **18**, 524 (2002).
19. S. G. Romanov, A. S. Susha, C. M. Sotomayor Torres, Z. Liang, and F. Caruso, J. Appl. Phys. **97**, 086103 (2005); Y. Jiang, C. Whitehouse, J. Li, W. Y. Tam, C. T. Chan, and P. Sheng, J. Phys. Condens. Matter **15**, 5871 (2003).
20. B. Rodrıguez-Gonzalez, V. Salgueirino-Maceira, F. Garcıa-Santamarıa, and L. M. Liz-Marzan, Nano Lett. **2**, 471 (2002); Y. Lu, Y. Yin, Z.-Y. Li, and Y. Xia, *ibid*. **2**, 785 (2002).
21. A. Moroz, Phys. Rev B **66**, 115109 (2002).
22. N. Eradat, J. D. Huang, Z. V. Vardeny, A. A. Zakhidov, I. Khayrullin, I. Udod, and R. H. Baughman, Synth. Met. **116**, 501 (2001); D. Wang, J. Li, C. T. Chan, V. A. SalgueiriÇo-Maceira, L. M. Liz-Marzan, S. Romanov, and F. Caruso, Small **1**, 122 (2005); N. Perez, A. Huls, D. Puente, W. Gonzalez-Vinas, E. Castano, and S. M. Olaizola, Sens. Actuators B **126**, 86 (2007); M. E. Kozlov, N. S. Murthy, I. Udod, I. I. Khayrullin, R. H. Baughman, and A. A. Zakhidov, Appl. Phys. A **86**, 421 (2007); J. C. Lytle and A. Stein, in *Annual Reviews of Nano Research*, edited by G. Cao and C. J. Brinker (World Scientific Publishing Co., New Jersey, 2006), Vol. 1, Chapter 1 .
23. F. Le, D. W. Brandl, Y. A. Urzhumov, H. Wang, J. Kundu, N. J. Halas, J. Aizpurua, and P. Nordlander, ACS Nano **2**, 707 (2008); R. C. Pennington, G. D'Alessandro, J. J. Baumberg, and M. Kaczmarek, Phys. Rev. A **79**, 043822 (2009).
24. S. A. Maier, *Plasmonics: Fundamentals and Applications* (Springer, New York, 2007); R. D. Meade, K. D. Brommer, A. M. Rappe, and J. D. Joannopoulos, Phys. Rev. B **44**, 10961 (1991); K. Ishizaki and S. Noda, Nature **460**, 367 (2009).
25. C. D. Dushkin, K. Nagayama, T. Miwa, and P. A. Kralchevsky, Langmuir **9**, 3695 (1993); L. Shi, X. Liu, H. Yina, and J. Zi, Phys. Lett. A **374** 1059 (2010); X. Yu, L. Shi, D. Han, J. Zi, and P. V.

Braun, Adv. Funct. Mater. **20**, 1 (2010); M. Lopez-Garcıa, J. F. Galisteo-Lopez, A. Blanco, J. Sanchez-Marcos, C. Lopez, and A. Garcıa-Martın, Small **6**, 1757 (2010).
26. P. B. Jonson and R. W. Christy, Phys. Rev. B **6**, 4370 (1972).
27. M. Inoue, Phys. Rev. B **36**, 2852 (1987); Y. Kurokawa, H. Miyazaki, and Y. Jimba, Phys. Rev. B **69**, 155117 (2004).
28. S. G. Romanov, M. Bardosova, M. Pemble, and C. M. Sotomayor Torres, Appl. Phys. Lett. **89**, 43105 (2006); S. G. Romanov, M. Bardosova, I. Povey, M. Pemble, and C. M. Sotomayor Torres, *ibid*. **92**, 191106 (2008).
29. A. Regensburger, A. V. Korovin, S. G. Romanov, and U. Peschel, submitted.
30. K. G. Mueller, M. Veith, S. Mittler-Neher, and W. Knoll, J. Appl. Phys. **82**, 4172 (1997); X. Li, D. Han, F. Wu, C. Xu, and X. Liu, J. Zi, J. Phys. Condens. Matter **20**, 485001 (2008).
31. M. López-García, J. F. Galisteo-López, Á. Blanco, C. López, and A. García-Martín, Adv. Funct. Mater. **20**, 4338 (2010).
32. V. N. Bogomolov, Sov. Phys. Usp. **21**, 77 (1978); S. G. Romanov and C. M. Sotomayor Torres, in *Handbook of Nanostructured Materials and Technology*, edited by H. S. Nalwa (Academic Press, San Diego, 2000), Vol. 4, Chapter 4, pp. 231–323.
33. H. Takei, J. Vac. Sci. Technol. B **17**, 1906 (1999); M. Himmelhaus and H. Takei, Sens. Actuators B**63**, 24 (2000).
34. H. J. Lezec and T. Thio, Opt. Express **12**, 3629 (2004).
35. P. Zhan, Z. Wang, H. Dong, J. Sun, J. Wu, H.-T. Wang, S. Zhu, N. Ming, and J. Zi, Adv. Mater. **18**, 1612 (2006); L. Landström, D. Brodoceanu, K. Piglmayer, and D. Bäuerle, Appl. Phys. A, **84**, 373 (2006); C. Farcau and S. Astilean, J. Opt. A: Pure Appl. Opt. **9**, S345 (2007).
36. L. Martin-Moreno, F. J. Garcia-Vidal, H. J. Lezec, K. M. Pellerin, T. Thio, J. B. Pendry, and T. W. Ebbesen, Phys. Rev. Lett. **86**, 1114 (2001).
37. N. Bonod, S. Enoch, L. Li, E. Popov, and M. Neviere, Opt. Express **11**, 482 (2003).
38. L. Landstrom, D. Brodoceanu, D. Bauerle, F. J. Garcia-Vidal, S. G. Rodrigo, and L. Martin-Moreno, Opt. Express **17**, 761 (2009).
39. I. Ursu, I. N. Mihailescu, A. M. Prokhorov, V. I. Konov, and V. N. Tokarev, Physica B + C **132**, 395 (1985).
40. C. Charnay, A. Lee, S.-Q. Man, C. E. Moran, C. Radloff, R. K. Bradley, and N. J. Halas, J. Phys. Chem. B **107**, 7327 (2003).
41. A. L. Aden and M. Kerker, J. Appl. Phys. **22**, 1242 (1951); E. Prodan, A. Lee, and P. Nordlander, Chem. Phys. Lett. **360**, 325 (2002).
42. A. I. Maaroof, M. B. Cortie, N. Harris, and L. Wieczorek, Small, **4**, 2292 (2008).
43. J. V. Sanders, Nature **4964**, 1151 (1964); V. G. Balakirev, V. N. Bogomolov, V. V. Zhuravlev, Y. A. Kumzerov, V. P. Petranovskii, S. G. Romanov, and L. A. Samoilovich, Crystallogr. Rep. **38**, 348 (1993).
44. Y. Xia, B. Gates, Y. Yin, and Y. Lu, Adv. Mater. **12**, 693 (2000); W. Khunsin, G. Kocher, S. G. Romanov, and C. M. Sotomayor Torres, Adv. Funct. Mater. **18**, 2471 (2008).
45. S. G. Romanov, Phys. Solid State **49**, 536 (2007); S. G. Romanov, U. Peschel, M. Bardosova, S. Essig, and K. Busch, Phys. Rev. B **82**, 115403 (2010).
46. E. Istrate and E. H. Sargent, Appl. Phys. Lett. **86**, 151112 (2005).
47. K. Wostyn, Y. Zhao, G. de Schaetzen, L. Hellemans, N. Matsuda, K. Clays, and A. Persoons, Langmuir **19**, 4465 (2003); E. Palacios-Lidón, J. F. Galisteo-López, B. H. Juárez, and C. López, Adv. Mater. **16**, 341 (2004); F. Fleischhaker, A. C. Arsenault, J. Schmidtke, R. Zentel, and G. A. Ozin, Chem. Mater. **18**, 5640 (2006).
48. K. Ishizaki, and S. Noda, Nature **460**, 367 (2009); R. Moussa, B. Wang, G. Tuttle, Th. Koschny, and C. M. Soukoulis, Phys. Rev. B **76**, 235417 (2007); A. Tikhonov, J. Bohn, and S. A. Asher, *ibid*. **80**, 235125 (2009).
49. B. Ding, M. E. Pemble, A. V. Korovin, U. Peschel, and S. G. Romanov, Phys. Rev. B, **82**, 035119 (2010).
50. I. R. Hooper and J. R. Sambles, Phys. Rev. B **70**, 045421 (2004).

51. M. V. Rybin, A. B. Khanikaev, M. Inoue, K. B. Samusev, M. J. Steel, G. Yushin, and M. F. Limonov, Phys. Rev. Lett. **101**, 023901 (2009).
52. B. R. Cooper, H. Ehrenreich, and H. R. Philipp, Phys. Rev. **138**, A494 (1965).
53. B. Ding, M. E. Pemble, A. V. Korovin, U. Peschel and S. G. Romanov, Appl. Phys. A **103**, 889 (2011).
54. S. G. Romanov, H. M. Yates, M. E. Pemble, and R. M. De La Rue, J. Phys.: Condens. Matter **12**, 8221 (2000); E. Istrate and E. H. Sargent, Rev. Mod. Phys. **78**, 455 (2006).
55. A. S. Romanova, A. V. Korovin, and S. G. Romanov, Phys. Solid State **53**, 7 (2011).
56. S. G. Romanov, C. M. Sotomayor Torres, J. Ye, and R. Zentel, Progr. Solid State Chem. **33**, 279 (2005).
57. B. Ding, M. Bardosova, M. E. Pemble, A. V. Korovin, U. Peschel, S. G. Romanov, Adv. Funct. Mater. **21**, 4182–4192 (2011).

4.9

Light Propagation in Photonic Crystals Infiltrated with Fluorescent Quantum Dots or Liquid Crystal: Different Dimensionality Analysis

Rabia Moussa and Alexander Kuznetsov
The University of Texas at Dallas

Ryotaro Ozaki
National Defense Academy

Anvar A. Zakhidov
The University of Texas at Dallas

CONTENTS

4.9.1 Introduction

Photonic crystals (PCs) attracted a lot of attention in the last two decades because of their unique electromagnetic features and potential applications.[1,2] One of the main features of PCs is that electromagnetic waves with frequencies within the photonic band gap (PBG) are prohibited to propagate, regardless of their polarization and propagation direction. On the basis of this, many interesting physical phenomena have been discovered.[2,3] However, in some cases the nanoscale of the structure and the complicated physics behind it make it too difficult to study these materials using the well-established techniques. Therefore, the need to think about other ways to approach this kind of problems is essential. Usage of nanoscale materials such as quantum dots (QDs) and liquid crystals are emerging techniques for studying new and unique properties that the PCs might have.

The idea is to infiltrate with QDs and liquid crystals all different PCs ranging from a simple multilayer porous silicon one-dimensional (1D) structure to a two-dimensional (2D) GaAs PC and direct and inverse opals. Studying these structures before and after infiltration provides information about the properties of PCs. QDs (which are sometimes called artificial atoms) fall into the category of nanocrystals. They have some exceptional advantages including being easily excited using excitation sources hundreds of nanometers away from the emission wavelength. Also they are highly photoresistant with long fluorescence lifetimes. Their intense fluorescence is usually used to track individual molecules. Finally, their broad excitation spectra allow the use of existing excitation sources, making imaging easier.

Similarly to QDs, liquid crystals[4] can induce dramatic changes in the properties of the PC; for example, the ability of PCs to switch the effective refractive index from positive to negative. Motivated by these interesting features of liquid crystals, we opted to infiltrate our PCs with nematic liquid crystals. In fact, optical switch based on PCs that exhibit very different optical properties and possess the ability to tune photonic band structures could be realized by taking advantage of the orientation transitions of the liquid crystal. Such switchable PCs provide a method by which light can be routed.

Without a doubt, one of the most interesting features that we investigating here in PCs is the possibility that these PCs can be regarded as negative refractive index materials[5,6] under certain conditions. Notomi[5] suggested that PCs, which are periodic optical nanostructures, also could have an effective negative refractive index.[3,4] Many recently published papers[6,7] have studied negative refraction and focused on the superlensing phenomenon or on the diffraction limit problem. The key feature of these phenomena is the excitation of surface waves and the amplification of the evanescent waves inside the negative index slab. However, applications in nanoscale technology require another procedure, especially when it comes to extracting light from optical devices. QDs seem to be a good solution.[8] In fact, by using the infiltration techniques with QDs or with liquid crystals we are looking for ways to visualize this negative refraction in such complicated structures such as direct and inverse opals, and to create the tunability of the band structures and the switching ability that the 2D PC may possess. In this case, infiltrating these structures with certain QDs is a key for detection of negative refraction.

This chapter is divided into three parts. In the first part, we focus on the infiltration of 1D porous silicon with QDs and analyze different infiltration methods to lay the ground for future work. In the second part, we focus on the infiltration of 2D PCs with liquid crystals and study their switching ability by taking advantage of the unique features of the liquid crystals. In the third part, we analyze the infiltration of QDs inside direct and inverse opals and track them down for a possible focusing inside the opals, which would indicate left-handed behavior of the opals.

4.9.2 One-Dimensional Photonic Crystals

The first part of this chapter studies the infiltration techniques of the fluorescent QDs inside 1D PCs. Our goal in this part is to identify the best technique to infiltrate QDs inside PCs and use this tool to investigate optical properties of PCs.

Our 1D PC structure is made of alternating layers of porous silicon with different refractive indices. Each layer is done by etching a silicon wafer in a solution of hydrofluoric acid, water, and ethanol.[9] For this study, the 1D PC is made of 20 alternating layers (AB) with layer A having 50% porosity and 80 nm thickness and layer B having 75% porosity and 120 nm thickness. Refractive index of porous silicon layers can be calculated using Bruggeman approximation[10] as pores are of nanometer range. The layers with 50%

and 75% porosity have refractive indices of about 2.3 and 1.5 (at 750 nm), respectively. Experimentally, the index and the thickness of each layer are tuned separately by using different etching conditions which leads to a better control of the position and the width of the PBG.

A scanning electron microscope (SEM) image of the top layer of a porous silicon PC is shown in Figure 4.9.1. Pores are distributed uniformly through the crystal and have an average size of about 15 nm.

Nano-sized objects (QDs) may be infiltrated inside porous silicon multilayered PCs by different methods. Solution soaking, sonication, and thermal- or vacuum-assisted infiltrations are some methods that have been used to infiltrate QDs inside PCs. Our purpose was to choose the optimal method that leads to the best infiltration results. The QDs used in this study were made of CdSe with an emission peak at 615 nm and a diameter of approximately 5.2 nm, which is few times smaller than the size of the pores in silicon. The emission of QDs in visible range allows for easier spectroscopic characterization.

The infiltration of QDs into our 1D porous silicon PC was carried out by four different methods. The first sample (later referred to as sample 1) was left for 20 h in a solution of QDs in toluene. The second sample (sample 2) was placed in the solution and heated for 20 h at 40°C. The third sample (sample 3) was placed in the solution and sonicated for 14 h. Finally, the last sample (sample 4) was placed in vacuum and the solution of QDs was immersed into the PC and left for 16 h.

Emission spectra measurements were carried out before and after QD infiltration and are shown in Figure 4.9.2. The spectra were measured using 500 nm excitation beam. At this excitation wavelength, bulk silicon has relatively strong emission around 650 nm. The peak is clearly observed in all samples before the infiltration of QDs (Figure 4.9.2). After the infiltration of QDs, another peak at 615 nm appeared. The height of this peak relative to the 650 nm silicon one is proportional to the concentration of infiltrated QDs. Figure 4.9.2 indicates that the infiltration used for sample 1 results in a prominent peak at 615 nm of the QDs. The second method showed quite similar results and the intensity of the peak could be enhanced by adjusting the heating temperature. The third method (sample 3) was not as successful as the first and the second samples because the first peak at 615 nm was relatively weak. Finally, method 4 was not successful at all. It is important to mention that all the infiltration methods do not provide any information about the percentage of QDs

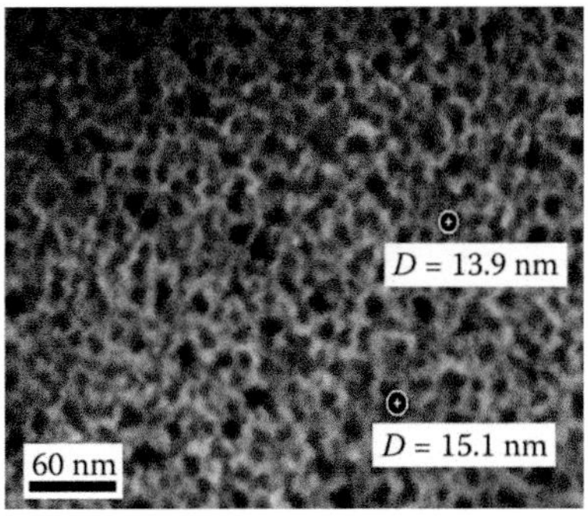

FIGURE 4.9.1
Scanning electron microscope image of the top layer of porous silicon multilayered one-dimensional photonic crystal showing an average pore size of 15 nm.

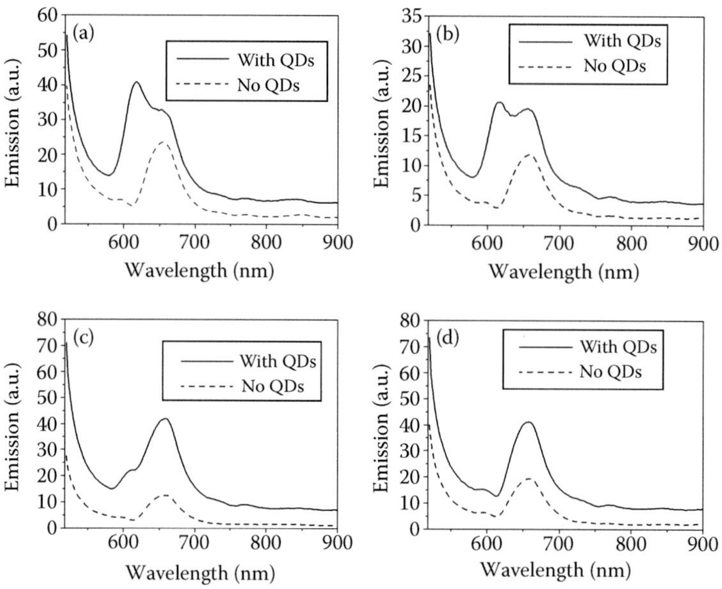

FIGURE 4.9.2
Emission spectra of one-dimensional porous silicon multilayered photonic crystals before (dashed line) and after (solid line) infiltration of quantum dots. (a) Sample 1, (b) sample 2, (c) sample 3, and (d) sample 4.

that were trapped on the surface and the percentage of them that went deeper into the sample. So we looked at other methods to characterize the distribution of QDs inside the porous silicon PC.

Reflectance spectroscopy was used. It is well known that a band gap of a PC depends on the refractive indices of the material it is made of. So infiltration of materials with different refractive indices into the PC would affect the width and the position of the band gap. Difference in refractive indices of porous silicon (2.3 for 50% porosity and 1.5 for 75% porosity) and CdSe (2.5) would result in significant shift of the band gap when CdSe QDs are successfully infiltrated.

Results of reflectance measurements before and after QD infiltration are shown in Figure 4.9.3. Samples 1, 2, and 3 show significant shift of the band gap, so we can conclude that in all these cases QDs are successfully infiltrated inside the PC. Again, sample 4 shows almost no change in the band gap, proving that infiltration by the fourth method was not successful.

Another important feature of these plots shows that maximum reflectance is decreased by about 8% for sample 1, 12% for sample 2, and only 2% for sample 3. The drop in maximum reflectance occurs because of the nonuniform distribution of QDs inside the sample. Once the QDs are infiltrated into the structure, the effective refractive index of the whole PC will change. However, when the QDs are concentrated in the first few layers and do not affect the whole structure in a uniform way, the quality of PBG is significantly decreased and the PBG peak reflection is lowered. As Figure 4.9.3 indicates, method 3 provides the most uniform distribution of QDs inside PCs whereas methods 1 and 2 result in gradient concentrations of QDs. This nonuniformity of QDs inside the structure also explains the higher emission peaks of QDs for samples 1 and 2 as compared to sample 3 (see Figure 4.9.2). So, even though the intensity of the 615 nm peak for sample 3 was very modest, the QDs

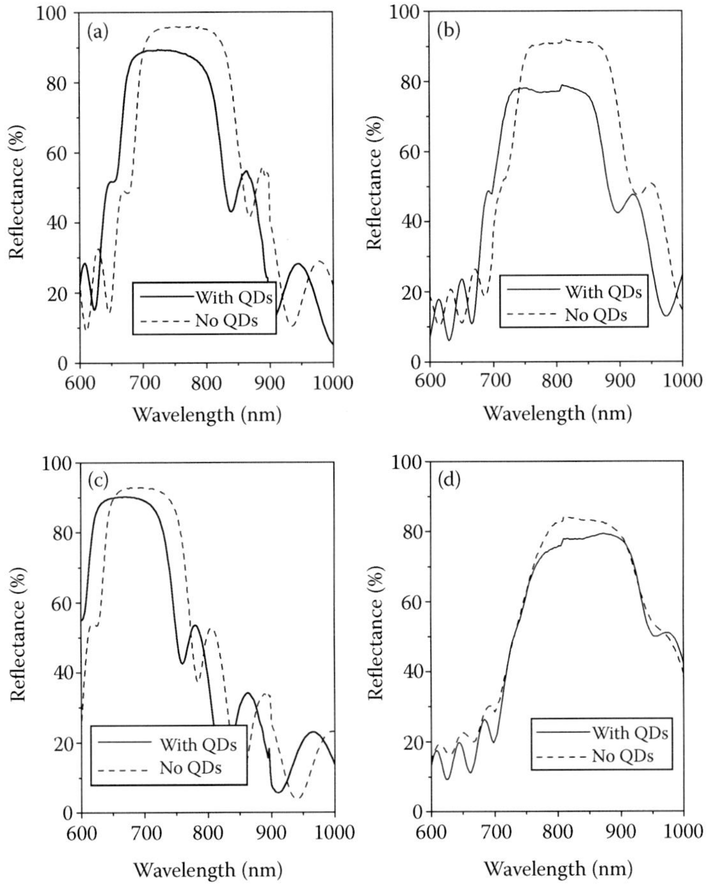

FIGURE 4.9.3
Emission spectra of one-dimensional porous silicon photonic crystals before (dashed line) and after (solid line) infiltration of quantum dots. (a) Sample 1, (b) sample 2, (c) sample 3, and (d) sample 4.

are distributed much more uniformly. The easy way to overcome this problem is to increase the concentration and the infiltration time. This will guarantee both higher concentration and better uniformity of QDs inside the PC.

Summarizing all the results, we can conclude that the best infiltration method is the sonication of the sample immersed into the solution of the dots. Although the soaking and thermally assisted sonication yields to high amounts of infiltrated QDs, sonication also allows achieving much more uniform distribution of QDs inside the porous silicon.

As seen in Figure 4.9.3, the reflectance spectra were shifted to the lower wavelengths on QDs infiltration. This so-called blue-shift is unexpected because the presence of QDs should increase the effective refractive index of the medium and would in fact lead to a red-shift (to higher wavelengths) of the band gap.

The observed blue-shift of the band gap indicates that the effective refractive index of the porous silicon PC was reduced on QD infiltration and we believe it is due to the oxidation of silicon (Si) to silicon dioxide (SiO_2). There are two main factors to support this idea. First, silicon dioxide has much smaller refractive index ($n = 1.45$) than silicon ($n = 3.87$).

Second, owing to high porosity of the samples their surface area (per unit volume) is extremely high. Hence, even a few-nanometer-thick oxide layer will cause a significant decrease in the effective refractive index.

To confirm that assumption, we intentionally oxidized a porous silicon film by putting it in the oven and heating it to 600°C in air for 2 h. Such a process should result in a 1–2-nm-thick oxide film.[11] On average, a blue-shift of about 30 nm was observed (see Figure 4.9.4a). We also observed rather slow oxidation of porous silicon samples when they were left in air at room temperature for a long time (Figure 4.9.4b). Thus, the total blue-shift of spectra is the result of the overlapping of two effects: the red-shift due to QD infiltration and the much stronger blue-shift due to the oxidation.

To get a better insight of the infiltrated QDs, we used confocal microscopy that has been proven to be one of the best techniques for three-dimensional (3D) visualization of embedded light sources.[12] By collecting only the focused light, it is possible to reconstruct the 3D structure of an object. Confocal microscopes scan through the sample by changing the focus distance. Then, special software is used to convert consecutive series of scans into 3D images.

Figure 4.9.5 shows the grayscale confocal image of porous silicon 1D PCs before and after infiltration of QDs. It is seen that after the infiltration the dots are distributed uniformly throughout the surface of the sample. Brighter spots indicate the dots that are located closer to the surface whereas darker spots correspond to the deep dots infiltrated inside the sample. Also, because the pore size is just few times larger than the size of the QDs, there is no agglomeration of the dots.

By taking a series of scans focusing inside the sample, a 3D reconstructed image of the sample was created. Figure 4.9.5c shows the side view of the image that indicates that QDs are also infiltrated inside the sample and not only at the surface. Unfortunately, the maximum resolution of the confocal microscope (about 2 μm) does not allow one to investigate the distribution of QDs through the 10-μm-thick sample in more detail.

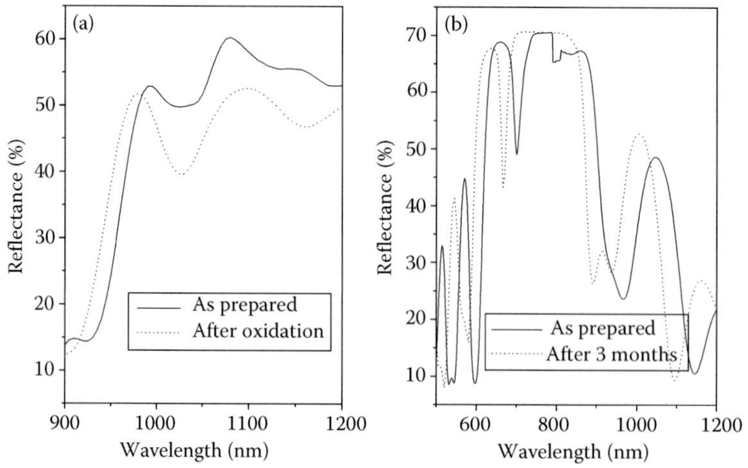

FIGURE 4.9.4
Reflectance spectra of (a) porous silicon film before (solid line) and after (dotted line) oxidation in air at 600°C for 2 h and (b) porous silicon photonic crystal before (solid line) and after (dotted line) air exposure at room temperature.

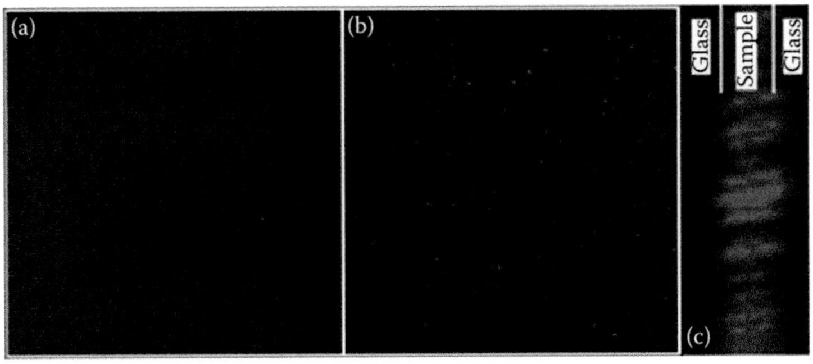

FIGURE 4.9.5
(**See color insert.**) Grayscale confocal image of porous silicon (a) before infiltration and (b) after infiltration with quantum dots. (c) Side view of reconstructed three-dimensional confocal image showing porous silicon sample infiltrated with quantum dots. Image size is 0.75 mm × 0.75 mm.

4.9.3 Two-Dimensional Photonic Crystal

In the second part of this chapter, we focus on the 2D PC infiltrated by another material, liquid crystals. Our goal in this part of the chapter is to show how PC properties can be dramatically changed simply by infiltrating it with certain materials. We numerically study light propagation in a liquid-crystal-infiltrated 2D PC at a high-order photonic band. The use of liquid crystals is widely considered for tunable PC investigations. Using liquid crystals, tunable superprism effect and tunable negative refraction have also been studied.[13] However, previous studies focus only on tunable refraction and not on tunable refractive index. Moreover, the equifrequency surfaces (EFSs) presented were not perfectly circular, which assumes the effective refractive index is inaccurate. An effective refractive index of a PC can be defined only when its EFS is a circle.[5] Here, we discuss a simple configuration to obtain a circular EFS and its photonic band structure.

We considered the 2D triangular lattice PC of GaAs infiltrated with a nematic liquid crystal. The reason behind the choice of GaAs, with a refractive index of 3.6, is the large index contrast that assures the notion of an effective refractive index in a PC. Since GaAs is not transparent for visible light, we assume that an infrared beam propagates into the PC. The R/a ratio of the PC is 0.42, where a is the lattice constant and R is the radius of the holes. For light at a wavelength of 1.55 μm, a and R are 845 and 355 nm, respectively.

Recently, there was a report that a dual-frequency nematic liquid crystal has a large birefringence ($\Delta n > 0.3$) in the infrared region.[14] In our calculation, we examined the GaAs PC with a dual-frequency nematic liquid crystal having a large birefringence. The ordinary and extraordinary refractive indixes of the liquid crystal are set to be 1.5 and 1.8, respectively. Figure 4.9.6a and b shows liquid crystal molecular alignment in the PC. We assume that liquid crystal molecules align parallel to the z-direction when a low-frequency electric field is applied along the z-direction. As a result of the use of a dual-frequency nematic liquid crystal, the liquid crystal molecules should be reoriented perpendicular to the z-direction when a high-frequency electric field is applied along the z-direction.

Let us consider transverse magnetic mode light propagation consisting of E_z, H_x, and H_y. The electric field E_z is affected by only the extraordinary index when the molecular directors align parallel to the z-direction. In contrast, E_z is affected by only the ordinary index

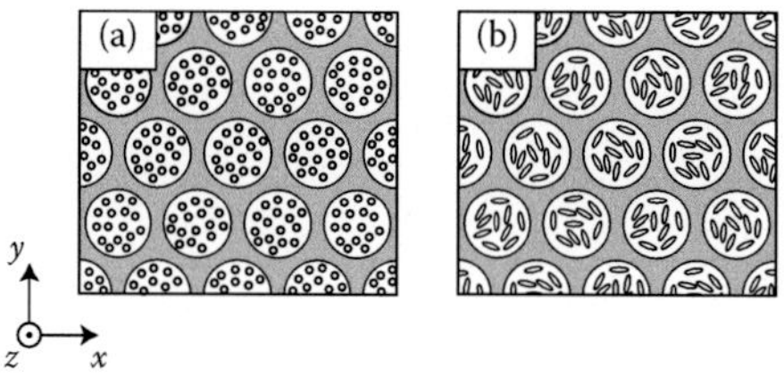

FIGURE 4.9.6
A GaAs two-dimensional triangular lattice photonic crystal infiltrated with a dual-frequency nematic liquid crystal. (a) The liquid crystal directors are parallel to the z-axis when a low-frequency electric field is applied along the z-axis. (b) The directors are perpendicular to the z-axis when a high-frequency electric field is applied along the z-axis.

when the molecular directors are perpendicular to the z-direction. In this configuration, the propagating light is not affected by both the ordinary and the extraordinary indices simultaneously. Thus, we could avoid the complicated propagation owing to the optical anisotropy of the liquid crystal because the light is affected by only the ordinary or the extraordinary index. This is the important key to obtain circular EFS in *k*-space.

Figure 4.9.7a shows the photonic band diagram of the structure at higher frequencies and calculated by a plane-wave method.[15] The solid and broken lines are for the ordinary and extraordinary refractive indices, respectively. At lower frequencies, there is not much difference between the solid and broken lines. However, the difference between the two lines becomes larger at higher frequencies. This is caused by the fact that the light effectively feels a larger liquid crystal domain at higher frequencies because the wavelength shortens. Therefore, the change in the refractive index can be utilized efficiently at a higher photonic band. An effective refractive index of a PC can be defined when a photonic band forms a bell shape near the Γ point.[5] The sign of the effective index is determined by the upward or downward direction of the bell-shaped band. An upward bell-shaped band means the PC could have positive refractive indices, whereas a downward bell-shaped band means that the PC could have negative refractive indices.

The fifth band for $n_{lc} = 1.5$ and the sixth and seventh bands for $n_{lc} = 1.8$ form a bell shape near the Γ point. The effective refractive index of the PC is determined by $|n_{eff}| = ck/\omega$

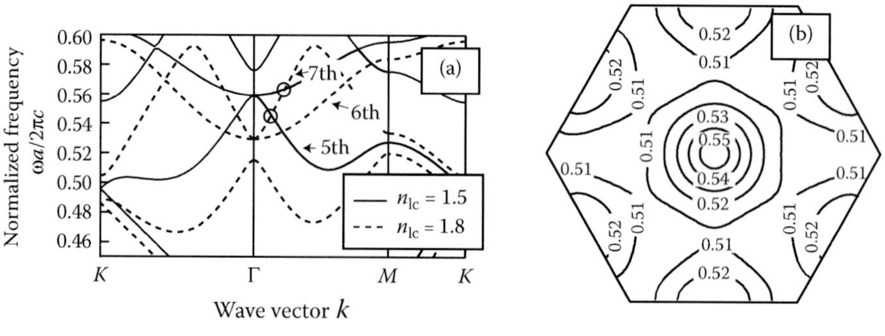

FIGURE 4.9.7
Photonic band diagram of the tunable photonic crystal calculated by the plane-wave method (a). The equifrequency surface of the fifth band for $n_{lc} = 1.5$ (b).

that is definable when an EFS is circular. Figure 4.9.7b shows the EFS of the fifth band for $n_{lc} = 1.5$. The EFS corresponds to a contour plot of the photonic band, and a hexagon represents the first Brillouin zone. It is clear that the contours become circular above the normalized frequency $\omega a/2\pi c$ of 0.52 at the center. These calculations indicate that the PC could have an effective refractive index between 0.52 and 0.55. Here, we emphasize again that the effective index is determined by $|n_{eff}| = ck/\omega$. Note that there are some intersection points in Figure 4.9.7a. The marked intersections mean that both bands for $n_{lc} = 1.5$ and $n_{lc} = 1.8$ have the same k and ω. That is, PCs with $n_{lc} = 1.5$ and 1.8 have the same $|n_{eff}|$ under these conditions. In contrast, the sign of the effective index is not the same because the directions of the bell-shaped bands are opposite between $n_{lc} = 1.5$ and 1.8. Therefore, the tunable PC can change the sign of the effective refractive index by using the liquid crystal.

Finite-difference time-domain (FDTD) simulation is performed to investigate light propagation in the tunable PC described earlier. We first investigate whether an incident light is coupled into the PC. It is known that an uncoupled mode exists at some high-order photonic bands. Coupling is very important because if the incident light is uncoupled, the light cannot enter the PC.

The transmission spectrum and the photonic band diagram for $n_{lc} = 1.5$ are shown in Figure 4.9.8. To compare with the transmission spectrum, the band diagram in the Γ–*M* direction shown in Figure 4.9.7a is again plotted. The transmission spectrum and band diagram are calculated by the FDTD method and plane-wave method, respectively.

In the FDTD simulation, the transmitted light is detected at the bottom of the PC. As evident from Figure 4.9.8, the transmittance significantly decreases at some frequencies. The gray backgrounds indicate strong reflection, and almost all of the reflection frequencies correspond to PBGs in the Γ–*M* direction. However, despite the existence of the third, fifth, and sixth bands, the transmittance is quite low at some frequencies in the bands. This means that the incident light is uncoupled to the PC at these frequencies.

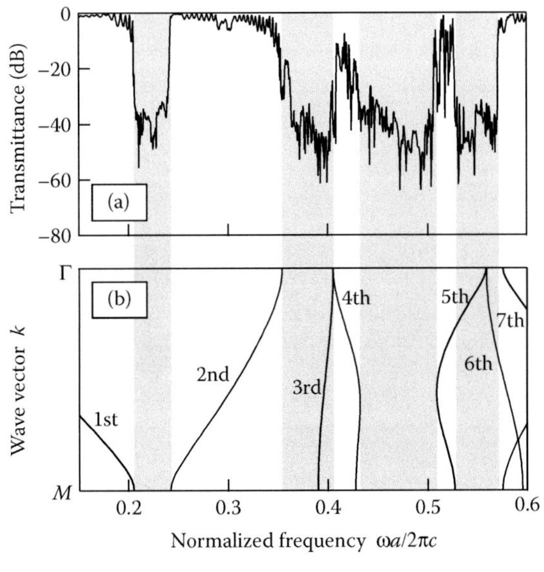

FIGURE 4.9.8
Transmission spectrum (a) and the photonic band diagram (b) for $n_{lc} = 1.5$.

The uncoupled mode is caused by a field pattern of an electromagnetic wave at a high-order band. Figure 4.9.9 shows the field patterns of the E_z component at the fifth and sixth bands.

In the field patterns, blue and red represent negative and positive strengths, respectively. The important thing here is that 2D electric field patterns have electric dipoles along the x- and y-directions. When an incident light is a plane wave propagating along the y-direction, that is, the incident light is perpendicular to the x-direction, the incident light cannot excite an electric dipole along the x-direction because the incident light does not have any x-component. In other words, a plane wave that has only the y-direction dipole hardly generates a 2D electric field pattern shown in Figure 4.9.9; that is why transmittances at the fifth and sixth bands are quite low. To resolve such an uncoupled mode, an antireflection structure has been studied.[16] Here, we consider a PC structure without the top and bottom edges. We indeed have some part of the rods of the structure being cut at the beginning and the end of the structure; in other words, we have a specific surface termination of the structure to allow some mode coupling that otherwise would be forbidden. The aim of the structure is to disrupt the phase of the incident plane wave at the interface. The phase disruption at the interface gives rise to the x-direction dipole, and that will improve the coupling efficiency. Figure 4.9.10 shows the transmission spectrum from the structure. It is clear that the transmittance of the fifth band is dramatically improved. The improvement of coupling efficiency is due to the fact that the incident plane wave is not in-phase at the interface. On the other hand, the sixth band is hardly improved and is still uncoupled. As the fifth and seventh bands are available in the structure, we could use the frequency at the intersection between the fifth and seventh bands in Figure 4.9.7a in order to control the sign of an effective index.

Let us examine now light propagation into the tunable PC having a positive or negative refractive index. Figure 4.9.11 shows the FDTD simulated propagation from the liquid crystal into the tunable PC at $\omega a/2\pi c = 0.545$ with an incident angle of 3°. In Figure 4.9.11, the PC is placed at the lower end. That is, the upper medium is liquid crystal and the lower medium is the tunable PC. Figure 4.9.11a and b represents $n_{lc} = 1.5$ and 1.8, respectively. As evident in the figure, the refraction direction in the PC depends on the refractive index of the liquid crystal. This indicates that we could control refraction direction between positive and negative by reorienting the liquid crystal. The effective refractive indices of the PC with $n_{lc} = 1.5$ and 1.8 are −0.132 and 0.132, respectively, which are determined by $|n_{eff}| = ck/\omega$. Although the value of the effective refractive index is less than the index of air $n_{air} = 1$, it is

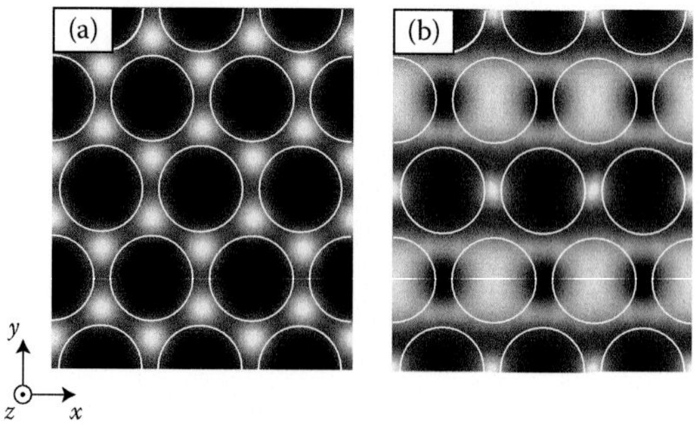

FIGURE 4.9.9
Electric field patterns of the E_z component at the fifth band (a) and the sixth band (b).

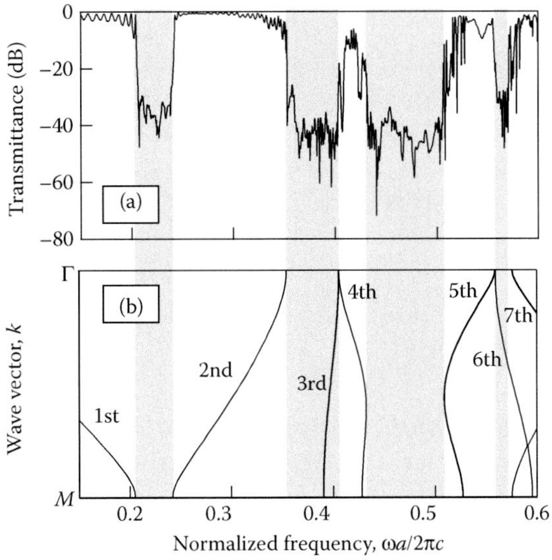

FIGURE 4.9.10
Transmission spectrum (a) and the photonic band diagram (b) of the tunable photonic crystal without the top and bottom edges. The refractive index of the liquid crystal is 1.5.

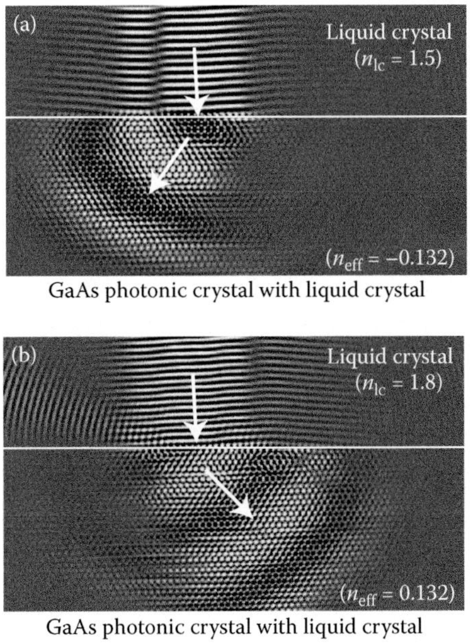

FIGURE 4.9.11
Finite-difference time-domain simulated propagation from the liquid crystal into the photonic crystal at $\omega a/2\pi c = 0.545$ with incident angle of 3°: (a) $n_{lc} = 1.5$ and (b) $n_{lc} = 1.8$. The tunable photonic crystal is placed at the lower part.

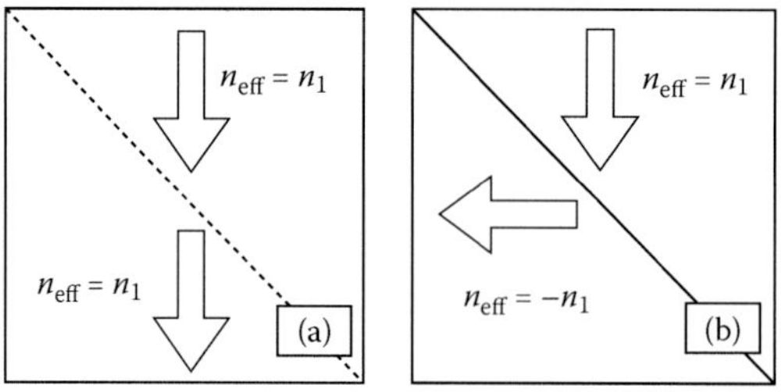

FIGURE 4.9.12
Optical switch using two triangular photonic crystals having positive or negative refractive index: (a) positive and positive and (b) positive and negative.

not wrong. This is because the effective refractive index is for phase velocity, not for group velocity. According to Snell's law, the refractive angles are –36.5° and 45.5° for $n_{lc}/n_{eff} = -1.5/0.132$ and 1.8/0.132, respectively. The arrows in Figure 4.9.11 are drawn on the basis of the angles determined by Snell's law. It is clear that the arrow directions agree well with FDTD simulated propagation.

Light propagation through a combination of positive and negative index media is also examined. Several groups have studied the optical splitter using a combination of triangular PCs.[17] Here, we consider an optical switch using two triangular PCs shown in Figure 4.9.12. The greatest advantage in our PC system is the controllable index sign.

When the two triangular PCs have the same effective index, an incident light propagates straight. However, if we use two triangular PCs having the same absolute value of effective index but opposite signs, an incident light could turn 90° at the interface as a result of negative refraction. Figure 4.9.13 shows the propagating electric field passing through the combined system of two triangular shapes. In Figure 4.9.13a, the light propagates straight because they have the same effective index. In contrast, when the two triangular PCs have positive and negative indices, the light turns 90° to the left. Note that light is deflected at right angles without a waveguide structure. However, in this calculation, the reflection at the interface between two triangular PCs is not zero. The interface should be optimized to reduce the reflection. We continue to work toward resolving the problem. Aside from the problem of internal reflection, FDTD simulation suggests that light propagation can be controlled with the tunable effective refractive index of a PC.

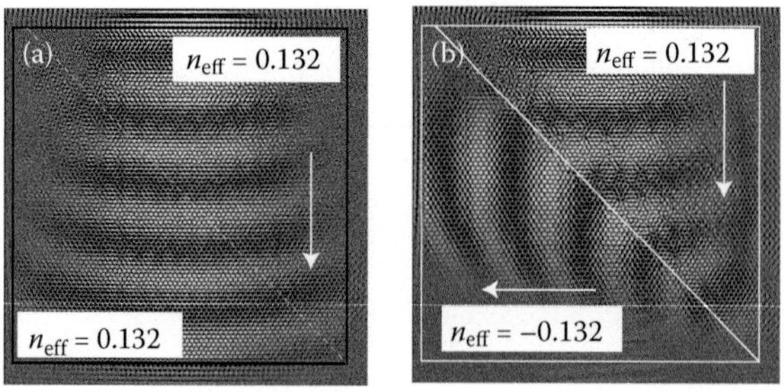

FIGURE 4.9.13
Finite-difference time-domain simulated propagation in the combination of two triangular photonic crystals. (a) Photonic crystals with the same effective index. (b) Photonic crystals with opposite effective refractive index signs.

In this section, we were able to numerically examine light propagation into the PC infiltrated with a nematic liquid crystal at a high-order photonic band. The photonic band calculation indicated that the PC could have a positive or negative effective refractive index depending on the liquid crystal refractive index. The coupling efficiency at a high-order photonic band was improved by changing the interface of the PC. The FDTD simulation showed that positive or negative refraction could be controlled by reorienting the liquid crystal. Further, by combining two triangular PCs having opposite index signs, an incident light could be turned 90° without a waveguide structure.

4.9.4 Three-Dimensional Photonic Crystal

It is important to notice that up to date there are no PCs operating as left-handed materials in the optical range and all attempts of building artificial material and metamaterial structures face the same major problems: namely, the high absorption and fragility of materials in the absence of strong and robust fabrication techniques that build structures in the nanoscale. Three-dimensional PCs are usually fabricated by the self-assembly of silica or polystyrene spheres.[18] As the PCs are self-assembled, many of the problems encountered with more traditional nanoscale fabrication techniques are bypassed. By using these self-assembled arrays infiltrated with QDs, one can easily track the propagation of light and demonstrate the material's ability to have negative refraction.

Although many studies have been done using external sources of light to investigate light propagation in PCs, very little work has been done using embedded sources.[19] Our approach involves using CdSe QDs that act as quasipoint sources that can be inserted into the PC through infiltration. Some groups have already studied the QD effect in PCs.[20] However, left-handed behavior has not been examined using this technique. The theoretical part of an embedded point source inside left-handed materials was extensively studied in a recent paper.[21]

In this section, we will experimentally demonstrate that a point-source QD embedded in a PC can be used to study the propagation of scattered light and to determine the rightness (whether it is right- or left-handed material) of the sample.

The 3D opal PCs used in this experiment are prepared by self-assembly of monodispersed silica particles in a close-packed arrangement of silicon dioxide spheres. To prepare the samples we use sedimentation of silica particles from solution. Monodispersed suspension of the particles is deposited by sedimentation on a polished silicon substrate over several weeks. After removal from the liquid, the deposit is sintered in order to reinforce it. Then it is polished with fine sandpaper. Grown PCs show perfect short-range order on a length scale of tens of microns whereas the domain structure is prevalent over a long range. An SEM image of the PC made of 820 nm silica spheres is shown in Figure 4.9.14a.

In general, inverse structures are prepared in the following way. First, a template is infiltrated with a solid material; then, the template is removed by etching. To make the inverse opals we used direct silica opals as a template. The direct opals were heated to 100°C in a vacuum oven to remove excess oxygen. Then they were infiltrated with optical glue with a refractive index $n = 1.6$. To ensure that all air voids are filled with the material, the infiltration was done in a vacuum oven heated to 100°C for 24 h. After the infiltration, the samples were etched in 5% hydrofluoric acid solution. An SEM image of the direct opal made of silica and inverse opal made of optical glue are shown Figure 4.9.14.[22]

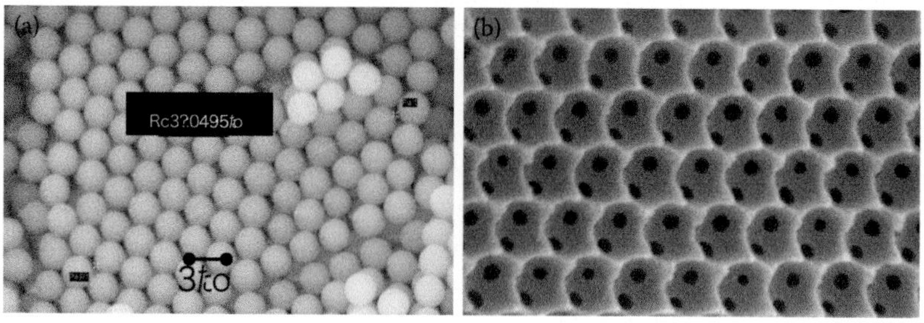

FIGURE 4.9.14
Scanning electron microscope (SEM) image of the direct opal made of 820 nm silica spheres (a) and SEM image of the inverse opal (b). (With kind permission from Springer Science+Business Media: *Nanoscale Photonics and Optoelectronics*, Vol. 9, 2010, p. 65, R. Moussa et al., Figures 4.1–4.8.)

Three microscopes were used to view our opal PC. First, the Cytoviva microscope was used; it allowed us to see real-time images of QDs simultaneously in fluorescent and dark-field modes. The Cytoviva microscope is the first step because it is a good indicator of where to locate aggregations of the QDs inside the sample. In order to see inside the PC, the sample needs to be transparent. Drops of specific index-matching oil were used not just to make our samples transparent but also to spread the light equally on the surface. Images were viewed under different light intensities to get glimpses of the structure of the opal and the QDs.

The second microscope used was the Leica microscope. It was possible to investigate the propagation of light inside our samples by using confocal imaging allowed by this microscope. By changing the focus distance and collecting only the focused light, confocal microscopes scan through the sample. Special software is then used to convert consecutive series of images into a 3D image. This technique requires the samples to be transparent so the PCs were immersed in oil as for Cytoviva imaging. Clear images of the light distribution can be determined even by scanning 100 μm into the sample.

The third microscope used was the Nikon confocal microscope. The illumination in the Nikon microscope is achieved by scanning one or more focused beams of light, usually from a laser or arc-discharge source, across the sample. The sequences of points of light from the sample are detected by a photomultiplier tube through a pinhole, and the output from the tube is built into an image and displayed by the computer. Regardless of the sample preparation, a primary benefit of the manner in which confocal microscopy is carried out is the flexibility in image display and analysis that results from the simultaneous collection of multiple images, in digital form, into a computer.

As we know what to expect from the simulation results,[21] we started investigating experimentally the opal PC system. One of the main difficulties that we faced while carrying out the experimental measurement was determining how deep the infiltrated QDs went inside the opal. Indeed, in the first samples we noticed that most of the QDs were concentrated at the interface; our conclusion concerning the nature of the focus thus was not as conclusive because it could be just a result of a rough interface. Therefore, we decided first to investigate several samples with different concentrations of QDs and analyze them. In the result plotted in Figure 4.9.15, three different samples were examined.

The three samples were treated the same way and were cut so that the [111] face was the main surface. The samples were then polished to remove any large scratches before being soaked in solutions of CdSe QDs in toluene. Several concentrations of QD solutions were

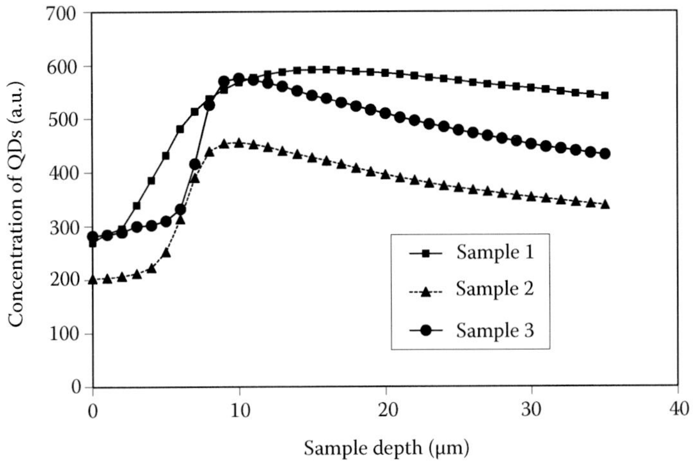

FIGURE 4.9.15
Concentration of quantum dots in arbitrary units versus the sample depth for three different samples. (With kind permission from Springer Science+Business Media: *Nanoscale Photonics and Optoelectronics*, Vol. 9, 2010, p. 69, R. Moussa et al., Figures 4.1–4.8.)

used to explore the effects of solution concentration on sample infiltration. Two concentrations that gave the best results were solutions with concentrations of 8×10^{-4} mg CdSe QD/mL toluene, which will be referred to as the low high concentration (LH), and 8×10^{-5} mg QD/mL toluene, which will be referred to as the medium concentration (M). The samples were soaked in these solutions for two and a half days and then imaged within a day of being removed from the solutions. Three samples were studied. The first sample consisted of 600 nm opal size infiltrated in an LH QD solution of 599 nm wavelength. The second consisted of 800 nm opal size infiltrated in an LH QD solution of 599 nm wavelength. The third consisted of 800 nm opal size infiltrated in an M QD solution of 612 nm wavelength. Before imaging, the samples were polished to remove large surface scratches and excess QD buildup on the surface of the opal with 1.0 μm polishing paper. After polishing, the samples were placed on a glass slide and then immersed in ethanol to make the opal transparent to allow imaging inside the opal. An Olympus FluoView 300 microscope with a 60× water immersion objective lens was used. Imaging started at the bottom of the opal, the top of the glass slide, and proceeded into the opal. The images were taken in 1.0 μm steps for 35 μm.

Figure 4.9.15 shows the concentration of the QDs versus the sample depth for the three samples. The opal–glass interface is shown to be between 0 and 10 μm. One possibility for the uncertainty in determining the interface is the change in the focus distance when switching from optical to computer scanning. The initial increase in these curves is mainly because of the high concentration of QDs close to the interface. In the case of these samples, this interface is roughly around 10 μm. The concentration starts to decrease afterwards, except for sample 1 in which it stays quite high deep inside the sample. As Figure 4.9.15 shows clearly, the concentration of QDs remains considerable even after 35 μm, and actually the depth would be greater if the ethanol had not dried out leaving the sample opaque and thus too dark to view. It is important to remind the reader that despite the differing size of the opals used in the three samples, the wavelength of the infiltrated QDs, and the concentration of the QD solution, the tendency for a high concentration of

QDs close to the interface and a gradual decrease away from the interface remains the same for all three samples.

Second, the knowledge of how deep the QDs can go inside the opal is precious information as scientists in the field are currently trying to determine the optical properties of materials by the different responses they get from the QDs embedded inside these materials. So, the more these QDs travel inside the samples, the better information we can collect from the sample. Therefore, results of Figure 4.9.15 are important for many fields.

Now that we are more confident that the QDs can indeed go as deep as 30 or 40 μm inside the opal, it is equally important to track the path of these QDs. Before proceeding to the result that tracks the propagation of QDs inside the opal and close to the interface with the glass, we ensure with the dark field/fluorescent Cytoviva microscope that this aggregation of QDs is present and localize its position for better confocal imaging.

The sample examined in this case is a red opal with blue CdSe QDs (481 nm) with high (H) concentration (8×10^{-2} mg/mL). QDs have a bright luminescence when seen with this kind of microscope. Figure 4.9.16 is a Cytoviva image of the opal infiltrated with blue QDs made using the 10× magnifying objective lens. Figure 4.9.16a is an optical image of the sample and Figure 4.9.16b is an image in florescent mode. As clearly seen from Figure 4.9.16a, the sample structure is in the [111] direction and this is shown through the honeycomb shape. Figure 4.9.16b shows the bright light emanating from the aggregation of QDs under the florescent light. Notice that Figure 4.9.16b is the top left corner of Figure 4.9.16a in which we found this high concentration of QDs. The real-time images from the Cytoviva microscope are good indicators of where to focus during the confocal experiment.

Indeed, the sample shown in Figure 4.9.16 was examined under the Leica confocal microscope exactly at the top left side. Two frames of multiple scans done on this sample are shown in Figure 4.9.17. Figure 4.9.17a displays a snapshot of the repeated scanning process for its entire imaged surface and Figure 4.9.17b shows the zoomed part of the top left of the imaged surface. It is important to mention that with this kind of confocal microscope setup, the sample is kept transparent by using specific Leica oil for the duration of imaging. A 10× magnifying objective lens was used. The excitation laser was chosen to be 456 nm. The multiple scans were generated using the z-scan feature with 1.5 μm step. Thus, the main experimental result is shown in the frames in which the high concentration of QDs close to the opal–glass interface is observed; however, the most important

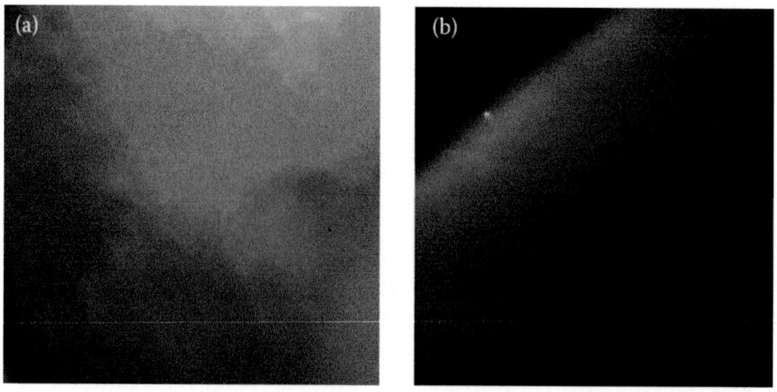

FIGURE 4.9.16
(**See color insert.**) Cytoviva images of the red opal with blue quantum dots. In (a) the opal is seen under the fully optical light and in (b) it is seen under fully florescent light. (With kind permission from Springer Science+Business Media: *Nanoscale Photonics and Optoelectronics*, Vol. 9, 2010, p. 70, R. Moussa et al., Figures 4.1–4.8.)

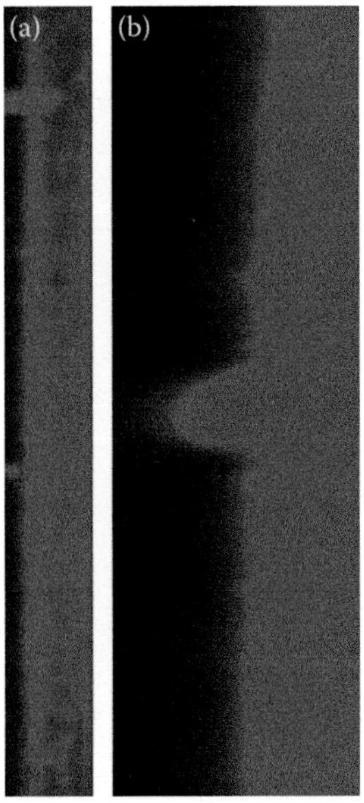

FIGURE 4.9.17
(**See color insert.**) Two snapshots of the multiple generated scans. (a) Snapshot for the whole surface of the imaged red opal with blue quantum dots and (b) snapshot for the top left part of the surface of the imaged red opal with blue quantum dots. (With kind permission from Springer Science+Business Media: *Nanoscale Photonics and Optoelectronics*, Vol. 9, 2010, p. 71, R. Moussa et al., Figures 4.1–4.8.)

aggregation is situated at the top left as indicated by the Cytoviva image as well as the Leica confocal microscope (Figures 4.9.16b and 4.9.17b). It is well known that if light is coming from a negative medium and propagating into a positive one it gets bent in the negative direction giving rise to a focus, as discussed theoretically in the study by Moussa et al.[21]

In Figure 4.9.17a, different aggregations of QDs close to the opal–glass interface are seen but the most important aggregation is close to the top left. Therefore, we focused more on that huge aggregation in Figure 4.9.17b that gives rise to the focus observed in the snapshot.

What we have seen in this result is what is expected from blue QDs having an excitation frequency right on the band that leads to the negative refraction. A focus forms above the interface. Such a result is very important because it leads us to the conclusion that this focus might be due to the negative refraction in nanoscale materials, namely, the opal PCs.

Another red opal was infiltrated with H concentration blue QDs and the results are shown in Figure 4.9.18. The images were obtained using the Leica confocal microscope. A sequence of 60 consecutive scans (2 μm step) was taken and processed into 3D pictures. The top view shows the honeycomb structure of the opal infiltrated with 481 nm CdSe QDs. Owing to the existence of several defects and emptiness between the domains of opals, most of the QDs are concentrated on boundaries of these domains (see Figure 4.9.18a). The 30°-inclined top-view image shows the focusing and the imaging on the glass side and particularly at the top right of the picture (Figure 4.9.18b). The side view (Figure 4.9.18c) is clearer and shows the focusing and the imaging on the glass side. The opal–glass interface is shown with the dotted line in Figure 4.9.18c.

In contrast with the results shown in the previous figure, this sample shows more homogeneous concentration of QDs at the opal–glass interface that leads to a well-focused

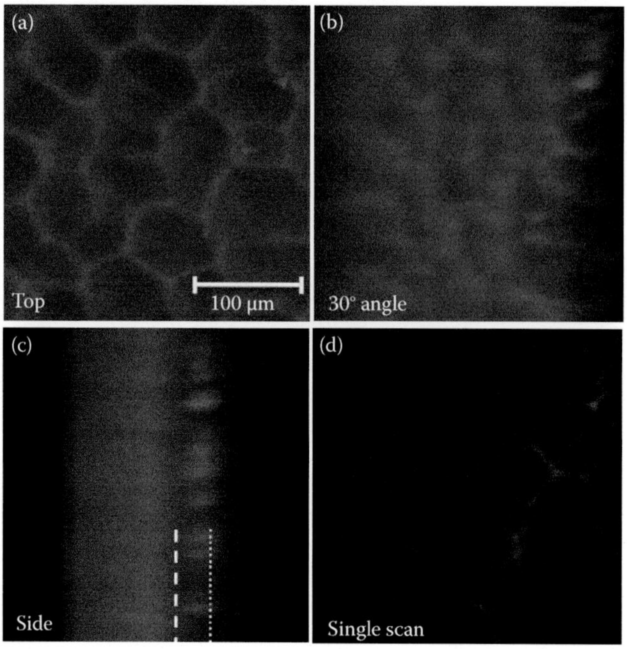

FIGURE 4.9.18
(**See color insert.**) Three-dimensional reconstructed image of light propagation in red opal infiltrated with 481 nm quantum dots. (a) Top view of the opal; (b) 30° inclined from top view; (c) side view of the opal where the dashed line shows opal–glass interface with opal being on the left side and dotted line is where the single scan was taken. (d) Single scan at 30 μm above the surface of the opal. (With kind permission from Springer Science+Business Media: *Nanoscale Photonics and Optoelectronics*, Vol. 9, 2010, p. 72, R. Moussa et al., Figures 4.1–4.8.)

image on the other side. We succeeded in obtaining a single scan of the top view of the glass side plotted in Figure 4.9.18d. Despite the fact that this image was taken 30 μm above the surface of the opal (the dotted line in Figure 4.9.18c) an image is clearly seen to form on the glass side. Focusing occurs at the opal–glass interface. Again, on the glass side we could recognize the honeycomb shape of the QDs localized primarily on the boundaries of the opal's domains that were translated and refocused in the glass.

It has been shown that the phenomenon of the negative refraction can be seen easily in strong modulation PC structures in which an effective refractive index can be attributed. Such a strong modulation has a chance to be present not in regular opals but in inverted opals in which the infiltrating material can play an important role and be a key factor for the negative refraction phenomenon. However, one has to choose carefully the infiltrating materials that lead to high refractive index contrast as well as the transparency needed for the material. It is equally important that the material should be nonabsorbing at the band-gap frequencies to avoid complications arising from electronic transitions. With such criteria in mind we infiltrated the opal with optical glue with a corresponding refractive index of $n = 1.6$. The refractive index was not as high as we would have liked it to be but with this material we succeeded in obtaining inverted transparent opals with a very successful etching phase process.

The inverted red opals were infiltrated later with blue QDs (481 nm wavelength) in a solution of 8×10^{-4} mg CdSe QD/mL toluene. The samples were soaked in these solutions for a day and then imaged within the first few days of being removed from the solutions.

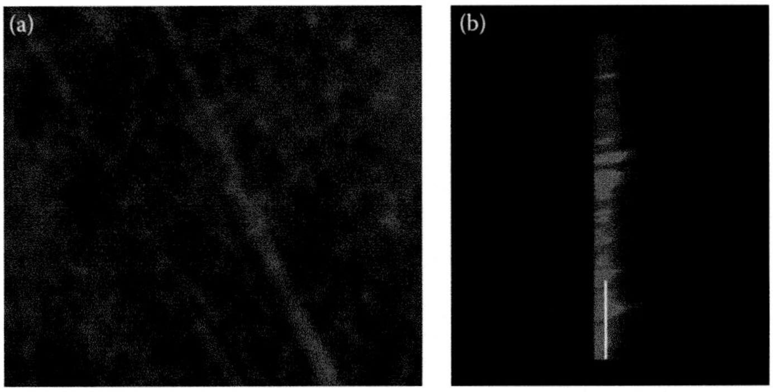

FIGURE 4.9.19
(**See color insert.**) Three-dimensional reconstructed image of light propagation in red inverted opal with optical glue and infiltrated with 481 nm blue quantum dots. (a) Image seen from the top; (b) image seen from the side. The solid line indicates roughly the interface between the inverted opal and the glass. (With kind permission from Springer Science+Business Media: *Nanoscale Photonics and Optoelectronics*, Vol. 9, 2010, p. 73, R. Moussa et al., Figures 4.1–4.8.)

Contrary to the regular opals, inverted opals and especially the one infiltrated with optical glue were very soft and curved. Therefore, we escaped the sanding part owing to the softness and fragility of the samples. The samples were imaged using the Leica confocal microscope. Figure 4.9.19a shows the imaged inverted red opal in the [111] direction with its honeycomb structure. The size of the sample was 1 mm by 1 mm and the magnifying lens used was 10×. The sample was not polished; therefore, large surface scratches and excess QD buildup on the surface of the inverted opal were seen, including the two diagonal lines due to the cuts operated on the sample. In comparison with regular opals the inverted ones attract more QDs around the domains, and actually those domains of 30–40 μm look beautiful in inverted opals because the QDs surround them and shape them beautifully. Thus, the QD infiltration technique can be used as a test to measure the quality of infiltration of the inverted opals, especially if we know that dissolving the silica spheres and filling the surrounding space with high refractive index material is never 100% complete.

The second step was to check with these samples the possibility of focusing. For that, several samples were checked and the result in Figure 4.9.19b corresponds to the red inverted opal infiltrated with optical glue and later on with 481 nm CdSe blue QDs at a concentration of 8×10^{-4} mg CdSe QD/mL toluene. Left for a day and removed and imaged right after, the image shows some indication of a converging and close-shape form of the focus. The dashed line in the figure indicates roughly the interface between the inverted opal and the glass, and as seen along this entire interface the focus of many close shapes are noticed and they all appear in the glass part. It is clear from previous works[23] and from the pioneering paper of Notomi[5] that achieving a focus due to a negative refraction can occur in strong modulation structures, but in our case this occurs for weak modulation. Thus, the real question that should be addressed in future work is whether this focus inside the opal and inverted opal occurs because of a negative refraction or because of some complicated mechanism involving the anisotropy of the system and what is the complicated physics behind it.

Thus, we have shown in this chapter that infiltrating PCs with QDs can be very useful not only to examine optical properties of PCs in the photonic band region, as some groups do, but also to follow the dispersion of these photons emitted by fluorescent QDs embedded inside PC structures and to determine whether the right-handed or the left-handed material electromagnetic properties take place in electromagnetic wave propagation. Although our experiments have not undoubtedly proved the existence of negative refraction in the upper bands in opal PCs, owing to the uncertainty of the photon path ways in defected opal structures, we have demonstrated here that there is a clear imaging of QD aggregates accumulated in defected parts with honeycomb shapes. To assign this type of focusing to focusing by negatively refracting beam we are creating defect-free opal films for future experiments.

4.9.5 Summary

In conclusion, using the techniques of infiltration of nano-sized materials (QDs and liquid crystals) into PCs of different dimensionality and shapes we tailored their optical properties. For 1D PCs, the distribution and the amount of QDs infiltrated inside them strongly affect their optical properties, namely, the refractive index and the transmission power. Also, we demonstrated how to change these properties by mastering different ways and methods by which these QDs are infiltrated. For 2D PCs, infiltration with liquid crystal does not just change their properties but creates a new device based on these PCs. Infiltrating a 2D PC with nematic liquid crystals showed the possibility for the same PC to have positive or negative refraction that can be controlled by reorienting the liquid crystals. Thus, electromagnetic waves traveling through two wedges of PCs with opposite indices could be turned 90° without a waveguide structure. For 3D PCs (both direct and inverse opals), infiltrating them with QDs was key to tracking down the luminescence path for a possible focus that occurs as a result of the left-handed behavior of these opals at certain frequencies. All these optical properties revealed in PCs of different dimensionality are the result of using powerful techniques of nano-sized particle infiltration. Finally, mastering the infiltration techniques will open avenues to investigate newly discovered materials and to reveal unique optical properties that these materials might have.

Acknowledgments

We thank Prof. Aleksey Efros for sharing his ideas and results[24] on optical routing in 2D silicon PCs controlled by temperature-induced changes in refractive index. We also acknowledge financial support of Air Force Office of Scientific Research under grants FA-9550-05-1-09-09 on "Negative Refraction in Visible and IR" and FA9550-08-1-0261 on "Tunable and Nonlinear Negative Refraction in Inverted Opal Photonic Crystals for Visible and Infrared."

References

1. S. Noda, M. Fujita, and T. Asanol, Nat. Photonics **1**, 449 (2007); J. D. Joannopoulos, P. R. Villeneuve, and S. Fan, Nature **386**, 143 (1997).
2. S. Foteinopoulou, E. N. Economou, and C. M. Soukoulis Phys. Rev. Lett. **90**, 107402 (2003).

3. S. Fan, P. R. Villeneuve, J. D. Joannopoulos, and E. F. Schubert, Phys. Rev. Lett. **78**, 3294 (1997).
4. S. Kubo, Z. Z. Gu, K. Takahashi, Y. Ohko, O. Sato, and A. Fujishima J. Am. Chem. Soc. **124**, 10950 (2002).
5. M. Notomi, Phys. Rev. B **62**, 10696 (2000).
6. C. Luo, S. G. Johnson, J. D. Joannopoulos, and J. B. Pendry, Phys. Rev. B **65**, 201104® (2002).
7. E. Cubukcu, K. Aydin, E. Ozbay, S. Foteinopoulou, and C. M. Soukoulis, Phys. Rev. Lett. **91**, 207401 (2003); Nature **423**, 604 (2003); P. V. Parimi, W. T. Lu, P. Vodo, and S. Shridar, Nature **426**, 404 (2003).
8. R. Schuster, M. Barth, A. Gruber, and F. Cichos, Chem. Phys. Lett. **413**, 280 (2005).
9. S. M. Weiss and Ph. M. Fauchet, Proc. SPIE **36**, 4654 (2002).
10. W. R. Tinga, W. A. G. Voss, and D. F. Blossey, J. Appl. Phys. **44**, 3897 (1973).
11. K. Keunyoo, H. L. Young, and H. Myung, Semicond. Sci. Technol. **11**, 1059 (1996).
12. D. Semwogerere and E. R. Weeks, in *Encyclopedia of Biomaterials and Biomedical Engineering*, edited by G. E. Wnek and G. L. Bowlin (Taylor & Francis, New York, 2005).
13. G. Alagappan, X. W. Sun, and P. Shum, Opt. Lett. **31**, 1109 (2006); H. Takeda and K. Yoshino, Phys. Rev. E **67**, 056607 (2003); Y-Y. Wang and L-W. Chen, Opt. Express **14**, 10580 (2006).
14. J. Sun, H. Xianyu, S. Gauza, and S-T. Wu, Liq. Cryst. **36**, 1401 (2009).
15. M. Plihal and A. A. Maradudin, Phys. Rev. B **44**, 8586 (1991).
16. T. Baba and D. Ohsaki, Jpn. J. Appl. Phys. **40**, 5920 (2001).
17. D. M. Pustai, S. Shi, C. Chen, A. Sharkawy, and D. W. Prather, Opt. Express **12**, 1823 (2004); V. Zabelin, L. A. Dunbar, N. Le Thomas, R. Houdre, M. V. Kotlyar, L. O'Faolain, and T. F. Krauss, Opt. Lett. **32**, 530 (2007).
18. M. N. Shkunov, Z. V. Vardeny, M. C. DeLong, R. C. Polson, A. A. Zakhidov, and R. H. Baughman, Adv. Funct. Mater. **12**, 21 (2002).
19. A. Berrier, M. Mulot, M. Swillo, M. Qiu, L. Thylen, A. Talneau, and S. Anand, Phys. Rev. Lett. **93**, 073902 (2004).
20. M. Barth, R. Schuster, A. Gruber, and F. Cichos, Phys. Rev. Lett. **96**, 243902 (2006); S. Strauf, Phys. Rev. Lett **96**, 127404 (2006).
21. R. Moussa, A. Kuznetsov, E. Neiser, A. L. Roberson, and A. A. Zakhidov, J. Nanophoton. **4**, 043503 (2010).
22. R. Moussa, A. Kuznetsov, E. Neiser, and A. A. Zakhidov, in *Nanoscale Photonics and Optoelectronics*, edited by Z. M. Wang and A. Neogi (Springer Science + Business Media, LLC, New York, 2010), Vol. 9, p. 65, Figures 4.1–4.8; Vol. 9, p. 69, Figures 4.1–4.8; Vol. 9, p. 70, Figures 4.1–4.8; Vol. 9, p. 71, Figures 4.1–4.8; Vol. 9, p. 72, Figures 4.1–4.8; Vol. 9, p. 73, Figures 4.1–4.8.
23. K. Ren, Z. Y. Li, X. Ren, S. Feng, B. Cheng, and D. Zhang, Phys. Rev. B **75,** 115108 (2007).
24. A. Efros, private communication.

Index

D

E

F

G

S